Multi-Phase Flow and Heat Transfer III

PART A – FUNDAMENTALS

Process Technology Proceedings, 1

Multi-Phase Flow and Heat Transfer III

Part A: Fundamentals

Proceedings of the Third Multi-Phase Flow and Heat Transfer Symposium — Workshop, Miami Beach, Florida, U.S.A., April 18–20, 1983

Edited by

T. Nejat Veziroğlu

University of Miami, Coral Gables, Florida, U.S.A.

and

Arthur E. Bergles

Iowa State University, Ames, Iowa, U.S.A.

ELSEVIER – Amsterdam – Oxford – New York – Tokyo 1984

ELSEVIER SCIENCE PUBLISHERS B.V.
Molenwerf 1
P.O. Box 211, 1000 AE Amsterdam, The Netherlands

Distributors for the United States and Canada:

ELSEVIER SCIENCE PUBLISHING COMPANY INC.
52, Vanderbilt Avenue
New York, NY 10017

Library of Congress Cataloging in Publication Data

Multi-phase Flow and Heat Transfer Symposium Workshop
(3rd : 1983 : Miami Beach, Fla.)
Multi-phase flow and heat transfer, III.

(Process technology proceedings ; 1)
Contents: v. 1. Fundamentals -- v. 2. Applications.
Includes bibliographies and indexes.
1. Multiphase flow--Congresses. 2. Heat--Transmission--Congresses. I. Veziroğlu, T. Nejat, 1924- . II. Bergles, A. E., 1935- .
III. Title. IV. Series.
TA357.M83 1983 620.1'064 84-10237
ISBN 0-444-42379-6 (U.S. : v. 1)
ISBN 0-444-42380-X (U.S. : v. 2)
ISBN 0-444-42381-8 (U.S. : set)

ISBN 0-444-42379-6 (Part A)
ISBN 0-444-42381-8 (Set)
ISBN 0-444-42382-6 (Series)

Printed in The Netherlands

MULTI-PHASE FLOW & HEAT TRANSFER III

PART A FUNDAMENTALS

PART B APPLICATIONS

Proceedings of the Third Multi-Phase Flow and Heat Transfer Symposium-Workshop, held in Miami Beach, Florida, U.S.A., on 18-20 April 1983, and presented by the Clean Energy Research Institute, College of Engineering, University of Miami, Coral Gables, Florida, U.S.A.; sponsored by the National Science Foundation, Washington, D.C., U.S.A.; in cooperation with the International Association for Hydrogen Energy, International Journal of Heat and Mass Transfer, and the Department of Mechanical Engineering, University of Miami, Coral Gables, Florida, U.S.A.

EDITORS

T. Nejat Veziroğlu
Clean Energy Research Institute
University of Miami
Coral Gables, Florida, U.S.A.

Arthur E. Bergles
Department of Mechanical Engineering
Iowa State University
Ames, Iowa, U.S.A.

MANUSCRIPT EDITOR

Sheila M. Puryear
Clean Energy Research Institute
University of Miami
Coral Gables, Florida, U.S.A.

EDITORIAL ASSISTANTS

İlker Gürkan
University of Miami
Coral Gables, Florida, U.S.A.

Aykut Menteş
University of Miami
Coral Gables, Florida, U.S.A.

ACKNOWLEDGEMENTS

The Symposium Organizing Committee gratefully acknowledges the sponsorship of the National Science Foundation and the assistance and cooperation of the International Association for Hydrogen Energy, International Journal of Heat and Mass Transfer, and the Mechanical Engineering Department of the University of Miami.

We wish to extend our sincere appreciation to the Invited Speakers: Dr. Win Aung, Director, Heat Transfer Program, National Science Foundation; Dr. Adrian Bejan, Associate Professor, Department of Mechanical Engineering, University of Colorado, Boulder, Colorado; and Dr. Jack T. Sanderson, Assistant Director for Engineering, National Science Foundation. We also thank Dr. Arthur E. Bergles, Distinguished Professor, Department of Mechanical Engineering, Iowa State University, Ames, Iowa, for his efforts in organizing and directing the workshop sessions.

Special thanks are due to our authors and lecturers, who have provided the substance of the symposium as published in the present proceedings.

And last, but not least, our debt of gratitude is owed to the Session Chairpersons and Co-Chairpersons, and to the Workshop Group Leaders, for the organization and execution of the technical and workshop sessions.

The Organizing Committee
Third Multi-Phase Flow and Heat
Transfer Symposium-Workshop

FOREWORD

The Third Multi-Phase Flow and Heat Transfer Symposium-Workshop continued the tradition established by its predecessors, the Two-Phase Flow and Heat Transfer Symposium-Workshop of October 1976 and the Second Multi-Phase Flow and Heat Transfer Symposium-Workshop of April 1979. It provided an event worthy of this important topic, including the latest information on the status of multi-phase flow and heat transfer research, development and applications. It also sought to identify important areas of multi-phase flow and heat transfer in urgent need of further research and development. In keeping with this latter objective, workshop discussions were held among the 188 attendees who represented universities, research establishments and industrial organizations. The Workshop Summaries are presented in a separate report.

It is particularly significant that this Third Symposium-Workshop included participants from 28 countries. This assured the international character of the information transfer, in keeping with the widespread need for better understanding of the behavior of multi-phase flows.

This two-volume set of proceedings includes 119 papers from the Symposium program. The great majority of the papers addresses gas-liquid flow; however, studies of gas-solid and liquid-solid flows are also included. While the heat transfer papers stress boiling (evaporation) or condensation, several papers consider phase change via freezing or melting. The papers have been divided into those that relate to fundamentals (Part A) and those that emphasize applications (Part B).

Included in Part A are papers dealing with the formulation of the multi-phase flow equations and solutions of these equations, flow regime transitions for gas-liquid flows, flow structure details and pressure drop of gas-liquid flows, steam separation and distribution, pressure wave propagation in gas-liquid and gas-solid flows, flashing and critical flow, heat transfer in pool boiling and forced convection boiling, heat transfer in dispersed gas-liquid flows, heat transfer in condensing, and thermal hydraulic instabilities.

Part B includes a large collection of papers addressing safety issues in nuclear reactor technology: thermal hydraulic code development, experiments to assess performance and validate codes and modelling of selected accident phenomena. Other papers in this part consider gas-solid flows, fluidized beds, liquid-solid flows, freezing and melting, mass transfer and chemical reactions, porous media, measurement techniques, and miscellaneous industrial applications.

The papers included in these two volumes are of both current and archival interest. The volumes should, therefore, serve as both a proceedings of the Symposium-Workshop and as a reference covering important topics in the field of multi-phase flow and heat transfer.

T. Nejat Veziroğlu
Arthur E. Bergles

PART A — FUNDAMENTALS

CONTENTS

5. STEAM SEPARATION AND DISTRIBUTION

6. PRESSURE WAVE PROPAGATION IN MULTI-PHASE FLOWS

7. FLASHING AND CRITICAL FLOW

8. BOILING HEAT TRANSFER

SYMPOSIUM COMMITTEE AND STAFF

SYMPOSIUM COMMITTEE

Arthur E. Bergles
Iowa State University, U.S.A.

Xuejun Chen
Xian Jiaotong University, China

Dimitri Gidaspow
Illinois Institute of Technology, U.S.A.

Sadik Kakaç
University of Miami, U.S.A.

R. T. Lahey, Jr.
Rensselaer Polytechnic Institute, U.S.A.

Samuel S. Lee
University of Miami, U.S.A.

Robert Lyczkowski
Argonne National Laboratory, U.S.A.

Samuel Sideman
Technion-Israel Institute of Technology, Israel

K. Stephan
Universität Stuttgart, F.R.G.

T. Nejat Veziroḡlu (Chairperson)
University of Miami, U.S.A.

STAFF

Symposium Coordinators:	Sheila M. Puryear Lucille J. Walter
Assistant Coordinators:	Barbara Berman Marina M. Blanco Ann Raffle
Manuscript Editor:	Sheila M. Puryear
Graduate Assistants:	O. Eser Ateşoglu İ. Gürkan A. Menteş O. T. Yıldırım
Undergraduate Assistants:	Mark Drews Clarence Mackey Cristina Robu

SESSION OFFICIALS

PLENARY SESSION	SYMPOSIUM OPENING
Chairperson:	A. E. Bergles, Iowa State University, Ames, Iowa, U.S.A.
Co-Chairperson:	S. S. Lee, University of Miami, Coral Gables, Florida, U.S.A.
SESSION 1A	MULTI-PHASE FLOW FUNDAMENTALS
Chairperson:	R. W. Lyczkowski, Argonne National Laboratory, Argonne, Illinois, U.S.A.
Co-Chairperson:	G. Kvajic, U-MAXS, Electromag Inc., Miami, Florida, U.S.A.
SESSION 1B	TWO-PHASE FLOW INSTABILITIES
Chairperson:	S. Kakaç, University of Miami, Coral Gables, Florida, U.S.A.
Co-Chairperson:	A. Awad, University of Miami, Coral Gables, Florida, U.S.A.
SESSION 1C	FLUIDIZED BEDS
Chairperson:	H. K. Fauske, Fauske and Associates, Inc., Burr Ridge, Illinois, U.S.A.
Co-Chairperson:	S. Sideman, Technion-Israel Institute of Technology, Technion City - Haifa, Israel
SESSION 2A	MULTI-PHASE FLOW REGIMES
Chairperson:	M. E. Salcudean, University of Ottawa, Ottawa, Ontario, Canada
Co-Chairperson:	E. E. Michaelides, University of Delaware, Newark, Delaware, U.S.A.
SESSION 2B	REACTOR SAFETY
Chairperson:	R. W. Lyczkowski, Argonne National Laboratory, Argonne, Illinois, U.S.A.
Co-Chairperson:	J. S. Chang, McMaster University, Hamilton, Ontario, Canada
SESSION 2C	MEASUREMENTS / SUSPENSIONS
Chairperson:	S. C. Kranc, University cf South Florida, Tampa, Florida, U.S.A.
Co-Chairperson:	S. M. Sami, Ecole Polytechnique de Montreal, Montreal, Quebec, Canada

SESSION 3A	MULTI-PHASE FLOW PRESSURE DROP AND HEAT TRANSFER
Chairperson:	S. G. Bankoff, Northwestern University, Evanston, Illinois, U.S.A.
Co-Chairperson:	J. P. Adams, EG & G Idaho, Inc., Idaho Falls, Idaho, U.S.A.
SESSION 3B	MULTI-PHASE FLOW MODELLING
Chairperson:	D. Gidaspow, Illinois Institute of Technology, Chicago, Illinois, U.S.A.
Co-Chairperson:	T. S. Andreychek, Westinghouse Electric Corporation, Pittsburgh, Pennsylvania, U.S.A.
SESSION 3C	MULTI-PHASE FLOW APPLICATIONS
Chairperson:	T. M. Romberg, CSIRO Division of Mineral Physics Sutherland, NSW, Australia
Co-Chairperson:	A. M. Tentner, Argonne National Laboratory, Argonne, Illinois, U.S.A.
SESSION 4A	MASS TRANSFER AND PHASE CHANGE
Chairperson:	K. V. Wong, University of Miami, Coral Gables, Florida, U.S.A.
Co-Chairperson:	J. J. J. Chen, University of Hong Kong, Hong Kong
SESSION 4B	STEAM GENERATION AND DISTRIBUTION
Chairperson:	J. H. Kim, Electric Power Research Institute, Palo Alto, California, U.S.A.
Co-Chairperson:	F. Dobran, Stevens Institute of Technology, Hoboken, New Jersey, U.S.A.
SESSION 4C	DROPLET DYNAMICS
Chairperson:	K. Johannsen, Technische Universität Berlin, Berlin, Federal Republic of German
Co-Chairperson:	S. C. Yao, Carnegie-Mellon University, Pittsburgh, Pennsylvania, U.S.A.
WORKSHOP SESSION 5A	BOILING AND CONDENSATION
Workshop Leader:	J. Lienhard, University of Houston, Houston, Texas, U.S.A.
WORKSHOP SESSION 5B	TWO-PHASE GAS-LIQUID FLOW REGIMES AND PRESSURE DROP
Workshop Leader:	J. Weisman, University of Cincinnati, Cincinnati Ohio, U.S.A.

WORKSHOP SESSION 5C	TWO PHASE GAS-LIQUID FLOW ISSUES IN NUCLEAR REACTOR AND CHEMICAL PLANT SAFETY
Workshop Leaders:	S. G. Bankoff, Northwestern University, Evanston, Illinois, U.S.A. M. E. Salcudean, University of Ottawa, Ottawa, Ontario, Canada
WORKSHOP SESSION 5D	GAS-SOLID AND SOLID-LIQUID FLOW AND HEAT TRANSFER
Workshop Leader:	R. Lyczkowski, Argonne National Laboratory, Argonne, Illinois, U.S.A.
WORKSHOP SESSION 5E	TWO-PHASE EQUIPMENT FOR THE POWER AND PROCESS INDUSTRIES
Workshop Leader:	G. Breber, Heat Transfer Research, Inc., Alhambra, California, U.S.A.
WORKSHOP SESSION 5F	OPEN FORUM (Subjects Not Covered in Other Sessions)
Workshop Leader:	W. M. Rohsenow, Massachusetts Institute of Technology, Cambridge, Massachusetts, U.S.A.
SESSION 6A	BOILING AND CONDENSATION
Chairperson:	P. Saha, Brookhaven National Laboratory, Upton, New York, U.S.A.
Co-Chairperson:	H. Auracher, Institute fur Techn. Thermodynamik, Stuttgart, Federal Republic of Germany
SESSION 6B	TRANSIENTS AND WAVE PROPAGATION
Chairperson:	T. M. Romberg, CSIRO Division of Mineral Physics, Sutherland, NSW, Australia
Co-Chairperson:	R. Taleyarkhan, Rensselaer Polytechnic Institute, Troy, New York, U.S.A.
SESSION 6C	HEAT TRANSFER / ENERGY CONVERSION
Chairperson:	T. W. Fogwell, Texas A & M University, College Station, Texas, U.S.A.
FINAL PLENARY SESSION	WORKSHOP REPORTS
Chairperson:	A. E. Bergles, Iowa State University, Ames, Iowa, U.S.A.

Multi-Phase Flow and Heat Transfer III. Part A: Fundamentals
edited by T.N. Veziroğlu and A.E. Bergles
Elsevier Science Publishers B.V., Amsterdam, 1984 — Printed in The Netherlands

SECOND-LAW ASPECTS OF HEAT TRANSFER ENGINEERING

Adrian Bejan
Department of Mechanical Engineering
University of Colorado
Boulder, Colorado 80309, U.S.A.

ABSTRACT

Used together, the first and second laws of thermodynamics indicate that any engineering device destroys available (useful) work at a rate proportional to its rate of entropy generation. Heat exchange apparatuses always generate entropy, due to the inherent irreversibility of heat transfer across a finite temperature difference and due to the irreversibility of fluid flow against resistive friction forces. On thermodynamic grounds, then, in this lecture it, is argued that the design philosophy that insures the conservation of available work ("useful energy") is the Minimization of Entropy Generation in the design of heat exchange devices and in the design of components and subcomponents for such devices. The ability to optimally select the geometry of a device such that its entropy generation rate is minimized, is illustrated based on examples drawn from duct design, fin design and heat exchanger design.

1. INTRODUCTION

I want to thank the organizers for the invitation to present this lecture at the Third Multi-Phase Flow and Heat Transfer Symposium-Workshop. I feel very honored by this invitation.

In this lecture I will rely on the classical statements of the first and second laws of thermodynamics to show the intimate connection between heat transfer processes and the destruction of available work in installations that employ heat transfer processes. I will then argue that one design philosophy that insures the reduction of available work destruction (i.e. the conservation of "useful" energy) is the minimization of entropy generation in each of the components of an engineering installation [1,2]. Finally, I will illustrate this design philosophy by means of examples ranging from the design of individual fins to that of heat exchangers processing two-phase streams.

2. THE NEED TO MINIMIZE ENTROPY GENERATION

Consider the steady (cyclic) operation of a power plant, as shown schematically in Fig. 1. The power plant receives its heat input at a rate $\dot{Q}_H$ from a high temperature reservoir of absolute temperature T_H. Shaft work is being produced at a rate $\dot{W}$, and heat is being rejected at a rate $\dot{Q}_L$ to a low temperature reservoir T_L. The first and second laws of thermodynamics for the power plant as a thermodynamic system, state that [3]

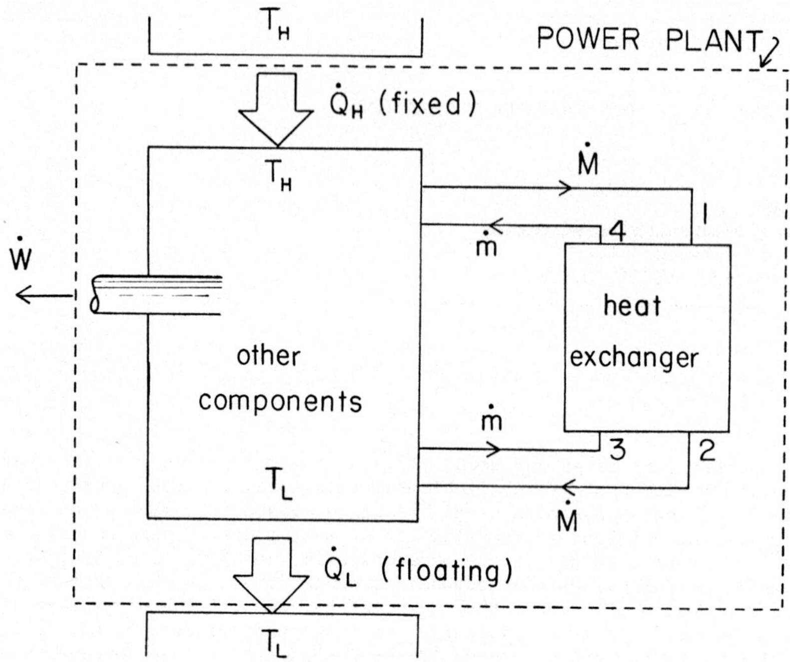

Fig. 1. Schematic of Power Plant Operation, with a Heat Exchanger as a Component

$$\dot{Q}_H - \dot{Q}_L - \dot{W} - 0 \tag{2.1}$$

$$\frac{\dot{Q}_H}{T_H} - \frac{\dot{Q}_L}{T_L} \leq 0 \tag{2.2}$$

The second law of thermodynamics, eq. (2.2), can be rewritten as

$$\dot{S}_{gen} = \frac{\dot{Q}_L}{T_L} - \frac{\dot{Q}_H}{T_H} \geq 0 , \tag{2.3}$$

which means that the power plant rate of entropy generation $\dot{S}_{gen}$ is never negative. The inequality sign appearing in the second law is the basis for distinguishing between real power plants that operate irreversibly,

$$\dot{S}_{gen} > 0, \tag{2.4}$$

and highly idealized power plants that, like the famous Carnot cycle, would operate reversibly,

$$\dot{S}_{gen} = 0. \tag{2.5}$$

A critical examination of the quality of the problems proposed to the student by most thermodynamics textbooks would show that young engineers are taught to be precise and, naturally, to achieve this precision using the equations made available to them by engineering thermodynamics. In the case of a power plant, these "equations" are (2.1) and (2.5): I believe this is the source of the "mystery" that surrounds the second law among engineers of all orientations. It is true that many engineering schools teach the first law and the second law of thermodyanmics, as the analysis listed here between equations (2.1) and (2.5) is present in most textbooks. However, the young engineer is not taught to use the second law to solve real engineering problems. In the above example, the second law is best represented by the inequality (2.4): as shown below, the concept of entropy generation or the severity of the inequality sign in (2.4) has a very important practical meaning.

Eliminating $\dot{Q}_L$ between equations (2.1) and (2.2) we find

$$\dot{W} \leq \dot{Q}_H \left(1 - \frac{T_L}{T_H}\right) \tag{2.6}$$

This is the well-known conclusion that the work produced by the power plant can never exceed the maximum that would be produced by a reversible power plant using the same heat input $\dot{Q}_H$ and the same heat reservoirs (T_H, T_L). In a different notation, equation (2.6) states that

$$\dot{W} \leq \dot{W}_{max} \tag{2.7}$$

where

$$\dot{W}_{max} = \dot{Q}_H \left(1 - \frac{T_L}{T_H}\right) \tag{2.8}$$

If the work output of the power plant is always lower than the ceiling value represented by $\dot{W}_{max}$, then it seems that a top-priority question for the engineer must be "How much lower is $\dot{W}$, and why?". Referring to the difference between maximum and actual work outputs as lost work,

$$\dot{W}_{lost} = \dot{W}_{max} - \dot{W}, \tag{2.9}$$

and using equations (2.1) - (2.3), we find that the lost work is proportional to the generated entropy,

$$\dot{W}_{lost} = T_L \dot{S}_{gen} \geq 0. \tag{2.10}$$

In conclusion, the severity of the inequality sign in the second law of thermodynamics, eq. (2.4), describes the degree to which the thermodynamic system acts as a destroyer of otherwise available work. The degree of system irreversibility is measured quantitatively as $\dot{S}_{gen}$. Therefore, in order to minimize the lost work, the power plant designer must concentrate on minimizing the rate of entropy generation inside each of the engineering components that make up the power plant. For example, if one of the components is the counterflow heat exchanger shown in Fig. 1, then we can prove that

$$\dot{S}_{gen} = (\dot{S}_{gen})_{\substack{\text{heat}\\ \text{exchanger}}} + (\dot{S}_{gen})_{\substack{\text{other}\\ \text{components}}} , \tag{2.11}$$

in other words,

$$\dot{S}_{gen} = \sum_k (\dot{S}_{gen})_k \ , \tag{2.12}$$

where k represents all the components of the system. To prove eq. (2.11), we start with the general statement of the second law for any flow system (see Ref. [1], p. 9)

$$\dot{S}_{gen} = \frac{\partial S}{\partial t} - \sum_i \frac{\dot{Q}_i}{T_i} + \sum_{\substack{\text{outlet} \\ \text{ports}}} \dot{m}s - \sum_{\substack{\text{inlet} \\ \text{ports}}} \dot{m}s \geq 0 \tag{2.13}$$

with the following notation:

$\frac{\partial S}{\partial t}$ = the time rate of entropy accumulation inside the system (S is the entropy of the system; S is a thermodynamic property that must not be confused with $\dot{S}_{gen}$, which is not a thermodynamic property)

$\frac{\dot{Q}_i}{T_i}$ = the entropy transfer; note that $\dot{Q}_i$ is the heat transfer interaction defined positive when directed into the system, and T_i is the absolute temperature of the system boundary (control surface) crossed by $\dot{Q}_i$.

$\dot{m}s$ = the entropy flowrate through a port in the control surface that defines the system ($\dot{m}$ is the mass flowrate and s is the specific entropy (a property) of the substance that just crosses the control surface).

Applying the $\dot{S}_{gen}$ definition (2.13) to the two subsystems of the power plant system, in the steady state we have

$$(\dot{S}_{gen})_{\substack{\text{heat} \\ \text{exchanger}}} = \dot{M}\,(s_2 - s_1) + \dot{m}\,(s_4 - s_3) \tag{2.14}$$

$$(\dot{S}_{gen})_{\substack{\text{other} \\ \text{components}}} = \frac{\dot{Q}_L}{T_L} - \frac{\dot{Q}_H}{T_H} + \dot{M}\,(s_1 - s_2) + \dot{m}\,(s_3 - s_4) \tag{2.15}$$

Adding equations (2.14) and (2.15) side-by-side, and recognizing that the overall entropy generation rate of the power plant is given by eq. (2.3), we obtain equation (2.11): the message of equation (2.11) is that the entropy generation rate of the heat exchanger must be minimized, if the overall $\dot{W}_{lost}$ of the power plant is to be reduced.

The above argument can be repeated by focusing on a component smaller than the heat exchanger of Fig. 1: for example, a fin or a flow passage in the heat exchanger. The conclusion remains the same, namely

$$\dot{W}_{lost} = T_L \sum_k (\dot{S}_{gen})_k \tag{2.16}$$

where k represents any system component or subcomponent. Equation (2.16) is known as the Gouy-Stodola theorem; it applies in exactly the same form to thermodynamic systems considerably more complicated than the closed power plant system shown in Fig. 1. A more general derivation of equation (2.16) is the subject of chapter 2 in Ref. [1], where it is shown that T_L generally represents the absolute temperature of the ambient reservoir to which the system rejects heat.

3. THE IRREVERSIBILITY OF HEAT TRANSFER AND FLUID FLOW

The proportionality between lost available work and entropy generation, eq. (2.16), should occupy a central position in heat transfer engineering, because heat transfer processes are inherently irreversible. In particular, the irreversibility of convection heat transfer processes is due to two characteristically irreversible phenomena:

i) the transfer of heat in the direction of a finite temperature gradient, or across a finite temperature difference;

ii) the flow of fluid against the restraining forces posed by friction, i.e. against duct skin friction or against the drag force experienced by a body bathed by the flow.

The importance of avoiding irreversibilities of type (i) and (ii) is, of course, part of the present-day heat transfer engineering methdology. The reduction of heat transfer irreversibility (i) is in fact the thermodynamic foundation for the field of Heat Transfer Augmentation [4], i.e. for the development of techniques aimed at reducing the temperature difference between walls and energy-carrying fluid. On the other hand, the reduction of fluid flow irreversibility (ii) is the thermodynamic reason for the desire to reduce pressure drop in heat exchanger design [5]. However, it is worth noting that these two activities, heat transfer augmentation and pressure drop reduction, are presently being discussed as two <u>different</u> items in the approach to good heat transfer engineering [6]. The concept of entropy generation and its design-related philosophy of entropy generation minimization constitute the framework in which to unite these activities into a direct approach to optimum thermodynamic design.

Before showing how irreversibilities (i) and (ii) can be minimized together in a certain heat transfer device, it is instructive to prove that heat transfer and fluid flow are indeed sources of entropy generation. Consider first the heat transfer rate $\dot{Q}$ between two entities separated by a finite temperature difference ΔT. As shown in Fig. 2a, the entropy generation rate becomes visible if we recognize the locus of irreversibility, that is, the "temperature gap system" shown by the dashed contour. According to the general expression (2.13), the entropy generation rate in the temperature gap system is

$$\dot{S}_{gen} = - \frac{\dot{Q}}{T+\Delta T} + \frac{\dot{Q}}{T} = \frac{\dot{Q}\Delta T}{T(T+\Delta T)} \tag{3.1}$$

or, when $\Delta T << T$ (as in most heat exchanger applications),

$$\dot{S}_{gen} = \frac{\dot{Q}\Delta T}{T^2} . \tag{3.2}$$

The heat transfer irreversibility is therefore proportional to both heat transfer rate and temperature difference. This relationship explains the coexistence of two apparently contradictory activities in heat transfer engineering:

1. heat transfer augmentation, whose purpose is to enhance thermal contact,
2. thermal insulation, whose purpose is to inhibit thermal contact.

In the case of heat transfer augmentation, the object is to decrease the

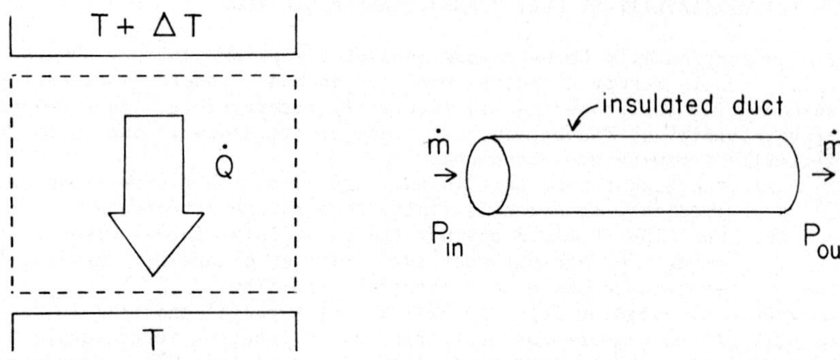

Fig. 2. The Irreversibility of Pure Heat Transfer and Pure Fluid Flow

temperature difference ΔT when the heat transfer rate $\dot{Q}$ is fixed (e.g. the design of cooling passages for the core of a nuclear reactor): as a result, the product $\dot{Q}\Delta T$ and the entropy generation rate are minimized. In the case of thermal insulation, the object of the design activity is to minimize the heat leak $\dot{Q}$ when ΔT is fixed (e.g. the insulation around a house, or that around a helium liquifier): as a result, the product $\dot{Q}\Delta T$ and the entropy generation rate are again minimized. The thermodynamic equivalence of these seemingly opposite practices in heat transfer engineering, heat transfer augmentation and thermal insulation, demonstrate once more the universality of "entropy generation minimization" as a design philosophy [1,2].

The irreversibility of fluid flow can be evaluated by applying the entropy generation formula (2.16) to the flow configuration of interest: in the case of steady flow through an adiabatic duct, Fig. 2b, eq. (2.16) yields

$$\dot{S}_{gen} = \dot{m}\,(s_{out} - s_{in}) \tag{3.3}$$

We obtain a more meaningful substitute for eq. (3.3) if we first recognize the canonical relationship for enthalpy [1],

$$dh = Tds + vdP \tag{3.4}$$

and the first law of thermodynamics for the duct as an open (steady flow) system

$$0 = h_{out} - h_{in} \tag{3.5}$$

Combining eqs. (3.3) - (3.5) we conclude that entropy is being generated as soon as a pressure drop can be measured,

$$\dot{S}_{gen} = \dot{m} \int_{P_{out}}^{P_{in}} \left(\frac{v}{T}\right)_{h=constant} dP, \tag{3.6}$$

Two special forms of eq. (3.6) are [1]

$$\dot{S}_{gen} \simeq \dot{m}R \frac{\Delta P}{P_{in}}, \text{ for ideal gas flow } (\Delta P << P_{in}) \tag{3.7}$$

$$\dot{S}_{gen} \simeq \dot{m} \frac{\Delta P}{\rho T_{in}}, \text{ for incompressible fluid flow } (\Delta P << \rho c T_{in}) \tag{3.8}$$

The entropy generation rate is proportional to both $\dot{m}$ and ΔP, hence, to the pump power required to sustain the flow.

In Figs. 2a and 2b, the irreversibilities of heat transfer and fluid flow have been considered independently; in any actual heat transfer device, however, these irreversibilities are strongly coupled, as they are both influenced sharply by the geometry and the size of the device. This strong coupling is illustrated in the remainder of this lecture by means of elementary examples of heat transfer apparatuses.

4. INTERNAL FLOW

As a first example, consider the general heat exchanger passage of hydraulic diameter $D_h = 4A/p$ shown in Fig. 3. Since, in general, the passage is characterized by both pressure drop and wall-fluid temperature difference, the system defined by the solid lines on Fig. 3 will generate entropy by fluid flow and by heat transfer. Again, the way to calculate the entropy generation rate for this system is via eq. (2.13); the result is

$$d\dot{S}_{gen} = \dot{m}\ (s + ds - s) - \frac{q''pdx}{T+\Delta T} \tag{4.1}$$

where s is the bulk specific entropy of the stream at location x along the duct. Similarly, the first law of thermodynamics requires

$$q''pdx = \dot{m}\ (h + dh - h) \tag{4.2}$$

or, using the canonical relation (3.4),

$$q''pdx = \dot{m}\ (Tds + \frac{1}{\rho}\ dP). \tag{4.3}$$

Eliminating ds between eqs. (4.1) and (4.3) we obtain the general result

$$\frac{d\dot{S}_{gen}}{dx} = \underbrace{\frac{q''p\Delta T}{T^2}}_{\substack{\text{heat transfer}\\ \text{irreversibility}\\ \text{per unit duct length}}} + \underbrace{\frac{\dot{m}}{\rho T}\left(-\frac{dP}{dx}\right)}_{\substack{\text{fluid flow}\\ \text{irreversibility}\\ \text{per unit duct length}}}, \tag{4.4}$$

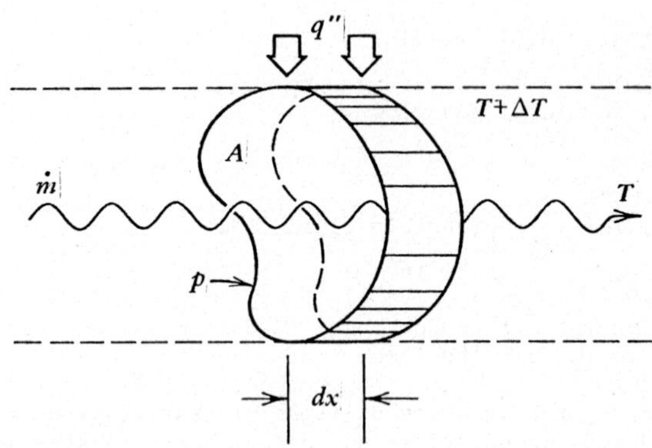

Fig. 3. Forced Convection Heat Transfer in a Duct of Arbitrary Geometry [1]

where the wall-fluid ΔT was assumed much smaller than the bulk absolute temperature of the stream. Therefore, on the basis of entropy generation analysis, we reach the important conclusion that the irreversibility of any heat exchanger passage is ruled by the sum of heat transfer and fluid flow irreversibilities, as indicated in eq. (4.4).

The geometric coupling between the two terms appearing on the right-hand side of eq. (4.4) becomes evident if we think of varying the hydraulic diameter while keeping (pq") and ($\dot{m}$) constant. As D_h decreases, the wall-fluid ΔT decreases and the longitudinal pressure gradient (-dP/dx) increases. Therefore, a given change in duct size (D_h) has opposing effects on the two terms in the $d\dot{S}_{gen}/dx$ formula (4.4). In the small-D_h limit the two-term expression (4.4) is dominated by fluid flow irreversibility, and in the large-D_h limit by heat transfer irreversibility. An optimum hydraulic diameter exists, such that the irreversibility per unit duct length is minimum.

To illustrate the existence of this optimum duct size for minimum entropy generation, consider the case of a tube of diameter D, for which eq. (4.4) assumes the special form

$$\frac{d\dot{S}_{gen}}{dx} = \frac{(q')^2}{\pi k T^2 Nu} + \frac{32(\dot{m})^3 f}{\pi^2 \rho^2 T D^5} \qquad (4.5)$$

Note that q', Nu and f are the heat transfer per unit duct length (pq"), the

Nusselt number based on tube diameter, and the friction factor. Relying on well-known engineering correlations for Nu (Re_D, Pr) and $f(Re_D)$, correlations appropriate for fully-developed turbulent flow in the range $10^4 < Re_D < 10^7$, it is easy to show that the optimum tube diameter D_{opt} for minimum total irreversibility obeys the dimensionless relationship [1]

$$Re_{D,opt} = 2.023\ Pr^{-0.071}\ B_0^{0.358}, \tag{4.6}$$

where B_0 is a "duty" parameter assumed known,

$$B_0 = \frac{\rho \dot{m} q'}{\mu^{5/2} (kT)^{1/2}} . \tag{4.7}$$

The sharpness of the irreversibility minimum that corresponds to D_{opt} is illustrated in Fig. 4: on this figure, $S'_{\Delta T}$ and $S'_{\Delta P}$ represent the first and second terms appearing on the right-hand side of eq. (4.4). Symbol S' stands for $d\dot{S}_{gen}/dx$, and S'_{min} for the minimum entropy generation rate per unit tube length. Marked along the curve of Fig. 4 is the irreversibility distribution ratio ϕ, defined as $\phi = S'_{\Delta P}/S'_{\Delta T}$. The fact that the minimum shown in Fig. 4 is sharp, and that right at the optimum the value of ϕ is 0.168, suggests that a precise balance between heat transfer and fluid flow irreversibilities must be preserved if the overall irreversibility is to be minimized.

It is worth keeping in mind that the optimum tube diameter given by eq. (4.6) is not a general result, since it is based on the assumption that certain parameters (contained in B_0) are fixed. A different optimum will result when

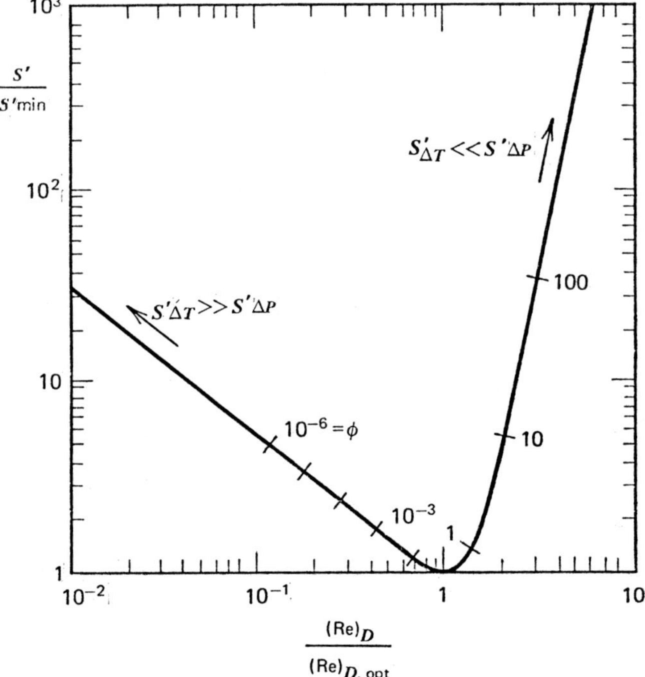

Fig. 4. Relative Entropy Generation Rate in a Smooth Tube [1]

$d\dot{S}_{gen}/dx$ is minimized subject to a different set of constraints. Regardless of constraints, however, the thermodynamic optimization is based on minimizing expressions (4.4) and (4.5), which are generally valid.

5. EXTERNAL FLOW

Similar trade-offs between heat transfer and fluid flow irreversibilities exist in the thermodynamic performance of convection devices that operate in external flow. One frequent example of external convection is the heat transfer from fins (extended surfaces). Along with the dx-long duct of Fig. 3, the fin is perhaps the most elementary subcomponent in the constitution of a heat exchanger: as demonstrated in section 2, the irreversibility caused by each fin is destined to contribute to the overall irreversibility of the heat exchanger (eq. 2.12).

The entropy generated by a single fin can be evaluated based on the schematic shown in Fig. 5. A fin with base temperature and heat transfer rate, T_B and $\dot{Q}_B$, is bathed by a fluid with uniform temperature and velocity, T_∞ and V_∞. As "thermodynamic system" we select the infinitely large heat reservoir represented by the T_∞ fluid. Relative to this system, which is shown by the dashed line in Fig. 5, the fin moves with the velocity V_∞. In the cut made by the system boundary through the fin, we distinguish the drag force F_D: since, as observers we are riding on the fluid, the fin performs work on the system, at a rate $F_D V_\infty$.

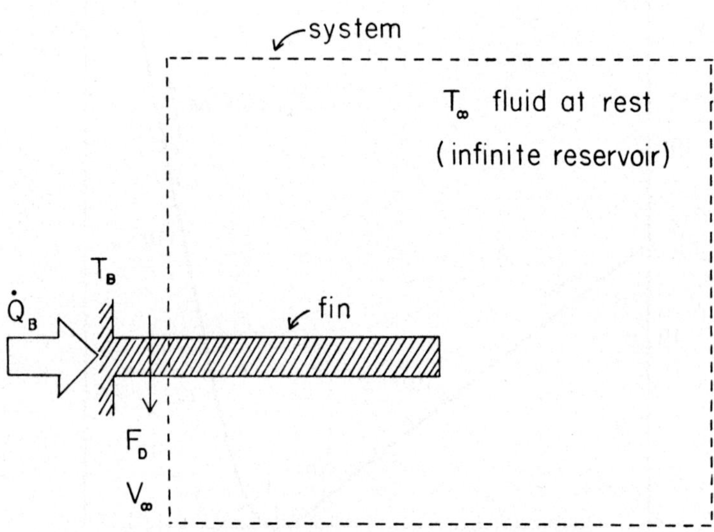

Fig. 5. Convection Heat Transfer from a Solitary Fin

The first and second laws of thermodynamics for the system of Fig. 5 are, respectively,

$$\underset{\text{heat transfer}}{\dot{Q}_B} + \underset{\text{work transfer}}{F_D V_\infty} = \underset{\text{energy change}}{\frac{dU_\infty}{dt}} \tag{5.1}$$

$$\underset{\text{entropy generation}}{\dot{S}_{gen}} = \underset{\text{entropy change}}{\frac{\partial S_\infty}{\partial t}} - \underset{\text{entropy transfer}}{\frac{\dot{Q}_B}{T_B}}$$

where U_∞ and S_∞ represent the instantaneous internal energy and entropy inventories of the system (the T_∞ heat reservoir). Recalling the canonical relation $du = Tds + Pdv$, and recognizing that the T_∞ heat reservoir is incapable of Pdv work transfer with its environment (i.e. with the fin), we conclude that

$$\frac{dU_\infty}{dt} = T_\infty \frac{dS_\infty}{dt} \tag{5.3}$$

The entropy generation rate follows from eqs. (5.1) - (5.3) through the elimination of U_∞ and S_∞,

$$\dot{S}_{gen} = \frac{(T_B - T_\infty)\dot{Q}_B}{T_\infty T_B} + \frac{F_D V_\infty}{T_\infty} \tag{5.4}$$

or, assuming $(T_B - T_\infty) << T_\infty$,

$$\dot{S}_{gen} = \underset{\substack{\text{heat transfer}\\ \text{irreversibility}\\ \text{due to a single fin}}}{\frac{(T_B - T_\infty)\dot{Q}_B}{T_\infty^2}} + \underset{\substack{\text{fluid flow}\\ \text{irreversibility}\\ \text{due to a single fin}}}{\frac{F_D V_\infty}{T_\infty}} \tag{5.5}$$

Equations (5.4) and (5.5) are of considerable generality and simplicity; they were derived first in Refs. [7] and [8] based on two alternative thermodynamic analyses, both different than the analysis presented here based on Fig. 5. Once again, the entropy generated by the heat transfer device (the fin in external flow) is equal to the sum of two contributions: imperfect thermal contact plus fluid friction. The interesting aspect of the additive nature of these two contributions is that changes in fin size affect the two contributions differently. For example, increasing the fin surface area will consistently reduce the heat transfer contribution to $\dot{S}_{gen}$ and, at the same time, will increase the fluid friction contribution. There exists a definite optimum fin size for which the total entropy generation rate of the fin is minimum.

The minimization of fin irreversibility is illustrated on a case-by-case basis in Refs. [7,8]. The procedure consists of combining the objective function (5.5) with the fin conduction solution for the relationship between $\dot{Q}_B$ and $(T_B - T_\infty)$, and with the usually empirical relationship between F_D and V_∞. For example, in the case of a cylindrical fin of length L and diameter D, the $\dot{S}_{gen}$ formula (5.5) becomes

$$\underbrace{\dot{S}_{gen}\frac{k\nu T_{\infty}^{2}}{\dot{Q}_{B}^{2}V_{\infty}}}_{N_S} = \frac{\frac{2}{\pi}(k/\lambda)^{1/2}}{Nu^{1/2}Re_D \tanh\left[2Nu^{1/2}\left(\frac{\lambda}{k}\right)^{1/2}\frac{Re_L}{Re_D}\right]} + \frac{B}{2}C_D\, Re_L\, Re_D \tag{5.6}$$

(the entropy generation number)

where Re_L and Re_D are $V_\infty L/\nu$ and $V_\infty D/\nu$, respectively. In general, the Nusselt number Nu is a known function of Pr and Re_D, and the drag coefficient C_D is a known function of Re_D. Finally, parameter B is dimensionless group associated with the heat transfer function prescribed for the fin,

$$B = \frac{\rho\nu^{3}kT_{\infty}}{\dot{Q}_{B}^{2}} \tag{5.7}$$

Assuming that the fin heat transfer duty ($\dot{Q}_B$) and the flow characteristics (V_∞, T_∞, fluid properties) are fixed, the entropy generation rate emerges as a function of four parameters,

$$N_S = \text{function}\ (Re_L, Re_D, M, B) \tag{5.8}$$

with $M = (k/\lambda)^{1/2}\ Pr^{-1/6}$. Solving simultaneously

$$\frac{\partial N_S}{\partial Re_L} = 0 \quad \text{and} \quad \frac{\partial N_S}{\partial Re_D} = 0 \tag{5.9}$$

it is possible to calculate once and for all the thermodynamic optimum fin geometry ($Re_{L_{opt}}$ and $Re_{D_{opt}}$) in terms of design parameters M and B. Figure 6 shows in chart form the result of this numerical calculation for M = 100: both $Re_{L_{opt}}$ and $Re_{D_{opt}}$ decrease as B increases, i.e. as the heat transfer duty of the fin decreases (note that the optimum slenderness ratio L_{opt}/D_{opt} is of order 10, and is relatively independent of B). Further examples of entropy generation minimization in fin design are given in Refs. [7,8]. Although the actual optimization procedure depends on the particular constraints present in the design, in all cases the starting point is eq. (5.4), which has general applicability.

6. HEAT EXCHANGERS

Perhaps, the most challenging application of the entropy generation minimization philosophy is in the area of heat exchanger design. Here, as in the basic internal and external convection problems illustrated earlier, the heat transfer and fluid friction irreversibilities compete once again towards a unique thermodynamic optimum. The heat exchanger entropy generation minimization problem was first studied in the context of counterflow heat exchangers for gas-to-gas applications [9,10]: however, instead of recounting the optimization work already published, I will use this opportunity to generalize this approach to heat exchangers processing any pure substance, in particular, to heat exchangers processing incompressible fluids and two-phase mixtures.

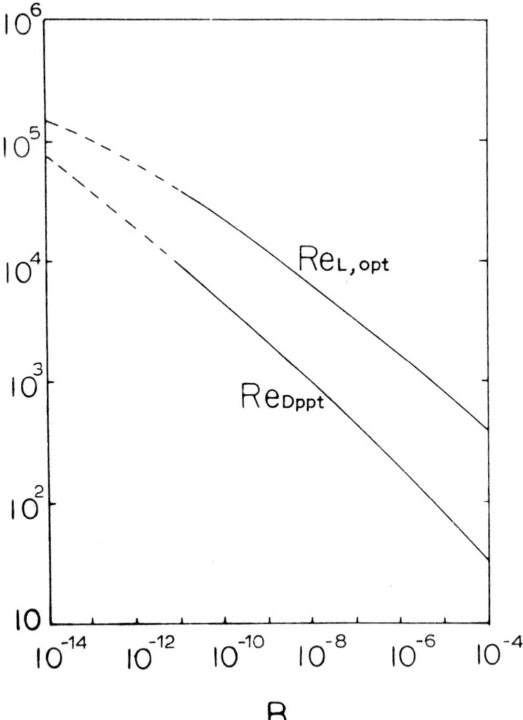

Fig. 6. Optimum Pin Fin Diameter and Optimum Length for Minimum Entropy Generation Rate (M = 100) [8].

Consider the two-stream heat exchanger shown schematically on the left side of Fig. 7, and disregard for a moment the counterflow relationship between the two streams (the counterflow heat exchanger is used as working example later in this section). Regardless of the relative sense of the two streams, the first and second laws of thermodynamics for steady flow are

$$[\dot{m}(h_{out}-h_{in})]_H + [\dot{m}(h_{out}-h_{in})]_C = 0 \qquad (6.1)$$

$$\dot{S}_{gen} = [\dot{m}(s_{out}-s_{in})]_H + [\dot{m}(s_{out}-s_{in})]_C = 0 \qquad (6.1)$$

where subscripts $(\ )_H$ and $(\ )_C$ denote the hot and cold streams, respectively. Equations (6.1) and (6.2) are based on the assumption that the heat exchanger enclosure is adiabatic.

The $\dot{S}_{gen}$ expression (6.2) shows that the entropy generation rate can be calculated as soon as the inlet and outlet thermodynamic states have been determined: according to classical heat exchanger analysis, the peripheral conditions can be determined from the first law (6.1) in conjunction with the rate equations for heat transfer (the effectiveness - N_{tu} relationship) and fluid flow (the pressure drop formula) [5]. The dependence of $\dot{S}_{gen}$ on measurable peripheral conditions such as temperature and pressure becomes clear if we focus on the following special models of pure substance behavior.

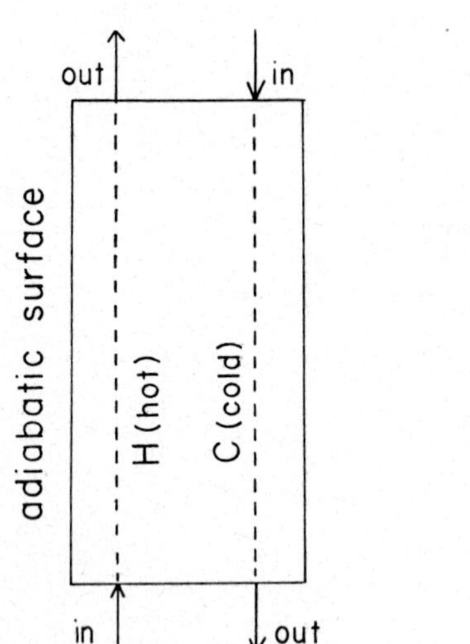

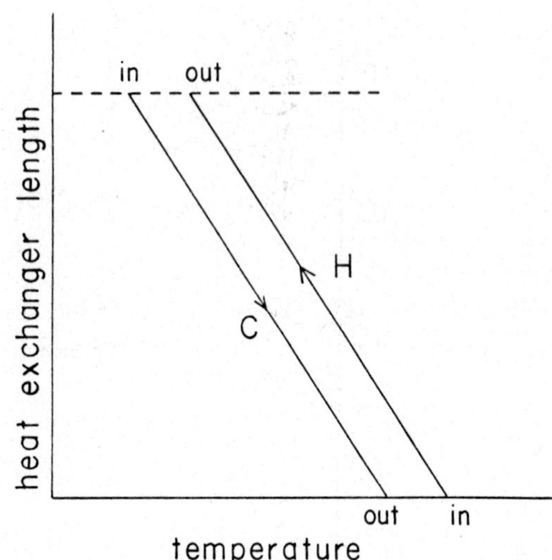

Fig. 7. Left Side: General Two-Stream Heat Exchanger. Right Side: Temperature Distribution Along a Nearly Ideal Balanced Counterflow Heat Exchanger.

6.1. Ideal Gases

Recognizing the ideal gas constitutive relations $dh = c_p dT$ and $ds = c_p\, dT/T - R\, dP/P$, the laws of thermodynamics (6.1) and (6.2) become

$$[\dot{m}c_p(T_{out}-T_{in})]_H + [\dot{m}c_p(T_{out}-T_{in})]_C = 0 \tag{6.3}$$

$$\dot{S}_{gen} = \underbrace{\left(\dot{m}c_p \ln\frac{T_{out}}{T_{in}}\right)_H + \left(\dot{m}c_p \ln\frac{T_{out}}{T_{in}}\right)_C}_{\text{heat transfer irreversibility (due to imperfect stream-to-stream thermal contact)}} + \underbrace{\left(R\frac{\Delta P}{P_{in}}\right)_H + \left(R\frac{\Delta P}{P_{in}}\right)_C}_{\text{fluid flow irreversibility (due to duct friction)}} \geqq 0 \tag{6.4}$$

where the pressure drop on each side of the heat exchanger surface has been assumed much smaller than the local absolute pressure,

$$\Delta P = P_{in} - P_{out} << P_{in}. \tag{6.5}$$

The contribution of the $\dot{S}_{gen}$ expression (6.4) is that it combines into the same objective function the issues of "thermal contact improvement" and "pressure drop reduction": traditionally, these issues have been pursued separately (i.e. as two distinct objectives) in classical heat exchanger design.

The minimization of the entropy generation rate $\dot{S}_{gen}$ relies once again on the proper selection of flow passage geometry: this trade-off is suggested by the similarities between eq. (6.4) and the expressions valid for duct flow, eq. (4.4), and for external flow, eq. (5.4). The first step towards minimizing $\dot{S}_{gen}$ is the selection of relative position between the two streams inside the heat exchanger enclosure shown on the left side of Fig. 7; for example, in a <u>balanced counterflow</u> heat exchanger with enough heat transfer area so that $(T_H-T_C) << T_C$, the entropy generation rate (6.4) becomes [1,9]

$$\frac{\dot{S}_{gen}}{\dot{m}c_p} \cong \left(\frac{\tau^2}{N_{tu}} + \frac{R}{c_p}\frac{\Delta P}{P_{in}}\right)_H + \left(\frac{\tau^2}{N_{tu}} + \frac{R}{c_p}\frac{\Delta P}{P_{in}}\right)_C \tag{6.6}$$

where

$$\tau = \frac{T_{in,H} - T_{in,C}}{(T_{in,H}T_{in,C})^{1/2}} \tag{6.7}$$

$$(N_{tu})_{H,C} = \frac{(hA)_{H,C}}{\dot{m}c_p} \tag{6.8}$$

$$\dot{m}c_p = (\dot{m}c_p)_H = (\dot{m}c_p)_C \tag{6.9}$$

According to expression (6.6), the overall irreversibility of the heat exchanger breaks up into identical terms corresponding to each side of the heat transfer surface. The irreversibility of each side is caused by a combination of heat transfer effect τ^2/N_{tu}, and fluid friction, $(R/c_p)(\Delta P/P_{in})$: this combination can be minimized for one side at a time, by first recognizing the identities

$$N_{tu} = \frac{4L}{D_h} St \tag{6.10}$$

$$\frac{\Delta P}{P_{in}} = f \frac{4L}{D_h} \frac{G^2}{2\rho P_{in}} \tag{6.11}$$

Therefore, the irreversibility contributed by each side of the surface has the same analytical form

$$\left(\frac{\dot{S}_{gen}}{\dot{m}c_p}\right)_{\substack{\text{one side only} \\ \text{(H or C)}}} = \frac{\tau^2}{St}\left(\frac{D_h}{4L}\right) + \frac{R}{c_p} f \frac{G^2}{2\rho P_{in}} \left(\frac{4L}{D_h}\right) \tag{6.12}$$

which shows very clearly the trade-off role played by the slenderness ratio of the heat exchanger passage, $4L/D_h$. Thus, the geometry of the duct can be selected optimally, for one side of the surface at a time: the general results of this thermodynamic optimization, and the impact of possible constraints such as fixed heat transfer surface area or fixed heat exchanger volume, have been developed in

Ref. [1]. For heat exchanger configurations other than "nearly ideal balanced counterflow", the general $\dot{S}_{gen}$ expression (6.4) must be minimized numerically in conjunction with eqs. (6.3), (6.10), (6.11) and the effectiveness - N_{tu} relation appropriate for the flow configuration of interest.

6.2 Incompressible Fluids

Substituting the incompressible-fluid constitutive relations $dh = cdT + \frac{1}{\rho}\, dP$ and $ds = cdT/T$ into the first and second laws, eqs. (6.1) and (6.2), yields

$$\dot{m}_H \left[c(T_{out}-T_{in}) - \frac{\Delta P}{\rho}\right]_H + \dot{m}_C \left[c(T_{out}-T_{in}) - \frac{\Delta P}{\rho}\right]_C = 0 \tag{6.13}$$

$$\dot{S}_{gen} = \left(\dot{m}c \ln \frac{T_{out}}{T_{in}}\right)_H + \left(\dot{m}c \ln \frac{T_{out}}{T_{in}}\right)_C \geq 0 \tag{6.14}$$

The entropy generation rate can be minimized as in the preceding example, by first substituting into eq. (6.14) the constraints placed by the first law, eq. (6.13), the effectiveness - N_{tu} relation and the pressure drop - flowrate relation. For example, in a balanced counterflow configuration in the limit $(T_H-T_C) << T_C$, eq. (6.14) reduces to

$$\frac{\dot{S}_{gen}}{\dot{m}c} = \left(\frac{\tau^2}{N_{tu}} + \frac{\Delta P}{\rho c T_{in}}\right)_H + \left(\frac{\tau^2}{N_{tu}} + \frac{\Delta P}{\rho c T_{in}}\right)_C \tag{6.15}$$

This expression is of the same form as eq. (6.6), as $\rho c T_{in}$ now replaces $c_p P_{in}/R$. The conclusions reached earlier in the case of gas-to-gas balanced counterflow, apply unchanged to the case of incompressible fluids. Therefore, the geometry of each side of the heat transfer surface can be selected optimally by properly balancing the heat transfer and fluid flow irreversibilities of each side.

6.3. Two-Phase Mixtures

Consider the steady flow of a two-phase mixture through a duct in thermal contact with a heat reservoir of temperature T_0 (Fig. 8). To understand the functioning of this heat exchanger it helps to think of the condenser in a Rankine-cycle heat engine, where T_0 is the absolute temperature of the ambient; however, as shown later in this subsection, the result of the entropy generation analysis is quite general. The first and second laws of thermodynamics, written for the system defined by the dashed-line boundary in Fig. 8, state that

$$\dot{m}\,(h_{in}-h_{out}) - \dot{Q} = 0 \tag{6.16}$$

$$\dot{S}_{gen} = \dot{m}\,(s_{out}-s_{in}) + \frac{\dot{Q}}{T_0} \geq 0 \tag{6.17}$$

From a heat-exchanger-design standpoint, we are interested in how the pressure drop ($\Delta P = P_{in}-P_{out}$) and the fluid-ambient temperature difference ($T_{in}-T_0$) affect the overall irreversibility of the heat exchanger ($\dot{S}_{gen}$). In order to express $\dot{S}_{gen}$ in terms of ΔP and $T_{in}-T_0$, we consult the T-s diagram shown on the right side of Fig. 8 and observe that we can write

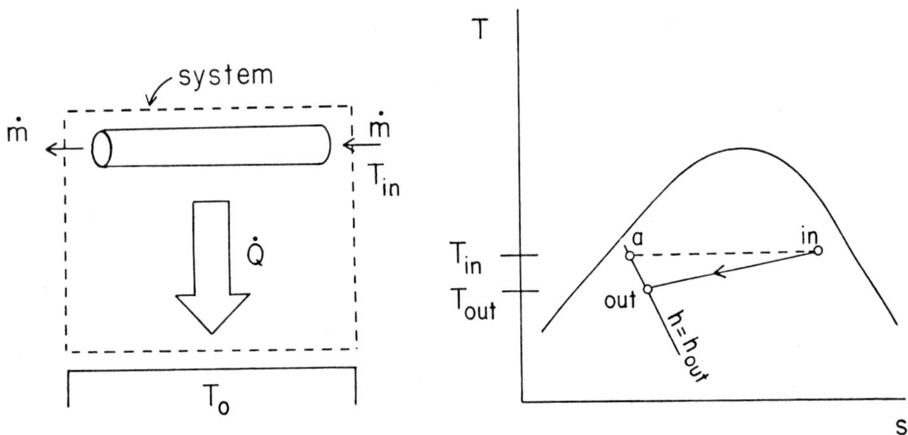

Fig. 8. Entropy Generation Analysis of Two-Phase Flow Through a Heat Exchanger Duct

$$s_{out} - s_{in} = (s_a - s_{in}) + (s_{out} - s_a) \tag{6.18}$$

where state "a" is defined as

$$T_a = T_{in}, \text{ and } h_a = h_{out}. \tag{6.19}$$

In other words, state "a" corresponds to the outlet state "out" in the theoretical limit where the pressure drop ΔP is zero. We note further that

$$T_{in}\,(s_a - s_{in}) = h_a - h_{in}$$

$$= h_{out} - h_{in} \tag{6.20}$$

Combining this result with eqs. (6.16)-(6.18), we obtain

$$\dot{S}_{gen} = \underbrace{\dot{m}\,(h_{in} - h_{out})}_{\dot{Q}}\left(\frac{1}{T_0} - \frac{1}{T_{in}}\right) + \dot{m}\,(s_{out} - s_a) \tag{6.21}$$

where, quite visibly, the first term on the right-hand side represents the contribution due to imperfect thermal contact, while the second term represents the irreversibility associated with the pressure drop (see the T-s diagram of Fig. 8). A more meaningful expression for the second term can be written by invoking once more the canonical relation $dh = Tds + \frac{1}{\rho}dP$, and recognizing that from "a" to "out" the enthalpy does not change. Therefore, for sufficiently small pressure drops, i.e. for $(T_{in}-T_{out}) \ll T_{in}$, we have

$$s_{out} - s_a \cong \frac{\Delta P}{\rho_{avg} T_{in}} \tag{6.22}$$

In the same limit, the entropy generation rate (6.21) becomes

$$\dot{S}_{gen} = \underbrace{\dot{Q}\,\frac{T_{in} - T_0}{T_{in}T_0}}_{\text{heat transfer irreversibility}} + \underbrace{\frac{\dot{m}\Delta P}{\rho_{avg} T_{in}}}_{\text{fluid flow irreversibility}} \tag{6.23}$$

Repeating the above analysis for a duct of length dx, where x is measured in the direction of the flowrate $\dot{m}$, we obtain the per-unit-length result

$$\frac{d\dot{S}_{gen}}{dx} = \frac{T-T_0}{T^2}\frac{d\dot{Q}}{dx} + \frac{\dot{m}}{\rho T}\left(-\frac{dP}{dx}\right) \tag{6.24}$$

where ρ and T are local bulk properties of the stream. Note that eq. (6.24) contains also the assumption that $(T-T_0) \ll T$. The structure of eq. (6.24) is the same as that of the entropy generation formulas encountered earlier. This means that by combining eq. (6.24) with appropriate correlations for heat transfer coefficient and friction factor, it is possible to select the duct diameter such that $d\dot{S}_{gen}/dx$ is minimum.

A final observation concerns the meaning of the temperature difference $T-T_0$ appearing in eq. (6.24). If the duct is surrounded by still air, then

$$T - T_0 = (T-T_w) + (T_w-T_0) \tag{6.25}$$

where T_w is the absolute temperature of the duct wall. In this case, the local entropy generation rate becomes

$$\frac{d\dot{S}_{gen}}{dx} = \left(\frac{d\dot{Q}/dx}{T}\right)^2 \left[\frac{1}{(ph)_i} + \frac{1}{(ph)_o}\right] + \frac{\dot{m}}{\rho T}\left(-\frac{dP}{dx}\right) \tag{6.26}$$

where p and h denote wetted perimeter and heat transfer coefficient, and subscripts $(\ )_i$ and $(\ )_o$ stand for the inside and outside of the duct, respectively. In writing eq. (6.26) it is assumed that the flow on the outside of the duct is driven by buoyancy (natural convection); thus, the entropy generation rate (6.26) is due to three effects, the imperfect thermal contact between fluid and wall, the imperfect thermal contact between wall and ambient, and the flow with friction through the duct.

If the condensation tube of Fig. 8 is surrounded not by a fluid reservoir, but by an evaporating stream at a lower temperature, then the entropy generation rate for this two-stream heat exchanger is

$$\frac{d\dot{S}_{gen}}{dx} = \left[\frac{1}{T^2 ph}\left(\frac{d\dot{Q}}{dx}\right)^2 + \frac{\dot{m}}{\rho T}\left(-\frac{dP}{dx}\right)\right]_H + \left[\frac{1}{T^2 ph}\left(\frac{d\dot{Q}}{dx}\right)^2 + \frac{\dot{m}}{\rho T}\left(-\frac{dP}{dx}\right)\right]_C \tag{6.27}$$

In this expression subscripts $(\ )_H$ and $(\ )_C$ stand for the condensing (hot) and evaporating (cold) sides of the heat exchanger surface, respectively. Equation (6.27) follows directly from eq. (6.24), by writing eq. (6.24) for each side of the surface (T_0 being the wall temperature in this case) and then adding the two expressions side by side. Note also that the first law of thermodynamics requires $(d\dot{Q}/dx)_H = (d\dot{Q}/dx)_C$. The competition between heat transfer and fluid flow irreversibilities, hence, the ability to achieve a thermodynamic optimum based on the proper selection of duct geometry, is evident in both terms on the right-hand side of eq. (6.27).

7. CONCLUSIONS

The entropy generation analysis of heat exchange devices provides an effective, often straightforward, approach to the optimum thermodynamic design of such devices. To summarize, in this lecture I showed that:

I. The destruction of available work in a heat exchange apparatus proceeds at a rate directly proportional to the rate of entropy generation of the apparatus.

II. Entropy is always generated by processes such as heat transfer across a finite temperature difference, and fluid flow against friction.

III. It is important to minimize the heat transfer and fluid flow irreversibilities together, as only a critical balance between them insures the minimum entropy generation rate possible.

IV. It is possible to optimally select the geometry of a heat transfer device so that its entropy generation rate is forever minimum.

The usefulness and general applicability of the entropy generation minimization philosophy is now being realized, as this methodology is being applied not only in the field of Heat Transfer, but also in Fluid Mechanics, Solar Energy, Cryogenics and Energy Policy: the recent advances in these fields are also reviewed in Ref. [1]. For a pedagogical standpoint, the identification, discussion and minimization of entropy generation in heat transfer engineering offers a unique opportunity to finally bridge the gap between three isolated cornerstones of engineering education: fluid mechanics, thermodynamics and heat transfer.

ACKNOWLEDGEMENTS

The work reviewed in this lecture has been sponsored by the Office of Naval Research. The author wishes to acknowledge the encouragement received early in this line of thinking from Mr. M.K. Ellingsworth, from the Office of Naval Research.

REFERENCES

1. Bejan, A., Entropy Generation Through Heat and Fluid Flow, Wiley, New York, 1982.

2. Bejan, A., "Second-Law Analysis in Heat Transfer and Thermal Design", Advances in Heat Transfer, Vol. 15, 1982, pp. 1-58.

3. Cravalho, E.G., and Smith, J.L., Jr., Engineering Thermodynamics, Pitman, Boston, 1981, chapter 6.

4. Bergles, A.E., "Principles of Heat Transfer Augmentation", in Heat Exchangers, Thermal-Hydraulic Fundamentals and Design, Hemisphere, Washington, 1981, pp. 819-842 and 857-881.

5. Kays, W.M., and London, A.L., Compact Heat Exchangers, McGraw-Hill, New York, 1964.

6. Bejan, A., Discussion of a previously published paper, Journal of Heat Transfer, Vol. 102, 1980, pp. 586, 587.

7. Poulikakos, D., "Fin Geometry for Minimum Entropy Generation", M.S. Thesis, Dept. of Mechanical Eng., University of Colorado, Boulder, December 1980.

8. Poulikakos, D., and Bejan, A., "Fin Geometry for Minimum Entropy Generation in Forced Convection", Journal of Heat Transfer, Vol. 104, 1982, pp. 616-623.

9. Bejan, A., "The Concept of Irreversibility in Heat Exchanger Design: Counterflow Heat Exchangers for Gas-to-Gas Applications", Journal of Heat Transfer, Vol. 99, 1977, pp. 374-380.

10. Sarangi, S., and Chowdhury, K.,"On the Generation of Entropy in a Counterflow Heat Exchanger," Cryogenics, Vol. 22, 1982, pp. 63-67.

LIST OF SYMBOLS

A heat transfer surface area

B dimensionless group, eq. (5.7)

B_0 duty parameter, eq. (4.7)

c specific heat of incompressible fluid

c_p specific heat at constant pressure

C_D drag coefficient

D tube diameter

D_h hydraulic diameter

f fanning friction factor

F_D drag force

G mass velocity

h specific enthalpy or heat transfer coefficient

k thermal conductivity

L	length
$\dot{m}$, $\dot{M}$	mass flowrates
Nu	Nusselt number
N_S	entropy generation number
N_{tu}	number of heat transfer units
p	wetted perimeter
P	pressure
ΔP	pressure drop
Pr	Prandtl number
q'	heat transfer rate per unit length
q''	heat flux
$\dot{Q}$	heat transfer rate
R	ideal gas constant
$Re_{D,L}$	Reynolds number based on D or L
s, S	entropy change (property)
$\dot{S}_{gen}$	entropy generation rate (non-property)
$S'_{\Delta P}$	entropy generation rate due to duct friction
$S'_{\Delta T}$	entropy generation rate due to imperfect thermal contact
St	Stanton number
t	time
T	absolute temperature
ΔT	temperature difference
u, U	internal energy
v	specific volume, $1/\rho$
V	velocity
$\dot{W}$	work transfer rate
$\dot{W}_{lost}$	lost work rate
$\dot{W}_{max}$	maximum work transfer rate
x	coordinate in the direction of flow

λ	fin material conductivity
ν	viscosity
μ	kinematic viscosity
ρ	density
τ	dimensionless temperature difference, eq. (6.7)
ϕ	irreversibility distribution ratio, $S'_{\Delta P}/S'_{\Delta T}$
$(\ )_{in}$	inlet conditions
$(\ )_{out}$	outlet conditions
$(\ )_B$	fin base conditions
$(\ )_\infty$	fluid reservior conditions
$(\ )_C$	cold
$(\ )_H$	hot, or high temperature
$(\ )_L$	low temperature

Multi-Phase Flow and Heat Transfer III. Part A: Fundamentals
edited by T.N. Veziroğlu and A.E. Bergles
Elsevier Science Publishers B.V., Amsterdam, 1984 — Printed in The Netherlands

ON THE FORMULATION OF CONSERVATION, BALANCE AND CONSTITUTIVE EQUATIONS FOR MULTIPHASE FLOWS

Flavio Dobran
Department of Mechanical Engineering
Stevens Institute of Technology
Hoboken, New Jersey 07030, U.S.A.

ABSTRACT

The formulation of conservation, balance and constitutive equations for multiphase flows requires the considerations of the levels of description of the theory and of the existence and consistency of the theoretical structure in a physical world. A theory of multiphase mixtures should be formulated on a level that avoids the solution of the multiboundary problem, and it should be consistent with proper and rigorous averaging procedures applied to the macroscopic field equations of each phase. The averaging or reductionistic procedure cannot, however, yield a rigorous theory, for, eventually, this must be abandoned in order to define rigorously kinematic concepts. The degree to which a reductionistic or a holistic theoretical structure is accepted plays a crucial role in the development of constitutive equations, and both of these approaches are discussed in this paper. Based on these investigations, results for constitutive equations are reviewed, and it is shown that although the holistic or continuum mechanics approach yields a self-consistent theory and many useful results, it is not very clear that this procedure will yield a theory of multiphase mixtures that is fully consistent with the physical world.

1. INTRODUCTION

While a considerable advance has been made in the development of useful models for two-phase flows, this development has not been as rapid for flows involving three or more phases. The main reasons for this are: 1) two-phase flows are easier to study analytically and experimentally, and 2) the practical need to understand two-phase flows has been greater. Modeling of transient and nonequilibrium flow phenomena places severe demands on any model of multiphase flows, and for this reason it is, therefore, necessary that the conservation, balance and constitutive equations have a strong foundation: pragmatic in order to conform with the physical reality and idealistic in order to possess a sound theoretical structure. The degree of pragmatism or idealism in a model is always disputable which, of course, must be in order to advance the present state of knowledge.

The analysis of multiphase flows is difficult due to their finite interfacial area. Most, if not all, of the present models reflect a particular special case of this area or flow regime. Flow regime modeling is very useful for steady flows; however, many practical situations involve unsteady multiphase flows for which the current modeling practices are highly unsatisfactory due to the inadequate knowledge of the flow regime transition physics. Although it is conceptually possible to describe a multiphase mixture by the well-established macroscopic conservation and balance equations of each phase separately and by the interfacial boundary conditions, it is, however, not possible practically to follow the usually very large number of interfaces, and, as a consequence, the analytic problem becomes intractable. The recognition of this fact has led to the diverse modeling practices - all of which are based on a continuum approach. The conservation and balance equations of multiphase flows constructed in this manner are either postulated [1-2] or they are obtained by an application of suitable averaging of the macroscopic field equations of each phase [3-8]. Both approaches have deficiencies: the first, because it lacks a strong physical justification, and the second, because it lacks a reconciliation between different levels of description. The above-mentioned idealistic or pragmatic approaches which are followed in the construction of conservation and balance equations for multiphase flows also play a crucial role in the subsequent development of constitutive equations, as discussed below.

In view of the fundamental issues which surround the formulation of conservation, balance and constitutive equations for multiphase mixtures, it is important that these issues be discussed in order that future theoretical developments be built on a stronger foundation. This discussion is important since all too often many proposed models tend to become personalized theories adopted by those who propose them and by the users of these theories who become familiar with them. Wallis [9], in a recent review article, concludes that "The ad hoc approach is likely to retain its preeminence as the dominant method of analysis for the foreseeable future." This conclusion, while being pessimistic, is nevertheless well-founded since it is based on extensive experiences in two-phase flow modeling. With a considerable amount of knowledge already existing and with the availability of computational techniques and advanced instrumentation, the future of multiphase flow modeling appears more optimistic to this author.

The formulation of conservation and balance equations for multiphase flows is discussed in Section 2 where it is shown that despite a considerable progress many problems still remain. In Section 3, the discussion is carried out on the formulation of constitutive equations where different methods of analysis are reviewed. It is shown there that it is not very clear which approach to utilize but that the holistic or continuum mechanics procedure provides very useful results.

2. ON THE FORMULATION OF CONSERVATION AND BALANCE EQUATIONS FOR MULTIPHASE FLOWS

All the formulations of conservation and balance equations for multiphase flows utilize continuum modeling which are based on either of the two following approaches:

a) Extension of single-phase multicomponent mixture equations to multiphase mixtures.

b) Averaging of the instantaneous field equations of each phase over the appropriate time or space segments.

In the first approach, the theory of single-phase multicomponent mixtures [10,11] is extended to multiphase mixtures [1,12]. The fundamental assumptions in this theory are the coexistence of all phases at each point in space and that each coexisting phasic continuum can be assigned unique properties such as density, velocity, energy, etc. Additional important characteristics of this theory are as follows:

1) No attempt is made in the theory to justify the field equations from any molecular or local macroscopic level for each phase.

2) The conservation and balance equations for the mixture as a whole do not allow for source terms.

3) In the absence of external torques or couple stresses, the stress tensor for the mixture as a whole is symmetric.

4) The entropy equation for the mixture as a whole is assumed to be valid unless the nonequality of entropy and heat fluxes is assumed in which case an entropy equation for each constituent in the mixture is also postulated to be valid.

While in the theory of single-phase multicomponent mixtures the last three conditions given above appear to be justifiable in so far as no contradictions with classical equilibrium thermodynamics have been found, it is not clear that these conditions should also be valid for a multiphase mixture - at least as long as we accept the validity of averaging procedures discussed below.

The averaging approach followed in the construction of conservation and balance equations for multiphase mixtures is based on the concept that a proper and clear way to obtain multiphase field equations is to perform averaging of the local macroscopic field equations for each phase over a suitable time or space domain. Thus, Drew [3] carried out time and volume averaging, Ishii [4] performed time averaging, Delhaye and Achard [5] derived one-dimensional volume-, time-, area- and space-averaged two-phase flow equations, Dobran [7] derived multidimensional volume-averaged equations, Bataille and Kestin [6] and Marle [8] utilized the theory of distributions to obtain volume and volume-time-averaged multiphase field equations, respectively.

With all these averaging procedures proliferating through the literature, some doubts arise whether any of them should be accepted as yielding the correct form of multiphase field equations. While ad hoc models can be justified on the basis of a particular averaging procedure, it is important to note that the objective of averaging is to produce a rigorous set of conservation and balance equations for multiphase mixtures. This being the case, then, is there any relative merit between different averaging approaches? In the following, we shall attempt to reconcile the relative advantages and disadvantages of various averaging formulations.

The time-, area- and segment-averaged equations have the disadvantage of being singular under certain flow conditions (for the case of time-averaged formulation when the interface becomes stationary in the flow field). This disadvantage is, clearly, bothersome and to smooth out the singularities, further averaging is usually performed. The smoothened field equations are all of space-time variety and immediately bring about the following concerns:

1) The field equations do not lend themselves to the investigation of constitutive equations by utilizing the modern foundations of continuum mechanics since the principle of objectivity or material frame invariance cannot be <u>consistently</u> utilized on the time-averaged equations unless they are also assumed to be ergodic.

2) Averaging over a restricted space domain to obtain one-dimensional field equations leads to difficulty when studying the constitutive

equations since the conclusions about the invariance properties can only be obtained from the general three-dimensional equations and not from any restricted class of them.

3) The time-averaged turbulent field equations are different from the space-averaged equations in the sense that the latter equations contain additional turbulent correlations, as discussed below.

The nature and subtleties involved in different averaging procedures discussed above can be expounded by selecting one such procedure for discussion and then by pointing out the advantages and disadvantages of this procedure and its relation to other averaging approaches. For this purpose, the multidimensional volume averaging procedure developed by Dobran [7] is selected.

The formulation of conservation and balance equations for multiphase flows utilizing the volume averaging procedure follows by considering a multiphase flow field such as that shown in Fig. 1. The volume pertaining to phase α, $\alpha = 1, \ldots, \gamma$, in the arbitrary and fixed averaging volume V is $V_\alpha = \sum_\delta V_{\alpha\delta}$ such that $V = \sum_{\alpha=1}^{\gamma} V_\alpha$. For each phase α and volume $V_{\alpha\delta}$ the macroscopic field equations representing the conservation of mass and balance of linear momentum, angular momentum, energy and entropy can be written as follows [7,13]:

$$\frac{\partial}{\partial t}(\rho_{\alpha\delta}\Psi_{\alpha\delta}) + \nabla\cdot(\rho_{\alpha\delta}\Psi_{\alpha\delta}\,\mathbf{v}_{\alpha\delta}) + \nabla\cdot J_{\alpha\delta} - \rho_{\alpha\delta}\Phi_{\alpha\delta} - \rho_{\alpha\delta}\,B_{\alpha\delta} = 0 \tag{1}$$

where $\Psi_{\alpha\delta}$, $J_{\alpha\delta}$, $\Phi_{\alpha\delta}$ and $B_{\alpha\delta}$ are given in Table 1. At the interface between phases α and β we have the jump conditions [7,13]:

$$(m_{\alpha\delta}\,\Psi_{\alpha\delta} + J_{\alpha\delta}\,\mathbf{n}_{\alpha\delta}) + (m_{\beta\eta}\Psi_{\beta\eta} + J_{\beta\eta}\,\mathbf{n}_{\beta\eta}) = \Delta_{\alpha\delta} \tag{2}$$

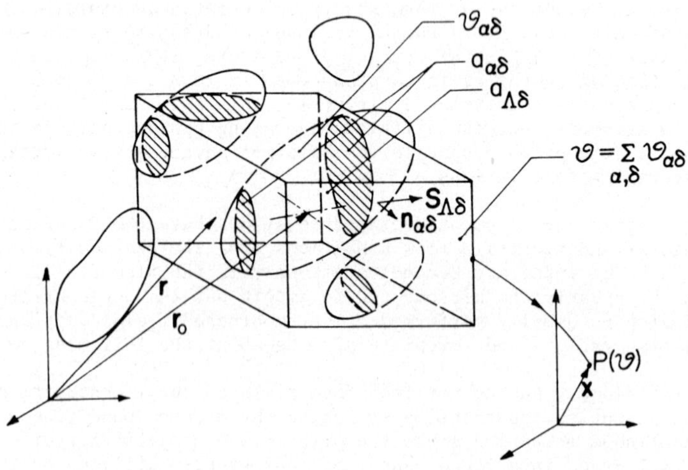

Figure 1. Definition of the Averaging Volume in a Multiphase Flow Field

CONSERVATION OF	$\Psi_{\alpha\delta}$	$J_{\alpha\delta}$	$\Phi_{\alpha\delta}$	$B_{\alpha\delta}$	$\Delta_{\alpha\delta}$
MASS	1	0	0	0	0
LINEAR MOMENTUM	$\mathbf{v}_{\alpha\delta}$	$-\mathbf{T}_{\alpha\delta}$	$\mathbf{b}_{\alpha\delta}$	$\mathbf{0}$	$(\boldsymbol{\Delta}_m)_{\alpha\delta} = (2H\nu\mathbf{n} + \nabla_s\nu)_{\alpha\delta}$
ANGULAR MOMENTUM	$\mathbf{r}\wedge\mathbf{v}_{\alpha\delta}$	$-\mathbf{r}\wedge\mathbf{T}_{\alpha\delta}$	$\mathbf{r}\wedge\mathbf{b}_{\alpha\delta}$	$\mathbf{0}$	$\mathbf{r}\wedge(\boldsymbol{\Delta}_m)_{\alpha\delta}$
ENERGY	$\varepsilon_{\alpha\delta}+\frac{v_{\alpha\delta}^2}{2}$	$\mathbf{q}_{\alpha\delta}-\mathbf{T}_{\alpha\delta}^T\mathbf{v}_{\alpha\delta}$	$\mathbf{b}_{\alpha\delta}\cdot\mathbf{v}_{\alpha\delta}$	$r_{\alpha\delta}$	$(\Delta_\epsilon)_{\alpha\delta} = (2H\nu\mathbf{n}\cdot\mathbf{S}+\nabla_s\nu\cdot\mathbf{S}+\nu\nabla_s\cdot\mathbf{S})_{\alpha\delta}$
ENTROPY	$s_{\alpha\delta}$	$\frac{\mathbf{h}_{\alpha\delta}}{\Theta_{\alpha\delta}}$	$\frac{r_{\alpha\delta}}{\Theta_{\alpha\delta}}$	$\xi_{\alpha\delta}\geq 0$	$(\Delta_s)_{\alpha\delta}\geq 0$

Table 1. Coefficients of the Conservation Equations (1) and (2)

where

$$m_{\alpha\delta} = \rho_{\alpha\delta}(\mathbf{v}_{\alpha\delta} - \mathbf{S}_{\Lambda\delta})\cdot\mathbf{n}_{\alpha\delta} \quad . \tag{3}$$

$\mathbf{S}_{\Lambda\delta}$ is the interface velocity, $\mathbf{n}_{\alpha\delta}$ is the unit normal vector which points outwards from the volume $V_{\alpha\delta}$, $m_{\alpha\delta}$ is the interphase mass transfer rate which arises due to the phase change, and $\Delta_{\alpha\delta}$ is the surface tension term (or some other interfacial source density) and is defined in Table 1. H is the mean interface curvature, ∇_S is the surface gradient vector and ν is the surface tension coefficient. Clearly, $\alpha \neq \beta$, $\mathbf{n}_{\alpha\delta} = -\mathbf{n}_{\beta\eta}$ and $\mathbf{S}_{\Lambda\delta} = \mathbf{S}_{\beta\eta}$. Furthermore in Table 1, $\mathbf{v}_{\alpha\delta}$ is the velocity, $\rho_{\alpha\delta}$ is the mass density, $\mathbf{T}_{\alpha\delta}$ is the stress tensor, $\mathbf{b}_{\alpha\delta}$ is the external body force per unit mass, $\varepsilon_{\alpha\delta}$ is the internal energy per unit mass, $\mathbf{q}_{\alpha\delta}$ is the heat flux vector, $s_{\alpha\delta}$ is the entropy per unit mass, $\mathbf{h}_{\alpha\delta}$ is the entropy flux vector, $\Theta_{\alpha\delta}$ is the absolute temperature, $r_{\alpha\delta}$ is the heat generation rate per unit volume, and $\xi_{\alpha\delta}$ and Δ_s are the entropy production rates in the continuum and at the interface, respectively.

The volume averaging process consists of volume averaging of Eq. (1) over the portion of volume V which the phase α occupies at time t, i.e.

$$\sum_\delta \int_{V_{\alpha\delta}} [\text{Eq. (1)}]dV = 0 \quad . \tag{4}$$

Utilizing Leibnitz's rule and Divergence Theorem [7], Eq. (4) is reduced to:

$$\frac{\partial}{\partial t} V_\alpha\langle\rho_\alpha\Psi_\alpha\rangle + \nabla\cdot V_\alpha\langle\rho_\alpha\Psi_\alpha\,\mathbf{v}_\alpha\rangle + \nabla\cdot V_\alpha\langle J_\alpha\rangle - V_\alpha\langle\rho_\alpha\Phi_\alpha\rangle - V_\alpha\langle\rho_\alpha B_\alpha\rangle =$$

$$- \sum_\delta \int_{a_{\Lambda\delta}} (m_{\alpha\delta}\,\Psi_{\alpha\delta} + J_{\alpha\delta}\,\mathbf{n}_{\alpha\delta})da, \tag{5}$$

where the volume-averaged quantities are defined as follows:

$$\langle F_\alpha\rangle \equiv \frac{1}{V_\alpha}\sum_\delta \int_{V_{\alpha\delta}} F_{\alpha\delta}\;dV \; . \tag{6}$$

Equation (5) is the general transport equation of phase α and will be identified as the macro-macroscopic transport equation of phase α as opposed to Eq. (1) which is the macroscopic transport equation of phase α. The volume averaging procedure constructed above induces a mapping such that Eq. (5) represents the superposition of γ continua, i.e., at each point of space and at each instant of time a continuum of phase α can be identified and in this report the theory of multiphase mixtures constructed in this manner is identical to the theory of single-phase multicomponent mixtures.

All space and time averaging procedures lead to an equation similar to Eq. (5); the difference between them, however, is in the interpretation of the averaged quantities. Furthermore, Eq. (5) is not useful until a connection is made between the variables defined in the macroscopic field Eq. (1) (ρ_α, Ψ_α, $\mathbf{v}_\alpha$, etc.) and the averaged variables in the macro-macroscopic field Eq. (5). In the next step of the analysis, therefore, two kinds of averaged variables are introduced. These are

i) The density-weighted averages, $\tilde{H}_\alpha$

$$\tilde{H}_\alpha \equiv \frac{\langle \rho_\alpha H_\alpha \rangle}{\langle \rho_\alpha \rangle} \tag{7}$$

ii) The phase averages, $\overline{H}_\alpha$ and $\overline{\overline{H}}_\alpha$

$$\overline{H}_\alpha \equiv \frac{V_\alpha}{V} \langle H_\alpha \rangle \tag{8}$$

$$\overline{\overline{H}}_\alpha \equiv \langle H_\alpha \rangle \tag{9}$$

The density-weighted variables are defined for all those macroscopic variables on a unit mass basis. In addition, the macro-macroscopic velocity of phase α, $\tilde{\mathbf{v}}_\alpha$, is also defined as the density-weighted average for the reasons that a finite size instrument can only measure this velocity and, when substituting for Ψ_α, J_α, Φ_α and B_α from Table 1 for the conservation of mass in Eq. (5), the conservation of mass equation of phase α,

$$\frac{\partial}{\partial t}\, \overline{\rho}_\alpha + \nabla \cdot \overline{\rho}_\alpha \, \tilde{\mathbf{v}}_\alpha = -\sum_\delta \int_{a_{\Lambda\delta}} \dot{m}_{\alpha\delta} \, da \quad , \tag{10}$$

becomes similar in appearance to the macroscopic continuity equation of the same phase. In the time-averaged formulation, the velocity $\tilde{\mathbf{v}}_\alpha$ is defined as the time-averaged velocity and in the volume-time-averaged formulation it is defined over a space-time domain.

All space and time averaging approaches lead to the conservation of mass equation that is similar to Eq. (10); however, the balance equations of momentum, energy and entropy differ from one formulation to another in the interpretation of the following term:

$$V_\alpha \langle \rho_\alpha \Psi_\alpha \mathbf{v}_\alpha \rangle \quad . \tag{11}$$

In the time-averaged formulation, the macroscopic field variable Ψ_α is decomposed into a density-weighted time average $\tilde{\Psi}_\alpha$ and into a fluctuating or turbulent field Ψ_α', i.e.

$$\Psi_\alpha = \tilde{\Psi}_\alpha + \Psi_\alpha' \quad , \tag{12}$$

whereas in the volume and time-averaged formulation $\tilde{\Psi}_\alpha$ in Eq. (12) must be interpreted as the density-weighted space-time average and Ψ'_α as the space time fluctuation. In the volume-averaged formulation, no splitting up of the field variables as defined by Eq. (12) is performed; instead, the covariance coefficients are introduced as the differences between the average of the product of macroscopic variables and the product of the macro-macroscopic field varibles, i.e.,

$$V_\alpha \langle \rho_\alpha \Psi_\alpha \mathbf{v}_\alpha \rangle - V \bar{\rho}_\alpha \tilde{\Psi}_\alpha \tilde{\mathbf{v}}_\alpha \quad . \tag{13}$$

From the above discussion, it is clear that depending on the averaging procedure used in the formulation of conservation equation of mass and balance equations of momentum, energy and entropy for multiphase flows, the resulting field equations can be quite different. Using Eq. (12) in the time- and space and time-averaged formulations gives rise to the macro-macroscopic field equations in terms of mean and fluctuating quantities where the latter quantities are interpreted as the turbulent contributions. The volume-averaged macro-macroscopic field equations [7] contain instead the covariance coefficients expressed by Eq. (13) where they are interpreted as the nonlocal effect in the theory. This interpretation arises from the effect of the size of averaging volume V on these coefficients. When the multiphase mixture becomes dispersed the effect of nonlocality disappears.

Another significant difference between the time-averaged and volume-averaged field equations is in the formulation of turbulence characteristics. Time averaging of the volume-averaged linear momentum equation, for example, gives rise to velocity-velocity, partial density-velocity, partial density-velocity-velocity and partial density-pressure correlations while the time-averaged linear momentum equation gives rise to the velocity-velocity correlations [4] only.

In view of the discussion above, it is my conclusion that the volume-averaged conservation and balance equations for multiphase mixtures have the most desirable properties to motivate the construction of a theory of multiphase flows since:

1. The equations are not singular.
2. The equations are amenable to the study of invariance properties using the principle of objectivity or frame indifference without giving rise to any inconsistencies.
3. The turbulent field equations are richer.

In the construction of a theory of multiphase mixtures utilizing the averaging approaches described above, one follows a reductionistic or pragmatic approach. This means that the theory should be formulated by piecing together its fundamental building blocks (the macroscopic field equations of each phase) with the pieces which describe how these blocks should be put together (the macroscopic interface or boundary conditions). Clearly, different averaging procedures lead to different interpretations of the resulting field equations, as discussed above, and some procedures are more desirable than others (for example, the volume averaging) since the multiphase equations constructed in this way insure the proper existence and consistency properties of the theory. This, however, should not be viewed necessarily as a proper way of constructing a rigorous theory, for if this were the case, then a reconciliation would have to be made between the macroscopic and macro-macroscopic field variables. How this can be done in a physically convincing manner is at the present time open to speculations.

The approach that has been followed in constructing a theory of multiphase mixtures in Reference 7 involves utilizing the volume averaging procedure not in the hope of deriving the multiphase field equations but to motivate the construction of the theory since, as discussed above, the volume-averaged equations have more desirable properties than the equations from other averaging approaches. Having established the macro-macroscopic transport Eq. (5) and definitions expressed by Eqs. (6)-(9), it is assumed that to each coexisting or superimposed continuum of phase α can be assigned particles X_α and configurations $\boldsymbol{\chi}_\alpha$ of these particles at time t such that at the current space position $\mathbf{x}$ we can write

$$\mathbf{x} \equiv \boldsymbol{\chi}_\alpha(X_\alpha, t) \quad . \tag{14}$$

Furthermore, to each particle of phase α can be assigned a reference configuration $\boldsymbol{\kappa}_\alpha$ at some time $to < t$ such that its reference position is expressed by the following equation:

$$\mathbf{X}_\alpha \equiv \boldsymbol{\kappa}_\alpha (X_\alpha, to) \quad . \tag{15}$$

Combining Eqs. (14) and (15) we obtain:

$$\mathbf{x} = \boldsymbol{\chi}_\alpha(\boldsymbol{\kappa}_\alpha^{-1}(\mathbf{X}_\alpha, to), t) \equiv \boldsymbol{\chi}_{\alpha\boldsymbol{\kappa}} (\mathbf{X}_\alpha, t) \quad , \tag{16}$$

where $\boldsymbol{\chi}_{\alpha\boldsymbol{\kappa}}$ is the deformation function of a particle of phase α. The velocity and acceleration of particle X_α follow by differentiating Eq. (16) with respect to time while keeping the identity of particle constant. In the next crucial step of the analysis, the velocity and acceleration obtained from Eq. (16) are identified with the density-weighted velocity and acceleration defined by Eq. (7), i.e.,

$$\tilde{\mathbf{v}}_\alpha \equiv \frac{\partial \boldsymbol{\chi}_{\alpha\boldsymbol{\kappa}}(\mathbf{X}_\alpha, t)}{\partial t} \tag{17}$$

$$\grave{\tilde{\mathbf{v}}}_\alpha \equiv \frac{\partial^2 \boldsymbol{\chi}_{\alpha\boldsymbol{\kappa}} (\mathbf{X}_\alpha, t)}{\partial t^2} \tag{18}$$

It should be noted in Eq. (18) and in the equations that follow that a backward prime affixed to the subscripted symbols with α indicates the material derivative 'following' the particle of the phase α.

The density of a multiphase mixture, ρ, and the center of mass velocity, $\mathbf{v}$, are defined respectively as

$$\rho \equiv \sum_{\alpha=1}^{\gamma} \bar{\rho}_\alpha \quad , \quad \rho \mathbf{v} \equiv \sum_{\alpha=1}^{\gamma} \bar{\rho}_\alpha \tilde{\mathbf{v}}_\alpha \tag{19}$$

while the diffusion velocity of phase α, $\mathbf{u}_\alpha$, is defined by

$$\mathbf{u}_\alpha \equiv \tilde{\mathbf{v}}_\alpha - \mathbf{v} \, . \tag{20}$$

If $\tilde{\Gamma}_\alpha$ and Γ, where $\rho\Gamma \equiv \sum_{\alpha=1}^{\gamma} \bar{\rho}_\alpha \tilde{\Gamma}_\alpha$, are differentiable functions of $\mathbf{x}$ and t then by utilizing the definitions of material derivatives

$$\grave{\tilde{\Gamma}}_\alpha \equiv \frac{\partial \tilde{\Gamma}_\alpha}{\partial t} + (\nabla \tilde{\Gamma}_\alpha)\, \tilde{\mathbf{v}}_\alpha \,, \qquad \dot{\Gamma} \equiv \frac{\partial \Gamma}{\partial t} + (\nabla \Gamma)\mathbf{v} \tag{21}$$

we have:

$$\rho \dot{\Gamma} = \sum_{\alpha=1}^{\gamma} [\bar{\rho}_\alpha \grave{\tilde{\Gamma}}_\alpha - \nabla \cdot \bar{\rho}_\alpha \tilde{\Gamma}_\alpha \mathbf{u}_\alpha + (\frac{\partial \bar{\rho}_\alpha}{\partial t} + \nabla \cdot \rho_\alpha \tilde{\mathbf{v}}_\alpha)\, \tilde{\Gamma}_\alpha - (\frac{\partial \rho}{\partial t} + \nabla \cdot \rho \mathbf{v})\tilde{\Gamma}_\alpha], \tag{22}$$

where $\partial\rho/\partial t + \nabla \cdot \rho \mathbf{v} = 0$ by virtue of the conservation of mass equation for the mixture as a whole (Eq. (A.1)).

Utilizing Eqs. (6) - (9) and (17) - (21) in Eq. (5) results in the following conservation and balance equations for phase α:

Mass: $$\grave{\bar{\rho}}_\alpha + \bar{\rho}_\alpha \nabla \cdot \tilde{\mathbf{v}}_\alpha = \hat{c}_\alpha \tag{23}$$

Linear Momentum: $$\bar{\rho}_\alpha \grave{\tilde{\mathbf{v}}}_\alpha = \nabla \cdot \bar{\mathbf{T}}_\alpha + \bar{\rho}_\alpha \tilde{\mathbf{b}}_\alpha + \hat{\mathbf{p}}_\alpha \tag{24}$$

Angular Momentum: $$\hat{\mathbf{M}}_\alpha = \bar{\mathbf{T}}_\alpha - \bar{\mathbf{T}}_\alpha^T \tag{25}$$

Energy: $$\bar{\rho}_\alpha \grave{\tilde{\varepsilon}}_\alpha = \mathrm{tr}(\bar{\mathbf{T}}_\alpha^T \nabla \tilde{\mathbf{v}}_\alpha) - \nabla \cdot \bar{\mathbf{q}}_\alpha + \bar{\rho}_\alpha \tilde{r}_\alpha + \hat{\varepsilon}_\alpha \tag{26}$$

Entropy: $$\bar{\rho}_\alpha \grave{\tilde{s}}_\alpha + \nabla \cdot (\frac{\bar{\mathbf{q}}_\alpha}{\bar{\bar{\Theta}}_\alpha}) - \frac{\bar{\rho}_\alpha \tilde{r}_\alpha}{\bar{\bar{\Theta}}_\alpha} + \hat{c}_\alpha \tilde{s}_\alpha + \hat{s}_\alpha \geq 0 \tag{27}$$

In Eqs. (23) - (27), $\hat{c}_\alpha$ is the mass supply of phase α or the rate of phase change per unit volume of the mixture, $\bar{\mathbf{T}}_\alpha$ is the (partial or phase average) stress tensor, $\tilde{\mathbf{b}}_\alpha$ is the external body force, $\hat{\mathbf{p}}_\alpha$ is the linear momentum supply, $\hat{\mathbf{M}}_\alpha$ is the angular momentum supply, $\tilde{\varepsilon}_\alpha$ is the internal energy, $\bar{\mathbf{q}}_\alpha$ is the (partial or phase average) heat flux vector, $\tilde{r}_\alpha$ is the heat generation rate, $\hat{\varepsilon}_\alpha$ is the energy supply, $\tilde{s}_\alpha$ is the entropy, $\hat{s}_\alpha$ is the entropy supply and $\bar{\bar{\Theta}}_\alpha$ is the absolute temperature. $\bar{\mathbf{T}}_\alpha^T$ denotes the transpose of the second order tensor $\bar{\mathbf{T}}_\alpha$ (a linear transformation), and $\mathrm{tr}(\bar{\mathbf{T}}_\alpha^T \nabla \tilde{\mathbf{v}}_\alpha)$ denotes the trace of the product of two linear transformations $\bar{\mathbf{T}}_\alpha^T$ and $\nabla \tilde{\mathbf{v}}_\alpha$, i.e. $\mathrm{tr}(\bar{\mathbf{T}}_\alpha^T \nabla \tilde{\mathbf{v}}_\alpha) = \bar{T}_{\alpha ij}\ \partial \tilde{v}_{\alpha i}/\partial x_j$. The complete expressions for the supply terms $\hat{c}_\alpha$, $\hat{\mathbf{p}}_\alpha$, $\hat{\mathbf{M}}_\alpha$, $\hat{\varepsilon}_\alpha$ and $\hat{s}_\alpha$ are presented in Reference 7. Here, only the expressions for mass supply $\hat{c}_\alpha$ and linear momentum supply $\hat{\mathbf{p}}_\alpha$ are shown. Thus:

$$\hat{c}_\alpha \equiv -\frac{1}{V} \sum_\delta \int_{a_{\Lambda\delta}} m_{\alpha\delta}\, da \tag{28}$$

$$\hat{\mathbf{p}}_\alpha \equiv -\hat{c}_\alpha \tilde{\mathbf{v}}_\alpha - \frac{1}{V} \sum_\delta \int_{a_{\Lambda\delta}} (m_{\alpha\delta} \mathbf{v}_{\alpha\delta} - \mathbf{T}_{\alpha\delta} \mathbf{n}_{\alpha\delta})\, da - \nabla \cdot \mathbf{C}_{1\alpha}\,, \tag{29}$$

where

$$\mathbf{C}_{1\alpha} \equiv \frac{V_\alpha}{V} < \rho_\alpha \mathbf{v}_\alpha \otimes \mathbf{v}_\alpha > - \bar{\rho}_\alpha \tilde{\mathbf{v}}_\alpha \otimes \tilde{\mathbf{v}}_\alpha \tag{30}$$

is the covariance coefficient that represents the nonlocal effect and the symbol $\otimes$ denotes the tensor product.

The conservation and balance equations for a multiphase mixture are obtained by summing up from $\alpha = 1$ to $\alpha = \gamma$ each of the Eqs. (23) - (27). The resulting field equations as well as the compatibility conditions between the phasic and mixture equations are reproduced from Reference 7 in Appendix A of this paper for completeness.

In the volume-averaged equations presented above, the covariance coefficients in the momentum, energy and entropy supplies or interaction terms of phase α can be viewed as accounting in the theory for nonlocal effects. When the multiphase mixture becomes dispersed and the averaging volume approaches zero the nonlocal effects should be negligible. With a finite interfacial area the region of nonlocality can be associated with the size of the averaging volume V and it is through such a construction that V can reenter into the framework of the theory.

The macro-macroscopic multiphase field equations (Eqs. (23) - (27)) are similar in appearance to the single-phase multicomponent mixture equations [10,11] except for the following features:

1) In the multiphase field equations above, the linear momentum source term $\hat{\mathbf{p}}_m$, the angular momentum source term $\hat{\mathbf{M}}_m$, the energy source term $\hat{e}_m$ and the entropy source term $\hat{s}_m$ are not necessarily equal to zero since they account for two effects: a) for the interfacial sources such as surface tension, and b) for the nonlocal effects.

2) The stress tensor for the mixture as a whole is not necessarily symmetric due to a) and b) in 1) above even if the mixture is not acted upon by the body couples and couple stresses.

3) In the above theory of multiphase mixtures, the entropy Eq. (27) for phase α appears naturally while in the theory of multicomponent mixtures [10,11] an entropy equation for the mixture as a whole can only be used in order to achieve consistency of the theory with the classical equilibrium thermodynamics. From this, it follows that if each phase is composed of two or more chemical constituents, then an entropy equation for each constituent in the phase is not allowed.

As the above discussion shows, it should be clear that the final set of field equations for multiphase flows, Eqs. (23) - (27), are not entirely formulated utilizing a reductionistic approach, for this would not have led to the rigorous theory as brought about by the definition of kinematic concepts in Eqs. (14) - (22). The final set of field equations reflect a blend of pragmatism as well as idealism: pragmatism in the sense that these equations are consistent with the volume averaging approach and idealism in the sense that the equations are given a theoretical foothold by the definition of kinematic concepts. We will see in the next section that the degree to which reductionistic or holistic belief is placed on the models has a great bearing on the investigation of constitutive equations for multiphase mixtures.

3. ON CONSTITUTIVE EQUATIONS FOR MULTIPHASE FLOWS

In the previous section, we have encountered some methods for formulating field equations for multiphase mixtures although, strictly speaking, there are only two such methods that are exclusively utilized: extension of the single-phase multicomponent mixture field equations to the multiphase mixtures and averaging of the macroscopic field equations of each phase over the appropriate time and space segments. A similar situation also exists in the construction of constitutive equations for multiphase flows in the sense that use is made of one of the following two general approaches:

a) Formulation of constitutive equations from tne macroscopic modeling considerations (reductionistic approach).

b) Formulation of constitutive equations from macro-macroscopic modeling considerations (holistic approach).

In the first approach, where the constitutive equations are deduced through the macroscopic modeling, two procedures can be distinguished: 1) averaging of the macroscopic constitutive equations of each phase over the appropriate time or space segments [4], and 2) phenomenological approach inspired by the flow-pattern modeling [6,9, workers carrying out ad hoc modeling of multiphase flows].

An attempt to deduce the constitutive equations for two-phase flows by time averaging of the macroscopic constitutive equations of each phase was made by Ishii [4]. This reductionistic approach can be criticized for the following deficiencies:

1) Failure to account in the constitutive equations for the organized structure of the multiphase flow field. The degree of this deficiency can, of course, be disputed but to justify it, consider a bubbly two-phase flow. At the macroscopic level of individual bubbles or parts of bubbles no explicit insight can be gained as to why individual bubbles are in a given global or macro-macroscopic state. Averaging of the macroscopic constitutive equations of each phase leads to a relation between the macroscopic and macro-macroscopic states. However, this procedure is incomplete, because the constitutive equations averaged in this manner lack the complete global information which is necessary for the proper formulation of macro-macroscopic constitutive equations. In a bubbly two-phase flow, the agglomeration or disintegration of bubbles is triggered by the global information produced from the organized character of large number of bubbles or by the conditions specified by the macro-macroscopic state of the flow field.

2) Failure to provide a closed system of equations since averaging of the macroscopic constitutive equations leads again to covariance coefficients or turbulent correlation coefficients that need modeling through additional constitutive equations.

The easiest way to construct constitutive equations for multiphase flows is through a phenomenological modeling of individual flow regimes (bubbly flow, annular flow, etc.). This approach has been widely used for the practical design of multiphase flow equipment, and it will continue to be used primarily because of its simplicity and established confidence [9]. For mixtures of three or more phases this approach has not been adequately explored.

In the second approach in the formulation of constitutive equations for multiphase mixtures, where the constitutive equations are deduced from the holistic or macro-macroscopic modeling considerations, two modeling procedures can be also identified. These are: 1) the procedure followed in the classical thermodynamics of irreversible processes [16], and 2) the procedure of single-phase continuum mechanics [13].

As pointed out by Bataille and Kestin [6], the constitutive equations for multiphase mixtures derived on the basis of an approach from the classical thermodynamics of irreversible processes require the existence of a macro-macroscopic fundamental equation of state or of a Gibbs equation. Furthermore, it is also required that the resulting entropy equation exhibits the bilinear form in terms of fluxes and forces. Marle [8] and Hung [17] utilized this procedure by making a highly questionable assumption on the equality of macroscopic and macro-macroscopic fundamental equations and derived some special form of constitutive equations.

The continuum mechanics approach to study the constitutive equations is refuted by some and praised by others. Those who disapprove this approach find the method "...sufficiently arbitrary and general to account for nearly any kind of situation" [6], or unjustifiable because it is based on questionable principles of material frame invariance, equipresence, positive entropy production and on a number of other points [18]. While it is true that some workers in single-phase continuum mechanics have used the approach beyond physical justification, it still remains to be proved that the general procedure is incorrect. Those who refute the foundational basis of continuum mechanics start from special models of the physical world and argue to prove the trouble spots. While their (reductionistic) argument might be correct, is it not also safe to argue that it is just possible for some information in this approach, perhaps inconsequential at a first glance, to have been neglected which leads to false conclusions? As far as the linear theory of constitutive equations is concerned, there is, however, essentially no opposition to the unsoundness of the continuum mechanics approach, and it is, therefore, in this approximation that certain results for constitutive equations of multiphase flows will be summarized below.

Constitutive equations for multiphase mixtures using the continuum mechanics approach, have been studied to various degrees in References 1, 12, 14, 15 and 19. The important feature in the continuum mechanics approach is that the entropy equation is utilized to obtain restrictions on the constitutive assumption. Based on the field equations extended from the theory of multicomponent mixtures to multiphase mixtures, Drumheller and Bedford [1] and Bowen [12] utilized the entropy equation for the mixture as a whole, while Dobran [14,15] used the single-phase entropy Eq. (27) to restrict the constitutive assumption. Drew and Lahey [19] considered only some general constitutive principles for two-phase flows using the time-averaged model of Ishii [4] and did not use the entropy equation to restrict the form of constitutive assumption. As discussed in Section 2, an entropy equation for each phase is consistent with all averaging approaches while the entropy equation for the mixture as a whole does not appear to be too restrictive and the constitutive equations based on it should then be more general.

In the remaining part of this section, certain results for constitutive equations for a multiphase mixture of fluids from the studies in References 14 and 15 will be presented. For this purpose, it is convenient to eliminate the heat generation rate $\tilde{r}_\alpha$ in the entropy Eq. (27) by using the energy Eq. (26) to obtain

$$-\bar{\rho}_\alpha(\grave{\tilde{\psi}}_\alpha + \grave{\bar{\bar{\theta}}}_\alpha \tilde{s}_\alpha) + \mathrm{tr}(\bar{\mathbf{T}}^T_\alpha \nabla \tilde{\mathbf{v}}_\alpha) - \frac{\bar{\mathbf{q}}_\alpha \cdot \nabla \bar{\bar{\theta}}_\alpha}{\bar{\bar{\theta}}_\alpha} + \hat{\varepsilon}_\alpha + \bar{\bar{\theta}}_\alpha \hat{s}_\alpha + \hat{c}_\alpha \tilde{s}_\alpha \bar{\bar{\theta}}_\alpha \geq 0 , \tag{31}$$

where the Helmholtz potential $\tilde{\psi}_\alpha$ is defined as follows:

$$\tilde{\psi}_\alpha \equiv \tilde{\varepsilon}_\alpha - \bar{\bar{\theta}}_\alpha \tilde{s}_\alpha \quad . \tag{32}$$

Equation (31) together with Eqs. (23)-(26) were used by Dobran [14] to study constitutive equations for a dispersed multiphase fluid mixture with a single but nonuniform temperature and without the possibility of phase change, i.e.,

$\overline{\overline{\Theta}}_\alpha = \Theta$ and $\hat{c}_\alpha = 0$. The constitutive assumption studied in Reference 14,

$$(\overline{\mathbf{T}}_\alpha, \hat{\mathbf{p}}_\alpha, \hat{\mathbf{M}}_\alpha, \tilde{\Psi}_\alpha, \overline{\mathbf{q}}_\alpha, \hat{\varepsilon}_\alpha, \tilde{s}_\alpha, \hat{s}_\alpha) = f[\Theta, \nabla\Theta, \overline{\rho}_\beta, \nabla\overline{\rho}_\beta, \mathbf{D}_\beta, \mathbf{W}_\beta - \mathbf{W}_\gamma, \tilde{\mathbf{v}}_\beta - \tilde{\mathbf{v}}_\gamma, \grave{\tilde{\mathbf{v}}}_\beta - \grave{\tilde{\mathbf{v}}}_\gamma - 2\,\mathbf{W}_\gamma(\tilde{\mathbf{v}}_\beta - \tilde{\mathbf{v}}_\gamma)], \tag{33}$$

also ignores the memory effects but accounts for the nonuniform temperature and densities; for heat conduction, $\nabla\Theta$; for density gradients, $\nabla\overline{\rho}_\beta$; for viscous effects, $\mathbf{F}_\beta$; for viscous drag, $\tilde{\mathbf{v}}_\beta$; and for virtual mass effects, $\grave{\tilde{\mathbf{v}}}_\beta$. Note that in the notation of Eq. (33) each consitutive variable of the phase α, $\alpha = 1, \ldots, \gamma$, depends on all the other variables, $\beta = 1, \ldots, \gamma$, in the functional expression for f. Using the principle of objectivity or the material frame indifference [7,14,15]: 1) $\mathbf{F}_\beta$ must be replaced by $\mathbf{D}_\beta$ and $\mathbf{W}_\beta - \mathbf{W}_\gamma$) where $\mathbf{D}_\beta = \frac{1}{2}(\mathbf{L}_\beta + \mathbf{L}_\beta^T)$, $\mathbf{W}_\beta = \frac{1}{2}(\mathbf{L}_\beta - \mathbf{L}_\beta^T)$ and $\mathbf{L}_\beta = \mathbf{D}_\beta + \mathbf{W}_\beta$ is the velocity gradient, $\nabla\tilde{\mathbf{v}}_\beta$, 2) $\tilde{\mathbf{v}}_\beta$ must be replaced by $\tilde{\mathbf{v}}_\beta - \tilde{\mathbf{v}}_\gamma$, and 3) $\grave{\tilde{\mathbf{v}}}_\beta$ must appear in the frame invariant form $\grave{\tilde{\mathbf{v}}}_\beta - \grave{\tilde{\mathbf{v}}}_\gamma - 2\,\mathbf{W}_\gamma(\tilde{\mathbf{v}}_\beta - \tilde{\mathbf{v}}_\gamma)$. By substituting Eq. (33) into Eq. (31) the following is established:

1. The equilibrium state of the multiphase mixture specified by the conditions $\nabla\Theta = \mathbf{0}$, $\mathbf{D}_\beta = \mathbf{O}$, $\mathbf{W}_\beta = \mathbf{O}$, $\tilde{\mathbf{v}}_\beta - \tilde{\mathbf{v}}_\gamma = \mathbf{0}$, $\grave{\tilde{\mathbf{v}}}_\beta = \mathbf{0}$, for $\beta = 1, \ldots, \gamma$.

2. The restrictions on the equilibrium state coefficients in the linearized constitutive equations.

The above procedure is long and it leads to the following linearized constitutive equations:

$$\tilde{\Psi}_\alpha = \tilde{\Psi}_\alpha(\Theta, \overline{\rho}_\alpha) \quad , \qquad \tilde{s}_\alpha = -\frac{\partial\tilde{\Psi}_\alpha(\Theta, \overline{\rho}_\alpha)}{\partial\Theta} \tag{34}$$

$$\hat{s}_\alpha = 0 \quad , \qquad \hat{\varepsilon}_\alpha = 0 \tag{35}$$

$$\overline{\pi}_\alpha \mathbf{I} \equiv -\overline{\mathbf{T}}_\alpha[\Theta, \nabla\Theta = \mathbf{0}, \overline{\rho}_\beta, \nabla\overline{\rho}_\beta = \mathbf{0}, \quad \mathbf{D}_\beta = \mathbf{O}, \mathbf{W}_\beta - \mathbf{W}_\gamma = \mathbf{O}, \tilde{\mathbf{v}}_\beta - \tilde{\mathbf{v}}_\gamma = \mathbf{0}, \grave{\tilde{\mathbf{v}}}_\beta - \grave{\tilde{\mathbf{v}}}_\gamma - 2\mathbf{W}_\gamma(\tilde{\mathbf{v}}_\beta - \tilde{\mathbf{v}}_\gamma) = \mathbf{0}] \tag{36}$$

$$\hat{\mathbf{p}}_\alpha = -\gamma_\alpha \nabla\Theta - \sum_{\beta=1}^{\gamma-1} \xi_{\alpha\beta}(\tilde{\mathbf{v}}_\beta - \tilde{\mathbf{v}}_\gamma) - \sum_{\beta=1}^{\gamma-1} \Delta_{\alpha\beta}(\grave{\tilde{\mathbf{v}}}_\beta - \grave{\tilde{\mathbf{v}}}_\gamma) \tag{37}$$

$$\hat{\mathbf{M}}_\alpha = \overline{\mathbf{T}}_\alpha - \overline{\mathbf{T}}_\alpha^T = 4 \sum_{\beta=1}^{\gamma-1} \phi_{\alpha\beta}(\mathbf{W}_\beta - \mathbf{W}_\gamma) \tag{38}$$

$$\overline{\mathbf{T}}_\alpha = -\overline{\pi}_\alpha \mathbf{I} + \sum_{\beta=1}^{\gamma} [\lambda_{\alpha\beta}(\mathrm{tr}\,\mathbf{D}_\beta)\mathbf{I} + 2\mu_{\alpha\beta}\mathbf{D}_\beta] + 2\sum_{\beta=1}^{\gamma-1} \phi_{\alpha\beta}(\mathbf{W}_\beta - \mathbf{W}_\gamma) \tag{39}$$

$$\overline{\mathbf{q}}_\alpha = -\kappa_\alpha \nabla\Theta - \sum_{\beta=1}^{\gamma-1} \zeta_{\alpha\beta}(\tilde{\mathbf{v}}_\beta - \tilde{\mathbf{v}}_\gamma) - \sum_{\beta=1}^{\gamma-1} \nu_{\alpha\beta}(\grave{\tilde{\mathbf{v}}}_\beta - \grave{\tilde{\mathbf{v}}}_\gamma) - \overline{\rho}_\alpha \tilde{s}_\alpha \Theta \mathbf{u}_\alpha , \tag{40}$$

where the coefficients γ_α, $\xi_{\alpha\beta}$, $\Delta_{\alpha\beta}$, $\phi_{\alpha\beta}$, $\lambda_{\alpha\beta}$, $\mu_{\alpha\beta}$, κ_α, $\zeta_{\alpha\beta}$ and $\nu_{\alpha\beta}$ depend on Θ, $\overline{\rho}_1, \ldots, \overline{\rho}_\gamma$. These coefficients are restricted by the entropy Eq. (31), and for a two-phase fluid mixture, it is shown by Dobran [14,15] that the results are in accord with the available experimental data.

For a two-phase highly dispersed fluid mixture where the linear momentum supply for the mixture $\hat{\mathbf{p}}_m = \mathbf{0}$, the angular momentum supply for the mixture $\hat{\mathbf{M}}_m = \mathbf{O}$, the energy supply for the mixture $\hat{\varepsilon}_m = 0$ and the entropy supply for the mixture $\hat{s}_m = 0$, it is shown by Dobran [15] that

1) $\hat{\mathbf{M}}_\alpha = \mathbf{O}$ and thus from Eq. (38) we have the result that the stress tensor for phase α is symmetric, i.e.,

$$\overline{\mathbf{T}}_\alpha = \overline{\mathbf{T}}_\alpha^T \quad , \qquad \phi_{\alpha 1} = 0 \tag{41}$$

2) Using Eq. (41) in Eq. (39) the stress tensor for phase α thus becomes:

$$\overline{\mathbf{T}}_\alpha = -\overline{\pi}_\alpha \mathbf{I} + \lambda_{\alpha\alpha}(\mathrm{tr}\,\mathbf{D}_\alpha)\,\mathbf{I} + 2\mu_{\alpha\alpha}\,\mathbf{D}_\alpha \tag{42}$$

The stress tensor for phase α, $\overline{\mathbf{T}}_\alpha$, is thus shown to be symmetric only in the special case of disperse flow and it does not depend on the variables from the other phase, ($\mathbf{W}_1$ - $\mathbf{W}_2$). In their studies of two-phase flow constitutive equations, Ishii [4] and Bataille and Kestin [6] assumed a priori the symmetry of $\overline{\mathbf{T}}_\alpha$, while Drew [20] invoked the principle of phase separation in which the bulk phase variables (such as $\overline{\mathbf{T}}_\alpha$ and $\overline{\mathbf{q}}_\alpha$) are allowed to depend only on the variables from the same phase. Although Drew's principle of phase separation might be correct in special circumstances (cit. the result above expressed by Eq. (42)), within the present theory, however, this principle remains questionable since it is in conflict with Eqs. (38) - (40). In the time averaging approach [4]: 1) the coefficient $\lambda_{\alpha\alpha} = 0$ in Eq. (42); 2) no account is taken in the theory for virtual mass effects; and 3) the Dufour effect in the expression for the heat flux vector is associated with the concentration gradient and not with the diffusion of phases as represented by Eq. (40) and also agreed upon in the theory of single-phase multicomponent mixtures [10].

The Helmholtz potential in Eq. (34), $\bar{\psi}_\alpha = \bar{\psi}_\alpha(\Theta, \bar{\rho}_\alpha)$, represents a fundamental relation of the macro-macroscopic state with the assumptions which lead to the constitutive Eqs. (34) - (40). The knowledge of such a relationship and the coefficients γ_α, $\xi_{\alpha\beta}$, $\Delta_{\alpha\beta}$, $\phi_{\alpha\beta}$, $\underline{\lambda}_{\alpha\beta}$, $\mu_{\alpha\beta}$, κ_α, $\zeta_{\alpha\beta}$ and $\nu_{\alpha\beta}$ for $\alpha, \beta = 1, \ldots, \gamma$ as functions of Θ, $\bar{\rho}_1, \ldots, \bar{\rho}_\gamma$ are sufficient to close the multiphase field Eqs. (23) - (27) when the constitutive Eqs. (37) - (40) are utilized. Unfortunately, this macro-macroscopic fundamental equation is not known at the present time and this is where the holistic or continuum mechanics approach presented in the paper becomes inadequate. To get around this problem, the constitutive assumption can be enlarged and additional transport equations can be assumed [12,21].

From the above, it is clear that the linearized theory of constitutive equations yields many results that are in accord with past modeling practices in two-phase flows, and it also gives more general results than are predicted by these models. The fundamental problem thus remains: should there be a meaningful physical reconciliation between the macroscopic and macro-macroscopic states or should there exist additional transport equations [2]. This problem is, at the present time, the main stumbling block towards the construction of a viable theory for multiphase flows.

4. SUMMARY AND CONCLUSIONS

The theory of multiphase flows is in an infancy of development. While a general accord exists on the form of conservation and balance equations for multiphase flows, different averaging procedures lead to the definition of different multiphase flow variables which are still contraversial. Averaging over time and space domains gives rise to different models and especially manifests itself in the turbulent field equations. The volume-averaged multiphase field equations appear to be the most desirable to motivate the construction of a physically meaningful and mathematically rigorous set of conservation and balance equations. The field equations based on this approach were presented in the paper where it is also shown that the averaging or reductionistic approach utilized to construct conservation, balance and constitutive equations for multiphase mixtures needs reconciliations between two different levels of description: the macroscopic and macro-macroscopic levels. To avoid this, a holistic or continuum mechanics approach is more desirable since it yields a more consistent theory.

The results for constitutive equations, using the continuum mechanics approach and discussed in the paper, should, through fundamental experiments, lead to better modeling practices. Other procedures such as flow regime modeling will continue to provide special forms of the constitutive equations and, therefore, the verification of various theoretical models. The approach, based on irreversible thermodynamics, lacks the appropriate macro-macroscopic fundamental equation and, for this reason, it has not to date found a wide support. Clearly, while considerable progress has been made in the formulation of conservation, balance and constitutive equations for multiphase flows, much more work remains to be done in the future through both analysis and fundamental experiments.

REFERENCES

1. Drumheller, D.S., and Bedford, A., "A Thermomechanical Theory of Reacting Immiscible Mixtures," Arch. Rational Mech. Anal., 1980, Vol. 73, pp. 257-284.

2. Sha, W.T., and Soo, S.L., "Multidomain Multiphase Fluid Mechanics," Int. J. Heat Mass Transfer, 1978, Vol. 21, pp. 1581-1595.

3. Drew, D.A., "Averaged Field Equations for Two-Phase Media," Studies in Appl. Math., 1971, Vol. L, pp. 133-166.

4. Ishii, M., Thermo-Fluid Dynamic Theory of Two-Phase Flows, Eyrolles, Paris, 1975.

5. Delhaye, J.M. and Achard, J.L., "On the Use of Averaging Operators in Two-Phase Flow Modeling," Symp. Thermal and Hydraulic Aspects of Nuclear Reactor Safety, ASME, New York, 1977, Vol. 1, pp. 289-332.

6. Bataille, J., and Kestin, J., "Continuum Modeling of Two-Phase Flows," Brown University Report No. GEOFLO/12, December, 1981.

7. Dobran, F., "Theory of Multiphase Mixtures," Stevens Institute of Technology Report No. ME-RT-81015, March, 1982.

8. Marle, C.M.,"On Macroscopic Equations Governing Multiphase Flow with Diffusion and Chemical Reactions in Porous Media," Int. J. Engng. Science, 1982, Vol. 20, pp. 643-662.

9. Wallis, G.B., "Review-Theoretical Models of Gas-Liquid Flows," J. Fluids

Engng., 1982, Vol. 104, pp. 279-283.

10. Bowen, R.M., "Theory of Mixtures," In Continuum Physics, Vol. III (ed. A.C. Eringen), Academic Press, New York, 1976.

11. Atkin, R.J., and Craine, R.E., "Continuum Theories of Mixtures: Basic Theory and Historical Development," Q.J. Mech. Appl. Math., 1976, Vol. 29, pp. 209-243.

12. Bowen, R.M., "Compressible Porous Media Models by Use of the Theory of Mixtures," Int. J. Engng. Science, 1982, Vol. 6, pp. 667-735.

13. Eringen, A.C., Continuum Physics, Vol. II, Academic Press, New York, 1975.

14. Dobran, F., "Constitutive Equations for Multiphase Mixtures of Fluids," Stevens Institute of Technology Report ME-RT-82002, March, 1982.

15. Dobran, F., "A Two-Phase Fluid Model Based on the Linearized Constitutive Equations," Advances in Two-Phase Flows and Heat Transfer, NATO Advanced Research Workshop, Spitzingsee, Schliersee, F.R. Germany, Aug. 31 - Sept. 3, 1982.

16. DeGroot, S.R., and Mazur, P., Nonequilibrium Thermodynamics, North-Holland, Amsterdam, 1962.

17. Hung, N., "On the Formulation of Constitutive Laws Required to Describe Two-Phase Flow Models," Letters in Heat and Mass Transfer, 1979, Vol. 6, pp. 513-518.

18. Woods, L.C., "The Bogus Axioms of Continuum Mechanics," Bulletin of the Institute of Math. and its Appl. 1981, Vol. 17, pp. 98-102.

19. Drew, D.A., and Lahey, R.T., "Application of General Constitutive Principles to the Derivation of Multidimensional Two-Phase Flow Equations," Int. J. Multiphase Flow, 1979, Vol. 5, pp. 243-264.

20. Drew, D.A., "Two-Phase Flows: Constitutive Equations for Lift and Brownian Motion of Some Basic Flows," Arch. Rational Mech. Anal., 1976, Vol. 62, pp. 149-163.

21. Gibeling, H.J., and McDonald, H., "An Implicit Numerical Analysis for Two-Dimensional Two-Phase Turbulent Interior Ballistic Flows," AIAA 21st Aerospace Sciences Meeting, Paper AIAA-83-0561, 1983, Reno, Nevada.

APPENDIX A

The conservation and balance equations for a multiphase mixture are obtained by summing up each of the Eqs. (23) - (27) and by utilizing Eq. (22), i.e.

Mass $$\dot{\rho} + \rho \nabla \cdot \mathbf{v} = 0 \tag{A.1}$$

Linear Momentum: $$\rho \dot{\mathbf{v}} = \nabla \cdot \mathbf{T} + \rho \mathbf{b} + \hat{\mathbf{p}}_m \tag{A.2}$$

Angular Momentum: $$\hat{\mathbf{M}}_m = \mathbf{T} - \mathbf{T}^T \tag{A.3}$$

Energy: $$\rho \dot{\varepsilon} = \mathrm{tr}(\mathbf{T}^T \nabla \mathbf{v}) - \nabla \cdot \mathbf{q} + \rho r + (\hat{e}_m - \mathbf{v} \cdot \hat{\mathbf{p}}_m) \tag{A.4}$$

Entropy:
$$\rho\dot{s} + \nabla\cdot\sum_{\alpha=1}^{\gamma}\left(\frac{\bar{\mathbf{q}}_\alpha}{\bar{\bar{\Theta}}_\alpha} + \bar{\rho}_\alpha\tilde{s}_\alpha\mathbf{u}_\alpha\right) - \sum_{\alpha=1}^{\gamma}\frac{\bar{\rho}_\alpha\tilde{r}_\alpha}{\bar{\bar{\Theta}}_\alpha} + \hat{s}_m \geq 0 \quad , \tag{A.5}$$

where the compatibility conditions between the phasic Eqs. (23) - (27) and mixture Eqs. (A.1) - (A.5) are expressed by the following relations:

$$\sum_{\alpha=1}^{\gamma}\hat{c}_\alpha = 0 \tag{A.6}$$

$$\mathbf{T} \equiv \sum_{\alpha=1}^{\gamma}(\bar{\mathbf{T}}_\alpha - \bar{\rho}_\alpha\,\mathbf{u}_\alpha\otimes\mathbf{u}_\alpha) \tag{A.7}$$

$$\rho\mathbf{b} \equiv \sum_{\alpha=1}^{\gamma}\bar{\rho}_\alpha\,\tilde{\mathbf{b}}_\alpha \tag{A.8}$$

$$\hat{\mathbf{p}}_m \equiv \sum_{\alpha=1}^{\gamma}(\hat{c}_\alpha\,\mathbf{u}_\alpha + \hat{\mathbf{p}}_\alpha) \tag{A.9}$$

$$\hat{\mathbf{M}}_m \equiv \sum_{\alpha=1}^{\gamma}\hat{\mathbf{M}}_\alpha \tag{A.10}$$

$$\rho\varepsilon \equiv \sum_{\alpha=1}^{\gamma}(\bar{\rho}_\alpha\tilde{\varepsilon}_\alpha + \tfrac{1}{2}\bar{\rho}_\alpha\,\mathbf{u}_\alpha\cdot\mathbf{u}_\alpha) \tag{A.11}$$

$$\rho r \equiv \sum_{\alpha=1}^{\gamma}(\bar{\rho}_\alpha\tilde{r}_\alpha + \bar{\rho}_\alpha\,\tilde{\mathbf{b}}_\alpha\cdot\mathbf{u}_\alpha) \tag{A.12}$$

$$\mathbf{q} \equiv \sum_{\alpha=1}^{\gamma}[\bar{\mathbf{q}}_\alpha - \bar{\mathbf{T}}_\alpha^T\,\mathbf{u}_\alpha + \bar{\rho}_\alpha\tilde{\varepsilon}_\alpha\,\mathbf{u}_\alpha + \tfrac{1}{2}\bar{\rho}_\alpha(\mathbf{u}_\alpha\cdot\mathbf{u}_\alpha)\,\mathbf{u}_\alpha] \tag{A.13}$$

$$\hat{e}_m - \mathbf{v}\cdot\hat{\mathbf{p}}_m \equiv \sum_{\alpha=1}^{\gamma}[\hat{\varepsilon}_\alpha + \mathbf{u}_\alpha\cdot\hat{\mathbf{p}}_\alpha + \hat{c}_\alpha(\tilde{\varepsilon}_\alpha + \tfrac{1}{2}\,\mathbf{u}_\alpha\cdot\mathbf{u}_\alpha)] \tag{A.14}$$

$$\rho s \equiv \sum_{\alpha=1}^{\gamma}\bar{\rho}_\alpha\,\tilde{s}_\alpha \tag{A.15}$$

$$\hat{s}_m \equiv \sum_{\alpha=1}^{\gamma}\hat{s}_\alpha \tag{A.16}$$

The mixture source terms $\hat{\mathbf{p}}_m$, $\hat{\mathbf{M}}_m$, $\hat{e}_m$ and $\hat{s}_m$ account for two effects [7]: 1) for the interfacial forces between the phases in the averaging volume V; and 2) for the covariance effect or the nonlocal material response. Notice also that Eqs. (23) - (27) and (A.1) - (A.5) can be time-averaged to obtain turbulent field equations.

Multi-Phase Flow and Heat Transfer III. Part A: Fundamentals
edited by T.N. Veziroğlu and A.E. Bergles
Elsevier Science Publishers B.V., Amsterdam, 1984 — Printed in The Netherlands

SOME REMARKS ON THE CONSTITUTIVE EQUATIONS OF STRATIFIED GAS-LIQUID FLOW

J. FABRE, L. MASBERNAT and C. SUZANNE
Institut National Polytechnique de Toulouse
31071 Toulouse, France

ABSTRACT

This paper deals with two phase gas-liquid flow in stratified configuration. In a first part some fundamental results about the kinematic structure in both phases are recalled and new results are presented. In a second part, these results are employed to determine interfacial and wall shear stress. In a third part, the influence of the flow behaviour is discussed and a simple idea for the flow modeling is propounded.

1. INTRODUCTION

In two phase pipe flow, prediction of averaged values, like pressure drop and phase fraction, are of practical interest in engineering applications. Space-averaged equations expressing the conservation of mass and momentum in each phase, constitute the logical basis for their determination. The set of partial differential equations introduces additional unknown quantities by space integration : space distribution coefficients, mass and momentum fluxes between the phases and between each phase and the wall. In stationary flows, when interfacial mass transfer is negligible, the constitutive laws for fluid-fluid and fluid-wall interactions are the keystones of the flow modeling.

Although stratified two phase flow is the simplest two phase regime, wall and interfacial momentum transfer problem has not yet been perfectly solved. As a rule, single phase results were used to express the friction factor equations, calibrated indirectly by comparison with hold-up and pressure drop experimental data. This approach ignores the specific effect of the waves and the behaviour of the flow; its predictive capability, limited to the calibration range, has to be improved.

The formulation of more realistic transfer models requires a knowledge of the kinematic structure of the flow in each phase, and specially of the turbulent motions in the immediate vicinity of the disturbed interface. Great advantages are provided by a detailled analysis of the mean and turbulent flow. First, the experimental determination of the interfacial shear stress, which cannot be deduced directly from the phase flow rates, the pressure drop, and the phase fraction, is possible. Secondly, experimental data of local velocity and Reynolds stress become available for numerical modeling of turbulent-turbulent two-phase flows.

The present work starts with an experimental study of the structure of stratified, near horizontal, air-water flow in a channel of rectangular cross-section. It follows a series of studies carried out, at the "Institut de Mécanique des Fluides de Toulouse" by MASBERNAT, SUZANNE, REFEROWSKI [1], SUZANNE [2], GAYRAL, MASBERNAT, SUZANNE [3], HADJ-FREDJ [4], GAYRAL [5], FABRE, MIRE, SUZANNE [6], and FABRE, MASBERNAT, SUZANNE [7]. From experimental measurements of mean velocity and Reynolds stress on a vertical axis of the channel, an apparent anomalous behaviour of the flow in each phase was shown: it was suggested that, under certain conditions, secondary flow may occur and play an important role in momentum and mass transfer through the interfacial wavy region. Although these experiments were conducted in a channel with a 2:1 aspect ratio, it does not appear that this behaviour is linked to the lateral proximity of the wall boundaries: some other studies, too, have pointed out similar characteristics in channels with larger aspect ratio [8-11].

The work presented in this paper was intended to give a more complete description of the flow and to discuss the influence exerted by secondary flows on the interfacial momentum transfer.

In the first part, some fundamental results about the kinematic structure in both phases are recalled, and new results on lateral mean velocity and shear stress distribution, are presented. In the second part, these results are used to compare different experimental methods to determine interfacial and wall shear stress. Finally, in the third part, the influence of the flow behaviour on the interfacial transfer is discussed.

2. KINEMATIC STRUCTURE OF STRATIFIED WAVY GAS-LIQUID FLOW

Few previous studies have investigated the local structure of stratified gas-liquid flow. The work of JEFFRIES et al. [8], [9], THEOFANOUS et al. [10], [11], AKAI et al. [12], [13], [14] may be cited for their attempts to give a description of the turbulent structure in each phase. Their experimental conditions are given in Table 1.

The experimental results presented in this paper, were obtained from a rectangular channel, 200 mm wide, 100 mm high, 12 m long fitted with gas and liquid recirculation loops and with equipment for measuring gas and liquid flow rates and pressure drops. The bottom slope was adjusted precisely, using a leveling technique. All the velocity measurements were performed in a test cross-section, located at a distance of 9 m from the inlet of the channel, so as to ensure a fully developed flow in both phases and a homogeneous wave field in the direction of the main flow.

Table 1. Flow conditions in other studies

Ref.	Fluids	Channel Aspect ratio	R_{e_L} max	R_{e_G} max	Measured Components
8-9	air-water	2:1	∿ 20,000	∿ 20,000	$\overline{U}$, $\overline{u'^2}$
10-11	air-water	12:1	∿ 20,000	∿ 60,000	$\overline{U}$, $\overline{u'^2}$, $\overline{u'v'}$
12	air-water	2.7:1	∿ 1,000	∿ 8,000	$\overline{U}$, $\overline{u'^2}$
13	air-mercury	2.7:1	∿ 8,000	∿ 8,000	$\overline{U}$, $\overline{u'^2}$
7	air-water	2:1	∿ 14,000	∿ 40,000	$\overline{U}$, $\overline{V}$, $\overline{u'^2}$, $\overline{v'^2}$, $\overline{w'^2}$, $\overline{u'v'}$.

In the gas phase, all velocity data were obtained by linearized constant-temperature hot-wire anemometry system, with DISA x probes. The velocity and Reynolds stress components $\overline{U}$, $\overline{V}$, $\overline{u'^2}$, $\overline{v'^2}$, $\overline{u'v'}$ were determined by analogical treatment.

In the liquid phase, instantaneous velocities were measured by a laser Doppler velocimeter using a monochromatic optical system, fitted with a Bragg cell, a frequency counter, and either a 2 W Argon laser or a 35 mW He-Ne laser. The forward scattering technique was used systematically to improve the accuracy of the turbulence intensity measurement:

1. $\overline{U}$, $\overline{V}$, $\overline{u'^2}$, $\overline{v'^2}$ data were determined directly from instantaneous output signals: special care was taken for the y-components.

2. $\overline{u'v'}$ data were obtained from smoothed values of the turbulence intensities of the velocity components at ± 45 ° in the vertical plane.

3. $\overline{W}$, $\overline{w'^2}$ data were obtained by setting the axis of the laser at ± 45° in the horizontal plane.

In the two-phase wavy region, the conditional averages $\overline{U}$ and $\overline{u'^2}$ were determined in the liquid phase by numerical acquisition and treatment of the digital samples from the frequency counter.

To measure the instantaneous liquid height, a capacitive probe was used. It was thus possible to characterize the structure of the liquid interface by the root-mean square wave height $\overline{h'^2}$ and by the y-distribution of the local void fraction α.

What the results on the y-axis show. Four runs were conducted [7] with a constant liquid flow rate and a mean velocity in the gas phase warying between 0 and 7.5 m/s. The flow conditions are given in Table 2. The bottom slope was determined from the shear stress gradient in the free surface flow corresponding to the run 10. Only small ripples of negligible amplitude were present at the interface in the run 7. In the two other runs developed wind waves were present.

The y-distribution of the mean velocity components $\overline{U}$ and $\overline{V}$, of the turbulence shear stress $-\rho\overline{u'v'}$, and of the turbulence energy e, are shown:

1. in figure 1 for the liquid phase.

2. in figure 2 for the gas phase.

They are in non-dimensional form using, as a length scale the mean height h_k of the phase k, and as a velocity scale, either the friction velocity u_{WL} at the lower wall, or the interfacial friction velocity u_{IG}.

These results exhibit some important features of the wavy flow corresponding to the runs 8 and 9, in comparison with the classical parallel shear flow corresponding to the run 7:

1. The mean flow is non-parallel in both phases. On the axis, we found a non-zero y-component, indicating a secondary mean motion of the two fluids from the interface towards the walls; this component is of the same order as the wall friction velocity. These results suggest that, in each phase, the streamline pattern of the secondary flows consists of cells, symmetrical with respect to the y-axis [7]. Therefore, the fluid motion is assumed to be helicoidal.

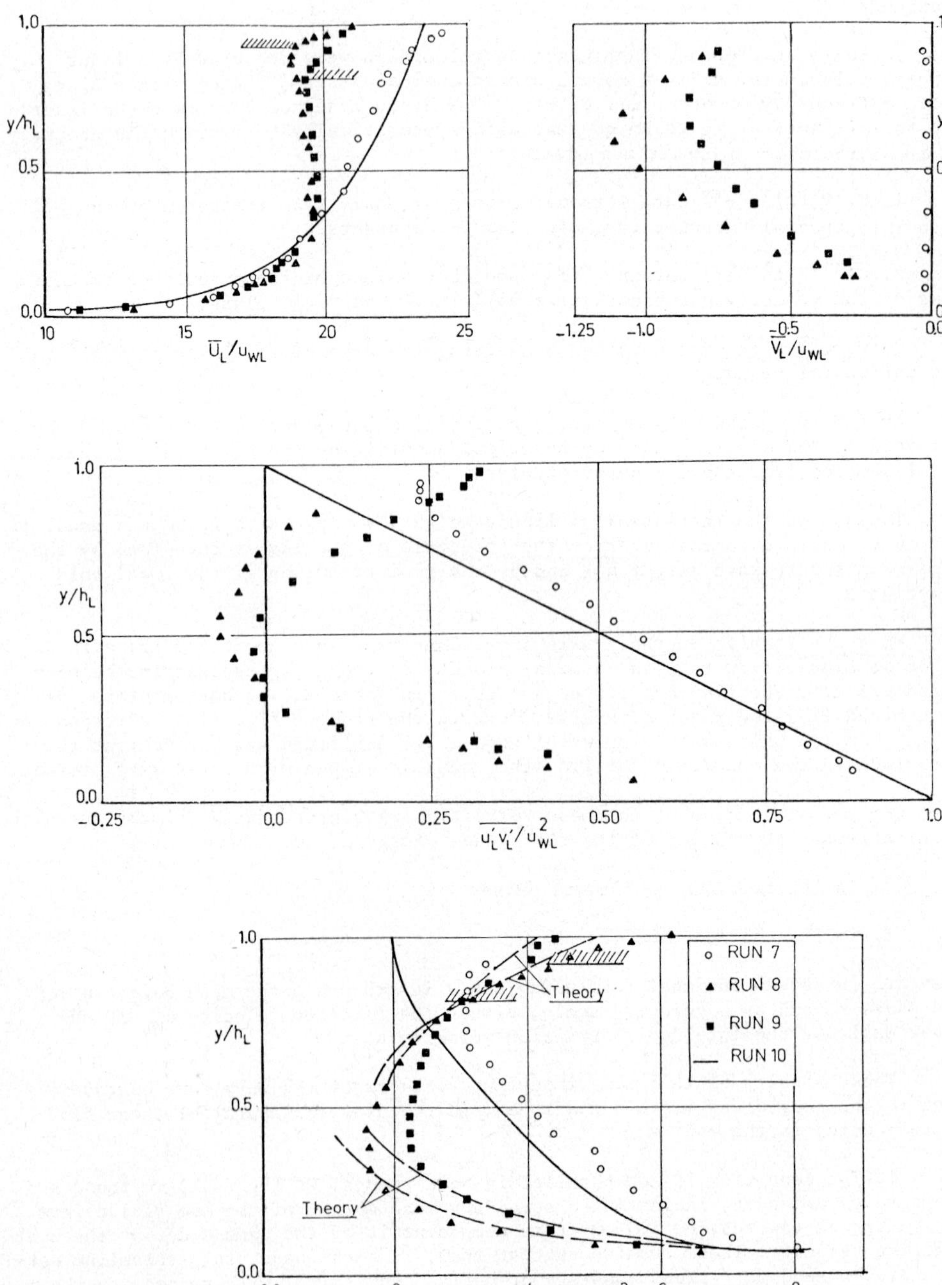

1.0
0.5
0.0
y/h_L
10
15
20
25
$\overline{U}_L/u_{WL}$
-1.25
-1.0
-0.5
0.0
$\overline{V}_L/u_{WL}$
-0.25
0.0
0.25
0.50
0.75
1.0
$\overline{u'_L v'_L}/u_{WL}^2$
0
2
4
6
8
e_L/u_{WL}^2
Theory
RUN 7
RUN 8
RUN 9
RUN 10

Fig. 1

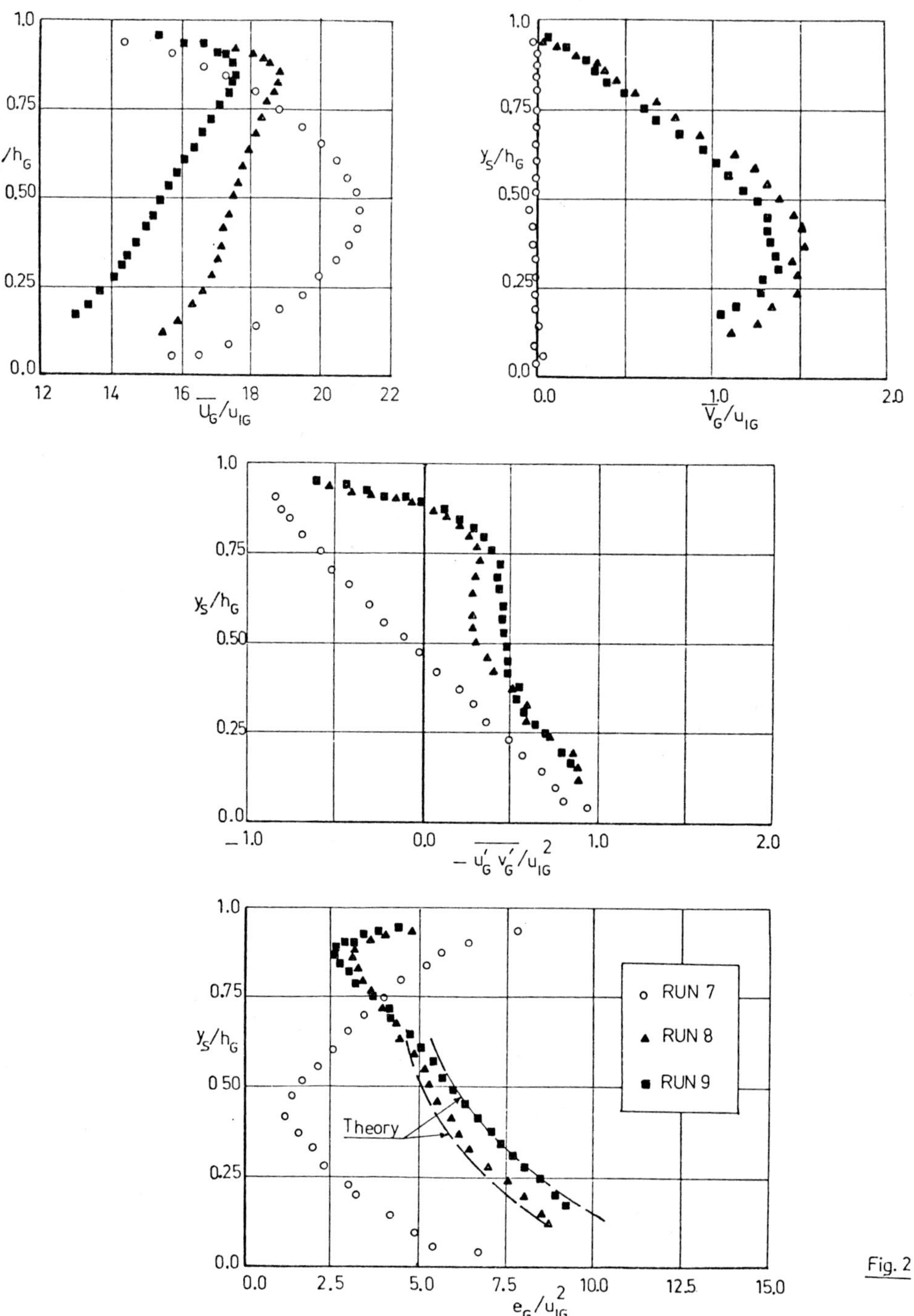

1.0
0.75
y_S/h_G
0.50
0.25
0.0
12
14
16
18
20
22
$\overline{U_G}/u_{IG}$
0.0
1.0
2.0
$\overline{V_G}/u_{IG}$
−1.0
$-\overline{u'_G v'_G}/u_{IG}^2$
2.5
5.0
7.5
10.0
12.5
15.0
e_G/u_{IG}^2
Theory
RUN 7
RUN 8
RUN 9

Fig. 2

Table 2. Flow conditions in this study

RUN	7	8	9	10	12
$I \cdot 10^3$	1.05	1.05	1.05	1.05	1.38
R_{e_L}	13,600	13,600	13,600	13,600	11,300
Fr_L	0.59	0.77	1.37	0.57	0.76
R_{e_G}	15,100	25,100	39,600	0	26,500

2. The shear stress distribution deviates from linearity in each phase, and takes, in the liquid phase, unexpected negative values. Near the waves, the productions of turbulence is negligible. Positive shear stress is associated with negative velocity gradients, therefore the effective viscosity concept does not apply for momentum transport, and "turbulent-viscosity" models fail to predict velocity profiles. This is certainly one of the most noticeable results,showing a particular rheological behaviour under the waves. Quantitative interpretation of the experimental results show that near the wall, the shear stress gradient balances the momentum advective transport. Near the waves, the contribution of the non-measured component of the Reynolds stress $\overline{u'w'}$ cannot be neglected.

3. In the gas phase, although the secondary flow influences the TKE distribution, the results exhibit a classical structure of wall turbulence. It was shown, in a previous study, that the TKE balance is controlled, mainly, by production, dissipation, and advective transport. In the liquid phase, the same conclusion holds for the flow near the lower wall. Under the wave region, all the Reynolds stress components, and especially the shear stress, present a steep gradient, and advective and turbulence transport play a dominant role. The tendency of the experimental results are explained by an asymptotic TKE balance, neglecting turbulence production. This crude evaluation shows the importance of the Reynolds stress transport from the waves.

What the new results show. From the y-distribution of the mean velocity and Reynolds stress, strong three dimensional effects are expected in each phase. In particular, the shear-stress may be non-uniform at the interface and at the wall. The analysis of these effects was made in the run 12, where the flow conditions were roughly the same as in the run 8. Only the bottom slope and the viscosity were slightly different, as shown in Table 2.

The measurements were performed, in the liquid phase, on four vertical axes, located at z/L = 0, 0.33, 0.60, 0.93, from the symmetrical axis, (L being the half-width of the channel). The experimental data are presented in Figure 3 in non-dimensional form, using as a velocity scale, either the bulk velocity or the wall friction velocity at z =0.

It appears that, on the symmetrical axis, the results are different from those corresponding to the run 8, for approximately the same gas and liquid flow rates. In fact, it was difficult, with these flow conditions, to reproduce exactly the same flow structure in the liquid phase. However, all the measured distributions are in qualitative agreement with previous conclusions.

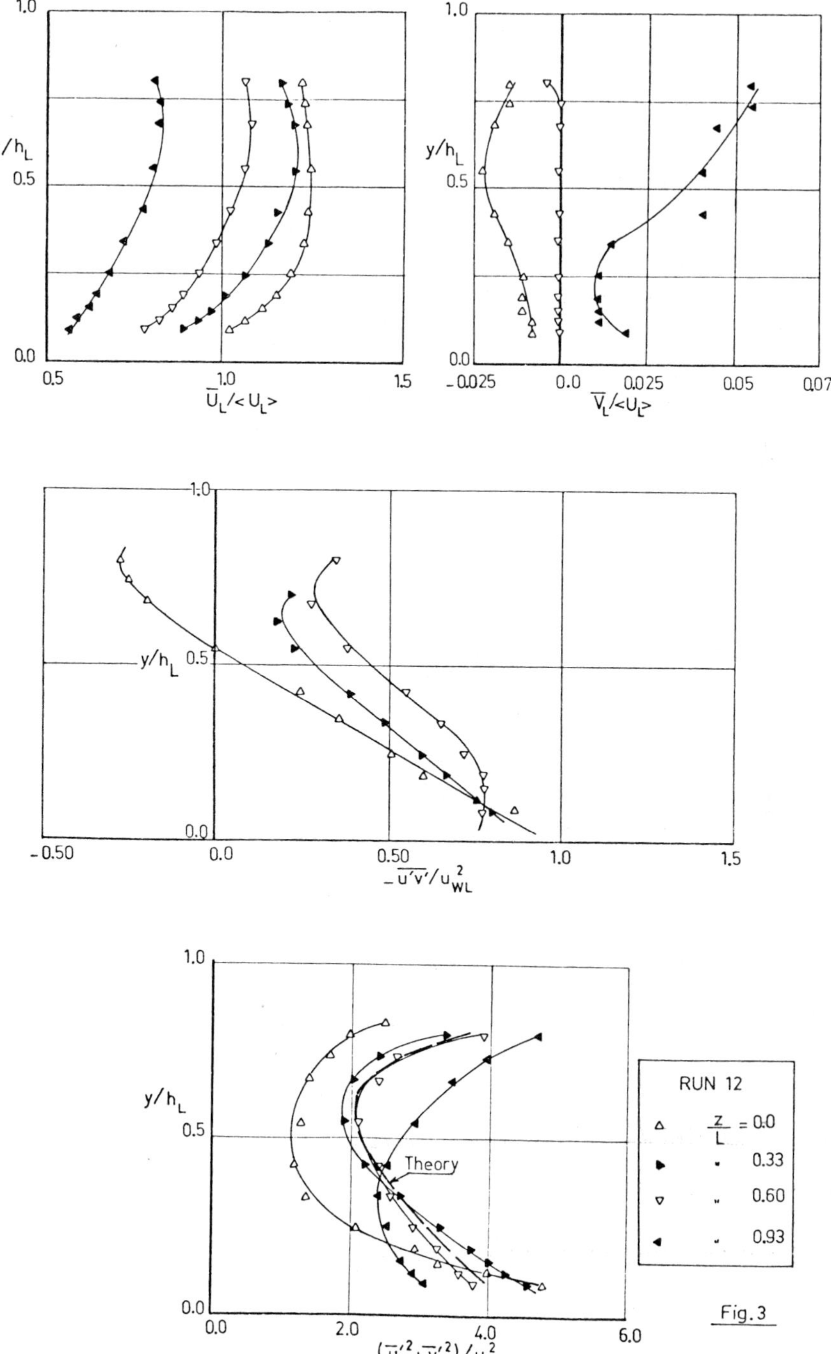

1.0
y/h_L
0.5
0.0
0.5
1.0
1.5
U_L/<U_L>
1.0
y/h_L
0.5
0.0
-0.025
0.0
0.025
0.05
0.075
V_L/<U_L>
1.0
y/h_L 0.5
0.0
-0.50
0.0
0.50
1.0
1.5
-u'v'/u_WL^2
1.0
y/h_L
0.5
0.0
0.0
2.0
4.0
6.0
(u'^2+v'^2)/u_WL^2
Theory
RUN 12
z/L = 0.0
" 0.33
" 0.60
" 0.93

Fig. 3

The occurence of secondary flow is confirmed by positive values of the y-component of the mean velocity at z/L = 0.93. Due to the fact that the $\overline{V}$ data collected outside two layers close to the lateral wall and to the axis were roughly zero, one can suggest that the secondary flow pattern is made up of two cells, symmetrical with respect to the y-axis. Thus, all the Reynolds stress components are influenced by the perturbative effect of the advective transport, near the y-axis and the lateral wall, as it can be shown on the shear stress and the turbulence intensity distributions.

Another characteristic feature of the flow in the liquid concerns the mean velocity distribution, where transverse gradients are more important than in parallel flow. But, the most noticeable result is in relation to the y-distribution of the velocity, where a maximum value appears. This fact seems to be surprising for a free surface flow subjected to a positive interfacial friction. However the analysis of the shear stress profiles under the wavy region, shows that positive values are associated with negative values of the velocity gradient at z/L = 0.33 and 0.60. Turbulence production has a weak or negative value. This confirms that Boussinesq's eddy viscosity concept does not apply in the upper region of the liquid flow. As a general rule, in wall turbulence, one expects the shear stress to be fixed by the momentum equation, and the constitutive rheological law to result from a crude balance neglecting turbulence diffusion in the Reynolds stress equation. Here, the shear stress, like the turbulence intensity, has a positive gradient under the waves. In fact, the shear stress is mainly controled by the turbulence fluxes and the pressure-strain correlation. Thus, positive values of the shear stress gradient, which appear as a consequence of the turbulence transport, do not agree for parallel flow. The secondary flow must be considered as a perturbation linked to the particular rheological behaviour near the waves.

The liquid flow, submitted to the wind shear, may be conveniently represented by a two-layer pattern as follows:

1. a lower layer, where the velocity gradients are important, as well as the turbulence production which results from the wall friction. The eddy viscosity concepts applies for the momentum equation.

2. an upper layer, where y-velocity distribution may be considered as roughly constant; here, the turbulence results from the transport of Reynolds stress produced in the wavy region and the eddy viscosity concept does not apply for the momentum equation.

This crude description will be used, further, to give new ideas on the constitutive equations in stratified two-phase flow.

3. THE DETERMINATION OF THE WALL AND INTERFACIAL SHEAR STRESS

In separated flow models, constitutive equations for the interfacial and wall shear stress have to be calibrated with experimental data. Unfortunately, few measurements are available today. It is a well-known problem that the three averaged quantities $< \tau_I >$, $< \tau_G >$, $< \tau_L >$ cannot be deduced directly from the measurement of the liquid and gas flow rates, of the pressure gradient and of the phase fraction.

Consequently, local measurements are needed for the experimental determinations. If one excludes the electrochemical techniques allowing a direct measurement of the wall shear stress in the liquid phase, the method developed by HANRATTY and ENGEN [15] using the location of the maximum in the velocity profile, is certainly the most convenient for two-dimensional flow. This method is based

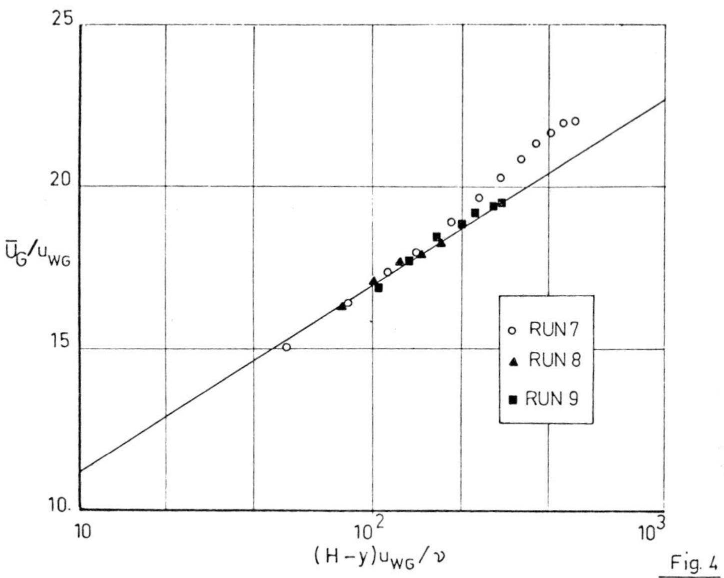
25
20
15
10.
$\overline{U}_G/u_{WG}$
10
10^2
10^3
$(H-y)u_{WG}/\nu$
o RUN 7
▲ RUN 8
■ RUN 9

Fig. 4

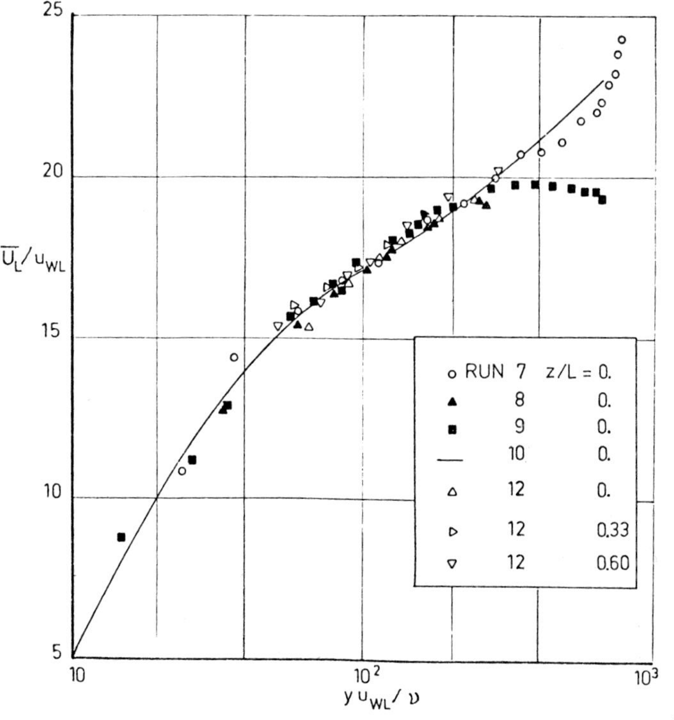
25
20
15
10
5
$\overline{U}_L/u_{WL}$
10
10^2
10^3
$y u_{WL}/\nu$
o RUN 7 z/L = 0.
▲ 8 0.
■ 9 0.
— 10 0.
△ 12 0.
▷ 12 0.33
▽ 12 0.60

Fig. 5

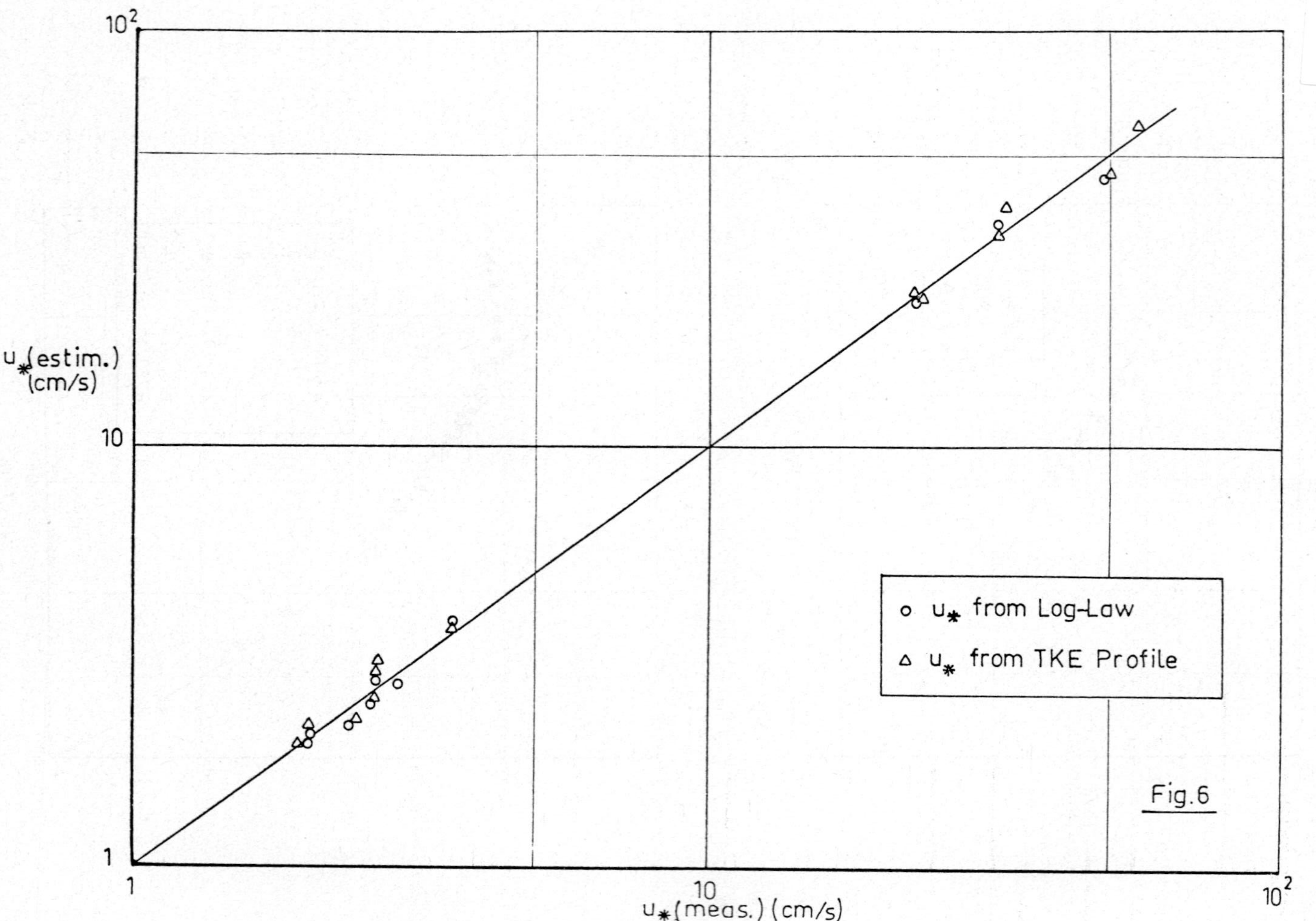
u_*(estim.) (cm/s)
10^2
10
1
1
10
10^2
u_*(meas.) (cm/s)
o u_* from Log-Law
△ u_* from TKE Profile
Fig.6

on two assumptions: the flow is parallel and the shear-stress gradient may be known from the pressure drop and the bottom slope; the eddy viscosity concept is applicable and the maximum in the velocity profile corresponds to zero stress. When secondary flow occurs, the shear stress is not linear and the method is not consistent.

We compared three other methods to obtain a good estimate of the wall and interfacial shear stress. The first is founded on the validity of the logarithmic velocity profile in wall turbulent flows which may be described by an equation of the form:

$$\frac{\overline{U}}{u_*} = \frac{1}{\kappa} \log \frac{y}{K_{s*}} + A \qquad (1)$$

where κ the Von Karman constant is assumed to be 0.4

A is a constant assumed to be 8.5

K_{s*} is an apparent roughness given by the relation [2]

$$K_{s*} = K_s \left(1 + B \frac{\nu}{u_* K_s}\right) \qquad (2)$$

with $B \simeq 3.32$

K_s is the roughness size

When $\frac{u_* K_s}{\nu} \ll 1$, the length scale K_s disappears in equation (1) and the logarithmic law for the "turbulent smooth" regime is given by:

$$\frac{\overline{U}}{u_*} = \frac{1}{\kappa} \log \frac{y u_*}{\nu} + A' \qquad (3)$$

where A' is roughly equal to 5.5.

When production balances dissipation, equations (1) and (2) agree well with experimental data, except in the viscous sub-layer. The friction velocity u_* and the roughness length scale K_s can be determined from $\overline{U}$ measurements. A good estimate is obtained in the "turbulent smooth" regime where the single quantity u_* is needed. The method generally used in our work, consists of fitting the experimental results with equation (3) for $\frac{y u_*}{\nu} \sim 100$.

The second method consists of an extrapolation of the shear stress data at the wall or the interface.

The third method is relative to the TKE profile. When production and dissipation balance, a rough approximation neglecting the turbulence transport in the TKE equation, leads to the following relation:

$$- \overline{u'v'} \frac{\partial \overline{U}}{\partial y} = \varepsilon \qquad (4)$$

Assuming the constitutive equation between shear and mean strain rate to be given by RODI [16]:

$$- \overline{u'v'} = C_\mu \frac{e^2}{4\varepsilon} \frac{\partial \overline{U}}{\partial y} \qquad (5)$$

One can obtain:

$$\frac{4}{C_\mu} \left|-\overline{u'v'}\right| = e \qquad (6)$$

With a recommended value of $C_\mu = 0.09$, the proportionality constant between the TKE and the shear stress is roughly equal to 6.7. Equation (6) may be used to determine the friction velocity, by extrapolation of the energy data at the wall and the interface. Special care must be taken to exclude the data which are located in the immediate vicinity of the wall.

A comparison of these methods was made for the different runs. In the Figure 4, the experimental data of the mean velocity $\overline{U}$ are closely grouped by a careful choice of the friction velocity. Due to the lack of precision in determining the distance to the interface, this method was not applied to the determination of the interfacial friction velocity in the gas phase.

Table 3. Estimated wall friction velocity in the liquid (cm/s)

RUN	7	8	9	10	12	12	12
z/L	0	0	0	0	0	0.33	0.60
Log law	<u>2.08</u>	<u>2.80</u>	<u>3.85</u>	<u>2.00</u>	<u>2.75</u>	2.45	2.20
Shear stress meas.	2.02	∿2.65	∿3.61	2.00	2.68	2.61	2.38
TKE meas.	2.19	∿2.90	∿3.71	2.00	∿(3.12)	∿(2.50)	∿(2.22)

Estimated wall and interfacial velocity in the gas (cm/s)

RUN	7	8	9	7	8	9
(z/L = 0)	wall	wall	wall	interfacial	interfacial	interfacial
Log law	22	33.7	44	–	–	–
Shear stress Meas.	23	32	49	<u>23</u>	<u>33</u>	<u>56</u>
TKE meas.	23	32	44	22	38	60

___ value used as a velocity scale
( ) estimated from $\overline{u'^2} + \overline{v'^2} = 3/4\ e$

All the results are summarized in Table 3. The Figure 5 gives an idea of the accuracy when the shear stress is obtained from the mean velocity or the TKE profiles. It shows clearly that no general trend appears in the differences between measured and estimated shear stress. The relative error is not greater than 16 %. The maximum values occur when secondary flows are present. We give great confidence to the estimations from the mean velocity profile which can be measured very accurately. The two other methods are very similar: extrapolation of TKE is inaccurate due to the non linearity of the turbulence energy distribution, and shear stress cannot be measured with great precision. In particular, some difficulties may arise when steep gradients, linked to the secondary flow are present; this is the case for the determination of quantities, like shear stress and energy, that need several measurements.

4. SOME COMMENTS ON THE FLOW MODELING

One of the questions arising now is how the secondary flows influence the momentum transfer. A discussion on the momentum equation averaged on the z-axis shows the dominant effect of the structure of the flow. Introducing the average operator:

$$< f >_z = \frac{1}{2L} \int_{-L}^{L} f.dz$$

and assuming the symmetry of the flow in respect to z, we can write the x-momentum equation:

$$\frac{1}{\rho_k} \frac{\partial}{\partial y} < \tau_{xy} >_z = - g(I_k + I) + \frac{\partial}{\partial y} < \overline{U}_k \overline{V}_k >_z - \frac{1}{\rho k} \frac{\tau^\circ_{xy}}{L} \tag{7}$$

Where τ°_{xz} is the shear stress on the lateral wall for z = L

and $I_k = - \dfrac{1}{\rho_k g} \dfrac{dP_g}{dx}$

The velocity $\overline{U}_k$ can be expressed as:

$$\overline{U}_k = < \overline{U}_k >_z + \hat{U}_k \tag{8}$$

Taking into account the equation of conservation of mass, leads to a new formulation of equation (7):

$$\frac{1}{\rho_k} \cdot \frac{\partial}{\partial y} < \tau_{xy} >_z = - g(I_k + I) + \frac{\partial}{\partial y} < \hat{U}_k \overline{V}_k >_z + \frac{1}{\rho_k} \frac{|\tau^\circ_{xz}|}{L} \tag{9}$$

In two-dimensional parallel flow, the two last terms disappear in the R.H.S. and the mean shear stress is linear. When the aspect ratio is not large, the influence of the proximity of the lateral walls is introduced by the friction $|\tau^\circ_{xz}|$ which is opposed to the gravity in downward flow. The second term in the R.H.S. takes into account the mean effect of the secondary flow on the momentum equation.

In the liquid, $\hat{U}_k \overline{V}_k$ has no significant value, except near the y-axis and the lateral wall, where it takes negative values. Consequently, the average quantity $< \hat{U}_k \overline{V}_k >$, which is zero at the wall and the interface, first increases, then decreases, when y increases. The mean shear stress profile deviates from linearity by an "apparent slope" effect of the secondary motion. In the upper layer, near the interface, the R.H.S. may be positive, like in the RUNS 7, 8, 9, 12, and the fluid motion may appear as an upward flow; in the lower layer, the slope seems to be increased. In the gas, an inverse effect of the correlation $< \hat{U}_k \overline{V}_k >$ is expected on the mean shear stress.

An idea for a simple solution of the flow in the liquid may be given by a two-layer model. In the lower layer, where production and dissipation balance, the "eddy viscosity" concept may be applied to relate the mean shear stress $< \tau_{xy} >$ to the mean gradient. In the upper layer, a constant velocity core may be assumed. From the knowledge of the perturbation $< \hat{U}_k \overline{V}k >$ and of the layer height, the problem can be solved. Therefore we need a length scale and a velocity scale, which result from the competition between the wall turbulence and the wave turbulence. At this stage, no quantitative prediction founded on experimental analysis can be given. However the length scale may be obtained from the following consideration.

Let us assume a two-layer flow as defined in the previous section. In the upper layer if one excludes the two regions near the y-axis and the lateral wall, the advection can be neglected. A diffusion balance leads to the following modeled equations:

$$C_\mu \frac{\partial}{\partial y} \left(\frac{e_L^2}{\varepsilon_L} \cdot \frac{\partial e_L}{\partial y} \right) = 8\varepsilon_L \qquad (10)$$

$$C_\mu \frac{\partial}{\partial y} \left(\frac{e_L^2}{\varepsilon_L} \cdot \frac{\partial \varepsilon L}{\partial y} \right) = 8C_2 \frac{\varepsilon_L^2}{e_L} \qquad (11)$$

in which C_μ, C_2 are classical constants in $k - \varepsilon$ models.

These equations have the solutions (MARODON [17]):

$$e_L = e_{IL} \quad \exp \left(- \frac{y_o - y}{1} \right) \qquad (12)$$

$$\varepsilon_L = \varepsilon_{IL} \quad \exp \left(- \frac{3}{2} \cdot \frac{y_o - y}{1} \right) \qquad (13)$$

with the condition $C_2 = 2$ (close to the value 1.9 used usually). The subscript o refers to the conditions near the waves. These equations introduce a local length scale 1, linked to the diffusion process under the waves.

An approximate general solution for e may be given by matching the two particular solutions in each layer. Taking into account the proportionality between τ and e in the lower layer, results in the following expression:

$$e_L = e_{IL} \exp\left(- \frac{y_o - y}{1}\right) + u_{WL}^2 \frac{4}{C_\mu} \left(\frac{y_o - y}{y_o}\right) \qquad (14)$$

Using $1/h = 0.15$ for the RUN 12, one obtains the solution shown in Figure 3. The experimental data, corresponding to the z-location where advection is negligible ($Z/L = 0.33$, 0.60), are in agreement with this solution. The turbulence energy minimum is located close to the mean velocity maximum. Therefore, one can predict the thickness of each layer from the local scales e_{IL} and 1 linked to the turbulence production in the waves. The length scale 1 is found to be well correlated to the wave length λ by MARODON [17]:

$$\frac{1}{\lambda} = 0.1 \qquad \text{when} \qquad \frac{u_{IG}^2}{g1} > 1 \qquad (15)$$

with the measured value 10 cm for λ, in RUN 12, we find $1/\lambda = 0.05$

Relating e_{IL} to the friction velocity u_{IG} gives the solution to the problem. Further experimental studies on the wave structure are needed to define constitutive equations for u_{IL} and l.

5. CONCLUSION

The experimental results presented in this paper show the particular structure of the flow in both phases. They confirm the presence of a perturbative secondary motion, with a streamline pattern, made up of two cells symmetrical with respect to the y-axis, in a cross section of the flow.

In such flow, the wall and interfacial shear stress are not determined easily. Based upon our experiences with the three methods tested, we suggest that the determinations founded on logarithmic profile be used whenever possible.

The turbulence in the liquid, suggests a two layer flow pattern: a lower layer, where the eddy viscosity concept applies and an upper layer, resulting from a dissipation-diffusion balance where the particular rheological behaviour is underlined.

From this description, we propose an idea for the flow modeling, requiring two local scales linked to the interfacial wavy structure. Further experiments are needed to confirm this idea.

ACKNOWLEGMENTS

The authors wish to acknowledge the assistance of Mrs. A. Miré and R. Fernandez Flores for their contributions to the experiments.

LIST OF SYMBOLS

$e = \overline{u_i' u_i'}$: twice the turbulent kinetic energy

h : flow height

I : bottom slope

I_G : pressure drop in gas

$\overline{P}$: mean pressure

$\overline{U}$: x-velocity component

u' : x-fluctuating velocity

$\overline{V}$: y-velocity component

v' : y-fluctuating velocity

$\overline{W}$: z-velocity component

u : velocity scale/friction velocity

x : longitudinal coordinate

y : vertical coordinate

z : lateral coordinate

α : local time fraction

ε : dissipation rate

ρ : density

τ_{ij} : shear stress component

ν_t : eddy viscosity

SUBSCRIPTS

I : relative to interface

G : relative to gas

L : relative to liquid

W : relative to wall

REFERENCES

[1] Masbernat, L., Suzanne, C., Referowski, V., Turbulent velocity measurements in stratified co-current two phase flow. Proceeding of the LDA Symposium, Copenhagen, 1975, pp. 593-598.

[2] Suzanne, C., Etude d'un écoulement stratifié gaz-liquide. Thesis, UPS Toulouse, 1977.

[3] Gayral, B., Masbernat, L., Suzanne, C., Mean velocities and Reynolds stresses in co-current gas-liquid stratified channel flow. Int. Sem., Momentum, Heat and Mass Transfer in two-phase energy and chemical systems, Dubrovnik, 1978.

[4] Hadj-Fredj, A., Structure cinématique de l'écoulement stratifié ondulé d'un gaz et d'un liquide. Thesis, INP Toulouse, 1978.

[5] Gayral, B., Bilan d'énergie turbulente en écoulement stratifié co-courant de gaz et de liquide. Thesis, INP Toulouse, 1979.

[6] Fabre, J., Miré, A., Suzanne, C., Structure locale de l'écoulement au voisinage d'un film d'huile épais, soumis au cisaillement du vent et du courant. Symposium IAHR : la mécanique des nappes d'hydrocarbures, Paris, 1981.

[7] Fabre, J., Masbernat, L., Suzanne, C., New results on the structure of stratified gas-liquid flow. Nato Advanced Research Workshop on two phase flows and heat transfer, Spitzingsee, 1982.

[8] Jeffries, R.B., Scott, D.S., Rhodes, E., The structure of turbulence close to the interface in the gas phase of a co-current stratified two-phase flow. Symp. on Mechanisms of two phase flow, AIChE, Portland, 1969.

[9] Jeffries, R.B., Scott, D.S., Rhodes, E., Structure of turbulence close to the interface in the liquid phase of a co-current stratified two-phase flow. Symp. on Fluid Mechanics and Measurements in Two-Phase Flow Systems, AIChE, 1969.

[10] Johns, D.M., Theofanous, T.G., Houze, R.N., Turbulent characteristics of two-phase, gas liquid channel flow. Proceeding of the Third Symposium on turbulence in liquids, 1973.

[11] Theofanous, T.G., Houze, R.N., Johns, D.M., Internal Report, School of Engineering, Purdue University, West Lafayette, 1975.

[12] Akai, M., Inoue, A., Aoki, S., Structure of a co-current stratified two-phase flow with wavy interface. Proceeding of the 25th National Congress for applied Mechanics, Tokyo, 1975.

[13] Akai, M., Inoue, A., Aoki, S., Endo, K., A co-current stratified air-mercury flow with wavy interface. Int. Journal of Multiphase Flow, 1980, Vol. 6, pp. 173-190.

[14] Akai, M., Inoue, A., Aoki, S., The prediction of stratified two phase flow with a two equation model of turbulence. Int. J. of Multiphase Flow, 1981, Vol. 7, pp. 21-39.

[15] Hanratty, T.J., Engen, J.M., Interaction between a turbulent air stream and a moving water surface. AIChE Journal, 1957, Vol. 3, pp. 299-304.

[16] Rodi, W., Turbulence models and their application in hydraulics, State-of-the art paper, IAHR, 1980.

[17] Marodon, D., Turbulence en écoulement d'eau cisaillé par le vent, Thesis, INP Toulouse, 1983.

Multi-Phase Flow and Heat Transfer III. Part A: Fundamentals
edited by T.N. Veziroğlu and A.E. Bergles
Elsevier Science Publishers B.V., Amsterdam, 1984 — Printed in The Netherlands

A STUDY OF A TWO-COMPONENT, TWO-PHASE FLOW SYSTEM IN ONE DIMENSION

Asim Ulke
Westinghouse R&D Center
Pittsburgh, PA 15235, U.S.A.

ABSTRACT

The conservation equations for one-dimensional, time-dependent, two-component, two-phase flows were developed using the two-fluid model. The final set of seven conservation equations in Eulerian coordinates, three equations of state, and the constitutive equations were solved by means of an explicit, space staggered finite difference scheme. The computer code which evolved from this effort was used to solve for the flow field in a piping system having a mixture of water, steam and noncondensable gas for its working fluid. The predictions of the gas and the water properties and steam concentrations along the flow path were made. The computed pressure distributions compared well with the experimental data. In spite of some localized fluctuations, the solutions were stable for time steps close to those predicted by the Courant-Friedrichs-Lewy criterion for the gas phase only.

1. INTRODUCTION

There are two fundamentally different formulations of the equations for two-phase flow systems, namely, the mixture models and the two-fluid model. The mixture models are formulated by considering a two-phase system as a whole. Highly transient two-phase flow situations often involve sudden accelerations of one phase with respect to the other and thus can not be appropriately described by the mixture models. In such cases each phase should be considered separately by use of the two-fluid model. This model requires two sets of conservation equations governing the balance of mass, momentum and energy. Constitutive equations for the transport of mass, momentum and energy across the interfaces and two equations of state complete the formulation of one-component, two-phase flow. The present study involves a second component which is a noncondensable gas. It was decided to treat the gas and vapor components as a homogeneous mixture. The resulting system requires third continuity and state equations. Since the noncondensable gas component was continuous while the vapor could appear or disappear, a complete set of conservation equations was assigned to the gas-vapor mixture and the vapor mass was computed from the third continuity equation.

2. THE CONSERVATION EQUATIONS

We consider a one-dimensional, compressible and inviscid flow except for interfacial friction and drag and friction at the boundaries. We assume the equilibrium of pressure between the phases. The Eulerian working equations which are derived from the integral forms of the conservation equations [1,2], after dropping the vector designators for convenience, are as follows:

Continuity:

$$\forall \frac{\partial D_a}{\partial t} + \sum F_a \cdot n\, A = \forall\, \dot{m}_{ab} + \dot{m}_{a,in} \tag{1}$$

Momentum:

$$\forall \frac{\partial F_a}{\partial t} + \sum F_a V_a \cdot n\, A = - \sum \alpha_a (g_c P + Q)\, n\, A$$

$$+ \forall\, (D_a\, g \cos\beta + F_{ca} + F_{wa} + \dot{m}_{ab} \hat{V} + F_{ab}) + \dot{m}_{a,in} V_{a,in} \tag{2}$$

Energy:

$$\forall \frac{\partial E_a}{\partial t} + \sum [E_a + \alpha_a (P + Q/g_c)]\, V_a \cdot nA = \dot{m}_{a,in} h_{a,in}$$

$$+ \forall\, [q_{wa} - q_{ai} + E_{ai} - (P + Q/g_c) \frac{\partial \alpha_a}{\partial t}] \tag{3}$$

where,

$$D_a \equiv \alpha_a \rho_a, \qquad F_a \equiv \alpha_a \rho_a V_a, \qquad E_a = \alpha_a \rho_a e_a .$$

The dummy subscripts a and b are replaced by g or ℓ to obtain the conservation equations for gas and liquid respectively. The first term on the right hand side represents the energy entering the control volume from an external source. The contribution of kinetic energy in this term has been neglected. The total specific energy, e, neglecting the potential energy is defined as:

$$e = u + V^2/2g_c \tag{4}$$

The last term in Equation (3) represents the interface work. In absolute terms its contribution to the energy equation is generally negligible. However, it can become large relative to the other terms when liquid is just entering a cell or disappearing from it. This drives the liquid energy equation to instability or forces the time steps to unacceptably small values. Consequently is was neglected.

The respective sums of the interaction terms must vanish when the conservation equations are written for the complete system. This leads to the so-called interface jump conditions [1].

$$\dot{m}_{g\ell} = - \dot{m}_{\ell g} = \dot{m} \tag{5}$$

$$\dot{m}\,\hat{V}_g - \dot{m}\,\hat{V}_\ell + F_{g\ell} + F_{\ell g} = 0 \tag{6}$$

$$- q_{gi} + E_{gi} - q_{\ell i} + E_{\ell i} = 0 \tag{7}$$

In Equation (5) the sign of evaporation has been taken positive. In Equation (6) $\hat{V}_g$ and $\hat{V}_\ell$ are the intrinsic velocities. They were taken equal to the liquid velocity for evaporation and vapor velocity for condensation for a given cell. As a result the first two terms cancel and Equation (6) becomes $F_{g\ell} = - F_{\ell g}$. The definitions of the terms in Equation (7) can be obtained from an energy balance at the interface which yields,

$$\dot{m}\,h_{vs} - \dot{m}h_{\ell s} = q_{gi} - q_{i\ell} \tag{8}$$

The terms on the left are the energy transfer rates from the interface to the gas and from the liquid to the interface due to mass transfer respectively. The terms on the right are the sensible heat transfer rates from the gas to the interface and from the interface to the liquid. The comparison of Equations (7) and (8) shows that, $q_{\ell i} = -q_{i\ell}$, $E_{gi} = \dot{m}\,h_{vs}$ and $E_{\ell i} = -\dot{m}\,h_{\ell s}$.

The final condition to be imposed on the solution is that the volume fractions must add to unity.

$$\alpha_g + \alpha_\ell = 1 \tag{9}$$

These equations define the two-phase flow. The continuity equation for the vapor component provides the information for the amount of vapor in the gas-vapor mixture.

$$\forall \frac{\partial D_v}{\partial t} + \sum \frac{D_v}{D_g} F_g \cdot n\,A = \forall\,\dot{m} \tag{10}$$

where, $D_v \equiv \alpha_v\,\rho_v$. The ratio, D_v/D_g, in the convection term comes from the assumption of homogenity of the gas components.

3. THE CONSTITUTIVE EQUATIONS

One of the difficulties encountered in the analyses of two-phase flows is that, in addition to the primary variables, the flow structure may also be changing in time and space. Each flow structure or "flow regime" has its own characteristics which must be represented by the constitutive equations. In the systems we were dealing with, the flow structure could be approximated by spherical droplets in a continuous gas flow, except early in the event when the flow regime changes from slug to churn and to annular with a droplet core in a very short time. The annular flow regime may even persist for a while. Thus the addition of an annular flow model might be an improvement but could not be justified at this time.

3.1 The Liquid Droplet Size

The droplet diameters at points of liquid injection were obtained using the Nukiyama and Tanasawa expression [3].

$$\bar{D}_d = \frac{585}{V_g - V_\ell} \left(\frac{\sigma}{\rho_j}\right)^{0.5} + 597 \left(\frac{1000\, A_j V_j}{A_g\, V_g}\right)^{1.5} \left[\frac{\mu_j}{(\sigma\, \rho_i)^{0.5}}\right]^{0.45} \tag{11}$$

The numerical constants in the above equation result from use of the CGS units with the liquid viscosity μ_j in poises (1 poise = 0.1 Pa-S).

The generally accepted procedure for the prediction of the droplet sizes in the flow stream is to use the Weber number criterion. The Weber number is defined by

$$W_e = D_d\, \rho_g (V_g - V_\ell)^2 / g_c \sigma \tag{12}$$

The suggested values for the critical Weber number which give the maximum stable droplet diameters range from 8 to 20. The mean droplet diameter is calculated from this maximum value and an assumed or empirical size spectrum. Moeck [4] recommends 13 for the critical Weber number and the following relationship between the maximum and the mean droplet diameters at a cross section.

$$\bar{D}_d = 0.06147\, (D_d)_{max} \tag{13}$$

It should be pointed out that the critical Weber number sets an upper limit on the droplet size in steady or slowly varying flow situations. In rapidly changing flows with significant heat transfer, droplet break-up and coagulation, it is not possible to predict the mean droplet size distributions with confidence. In this study, the Weber number criterion was used to determine the tendencies of droplets to grow or decrease in size along the flow path. The droplet diameters at any time were restricted to the range of 0.8 to 2.0 times the diameter at the last upstream liquid injection cell, computed from Equation (11).

3.2 Interface Momentum Exchange

The interfacial force can be obtained from [1],

$$F_{g\ell} = -A_{g\ell}\, \rho_g\, |V_g - V_\ell|\, (V_g - V_\ell)\, C_d / 8 \tag{14}$$

where, $A_{g\ell}$ is the total surface area of the droplets per unit volume and equal to $6\, \alpha_\ell / \bar{D}_d$. C_d is the drag coefficient given by,

$$C_d = \frac{24}{Re_d} (1 + 0.15\, Re_d^{0.687}) + \frac{0.42}{1 + 42500\, Re_d^{-1.16}} \tag{15}$$

and Re_d is the droplet Reynolds number.

$$Re_d = \rho_g\, \bar{D}_d\, |V_g - V_\ell| / \mu_g \tag{16}$$

3.3 Wall to Phase Momentum Transfer

The wall frictional force for the gas phase was calculated from [1],

$$F_{wg} = - f_{wg} \alpha_g \rho_g V_g |V_g| /2D_h \tag{17}$$

The liquid frictional force was obtained from the Martinelli-Nelson equation given by [5],

$$F_{w\ell} = - f_{w\ell} \rho_\ell V_\ell |V_\ell| \left(\frac{\dot{m}_\ell}{\dot{m}_g + \dot{m}_\ell}\right)^{1.75} /2D_h \tag{18}$$

The friction coefficients for both phases were calculated from,

$$f_{wa} = 64/Re_{wa}, \quad \text{for laminar flows,} \tag{19}$$

$$f_{wa} = 0.224/Re_{wa}^{0.2}, \quad \text{for transitional flows,} \tag{20}$$

$$f_{wa} = 1/[2.0 \log_{10} (\frac{D_h}{2k_s}) + 1.74]^2 \tag{21}$$

for fully turbulent flows, with the pipe Reynolds number defined by

$$Re_{wa} = D_h \rho_a |V_a| /\mu_a \tag{22}$$

Equations (19) and (20) give the same value for f_{wa} at Re_{wa} = 1175. This value was used as the critical Reynolds number to avoid the discontinuity at the expense of some accuracy. Equation (21) was used as the lower bound to f_{wa}.

3.4 Interface Mass and Energy Exchange

The net rate of mass transfer $\dot{m}$ is the sum of the contributions due to the temperature difference between the phases, and due to the pressure changes (flashing). Similarly, the heat transfer rates q_{gi} and q_{li} have two contributions. The heat transfer from the interface to the droplet was assumed to be by conduction only. The heat transfer coefficient was derived from the temperature distribution in a sphere with a constant surface temperature [6].

$$H_\ell = 6.58 \, k_\ell/\bar{D}_d \, [0.333 \left(\frac{D_g - D_v}{D_g}\right)^{-0.198}] \tag{23}$$

The term inside the brackets is a correction factor which accounts for the presence of the noncondensable gas [7]. The corresponding heat transfer rate is

$$q_{\ell,\Delta T} = H_\ell A_{g\ell} (T_s - T_\ell) \tag{24}$$

The heat transfer coefficient for convection from the vapor to the interface was obtained from the Frössling equation [8], modified to account for the vapor blanketing in evaporation [9].

$$H_g = (2.0 + 0.74\, Re_d^{0.5}\, Pr_g^{0.33})\, k_g/\bar{D}_d\, (1 + B)^{-0.6} \tag{25}$$

where, B is the Spalding parameter given by

$$B = C_{pv}\, (T_g - T_\ell)/(h_{vs} - h_{\ell s}) \tag{26}$$

The corresponding heat transfer rate is

$$q_{g,\Delta T} = H_g\, A_g\, (T_g - T_s) \tag{27}$$

The mass transfer rate due to temperature difference is

$$\dot{m}_{\Delta T} = (q_{g,\Delta T} - q_{\ell,\Delta T})/(h_{vs} - h_{\ell s}) \tag{28}$$

The energy transfer rates due to flashing are derived from the system energy equation [1].

$$q_{g,\Delta P} = - T_s\, F_{vs}\, \alpha_v\, \rho_{vs} \frac{\partial S_{vs}}{\partial P} \left(\frac{\partial P}{\partial t} + V_g \frac{\partial P}{\partial x}\right) \tag{29}$$

$$q_{\ell,\Delta P} = - T_s\, F_{\ell s} \alpha_\ell\, \rho_{\ell s} \frac{\partial S_{\ell s}}{\partial P} \left(\frac{\partial P}{\partial t} + V_\ell \frac{\partial P}{\partial x}\right) \tag{30}$$

The corresponding mass transfer rate is

$$\dot{m}_{\Delta P} = (q_{g,\Delta P} + q_{\ell,\Delta P})/(h_{vs} - h_{\ell s}) \tag{31}$$

The sum of Equations (28) and (31) is equivalent to the jump condition given by Equation (8) or,

$$q_{gi} = q_{g,\Delta T} + q_{g,\Delta P} \tag{32}$$

$$q_{i\ell} = q_{\ell,\Delta T} - q_{\ell,\Delta P} \tag{33}$$

Since flashing can only cause the condensation of the vapor or evaporation of the liquid, the reverse processes implied by the positive values of $q_{g,\Delta P}$ and the negative values of $q_{\ell,\Delta P}$ are not admissible.

The fraction of vapor at saturation F_{vs} and that of liquid $F_{\ell s}$ in Equations (29) and (30) were approximated by [10],

$$F_{vs} = \exp\left\{- (T_g - T_s)\ [1/(1 + T_s - T_\ell)]\right\} \tag{34}$$

$$F_{\ell s} = \exp\left\{- (T_s - T_\ell)\ [1/(1 + T_g - T_s)]\right\} \tag{35}$$

3.5 Wall to Phase Heat Transfer

The forced convection heat transfer coefficients were obtained from [11],

$$H_{ca} = 4.36\ k_a/D_h, \quad \text{laminar flows,} \tag{36}$$

$$H_{ca} = 0.023\ Re_{wa}^{0.8}\ Pr_a^{0.4}\ k_a/D_h, \quad \text{turbulent flows.} \tag{37}$$

The wall heat transfer rate is then given by,

$$q_{wa} = H_{ca}\ A_{wa}\ (T_w - T_a) \tag{38}$$

The above equation requires the wall surface temperature T_w. In processes of short duration the effects of heat transfer are mostly confined within a narrow region adjacent to the surface. Consequently, the surface temperature can be approximated by an analytical solution of the surface temperature for a semi-infinite solid subjected to a constant heat flux [6].

$$T_w = T_{wi} + \frac{2(\dot{q}_w/A_w)}{k_w}\left[\frac{\alpha_w(t - t_i)}{\pi}\right]^{0.5} \tag{39}$$

where, $\dot{q}_w/A_w$ is the net average heat flux during the time period $t - t_i$.

3.6 The Flow Losses Due to Area Change and Turning

Sudden changes in cross sectional areas, elbows, bends, valves, etc., along a flow stream cannot be described by a one-dimensional inviscid modeling. Such irregularities cause flow separations and secondary flows and lead to the conversion of kinetic energy to thermal. In most situations only a fraction of this thermal energy can be recovered and the remainder is considered a loss. It is possible to predict the magnitude of this loss in only a few special cases of compressible flow. In general these losses can only be approximated or guessed. In this work the loss of momentum is calculated as follows.

$$F_{ea} = -\frac{1}{2\forall}\ A\ \alpha_a\ \rho_a\ K\ V_a^2 \tag{40}$$

where the values of the loss coefficient K are obtained from hydraulic loss coefficient tables for elbows and bends, and calculated for a sudden area expansion from,

$$K = (1 - A_1/A_2)^2 \tag{41}$$

and for a sudden contraction from,

$$K = (1/C_c - 1)^2 \tag{42}$$

3.7 The Equations of State

The specific internal energy of the noncondensable gas mixture and that of the vapor (in this case steam) were computed by fitting least squares polynomials to Gas Tables [12] data. The gas and the steam thermal conductivities were computed by curve fits to the data in Kreith [13]. The saturation properties of water was obtained using the Steam Tables [14] data by interpolation. The system pressure was obtained from the ideal gas equation using the mixture gas constant, the density and the temperature. The liquid was assumed to be incompressible with a constant specific heat.

4. THE SOLUTION TECHNIQUE

The conservation equations were discretized using a space staggered, explicit finite difference scheme based on that suggested by Crowley [15]. The recommended upstream scheme was replaced by central differencing which is always unstable without some artificial damping. The damping was provided by the von Neumann artificial viscosity [16] represented by Q in the conservation equations and computed as follows for a cell at j-1/2.

$$Q_{j-1/2} = D_{j-1/2} \left\{ (C_Q \bar{S}_{j-1/2})^2 - C_L \, C_{j-1/2} \overset{+}{S}_{j-1/2} \right\} \tag{43}$$

$$\bar{S}_{j-1/2} = 1/2 \left\{ |V_j - V_{j-1}| - (V_j - V_{j-1}) \right\} \tag{44}$$

$$\overset{+}{S}_{j-1/2} = 1/2 \left\{ |V_j - V_{j-1}| + (V_j - V_{j-1}) \right\} \tag{45}$$

where, $C_{j-1/2}$ is the sound speed, C_Q is the coefficient of quadratic artificial viscosity with a value around 2.0 and C_L is the coefficient of linear viscosity with a value around 0.1. The artificial viscosity alone was inadequate for two-phase flows and a smoothing procedure was developed. The values of D_g, F_g, E_g and D_v were replaced every time step by their smoothed values, obtained from equations of type,

$$\tilde{D}_{j-1/2} = (1-\omega)\, D_{j-1/2} + 0.5\, \omega (D_{j-1} + D_j) \tag{46}$$

where the smoothing parameter ω was computed from

$$\omega = \omega_o \exp[-\varepsilon(N - N_o)] \Delta t/\Delta t_{ref} \tag{47}$$

$$\omega = \omega_o \Delta t/\Delta t_{ref} \quad , \quad N \leq N_o \tag{48}$$

$$\omega \geq \omega_{min} \Delta t/\Delta t_{ref} \tag{49}$$

The smoothing is nonconservative and leads to errors in the global mass and energy balances. The value of ω should be chosen as the smallest number which leads to a stable solution.

Originally the time steps were computed automatically from a Courant-Friedrichs-Lewy criterion. It proved to be of questionable value for two-phase flows and it was discarded to save computation time.

The boundary conditions are the pressure and the gas temperature at the inlet and the pressure at the exit. The forward and backward differencing over half cells at the inlet and exit produce the velocities at the inlet and exit and the temperatures at the exit. The finite difference equations and the mesh are given in Reference [2] and will not be repeated. The computational procedure starts with the solution of the momentum equations at the nodes, this is followed by the continuity equations at the cell centers. The gas phase energy equation and the equation of state are involved iteratively for the system pressure. The solution of the liquid energy and state equations complete the main computational algorithm.

It should be pointed out that the constitutive equations described in Section 3 represent a set which works. Since they do not affect the computational scheme, any or all of these equations can be replaced by different ones which may describe a particular system better.

5. THE RESULTS AND DISCUSSION

This work was undertaken to develop an analytical tool capable of predicting the flow field in a large piping system. The system consists of a gas supply section, a vertical inlet pipe, a nozzle and a discharge piping assembly. A partial schematic of a typical system is shown in Figure 1. Initially, the inlet pipe is filled with water and sealed with a diaphragm downstream of the nozzle. At time zero a gas mixture at high pressure and temperature starts entering the system. This punctures the water closure diapragm and forces the water into the discharge piping assembly. At the same time the cooling water is metered into the discharge pipe just below the nozzle. It takes about 0.040 seconds for continuous gas flow to set up. Therefore, 0.040 seconds was the starting time for the solution. The initial distributions of P, T_g, T_ℓ, V_g, V_ℓ, α_ℓ and the vapor mass fraction were estimated to start the solution. The following results were obtained using 50 nodes with the nodal spacing varying from 0.005 units at the nozzle to 0.05 units in the piping.

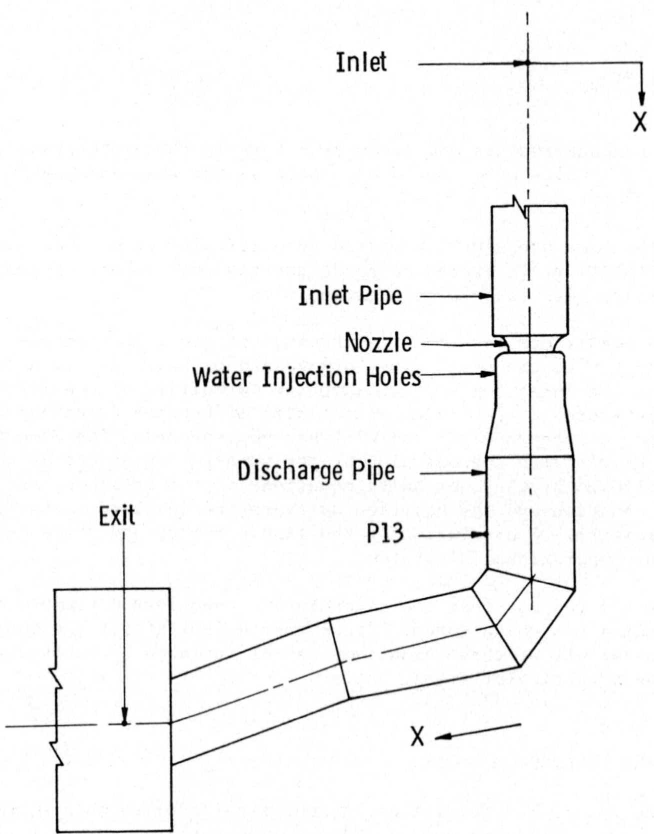

Fig. 1—System geometry

Figure 2 shows the system pressure distributions initially (0.040 S) and at 0.051 S along with the test data. The vapor mass fraction distributions initially and at 0.051 S are shown in Figure 3. The spike is at the water injection area. The water volume fractions are shown in Figure 4. In the earlier runs the initial distribution of water was extended upstream of the injection area. This produced much better results at the beginning of the event but required much smaller time steps for short periods near the drying out of each cell. Some accuracy was traded off for expediency and shorter computation time. The pressures, the gas temperatures and the vapor fractions at 0.196 S are shown in the next three figures. The temperature distribution of Figure 6 was typical for the entire run and was supported by the limited data

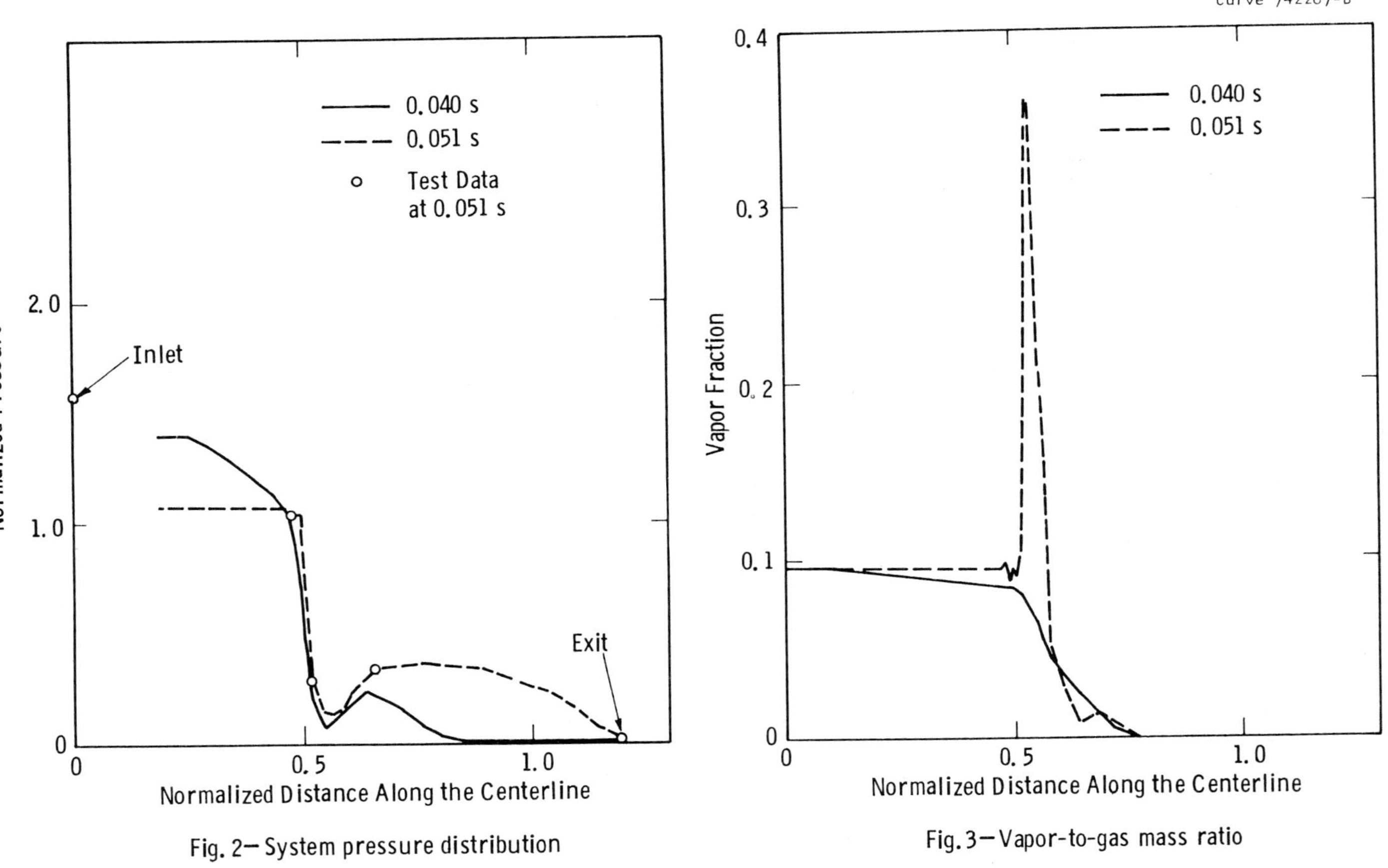

Fig. 2— System pressure distribution

Fig. 3—Vapor-to-gas mass ratio

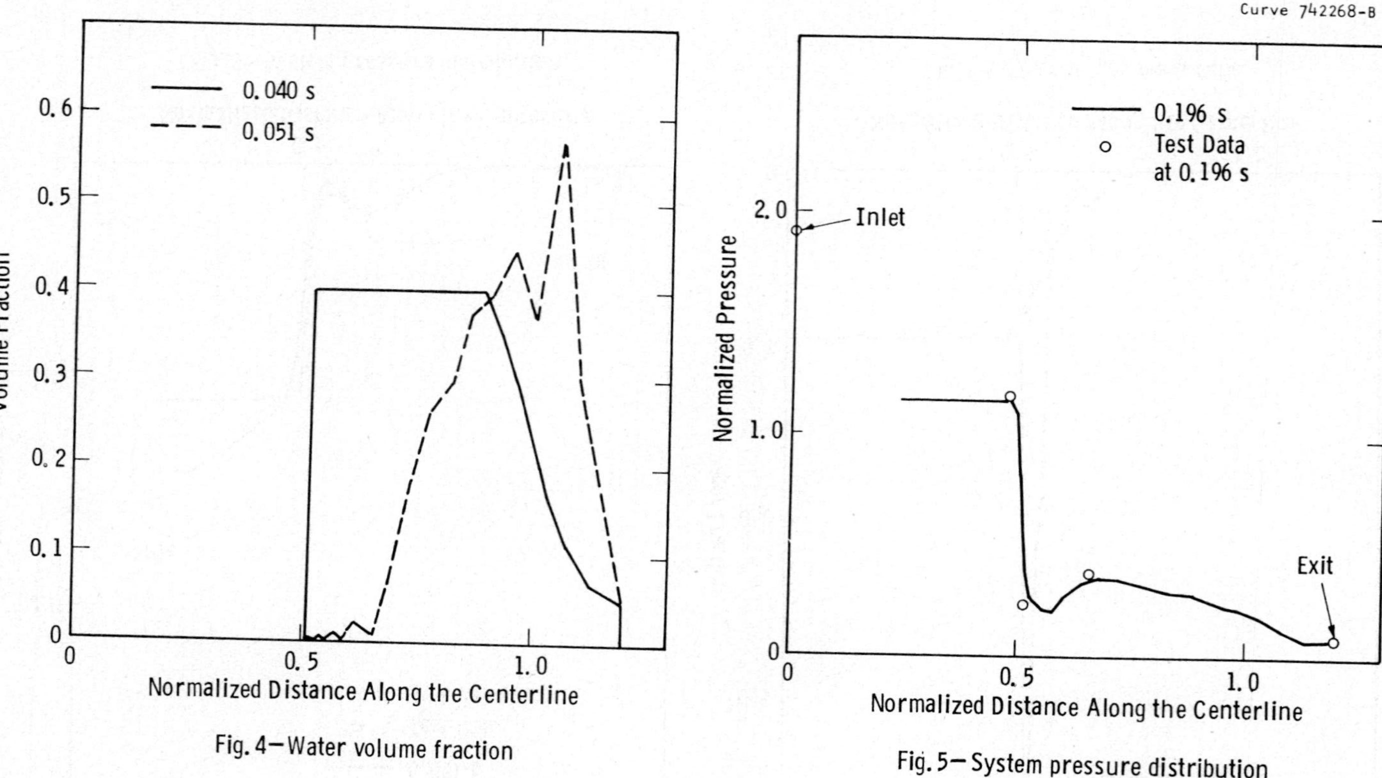

Fig. 4—Water volume fraction

Fig. 5—System pressure distribution

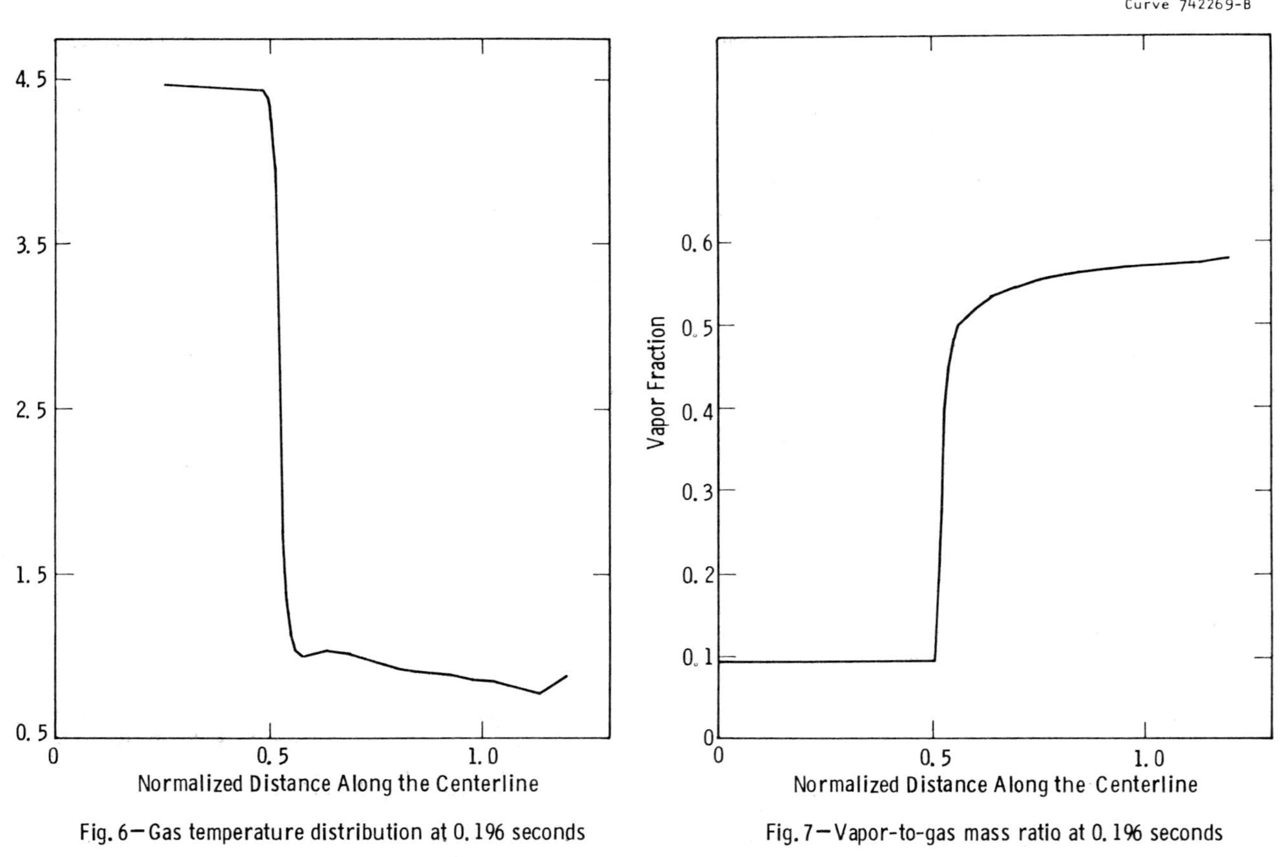

Fig. 6—Gas temperature distribution at 0.196 seconds

Fig. 7—Vapor-to-gas mass ratio at 0.196 seconds

we have. The vapor fraction distribution did not change appreciably during the remainder of the run from that in Figure 7. Figure 8 shows the pressure history at the inlet boundary and compares the predicted and the recorded inlet pipe pressures. The waviness in the data is due to a peculiar leakage flow in the gas supply system which was not accounted for in the solution. Figure 9 shows the pressure history at the exit boundary and at a transducer location about 0.15 units downstream of the nozzle throat. The discrepancy at 0.10 S is mostly due to the inaccuracies in the initial data. During the remainder of the run, the predictions are about 20.0 percent lower than the data. There are a number of contributors to this such as the differences between the actual and the modeled geometry, deficiencies in the flow regime, the droplet size, the interface heat transfer, the frictional and minor loss models, the grid structure, etc. The underprediction of the losses due to turning and area change is probably the most important among these. The oscillations of the predicted data coincided with the decrease of the water content of the piping system to a relatively small value, about 1.0 lbm (454.0 gm). The test data also show oscillations with somewhat smaller amplitude. These oscillations were not common to some of the modifications of the basic system. In fact, Figure 10 shows the pressure history at a comparable location for a similar system. In that run the loss coefficients for area change were modeled slightly higher. Besides having no oscillatory behavior, the pressure history shows a trend similar to those at the inlet pipe which is more likely the case.

6. CONCLUSIONS

The results of this work have shown that it is possible to obtain practical and reasonably accurate solutions to one-dimensional, transient, two-component two-phase flow problems using the two-fluid model with an explicit finite difference scheme. However, the effort is by no means complete. As a result of numerical smoothing the solution deviates from the full conservation law form which is somewhat disturbing. Several attempts to eliminate the need for it by changing the differencing scheme were unsuccessful, though, the possibilities may still exist.

The extension of the solution to the fully liquid and to the annular flow regimes would widen the range of applicability of the solution and would also improve its accuracy especially for the type of problems discussed here.

Our knowledge of the two phase flow phenomena is simply inadequate to describe it especially at the gas liquid interface. Furthermore, many of the existing two-phase flow correlations consider the flow as a whole. The need to describe the phenomena by considering each phase or component separately, which is unique to two-fluid models, often requires new interpretations and further approximations of these correlations. More work is needed in this area to extend the usefulness of two-fluid models.

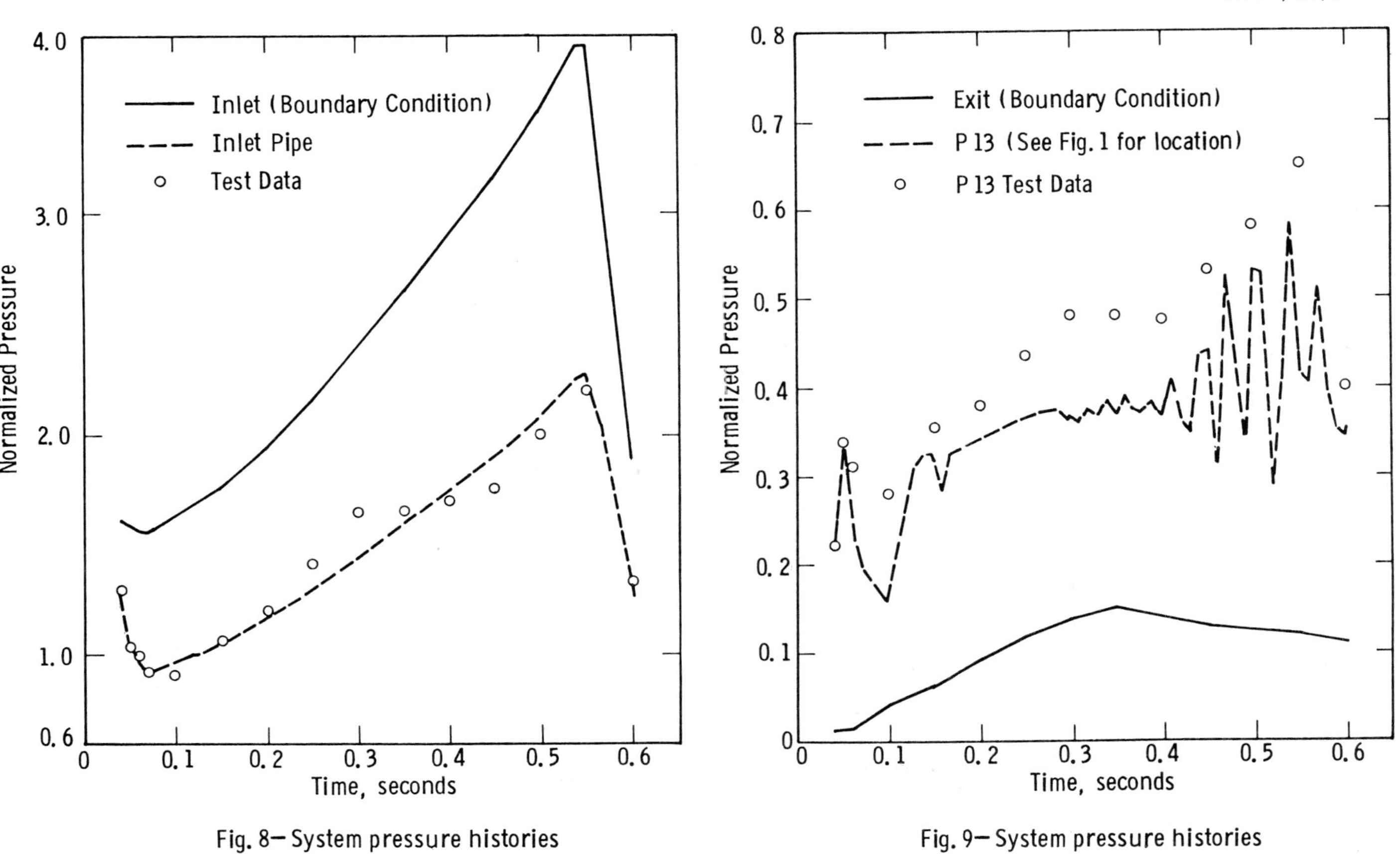

Fig. 8— System pressure histories

Fig. 9— System pressure histories

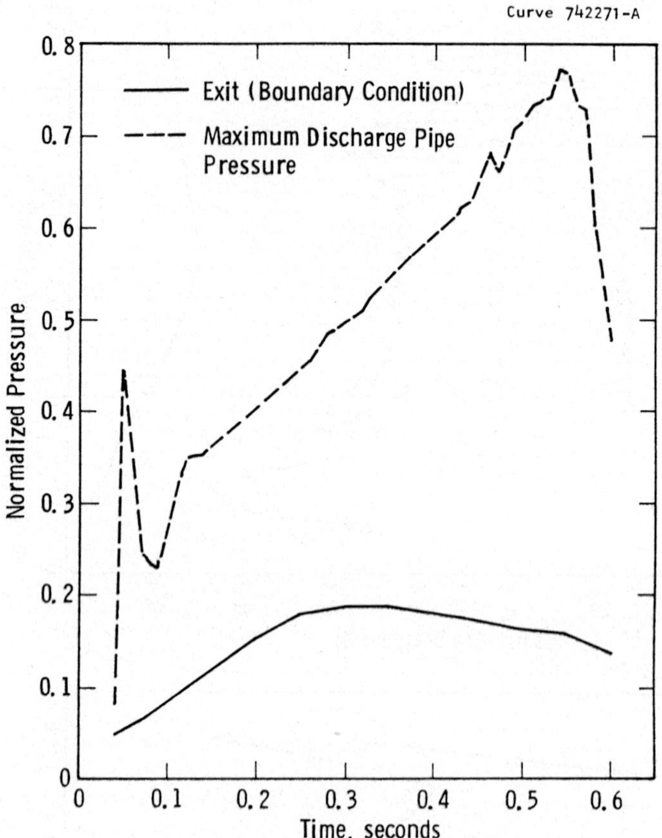

Fig. 10— System pressure histories

ACKNOWLEDGEMENTS

The work reported here was sponsored by the Westinghouse Marine Division, Sunnyvale, California.

REFERENCES

1. Hughes, E. D., Lyczkowski, R. W., and McFadden, J. H., An Evaluation of State-of-the-Art Two-Velocity Two-Phase Flow Models and Their Applicability to Nuclear Reactor Transient Analysis, EPRI NP-143, Vol. 2, Feb. 1976.

2. Ulke, A. "Time Dependent, Two-Component, Two-Phase Flow in a Duct," AIChE Fall Annual Meeting, Los Angeles, Nov. 14-18, 1982.

3. Nukiyama, S. and Tanasawa, Y., "Experiments on the Atomization of Liquids in an Air Stream," Transactions of the Society of Mechanical Engineers (Japan), Vol. 5, No. 18, Feb. 1939, pp. 62-67.

4. Moeck, E. O., Annular-Dispersed Two-Phase Flow and Critical Heat Flux, AECL-3650, 1970.

5. Tong, L. S., Boiling Heat Transfer and Two-Phase Flow, John Wiley & Sons, Inc., New York, 1966.

6. Carslaw, H. S., and Jaeger, J. C., Conduction of Heat in Solids, Oxford, London, 1959.

7. Chivate, H. S., and Agarwal, S. K., "Partial Condensation of a Vapor in the Presence of Noncondensable Gases," Indian Chemical Journal, Vol. 7, 1972, pp. 21-25.

8. Fuchs, N. A., Evaporation and Droplet Growth in Gaseous Media, Pergammon, 1959.

9. Hoffman, T. W., and Ross, L. L., "A Theoretical Investigation of the Effect of Mass Transfer on Heat Transfer to an Evaporating Droplet," Int. J. Heat Mass Transfer, Vol. 15, pp. 599-617, 1972.

10. Solbrig, C. W., Hocevar, C. J., and Hughes E. D., "A Model for a Heterogeneous Two-Phase Equal Temperature Liquid," Submitted through AIChE H. T. & E. C. Division for the 1977 Nat. H. T. Conference.

11. Rohsenow, W. M., and Choi, H., Heat, Mass, and Momentum Transfer, Prentice Hall, Englewood Cliffs, NJ 1961.

12. Keenan, J. H., and Kaye, J., Gas Tables, John Wiley & Sons, Inc., New York 1961.

13. Kreith, F., Principles of Heat Transfer, International Textbook Co., 1966.

14. Keenan, J. H., and Keyes, F. G., Thermodynamic Properties of Steam, John Wiley & Sons, Inc., New York 1963.

15. Crowley, W. P., Numerical Methods in Fluid Dynamics, Lawrence Livermore Lab., UCRL-51824, May 1975.

16. von Neumann, J., and Richtmyer, R. D. , "A Method for the Calculation of Aerodynamic Shocks," Journal of Applied Physics, Vol. 21, March 1950.

NOMENCLATURE

A   Area, ft^2 (m^2)[1]

C   Speed of sound, ft/s (m/s)

C_c Contraction coefficient

---

[1]Units in parentheses are the consistent SI equivalents

C_d	Drag coefficient
C_L	Coefficient of linear artificial viscosity
C_Q	Coefficient of quadratic artificial viscosity
D	Macroscopic density, $\alpha\rho$, lbm/ft^3 (kg/m^3)
D_h	Hydraulic diameter, ft (m)
D_p	Droplet diameter, ft (m)
e	Total specific energy, $ft\text{-}lb_f/lb_m$ (J/kg)
E	Macroscopic energy, $\alpha\rho e$, $ft\text{-}lb_f/ft^3$ (J/m^3), or energy exchange due to mass exchange when double subscripted, $ft\text{-}lb_f/ft^3\text{-}s$ (W/m^3)
F	Force per unit volume, $lb_m/ft^2\text{-}s^2$ (N/m^3), or fraction of a phase at saturation, or macroscopic momentum, $\alpha\rho V$, $lb_m/ft^2\text{-}s$ $(kg/m^2\text{-}s)$
g	Gravitational acceleration, ft/s^2 (m/s^2)
g_c	Gravitational constant, 32.174 $ft\text{-}lb_m/lb_f\text{-}s^2$ (1.0 $kg\text{-}m/n\text{-}s^2$)
h	Specific enthalpy, $ft\text{-}lb_f/lb_m$ (J/kg)
H	Heat transfer coefficient, $ft\text{-}lb_f/s\text{-}ft^2\text{-}°R$ $(W/m^2\text{-}°K)$
k	Thermal conductivity, $ft\text{-}lb_f/s\text{-}ft\text{-}°R$ (W/m-°K)
k_s	Surface roughness, ft (m)
K	Loss coefficient
$\dot{m}$	Mass flow rate, lb_m/s (kg/s), or mass exchange rate, $lb_m/ft^3\text{-}s$ $(kg/m^3\text{-}s)$
n	Unit vector along the outward normal to a control surface
N	Iteration count
P	Pressure, lb_f/ft^2 (Pa)
q	Heat transfer rate, $ft\text{-}lb_f/ft^3\text{-}s$ (W/m^3)
Q	Artificial viscosity, $lb_m/ft\text{-}s^2$ (Pa)
Re	Reynolds number
S	Specific entropy, $ft\text{-}lb_f/lb_m °F$ (J/kg-°K)
$\bar{S}$	Defined by Equation 44
$\overset{+}{S}$	Defined by Equation 45
T	Temperature, °R (°K)
t	Time, s (s)
u	Specific internal energy, $ft\text{-}lb_f/lb_m$ (J/kg)
V	Velocity, ft/s (m/S)
$\hat{V}$	Intrinsic velocity, ft/s (m/S)
¥	Volume, $ft^3 (m^3)$
W	Rate of work per unit volume, $ft\text{-}lb_f/ft^3\text{-}s$ (W/m^3)
We	Weber number
X	Flow direction, ft (m)

α	Volume fraction
β	Angle between the flow direction and the gravity vector
ρ	Density, lb_m/ft^3 (kg/m^3)
σ	Surface tension, lb_f/ft (N/m)
μ	Absolute viscosity, lb_m/ft-s (Pa-s)

Subscripts

1	Before expansion	in	Entering
2	After expansion	j	Jet
a,b	Dummy subscripts	ℓ	Liquid
c	Convective	s	Saturation
d	Droplet	v	Vapor
g	Gas	w	Wall
i	Interface, initial	ΔT	Temperature related
ref	Reference value	ΔP	Pressure related

Multi-Phase Flow and Heat Transfer III. Part A: Fundamentals
edited by T.N. Veziroğlu and A.E. Bergles
Elsevier Science Publishers B.V., Amsterdam, 1984 — Printed in The Netherlands

A STUDY OF MOMENTUM TRANSFER IN TWO-FLUID FORMULATION OF TWO-PHASE FLOW

George Egely* and Pradip Saha
Department of Nuclear Energy
Brookhaven National Laboratory
Upton, New York 11973, U.S.A.

ABSTRACT

Advanced nuclear safety codes such as TRAC and RELAP5 use two-fluid hydraulic models. However, there are uncertainties for the application of different correlations. This paper intends to show the effects and importance of a number of correlations for wall friction, interphase drag, and virtual mass. It has been shown that the homogeneous wall shear model yields good results up to the annular flow regime, the single-bubble drag correlation is acceptable, and the inclusion of virtual mass coefficient is helpful. (It has been found that the critical Weber number is not appropriate for bubble radius calculation; it predicts an opposing tendency when compared with the test data.) Also, a two-phase diffuser efficiency is required for diverging ducts and a correlation for the same has been proposed.

NOMENCLATURE

A	flow area
C	coefficient (drag, virtual mass)
D	diameter
F	force
f	friction coefficient
g	acceleration due to gravity
p	pressure
R	bubble radius
Re	Reynolds number
v	velocity
We	Weber number
x	quality
z	axial distance
α	void fraction
σ	surface tension
ρ	density
η	diffuser efficiency
μ	viscosity (kinematic)
ν	viscosity (dynamic)
θ	duct inclination angle with horizontal plane

* Permanent Address: Central Institute of Physical Research, Budapest, Hungary.

Subscripts

b bubble
d drag
g gas
gm gas diffusion
i interphase
ℓ liquid
m mixture
o initial condition
p particle
r relative
vm virtual mass
w wall

1. INTRODUCTION

Interfacial and wall-to-fluid momentum transfer are two important aspects of the generalized two-phase flow analysis. These two items play a major role in determining the pressure drop, void fraction, relative velocity and the interfacial heat and mass transfer in a two-phase flow system. Thus it is essential to have adequate models and correlations for these terms for both adiabatic and heated systems.

The wall-to-fluid momentum transfer has traditionally been accounted for by suitable modification of the single-phase friction factor at the wall. The most popular modification is, of course, the use of a two-phase multiplier. A comparative study of frictional pressure drops is given by Idsinga, et al [1]. However, the question of interfacial momentum transfer is still unresolved. Controversy exists even on the types of terms such as interfacial drag, virtual mass, lift force, Basset force, etc. that should be included in the interface force density of the phasic momentum equations. For example, although the advanced light water reactor thermal-hydraulic codes such as TRAC-PF1 [2] and RELAP5/MOD1 [3] are based on a two-fluid formulation, considerable differences still exist regarding the interfacial force term in these codes. TRAC-PF1 includes only the interfacial drag term, whereas RELAP5/MOD1 includes the interfacial drag and the virtual mass terms. Moreover, the models and correlations used for the interfacial drag term are also different in these two advanced codes.

The main purpose of the present paper is to study, in detail, the various terms and parameters of the interface force density in the one-dimensional two-fluid formulation of two phase flow. These include (i) flow pattern transition, (ii) interfacial drag, and (iii) virtual mass effects. Various correlations for the wall friction have also been studied. Finally, the effect of adverse pressure gradient and the limitation of one-dimensional formulation are discussed.

The study was undertaken with application to two-phase critical flow in mind. The present analysis was, therefore, compared with the high fluid velocity data from the Moby-Dick nitrogen-water tests [4] and the air-water tests of Wisman [5] as reported by Van der Welle [22]. Based on these comparisons, a number of recommendations have been made on the various aspects of momentum exchange for high velocity two-phase flows. It is believed that these recommendations will be useful in developing a general model for two-phase flashing flows where interfacial heat and mass transfer are coupled with the interfacial momentum transfer.

2. PRESENT FORMULATION

Since the interfacial and wall-to-fluid momentum transfer are the main topics of this study, the present analysis is based on the one-dimensional, two-fluid formulation of two-component, two-phase flow in an adiabatic channel with no interfacial heat and mass transfer. Moreover, the Moby-Dick nitrogen-water [4] and Wisman's air-water [5] tests, which are used for the model comparison, were steady-state experiments conducted in variable area channels. Therefore, the steady-state equations are sufficient for the present work. It should be kept in mind that the two-phase mixture did accelerate in the straight and converging part, and decelerate in the diverging part of the test channels in the above experiments. Similar acceleration and deceleration are experienced by a two-phase mixture during transient applications. Therefore, the results of this study should be valid for both steady-state and transient situations.

The one-dimensional, steady-state conservation equations for a variable area channel, but with no interfacial heat and mass transfer, can be written as:

Conservation of mass for gas:

$$\frac{1}{A}\frac{d}{dz}(\alpha\rho_g v_g A) = 0 \tag{1}$$

Conservation of mass for liquid:

$$\frac{1}{A}\frac{d}{dz}\left[(1-\alpha)\rho_\ell v_\ell A\right] = 0 \tag{2}$$

Conservation of momentum for gas:

$$\frac{1}{A}\frac{d}{dz}(\alpha\rho_g v_g^2 A) = -\alpha\frac{dp}{dz} - F_i - F_{wg} - \alpha\rho_g g \sin\theta \tag{3}$$

Conservation of momentum for liquid:

$$\frac{1}{A}\frac{d}{dz}\left[(1-\alpha)\rho_\ell v_\ell^2 A\right] = -(1-\alpha)\frac{dp}{dz} + F_i - F_{w\ell} - (1-\alpha)\rho_\ell g \sin\theta \,. \tag{4}$$

Notice that all the flow parameters are assumed to be uniform over a cross-section and the pressure difference between the gas and liquid phase is neglected. These assumptions are justified for the highly turbulent situations and bubble sizes ($R_b > 10^{-5}$ m) considered in this study.

The dependent variables in the above set of equations are: pressure, p, area averaged void fraction, α, liquid velocity, v_ℓ, and gas velocity, v_g. The phasic specific enthalpies are kept constant so that the phasic densities, ρ_g and ρ_ℓ, are functions of pressure only. Moreover, the gas phase is assumed to behave as a perfect gas. The angle of inclination, θ, and flow area, A, which may be a function of axial coordinates, z, are obtained from the test section orientation and geometry. However, constitutive relations for the interfacial force density, F_i, and wall friction to the gas and liquid phases, F_{wg} and $F_{w\ell}$, are needed to close the problem. It should be noted that the interfacial force density includes the interfacial drag as well as the resistance due to virtual mass.

For easier comparison of the present formulation with the homogeneous and drift-flux (or diffusion) models, the above phasic conservation equations (Eq. 1 through 4) were recast in a different form. With the introduction of the following variables:

Mixture density:

$$\rho_m = \alpha\rho_g + (1-\alpha)\rho_\ell \tag{5}$$

Mixture velocity:

$$v_m = \frac{\alpha\rho_g v_g + (1-\alpha)\rho_\ell v_\ell}{\rho_m} \tag{6}$$

Relative velocity:

$$v_r = v_g - v_\ell \tag{7}$$

Gas diffusion velocity:

$$v_{gm} = v_g - v_m = \frac{(1-\alpha)\rho_\ell v_r}{\rho_m} \tag{8}$$

one can write a new system of governing equations for

Gas mass:

$$\frac{d}{dz}(\alpha\rho_g v_m + \alpha\rho_g v_{gm}) + \frac{1}{A}(\alpha\rho_g v_m + \alpha\rho_g v_{gm})\frac{dA}{dz} = 0 \tag{9}$$

Mixture mass:

$$\frac{d}{dz}(\rho_m v_m) + \frac{1}{A}(\rho_m v_m)\frac{dA}{dz} = 0 \tag{10}$$

Mixture momentum:

$$\rho_m v_m \frac{dv_m}{dz} + \frac{d}{dz}(\alpha\rho_g v_{gm} v_r) + \frac{1}{A}\alpha\rho_g v_{gm} v_r \frac{dA}{dz} = -\frac{dp}{dz} - F_w - \rho_m g \,\mathrm{Sin}\,\theta \tag{11}$$

and, Difference of phasic momentum:

$$\begin{aligned} &\rho_m v_m \frac{d}{dz}\left[\frac{(\alpha\rho_g - (1-\alpha)\rho_\ell)}{\rho_m} v_m + \frac{4\alpha\rho_g}{\rho_m} v_{gm}\right] \\ &+ \frac{d}{dz}\left[(1-\frac{2\alpha\rho_g}{\rho_m})\alpha\rho_g v_{gm} v_r\right] + \frac{1}{A}\left[(1-\frac{2\alpha\rho_g}{\rho_m})\alpha\rho_g v_{gm} v_r\right]\frac{dA}{dz} \\ &= -(2\alpha-1)\frac{dp}{dz} - 2F_i - (F_{wg}-F_{w\ell}) - (\alpha\rho_g - (1-\alpha)\rho_\ell)\, g \,\mathrm{Sin}\,\theta\,. \end{aligned} \tag{12}$$

Note that the wall friction to the mixture, F_w, is the sum of F_{wg} and $F_{w\ell}$.

After further manipulation of the above set with the assumptions that $\rho_\ell = \rho_\ell(p)$ and $\rho_g = \rho_g(p)$, the following system of equations was obtained:

$$\begin{bmatrix} a_{11} & a_{12} & a_{13} & a_{14} \\ a_{21} & a_{22} & a_{23} & a_{24} \\ a_{31} & a_{32} & a_{33} & a_{34} \\ a_{41} & a_{42} & a_{43} & a_{44} \end{bmatrix} \begin{bmatrix} \frac{dp}{dz} \\ \frac{d\alpha}{dz} \\ \frac{dv_m}{dz} \\ \frac{dv_r}{dz} \end{bmatrix} = \begin{bmatrix} b_1 \\ b_2 \\ b_3 \\ b_4 \end{bmatrix} . \quad (13)$$

The elements of matrix [A] and vector B, i.e., a_{ij}'s and b's, are given in the Appendix. Note that the dependent variables are: pressure, p, area-averaged void fraction, α, mixture velocity, v_m, and the relative velocity, v_r. However, the phasic velocities, v_g and v_ℓ, can be calculated from the following identities:

$$v_g \equiv v_m + \frac{(1-\alpha)\rho_\ell}{\rho_m} v_r \quad (14)$$

and

$$v_\ell \equiv v_m - \frac{\alpha\rho_g}{\rho_m} v_r . \quad (15)$$

This completes the present formulation. The constitutive relations that were used for F_w, F_{wg}, $F_{w\ell}$ and F_i will be discussed later.

3. SOLUTION METHOD

A set of ordinary differential equations of the form:

$$\frac{d\tilde{Y}}{dz} = \tilde{F} \quad (16)$$

was first obtained from Equation (13) by using a matrix inversion routine with absolute error less than 10^{-12}. Here $\tilde{Y} = \tilde{Y}(p, \alpha, v_m, v_r)$ and $\tilde{F} = [A]^{-1}B$. Equation (16) was then solved by using an ordinary differential equation solver [6]. In a preliminary study, six methods [7], namely the (i) first-order Euler, (ii) Runge-Kutta-Merson, (iii) fourth-order Runge-Kutta, (iv) Adams-Moulton predictor-corrector, (v) Revised Gear method, and (vi) third-order

Callahan implicit method, were used for the solution of Equation (16). Based on the speed and accuracy, the Adams-Moulton predictor-corrector method was chosen for all the results presented in this paper. The maximum error during the integration was maintained under 1%, and the minimum and maximum step lengths were usually 6×10^{-4} m and 2×10^{-2} m, respectively.

4. EXPERIMENTS SIMULATED

A number of Moby-Dick nitrogen-water [4] and Wisman's air-water [5] tests conducted in vertical, variable area channels have been simulated, with the present model. The Moby-Dick test section consisted of a straight round pipe of 0.014 m inside diameter followed by a diffuser with 7° total included angle. For the tests simulated inlet void fraction and mass-flux ranged from 0.02 to 0.61 and 8100 kg/m^2s to 13000 kg/m^2s, respectively. The entrance pressure was in the range of 1.2 to 2.6 bar. The axial pressure distribution and area-averaged void fraction at 0.34 m upstream of the throat were measured. The void fraction measurement station was considered to be the starting point for the present calculations and the experimental values of pressure, void fraction and mixture velocity (from mass-flux and void fraction) were specified. There was no measurement of phasic or relative velocity. Therefore, some trials and errors for the initial value of relative velocity were needed. It was found that the results were not sensitive if a relative velocity in the range of 0.1 m/s to 0.4 m/s was used. The predicted pressure distributions for various constitutive relations were compared with the measured values. These will be discussed later.

The test section for the Wisman's air-water experiment had a converging, straight and diverging part. The inside diameter at the straight section was 0.005 m, whereas the total included angles for the converging and diverging parts were 12° and 26°, respectively. Axial distribution of pressure and diametrical void fraction were measured. However, no measurements were taken for the area-averaged void fraction. Therefore, at the starting point of the calculation, the diametrical void fraction was assumed to be equal to the area-averaged void fraction. The inlet void fraction and pressure ranged from 0.38 to 0.62 and 1.8 to 3.5 bar, respectively. The mass-flux at the constant diameter section ranged from 5500 to 12000 kg/m^2s. Comparisons between the test data and model predictions are discussed in the following sections.

5. CONSTITUTIVE RELATIONS CONSIDERED

Constitutive relations for the wall friction, i.e., F_w, F_{wg} and $F_{w\ell}$, and interfacial force density, F_i, are needed for the solution of Equation (13). It is well-known that these constitutive relations, particularly those for the interfacial force density, depend strongly on two-phase flow regime. Therefore, a flow regime map must be selected first.

5.1 Flow Regime Map

Since only the high mass-flux two-phase flow is considered in this study, the following flow regime map as a function of void fraction has been selected:

Bubbly flow for $0 < \alpha < 0.4$
Churn-turbulent flow for $0.4 < \alpha < 0.75$
Annular flow for $\alpha > 0.75$.

The above flow regime boundaries are somewhat arbitrary. However, no significant difference in the result was found if the boundary between the bubbly and the churn-turbulent flow was assumed to be at $\alpha = 0.5$ instead of $\alpha = 0.4$. Similar insensitivity is also expected for the boundary between the churn-turbulent and annular flow.

Based on the work of Thang and Davis [8] it is assumed that the bubbles are of uniform size at a particular cross-section. Also, the bubbles are spherical, at least in the bubbly flow regime. In the churn-turbulent regime, the bubbles are distorted and some bubble coalescence is expected. It should be noted that an equivalent bubble radius is required for the calculation of the interfacial drag, the most important part of the interfacial force density, F_i, in the bubbly and churn-turbulent flows. The details of the bubble diameter calculation are discussed later.

5.2 Wall Friction

Many correlations have been proposed for the calculation of frictional pressure drop in two-phase systems. An assessment of various correlations for steam-water system is given by Idsinga, et al [1]. These correlations were developed based on a mixture model where only the frictional force between the wall and the two-phase mixture, i.e., the term F_w in Equation (11), was needed. However, in a two-fluid model, a partitioning of the total wall friction, F_w, to the individual phases, i.e., F_{wg} and $F_{w\ell}$, is also needed. This introduces another area of uncertainty in the two-fluid model.

For the bubbly and churn-turbulent flow, i.e., $0 < \alpha < 0.75$, the following correlations were used in this study:

1. Homogeneous model, i.e.,

$$F_w = \frac{2 f_m \rho_m v_m^2}{D} \tag{14}$$

where f_m is the two-phase friction factor. For the turbulent flow, f_m is determined from:

$$f_m = 0.046 \, Re_m^{-0.2} \tag{15}$$

where

$$Re_m = \frac{\rho_m v_m D}{\mu_m} \tag{16}$$

$$\frac{1}{\mu_m} = \frac{x}{\mu_g} + \frac{(1-x)}{\mu_\ell} \tag{17}$$

and

$$x = \frac{\alpha \rho_g v_g}{\rho_m v_m} \quad . \tag{18}$$

For the partitioning of the wall friction to the individual phases, two different models were employed. They were:

(i) $F_{w\ell} = (1-x)\, F_w$ and $F_{wg} = x\, F_w$, (19)

and

(ii) $F_{w\ell} = (1-\alpha)\, F_w$ and $F_{wg} = \alpha\, F_w$. (20)

Preliminary calculations showed that the partitioning according to Equation (20) yielded unreasonable results. Therefore only the partitioning according to Equation (19) was used in the subsequent calculations. It should be noted that for the cases considered, the flow quality, x, was very small (x < 0.02). Therefore, the entire wall friction was essentially imposed on the liquid. This is reasonable since only the liquid was in contact with the wall.

2. Separated flow model, i.e.,

$$F_{w\ell} = \frac{2(1-\alpha) f_\ell \rho_\ell v_\ell^2}{D} \quad \text{and} \quad F_{wg} = \frac{2\alpha f_g \rho_g v_g^2}{D} \qquad (21)$$

where f_ℓ and f_g are single-phase friction factors which were calculated by using Equations (15) and (16), but replacing the mixture properties with the appropriate phasic properties.

For the <u>annular flow regime</u>, i.e., $0.75 < \alpha < 1$, the annular flow friction factor model as employed in the TRAC-PF1 code [2] has been used. However, during the present study, void fractions greater than 0.75 were rarely reached. The details of the annular friction factor model may be found elsewhere [2, 9, 10].

5.3 Interfacial Force Density

The interfacial force density, F_i, is the sum of a number of terms, namely the interfacial drag, virtual mass, lift and diffusive terms [11]. The interfacial drag term is believed to be the most dominant term, and the virtual mass term is likely to be important for highly accelerating and/or decelerating two-phase flows. Quantitative estimates for the last two terms, i.e., lift and diffusive terms, are not available yet, although it is believed that they are of secondary importance. Therefore, only the interfacial drag and the virtual mass terms have been included in the present study.

<u>Interfacial drag</u>. When a phase (liquid or gas) attempts to move faster than the other phase, an interfacial drag force develops which tries to retard the faster moving phase. The magnitude of this force depends on the shape of the gas-liquid interface, the relative velocity and the phasic properties. The specific models and correlations that were used for the interfacial drag in this study are described below.

For the <u>bubbly flow</u>, i.e., $0 < \alpha < 0.4$, the interfacial drag force per unit mixture volume is given by:

$$F_{i,d} = \frac{3\alpha}{8R_b} \rho_\ell v_r |v_r| C_{d,b} \tag{22}$$

where α, R_b, ρ_ℓ, v_r and $C_{d,b}$ are the void fraction, bubble radius, liquid density, relative velocty and drag coefficient for bubbles, respectively. Notice that although the void fraction, α, and relative velocity, v_r, may be obtained from the solution of Equation (13), yet the bubble radius, R_b, and the drag coefficient, $C_{d,b}$, must be determined separately.

In some calculations [2,3], the bubble radius is computed from a constant value of critical Weber number, i.e.,

$$R_b = \frac{We_{b,crit}\sigma}{2\rho_\ell v_r^2} \tag{23}$$

which yields larger bubbles for lower relative velocity, and vice-versa. This is in contradiction with test results of Thang and Davis [8].

In the present study, a bubble radius was assumed at the entrance of the two-phase mixture, and the bubble radius at a downstream location was computed from:

$$R_b = R_{b,o} (\rho_{g,o}/\rho_g)^{1/3} \tag{24}$$

where subscript o refers to the entrance condition. The above expression was obtained under the assumptions of (i) no mass transfer between the gas and the liquid, (ii) no bubble coalescence and (iii) no bubble break-up. The bubble Weber number, We_b, was calculated and monitored, but was not used to determine the bubble radius. However, the effect of bubble radius was studied and the results will be discussed later.

Drag coefficients for a single bubble as well as a group of bubbles were studied. For a single bubble, the following correlations were used, as suggested by Wallis [9] and used by Kuo, Wallis and Richter [12]:

$$C_{d,b} = \begin{cases} \frac{24}{Re_b} (1 + 0.15\, Re_b^{0.687}) & \text{for } Re_b < 10^3 \\ 0.44 & \text{for } Re_b \geq 10^3 \end{cases} \tag{25}$$

where the bubble Reynolds number, Re_b, is defined by $Re_b \equiv 2\rho_\ell v_r R/\mu_\ell$.

It has been suggested that for a group of bubbles, the drag coefficient might be larger than that for a single bubble. Based on Rowe's experiment [13] with a multilayered grid of solid spheres, Wallis [9] suggested the following correlations:

$$C_{d,b} = \begin{cases} \frac{24}{Re_b} (1 + 0.15\, Re_b^{0.687})(1-\alpha)^{-2.7} & \text{for } Re_b < 10^3 \\ 0.44\, (1-\alpha)^{-2.7} & \text{for } Re_b \geq 10^3 \; . \end{cases} \tag{26}$$

Effects of both correlations (25) and (26) were studied, and the results will be discussed later.

For the churn-turbulent flow, i.e., $0.4 < \alpha < 0.75$, three different approaches were attempted for the calculation of interfacial drag. They were:

(i) extension of the bubbly flow regime as discussed above,

(ii) extension of the bubbly flow regime for the calculation of an equivalent bubble radius, but use of the following drag coefficient as suggested by Ishii and Zuber [14]

$$C_{d,b} = \frac{8}{3} (1-\alpha)^2 \quad , \tag{27}$$

(iii) linear interpolation between the interfacial drag for the bubbly regime at $\alpha = 0.4$ and the interfacial drag (or shear) for the annular flow regime at $\alpha = 0.75$, i.e.,

$$F_{i,d} = \frac{(0.75-\alpha)}{0.35} F_{i,d}(\alpha = 0.4) + \frac{(\alpha - 0.4)}{0.35} F_{i,d}(\alpha = 0.75) \quad . \tag{28}$$

In the present study, all of the above approaches yielded approximately the same results.

For the annular flow, i.e., $\alpha > 0.75$, the following expression based on the work of Wallis [10] was used for the interfacial drag per unit mixture volume:

$$F_{i,d} = \frac{0.01 \left[1 + 75\,(1-\alpha)\right] \sqrt{\alpha} \quad \rho_g v_r |v_r|}{D} \quad . \tag{29}$$

This completes the description of the constitutive relation used for the interfacial drag.

Virtual mass. When a dispersed phase, particularly gas or vapor bubble, attempts to accelerate in a two-phase medium, it experiences an additional resisting force because a part of the neighboring fluid ahead of the accelerating phase must also be accelerated. This resisting force is called the virtual or apparent mass force. A general form of this force per unit mixture volume is given by Drew, et al [15] as

$$F_{i,vm} = \alpha \rho_\ell C_{vm} \left[\frac{\partial}{\partial t} (v_g - v_\ell) + v_g \frac{\partial}{\partial z} (v_g - v_\ell) + (\lambda - 2)(v_g - v_\ell) \frac{\partial v_g}{\partial z} + (1-\lambda) \frac{\partial v_\ell}{\partial z} \right] \tag{30}$$

where C_{vm} is the virtual mass coefficient and λ is a parameter to be determined experimentally. Drew, et al., have also suggested that the parameter λ should be a function of void fraction and its value is in the range of 0 to 2. With λ = 1 and for steady-state, the above expression becomes

$$F_{i,vm} = \alpha\rho_\ell C_{vm} \left[v_\ell \frac{dv_g}{dz} - v_g \frac{dv_\ell}{dz} \right] . \tag{31}$$

For high velocity two-phase flow, as considered in this study, $(v_g - v_\ell) << v_g$ or v_ℓ. Therefore, Equation (31) may be approximated as

$$F_{i,vm} = \alpha\rho_\ell v_g C_{vm} \frac{d(v_g - v_\ell)}{dz} . \tag{32}$$

Equation (32) with various correlations for the coefficient, C_{vm}, has been used in this study. Many other researchers [16,17] have also used the above expression.

The specific correlations for C_{vm} that were used for both the <u>bubbly and the churn-turbulent flow</u> in this study are:

(i) C_{vm} = 0, i.e., no virtual mass effect, (33)

(ii) C_{vm} = 0.5, i.e., the value for a spherical bubble accelerating in an infinite still liquid, (34)

(iii) $C_{vm} = 0.5\,(1-\alpha)^{3/2}$ as proposed by Wisman [5], (35)

(iv) $C_{vm} = 0.5\,(1+2\alpha)/(1-\alpha)$ as proposed by Ishii and Mishima [24], (36)

(v) C_{vm} = from Figure 5 of Srebnyuk and Gorban [18]. (37)

Note that Correlations (35) through (37) yield $C_{vm} = 0.5$ at $\alpha = 0$. However, Equation (35) yields a decreasing value of C_{vm} with an increasing value of void fraction, whereas Correlations (36) and (37) predict the opposite trend.

For the annular flow regime, there is no virtual mass effect and C_{vm} is set equal to zero.

<u>Other effects</u>. Stuhmiller [16] has suggested a dynamic pressure term of the following form to be included in the interfacial force density:

$$F_{i,pressure} = 0.37\, C_{d,b}\, \rho_\ell v_r^2 \frac{d\alpha}{dz} . \tag{38}$$

Preliminary calculations showed that this term is two orders of magnitude smaller than even the virtual mass term which in itself is much smaller than the interfacial drag term. Therefore, the dynamic pressure term has been neglected in the final calculations.

The Basset force [9] has also been neglected since it is important only for creeping flows with very small bubble Reynolds number. In the present study, the bubble Reynolds number was always much higher.

6. CORRECTION FOR FLOW IN DIVERGING NOZZLE

Results of preliminary calculations showed that in some cases the pressure recovery in the diverging part of the Moby-Dick and Wisman's test sections was highly overpredicted. The discrepancy was larger with higher void fraction and higher Mach number at the throat, although reasonable agreement with data was obtained at the straight and converging parts upstream of the throat. Even large variations in the constitutive relations did not make any significant improvement in the pressure prediction at the diverging part. Therefore, a phenomenon which appears only in the diverging part and not in the straight or converging part must be responsible for this discrepancy.

After some investigation, it was concluded that stalling or boundary layer separation because of the positive (or adverse) pressure gradient in the diffuser was the most probable cause of the discrepancy. Stalling with backflow in the diffuser is known to cause poor pressure recovery even for incompressible single-phase flow [19]. To rectify this problem, a correction factor, η, was introduced in this study. Specifically, the calculated pressure gradient, dp/dz, in the diverging part was multiplied by a correction factor:

$$\eta = \eta_{1\phi}\ \eta_{\alpha}\ \eta_{M} \tag{39}$$

where $\eta_{1\phi}$ is the single-phase diffuser efficiency and η_{α} and η_M are corrections due to two-phase flow parameters namely, the void fraction and Mach number at the throat.

It is known that the single-phase diffuser efficiency or coefficient of pressure recovery decreases with increasing value of the total included angle of a diffuser. For the diverging part of the Moby-Dick test section, the single-phase diffuser efficiency, $\eta_{1\phi}$, was estimated to be approximately 0.95. This was obtained from [20] and [21] with the consideration that substantially long straight parts existed both at the upstream and downstream of the Moby-Dick diffuser section.

The corrections due to void fraction and Mach number were obtained from simulation of a number of Moby-Dick nitrogen-water experiments. Finally, the following correlations were developed:

$$\eta_{\alpha} = 1.05 - 1.95\ \alpha_{throat}^{2.7} \qquad \text{for } 0.02 < \alpha_{throat} < 0.8 \tag{40}$$

and

$$\eta_M = 1 - 0.137\ M_{throat}^2 \qquad \text{for } 0 < M_{throat} \leq 1 \ . \tag{41}$$

Note that the Mach number is defined as the ratio of the mixture velocity to the homogeneous equilibrium sound velocity. For cases with $M_{throat} > 1$, the measured axial pressure distribution showed a continuous drop in pressure downstream of the throat which was not calculated even with the correction factor. Two-dimensional analysis with boundary layer effects is clearly needed for these cases.

The above correlations, i.e., Equations (40) and (41), were also applied to the Wisman's air-water test simulations. Acceptable agreement between the data and the calculation was obtained even though the included angle for the Wisman's diffuser was much larger than that for the Moby-Dick test section. Incidentally, the single-phase diffuser efficiency for the Wisman's diffuser was approximately 0.5.

7. RESULTS AND DISCUSSION

In view of space limitation, only some selected comparisons between the data and the calculations will be shown. However, the effects of various constitutive relations will be made clear. For easier understanding, the base or recommended constitutive relations are presented in Table 1.

Figure 1 shows the effect of two different wall friction factor models, i.e., the base case and the separated flow model. Although the predicted pressure distributions were not significantly different, the homogeneous model has been recommended based on the better prediction of the throat pressure. Simulation of other tests confirmed this trend.

Figure 2 shows the result for a Moby-Dick test with very low void fraction (~ 0.02). Good agreement with the measured pressure distribution was obtained with an entrance bubble radius of 5×10^{-4} m. This resulted in a bubble Weber number of approximately 5 for most of the straight part of the test section. For this low value of void fraction, the results were not very sensitive to the variation of entrance bubble radius. Notice that the relative velocity changed sign at the throat, i.e., vapor slowed down just downstream of the throat and liquid moved faster than the vapor for a significant length in the diffuser. This agrees with physical intuition, experimental observation [12] and other calculations [22].

Figures 3 and 4 show the results for two Moby-Dick tests with entrance void fraction of 0.25. Very good agreement between the data and the predictions was obtained with the base correlations as presented in Table 1. Moreover, Figure 3 shows that inclusion of a diffuser efficiency is essential for a reasonable pressure prediction in the diverging part. Figure 4 shows the effect of various entrance bubble radius. It is clear that the bubble radius must be chosen carefully and the bubble Weber number should be in the range of 5 to 20 for as long as possible.

Figure 5 shows the results for a Moby-Dick test with entrance void fraction of 0.61. Good agreement was obtained with the recommended correlations. Once again, the importance of diffuser efficiency is quite clear.

Figure 6 shows the effect of various drag coefficients, i.e., one for a single bubble (Equation (25)) and the other for a group of bubbles (Equation (26)). With all other parameters or correlations being the same, the calculation with the single bubble drag coefficient produced better agreement with the data. It should be noted that by reducing the entrance bubble radius, the prediction with the drag coefficient for a group of bubbles could have been

Table 1. Base or Recommended Constitutive Relations

ITEM	BASE CORRELATIONS
1. Wall Friction	Homogeneous model, i.e., Equation (14) through (18) for F_w. Partitioning to phases according to Equation (19).
2. Interfacial Drag	
(a) Bubbly Flow	Equation (25) for drag coefficient. Equation (24) for bubble radius. Initial bubble radius, $R_{b,o}$ must be chosen such that the bubble Weber number, We_b, stays in the range of 5 to 20.
(c) Churn-Turbulent Flow	Equation (27) for drag coefficient and the equivalent bubble radius from the extension of bubbly flow regime.
3. Virtual Mass Coefficient	Equation (35) as proposed by Wisman [5].
4. Correction for Diffusers	Equations (39) through (41).

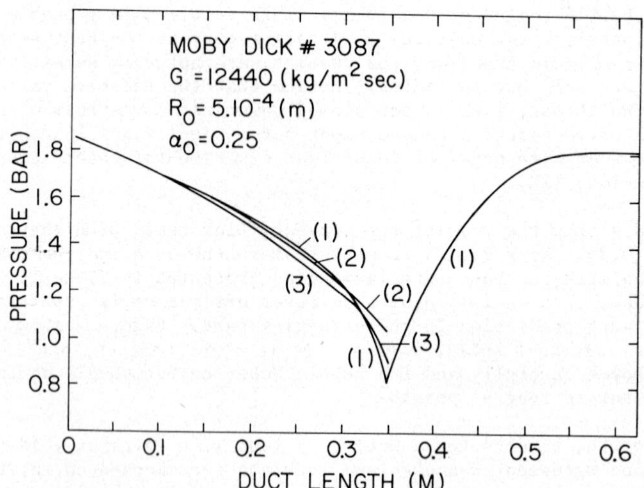

Figure 1. Comparison of Measured and Calculated Pressure for Different Wall Friction Models. (1) Measured Pressure. (2) Separated Wall Friction Model. (3) Homogeneous Wall Friction Model.

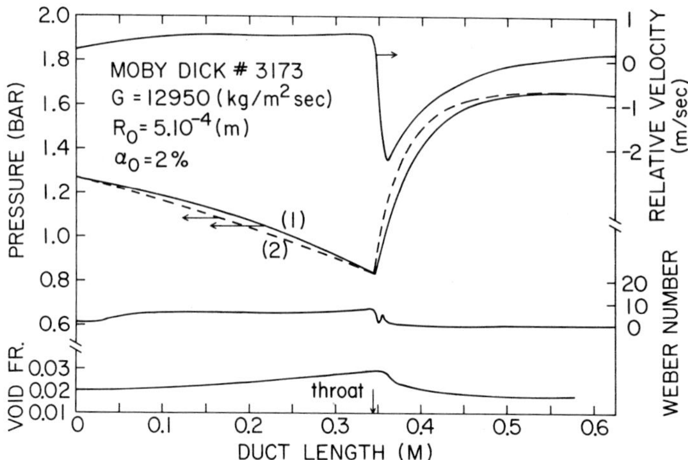

Figure 2. Comparison of Measured and Calculated Pressure for Low Void Fraction. (1) Measured Pressure. (2) Calculated Pressure.

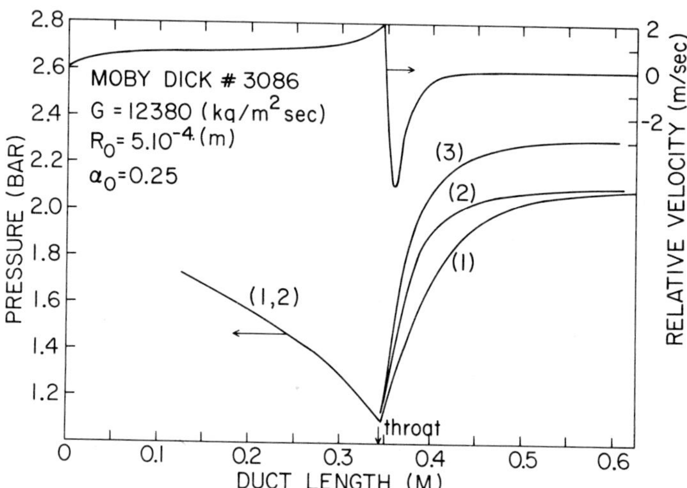

Figure 3. Comparison of Measured and Calculated Pressure for Medium Void Fraction. (1) Measured Pressure. (2) Calculated Pressure. (3) Calculated Pressure Without Diffuser Efficiency.

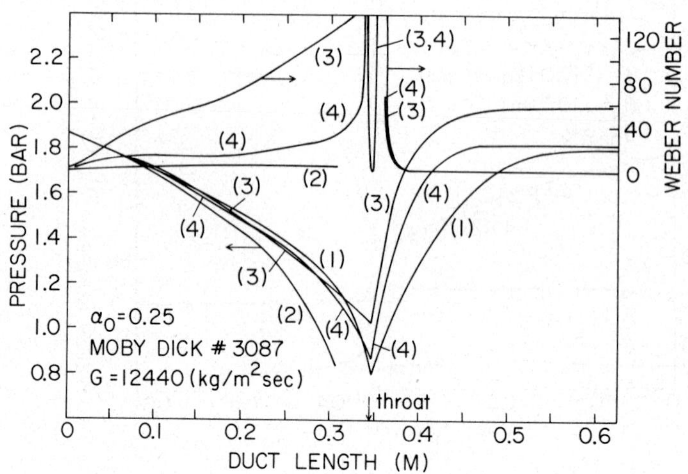

Figure 4. Comparison of Measured and Calculated Pressure for Different Initial Bubble Radius. (1) Measured Pressure. (2) Calculated Pressure, $R_o=10^{-4}$(m). (3) Calculated Pressure, $R_o=10^{-3}$(m). (4) Calculated pressure $R_o=5\times10^{-4}$(m).

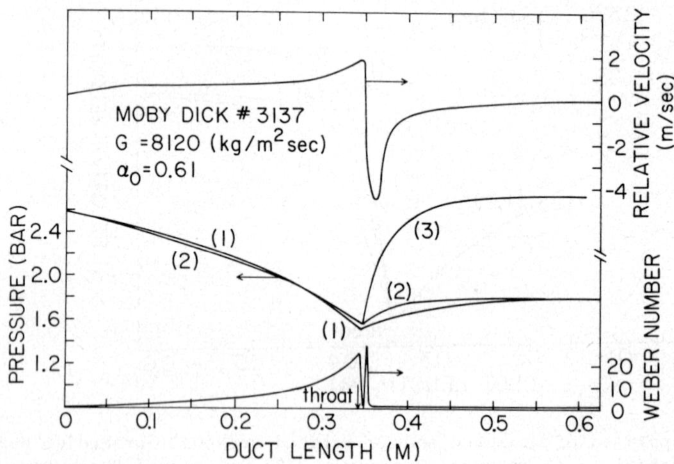

Figure 5. Comparison of Measured and Calculated Pressure for High Inlet Void Fraction, $R_o=2.5\times10^{-4}$(m). (1) Measured Pressure. (2) Calculated Pressure. (3) Calculated Pressure Without Diffuser Efficiency.

improved somewhat. However, after simulation of a number of tests, it was found that the drag coefficient for a single bubble produced better overall results, particularly near the throat. Therefore, the drag coefficient for a single bubble was selected as the base or recommended correlation instead of that for a group of bubbles.

The effect of various virtual mass coefficients is shown in Figure 7. From a quantitative viewpoint, the effect of virtual mass is rather small. However, inclusion of the virtual mass term did improve calculational stability and speed of computation as observed by other researchers [23]. After a number of test simulations, the correlation recommended by Wisman, i.e., Equation (35), was adopted as the base correlation.

Finally, Figures 8 and 9 depict the comparison between the predicted and the measured data for two air-water tests of Wisman as reported by Van der Welle [22]. The recommended correlations as presented in Table 1 were used for the calculations. Good agreement was obtained for both the tests. The importance and need of including a diffuser efficiency is once again shown in Figure 8.

8. SUMMARY AND CONCLUSIONS

Two-component, two-phase flow has been studied with a one-dimensional two-fluid formulation. Various terms of the interfacial force density have been studied in detail, and recommendations have been made based on the model comparison with the Moby-Dick nitrogen-water and Wisman's air-water data. It is hoped that these recommendations as presented in Table 1, will be useful in developing a general two-phase critical flow model.

The major conclusions of the present study are:

1. The homogeneous model for wall shear with partitioning based on flow quality is acceptable for bubbly and churn-turbulent flow.

2. The drag coefficient for a single bubble is adequate for the bubbly flow regime. Inclusion of a bubble interaction factor does not necessarily improve the results. For the churn-turbulent flow, the correlation due to Ishii and Zuber [14] is adequate.

3. The initial bubble radius must be selected very carefully. The bubble Weber number should be in the range of 5 to 20 for as long as possible. A more sophisticated model should have bubble coalescence and break-up criteria. However, use of a critical Weber number for calculation of bubble radius at all times contradicts reality.

4. The virtual mass effect is small compared to the interfacial drag effect, and its inclusion is not important from the quantitative viewpoint. However, its inclusion is justified for numerical stability and speed of computation.

5. A diffuser efficiency must be included if a one-dimensional model has to be applied to a diverging nozzle with significant total included angle. A preliminary correlation for the diffuser efficiency has been proposed.

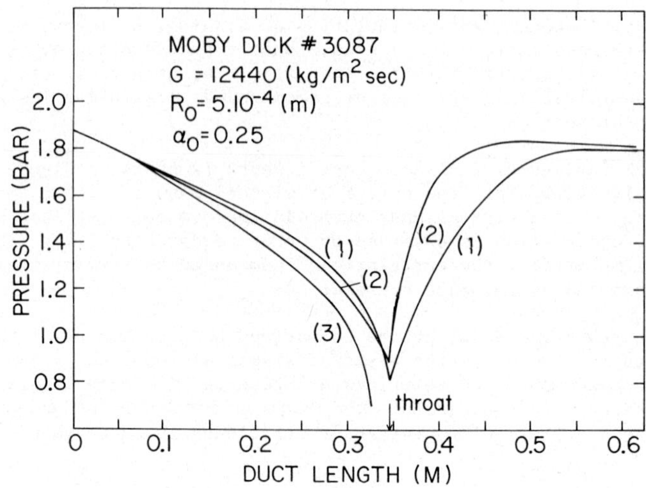

Figure 6. Comparison of Measured and Calculated Pressure, With Different Drag Correlations. (The Flow is Near Choking). (1) Measured Pressure. (2) Calculated Pressure with Single-Bubble Drag Correlation. (3) Calculated Pressure with Bubble Group Drag Correlations.

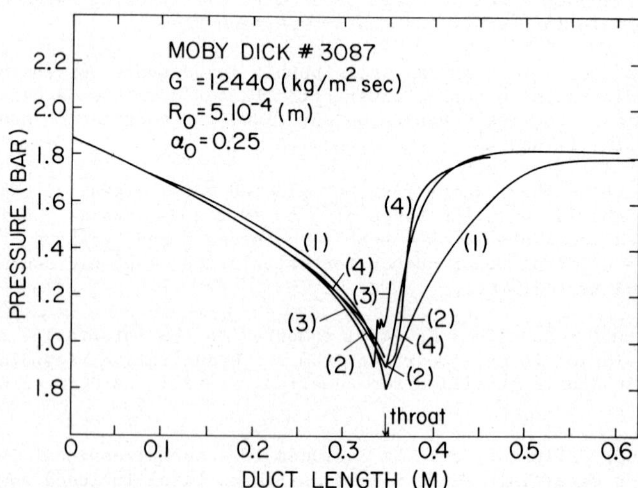

Figure 7. Comparison of Measured and Calculated Pressure for Different Virtual Mass Coefficients. (1) Measured Pressure. (2) Calculated Pressure with $C_{vm}=1/2(1-\alpha)^{3/2}$. (3) Calculated Pressure with $C_{vm}=1/2\ \frac{1+2\alpha}{1-\alpha}$. (4) Calculated Pressure With $C_{vm}=0$.

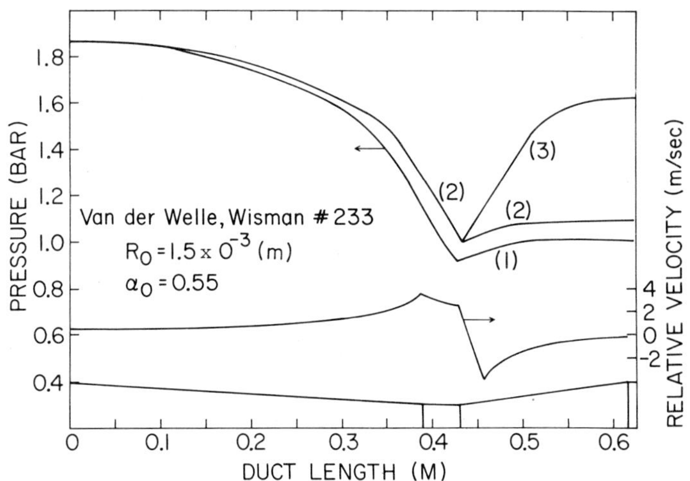

Figure 8. Comparison Between Measured and Calculated Pressure. (1) Measured Pressure. (2) Calculated Pressure. (3) Calculated Pressure Without Diffuser Efficiency.

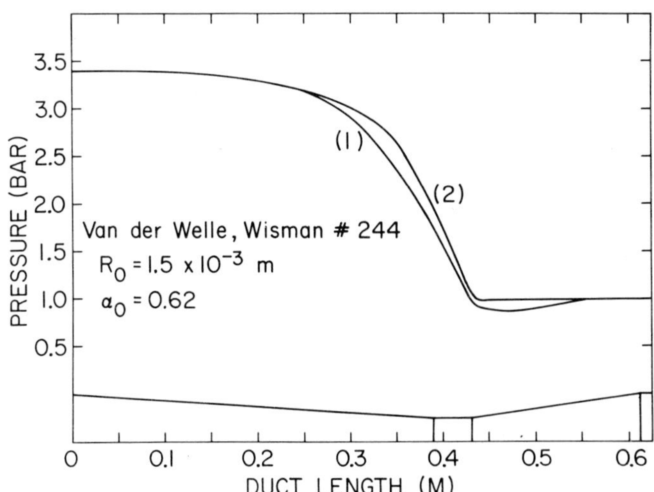

Figure 9. Comparison Between Measured and Calculated Pressure. (1) Measured Pressure. (2) Calculated Pressure.

APPENDIX

With the following definitions:

$$Q_1 = \alpha \frac{d\rho_g}{dp} + (1-\alpha) \frac{d\rho_\ell}{dp}$$

$$M = \frac{\alpha(1-\alpha)\, \rho_g \rho_\ell\, v_r}{\rho_m}$$

$$\frac{dM}{dz} = \underbrace{\frac{1}{\rho_m} \left[\alpha\, (1-\alpha) v_r (\rho_g \frac{d\rho_\ell}{dp} + \rho_\ell \frac{d\rho_g}{dp}) - MQ_1 \right]}_{Q_2} \frac{d\alpha}{dz}$$

$$+ \underbrace{\frac{1}{\rho_m} \left[\rho_\ell \rho_g v_r (1-2\alpha) M (\rho_\ell - \rho_g) \right]}_{Q_3} \frac{d\alpha}{dz} + \underbrace{\frac{1}{\rho_m} \alpha(1-\alpha) \rho_g \rho_\ell}_{Q_4} \frac{dv_r}{dz}$$

$$Q_5 = \frac{\rho_\ell - \rho_g}{\rho_m}$$

the coefficients of matrix A and vector B are:

$$a_{11} = v_m \alpha \frac{d\rho_g}{dp} + Q_2 \qquad b_1 = - \frac{\alpha \rho_g v_m + M}{A} \frac{dA}{dz}$$

$$a_{12} = \rho_g v_m + Q_3$$

$$a_{13} = \alpha \rho_g$$

$$a_{14} = Q_4$$

$$a_{21} = v_m Q_1 \qquad b_2 = - \frac{\rho_m v_m}{A} \frac{dA}{dz}$$

$$a_{22} = (\rho_g - \rho_\ell) v_m .$$

$$a_{23} = \rho_m$$

$$a_{24} = 0$$

$$a_{31} = 1 + Q_2 v_r \qquad b_3 = -\frac{Mv_r}{A}\frac{dA}{dz} - g\rho_m \operatorname{Sin}\theta - F_w$$

$$a_{32} = v_r Q_3$$

$$a_{33} = \rho_m v_m$$

$$a_{34} = M + v_r Q_4$$

$$a_{41} = v_m^2 (1-\alpha)\left[\frac{d\rho_\ell}{dp} - \frac{Q_1 \rho_\ell}{\rho_m}\right] + v_m^2 \alpha \left[\frac{\rho_g Q_1}{\rho_m} - \frac{d\rho_g}{dp}\right]$$

$$+ 4 v_m\left[\frac{MQ_1}{\rho_m} - Q_2\right] + \frac{2\alpha v_r}{\rho_m}\left[Q_2 \rho_g - \frac{MQ_1\rho_g}{\rho_m} + M\frac{d\rho_g}{dp}\right]$$

$$- v_r Q_2 + 1-2\alpha$$

$$a_{42} = v_m^2 \rho_\ell\left[Q_5 (1-\alpha)-1\right] - v_m^2 \rho_g (Q_5\alpha+1) - 4v_m(Q_5 M + Q_3)$$

$$+ \frac{2\rho_g v_r}{\rho_m}\left[Q_5 M\alpha + M + \alpha Q_3\right] - v_r Q_3$$

$$a_{43} = v_m (1-\alpha)\rho_\ell - v_m \alpha\rho_g$$

$$a_{44} = -4v_m Q_4 + (M+v_r Q_4)\left(2\alpha\frac{\rho_g}{\rho_m} - 1\right) - 2\alpha C_{vm}\rho_\ell v_g$$

$$b_4 = - g\left[(1-\alpha)\rho_\ell - \alpha\rho_g\right] \operatorname{Sin}\theta + \frac{1}{A} M v_r \left(1- \frac{2\alpha\rho_g}{\rho_m}\right)\frac{dA}{dz}$$

$$- F_{w\ell} + F_{wg} + 2F_i$$

ACKNOWLEDGEMENTS

The work reported in this paper was made possible due to the International Atomic Energy Agency (IAEA) Research Fellowship to G. Egely during his stay at Brookhaven National Laboratory. Also, the general support of the U. S. Nuclear Regulatory Commission is gratefully acknowledged.

REFERENCES

1. Idsinga, W., Todreas, N., and Bowring, R., "An Assessment of Two-Phase Pressure Drop Correlations for Steam-Water Systems," Int. J. Multiphase Flow, Vol. 3, pp. 401-413, 1979.

2. Safety Code Development Group, "TRAC-PF1: An Advanced Best Estimate Computer Program for Pressurized Water Reactor Analysis," Los Alamos National Laboratory Draft Report, 1981.

3. Ransom, V. H., et al., "RELAP5/MOD1 Code Manual," NUREG/CR-1826, EGG-2070, March 1982.

4. Jeandey, C., and Barriere, G., "Moby-Dick Nitrogen-Water Test Report," T. T. No. 599, 1979.

5. Wisman, R., "Fundamental Investigation on Interaction Forces in Bubble Swarms and its Application to the Design of Centrifugal Separators," Ph.D. Thesis, Delft University of Technology, 1979.

6. Wait, J. V., and Clarke, D., "Dare, P., User's Manual," Version 4.3, University of Arizona, Department of Electrical Engineering, 1978.

7. Lapidus, L., and Seinfeld, J.H., Numerical Solution of Ordinary Differential Equations, Academic Press, New York, 1971.

8. Thang, N. T., and Davis, M. R., "The Structure of Bubbly Flow Through Venturis," Int. J. Multiphase Flows, Vol. 5, pp. 17-37, 1979.

9. Wallis, G. B., One-Dimensional Two-Phase Flow, McGraw Hill, New York, 1969.

10. Wallis, G. B., "Annular Two-Phase Flow," ASME Journal of Basic Eng., Vol. 92D, No. 1, pp. 59-73, 1970.

11. Drew, D. A., and Lahey, R. T., Jr., "Interfacial Dissipation in Two-Phase Flow," in Basic Mechanisms in Two-Phase Flow and Heat Transfer, ASME Winter Annual Meeting, Chicago, Illinois, November 1980.

12. Kuo, J. T., Wallis, G. B., and Richter, H. J., "Interphase Momentum Transfer in the Flow of Bubbles Through Nozzles," EPRI Report NP-980, Project 443-2, 1979.

13. Rowe, P. N., and Henwood, G. A., "Drag Forces in a Hydraulic Model of a Fluidized Bed - Part 1," Trans. Institute of Chem. Eng., Vol. 39, pp. 43-59, 1961.

14. Zuber, N., and Ishii, M., "Drag Coefficient and Relative Velocity in Bubbly, Droplet or Particulate Flows," AIChE Journal, Vol. 20, No. 5, pp. 843-859, 1979.

15. Drew, D., Cheng, L., and Lahey, R. T., Jr., "The Analysis of Virtual Mass Effect in Two-Phase Flow," Int. J. Multiphase Flow, Vol. 5, pp. 233-242, 1979.

16. Stuhmiller, J. M., "The Influence of Interfacial Pressure Forces on the Character of Two-Phase Flow Model Equations," Int. J. Multiphase Flow, Vol. 3, pp. 551-560, 1977.

17. Richter, H. J., "Separated Two-Phase Flow Model: Application to Critical Two-Phase Flow," EPRI Report NP-1800, Project 443-2, 1981.

18. Srebnyuk, S. M., and Gorban, V. A., "Concerning the Induced Mass of a System of Bubbles," Fluid Mechanics-Soviet Research, Vol. 8, No. 3, pp. 93-102, 1979.

19. Ghose, S., and Kline, S. J., "The Computation of Optimum Pressure Recovery in Two-Dimensional Diffusers," ASME Journal of Fluids Eng., Vol. 100, pp. 113-130, 1981.

20. "Duct with Expanding Cross-Sections," General Electric Data Book on Heat Transfer and Fluid Flow, Section 406, 1969.

21. Klein, A., "REWIW: Effects of Inlet Conditions on Conical-Diffuser Performance," ASME Journal of Fluids Eng., Vol. 103, pp. 250-257, 1981.

22. Van der Welle, R., "A Numerical Procedure for Vertical One-Dimensional Two-Phase Flow," Nucl. Eng. and Design., Vol. 65, pp. 113-130, 1981.

23. Lahey, R. T., Cheng, L. Y., Drew, D. A., and Flaherty, J. E., "The Effect of Virtual Mass on the Numerical Stability of Accelerating Two-Phase Flows," Rensselaer Polytechnic Inst. Report, Troy, New York, 1980.

24. Ishii, M., and Mishima, K., "Study of Two-Fluid Model and Interfacial Area," NUREG/CR-1873, ANL-80-11, December 1980.

Multi-Phase Flow and Heat Transfer III. Part A: Fundamentals
edited by T.N. Veziroğlu and A.E. Bergles
Elsevier Science Publishers B.V., Amsterdam, 1984 — Printed in The Netherlands

A NUMERICAL METHOD FOR SOLVING STEADY MULTI-PHASE FLOW PROBLEMS

David L. Eddingfield, Ali Setork, and James L. Evers
Engineering Mechanics and Materials Department
Southern Illinois University
Carbondale, Illinois 62901, U.S.A.

ABSTRACT

A numerical method for solving steady, incompressible viscous multi-phase flow problems is introduced. This method which employs velocities, void fractions, and the pressure as dependent variables, is applicable to both homogeneous and unequal velocity models, and is equally applicable to problems in two and three space dimensions. Its basic principle lies in the introduction of an artificial compressibility δ and an artificial density ρ into the equations of motion, in such a way that the final results do not depend on δ. The procedure is to introduce an auxiliary time-dependent system of conservation equations in which the continuity and the momentum equations are coupled by means of an artificial equation of state. If the calculation progresses in the artificial time coordinate, then the solution of the auxiliary system of equations converges to the desired steady state solution and does not depend on δ. This approach is demonstrated in an example where the homogeneous model is solved for the case of particle injection into a two-dimensional gaseous flow.

1. INTRODUCTION

The numerical solution of the classical single phase flow equations presents major difficulties due to the special role of the pressure. In two space dimensions, one way to avoid excessive pressure iterations is to eliminate pressure terms from the system of equations by using the stream function and the vorticity formulation [1]. If, however, a solution in three space dimensions is desired, one is thrown back upon the primary variables.

In multi-phase flow equations, the pressure difficulty is magnified due to the closure problem of the equations [2]. In some cases of multi-phase flows the handling of the pressure term produces imaginary characteristics which lead to numerical instabilities when the method of lines is employed. For example, Setork [3] examines a model which could numerically be solved by the method of lines if the pressure term is eliminated from radial direction momentum equation, but would be unstable otherwise.

In this work a numerical method for solving steady, incompressible, viscous multi-phase flow problems is introduced. This method which employs velocities, void fractions, and the pressure as dependent variables, is applicable to both homogeneous and unequal velocity models, and is equally applicable to problems in two and three space dimensions.

The method was developed for single phase flows, by Chorin [4] and later used by Plows [5], and Setork [6]. The original method of Chorin cannot be applied to multi-phase flow problems due to the presence of void fraction variables in the conservation equations. This work provides proper modifications to the method, in order to make it applicable to multi-phase flow problems. It is believed that this method can be used to solve almost any two or three dimensional multi-phase flow problems and has substantial advantages in terms of the convergence rate and stability requirements over other widely used numerical methods.

In the present paper the step by step procedure for the method is given. As an example, the method has been applied to the calculation of the velocity and concentration fields in the case of injection of particles into a two dimensional flow. Promising results have been obtained. Another application of the method can be found in Setork [3].

2. DESCRIPTION OF METHOD

The basic principle of this method lies in the introduction of artificial time dependent conservation equations, an artificial compressibility δ, and an artificial equation of state. An explicit numerical method is then employed to solve this artificial time dependent system of equations by marching in time in such a way that the final result is the desired solution.

Consider the homogeneous model formulation [2] of a steady two phase flow problem. An auxiliary time-dependent system of conservation equations are introduced in which the artificial time is an auxiliary variable whose role is analogous to that of actual time in unsteady compressible flow problems. That is,

$$\frac{\partial}{\partial t}(\alpha_i \rho_i) + \nabla \cdot (\alpha_i \rho_i V_m) = 0 \qquad i = 1,2 \tag{1}$$

$$\rho_m \left[\frac{\partial}{\partial t}(V_m) + (V_m \cdot \nabla) V_m\right] = - \nabla P + f_w \tag{2}$$

$$\sum_{i=1}^{2} \alpha_i = 1 \tag{3}$$

$$\rho_m = \sum_{i=1}^{2} \alpha_i \rho_i \tag{4}$$

Equation 1 gives the continuity equations for each of the phases, and equation 2 is the mixture momentum equation. α_i is the void fraction for the phase i, subscript m denotes mixture variables, and f_w is the friction for which emperical correlation is needed.

The idea is to numerically solve the above system of equations by marching in the artificial time coordinate t. If both phases are incompressible, then the two continuity equations can be used to march for the void fractions and the mixture velocity is marched for using the momentum equation. However, there is no guarantee to satisfy equation 3, and there is no equation to march for the pressure. In order to overcome this difficulty, the continuity and momentum equations are coupled together by means of

$$P = \frac{\rho_j}{\delta} \qquad j = 1 \text{ or } 2 \tag{5}$$

which is called an artificial equation of state. That is, one of the phases is artificially assumed to be compressible with corresponding density being an unknown. This unknown density is marched for using the corresponding continuity equation and the other continuity equation is used to march for one of the void fractions. Equation 3 is then satisfied by using it to find the other void fraction at the new time level. Equation 5 is finally used to compute the pressure at the new time level. It is believed that if the calculation progresses in the artificial time direction then the solution of the auxiliary system of equations converges to the desired steady state solution and does not depend on δ. It now remains to show that

a) The finite difference approximation to the system is stable.

b) Steady state limit of the differenced system exists and it tends to the steady state solution of equations 1 to 4, as the mesh width tends to zero.

c) Find a value of δ and of any other parameter in the finite difference system such that the steady limit is reached as fast as possible, and show that the resulting procedure is indeed efficient.

No attempt has been made to carry out that method analytically, forcing heavy reliance on the numerical evidence. The method has been successfully used by Chorin [4] in 1967, Plows [5] in 1968, and Setork [6] in 1976, in the context of single phase fluid flow. The present work provides the first application of the method in two-phase flow area.

3. EXAMPLE PROBLEM

This example is designed to examine the applicability of the method to multi-phase flow problems, as well as to show the detail of the computational steps involved in the method. The problem is to numerically solve the case of the particle injection into the rectangular gaseous region shown in Figure 1.

The system of equations to be solved are two-dimensional versions of equations 1 to 5 with gas and solid being the two flowing components. The gas density ρ_g is taken to be the variable density in equation 5. Thus

$$P = \frac{\rho_g}{\delta} \tag{6}$$

will be used as the artificial equation of state. These equations can be discritized with various differencing schemes. In the one adopted here, the inertia and pressure terms are differenced by central differences, and the viscous dissipation terms are differenced according to the Dufort-Frankel pattern, in which a second derivative such as

$$\left.\frac{\partial^2\phi}{\partial x^2}\right|_{i,j}^{n} = \frac{1}{(\Delta x)^2}\left[\phi_{i+1,j}^{n} + \phi_{i-1,j}^{n} - \phi_{i,j}^{n+1} - \phi_{i,j}^{n-1}\right] \tag{7}$$

where $\phi_{i,j}^{n}$ is the nodal value of ϕ at location $(i\Delta x,\ j\Delta y)$ and at the time $n\Delta t$. Accordingly, the discritized system of equations will be

$$(\alpha_g\rho_g)_{i,j}^{n+1} = (\alpha_g\rho_g)_{i,j}^{n-1} - \text{FDT at n-th time level} \tag{8}$$

$$(\alpha_s)_{i,j}^{n+1} = (\alpha_s)_{i,j}^{n-1} - \text{FDT at n-th time level} \tag{9}$$

$$u_{i,j}^{n+1} = \text{FDT at both n-th and (n+1)-th time level} \tag{10}$$

$$v_{i,j}^{n+1} = \text{FDT at both n-th and (n+1)-th time level} \tag{11}$$

where FDT stands for Finite Differenced Terms. The subscript m is removed from the velocity components u_m and v_m for simplicity. Expressions for FDT in equations 8 to 11 can be easily derived by discritizing the gas continuity equation 1, the solid continuity equation 1, and the mixture momentum equation 2, respectively.

With regard to Figure 2, the computational procedure for all the interior nodes are as follows:

1. Specify the initial condition for the first two computational levels at $t = 0$ and at $t = \Delta t$.

2. March for u and v at all interior nodes using equations 10 and 11.

3. March for α_s at all the interior nodes using 9.

4. March for $\alpha_g\rho_g$ at all the interior nodes using 8.

5. Compute $\rho = \alpha_g\rho_g + \alpha_s\rho_s$ at all the nodes in the new time level.

6. Compute α_g at all the nodes in the new time level using $\alpha_g = 1 - \alpha_s$.

7. Compute ρ_g at all the nodes in the new time level using $\rho_g = \alpha_g\rho_g/\alpha_g$.

8. Compute P at all the nodes in the new time level using equation 6.

9. Repeat steps 2 to 8 to march to the next time level, and stop when convergence is maintained.

The boundary conditions used for this model are

$$u\ (x,o) = 120\ \text{ft/sec} \tag{12}$$

$$v\ (x,o) = 10\ \text{ft/sec} \tag{13}$$

$$\alpha_s\ (x,o) = .5 \tag{14}$$

$$u\ (x,1) = v\ (x,1) = \alpha_s\ (x,1) = 0 \tag{15}$$

$$\frac{\partial}{\partial x} u\ (o,y) = \frac{\partial}{\partial y} v\ (o,y) = \alpha_s\ (o,y) = 0 \tag{16}$$

The boundary condition at the out flow boundary is assumed to be assymptotic. The values of $\alpha_g\rho_g$ is not known at the boundary and have to be computed in the course of marching by using special equations derived by one sided differencing at the boundary. For example, at the inflow boundary the equation 8 is differenced as

$$(\alpha_g\rho_g)^{n+1}_{i,j} = (\alpha_g\rho_g)^{n-1}_{i,j} - \frac{2\Delta t}{\Delta x} [(\alpha_g\rho_g u)^n_{i+1,j} - (\alpha_g\rho_g u)^n_{i,j}]$$

$$- \frac{\Delta t}{\Delta y} [(\alpha_g\rho_g v)^n_{i,j+1} - (\alpha_g\rho_g v)^n_{i,j-1}] \tag{17}$$

Similar equations are needed at the upper and lower boundaries. The stability requirement severely restricts the step size Δt. According to Chorin [4], if the boundary conditions consist of prescribed velocities, the system is stable when

$$\Delta t \leq \frac{2}{\sqrt{2}\ (1 + \sqrt{5})} [\min\ (\Delta x,\ \Delta y)]\ \sqrt{\delta} \tag{18}$$

Note that the value of δ is independent of Δx or Δy, but is a factor related to rate of convergence. In this problem, δ was taken to be 1, and Δt was taken to be 0.0001. With 10 nodes in y direction and 50 nodes in x direction, 15 minutes of IBM 370 CPU time was needed for convergence.

The computed mixture velocity profile is shown in Figure 3, and the lines of constant α_s are plotted against coordinate axes in Figure 4. These results which look very reasonable, have also been produced by the methods lines [3]. The CPU time needed for the method of lines is almost three times more than the time required by our method. The reason is of course highly restricted step size enforced over the method of lines by the stability requirement.

4. CONCLUSIONS

The main conclusions to emerge from this study may be summarized as follows:

1. The method is explicit and relatively easy to program.

2. The pressure calculation is handled without any additional iteration.

3. The imaginary characteristics difficulty, which sometimes is associated with the method of lines, do not even appear in this method.

4. It is believed that the required computation time for this method is substantially less than that of the other numerical methods, due to a more relaxed stability condition.

5. The method exhibits a lot of promise in numerical solution of multi-phase flow problems. The future work includes the application of the method to more involved models such as unequal velocity models, and those models which involve temperature solution.

REFERENCES

1. Roache, P. J., Computational Fluid Dynamics, Hermosa Publishers, Albuquerque, New Mexico, 1972.

2. Ishii, M., Thermo-Fluid Dynamics Theory of Two-Phase Flow, Eyrolles, Paris, 1975.

3. Setork, A., "Mathematical Models and Their Numerical Solution for the Flow Field of High Velocity Water Jets," Ph.D. Dissertation, Southern Illinois University at Carbondale, Illinois.

4. Chorin, A. J., "A Numerical Method for Solving Incompressible Viscous Flow Problems," Journal of Computational Physics, 2, 12-26, 1967.

5. Plows, W. H., "Some Numerical Results for Two-Dimensional Steady Laminar Benard Convection," Physics of Fluids, Vol. 11, No. 8, pp. 1593-1599.

6. Setork, Ali, "Numerical Solution for a Two-Dimensional Flow with Temperature Dependent Viscosity," Master Thesis, Brown University, Providence, R.I., 1976.

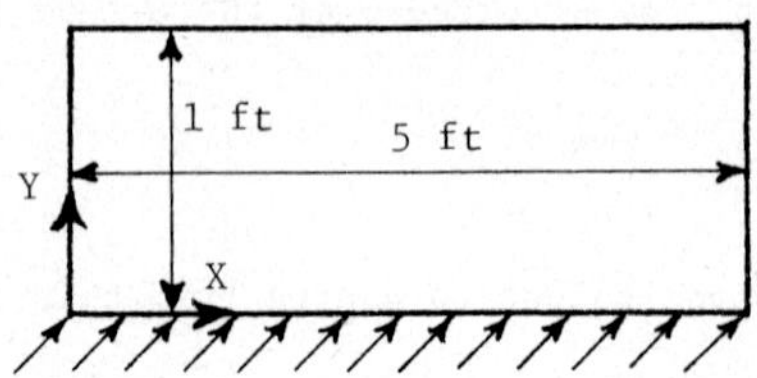

Figure 1. Particle injection into a gaseous region.

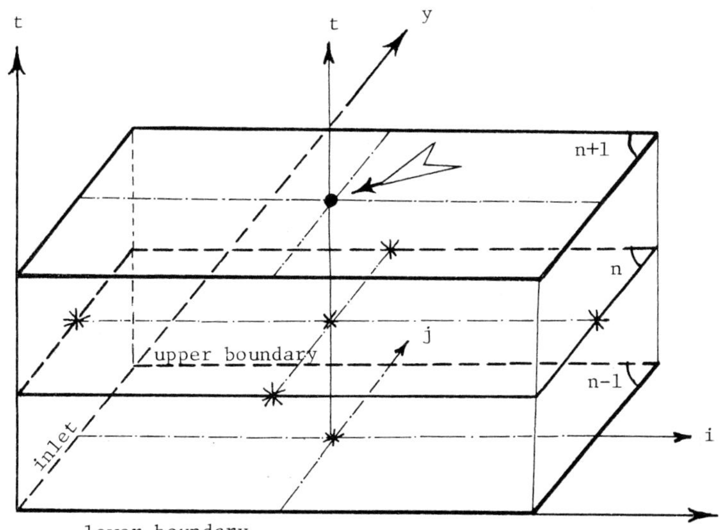

Figure 2. The finite difference molecule for the method. The value of any variable at a point shown by the arrow is found in terms of the values of the variables at the points shown by *.

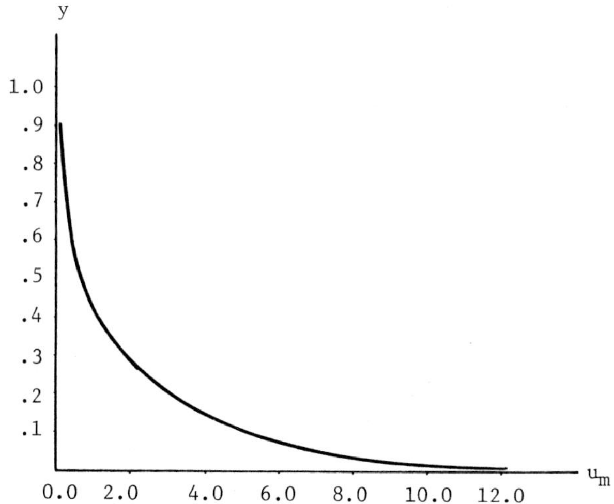

Figure 3. Axial component of mixture velocity (u_m) against vertical distance (y).

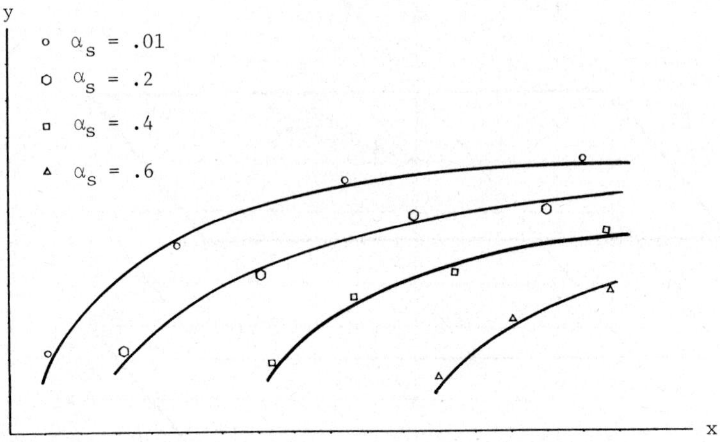

Figure 4. Lines of constant α_s against coordinate axes.

ACKNOWLEDGMENTS

This work was performed at Southern Illinois University under a contract for the U.S. Department of Energy, the Division of Surface Mining, Carbondale Mining Technology Center.

Multi-Phase Flow and Heat Transfer III. Part A: Fundamentals
edited by T.N. Veziroğlu and A.E. Bergles
Elsevier Science Publishers B.V., Amsterdam, 1984 — Printed in The Netherlands

TRANSIENT TWO-PHASE FLOW ANALYSIS USING HIGH-ORDER NUMERICAL SCHEMES

N.P. Kolev
Institute for Nuclear Research
and Nuclear Energy
1184 Sofia, Bulgaria

E.A. Katkovsky
VNIIAM, 125171 Moskow, USSR

ABSTRACT

Selected high-order methods are discussed to solve the equations for one-dimensional boiling flow. Characteristic based methods using directional schemes, and methods using cubic spline or generalized cubic spline approximations are considered. Implicit integration in time is possible in both the discrete and semidiscrete variants presented. The methods are applicable to flows ranging from homogeneous-equilibrium to separated-phase flow.

A two-phase benchmark problem, transients relevant to water cooled nuclear reactors and the Burger's equation shock problem are used to illustrate the methods. These problems include propagation phenomena as well as asymptotic approach to steady state. Comparisons with available experimental and numerical data demonstrate the significant improvement in accuracy and efficiency, offered by the high-order schemes. Discrete cubic spline (S_D) approximations are found effective and flexible, with good potential for extensions to adaptive grids in multi-phase flow problems.

1. INTRODUCTION

In the last few years a number of high-order methods have been proposed for numerical solution of hyperbolic and mixed hyperbolic-parabolic equations. Advantages and limitations of these algorithms have been extensively studied for simple model equations and in gas dynamics problems, see [1-11] and the literature cited there. Some conclusions are instructive.

An optimal spatial scheme minimizes both the errors in amplitude and phase velocity, and the spurious numerical oscillations [12-14]. Therefore a measure of the quality of numerical schemes must include the integrated absolute error (L_2 norm) from the time response in addition to the gain and phase error from the frequency response.

An acceptable balance between numerical diffusion and nonphysical oscillations requires that the schemes somehow take into account the directional nature of the solution.

Computationally most efficient scheme is a function of the error tolerance, and depends on the problem, but as a general rule schemes of second order or higher in time, and fourth or higher order in space, are to be preferred [1-4].

Significant gains in accuracy and efficiency can be expected when adaptive mesh refinement is used with certain higher-order methods to follow large moving gradients [15-17].

This research has stimulated the development of efficient algorithms for two-phase flows. Applications of higher order methods to transient thermal hydraulic analysis for nuclear reactors [18-22] have demonstrated a significant improvement in accuracy, compared to conventional methods. The improved accuracy and conservation permit the use of coarser grids and correspondingly improved computational efficiency (accuracy vs effort). First experience has been favourable enough to encourage further research and practical use of these methods.

Transient two-phase flows as encountered in nuclear engineering make particular demands on the solution algorithms [15,23]. Widely varying time scales, discontinuous propagation velocities in some cases, stationary and moving steep density gradients may be present. It should be possible to use a large number of segments and to simulate transients over long periods of time. The higher-order methods to be used in production codes should minimize the problems at the boundaries, the spurious oscillations and should be capable of handling flow reversals, propagation of shocks and discontinuities. Time step control and potential for extensions to adaptive spatial grids are desirable.

We will outline the development and discuss two-phase flow applications of selected higher-order methods, namely:

(a) extensions of the characteristic finite difference (CFD) method [24,25] to higher orders in space, using directional differences or discrete splines;

(b) pseudo-characteristic (PC) method of lines [10] using variable-order schemes in time and high-order upwind differences [20-22] or discrete splines in space;

(c) an adaptive grid method of lines using cubic spline or discrete cubic spline approximations and applicable to the flow equations in nonconservative form.

One-dimensional transients involving pipe blowdown [25] and heat addition to vertical flow channels are used to illustrate the methods. Results obtained using the homogeneous equilibrium model (HEM) and the equal velocity, unequal temperature (EVUT) model [26], are shown. The performance of an adaptive grid variant (c) is illustrated by solving the Burger's nonlinear equation shock problem [27].

2. BASIC EQUATIONS

The equations governing transient one-dimensional two-phase flow are used in their non-conservative formulation. The conservation equation set is transormed to the form:

$$\tilde{J}_1(\overline{\Psi})\overline{\Psi}_t + \tilde{J}_2(\overline{\Psi})\overline{\Psi}_x = \overline{d}(\Psi) \tag{1}$$

or

$$\overline{\Psi}_t + \tilde{A}\overline{\Psi}_x = \overline{f} \tag{2}$$

where $\tilde{A} = \tilde{J}_1^{-1}\tilde{J}_2$ and $\overline{f} = \tilde{J}_1^{-1}\overline{d}$; $\tilde{J}_1$ and $\tilde{J}_2$ are

square matrices of coefficients; $\overline{\Psi}$ is a vector of dependent variables and $\overline{f}$ is a vector of source/sink terms. Equation (2) is frequently refered to as the primitive form of the conservation equations. The eigenvalues $\lambda(A)$ represent the velocities of propagation of small amplitude, short-wavelength disturbances. For this kind of disturbances the characteristic analysis is adequate and the type of (2) is determined from $\lambda(A)$.

When the system is hyperbolic, the matrix A can be reduced to diagonal form Λ, and equation (2) written in canonical form, by a similarity transform $A = B^{-1}\Lambda B$, where Λ is the matrix of eigenvalues $\lambda(A)$ and the eigenvectors $\ell(A)$ are columns of B^T. Multiplying through by B gives the characteristic form:

$$\tilde{B}\overline{\Psi}_t + \tilde{\Lambda}\tilde{B}\overline{\Psi}_x = \overline{c} \tag{3}$$

or

$$\overline{\ell}^i(\overline{\Psi}_t+\lambda_i\overline{\Psi}_x) = c_i \qquad (\overline{\ell}^i = \tilde{B}_i^T) \quad i = 1,\dots,n \tag{4}$$

or

$$\tilde{B}\overline{\Psi}_t + \tilde{\Lambda}\overline{\Psi}^*_x = \overline{c}, \qquad (\overline{\Psi}^*_x = \tilde{B}\overline{\Psi}_x) \tag{5}$$

with

$$\Psi^{*i}_{x_i} = B_{i1}\Psi^i_{x_1} + B_{i2}\Psi^i_{x_2} + \dots + B_{in}\Psi^i_{x_n}, \quad i = 1,\dots,n \tag{6}$$

where superscript indicates which eigenvalue is used to govern the direction of differentiating according to $\mathrm{sign}(\lambda_i)$. See references [15,28].

The pseudo-characteristic equations can be obtained by solving (3) for the time derivative terms, or from (2) using $A = B^{-1}\Lambda B$:

$$\overline{\Psi}_t + \tilde{A}^*\overline{\Psi}^*_x = \overline{f} \tag{7}$$

where

$$\tilde{A}^* = \tilde{B}^{-1}\tilde{\Lambda} = \tilde{A}\tilde{B}^{-1}$$

See Carver [10], also Wang and Johnson [29].

Equation (2) is the basis for standard method of lines discretization, using finite differences or splines. For characteristic based algorithms, the equations to be discretized are (5) or (7).

Two hyperbolic sets of conservation equations were considered, the homogeneous equilibrium model and EVUT model. A dependent variable set $\Psi^T = (G,h,p)$, with $\rho = \rho(h,p)$ and speed of sound $c^{-2} = \rho_p + \rho_h/\rho$ were used for the HEM model. The appropriate matrices and vectors J_1, J_2, B, Λ, f and c are given in [23]. The EVUT model, $\Psi^T = (u,\alpha,p,h_f,h_g)$ is described in [26]. Expanded forms of equations (3) and (7) are given in the sections where the particular methods are discussed.

Burger's equation shock problem [27], with known analytical solution, was used to develop and test an adaptive grid method. For the present analysis the problem is:

$$
\begin{aligned}
u_t + uu_x &= \mu u_{xx} \qquad (8)\\
u(x,0) &= u_e(x,0)\\
u(0,t) &= u_e(0,t)\\
u(1,t) &= u_e(1,t)
\end{aligned}
$$

where

$$u_e(x,t) = \frac{0.1e^{-a} + 0.5e^{-b} + e^{-c}}{e^{-a} + e^{-b} + e^{-c}} \qquad (9)$$

$$
\begin{aligned}
a &= 0.05(x-0.5+4.95t)/\mu\\
b &= 0.25(x-0.5+0.75t)/\mu\\
c &= 0.5(x-0.375)/\mu
\end{aligned}
$$

$u_e(x,t)$ is also the exact solution. Equation (8) can be used as a model of mixed hyperbolic-parabolic systems, or can be regarded as a non-linear hyperbolic (advective) equation with the addition of an arbitrary dissipation term. Here $\mu = 0.003$ which gives the problem a strongly hyperbolic character. This establishes two near shock waves moving at different speeds and merging at $t = 0.5$. The dominance of the dissipation term increases with the number of mesh points unless μ is made function of the mesh spacing, and this should be taken into account when comparing numerical results.

3. HIGHER ORDER CFD METHODS

Given that a hyperbolic system can be established for a particular model, characteristic based numerical techniques can be used to solve it numerically. This approach makes it possible to combine elements of mathematical rigour with efficient discretization schemes.

The wave tracing method of characteristics (or grid of characteristics method) [30,31] is the only method which permits discontinuities to propagate without numerical diffusion, and yields highly accurate solutions regarded as benchmark references [25,26]. However, at least at this stage of development, wave tracing algorithms are expensive for routine calculations and more difficult to apply to complex geometries, compared to characteristic methods using specified number of grid points for representing the flow field [15,21,25,32]. The latter are, in contrast, relatively faster and flexible but introduce numerical diffusion, conservation errors and, in some cases, departures from monotonicity. Problems with the Jacobians used to derive the primitive and characteristic forms contribute additionally to the conservation problems related to discretization schemes. In the wave tracing method these effects are generally insignificant as the mesh spacing is dynamic and iteration is used. In the "fixed grid" characteristic methods special techniques are needed to minimize the above problems while preserving computational efficiency. Two approaches have proved particularly effective: the adaptive grid CFD method of first order [15,33], and higher order characteristic methods [10].

In the characteristic finite-difference method [24,33], first order differencing in space and time is used to discretize equation (3). The resulting equations are implicit in time and allow efficient integration. The method is used in the production code RAMA for transient thermalhydraulic analysis of nuclear reactors [34], and has been applied to a spectrum of flow-boiling models including two-fluid models [34,35]. Hancox and McDonald [15] reported a marked increase in accuracy with only a minimal increase in computing time, when using adaptive grid CFD strategies.

Applications of high-order pseudo characteristic semi-discretizations [10,20-22] demonstrated the significant improvement in conservation, accuracy and efficiency, offered by the high-order methods. This research indicated the potential of the CFD method for extensions to higher orders in space, combined with adaptive grid techniques. Here we present two fixed grid variants - a straightforward extension based on four and five-point biased upwind differences, and schemes based on discrete cubic splines. Discrete spline approximations are of particular interest as they have potential for practical use of adaptive grids and are easy to incorporate in production codes.

3.1. Biased Upwind Difference Approximations

For a uniform grid and $s_i = \text{sign}(\lambda_i)$, $\lambda = \lambda(A)$, the four- and five-point biased upwind difference schemes (4PBU, 5PBU) may be written as:

$$\frac{\partial}{\partial x}(u_i) = s_i(u_{i-2s_i} - 6u_{i-s_i} + 3u_i + 2u_{i+s_i})/6\Delta x + O(\Delta x^3) \tag{10}$$

$$\frac{\partial}{\partial x}u_i = s_i(-u_{i-3s_i} + 6u_{i-2s_i} - 18u_{i-s_i} + 10u_i + 3u_{i+s_i})/12\Delta x + O(\Delta x^4) \tag{11}$$

The schemes are based on piecewise Lagrange polynomial interpolation and have third and fourth order of accuracy, respectively. Numerical experiments with the PC method [10,20,21] have shown that these schemes give stable solutions with significantly improved wave front resolution. In fact, they are among the very few schemes, that can follow with minimal error a propagating triangular or square wave, see references [5,8].

If the five-point biased upwind scheme (11) is used in the characteristic form (3) to approximate the spatial derivatives at the end of each timestep, the following semi-implicit algorithm results (for $\overline{U} = \overline{\Psi}$, at mesh point k and time n):

$$\tilde{B}_k^n \frac{\overline{U}_k^{n+1} - \overline{U}_k^n}{\Delta t} + \tilde{\Lambda}_k^n \tilde{B}_k^n [\tilde{S}_k (-\overline{U}_{k-3} + 6\overline{U}_{n-2} - 18\overline{U}_{k-1} + 10\overline{U}_k + 3\overline{U}_{k+1}) +$$

$$+ T_k (\overline{U}_{k+3} - 6\overline{U}_{k+2} + 18\overline{U}_{k+1} - 10\overline{U}_k + 3\overline{U}_{k+1})]/12\Delta x = \overline{C}_k^n \quad (12)$$

where S and T are diagonal matrices with $S_{ii} = 1$ if λ_i is positive and zero otherwise; $T_{ii} = 1$ if λ_i is negative and zero otherwise.

Equation (12) can be rearanged to yield a banded system of linear algebraic equations:

$$\tilde{M}_{k,k-3} \overline{U}_{k-3}^{k+1} + \tilde{M}_{k,k-2} \overline{U}_{k-2}^{n+1} + \tilde{M}_{k,k-1} \overline{U}_{k-1}^{n+1} + \tilde{M}_{kk} \overline{U}_k^{n+1}$$

$$+ \tilde{M}_{k,k+1} \overline{U}_{k+1}^{n+1} + \tilde{M}_{k,k+2} \overline{U}_{k+2}^{n+1} + \tilde{M}_{k,k+3} \overline{U}_{k+3}^{n+1} = \overline{R}_k^n \quad (13)$$

where

$$\tilde{M}_{k,k-3} = -\tilde{S}_k \tilde{\Lambda}_k^n \tilde{B}_k^n$$

$$\tilde{M}_{k,k-2} = 6\tilde{S}_k \tilde{\Lambda}_k^n \tilde{B}_k^n$$

$$\tilde{M}_{k,k-1} = -(18\tilde{S}_k + 3\tilde{T}_k) \tilde{\Lambda}_k^n \tilde{B}_k^n$$

$$\tilde{M}_{kk} = \tilde{B}_k^n \frac{12\Delta x}{\Delta t} + 10(\tilde{S}_k - \tilde{T}_k) \tilde{\Lambda}_k^n \tilde{B}_k^n$$

$$\tilde{M}_{k,k+1} = (3\tilde{S}_k + 18\tilde{T}_k) \tilde{\Lambda}_k^n \tilde{B}_k^n;$$

$$\tilde{M}_{k,k+2} = -6\tilde{T}_k \tilde{\Lambda}_k^n \tilde{B}_k^n;$$

$$\tilde{M}_{k,k+3} = -\tilde{T}_k \tilde{\Lambda}_k^n \tilde{B}_k^n;$$

$$\overline{R}_k^n = \frac{12\Delta x}{\Delta t} (\overline{C}_k^n \Delta t + \tilde{B}_k^n \overline{U}_k^n).$$

When the system (13) is written in the form $A'y = b$ the only structural difference from the first-order CFD formulation in [24] is the band width. Here, at internal points, we have banded coefficient matrices A' with five (4PBU) or seven (5PBU) blocks M(nxn) instead of three, n being the number of dependent variables.

3.2. Discrete Cubic Splines

Discrete cubic spline (S_D) is defined using discrete analogues of the derivative continuity conditions in the form of divided differences [36]. It is regarded as a generalized cubic spline, with a free parameter ε used as increment to define discrete derivative approximations. In the limit as $\varepsilon \to 0$ standard cubic spline is obtained. In the particular case $\varepsilon_i = h_i$, where

$h_i = x_{i+1}-x_i$, directional spline approximations of the spatial derivatives at mesh point x_i can be derived in the form:

$$S_D'(x_i+) = \frac{f_{i+1} - f_i}{\Delta x_i} - \frac{\Delta x_i}{6}(2M_i + M_{i+1})$$
$$S_D'(x_i-) = \frac{f_i - f_{i-1}}{\Delta x_{i-1}} - \frac{\Delta x_{i-1}}{6}(M_{i-1} + 2M_i) \qquad (14)$$

where

$$M_i = S_D''(x_i), \qquad i = 1,\dots,N.$$

It can be shown that:

(i) A number of discretization schemes, including fourth order formulas, can be constructed using selected approximations for M_i. For example, (14) reduces to a simple two-point upwind scheme when M_i are set equal to zero, and to the 4PBU or a 6PBU scheme if M_i are approximated by three point, respectively five-point central differences;

(ii) The standard order of approximation,

$$\| S^{(r)}(x) - f^{(r)}(x) \|_\infty = O(\overline{\Delta x}^{4-r}) \quad r=1,2 \quad \overline{\Delta x} = \max_i \Delta x_i$$

can be increased at mesh points for uniform grids. In principal this can be done for non-uniform grids too, but the algorithm will be slower.

(iii) Increasing the order of the M_i approximation, using standard centered differences, generally changes the residual terms and the filtering properties, but not the order of the total scheme. The latter can be increased by particular choice of the M_i approximation. The effect in both cases is seen as increase in sharpness of the profile at discontinuities, and resembles the effect of combining a low-order monotone scheme and a high-order scheme in the flux-corrected (FCT) algorithm [9].

An example of a scheme, which gives fourth order of accuracy for $S_D'(x_i)$ over an uniform grid, at interior points, is:

$$M_i = \frac{1}{6h^2}(-f_{i-2}+10f_{i-1}-18f_i+10f_{i+1}-f_{i+2}) = f_i''+O(h^2)$$
$$i = 3,4,\dots,N-2. \qquad (15)$$

Then, following [36], one can obtain

$$S_D'(x_i) = f_i' + O(h^4). \qquad (16)$$

Implementation of approximations (14) requires a simple modification of the first order CFD algorithm, as follows:

(a) Evaluate numerically $M_k = S_D''(x_k)$ at time n, by suitable stand-alone routines and use the "high order" parts of schemes (14) as explicit dissipation terms in the right hand side:

$$\bar{R}_k^n = \frac{\Delta x}{\Delta t}\tilde{B}_k^n \bar{U}_k^n + \Delta x\{\bar{C}_k^n + \frac{\Delta x}{6}\tilde{\Lambda}_k^n \tilde{B}_k^n [\tilde{T}_k(2M_k + M_{k+1}) - \tilde{S}_k(M_{k-1} + 2M_k)]^n\} \tag{17}$$

(b) Use simple two-point upwind schemes to approximate the spatial derivatives in the characteristic form.

(c) Solve the resulting system, $A'y=b$, using the standard first-order CFD techniques with coefficients evaluated at mesh points k.

3.3. Boundaries

General boundary conditions are specified as required by the original differential equations. When the solution point is at a boundary, the nesessary and sufficient conditions are determined by considering the domain of dependence requirements.

Nodes near the boundaries cannot be treated in exactly the same way as the others. In case of fixed directional difference schemes a choice is required between higher order of accuracy or maintaining the directionality by using lower order formulas. As a general rule directionality should be maintained, and this appears to be more stringent when more detailed models and fast transients are involved. Numerical schemes based on discrete cubic splines are more flexible at the boundaries since they clearly separate a low-order and a higher-order component, the low-order part always maintaining directionality. If formulas (14) and (15) are used at internal points, the following schemes can be derived for M_i at near boundary points:

$$
\begin{aligned}
M_1 &= \frac{1}{6h^2}(17f_1-50f_2+54f_3-26f_4+5f_5) = f''+0(h^2) \\
M_2 &= \frac{1}{6h^2}(5f_1-8f_2+0.f_3+4f_4-f_5) = f''+0(h^2) \\
M_{N-1} &= \frac{1}{6h^2}(-f_{N-4}+4f_{N-3}+0.f_{N-2}-8f_{N-1}+5f_N) = f''+0(h^2) \\
M_N &= \frac{1}{6h^2}(5f_{N-4}-26f_{N-3}+54f_{N-2}-50f_{N-1}+17f_N) = f''+0(h^2)
\end{aligned}
\tag{18}
$$

which yield

$$S'_D(x_i) = f'(x_i) + 0(h^4). \tag{19}$$

For exploratory solution of particular models, use of fictitious point techniques at the boundaries is possible [22,35]. Upstream boundary conditions can be applied in a non-reflecting form [37], if needed.

Finite-difference algorithms, derived from the non-conservative forms of the flow equation, experience difficulties in the neighbourhood of phase transitions where large property gradients or discontinuous propagation velocities may occur. First order methods fail to conserve adequately in this region whereas higher order methods exhibit improved conservation [20,22] and do not require special techniques for tracing the transitions, at least for the conditions discussed here.

Pipe junctions, breaks, and piping networks can be treated as in the first order CFD algorithm [24]. Each pipe is usually taken as a separate segment and appropriate boundary conditions written. At each such boundary only the compatibility equations corresponding to inward directed characteristics are selected, and the missing equations are substituted by additional conditions. One-sided derivative approximations are used in the characteristic equations (3) which apply at boundary points. For example, abrupt area changes are considered as point discontinuities and assigned dual grid points. When the HEM model is used, and the discontinuous area change is between grid points k and k+1, there are six unknowns G_k, G_{k+1}, h_k, h_{k+1}, p_k, p_{k+1} and three equations from (3). The additional equations can be obtained from the steady state conservation laws:

$$\begin{aligned} &A_k G_k = A_{k+1} G_{k+1} \\ &\int_k^{k+1} \frac{dp}{\rho} + \frac{u_{k+1}^2 - u_k^2}{2} + F_{kk+1} = 0 \\ &h_k + \frac{u_k^2}{2} = h_{k+1} + \frac{u_{k+1}^2}{2} \end{aligned} \tag{20}$$

Following [24] these can be linearized

$$\begin{aligned} &A_k G_k^{n+1} - A_{k+1} G_{k+1}^{n+1} = 0 \\ &h_{k+1}^{n+1} - h_k^{n+1} + \frac{u_k^n + u_{k+1}^n}{2}(u_{k+1}^{n+1} - u_k^{n+1}) = 0 \\ &\frac{v_k^n + v_{k+1}^n}{2}(p_{k+1}^{n+1} - p_k^{n+1}) - h_{k+1}^{n+1} + h_k^{n+1} = -F_{kk+1}^n \end{aligned} \tag{21}$$

and incorporated in the overall problem. When critical flow occurs at the location of an abrupt area change, a procedure similar to that described in [31] can be used, and the resulting equations incorporated in the overall problem.

3.4. Numerical Results

Standard Problem 2, from reference [25], was used to test the higher order variants of the CFD method. This is a pipe blowdown problem, with geometry and initial conditions selected to agree with the experiment of Edwards and O'Brien [38]. Water at 243°C and 7 MPa is contained in a horizontal 4 meter pipe of diameter 0.032 m. At time t=0 one end is removed, and pressure wave propagation from the open to the closed end, in the early stage of the transient is considered. During the first 3.5 ms of the transient there are regions of both subcooled water and two-phase mixture. The HEM model predicts large discontinuous change in the sound speed at the boundaries between such regions, and this is convenient for testing purposes. In this paper the problem is referred to as Test Problem 1. The open end conditon was applied as follows: (1a) as in [25] (CFD, benchmark), or $u = c$ for the first several miliseconds (CFD 4PBU); (1b) as in [26].

The problem specifications are, correspondingly:

Initial Conditions

	Position	u(m/s)	h(MJ/kg)	(kg/m^3)	p(MPa)
	x = 4	70.1	1.03826	183.72	2.331
(1a)	0<x<4m	0	1.0459	815.09	7.0
(1b)	0<x⩽4m	0	1.0459	815.09	7.0

Boundary Conditions

(1a) x = 0m u = 0 , x = 4m p= 100kPa if u<c

(1b) x = 0m u = 0 , x = 4m

$$p=\begin{cases}3.4\text{MPa}, & t=0.0^{+}\text{s}\\ 0.1\text{MPa}, & t\geqslant 0.005\text{s}\\ \text{linearly interpolated}, & 0.0<t<0.005\text{s}\end{cases}$$

or u = c if u ⩾ c.

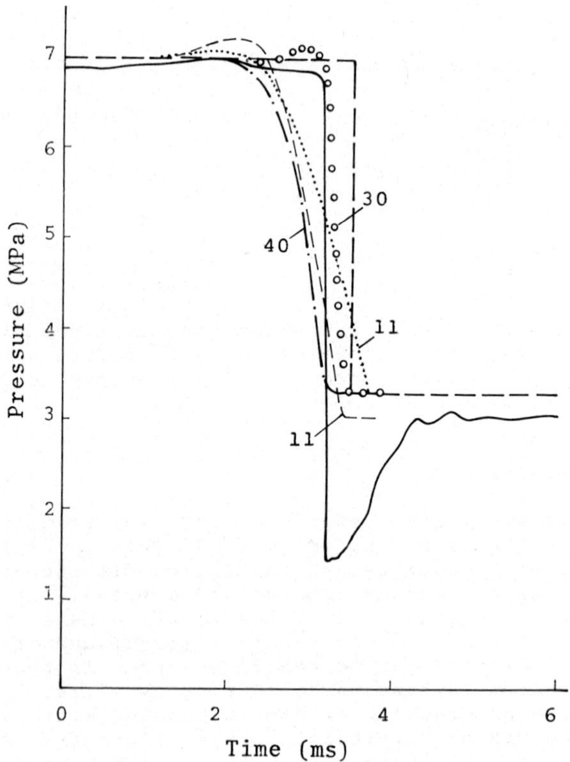

Fig.1. Closed end pressure, Test Problem 1. Experiment ——; Problem 1a: MOC benchmark— —; CFD two-point—·—; CFD 4PBU, with u(4,t)=c(4,t)– – –; Problem 1b: pseudo-characteristic, 4PBU o; discrete spline/CFD····· . Number of mesh points is indicated.

Additional constraints, derived directly from the equations, can be applied to compensate for the increased truncation errors in the schemes at a closed end. In the HEM case the following relations are obtained.

$$\frac{\partial p(0,t)}{\partial x} = 0 \tag{22}$$

$$\frac{\partial h(0,t)}{\partial t} = \left(\frac{1}{\rho}\frac{\partial p}{\partial t}\right)(0,t) \tag{23}$$

The former follows from the momentum equation and the latter from the energy equation, given G = 0 and T.dS = 0 = dH - vdP. Relation (23) was used for the solutions considered here.

Preliminary calculations have been completed using the HEM model, a fixed time step of 5 μs, and an 11 point grid. Figure 1 shows results of the CFD method using the four-point biased upwind scheme (10) in Test Problem 1a, and a discrete cubic spline appoximation in Test Problem 1b. The CFD/4PBU solution obtains approximately the same accuracy with 11 points as the first-order CFD did on a 40 point grid. When more points are used, for instance 30, the third order CFD solution will approximate the pseudo-characteristic solution shown in the same figure. The small overshoot in pressure does not become unstable on reflection [10] and can be minimized by selecting suitable boundary approximations, which may introduce additional damping. The approximations at points 1, 2, N-1, N were selected as follows: two-point upwind and 4PBU with an opposite direction of bias for u>0 at the closed end; two-point upwind schemes at the open end. For fixed grid applications, a more accurate and natural approach would be to use discrete spline approximations of FD type (14), (17) at the boundaries. Note that the 4PBU scheme can be derived from S_D approximations (14) using three point centered finite differences for $S_D''(x_i)$. The amplitude difference between the benchmark and the CFD/4PBU solution is attributed to the assumption u=c for the first several miliseconds of the transient.

When comparing with experimental data and the benchmark solution it should be noted that the model does not account for the initial tension release, propagating at the speed of sound in the pipe metal, for the pipe elasticity and the thermal nonequilibrium effects. The tension release results in about 1 percent lower initial pressure [38]. If the pipe elasticiy is taken into account following [39] a higher pulse propagation velocity is obtained and it can be shown that the benchmark front closely approximates the experimental one. In the example considered here, when finer grids are used, the numerical solution will approximate the benchmark solution and the wave front will be located to the right of the experimental curve. None of the HEM models predict the undershoot in pressure observed in the experiment and predicted by nonequilibrium models.

A CFD algorithm based on discrete cubic spline approximation (14), (17) with standard five-point centered differences for M_i (hence the notation DS5C), was tested on Problem 1b and produced the result shown in Fig. 1. Again, considerably less diffusion is seen, compared to the two point CFD results, and the spurious oscillations are minimized. Numerical experiments with the PC method indicate that the slight smearing in downstream direction can be minimized by optimizing the time and space discretization. Since the CFD/DS variant is of first order in time and antidiffusion corrections are calculated at time n, this algorithm is more sensitive to time step selection, and time step control as in [33] is required.

In the above examples the resulting system of algebraic equations was solved using an efficient banded matrix algorithm [40]. In case of hydraulic network simulation there may be nonzero elements outside the banded structure and sparse matrix solution techniques such as [41,42] are needed for efficient integration. More experimental evidence is needed for the quantitative assesment of computational efficiency. Available data, though limited, allow us to conclude that the computational overhead of the higher-order schemes is not large and significant gains in efficiency can be expected when a coarser mesh can be used. Moreover, experience with discrete spline semi-discretizations (Section 5) indicates that practical use of adaptive grid CFD methods is possible, with order of accuracy no less than third.

4. PC METHOD OF LINES SOLUTION

In the method of lines, partial differential equations are transformed into ordinary differential equations by means of piecewise approximations to the dependent variables. Thus, the original initial-boundary value problem for PDEs is approximated by an initial value problem for an ordinary differential equation system. If N spatial points are chosen each one-dimensional equation generates N ordinary differential equations. These are then integrated using high-quality software for stiff and non-stiff ODE systems.

The pseudo-characteristic method of lines proposed by Carver [10] combines the principal advantages of the method of characteristics and the method of lines. The PDEs to be discretized are written in the form of Eq. (7) amenable to solution by standard algorithms for ordinary differential equation integration in which time step is controlled by local convergence error. This permits practical use of higher order approximations in both time and space. Given a hyperbolic system, the numerical solution procedure requires the following steps:

(a) Transform the PDE system to pseudo-characteristic form, Eq. (7), with explicitly defined directions of differentiation in space according to Eq. (6) (see Section 2);

(b) Specify the required boundary conditions, following the characterictic rules, and discretize the spatial derivatives by directional approximations of suitable accuracy;

(c) Solve the resulting ODE system by an efficient error-controled algorithm.

Numerical eigenvalue and eigenvector evaluation, if required, may be an additional step in the solution procedure. When the elements of B and Λ are analytically expressible, only (b) and (c) are repeated during transient computations.

4.1. Discretization

In principal any directional approximation of the spatial derivatives can be used as a numerical differentiator. Here we discuss application of higher-order finite differences and generalized cubic splines.

Suitable finite-difference approximations for upwind derivatives are reviewed by Carver and Hinds [5], and Carver and Schiesser [8]. The method has been applied to several problems [10,20-22,29], including equilibrium and nonequilibrium two-phase flow models. Comparisons with numerical and experimental results [20-22] have shown that biased upwind FD schemes of third and fourth order yield stable PC solutions and minimize the numerical diffusion. An implicit in time PC method, using 4PBU and 5PBU schemes, performed very effectively on a reactor start up transient [20,22]. In fast transients, for relatively fine grids, the method was competitive with the two-step LW method when the time step was restricted by the accuracy condition, and faster in the case of low sonic velocities [22]. Higher-order approximations produced sharper profiles at discontinuities.

Fixed grid directional schemes of order higher than first, based on Lagrange and Hermite polynomial interpolation, diminish the spurious oscillation problem although they do not eliminate it. Small spurious peaks may be observed in the neighbourhood of discontinuities or large gradients. They do not become unstable on reflection [10], and can be minimized by adding numerical damping through appropriate boundary approximations [22]. The performance of schemes (10), (11) improves when the forcing function departs from a true discontinuity to a function with some degree of continuity [8].

Generalized cubic splines offer techniques for achieving a better balance between numerical diffusion and nonphysical oscillations. Use of discrete spline approximations was attempted. As previously discussed, formulas (14) could be a basis for constructing directional schemes with different filtering properties depending on the particular approximation of the second derivatives M_i. A variety of approximations can be used for M_i. Details on the resulting schemes will be presented elsewhere [43]. Here we refer to Section 3 and reference [36], and consider approximations obtained using standard three- and five-point centered differences (3PC, 5PC), or schemes (15) and (16) for M_i. Recall that the former yields approximations to Ψ_x in Eqs. (6),(7), which are generally $O(h^3)$ for nonuniform grids; the latter yields $O(h^4)$ for uniform grids. Increasing the order of M_i approximations normally changes the remainder and dissipation, but not the order of the total scheme. The order of $S_D'(x_i)$ can be increased using asymptotic expansions and special choice of M_i approximations, not necessarily of high order, such as (15),(17).

The approximations considered are summarized below.

	i-3	i-2	i-1	i	i+1	i+2	i+3
4PBU		x	x	x	x		
5PBU	x	x	x	x	x		
DS3C		x	x	x	x		
DS5C	x	x	x	x	x	x	

i is the grid point where the derivative Ψ_x in Eqs. (6),(7) is approximated, assuming $\lambda_i > 0$.

The direction of differentiation is governed by the characteristic slopes according to Eq. (6), Section 2. For a model with n dependent variables, $\overline{\Psi} = (\Psi_1, \Psi_2, \ldots, \Psi_n)^T$, Eq. (6) can be written in vector form as:

$$\overline{\Psi}^*_x = \begin{bmatrix} B_{11}(\frac{\partial\Psi 1}{\partial x})_{\lambda_1} + B_{12}(\frac{\partial\Psi 2}{\partial x})_{\lambda_1} + \ldots B_{1n}(\frac{\partial\Psi n}{\partial x})_{\lambda_1} \\ B_{21}(\frac{\partial\Psi 1}{\partial x})_{\lambda_2} + B_{22}(\frac{\partial\Psi 2}{\partial x})_{\lambda_2} + \ldots B_{2n}(\frac{\partial\Psi n}{\partial x})_{\lambda_2} \\ \vdots \\ B_{n1}(\frac{\partial\Psi 1}{\partial x})_{\lambda_n} + B_{n2}(\frac{\partial\Psi 2}{\partial x})_{\lambda_n} + \ldots B_{nn}(\frac{\partial\Psi n}{\partial x})_{\lambda_n} \end{bmatrix} \quad (24)$$

4.2. Boundary Nodes

Treatment of boundary nodes is analogous to that discussed in Section 3.3 for fully discrete high-order methods. Ordinary differential equations can be written for the unknowns at the junction, and the missing equations substituted by appropriate relations. When the additonal equations closing the system are expressible in differential form these can be incorporated in the overall problem.

Semi-discretizations by the method of lines offer some additional possibilities. For example, in order to ensure that the boundary condition at an outflow end does not cause false reflections, we can merely set $\Psi_{x_k} = \Psi_{x_{k-1}}$. In many cases it is easier to maintain order, using $\Psi_{x_1} = \Psi_{x_2}$ and $\Psi_{x_N} = \Psi_{x_{N-1}}$.

4.3. Numerical Results

Two transients are considered. The first is Test Problem 1b (pipe blowdown) described in Section 3.4. The second concerns flow with heat addition and phase transitions in vertical channels.

Test Problem 1b

The homogeneous equilibrium and EVUT models were used to produce the results shown in Figures 1 to 3. Expanded forms of

Eq. (7) for these models are given in [20-22]. The relevant matrices and vectors A, J_1, J_2, B, Λ, f, c and constitutive equations are given in [23],[26]. The EVUT PC form is illustrated below. Given

$$\Psi = \begin{bmatrix} u \\ \alpha \\ p \\ h_f \\ h_g \end{bmatrix} \qquad B = \begin{bmatrix} \rho c & 0 & 1 & 0 & 0 \\ 0 & 1 & b_o & 0 & 0 \\ -\rho c & 0 & 1 & 0 & 0 \\ 0 & 0 & -b_1 & 0 & 1 \\ 0 & 0 & -b_2 & 0 & 1 \end{bmatrix} \qquad \Lambda = \begin{bmatrix} u+c \\ u \\ u-c \\ u \\ u \end{bmatrix} \text{DIAG} \tag{25}$$

$$b_o = \alpha(1-\alpha)\left(\frac{1}{\rho_g c_f^2} - \frac{1}{\rho_f c_f^2}\right), \quad b_1 = \frac{1}{\rho_f}, \quad b_2 = \frac{1}{\rho_g}, \quad \rho = \alpha\rho_g + (1-\alpha)\rho_f,$$

$$\alpha = \alpha_g = 1-\alpha_f, \quad c^{-2} = \rho\left(\frac{\alpha}{\rho_g c_g^2} - \frac{1-\alpha}{\rho_f c_f^2}\right), \quad c_k^{-2} = (\rho_p + \rho_h/\rho)_k$$

and the matrix A in Eq. (2) from [26], the PC form is obtained either by solving analytically Eq. (3) for the time derivative terms, or by direct substitution in Eq. (7), yielding equations (26).

$$\frac{\partial p}{\partial t} = -[(u+c)\frac{\partial p}{\partial z_+} + (u-c)\frac{\partial p}{\partial z_-} + \rho c((u+c)\frac{\partial u}{\partial z_+} - (u-c)\frac{\partial u}{\partial z_-})]/2 + \rho c^2 f_1$$

$$\frac{\partial \alpha}{\partial t} = -b_o(\frac{\partial p}{\partial t} + u\frac{\partial p}{\partial z_.}) - u\frac{\partial \alpha}{\partial z_.} + f_3$$

$$\frac{\partial h_k}{\partial t} = b_k(\frac{\partial p}{\partial t} + u\frac{\partial p}{\partial z_.})_k - \frac{\partial h_k}{\partial z} + f_k \tag{26}$$

$$\frac{\partial u}{\partial t} = -[(u+c)\frac{\partial u}{\partial z_+} + (u-c)\frac{\partial u}{\partial z_-} + ((u+c)\frac{\partial p}{\partial z_+} - (u-c)\frac{\partial p}{\partial z_-})/\rho c]/2 + f_2/\rho$$

Predicted pressure histories at the closed end are compared in Figs. 1, 2 and 3 with HEM and EVUT benchmark solutions [25,26], with experimental data [38], and a first-order CFD solution [25]. Figure 1 illustrates the PC 4PBU solution with one sided four-point schemes at the boundaries. An implicit Hindmarsh-Gear method [44] with a local error tolerance of 0.001 was used to obtain the PC solution. A significant improvement in accuracy and a much closer agreement with the reference data is seen, compared to the first-order results.

Discrete spline approximations yield the results shown in Figs. 2,3. The EVUT model predicts the short-lived pressure trough, seen in the experiment and attributed to departure from thermal equilibrium. Duration of the trough and the rate of recovery are function of the rate constant t_{ik}^{-1} of interphase heat transfer [26], while the exact location of the minimum and of the rarefaction wave front vary according to exit boundary treatment, which affects the **overall solution.** Ferch [26] modelled the heat transfer by the relationship $q_{ik}=\alpha_f\alpha_g A\rho_k(h_{sk}-h_k)/t_{ik}$, with t_{ik}=5 ms, and used a linear pressure ramp over 5 ms to avoid numerical difficulties. The PC 4PBU method [21] produced stable solutions using exit pressure ramp time 5 to 1 ms or less, but smaller time steps were needed with faster ramp time. In the present study a value of 5 ms was specified, in order to enable comparison with the benchmark results [26], and t_{ik} was permitted to vary roughly according to the amount of disequilibrium, 1 ms$\leq t_{ik}\leq$3 ms. A local loss (abrupt expansion) coefficient, K_N=1, was used at the outlet.

General end conditions may cause some concern, since conditions analogous to those for usual cubic splines are required to construct S_D splines. In practice only f or f,f' values are needed for the approximations considered. In a more general case modified S_D splines can be considered which use the above approximations at the boundaries. The following conditions were specified in the EVUT case: (a) closed end 1) $u_1=0$ **(b) open end 2)** $P_N=P(t)$, or u=c if u>c.

Figure 2 shows two PC DS5C solutions which use different logic for assigning direction of differentiation. The standard PC algorithm in which the direction is dictated by each characteris-

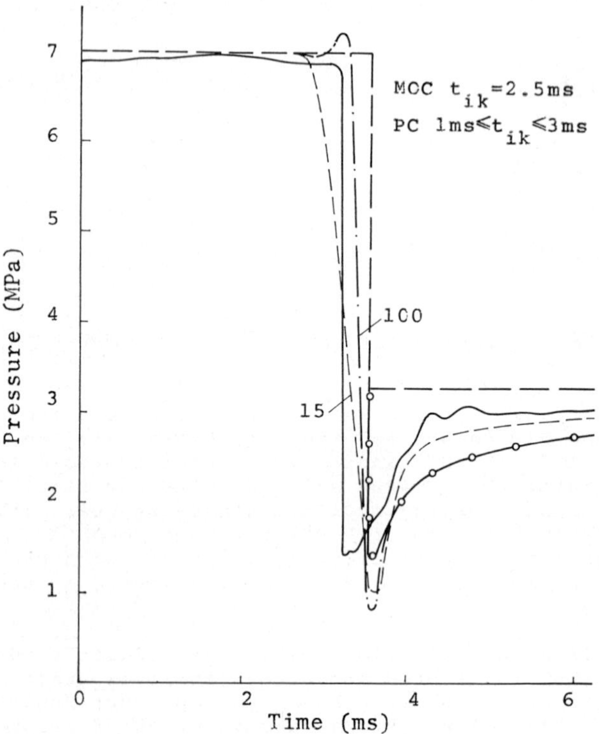

Fig.2. Closed end pressure, EVUT (Test Problem 1b). Experiment——— ; HEM benchmark — —; EVUT benchmark —o—; discrete spline (DS5C)– – – ; pseudo characteristic/DS5C —·—.

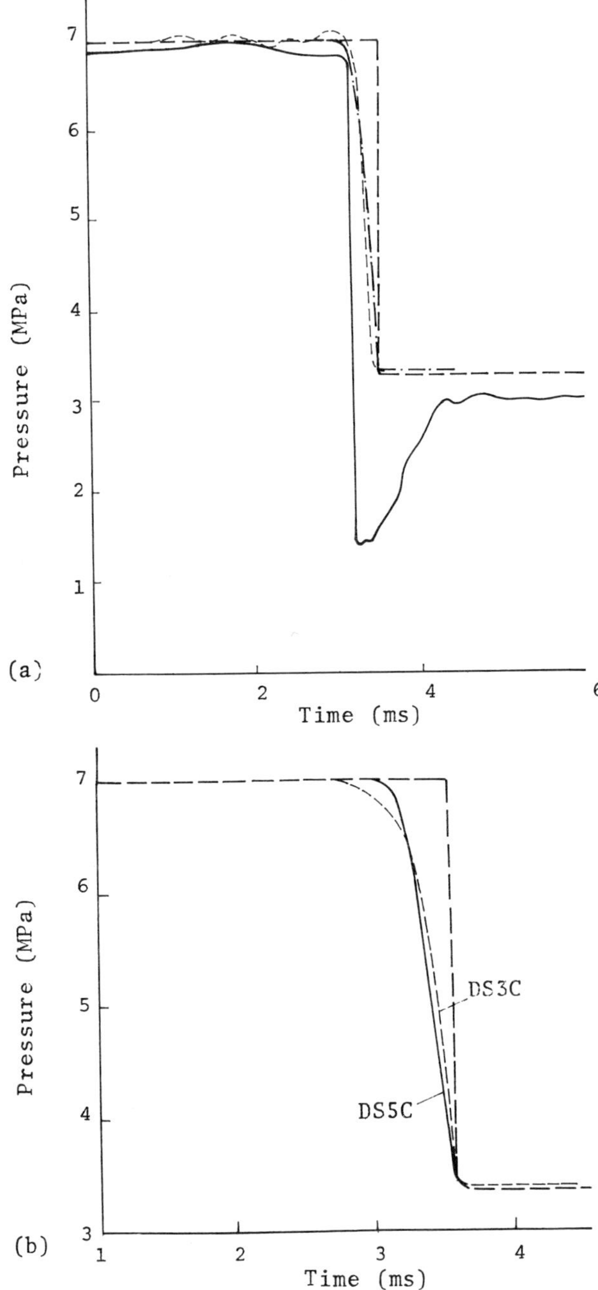

Fig.3. Closed end pressure, Test Problem 1b. Cubic spline and discrete cubic spline results on a 30 point grid. (a) experiment——; MOC benchmark— —; cubic spline– – –; discrete cubic spline (DS5C)—·—; (b) DS3C and DS5C results, enlarged.

tic slope (Eqs. (6),(7) produced the 100 point solution with a small spurious peak at the wave front. When only the "material" characteristic directions were used to govern the derivative biasing, that is Ψ_{x_+} instead of Ψ_{x_-} in the EVUT case, numerical oscillation was eliminated. This is illustrated by the solution on a 15 point grid. The same effect was observed in the HEM case, Fig. 3, using DS3C and DS5C approximations. The usual cubic spline solution is shown for comparison. As can be seen, monotonicity is achieved without introducing excessive dissipation. This is a desirable property and it is instructive to consider the special case above.

Note that in the modified algorithm the u-c compatibility equation is used with downwind biasing of Ψ_x for subsonic flow. Upstream S_D semi-discretization of the primitive form (2) also gave solutions free of spurious oscillations. A slight overshoot reappeared when the strictly characteristic algorithm was applied. Selected S_D variants, used in strictly characteristic sense and compared to 4PBU schemes, produced less oscillations but did not eliminate them (Figs. 1,3), Finally, S_D schemes produced some oscillations when applied to the advective equation with a step forcing function. Numerical results show that for the conditions considered here nonphysical oscillations are eliminated by simple upstream biased differentiation in the set of coupled conservation equations. Analysis of the differential groups $(u+c)\,\Psi_{z_+} + (u-c)\Psi_{z_-}$ in the PC form (26), using (14) for Ψ_z and $M_i=(f_{i-1}-2f_i+f_{i+1})/h^2$, indicates that numerical oscillations can be minimized or eliminated not only by selecting appropriate derivative approximations, but also by suitable combination of approximations in the differential groups. Substitution of (14) in the PC form yields a system which can be regarded as the primitive form, discretized by second order centered schemes, with explicitly defined dissipation terms added. For supersonic flow, and for the modification above the dissipation is sufficient to eliminate the oscillations, while for subsonic flow this would be the case only for particular choice of M_i or Ψ_z. Suitable approximations are under investigation. Rational splines (S_R), using two parameters to combine a first order spline and a cubic spline, may offer additional possibilities.

Discrete spline approximations (14) combine a first order scheme and a higher-order component to minimize both oscillation and dissipation. We have considered two particular cases: (a) $\varepsilon_i=0$ which reduces S_D to a C^2 cubic spline; and (b) $\varepsilon_i=h_i$ which generates local approximations of biased finite-difference type. The effect of $\varepsilon_i>h_i$ is yet to be determined. Increasing the order of M_i in (14) or the order of S'_D approximations, the sharpness of the profile at discontinuities increases (Fig. 3b). Jacobian map and eigenvalue analysis of the resulting DS5C ODE system reveals that the Jacobian matrix is not denser than the 5PBU one (Fig. 5), stiffness is lower, there are less eigenvalues with large imaginary parts and no positive real parts appear.

Test Problems 2a and 2b.

Problem 2a involves a transient initiated by rapid heat addition to a steady upward flow of subcooled water in a vertical channel. Problem 2b is the calculation of inital steady state for a dynamic problem, involving upward vapour-liquid flow with superheating in a part of the channel. Both problems include phase transitions and asymptotic approach to steady state and are used here to establish the conservation properties and certain computational characteristics of the solution algorithms. Problem specifications are given in Table I. Problem 2a is discussed in detail in [20,22]. Figures 4,5 show results of the Jacobian map and eigenvalue analysis of approximating ODE systems for selected schemes.

Table I

	Problem 2a	Problem 2b
Initial conditions	$u(z)=2.231$ m/s $h(z)=1.163$ MJ/kg $p(O)=12.28$ MPa $p(L)=12.20$ MPa $p(z)=p(O)-z/L[p(O)-p(L)]$ $O<z<L$	$G(z)=2014.$ kg/$(m^2 s)$ $h(O)=1.974$ MJ/kg $p(O)=16.912$ MPa $p(L)=16.4$ MPa $h(z)=h(O)-z/L[h(O)-h(L)]$ $p(z)=p(O)-z/L[p(O)-p(L)]$
Boundary conditions	$p(O)=12.28$ MPa $h(O)=1.163$ MJ/kg $p(L)=12.20$ MPa $\dot{q}''=2.38\bar{q}''\sin[\pi(z+0.07)/2.64]$ corresponding to fuel assembly power raised from 0 to 7.0 MW in 0.2 s, then to 7.32 MW in 5.8 s and then held constant [20].	$p(O)=16.912$ MPa $h(O)=1.974$ MJ/kg $p(L)=16.4$ MPa $q''=1.44\bar{q}''\sin[\pi(z+0.05)/0.75]$ corresponding to integral thermal power of 47.0 MW.
Parameters	$L=2.5$ m, $D_h=8.6$ mm $A=8840$ mm^2, $K/\ell=10.52$ m^{-1}	$L=0.65$ m, $D_h=2.462$ mm $A=5567$ mm^2, $K/\ell=22.0$ m^{-1}

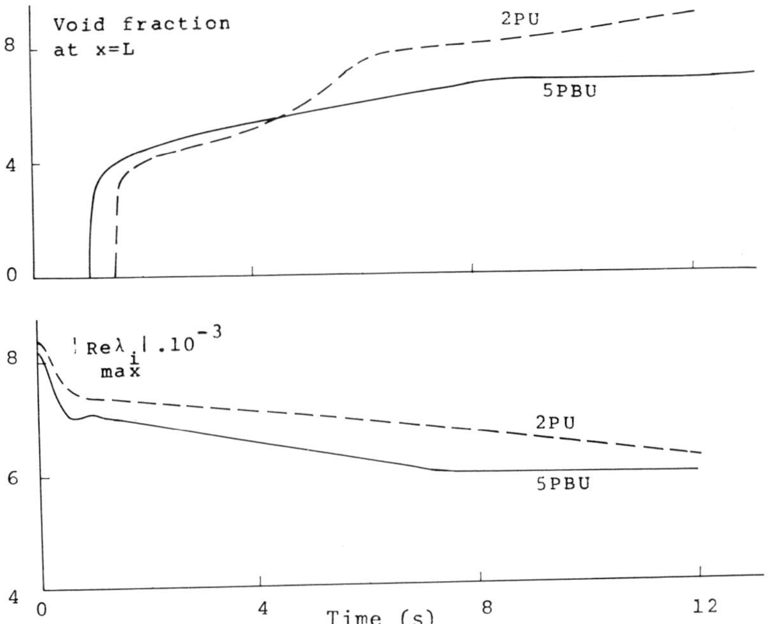

Fig. 4. Eigenvalue variation with problem time for Test Problem 2a, using two-point upwind and five-point biased upwind schemes.

```
7           31          67766
67766       32332       56665
889988      556655      566665
8999998     5566655     5663655
 8999998     5566655     5663655
  8999998     5566655     5563655
   8999998     5566655     5663655
    8999998     5566655     5663655
     889988      556655      566665
      66776       23443       56665
          7       334431      66776
```

```
99998       67766       34433
899998      566765      566665
8997998     5676665     5666665
 8997998     5676665     5666665
  8997998     5676665     5666665
   8996998     5676665     5666665
    8996998     5676665     5666665
     899998      567665      566665
      89999       66776       33443
          7       677774      34444
```

```
CCCCB       89998       66665
BCCCCB      899998      889988
BCCACCB     8897988     8999998
 BCCACCB     8898988     8999998
  BCC9CCB     8898988     8999998
   BCC9CCB     8898988     8999998
    BCC9CCB     8898988     8999998
     BCCCCB      899998      889988
      BCCCC       89998       56666
```

(a) finite differences, 5PBU

```
8           56655       77776
99987       56665       67766
899887      566654      665765
7899887     4666654     5675765
 7899887     4566654     5675765
  7899887     4566654     5675765
   7899987     4566654     5675765
    7899987     4566654     5675765
     789988      566665      567666
      88888       56665       66776
      89999       56666       67777
```

```
99987       78877       55665
899987      787764      566665
789998      6787765     5666665
 89A987      6787765     5667665
 799A997     6787865     5667665
  899A997     6787865     5667765
   89AAA97     6788875     5677765
    89AAA9      678887      567776
     9AAA9       78888       67766
     AABBA       88999       66776
```

```
BBBA9       89998       88888
BBBBA9      898987      889887
9BBBBB9     7897987     7899987
 9BBBBB9     7897987     7899987
  ABCBCB9     7897987     7899987
   ABCBCB9     7897987     7899987
    ABCBCBA     7897987     7899987
     ABCBCB      789998      789998
      BCCCB       89A99       89998
```

(b) discrete spline, DS5C

Fig. 5. (a) Jacobian map for Test Problem 2a, with HEM model in pseudo-characterictic form, on an 11 point grid; (b) Jacobian map for Test Problem 2b. The entries in the Jacobian map (m_{ij}) indicate the magnitude of the nonzero elements in the ODE system Jacobian (J_{ij}) according to $|J_{ij}| \sim 10^{**}(m_{ij}-6)$.

The results correspond to a (G,h,p) HEM model formulation, a grid of 11 points and constitutive equations as in [25]. Armand's two-phase multiplier is used in 2b. First-order PC method introduces large spatial discretization errors and the exit void fraction differs considerably from the 5PBU prediction. Dominating largest $Re\lambda_i$ are shown in Fig. 4. The 5PBU ODE set is not stiff enough ($|Re\lambda_i|_{max}/|Re\lambda_i|_{min} \sim 4.10^3$) to warrant the use of stiffly stable implicit methods and a 'no Jacobian' option of the Hindmarsh-Adams algorithm [44] performed best. Jacobian matrix shown in Fig. 5 can be made banded if the discretized PC form is ordered as in the CFD algorithm. When 2b is considered as a part of a reactor thermohydraulic problem, the stiffness of the coupled equation set increases and the system Jacobian is sparse. Although sparse matrix techniques and stiff-nonstiff partitioning can be used, experience has indicated that separate integration of the heat conduction equation by implicit methods is more effective. Since the channel is of constant area, mass velocity is expected to be constant over the channel length in the steady state. As shown in Table II for a DS5C scheme, mass velocity distribution at time 4 s in Problem 2b approximates a constant value.

Table II

Position (cm)	0	6.5	13	19.5	26	32.5	39	45.5	52	58.5	65
Void fraction	0.707	0.736	0.788	0.845	0.897	0.941	0.975	0.999	1	1	1
$G(kg/sm^2)$	2070	2070	2070	2071	2071	2072	2071	2074	2071	2073	2058

5. ADAPTIVE GRID METHODS

A number of fluiddynamic problems involving large moving gradients [15,45] can be solved more accurately and more economically using adaptive grid techniques that concentrate grid points dynamically across steep fronts. The implementation of adaptive grids requires: (a) development of nonuniform grid differentiators, (b) development of placement criteria, (c) procedure for initialization of the added equations. Placement criteria may be physical, using elements of Lagrangian techniques, or numerical, in terms of u_{xx}, u_x thresholds, L_2-norm etc. Spline approximations considered in the preceeding sections are particularly suitable for such algorithms since they permit economic evaluation of first and second derivatives, tracking the progress of large gradients between mesh points and spline interpolation for initialization of new grid points. The improved accuracy permits the use of coarser grids.

An adaptive grid variant of the semi-discretization methods considered was tested on the Burger's equation shock problem described in Section 2. Cubic spline (C^2) and discrete cubic spline approximations were considered. Figures 6,7 illustrate the effect of two adaptive points when a cubic spline differentiator is used. The integrated absolute error, $E=[\sum_i (U_i - U_{i,\,exact})]^{1/2}$ is shown in Fig. 7. The spatial grid is controlled dynamically by several user-defined parameters: u_{xx} and u_x thresholds, permissible number of grid points, of adative points and of subintervals. In the particular case above the number of grid points was restricted in order tominimize the change in the dominance of parameter μ in the equation (see Section 2). When a nonuniform grid, directional DS3C differentiator was used, the integrated absolute error was about an order of magnitude smaller. Spurious numerical effects were minimized in both cases. Variable grid differentiators were about 15-20 percent slower per mesh point, compared to the uniform grid variants. The results of these simple tests clearly indicate the advantages of the adaptive grid methods. The algorithms are directly applicable to the pseudo-characteristic and primitive forms of the flow equations.

6. CONCLUSIONS

Numerical methods for transient two-phase flow have been considered that combine accuracy and economy. Practical use of adaptive grid variants is possible. Discrete cubic spline approximations yield very good accuracy in case of functions with some degree of continuity. In case of true discontinuities the performance of local S_D approximations is comparable to or better than that of the well tested four- and five-point biased upwind approximations. Generalized cubic spline approximations, that are particularly well suited for treatment of discontinuities, are under investigation.

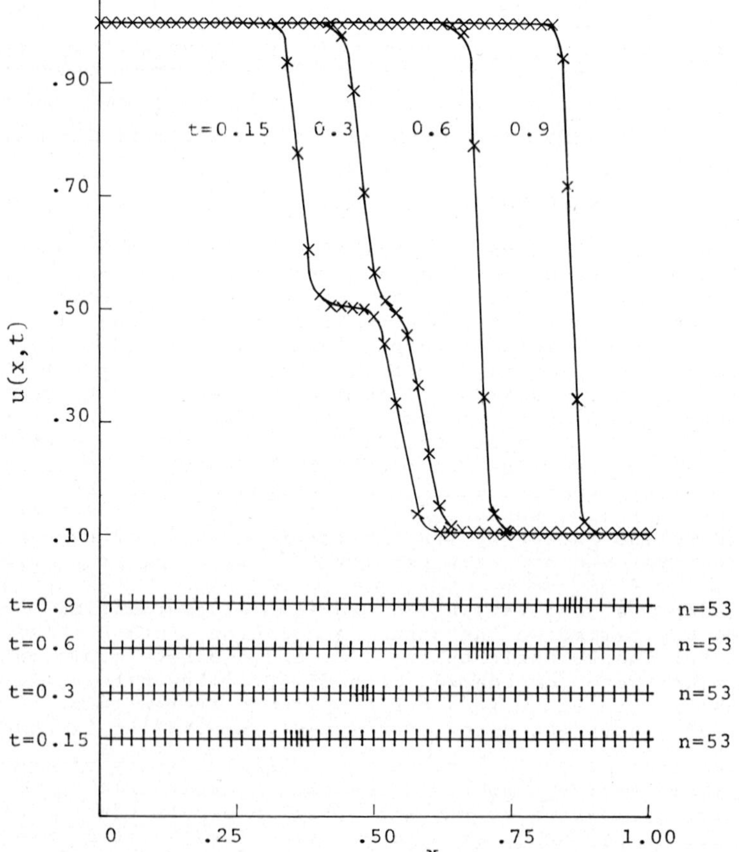

Fig. 6. Burger's equation solution: adaptive grid cubic spline, two moving points; line-analytical, points-numerical solution.

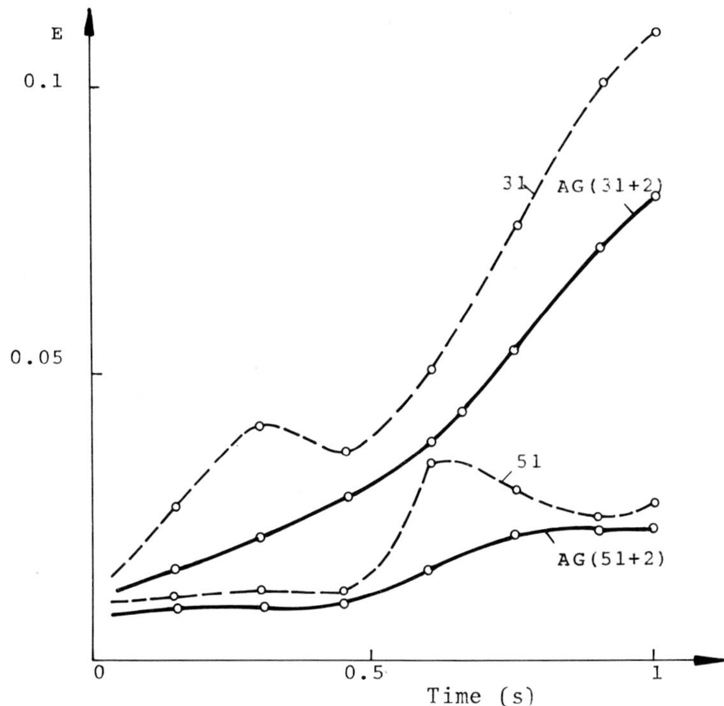

Fig. 7. Integrated absolute error, Burger's equation solution using fixed and adaptive grid cubic spline, with two moving points.

REFERENCES

1. Kreiss, H.O. and Oliger, J., Tellus 24 (1972), 199.

2. Swartz, B. and Wendroff, B., SIAM J. Numer. Anal. 11 (1974), 972.

3. Fornberg, E., SIAM J. Numer. Anal., 12, (1075), 509.

4. Gottlieb, D. and Orszag, S.A., "Numerical Analysis of Spectral Methods: Theory and Application", SIAM, Philadelphia, 1977.

5. Carver, M.B. and Hinds, H.W., "The Methods of Lines and the Advective Equation", Simulation, Sept. 1978.

6. Sod, G.A., J. Comp. Phys. 24 (1978), 402.

7. Carver, M.B., "Pseudo Characteristic Method of Lines Solution on the Conservation Equations", J.Comp.Phys., 35, 1 (1980) p. 57.

8. Carver, M.B. and Schiesser, W.E., "Biased Upwind Difference Approximations for First-Order Hyperbolic (Convective) Partial Differential Equations", 73rd Annual AIChE Mtg., Nov. 16-18, 1980.

9. Zalezak,S.T., in Advances in Computer Methods for PDEs, R. Vichnevetsky and R. Stepleman, eds., IMACS Press, 1981,pp.126-134.

10. Turkel, E., J. Comp. 35 (1980), 319.

11. Hughes, J.J.R., ed. "Finite Element Methods for Convection Dominated Flows", ASME, New York, 1979.

12. Vichnevetsky,K. and Tomalesky,A.W., "Spurious Wave Phenomena in Numerical Approximations of Hyperbolic Equations", Proc. Fifth An. Princeton Conf. on Information Systems Science,1971.

13. Vichnevetsky, R. and Pfeiffer, B., "Error Waves in Finite Element and FD Methods for Hyperbolic Equations", in Advances in Computer Methods for PDEs, AICA Press, 1975, pp. 52-57.

14. Vichnevetsky, R., "Group Velocity and Reflection Phenomena in Numerical Approximations of Hyperbolic Equations: a Survey ", DCS-TR-116, Rutgers Univ., New Brunswick, NJ, (1982).

15. Hancox, W.T. and McDonald, B.H., "Finite Difference Algorithms to Solve the One-Dimensional Flow Boiling Equations", Proc. ANS/ASME Topical Mtg. on Nucl. Reactor Thermal-Hydraulics, Saratoga, New York, R.T. Lahey, ed. 1980.

16. Hu, S.S. and Schiesser, W.E., "An Adaptive Grid Method in the Numerical Method of Lines", Advances in Computer Methods for PDEs, IMACS Press, 1981, pp. 305-311.

17. Gelinas, R.J. and Doss, S.R., in Advances in Computer Methods for PDEs, IMACS Press, 1981, pp. 230-238.

18. Romstedt, P. and Werner, W., Nucl. Sci. Eng., 64, (1977), 208.

19. Werner, W.F., "Weighted Residual Methods for the Solution of Fluid Dynamics Problems", Proc. Topl. Mtg. - Advances in Mathematical Methods for Nuclear Eng. Problems, Munich, 1981

20. Kolev, N.P., Manahilov, B. and Dimitrov, Ch., "Application of the PC Method of Lines to 1D Transient Thermohydraulic Analysis for Nuclear Reactors", Int. J. Modelling and Simul.,1,3(1981); Proc. SIMULATION'80 Symp., Interlaken, Switzerland, 1980.

21. Kolev, N.P. and Carver, M.B., "Pseudo Characteristic Method of Lines Solution of the Two-Phase Conservation Equations", Proc. III CSNI Specialists' Meeting on Transient Two-Phase Flow, Pasadena, March 1981.

22. Kolev, N.P. and Katkovsky, E.A., "A Comparative Study of the Pseudo Characteristic and LW Methods in Flow-Boiling Problems", Adv. in Comp. Methods for PDEs, IMACS 1981, pp. 391-398.

23. McDonald, B.H., Hancox, B.H. and Mathers, W.G., "Numerical Solution of the Transient Flow-Boiling Equations", II CSNI Spec. Mtg. on Transient Two-Phase Flow, Paris, June 1978.

24. Mathers, W.G., Zuzak, W.W., McDonald, B.H. and Hancox, W.T., Proc. CSNI Specialists Mtg., Toronto, Aug. 1976, vol. 1.

25. Hancox, W.T. and Banerjee, S., Nucl. Sci. Eng., 64,(1977),106.

26. Ferch, R.L., Int. J. Multiphase Flow, 5, (1979), 265.

27. Madsen, N.K. and Sincovec, in Numerical Methods for Differential Systems , L. Lapidus and W.E. Schiesser, eds., p. 229, Acad. Press, NY, 1975.

28. Rojdestvensky, B.L. and Yanenko, N.N., Quasi-Linear Equation Systems , Nauka, Moskow, 1978 (in Russian).

29. Wang, Y.Z. and Johnson, S.H., "Application of the Pseudo Characteristic Method of Lines", Proc. 10th IMACS Congress, Montreal, (1982) vol. 1, pp. 146-148.

30. Rudinger, G., Wave Diagrams for Non Steady Flow in Ducts , Van Nostrand, NY, 1955.

31. Weisman, J. and Tentner, A., Nucl. Sci. Eng. 78 (1981), 1.

32. Grillenberger, T., MRR-166,Technische Universität, München,1976.

33. McDonald, B.H., Richards, D.J. and Mills, P.J., "Appplication of Dynamic Grid Control in the Rama Code", Proc. 10th IMACS Congress, Montreal, 1982, vol. 1, pp. 162-163.

34. Rajan, V.S.V., Barclay, F.W., Hawley, E.H., McDonald, B.H. and Hanna, B.N., "RAMA: A Code for Analysis of the Thermohydraulics of Horizontal Heated Channels in a Pipe Network", CSNI Spec. Mtg. on Transient Two-Phase Flow, Pasadena 1981.

35. Roy, R.P. and Ho, S.A., Nucl. Sci. Eng. 81 (1982), 459.

36. Zavyalov, U.S., Kvassov, B.I. and Miroshnitchenko, V.L., Methods of Spline Functions , Nauka, Moscow, 1980 (in Russian).

37. Vichnevetsky, R., et al., Advances in Computer Methods for PDEs, IMACS Press, 1981, pp. 259-263.

38. Edwards, A.R. and O'Brien, T.P., J. Brit. Nucl. Energy Soc., 9, (1970), 125.

39. Katkovsky, E.A. and Poletaev, G.N., "Wave Propagation in Hydraulic Networks", IAE-2491 (1975) Moskow (in Russian).

40. Dongara,J. et al., "LINPACK-User's Guide",SIAM,Philadelphia,1979

41. Zlatev, Z., Wasniewski, J. and Schaumburg, K., "Y12M-Solution of Large and Sparse Systems of Linear Algebraic Equations", Lecture Notes in Computer Science, Springer, Berlin, 1981.

42. Curtis, A.R. and Reid, J.K., AERE R6844 (1971).

43. Katkovsky, E.A., and Kolev,N.P., "Generalized Cubic Spline Approximations for First-Order Hyperbolic PDEs", in preparation.

44. Hindmarsch,A.C., Gear ODE Solver,LLL Rept. UCID-30001-R3,1974.

45. Lahey, R.T., "Some Physical Problems Involving Large Moving Gradients", Proc. 10th IMACS Congress, Montreal, 1982.

Multi-Phase Flow and Heat Transfer III. Part A: Fundamentals
edited by T.N. Veziroğlu and A.E. Bergles
Elsevier Science Publishers B.V., Amsterdam, 1984 — Printed in The Netherlands

EFFECT OF THE GAS-PHASE DENSITY ON FLOW PATTERN AND FRICTIONAL PRESSURE DROP IN TWO-PHASE, TWO-COMPONENT VERTICAL FLOW

Mohamed A. Aggour
Exxon Company, U.S.A.
4550 Dacoma, Houston, TX 77092
and
Grant E. Sims
Department of Mechanical Engineering
University of Manitoba
Winnipeg, Manitoba R3T 2N2, Canada

ABSTRACT

Experimental data on the effect of the gas-phase density on flow-pattern transitions and frictional pressure drop were obtained during co-current gas-water flow in a vertical tube. Three gases (air, helium and Freon 12 vapor) were used allowing a change in the gas-phase density by a factor of approximately 31. The flow patterns were found to depend strongly on the relative superficial velocities of the phases; and certain flow-pattern transitions were found to depend on the gas-phase density. The gas density had an effect on the frictional pressure drop only at low water flow rates and moderate-to-high superficial gas velocities. The present data are used to evaluate some of the existing flow-pattern transition and pressure drop models and correlations.

1. INTRODUCTION

Two-phase, two-component flow has been the subject of intensive investigation in recent years because of its importance in many industrial applications, especially in the energy-related industries. Complete understanding of this subject, however, has not been achieved as yet. This is mainly because of the interdependence among the various flow parameters (flow patterns, pressure drop, etc.), and the dependence of these parameters on the many variables involved in this type of flow such as fluid properties, superficial velocities, flow geometry, etc.

Although a considerable amount of work has been published on the hydrodynamic aspects of two-phase, two-component flow, few data are found on the effect of fluid properties, particularly the gas-phase properties, on flow-pattern transitions and frictional pressure drop in two-component vertical flow. With the growing interest in modeling flow-pattern transitions, and the numerous existing correlations for predicting flow patterns and pressure drop, the need for additional data to examine the validity and generality of such models and correlations can not be overemphasized.

This paper presents experimental data on the effect of the gas-phase density on flow patterns and frictional pressure drop in two-component vertical flow. The present data were obtained from heat-transfer experiments with air-water, helium-water and Freon 12 vapor-water. Although the selection of these three gases and the experimental conditions described below allowed a large change in the gas-phase density, there were changes in the gas-phase viscosity as well (the viscosity of Freon 12 vapor compared to the air and helium viscosities which are approximately equal). For the purpose of this study, however, and until we have evidence of the effect of gas viscosity on flow-pattern transitions and frictional pressure drop, all observed effects are attributed to the gas-phase density. The present data, which cover a wide range of gas density and superficial velocities, are used to test some of the existing models and correlations with regard to their validity and generality.

2. EXPERIMENTAL APPARATUS

A schematic of the experimental apparatus is shown in Figure 1. The test section on the left end of Figure 1 was used for another investigation; the other test section (the heat-transfer test section) was used for the present study. It consisted of: (1) a 0.46 in. (1.17 cm) ID, 2.0 ft (60.96 cm) long, type 304 stainless steel tube which was used for heat-transfer and pressure drop measurements; and (2) a 1.0 ft (30.48 cm) long transparent cast acrylic section (having the same inside diameter as the stainless steel tube) connected immediately downstream of the stainless steel tube. A 5.0 ft (152 cm) long calming section was provided upstream of the test section. The mixer consisted of a cylindrical chamber with a concentric porous tube having the same inside diameter as the test section. Water flowed axially upward through the porous tube while the gas was forced through the pores of the tube to mix with the water.

The pressure drop was measured across the stainless steel tube by means of a water inverted-manometer or a mercury manometer depending on the range of the pressure drop. The flow patterns were observed, through the transparent section, visually unaided and with the aid of a stroboscope, and photographs were taken using a Pentax Spotmatic camera equipped with Super Takumar 50 mm - f/4 close-up lense and an electronic microflash unit with 0.5 sec flash duration.

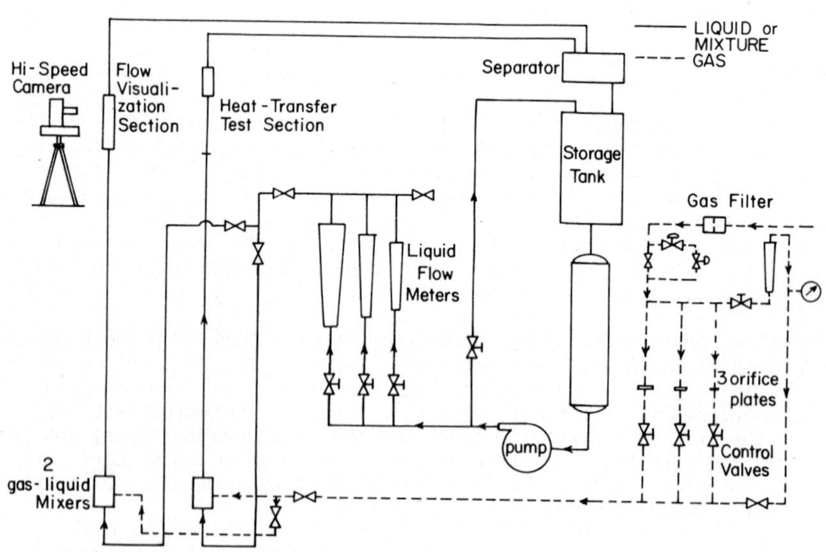

Fig. 1 Simplified Schematic of Experimental Apparatus

3. EXPERIMENTAL CONDITIONS

The present experiments were conducted using three gases (air, helium and Freon 12 vapor) with water as the liquid phase. These gases gave, under similar experimental conditions (same V_{SG} & V_{SL}), a density ratio of approximately 31 (the maximum density ratio for all experiments was about 55). As mentioned in the Introduction, the present data were obtained under heat-transfer conditions where the stainless steel tube was electrically heated to investigate the effect of the gas-phase density on heat-transfer coefficients. During the experiments, the difference between the wall and bulk temperatures was kept small (typically 15°F, i.e., 8°C), and the difference between the mixture temperatures at the outlet and inlet of the stainless steel tube varied between one and 20°F (0.6 and 11°C). further, the difference between the mixture's mean temperature in the heat-transfer section and in the transparent section was between 0.5 and 10°F (0.3 and 6°C). Under these conditions, the effect of heat transfer on flow patterns and pressure drop was deemed to be small.

The experimental conditions and the range of variables covered by the present experiments are summarized in Table 1, where V_{SL} is the superficial liquid velocity, V_{SG} is the superficial gas velocity, ρ_G is the gas density, P is the absolute mean pressure, and T is the mean temperature of the mixture. It should be mentioned here that the values of V_{SG} in Table 1 and on the figures presented below are the mean values calculated at the mixture's mean pressure and temperature. Since the difference between the mean V_{SG} and V_{SG} at the inlet to the transparent section was small (typically 2 to 3% and a maximum of 10%), no errors would be expected in the presentation of flow-pattern transitions discussed below.

Table 1. Experimental Conditions and Range of Variables

VARIABLE	RANGE COVERED BY EXPERIMENTS		
	HELIUM-WATER	AIR-WATER	FREON 12-WATER
V_{SL} ft/sec (m/sec)	1.03 - 34.7 (0.314 - 10.6)	1.03 - 34.7 (0.314 - 10.6)	1.03 - 13.9 (0.314 - 4.24)
V_{SG} ft/sec (m/sec)	0.423 - 484 (0.129 - 148)	0.240 - 315 (0.073 - 96.0)	0.515 - 118 (0.157 - 36.0)
ρ_G lbm/ft^3 (kg/m^3)	0.0109 - 0.0375 (0.1745 - 0.6003)	0.0772 - 0.3453 (1.236 - 5.527)	0.3367 - 0.6034 (5.390 - 9.660)
P Psia (N/Cm^2)	15.6 - 53.3 (10.8 - 36.7)	15.4 - 71.5 (10.6 - 49.3)	15.9 - 27.8 (11.0 - 19.2)
T °F (°C)	68 - 82 (20 - 28)	72 - 101 (22 - 38)	75 - 84 (24 - 29)

4. RESULTS AND DISCUSSION

4.1 Flow Patterns

Although different terminology of the flow patterns is found in the literature [1], some attempts have been made recently to standardize the description and terminology of flow patterns [2,3,4]. The basic flow patterns observed in the present study are described below.

1. Bubble Flow: The liquid flows as a continuous phase with the gas phase approximately uniformly distributed as discrete bubbles.

2. Slug Flow: The gas flows mostly as large bullet-shaped bubbles surrounded by a thin liquid film which flows downward. The large gas bubbles are separated by slugs of continuous liquid which may or may not contain small gas bubbles.

3. Annular Flow: The gas flows as a continuous phase along the core of the pipe and the liquid flows mainly as an annulus against the wall. The gas core may or may not contain liquid drops.

4. Froth Flow: This is also called emulsion flow. The gas and liquid phases flow in a highly turbulent, evenly dispersed mixture which has the appearance of milky emulsion. This pattern was observed only at high liquid flow rates and occurred directly following bubble flow.

5. Churn Flow: This pattern occurred only at low liquid rates. It has such a collapsing and pulsating nature. The gas flows in large distorted bubbles with the liquid falling downward forming a slug which is then lifted by the gas. It should be mentioned here that this churn pattern is similar to the froth flow defined by Govier and Aziz [2].

Detailed presentation of the flow-pattern photographs obtained in the present study can be found in [5]. In reference [5], however the churn flow described here was identified as a slug-annular transition.

Flow patterns are determined largely by the relative velocities of the phases. The fluid properties and pipe diameter are, in general, the other factors that affect flow-pattern transitions. Several investigators, e.g. [2,6,7,8,9], have studied the effect of liquid properties and pipe diameter on flow-pattern transitions. The following discussion deals only with the effect of the gas-phase density; and the comparison with existing correlations and models will be limited to those flow-pattern transitions which are affected by the gas density.

The present flow-pattern data are presented on Figure 2 in terms of V_{SL} versus V_{SG}. The general areas for bubble, slug, annular, and froth flow patterns are marked on the figure. Examination of Figure 2 shows that the superficial velocities are reasonably adequate for defining the bubble-slug, froth-slug and froth-annular transitions. The other transitions (slug-annular, slug-churn and churn-annular), however, can not be defined on Figure 2. This suggests that these transitions may be dependent on the gas-phase density as will be shown below.

The effect of the gas-phase density on flow-pattern transitions has been studied experimentally by Brown et al [10] and Weisman and Kang [8]. Brown et al [10] obtained data with air-water mixtures at pressures ranging from 18 to 110 psia (12.4 to 75.8 N/cm^2); i.e., a range of gas density of 0.092 to 0.552

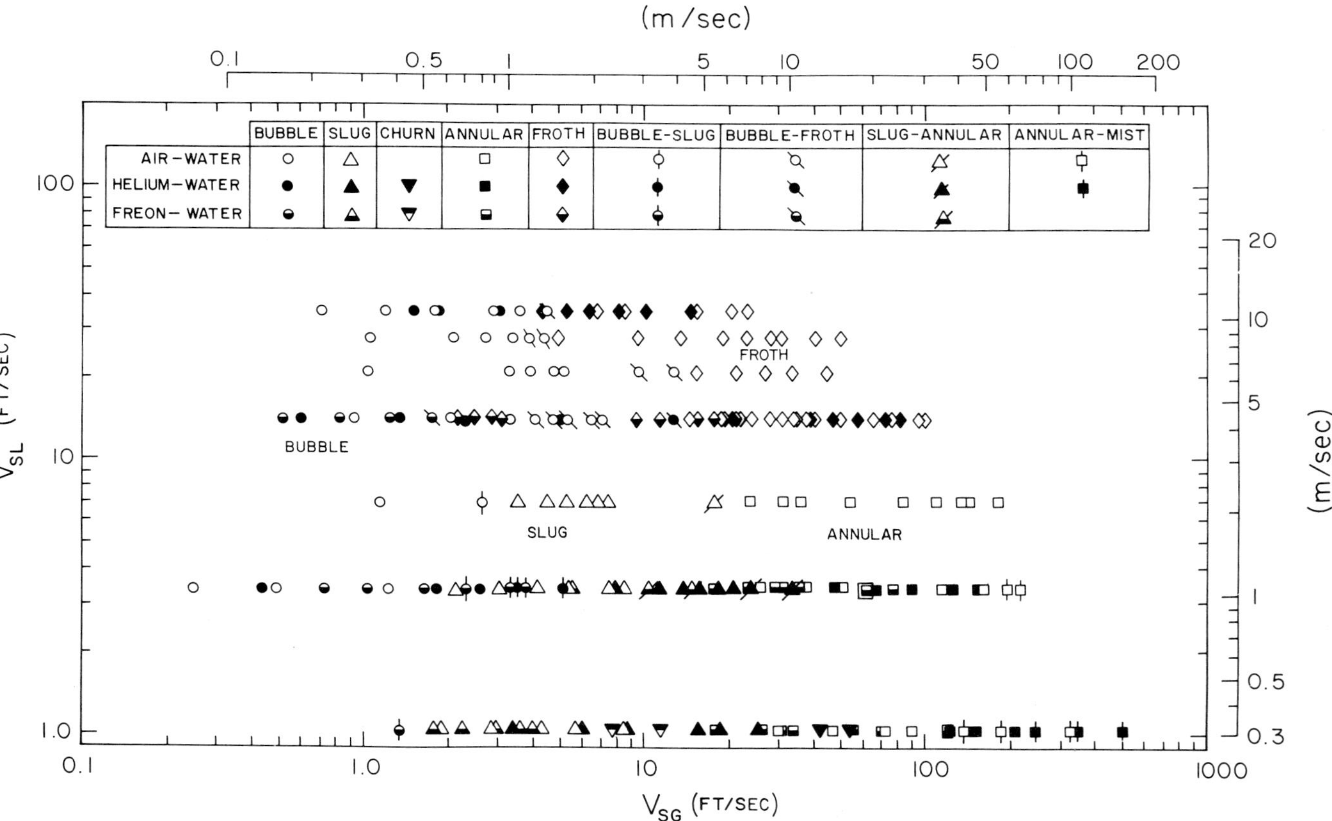

Fig. 2 Flow - Pattern Data

lb/ft^3 (1.47 to 8.84 kg/m^3). They concluded that the effect on flow-pattern transitions is proportional to the one-third root of the gas density. This effect was incorporated in the generalized flow-pattern map recommended by Govier and Aziz [2]. The map of [2] is tested against the present data in Fig. 3 where modified superficial liquid velocity Y^*V_{SL} and superficial gas velocity X^*V_{SG} are used as ordinate and abscissa respectively, where

$$Y^* = (\rho_L \sigma_{WA} / \rho_W \sigma_{LG})^{1/4},$$

$$X^* = Y^* (\rho_G / \rho_A)^{1/3},$$

ρ_L is the liquid density at flow conditions,

ρ_W is the density of water at standard conditions,

σ_{WA} is the surface tension of water-air system at standard conditions,

σ_{LG} is the surface tension of the liquid-gas sytem at flow conditions,

ρ_G is the density of the gas at flow conditions, and

ρ_A is the density of air at standard conditions.

Examination of Figs. 2 and 3 shows that the map of [2], provides great improvement in defining the slug-churn, slug-annular, and churn-annular transition boundaries over Fig. 2; the latter is better for the bubble-slug transition while no single conclusion can be drawn as to the bubble-froth transition.

Weisman and Kang [8] obtained data with Freon 113 vapor-liquid mixtures in a vertical tube at pressures between 14.7 and 43.5 psia (10 and 30 N/cm^2) and proposed an overall flow-pattern map with the coordinates being the superficial velocities each multiplied by a dimensionless factor to correct for the effect of fluid properties and pipe diameter; such factors being different for the different transitions. The transition criteria proposed by [8] suggested that the gas density has some effect on the intermittent-annular transition (intermittent flow includes both slug and churn flows). This transition was proposed to be determined by the equation:

$$1.9\ (V_{SG}/V_{SL})^{1/8} = Ku^{0.2}\ Fr^{0.18} \qquad (1)$$

where,

Ku = Kutataladze number = $V_{SG}\ \rho_G^{1/2} / [\sigma g(\ L-\ G)]^{1/4}$.

Fr = Froude number = $V_{SG}/\sqrt{g\ D}$,

D is the pipe diameter, and g is the acceleration of gravity.

The proposed criterion, eq. (1), is tested against the present data in Figure 4, where (V_{SG}/V_{SL}) is plotted against $(Ku^{0.2}\ Fr^{0.18})$. The solid line in Figure 4 represents Eq. (1), and the cross-hatched area represents the transition region between intermittent and annular flows. The figure shows that the proposed transition criterion works very well up to a value of (V_{SG}/V_{SL}) of approximately 25.

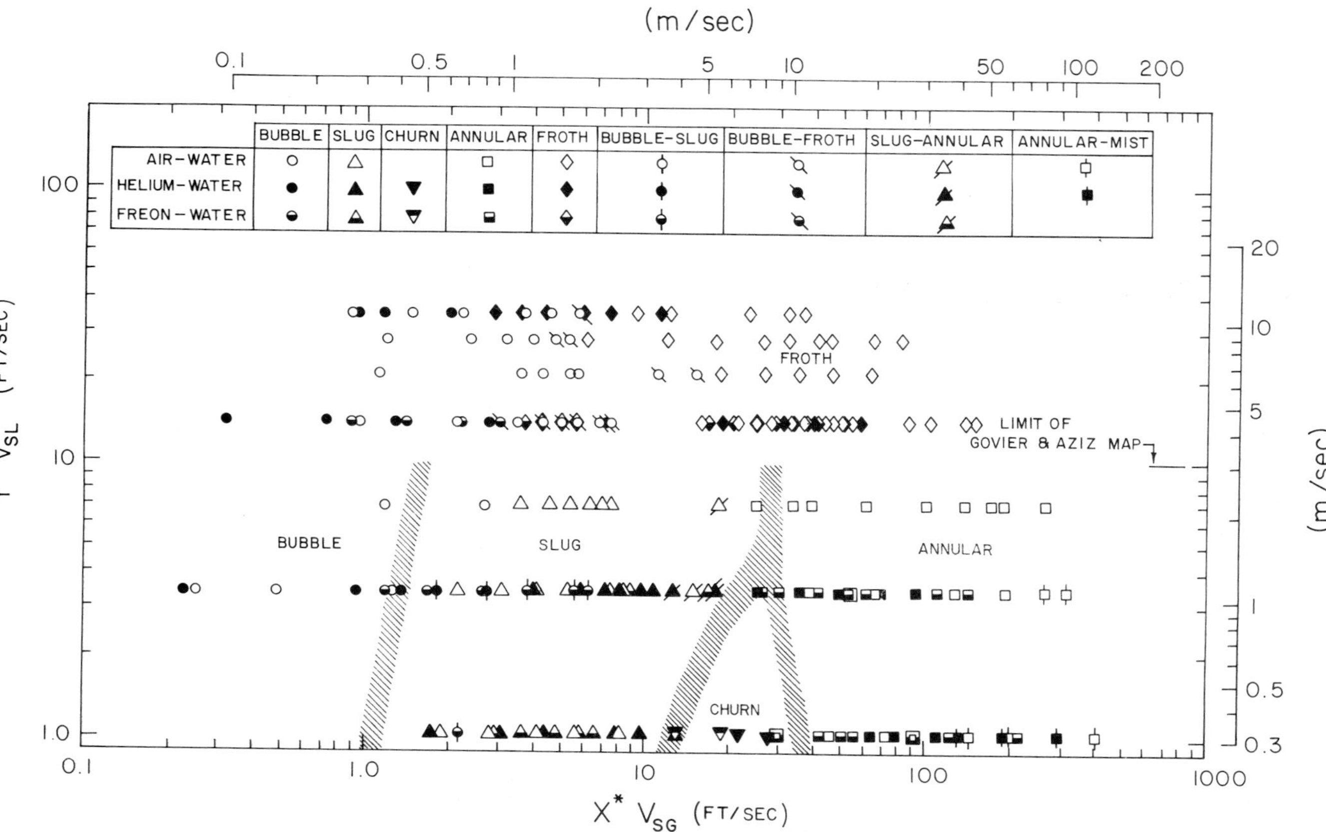

Fig. 3 Comparison of Data Against Govier & Aziz Map

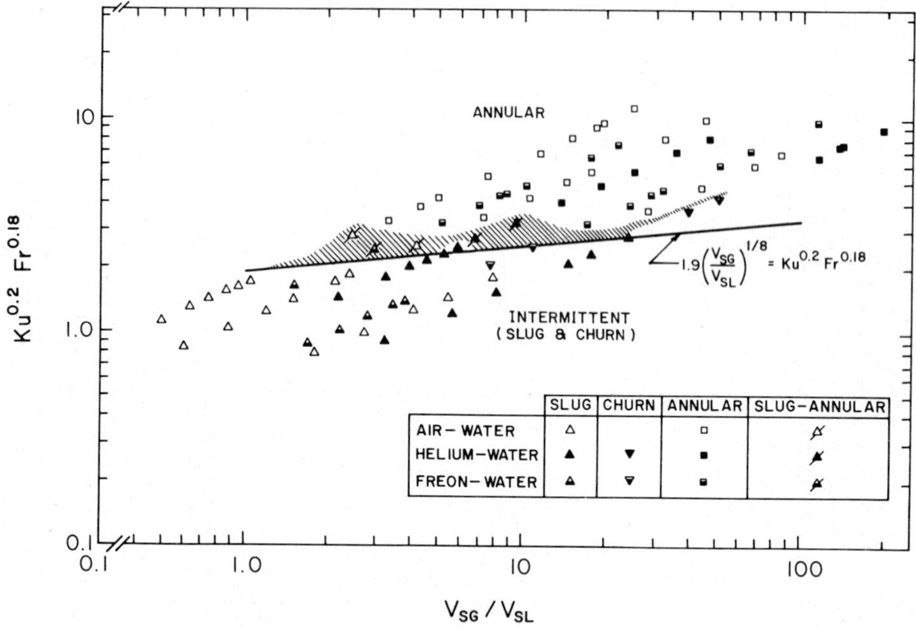

Fig. 4 Comparison of Data with Intermittent - Annular Transition of Weisman & Kang

Taitel et al [9] have recently proposed models for the various flow-pattern transitions in upward vertical flow and compared them against air-water data obtained in 2.5 and 5.1 cm diameter tubes. The models of [9] suggest that the gas-phase density has an effect on the churn- or slug-annular transition. Further, [9] showed that the transition to annular flow is independent of the liquid flow rate and tube diameter, and is determined by the equation.

$$Ku = 3.1 \tag{2}$$

This model [9] is tested against the present data in Figure 5, where Ku is plotted aginst V_{SL}. The numbers (1,2 and 3) identify the first observed annular flow for the three V_{SL} values shown on the figure. Examination of Figure 5 shows that Ku provides excellent separation for the flow-pattern regions, illustrating the success of the proposed model. The figure shows, however, that the value of Ku at which annular flow was observed in the present study is consistently lower than 3.1.

4.2 Frictional Pressure Drop

The present frictional pressure drop data were calculated from the measured total pressure drop (measured across the stainless steel tube) and the gravitational (hydraulic) pressure drop which was calculated using the void fraction correlation of Chisholm [11]. The accelerational pressure drop was neglected since it was very small compared to the total and gravitational pressure drops.

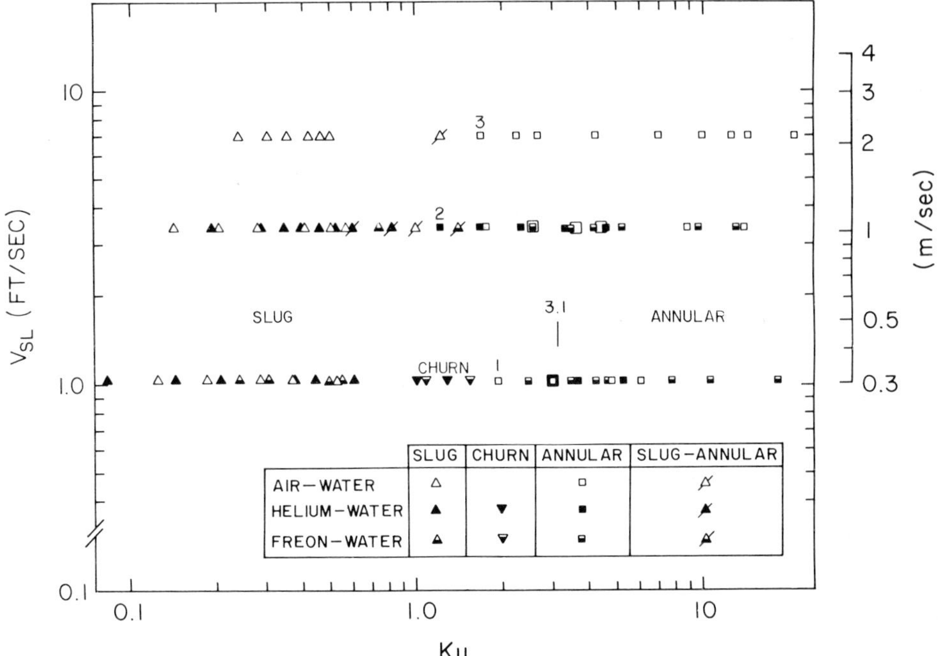

Fig. 5 Comparison of Data with Slug-or Churn-Annular Transition of Taitel et al

The present data are shown on Figure 6,7 and 8 for air-water, helium-water and Freon-water respectively. In these figures, the frictional pressure drop P_{TPF} is plotted against V_{SG} with V_{SL} as a parameter. The letters appearing next to the data points indicate the observed flow pattern, where, B is for bubble, S for slug, C for churn, A for annular, F for froth, and the double letters indicate transitional flow patterns. It should be mentioned here that the data for which the frictional pressure drop was less than 0.01 psi, equivalent of 0.2 in. (0.51 cm) of water, were excluded and are not shown on the figures. These data were obtained at the lowest V_{SL}; i.e. 1.03 ft/sec (0.314 m/sec), and some low values of V_{SG}, and were all in the slug flow pattern. Under these conditions, the fluctuations in the water levels in the inverted manometer used for measuring the pressure drop were too large to obtain accurate measurement of the pressure drop.

To illustrate the effect of the gas-phase density on frictional pressure drop, the data for all three gas-water mixtures are plotted on Figure 9. The following observations can be made by examining Figure 9:

1. At low liquid velocities and low gas velocities, and at high liquid rates and any gas velocity, the gas-phase density appears to have no effect on frictional pressure drop.

2. The gas density, however, has a significant effect on pressure drop at low liquid rates and moderate-to-high gas velocities; under these conditions, the frictional pressure drop increases with increasing gas density.

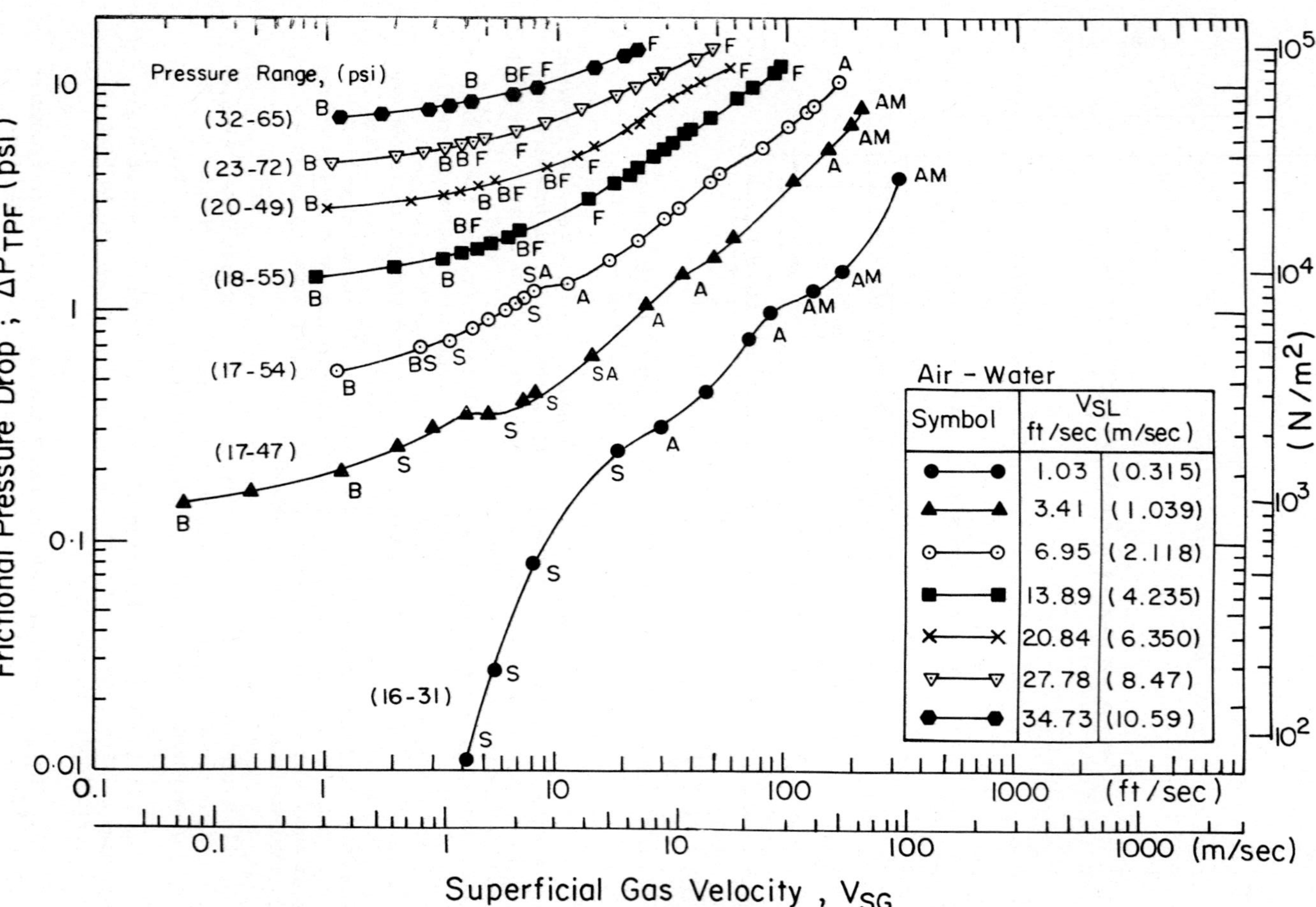

Fig. 6 Air-Water Frictional Pressure Drop Data

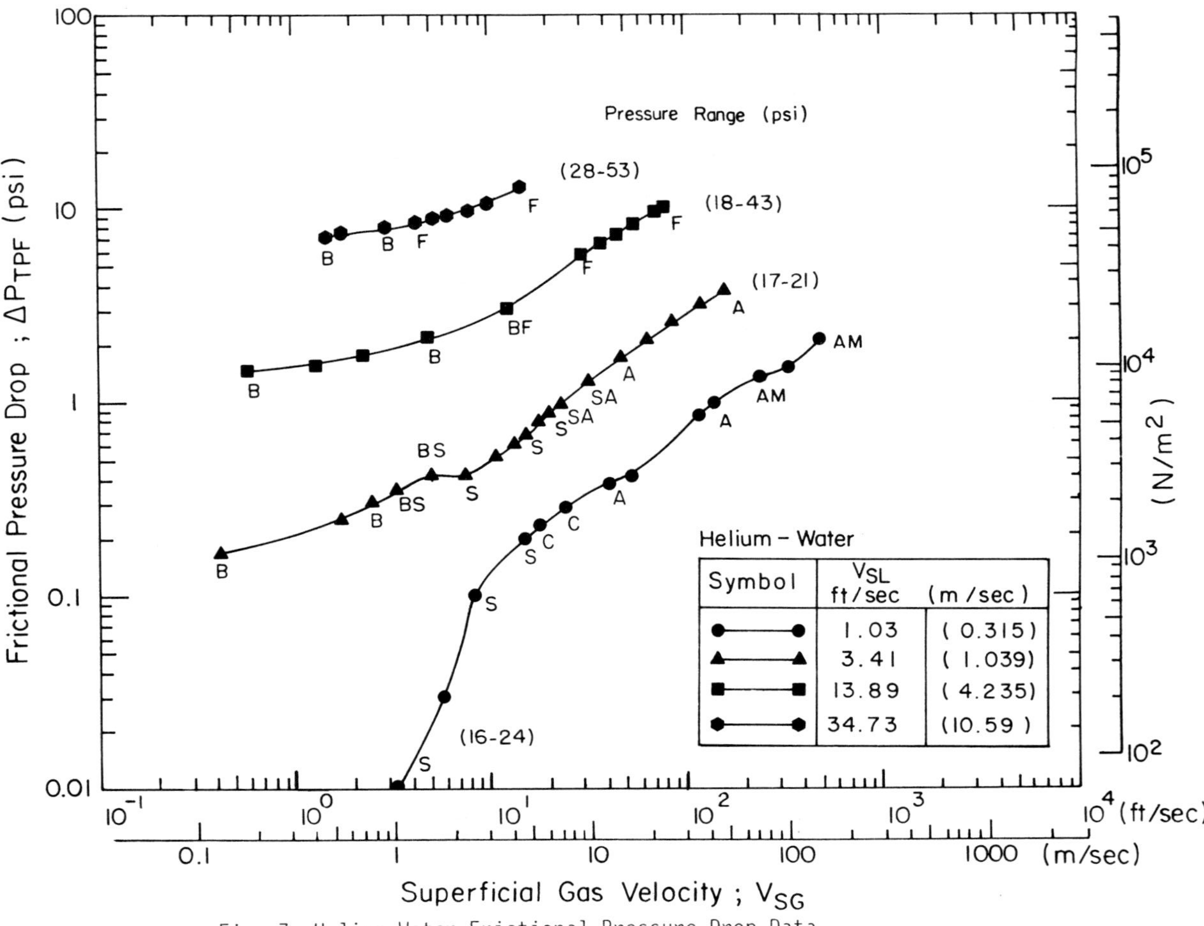

Fig. 7 Helium-Water Frictional Pressure Drop Data

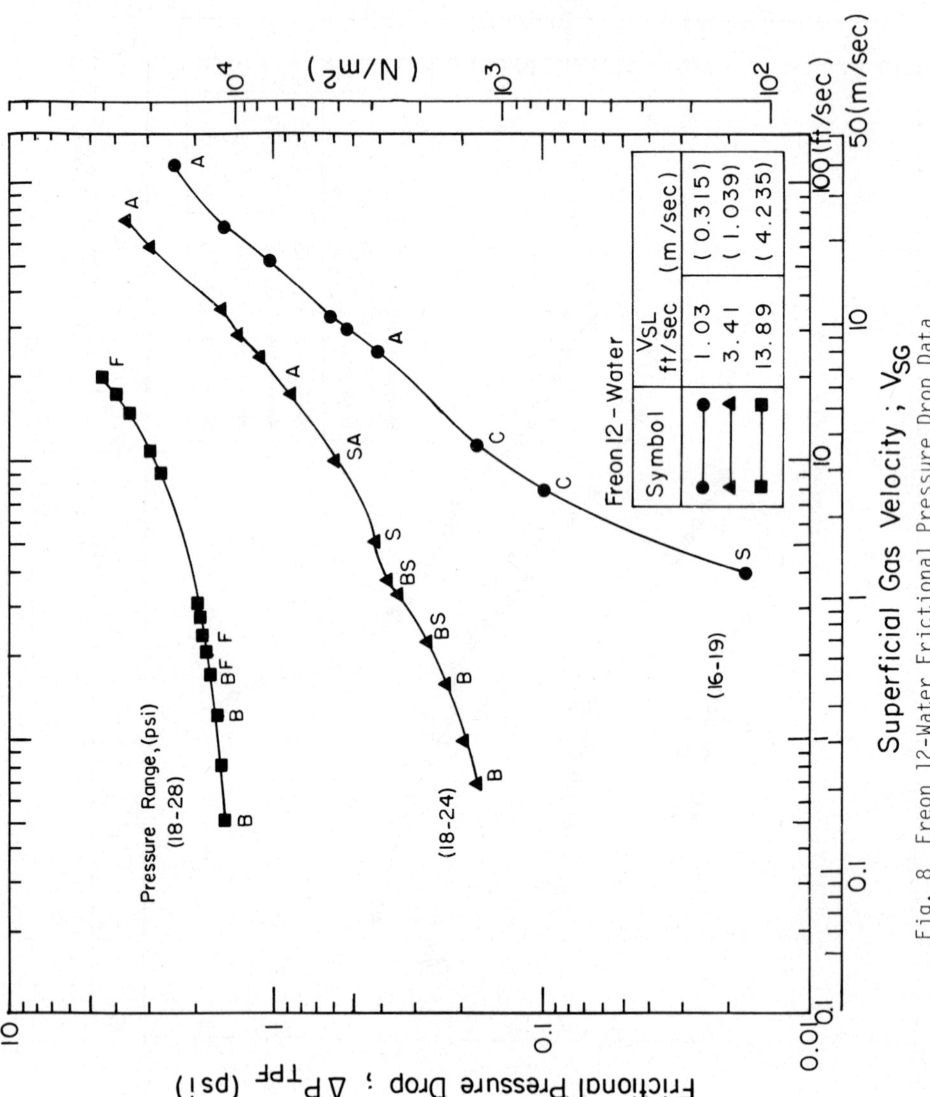

Fig. 8 Freon 12-Water Frictional Pressure Drop Data

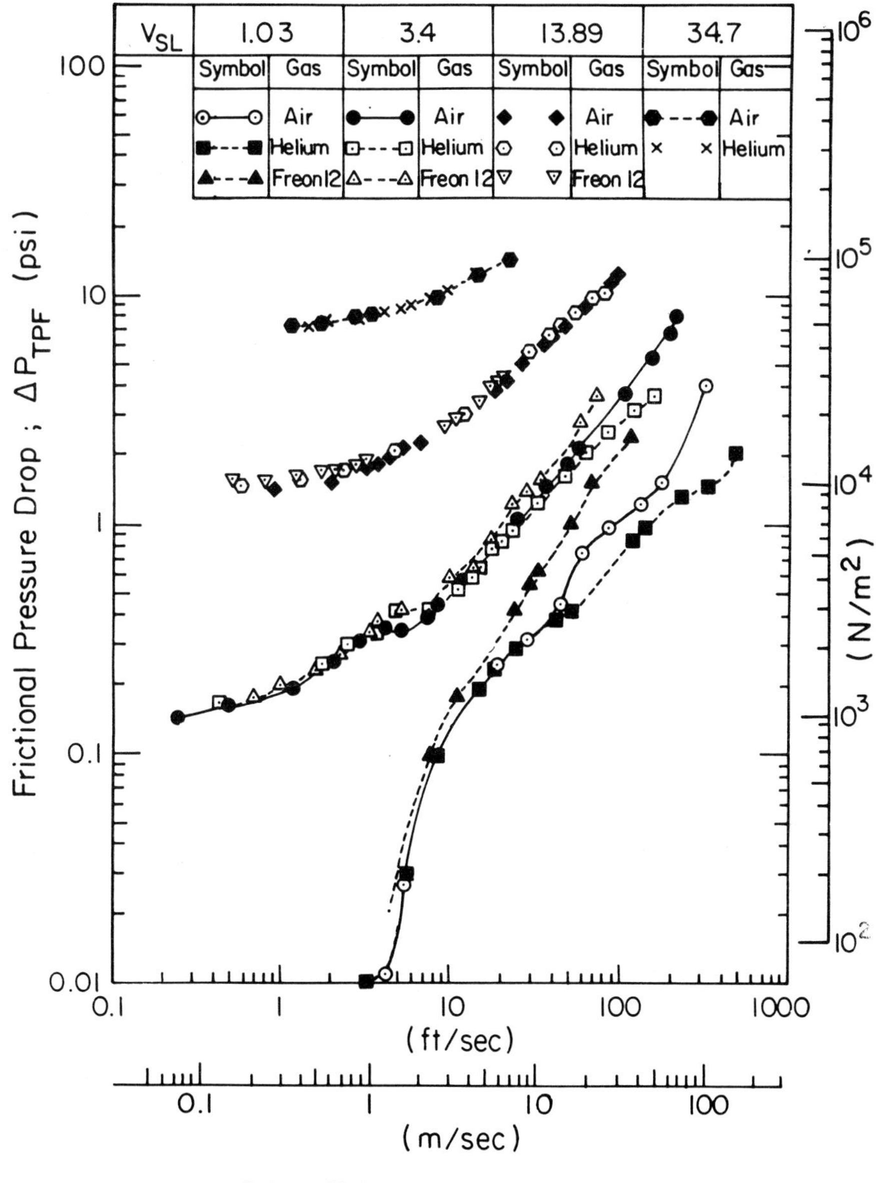

Fig. 9 Frictional Pressure Drop, All Data

3. The above effect is more pronounced at high gas velocities and decreases as the liquid velocity increases.

The above observations suggest that the effect of the gas density on pressure drop can not be isolated independently from the relative flow rates of the liquid and gas phases. When all data were examined, no such effect was observed for values of flow quality, which is a measure of the relative flow rates of the two phases, below 0.001; further, the effect increased with increasing quality. It is therefore, suggested that the variation of the pressure drop with the gas-phase density may be better described by relating this variation to the density of the two-phase mixture ρ_{MIX} defined as a function of the flow quality X; i.e.,

$$\rho_{MIX} = \left[\frac{X}{\rho_G} - \frac{(1-X)}{\rho_L}\right]^{-1} \tag{3}$$

To examine the validity of the above suggestion, the air-water mixture was considered as a reference; and the ratio of the frictional pressure drop of a particular gas-water mixture; i.e., helium-water or Freon 12-water, to that of air-water at the same V_{SL} and V_{SG} was plotted against the ratio of the gas-water mixture density to the air-water mixture density at the same conditions. The results are shown in Fig. 10 where the cross-hatched area represents the data at low V_{SL} and V_{SG}, and at high $V_{SL}$and all values of V_{SG}. For these data, the ratios of the mixture densities were equal or very close to one and so were the ratios of the frictional pressure drops. For the data at the lower V_{SL} and high values of V_{SG}, the density ratios differed from one and so did the pressure-drop ratios, being higher than one for Freon 12-water and lower than one for helium-water.

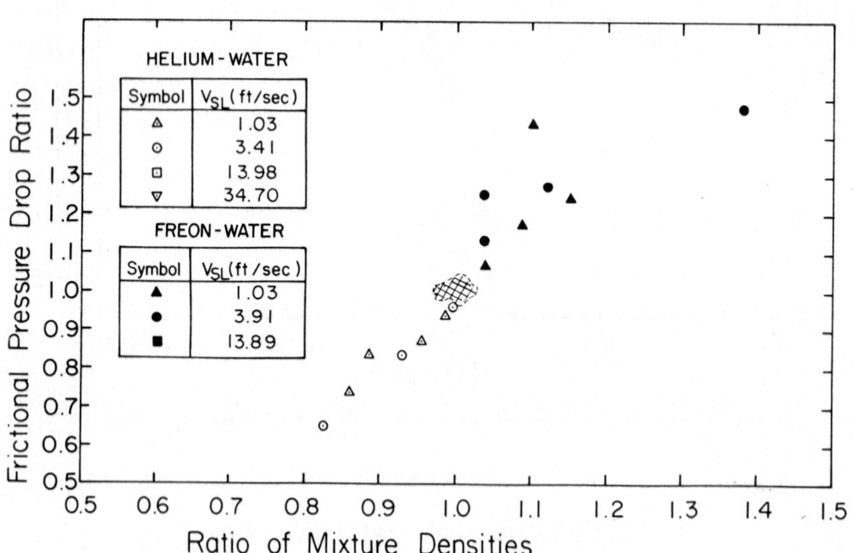

Fig. 10 Effect of Mixture Density on Frictional Pressure Drop

Brown et al. [10] studied the effect of the gas-phase density on pressure drop using air-water mixtures at different pressures (a range of gas density of 0.092 to 0.552 lb/ft^3). They developed a friction factor correlation which suggested that the gas density has no significant effect on pressure drop. This correlation, however, was limited to certain flow regimes and was not recommended for their regime IV where annular flow is found. Where, in the present work, significant effects of the gas density on pressure drop are found, it is especially in the annular flow and annular-mist transition.

Several correlations and schemes exist in the literature for predicting frictional pressure drop in two-phase vertical flow. The simplest three recommended correlations were tested against the present data. These are the homogeneous flow model [12], the Lockhart-Martinelli [13] correlation, and the Chisholm [14] correlation.

The homogeneous flow model was found to be the best among the three correlations. Excellent agreement was obtained between the predictions of the model and the present data (the mean deviation and the r.m.s. deviation were 6% and 23% respectively). The results of the comparison are shown on Figure 11.

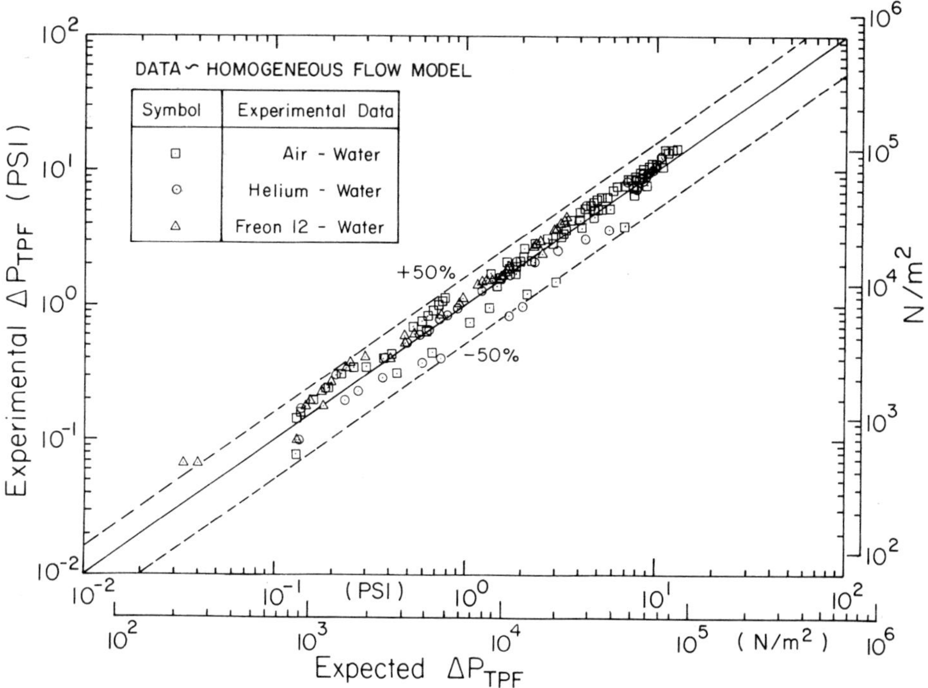

Fig. 11 Comparison of Frictional Pressure Drop Data with the Homogeneous Flow Model

5. CONCLUSIONS

1. New data are presented on the effect of the gas-phase density on flow-pattern transitions and frictional pressure drop in two-phase, two-component vertical flow.

2. The gas-phase density appears to have some effect only on the slug-annular, churn-annular, and slug-churn transitions.

3. The generalized flow-pattern map of Gavier and Aziz [2] provided good predictions of the present slug-churn, slug-annular and churn-annular transitions. This map did not appear to work as well for the other transitions.

4. The intermittent-annular transition criterion proposed by Weisman and Kang [8] was found to work well up to a value of V_{SG}/V_{SL} of about 25.

5. The churn- or slug-annular transition model proposed by Taitel et al [9] provided excellent separation of the flow patterns. The values of Ku at which annular flow boundary was observed were, however, consistently lower than the value proposed by [9], i.e. 3.1.

6. No effect was observed of the gas density on frictional pressure drop at low water rates and low superfical gas velocities, and at high water rates and all superficial gas velocities studied.

7. The gas density was found to have a significant effect on pressure drop at low water rates and moderate-to-high gas velocities. At a fixed water velocity, the pressure drop increased with increasing gas density. This effect was more pronounced at higher gas velocities and decreased with increasing liquid velocity.

8. It is suggested that the effect of gas density on pressure drop may be represented in terms of the density of the mixture defined as a function of flow quality and densities of the two phases.

9. Among the three correlations tested here, the homogeneous flow model was found to provide excellent predictions of pressure drop over the wide range of variables covered in the present study.

ACKNOWLEDGEMENT

The authors gratefully acknowledge the support of the National Research Council of Canada.

REFERENCES

1. Vohr, J. H., Flow Patterns of Two-Phase Flow - A Survey of Literature, USAEC Report TID-11514 (1960).

2. Govier, G. W, and Aziz, K., The Flow of Complex Mixtures in Pipes, Van Nostrand Reinhold Company, New York (1973).

3. Hewitt, G. F. and Semeria, R., Aspects of Two-Phase Gas-Liquid Flow, Chap. 12, in Heat Exchangers: Design and Theory Source Book, Eds., Afgan, N. and Schlünder, E.; Scripta Book Company, Washington, U.S.A. (1974).

4. Wallis, G.B., One Dimensional Two-Phase Flow, McGraw-Hill Book Company, N.Y. (1969).

5. Aggour, M. A., Hydrodynamics and Heat Transfer in Two-Phase, Two-Component Flows, Ph.D. Thesis, University of Manitoba, Manitoba, Canada (1978).

6. Dens, H., Jr., and Ros, N. C. J., Sixth World Petroleum Congress, Section 2, Paper No. 22, Frankfurt, Germany (1963).

7. Vijay, M. M., Ph.D. Thesis, University of Manitoba (1978).

8. Weisman, J., and Kang, S. Y., "Flow Pattern Transitions in Vertical and upwardly Inclined Lines", Int. J. Multiphase Flow, Vol. 7, 271-291 (1981).

9. Taitel, Y., Bornea, D., and Dukler, A. E., "Modelling Flow Pattern Transitions for Steady Upward Gas-Liquid Flow in Vertical Tubes", AIChE Journal, Vol. 26, No. 3, 345-354 (1980).

10. Brown, R. A. S., Sullivan, G. A., and Govier, G. W.," The Upward Vertical Flow of Air-Water Mixtures, III. Effect of Gas Phase Density on Flow Pattern, Holdup and Pressure Drop", CJChE, Vol. 38, No. 2, 62-66 (1960).

11. Chisholm, D., Research Note: Void Fraction During Two-Phase Flow, Jour. Mech. Eng. Sci. 15 (3), (1973).

12. Anon, The Frictional Component of Pressure Gradient for Two-Phase Gas or Vapour/Liquid Flow Through Straight Pipes, Engineering Sciences Data Item Number 76018, September (1976), Sponsored by I. Chem. E., and I. Mech. E., England, U.K.

13. Lockhart, R. W., and Martinelli, R. C., Proposed Correlation of Data for Isothermal Two-Phase Two-Component Flow in Pipes, Chem. Eng. Progr. 45, 39-48 (1949).

14. Chisholm, D., Pressure Gradients due to Friction During the Flow of Evaporating Two-Phase Mixtures in Smooth Tubes and Channels, Int. J. Heat Mass Transfer 16, 347-358 (1973).

Multi-Phase Flow and Heat Transfer III. Part A: Fundamentals
edited by T.N. Veziroğlu and A.E. Bergles
Elsevier Science Publishers B.V., Amsterdam, 1984 — Printed in The Netherlands

FLOW REGIME MAPS AND PRESSURE DROP IN TWO-PHASE GAS-LIQUID FLOW IN A RECTANGULAR CHANNEL

Leon Troniewski, Roman Ulbrich
Heat Technique and Chemical Engineering Department
Opole Technical University
ul.Oleska 114, 45-233 Opole, Poland

ABSTRACT

The results of investigations of hydraulics in a two-phase gas-liquid flow in vertical and horizontal rectangular channels have been presented in the paper. Flow regime maps have been worked out and a method for calculating the pressure drop in a two-phase mixture flow has been proposed. The methods applied in a two-phase flow in pipes have been used to describe the results of investigations after adopting the correction for the difference between the flow in rectangular channels and that in pipes.

1. INTRODUCTION

A two-phase gas-liquid flow in rectangular channels was investigated by many research workers (Hosler [1] , Richardson [2] , Hoglund et al. [3] , Janssen and Kervinen [4] , Iida and Takahashi [5] , Jones and Zuber [6] , Huhn and Wolf [7] , Friedel [8] , Sadatomi,Sato and Surawatari [9]) . It arises from the discussed survey of literature that the state of knowledge in the field of a two-phase flow in rectangular channels did not allow the influence of the ratio of channel sides as well as the properties of components of a two-phase mixture upon the range of occurence of flow patterns and a pressure drop to be determined. In consideration of little material which is concerned with the investigation of a two-phase flow in rectangular channels and in the presence of extensive material dealing with those of the flow in pipes ,the authors decided to take account of the correction including the geometry of a channel in the chosen methods that could be treated as universal ones in a two-phase flow through a pipe. As a result of the analysis of flow pattern maps published in the literature and the authors own investigations, the coordinates of universal flow regime maps were proposed :

vertical pipes [10]
$$\frac{w_{Gs}}{w_{Ls}}\sqrt{\frac{\rho_G}{\rho_a}\quad\frac{\rho_w}{\rho_L}} \quad - \quad w_{Ls}\sqrt{\frac{\rho_L}{\rho_w}} \tag{1}$$

horizontal pipes [11]
$$\frac{g_{Ls}}{g_{Gs}}\ \lambda_B\ \psi_B \quad - \quad \frac{g_{Gs}}{\lambda_B} \tag{2}$$

On the basis of the analysis of calculation methods of the pressure drop in a gas-liquid pipe flow, the Lockhart and Martinelli method [12] has been adopted as a universal one.

It has been assumed that :

the difference between the two-phase flow in a rectangular channel and that in a pipe results from the same relations which occur in a single-phase flow through a rectangular channel and through a pipe.

As a result of the theoretical considerations [13] for a steady and fully-developed laminar flow in a rectangular channel and numerical calculations for the turbulent flow, while applying k - ε model of turbulence [14] , the correction ω was suggested

$$\omega(k) = \frac{3\ k^2}{2\ (1+k)^2\ \left\{1-\frac{192}{\pi^5 k}\sum_{m=0}^{\infty}\left[\frac{1}{(2m+1)^{2m+1}}\ \mathrm{tgh}\left(\frac{\pi k}{2}(2m+1)\right)\right]\right\}} \tag{3}$$

which can be treated as the measure of difference of flow in the rectangular channel in relation to the flow in the pipe of the equivalent diameter D_e and for the same pressure gradient $\partial P/\partial z$. For the design calculations we can assume, with the accuracy to 2% (k=1÷20)

$$k = \left(\frac{k}{2}\right)^{0.16} \tag{4}$$

Pressure drop in a two-phase mixture flow through rectangular channels are suggested to be calculated by the Lockhart-Martinelli method including the difference between a single-phase flow both in pipes and in rectangular channels. Thus, the authors propose to calculate the pressure drops of a single-phase flow in rectangular channel from:

$$\frac{\Delta P}{\Delta L} = \lambda^* \frac{\rho \bar{w}^2}{2 D_e} \tag{5}$$

$$\lambda^* = \begin{cases} 64 \quad (Re^*)^{-1} & \text{for } Re^* \leq 2{,}100 \\ 0.3164 \, (Re^*)^{-0.25} & \text{for } Re^* > 2{,}100 \end{cases} \tag{6}$$

$$Re^* = \frac{\bar{w}^* D_e \rho}{\eta} \tag{7}$$

$$\bar{w}^* = \frac{\bar{w}}{\omega(k)} \tag{8}$$

Assuming that differences between the flow in rectangular channels and that in pipes are descirbed by the correction ω , it was necessary to decide where and in what form the correction should be used in the coordinates in the proper flow regime maps. On the basis of the considerations which are presented in [13] , the following coordinate system of flow regime maps for rectangular channels are proposed:

vertical
$$\frac{w_{Gs}}{w_{Ls}} \sqrt{\frac{\rho_G}{\rho_a} \frac{\rho_w}{\rho_L}} - w_{Ls} \, \omega \sqrt{\frac{\rho_L}{\rho_w}} \tag{9}$$

horizontal
$$\frac{g_{Ls}}{g_{Gs}} \psi_B \lambda_B - \frac{g_{Gs}}{\lambda_B \, \omega} \tag{10}$$

The given calculation method of the pressure drop and the usefulness of the proposed coordinate systems in the flow regime maps have been decided to be verified on the basis of experimental investigations.

2. EXPERIMENTS.

A two-phase gas-liquid flow in rectangular channels was studied at the test stand whose scheme , in its vertical position, is presented in Fig.1. The investigations were carried out in the range of 1÷12 and in the horizontal channels in the range of .1÷10 (k<1 when the shorter side was in its vertical position) , 1,100 mm long, the equivalent diameter from 7 up to 14 mm, the cross section area ca. 175 mm^2. An inlet mixer with the perimeter crevice between the air feeding channel and the test channel was applied. Pressure tappings for measuring the pressure drop

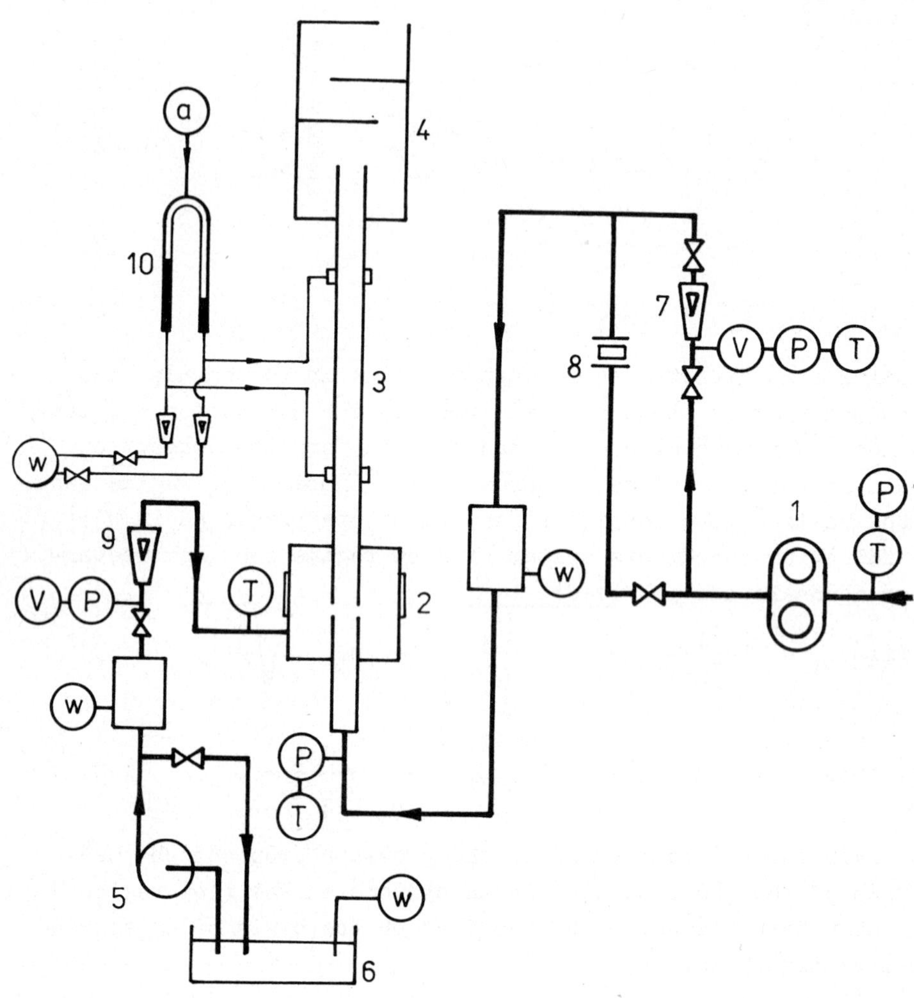

Fig.1. Scheme of installation for investigation
1-blower, 2-mixing chamber, 3-test section,
4-separator, 5-pump, 6-tank, 7-gas rotameter,
8-orfice, 9-liquid rotameter, 10-manometer
V - volume flux, P - pressure, T - temperature
a - air, w - water.

were placed in the test channel at the distance of 300 and 800mm from the inlet mixer. The patterns of the two-phase flow were observed along the whole lenth of the test channel. In order the determine the range of the occurrence of particular flow patterns a visual method was applied to observe the distribution of phases in the flowing mixture and to assign the transition point from one flow pattern into another one with constant liquid flux and varying gas flux. The pressure drop in the two-phase flow was measured by means of the system proposed by Hewitt [15] . Investigations of the occurrence of particular patterns of the two-phase flow were carried out by air-water and aqueous solution of sugar-air mixtures, whereas the pressure drop was investigated by the use of the water-air mixture. The superficial mass velocity varied in the range:

liquid $g_{Ls} = 20 \div 1{,}500 \text{ kg}/(\text{m}^2\text{s})$
gas $g_{Gs} = 0.1 \div 40 \text{ kg}/(\text{m}^2\text{s})$

3. RESULTS.

Flow pattern.

In the flow through vertical channels an excellent agreement of the existing flow patterns with those observed in a pipe was found. In the flow through horizontal channels it was found that apart from the patterns existing in the flow through a pipe there occurred a flow which was called the semi-annular flow, in particular for $k>2$, when the fluctuating froth-slug flow decreased and the liquid film had not occurred yet at the upper wall of channel. It was also observed, just as Iida and Takahashi [5] noticed, that the stratified and wavy flows did not occur in the water-air mixture flow for $k \leq 0.1$ and in the aqueous solution of sugar-air mixture flow for $k \leq 0.5$.

Flow regime maps.

The ranges of occurrence of the froth-slug flow for the air-aqueous solution of sugar mixture flow for different channels in the proposed coordinate system (9) , are given in Fig.2. The pattern map of the two-phase gas-liquid flow in vertical rectangular channels is given in Fig.3. The results of investigations of the range of occurrence of the particular patterns of the two-phase

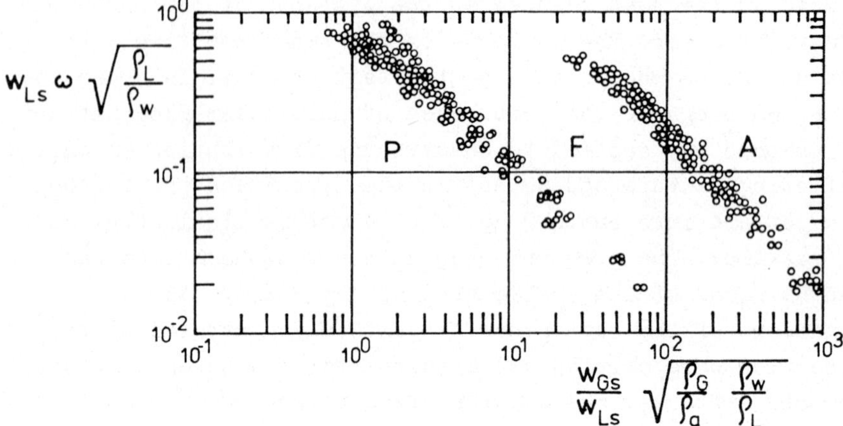

Fig.2. Result of investigations of range of occurrence of particular flow patterns in two-phase air-aqueous solution of sugar (η_L = 30 cP) flow in vertical rectangular channels with ratio of sides k = 1÷12.

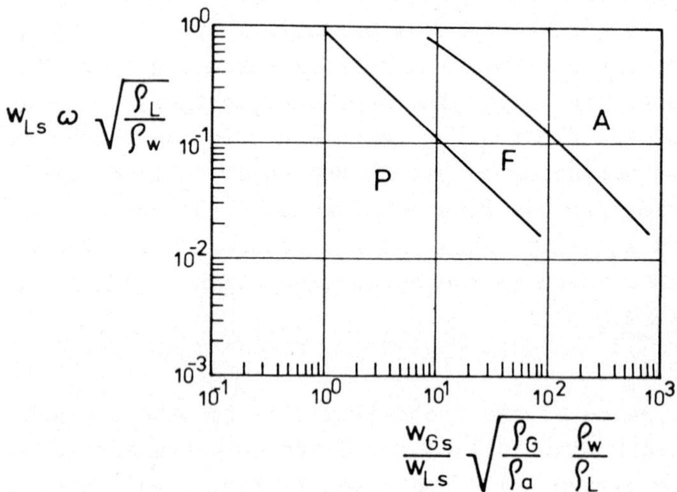

Fig.3. Flow regime map in two-phase gas-liquid flow in vertical rectangular channel.

flow in horizontal channels are presented, by way of example, in Fig.4 and Fig.5. While mapping the same flow patterns for the rest of rectangular channels it was found that the channels could be divided into two groups:

- rectangular channels for which the line dimension in the direction of the force of gravity effects considerably the formation of patterns of a stratified flow (Fig.4)
- rectangular crevices, when the dimension of a channel in the direction of the force of gravity is very small; for this group of channels the existance of stratified and wavy flows was not found (Fig.5) .

As a result of the carried out investigations, the channels with the ratio of sides: $k<0.1$ for the water-air mixture and $k\leq0.5$ for the aqueous solution of sugar-air mixture ($\eta_L = 30\cdot10^{-3}$ Pa·s) were included into the group of rectangular crevices. The pattern maps of the two-phase gas-liquid flow in horizontal rectangular channels are given in Fig.6.

Pressure drop.

The proposed manner of calculation of the pressure drop was verified by comparing the results of investigations with those in Fig.7 and Fig.8. Lines limiting the area in which 85% of points were marked, are given in diagrams. The obtained spread of points is of the same order as the accuracy of the Lockhart-Martinelli method while calculating the pressure drop of the two-phase mixture flow in a pipe.

4. CONCLUSIONS.

On the basis of the carried out considerations and experiments it can be assumed that in the two-phase flow through rectangular channels, the departures from dependences obligatory for the flow through pipes, result from differences in the single-phase flows through rectangular channels. The correction ω (equations (3) or (4)) , being a function of geometry, and which is derived from the analysis of the single-phase flow in rectangular channels, has proved to be entirely useful when the two-phase flow is described. In order to determine the range of occurrence of particular patterns of the two-phase flow in rectangular channels,

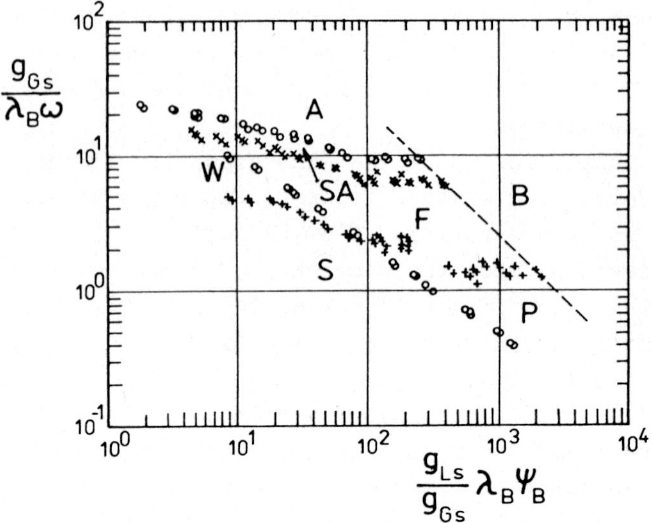

Fig.4. Results of investigation of range of occurrence of particular flow patterns in two-phase air-water flow in horizontal rectangular channel with ratio of sides k = 2.

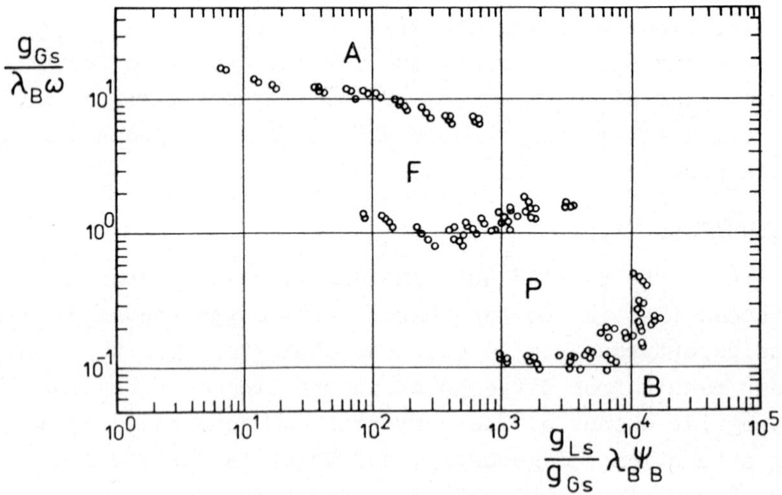

Fig.5. Results of investigation of range of occurrence of particular flow patterns in two-phase air- aqueous solution of sugar (η_L = 30 cP) in horizontal rectangular channel crevice with ratio of sides k = 0.1.

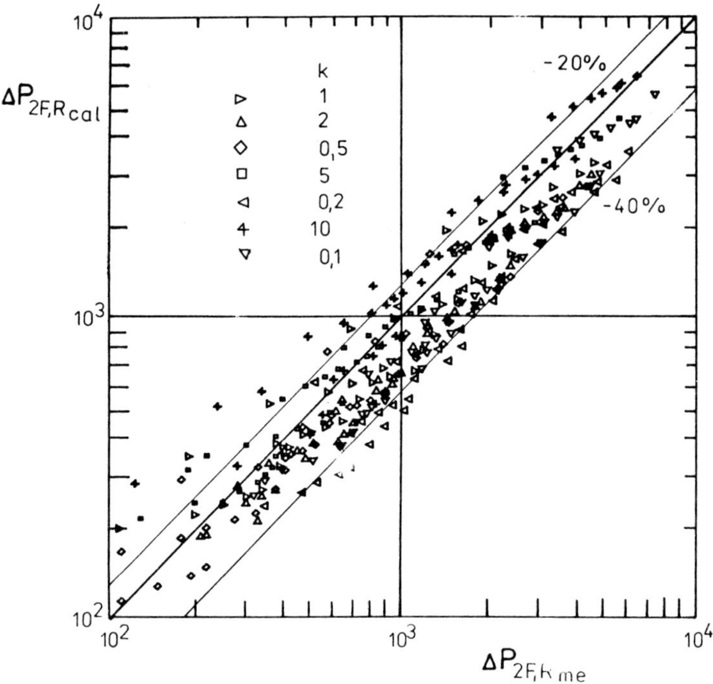

Fig.6. Flow regime maps in two-phase gas-liquid flow:
a - in horizontal rectangular channels
b - in horizontal crevices.

the authors propose to make use of Fig.3 for vertical channels and Fig.6 for horizontal channels and crevices, whereas the pressure drop of the two-phase mixture flows through rectangular channels ought to be calculated according to the modified Lockhart-Martinelli method.

5. NOTATION.

D	- diameter	m
g	- mass velocity	$kg/(m^2 \cdot s)$
k	- ratio of sides	
L	- lenth	m
P	- pressure	Pa
Re	- Reynolds number	
w	- velocity	m/s
η	- viscosity	Pa·s
λ	- friction factor	

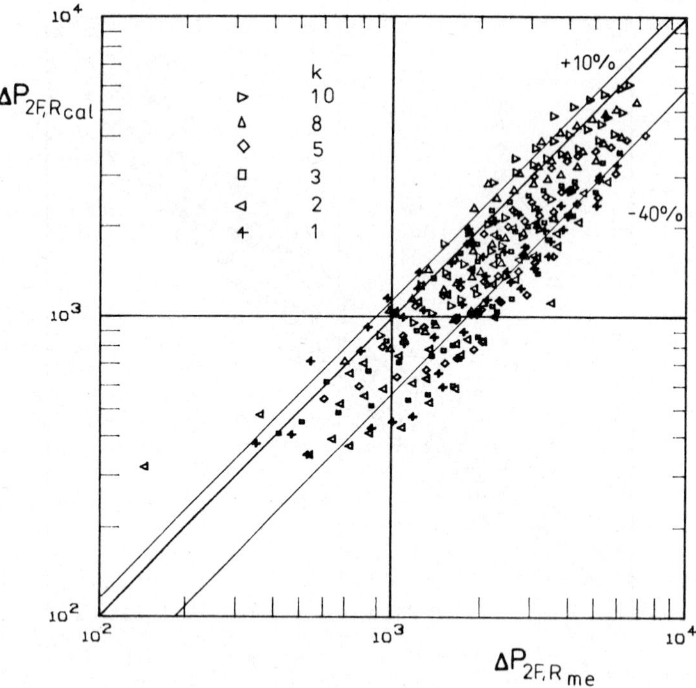

Fig.7. Comparison of measured pressure drops in two-phase air-water flow in vertical rectangular channels with those calculated by modified Lockhart-Martinelli method.

λ_B - Baker correction

ρ - density kg/m^3

Ψ_B - Baker correction

ω - correction for fifference between the flow in rectangular channel and that in a pipe

Subscripts

a - air (for t = 20°C, P = 0.1 MPa)

cal - calculated

e - equivalent

G - gas

L - liquid

me - measured

s - superficial

w - water (for t = 20°C)

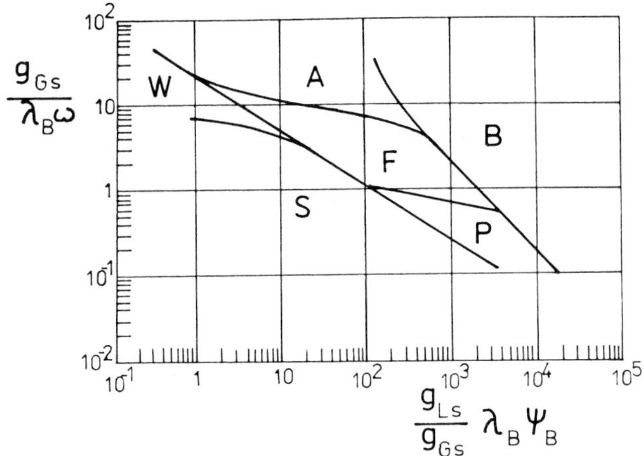

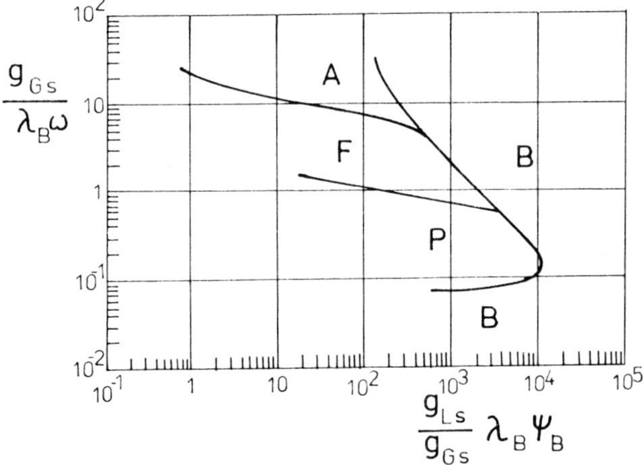

Fig.8. Comparison of measured pressure drops in two-phase air-water flow in horizontal rectangular channels with those calculated by modified Lockhart-Martinelli method.

$\overline{}$ - mean value

*- conventional value

6. REFERENCES.

[1] HOSLER E.R. 1967 Report WAPD-TM-658.

[2] RICHARDSON B.L. 1958 Report ANL-5949.

[3] HOGLUND B.M. and all 1961 Report ANL-5760.

[4] JANSSEN E., KERVINEN J.A. 1964 Report GEAP-4616.

[5] IIDA Y., TAKAHASHI K. 1976 Kagaku Kogaku Ronbunshu 2 3 228

[6] JONES O.C., ZUBER N. 1975 Int.J.Multiphase Flow 2 3 273.

[7] HUHN J., WOLF J. 1975 Zwiephasenstromung, Leipzig.

[8] FRIEDEL L. 1975 VDI-Forschungsheft 572.

[9] SADATOMI M., SATO Y., SARUWATARI S. 1982 Int.J. Multiphase Flow 8 6 641.

[10] TRONIEWSKI L., ULBRICH R. 1980 10th National Conference on Chemical and Process Engineering, Łódź.

[11] BAKER O. 1954 Oil and Gas J. 53 185.

[12] LOCKHART R.W., MARTINELLI R.C. 1949 Chem.Eng.Progr. 45 39.

[13] ULBRICH R. 1981 Ph.D.-thesis, Wrocław Technical University, Wrocław.

[14] LAUNDER B.E., JONES W.P. 1972 Int.J.Heat Mass Transfer 15 301.

[15] HEWITT G.F. 1978 Measurements of two-phase flow parameters Acadamic Press, London.

Multi-Phase Flow and Heat Transfer III. Part A: Fundamentals
edited by T.N. Veziroğlu and A.E. Bergles
Elsevier Science Publishers B.V., Amsterdam, 1984 — Printed in The Netherlands

EXPERIMENTS IN TWO PHASE FLOW:

BUBBLE SLUG TRANSITION IN A VERTICAL ANNULUS

M.M.Sorour
Mechanical Engineering Dept., Faculty of Engineering
Alexandria University, Alexandria, Egypt

ABSTRACT

This is an experimental study on the transition from bubbly to slug flow in a vertical annulus. The two phase gas liquid flow was simulated by air and water, and the radius ratio of the apparatus was varied to study its effect and to compare it with the extreme case of a long vertical cylinder. In addition the air introduced in the annulus was controlled such that various combinations of bubbles with different sizes and numbers were investigated.

This investigation indicates that the transition from bubbly to slug flow is not only a function of the gas and liquid flow rates as previously indicated but is also a strong function of the distribution and size of bubbles at their initial formation and the characteristic dimension of the annulus. For an annulus with small radius ratio large number of small bubbles delays the transition to slug flow, conversely for an annulus with large radius ratio this is only the trend up to a certain number of bubbles, where the reverse is true.

NOMENCLATURE

A1 = Cross section area of annulus
A2 = Area of holes of perforations
d = Diameter of hole
D_b = Average bubble diameter
D = Hydraulic diameter of the annulus.
P = Pitch between holes
N = Radius ratio
L = Aspect ratio
Q_g = Volumetric gas flow rate
Q_t = Volumetric liquid flow rate

ν_g = Kinematic viscosity of gas
ν_1 = Kinematic viscosity of liquid.
α = Void fraction
Re_g = Gas Reynolds number = $4\ Q_g/\pi D \nu_g$
Re_1 = Liquid Reynolds number = $4Q_l/\pi D \nu_l$
n = Number of holes

INTRODUCTION

Two phase gas liquid flow applications is extensively growing in many kinds of Engineering technology such as heat exchangers, pipe lines, and chemical mixing mechanisms. Consequently, many contributions have lately been made into the subject conserning the heat, mass and momentum transfer of two phase gas liquid flows in various channels and directions.

Two phase gas liquid flow is classified by the flow patterns which represent the way in which the gas and liquid phases are distributed in each other. There are four basic flow patterns for vertical two phase gas liquid flow; viz bubbly, slug, annular and froth flow. The bubbly flow is that flow where the gas phase is approximately uniformly distributed in the liquid phase as bubbles. Conversely, the slug flow is characterised by large bullet shapped bubbles which occupy the whole pipe diameter and other distributed small bubbles in the liquid phase; these large bubbles are called "Taylor bubbies". It is to be noted that, the exact knowledge of the hydrodynamic and thermal characteristics of the flow is a strong function of the flow regime. Therefore careful identification of various flow regimes in two phase flow is of vital importance; it is analogous to knowing whether the flow is laminar or turbulent in single phase flow.

There are many flow maps for classifying the flow regimes and identifying the boundaries of transition for two phase, gas liquid flow in a vertical cylinder. However, there is very limited knowledge concerning the transition in a vertical annulus. In addition, due to the lack of complete understanding of the transition mechanism, there is limited agreement between various investigations. Sternling[(1)] and Wallis[(2)] used superficial velocities as their criteria for transition, while Hewitt et al[(3)] and Govier et al[(4)] used superficial momentum flux. On the other hand, Duns et el[(5)] and Gould et al[(6)] used other dimensionless coordinates with little success.

The most complete model for transition in two phase gas liquid flow is that of Tait et al[(7)]. This model is based on physical mechanism suggested for each transition and accounts for the fluid properties and container size. However, the bubble size distribution was not accounted for in the transition and preliminary experiments by the auther indicated the importance of this parameter in the transition from bubbly to slug flow. Therefore, this investigation is devoted to experimentally study the effect of inlet bubble size and distribution in the transition from bubbly to slug flow in a vertical annulus to gain insight on the mechanism of transition and the stages of develcpment.

2. APPARATUS

The experimental set up is schematically shown in figure 1. In essence, it consists of a vertical annulus through which the mixture flows, compressor to provide the air under pressure, overhead tank to provide the motion of the water, control valves and pressure regulator to adjust each phase in the mixture, and flow meters for measuring the air and water flow rates. The inner cylinder of the annulus was made of light plastic material to ease its fixation and was painted black for photographic purposes; two such cylinders were used having 34 and 48 mm diameter. On the other hand, the outer cylinder was made transparent and of diameter equal to 75 mm thus giving three annuli of radius ratios N= 0.65, 0.45 & 0.00 and aspect ratios L = 88, 60 & 32 respectively.

The water flows from a constant head tank via control valve and a measuring orifice to the test section. To introduce and deliver the water, two 20 cm sections containing three circumferential holes set at 120° apart were connected to the bottom and top parts of the outer tube such that upwards or downwards flow of water was established through the cylinder.

The flow system allowed the air to be introduced into the annulus through a horizontal perforated plate located between a small air chamber and the annulus. Various perforated plates (Distributing plates) were used throughout this investigation. These plates were divided into two groups, one group have equal area of perforations with different number and size of holes and the other group have equal hole size but different number. The complete details of the distributing plates are presented in table 1.

Lastly, it is to be noted that this investigation is more concerned with the range where the air bubbles flow at a much higher velocity than the liquid and when the bubbles travel freely in the core of the liquid and not only near the walls.

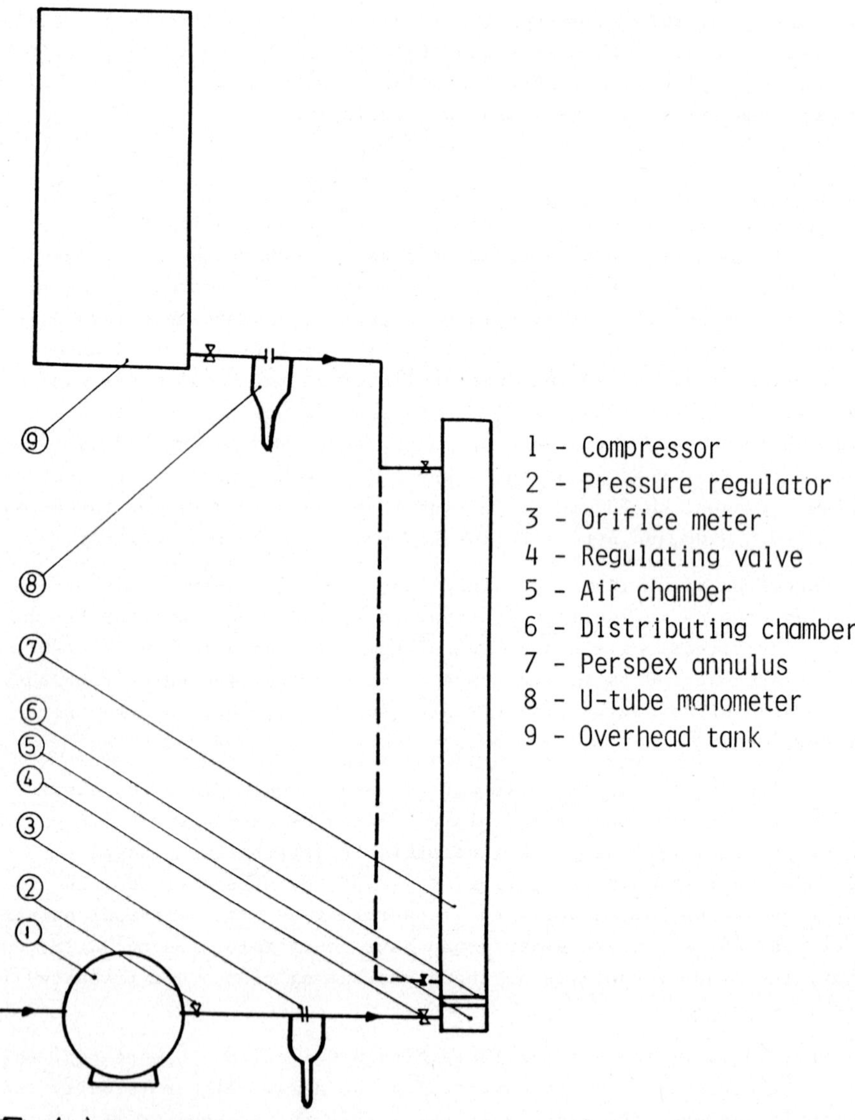

Fig.(1):Schematic diagram of the apparatus.

Table 1. Details of The distributing plates in SI units

$A1 = 26 \times 10^{-4}$, N = 0.65 , L = 88						
	Group I			Group II		
n	3	12	48	6	30	60
$d \times 10^3$	2	1	0.5	0.5	0.5	0.5
$P \times 10^3$	64	16	16	32	18	16
P/d	32	16	32	64	36	32
$A2 \times 10^6$	9.42	9.42	9.42	1.17	5.89	11.78
$10^3 \times A2/A1$	3.6	3.6	3.6	0.45	2.26	4.53
$A1 = 35 \times 10^{-4}$ N = 0.45 L = 66						
	Group I			Group II		
n	3	12	48	6	30	60
$d \times 10^3$	2	1	0.5	0.5	0.5	0.5
$P \times 10^3$	57	14	14	28	16	14
P/d	28.5	14	28	56	32	28
$A2x \ 10^6$	9.42	9.42	9.42	1.17	5.89	11.78
10^3 A2/A1	2.7	2.7	2.7	0.33	1.7	3.3
$A1 = 44 \times 10^{-4}$, N = 0.00 , L = 32						
	Group I			Group II		
n	3	12	48	6	30	60
$d \times 10^3$	2	1	0.5	0.5	0.5	0.5
$P \times 10^3$	40	15	15	20	17	15
P/d	20	15	30	40	34	30
$A2 \ x10^6$	9.42	9.42	9.42	1.17	5.89	11.78
10^3x A2/A1	2.1	2.1	2.1	0.26	1.3	2.6

3. RESULTS AND DISCUSSION

The transition from bubbly to slug flow was qualitatively investigated in various annular gaps and for many flow configurations. Visual observations aided by photographs indicated that the transition mechanism is not catastrophic, ie the bubbles do not instantaneously agglomerate to form large bullet shaped bubble having a diameter equal to that of the duct. In addition, in all investigated flow rates, distribution of bubbles, and radius ratios, no slugs were observed at the onset of transition. In the entrance section of the duct instead the early stages of transition is characterised by churning flow in the entrance section while slugs were observed downstream.

The stages of development to slug flow for large number of bubbles differ from that for small number of bubbles. It can be seen in plate 1 that for large number of small bubbles in the cylinder case the transition to slug flow is developed by an initial increase in the density of the bubbles to the extent that air bubbles become too close to each other, the second stage is characterised by circulation in the flow that increases in strength with the air flow rate, the last stage of development results in patches of water and patches of air in the entrance section that develop to slugs further downstream. On the other hand, the transition to slug flow for small number of large bubbles plate 2, occur by agglomeration and coalescence when the density of air bubbles increases and the probability of collision increases. The most distinctive illustration of the circulation in the flow is found with the 48 hole arrangement in the cylinder case. (plate 3). Lastly, a sample of the transition in a vertical annulus is presented in plate 4; it can be seen that the gap width significantly modifies the bubble distribution.

The transition boundaries, or the locus that divide the bubble regime from the slug regime are presented in dimensionless form as a gas Reynolds number, Re_g, versus liquid Reynolds number Re_l which indicate the inertia to the viscous forces in each phase at transition. In addition, the void fraction at transition a, which is the percentage by volume of the gas phase to the liquid phase, is given as a function of the liquid Reynolds number, Re_l. Both Reynolds number and void fraction completely identify the transition, and other parameters like the flow rate, momentum flux and superficial velocity can be deduced from these quantities.

Fig. 2 a presents the transition boundaries on a Re_g - Re_l plot for group A where equal area of perforations was distributed between large number of small holes and small number of large holes in three annuli of radius ratio, N = 0.00 , 0.45 & 0.65 respectively. It can be seen that as the liquid Reynolds number increases the corresponding gas Reynolds number increases as expected. However, the variation of Re_g with the distribution of holes is much more greater than with Re_l. As the number of holes increases and its size decreases the transition is significantly delayed. In addition, the transition is very much dependent on the gap size or the radius ratio.

On the other hand, fig. 2 b presents the transition boundaries on a Re_g-Re_l plot for group B, where equal hole size were used in the distributing plate with different numbers in the three studied annuli. It can be seen that there is no unique trend for the variation with the distributing plate. Re_g for n = 30 is always greater than Re_g for n = 6 in all the annuli. Conversely Re_g for n=60 is always smaller than Re_g

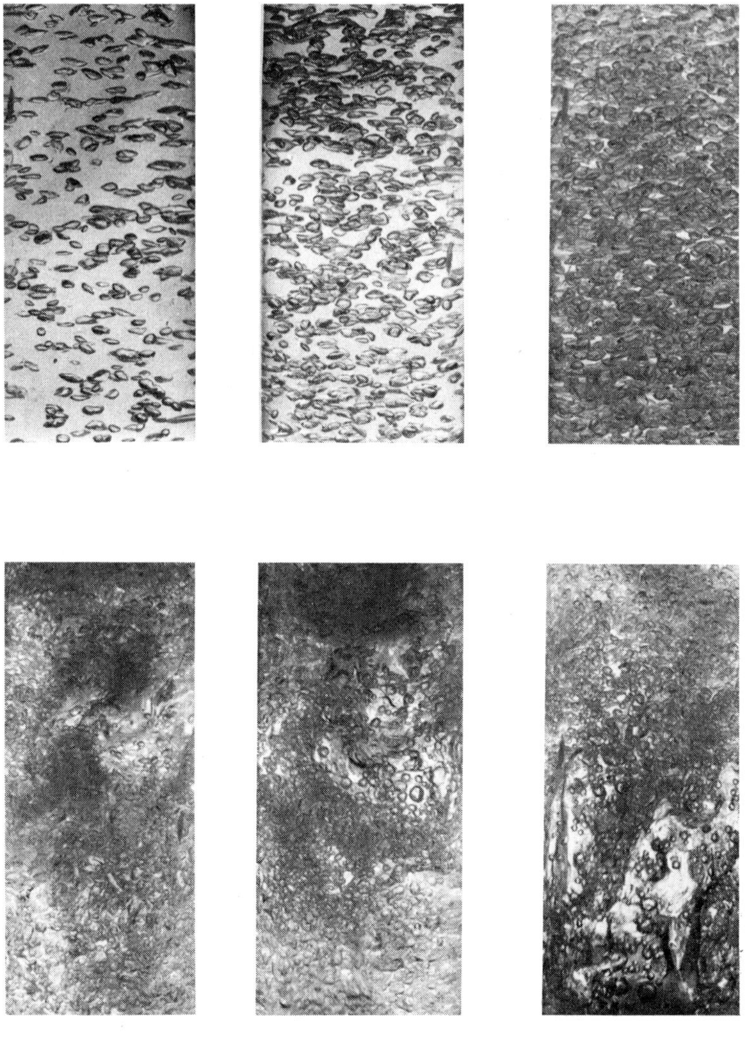

PLATE 1 $\Pi = 30$ $N = 0$

PLATE 2 $\Pi = 12$ $N = 0$

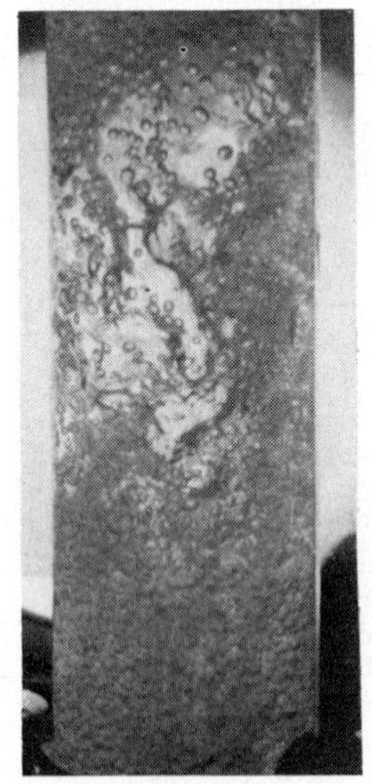

PLATE 3 $\Pi = 30$

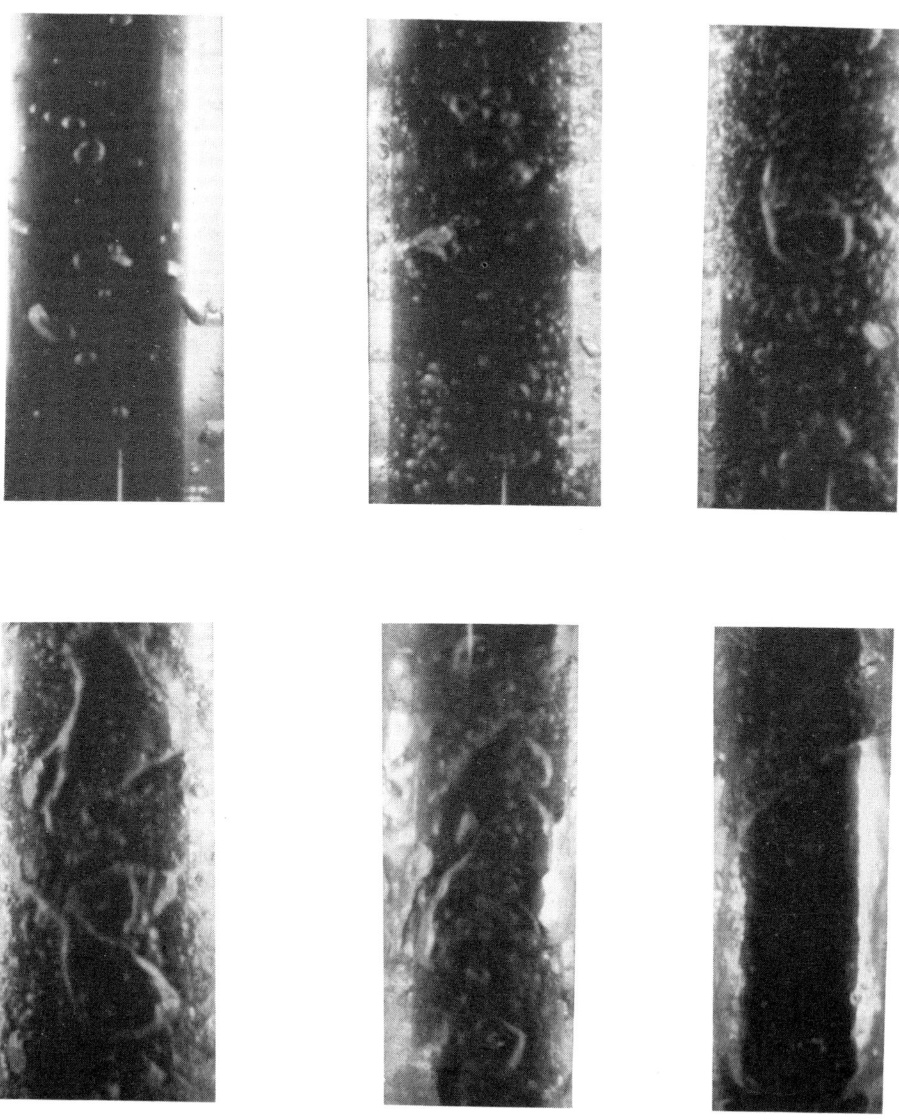

PLATE 4 $n = 12$ $N = 0{\cdot}65$

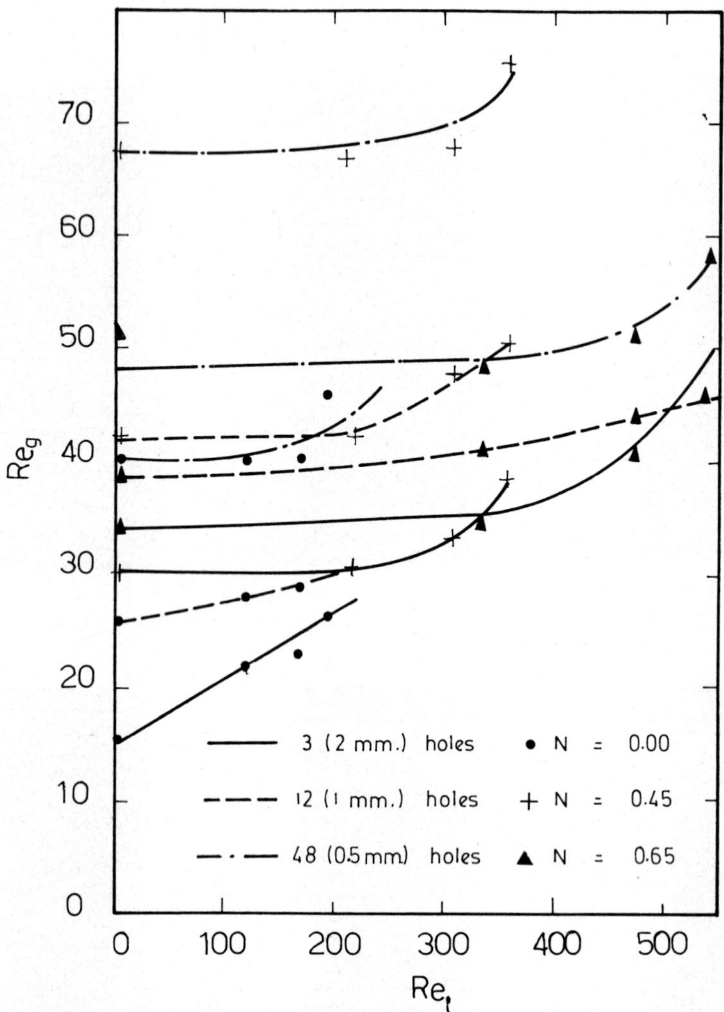

Fig. (2a): Transition boundaries for bubbly-slug flow.

for n = 30. In addition, the variation of Re_g with Re_l is similar to that in group A.

Figs. 3 a & 3 b present the average void fraction at transition versus the liquid Reynolds number, for group A and B respectively. It can be seen that the average void fraction is not a strong function of the Reynolds number, but is very sensitive to the distributing plate, and

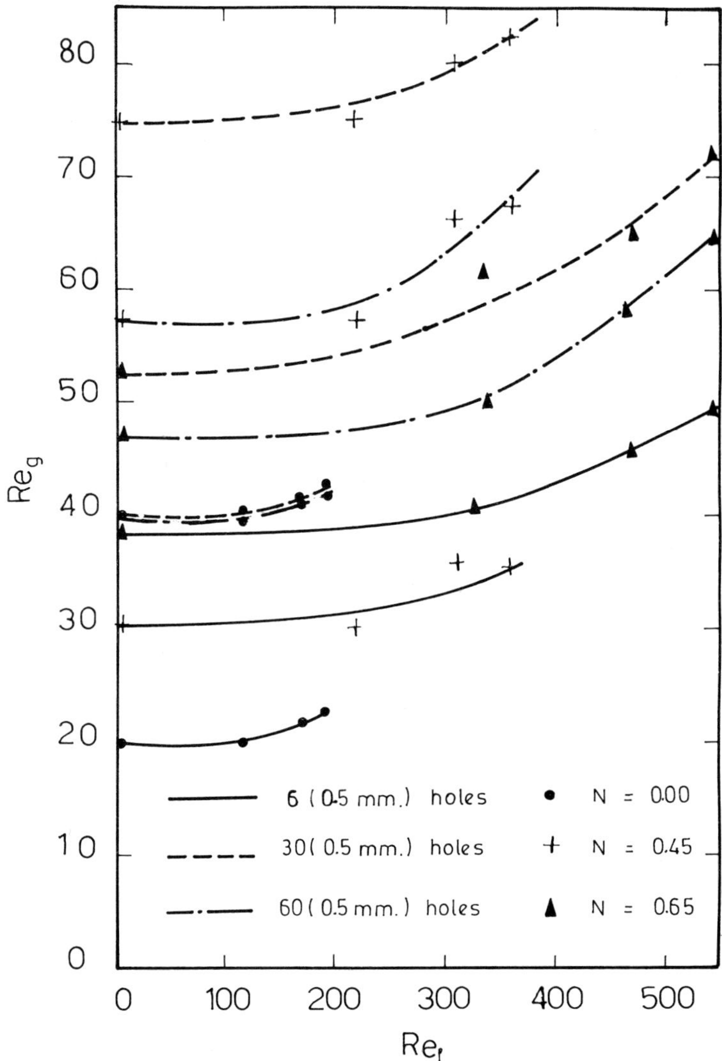

Fig.(2-b): Transition boundaries for bubbly-slug flow.

the radius ratio or the gap size. Large void fractions were found at transition when large number of holes were used in wide gaps. Conversely, small void fractions were observed with small number of holes at various

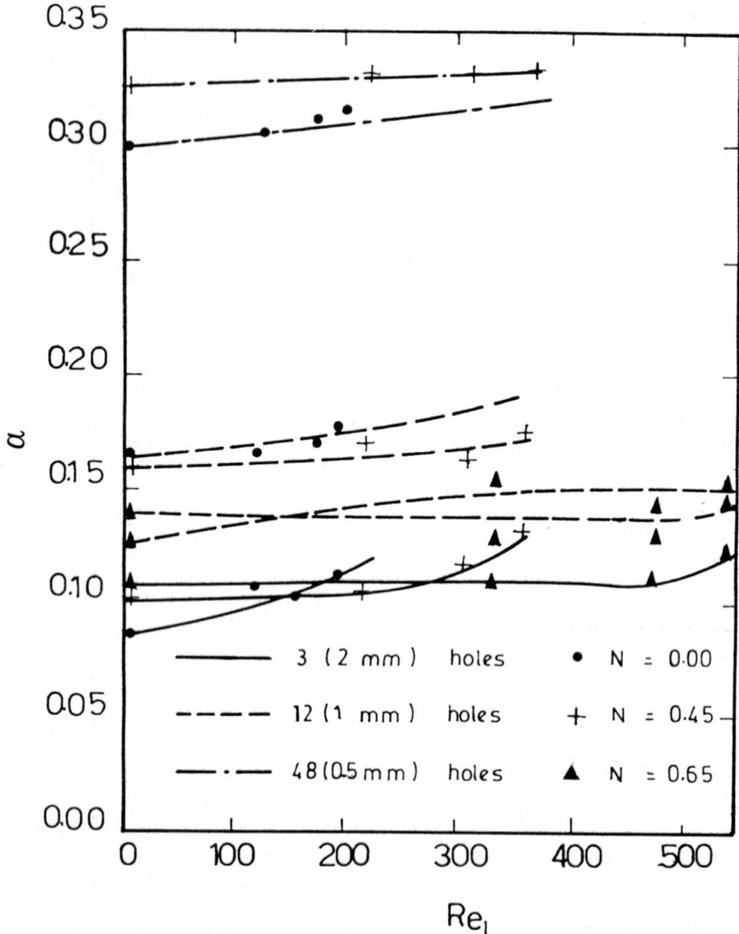

Fig.(3-a): Void fraction, α, at transition boundaries.

annuli. In addition, although for the wide gap case, there is significant variation of the void fractions between various configurations $0.1 \leqslant \alpha \leqslant 0.325$, the void fraction for the narrow gap case (N=0.65) was found in a very narrow range $0.1 \leqslant \alpha \leqslant 0.15$.

The reported transition boundaries in all these cases were obtained for upwards flow of water and air, although it was previously pointed out that provision in the apparatus for downwards water flow was available. Therefore, all the studied conditions (geometric and dynamic) were repeated when the liquid direction was reversed. A sample of the state points of the transition for upwards and downwards flow is presented in table 2 for comparison. It can be seen that both Reynolds number

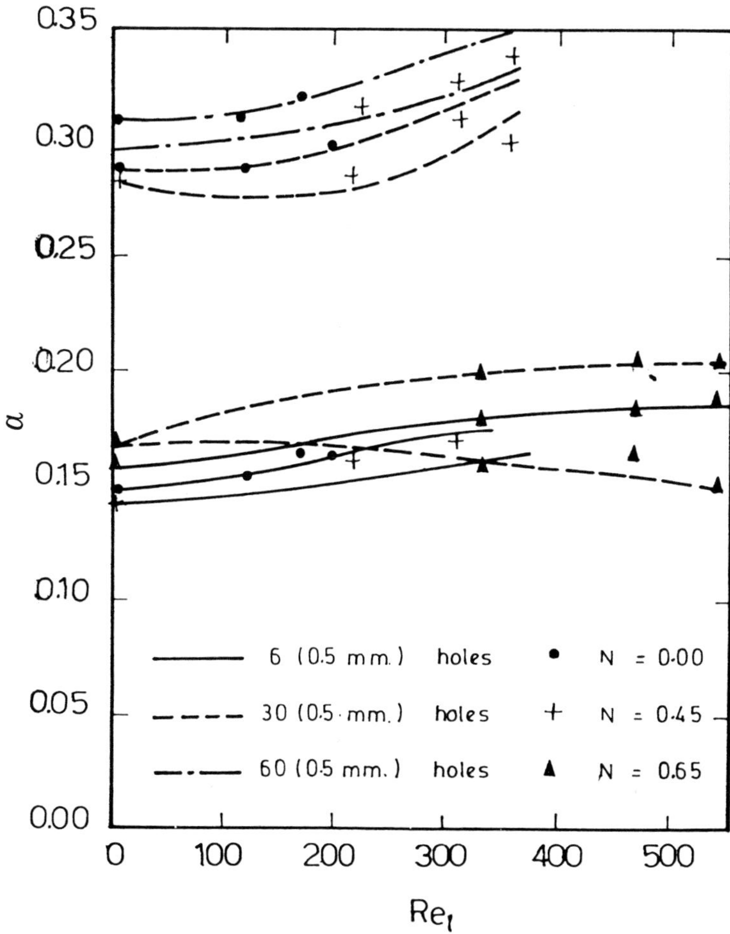

Fig.(3-b) : Void. fraction , α , at transition boundaries.

and void fraction vary with the liquid Reynolds number and the distributer plate in a very similar manner to these of upward flow. However, it is noticable that the magnitudes of Re_g and α are slightly smaller than their equivalents. These differences are too small to be analysed without an error of the results themselves. The auther feels that since the maximum flow rate of water was 1×10^{-5} m^3/s and the maximum flow rate of air was 3×10^{-4} m^3/s the similarity between upwards and downwards flow is not surprising.

Table 2. Comparison between upwards and downwards flow

	N = 0.45	Re_L	Re_g	$\tilde{Re}_g$	α	$\tilde{\alpha}$
Group A	3 holes(2mm)	0	30	30	0.100	0.100
		220	30.3	27	0.104	0.081
		311	33	30.1	0.114	0.080
		358	38	38	0.131	0.104
	12 holes(1mm)	0	42.5	42.5	0.157	0.157
		220	42.7	40.4	0.167	0.163
		311	48	46.96	0.165	0.166
		358	50.2	50.6	0.171	0.165
	48 holes(0.5)	0	67.5	67.5	0.325	0.325
		220	67.5	60.5	0.331	0.293
		311	67.5	60.7	0.330	0.300
		358	75.2	69	0.330	0.330
Group B	6 holes (0.5m)	0	30.3	30.3	0.145	0.145
		220	29.9	28.6	0.161	0.141
		311	35.7	30.3	0.167	0.130
		358	35.7	33	0.159	0.132
	30 holes(0.5 mm)	0	75.0	75.0	0.290	0.290
		220	75.4	77	0.283	0.325
		311	80.5	78.4	0.325	0.322
		358	82.6	82.1	0.302	0.310
	60 holes(0.5 mm)	0	57.0	57.0	0.300	0.300
		220	57	55.3	0.315	0.300
		311	66	60.2	0.395	0.310
		358	67.5	67.7	0.335	0.325

N.B $\tilde{Re}_g$ and $\tilde{\alpha}$ are the downwards flow Reynolds number and void fraction.

The presented results indicate that the transition from bubbly to slug flow depends greatly on the initial formation of bubbles. It is a function of the number of holes but not on the hole size since bubbles formed in chain or series bubbling are a function of the air flow rate per hole only, a function of the flow area but not on the interholes spacing since chain bubbling are quickly separated from the distributing plate. Therefore equal area of perforations and equal hole size does not offer any distinct classification for the transition, instead the number of released bubbles offers a much more sound criterion.

The average bubble diameter D_b at transition is plotted as a function of the number of released bubbles, or the number of holes, in fig 4. It can be seen that, as the number of bubbles increases; the

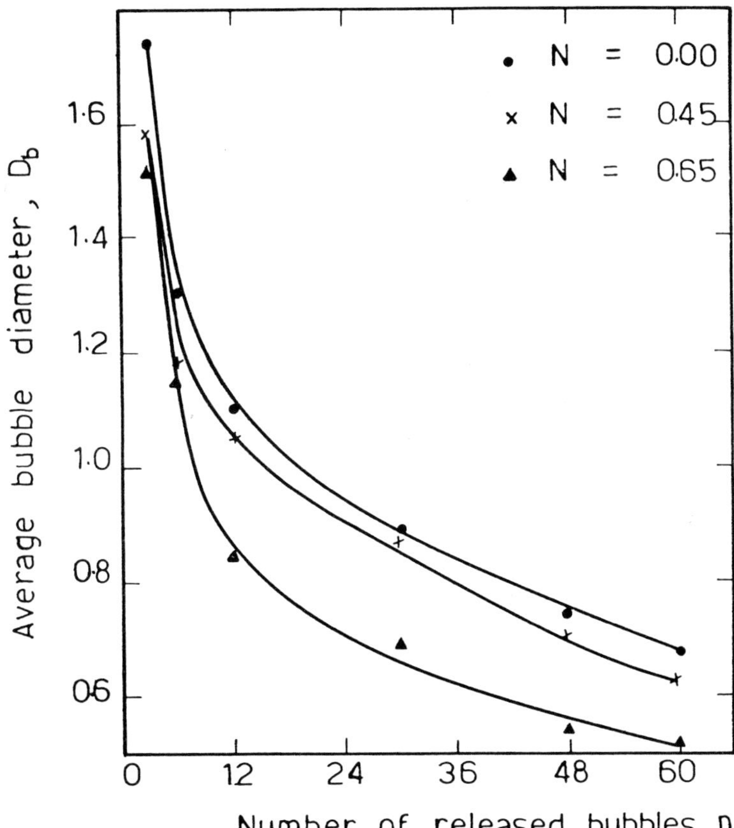

Fig.(4): Average bubble diameter at transition versus the number of released bubbles.

bubble size at transition decreases which indicates the significant decreases in the actual flow rate since the variation of the bubble size with its number for a constant flow rate is very small compared with the observed variation. Furthermore, as the radius ratio increases ie the gap size decreases, the bubble size at transition becomes smaller for the same number of bubbles. However, this variation is proportional to the area ratio in various annuli.

A sample of the transition boundaries as a function of the number of released bubbles for a constant liquid Reynolds number is plotted in Fig. 5. It can be seen that Re_g follows an interesting and characteristic behaviour. The gas Reynolds number at transition increases as the number of initial bubbles increases for $n<30$, conversely for $n>30$ the opposite is true; this being observed in the three annuli. Therefore, increasing the number of bubbles is in favour of bubbly flow pattern up to a certain limit where the opposite trend occur. Furthermore, it canbe seen that for $n<12$, as the radius ratio N increases the transition boundary is delayed, on the other hand, for $n>12$ this phenomenon seems to apply only in a limited manner.

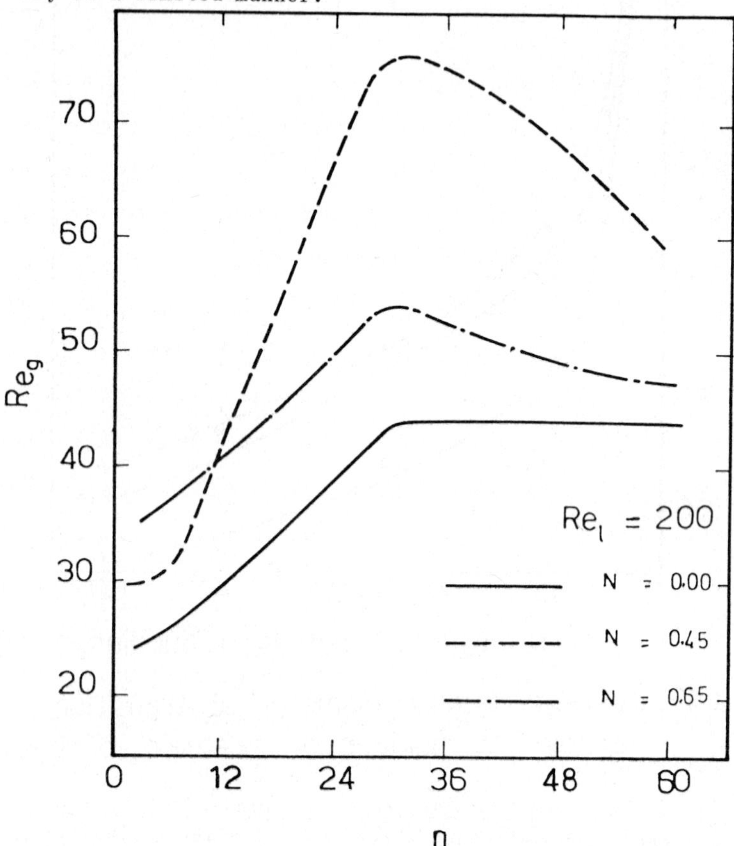

Fig.(5): Bubbly_slug transition as a function of the number of released bubbles

The average void fraction at transition is also presented as a function of the number of released bubbles in fig 6. It can be seen that as the number of released bubbles increases the void fraction initially increases, reaches a maximum, and then decreases again. In addition, the maximum void fraction and the corresponding number of released bubbles are two strong functions of the annular width or the radius ratio. Although, the maximum void fraction at transition for N=0.0 & N=0.45

are nearly equal, the corresponding number of bubbles are different (for N=0.00, n=60 & N=0.45, n = 48). Conversely the maximum void fraction for N=0.65 is much smaller than the other two annuli and occurs at smaller number of bubbles ie = 30.

For a constant air and liquid flow rates, as the number of holes

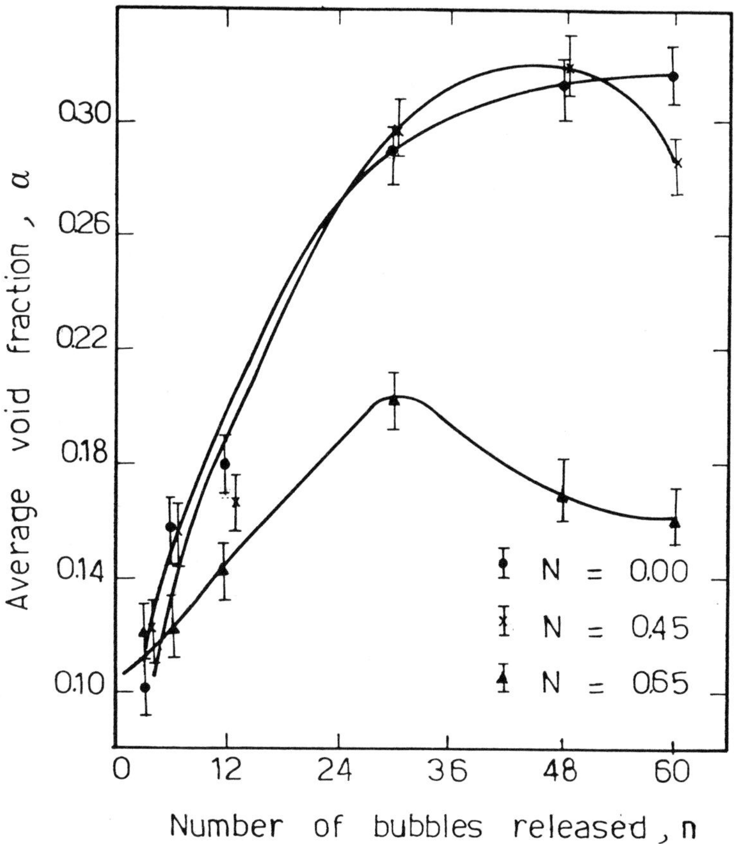

Fig.(6):The average void fraction at transition versus the number of released bubbles.

increases the generated bubbles become smaller in size, however the distance between two consecuitive and/or adjacent bubbles decreases. Simple calculations reveal that the decrease of the size of bubbles is less significant than the decrease in the distant between bubbles. ie increasing the number of holes increases the probability of collision between the generated bubbles. However, it appears that the more regular shape of the small bubbles and their uniform motion delay the process of agglomeration and hence the transition to slug flow. This trend continues as long as the gap size allows more bubbles to travel freely. Therefore, for smaller gaps or larger number of bubbles the bubbles become too close to each other and the transition is inhanced. That is being the reason for the significant influence of the gap size for large number of bubbles and its small effect for small number of bubbles.

4. CONCLUSION

This investigation indicates that the transition from bubbly to slug flow is a strong function of the potential energy of bubble surfaces or the bubble size distribution. In addition the area of flow determines the optimum size distribution of bubbles for the transition criteria. For an annulus with small radius ratio large number of small bubbles delay the transition to slug flow, conversely for an annulus with large radius ratio this is only the trend up to a certain number of bubbles where the reverse is true.

ACKNOLEDGMENT

The author acknowledges the help of Electronic Engineer Ashraf. A. Osman in obtaining the photographs.

REFERENCES

1. Strenling V.C." Two phase flow theory and engineering Decision" award lecture presented at AICHE Annual Meeting (December 1965).

2. Wallis. G.B, One dimensional two phase flow, McGraw Hill (1969).

3. Hewitt. G.F and Roberts: D.N " Studies of two phase flow patterns by simultaneous x-ray and flash photography "United Kingdom Atomic Energy Authority Report AERE-M 2159 (1969).

4. Govier. G.W and Aziz.K., The flow of complex mixtures in pipes, van Nostrand Reinhold Co (1972).

5. Duns, Jr.H, and Ras. N.C.J "Vertical flow of gas and liquid mixtures from Boreholes" Proc 6th World Petroleum Congress, Frankfurt June (1963).

6. Gould. T.L., Tek. M.R and Katz. D.L " Two-phase flow through vertical, inclined or curved pipe "J. Petroleum Tech, 26, 915 (1974).

7. Taitel Y. Bornea. D and Dukler. A.E. "Modelling flow pattern transition for Steady upward gas-liquid flow in vertical tubes " AIChE. Journal vol 26, No 3 pp 345-345, May (1980).

Multi-Phase Flow and Heat Transfer III. Part A: Fundamentals
edited by T.N. Veziroğlu and A.E. Bergles
Elsevier Science Publishers B.V., Amsterdam, 1984 — Printed in The Netherlands

AN INVESTIGATION ON FLOW PATTERN TRANSITIONS FOR GAS-LIQUID TWO-PHASE FLOW IN HELICAL COILS

Chen Xue-jun, Zhang Ming-yuan
Xi'an Jiaotong University
Xi'an, Shaanxi Province
The People's Republic of China

ABSTRACT

The air-water two-phase flow patterns and their transitions in helical coils are described. A systematical investigation of effects of coil diameter and helix angle on the flow pattern transitions is made. Using dimensional analysis method and based on the experimental data, the authors have proposed three criterion correlations for predicting the transitions under different flow conditions.

1. INTRODUCTION

Heat exchangers, steam generators and chemical reactors consisting of coiled tubes have been widely used in chemical industry, power industry and atomic energy techniques. Up to now, very little work has been done in investigating the characteristics of two-phase flow and heat transfer in helical coils, especially the transitions of two-phase flow patterns (1).

Mostly the investigations on gas-liquid two-phase flow patterns inside tubes have been concentrated on the horizontal and vertical flows. The work for investigating the gas-liquid two-phase flow patterns in helical coils has been limited only to observe the various flow regimes, both experimental data and theoretical analysis are scarce. The authors have investigated not only the flow patterns but also the transitions of flow patterns for gas-liquid two-phase flow in helical coils. The flow maps were obtained on the basis of the experimental results. A systematical investigation on effects of the coil diameter D and the helix angle α on the flow pattern transitions was made. Using the dimensional analysis and based on the experimental data, three correlations have been proposed for predicting the flow pattern transitions in helical coils under different flow conditions.

2. EXPERIMENTAL APPARATUS AND PROCEDURES

Schematic diagram of the experimental apparatus is shown in Fig. 1. By centrifugal pump 2, water was forced to flow from tank 1 through pressure stabilizator 3, Via flowmeter 4 or 5 to air-water mixer 12. Air was delivered from air compressor 6 through oil separator 7. The air was cooled in cooler 8 consisting of a small copper coil immersing in water, and then via flowmeter 10 or 11 entered into air-water mixer 12. After mixing, the air-water mixture flowed through an entrance length and test coil 13 to air-water separator. After separation, the air was discharged into atmosphere and the water was returned to tank 1. 9 is blow-off valve, 14 air-water separate bottle, 18 thermometer, 15 and 16 are mercury U-tube manometers.

The six test coils used in this experiment were all made of plexiglass tube, 22mm ID by 4mm wall thickness. Their structural parameters are listed in Table 1.

Table 1 The Structural parameters of the tested coils

No.	Coil diameter (m)	Helix angle (°)	turns
1	0.623	1.2	2
2	0.623	3	2
3	0.623	12	2
4	0.910	5	2
5	0.213	5	5
6	0.129	5	6

The observed range of air and water flowrates is as follows: The air superficial velocities were from 0.2 to 18 ms^{-1} , the water superficial velocities from 0.04 to 2.7 ms^{-1}.

At the beginning of the experiment, various flow patterns and their transitions were carefully observed in a horizontal straight glass tube, 18 mm ID and 1960 mm long. The results were compared with some previous flow pattern maps of horizontal flow, such as Baker's (2) map, Mandhane's (3) map and Weisman's (4) map. There were two purposes to do so: one was to be familiar with and to master the structure characteristics of individual flow regimes and the regularity of their transitions, the other was to check the accuracy of the equipment and instrumentation.

Then, the helical coils were tested in the test bed one by one. In order to find out the transition flows from one flow regime to another, the air flowrates were slowly varied at a selected water flowrate until the flow regime was changed. Then, after regulating the water flowrate to another value, the air flowrates were varied again.

3. THEORETICAL ANALYSIS OF FLOW PATTERN TRANSITIONS IN HELICAL COILS

The gas-liquid two-phase flow patterns difined in this experimental investigation are wavy stratified flow, plug flow, slug flow, annular flow and dispersed-bubble flow (see Fig. 2).

Wavy stratified flow: water flows at the bottom of tube with gas at the top. The interface continues to wave. In the whole period of the flow regime, sometimes the aerated water lumps sweeping through the tube can be observed.

Plug flow: The characteristic bullet-shaped bubbles move along in a position closer to the top of tube. Under the bubbles a water layer waving slightly flows along the buttom of tube. Bullet-shaped bubbles are separated by liquid lumps which fill the pipe.

Slug flow: The length of bullet-shaped bubbles in plug flow rapidly increases with the increase of air flowrates. When air flowrates increase to a certain value, more and more small air bubbles which usually concentrate on the front and top of liquid lumps will appear in the liquid lumps between bullet-shaped bubbles.

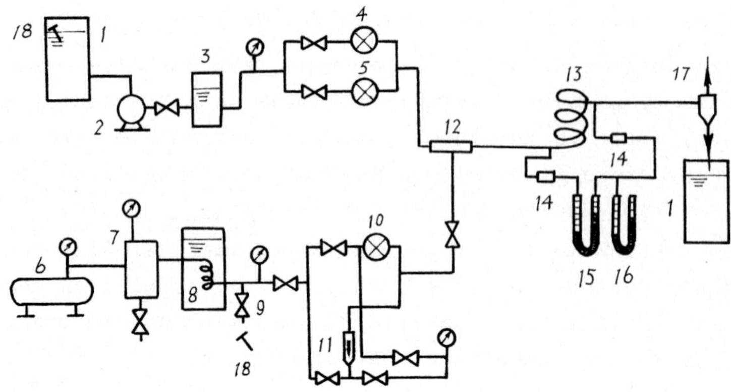

Fig.1 Schematic diagram of experimental apparatus

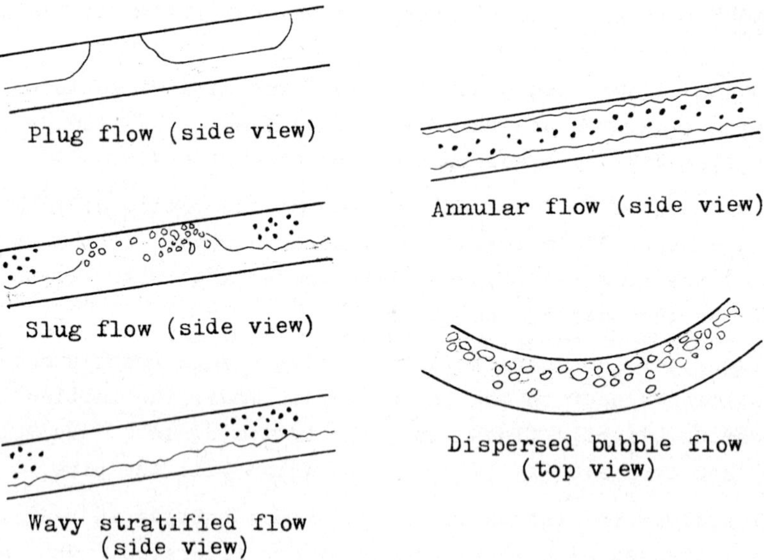

Fig.2 The flow regimes observed in helical coils

Then, the slug flow is entered. When bullet-shaped bubbles in slug flow become quite long, the water layers under bubbles begin to flow downward partially, but this downward flow will stopped as air velocity becomes sufficient high. It should be pointed out that pressure fluctuation is very strong during slug flow.

Annular flow: The liquid phase flows as a film around the tube wall, and the high velocity gas phase flows in the center of tube. The gas core always contains entrained liquid droplets, which is more evident at high liquid velocity. The liquid film is thicker at the bottom of tube than at the top.

Dispersed bubble flow: The gas phase is distributed as discrete small bubbles within a continuous liquid phase. Due to the effect of centrifugal force, liquid flows along the outer wall inside tube, i.e. far from coil axis, but the air bubbles concentrate near the inner wall, i.e. closer to coil axis.

The occurences and transitions of various flow regimes for gas-liquid two-phase flow are governed by the interaction and balance of the forces acting on the gas and liquid phases (5,6). The criterion numbers related to flow pattern transitions can be obtained by analyzing acting forces there, and then the criterion correlations for flow pattern transitions might be proposed. In the case of gas-liquid two-phase flow in a tube, the forces acting on the fluid are inertia force, gravitational force, viscous force and surface tension. Taking the inside diameter of tube as characteristic scale, the forces can be expressed as: inertia force, $1/2\ \rho u^2$; viscous force, $\mu u/d$; gravitational force, $\rho g d$; surface tension, σ/d. Dividing inertia force by viscous force, Reynolds number could be obtained, $Re=\rho du/\mu$; Weber number could be derived from the ratio of inertia force and surface tension, $We=\rho u^2 d/\sigma$; Froude number from the ratio of inertia and gravitational forces, $Fr=u^2/(gd)$. A few another criterion numbers could be gained from the numbers given above through various treatment, but the independent numbers are only three.

Three independent number, together with the ratio of volumetric flowrate u_{lo}/u_{go} , the density ratio ρ_l/ρ_g , viscosity ratio μ_l/μ_g , the coil structural parameters D/d and α may completely indicate the various transitions of flow regimes. In fact, the state of force acting for individual transitions are different. For a certain transition, some forces may play a vital part, others may not or even can be omitted. Therefore, when a certain individual transition is investigated, it is unnecessary to consider all these numbers, only a part of them is needed.

The authors' experimental investigation as well as others'[7] indicates that when the helix angle is small (the maximum value of α in available data is $6°$[8], the biggest value of the angle in the authors'experiment is $12°$), the flow patterns observed in helical coils are similar to those seen in horizontal flow, and the flow pattern maps for helical coils are similar to those for horizontal flow. On the other hand, there is something different in helical coils. The centrifugal force acting on the fluid results in a secondary flow: fluid moves from inner wall (i.e. the side close to the helix axis) to outer wall (i.e. the side far from the helix axis) in the centre section of tube, but on the wall, fluid draines towards helix axis. This feature of fluid motion in helical coils will affect flow pattern transitions.

The state of force acting for various pattern transitions in helical coils will be analyzed below. Here, plug and slug flow regimes will be considered as a identical flow regime, intermittent flow regime. Hence, the flow pattern transitions which should be considered here become three: intermittent-annular, intermittent-dispersed bubble, and wavy stratified-intermittent or annular.

The transition from wavy stratified to intermittent or annular flow regimes. For horizontal and near horizontal flow, this transition occurs when gas flows and accelerates along the liquid surface, the force due to Bernoulli effect above the wave is greater than the gravity force so that the wave grows continuously.

In the Taitel and Dukler semi-theoretical correlation, dimensionless group $F=\sqrt{\frac{\rho_g}{\rho_L-\rho_g}}\frac{u_{go}}{\sqrt{gd\cos\alpha}}$ is adopted. In fact, F is the square root of the ratio of gas inertia force to liquid gravity force. According to author's observation, the flow phenomena in helical coils and in horizontal pipe during this transition are very similar, the shape and the tendency of varing of the transition boundary in helical coil flow pattern maps are nearly same as that in inclined flow pattern map. Hence it is reasonable to think helical coils and the horizontal or near horizontal pipes are the same in transition mechnism. To consider the actions of both gas inertia force and liquid gravity, the criterion number $F_g=\sqrt{\frac{\rho_g}{\rho_L-\rho_g}}\frac{u_{go}}{\sqrt{gd}}$ is also used here, the influence of α is considered as another independent factor. The Bernoulli effect is also related to the liquid level in tube, but the liquid level of wavy flow in helical coils is controled by the factors such as u_{go}/u_{LO}, D/d and α. Therefore, the criterion correlation for this transition can be expressed as

$$f_1(F_g, u_{go}/u_{LO}, D/d, \alpha) = 0 \qquad (1)$$

The transition from intermittent to annular flow regimes. For the transition to annular flow regime in horizontal, near horizontal and vertical flows, many investigators have proposed a lot of theoretical models in different ways which explain the formation of annular flow regime (9, 10, 11). The consistent idea among them is that the formation of annular flow regime is due to the effects of high velocity gas stream, it overcomes the effect of liquid gravity force and makes the liquid to flow around the tube wall as a film. Therefore, for the transition to annular flow regime, gas inertia force and liquid gravity force should be considered as main affecting factors. The results of the observation in helical coils indicate that during the transition from intermittent to annular flow, the air superficial velocity maintains approx. constant at various liquid flowrates, which means that the transition to annular flow in helical coils is controled by gas inertia force as well. The experimental data for horizontal flow indicate that the transition is also affected by the surface tension (4), it is true in the case of helical coils. In order to express the interaction of gas inertia force, liquid gravity force and surface tension, the criterion numbers $F_g=\sqrt{\frac{\rho_g}{\rho_L-\rho_g}}\frac{u_{go}}{\sqrt{dg}}$ and $We_g=\rho_g u_{go}^2 d/\sigma$ are adopted here. From F_g and We_g, we could get the Kutadelaze number:

$$K_u=(We_g\cdot F_g^2)^{1/4}=\frac{u_{go}\,\rho_g^{1/2}}{(g(\rho_L-\rho_g)\sigma)^{1/4}}$$

In both Taitel's semi-theoretical correlation (10) for vertical tubes and Weisman's experimental correlation (4) for horizontal tubes, Ku is employed for the transition to annular flow. Replacing We_g with Ku , F_g and Ku are used here. Forthermore, it is discovered in the experimental observation that coil diameter D and water superficial velocity U_{Lo} have some influence on the transition, but the helix angle nearly has no effect on it. Hence, D/d and U_{go}/U_{Lo} should be also adopted. So the criterion correlation for transition to annular flow can be expressed by

$$f_2(Ku, F_g, D/d, U_{go}/U_{Lo}) = 0 \qquad (2)$$

The transition from intermittent to dispersed bubble flow regimes. In horizontal flow, the transition to dispersed flow occurs when the turbulent fluctuations are sufficient high so that it overcomes the buoyance [9]. The frictional pressure drop is related to the turbulent fluctuations, Taitel and Dukler propose that the data can be correlated in terms of a dimensionless ratio $T = \frac{|(dp/dx)_L^S|}{[(\rho_L - \rho_g)g]}$, where $|(dp/dx)_L^S|$ is the frictional pressure gradient when water flowing alone along the tube. It is more complex in helical coils where not only gravity force but centrifugal force acts on the fluid also. On one hand, this will impel the gas and liquid phases to be separated. On the other hand, in helical coils the secondary flow will occur when fluid velocity is very high, which, like turbulent fluctuation, will force the two phases to be mixed. As the results of mixing of two phases is greater than that of separating, the transition to dispersed bubble flow occurs. The frictional pressure drop is strongly connected with the secondary flow and hence the dimensionless ratio $\frac{|(dp/dx)_{LO}|}{[(\rho_L - \rho_g)g]}$ is also employed in helical coils, but $|(dp/dx)_{LO}|$ should be calculated from the correlations for helical coils. It is noticed in horizontal flow that surface tension and tube diameter affect the transition to dispersed bubble flow. To consider these effects, the dimensionless group $\frac{\sigma}{g(\rho_L - \rho_g)d^2}$ was adopted in the horizontal flow correlation [4]. This is supposed to be true also for helical coils. Besides, the experimental observation indicates that U_{go} and D have an effect on the transition, so U_{go}/U_{Lo} and D/d are needed. Considering all the factors mentioned above, the criterion correlation for the transition to dispersed bubble flow regime can be written as:

$$f_3\left(\frac{|(dP/dx)_{LO}|}{[(\rho_L-\rho_g)g]},\frac{\sigma}{g(\rho_L-\rho_g)d^2},\frac{u_{go}}{u_{LO}},\frac{D}{d}\right)=0 \qquad (3)$$

4. TESTING RESULTS AND ANALYSIS

The flow pattern maps obtained on the basis of experimental data for the six testing helical coils are illustrated in Fig. 3-Fig. 8 respectively. The gas and liquid superficial velocities are plotted as the axes. In figures: • expresses the experimental transition point, //// the experimental transition region, —— calculated results based on criterion correlations.

The transition boundaries distinguished in the experiment are: slug-plug transition boundary, slug-dispersed bubble transition boundary, slug-annular transition boundary and wavy stratified-slug or annular boundary. The shapes of the transition regions in the flow pattern maps are similar to each other. Taking Fig. 3 as an example, it can be seen from it that the boundaries of slug-plug transition and slug-annular transition appear to be a vertical belt-shaped region, i.e. the transition depend on air superficial velocity, but water superficial velocity only has a little effect on it, when U_{go} =0.3-0.5 ms^{-1} the transition to plug flow occurs, when U_{go} =11-13 ms^{-1} the transition to annular flow appears; The boundary of slug-dispersed bubble transition appears to be a near horizontal belt-shaped region, the water superficial velocity during the transition slightly increases with the increase of air superficial velocity; The boundary from wavy stratified to slug or annular flow transition forms a bell shape and the region of wavy stratified flow regime is much smaller than that in horizontal flow.

The effects of helix angle α on the transitions. The effects could be seen from Fig. 3-5. Helix angle nearly has no effect on the transitions to annular and to dispersed bubble flow. However, the boundary of wavy stratified-slug or annular transition is rather sensitive to the change of α . As the helix angle α increases, the wavy stratified flow region shrinks rapidly, when $\alpha=12^\circ$, wavy stratified flow regime will disappear for the observed range of sir and water flowrates. Compared with the experimental

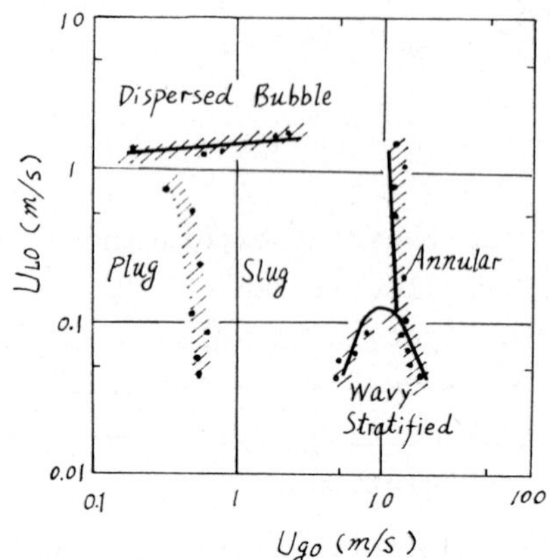

Fig.3 The flow pattern map in the helical coil of 0.623m D and 1.2° α

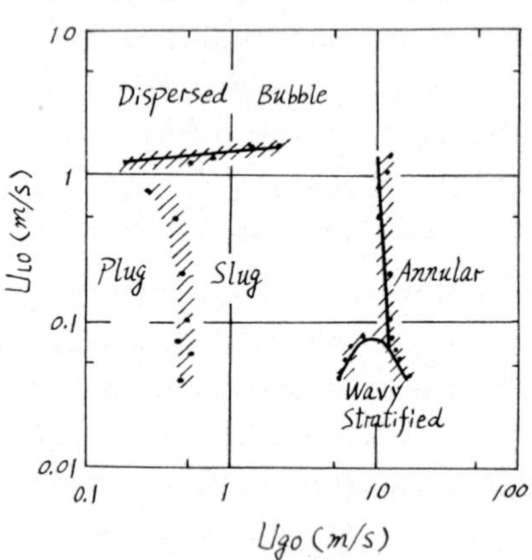

Fig.4 The flow pattern map in the helical coil of 0.623m D and 3° α

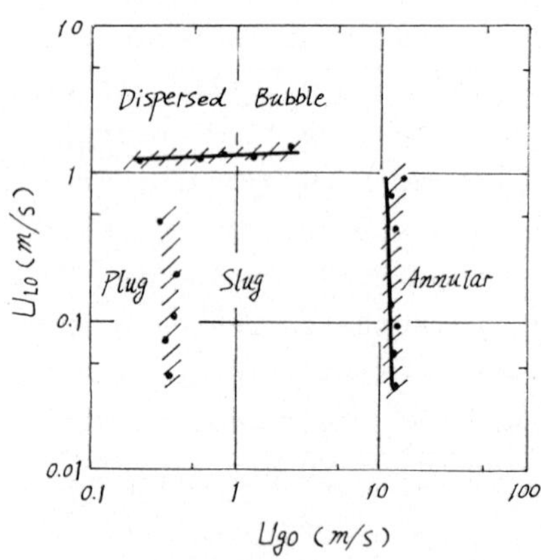

Fig.5 The flow pattern map in the helical coil of 0.623m D and 12° α

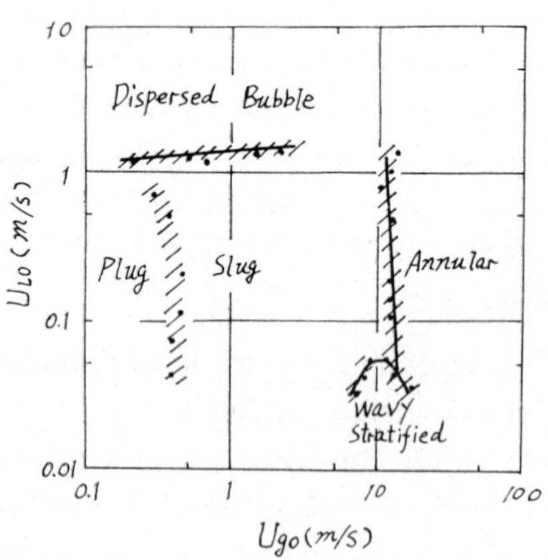

Fig.6 The flow pattern map in the helical coil of 0.910m D and 5° α

results for inclined pipe (upward flow) published by Barned et al.[12], if their so-called wavy annular flow regime is considered as slug flow regime, helical coils and inclined pipes are consistent in tendency of changing of transition boundaries with the change of α and with the change of inclination respectively, except that the reduction rate of wavy stratified flow region with the increase of α in helical coils is slower than that in inclined pipes. It can also be seen that the boundaries between plug and slug flow regimes move towards the direction where air flowrates are lower while increasing α .

The effects of coil diameter D on the transitions. The effects can be seen in Figs. 6 through 8. For the slug-annular and plug-slug transitions, the boundaries locate at the lower air flowrates with decreasing D; when D=0.129 m, plug flow regime will not be seen at all for the observed range of air and water flowrates, meanwhile, the region of wavy stratified flow regime shrinks slightly. At the same air flowrates, the water flowrates during the transition to dispersed bubble flow increase when D decreases, and the slope of the transition boundary in the map has the tendency to increase.

5. THE CRITERION CORRELATIONS FOR FLOW PATTERN TRANSITIONS IN HELICAL COILS

Using theoretical analysis, we got the three criterion correlations (1), (2), (3); and based on the experimental data, three concrete criterion correlations have been proposed as follows:

The correlation for the transition from wavy stratified to intermittent or annular flow regimes is

$$318\left(\frac{d}{D}\right)^{0.05}\cdot(\sin\alpha)^{0.4}\cdot(1.66F_g^2)^{0.526\ln(1.66F_g^2)+0.45}=\left(\frac{u_{go}}{u_{Lo}}\right)^{0.9} \qquad (4)$$

The correlation for the transition from intermittent to annular flow regimes is

$$K_u^{0.2}\cdot F_g^{0.36}\cdot\left(\frac{d}{D}\right)^{0.07}=0.816\left(\frac{u_{go}}{u_{Lo}}\right)^{0.03} \qquad (5)$$

The correlation for the transition from intermittent to dispersed bubble flow regime is

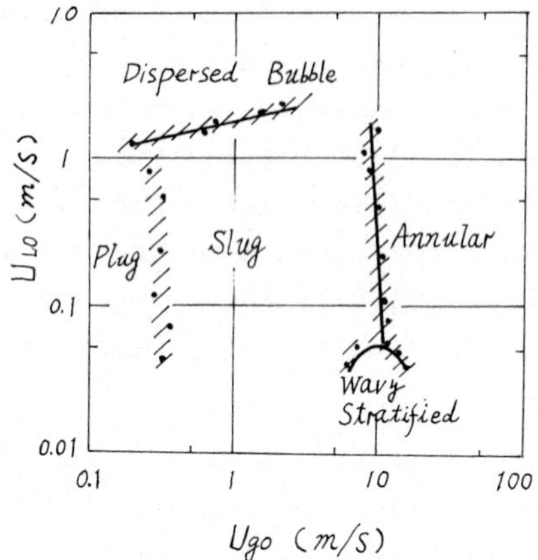

Fig.7 The flow pattern map in the helical coil of 0.213m D and 5° α

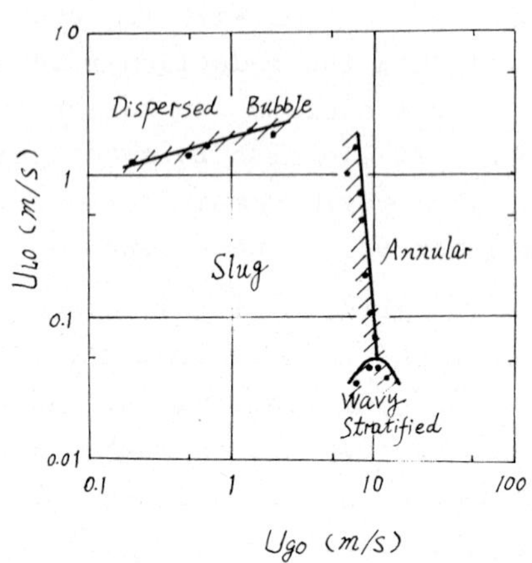

Fig.8 The flow pattern map in the helical coil of 0.129m D and 5° α

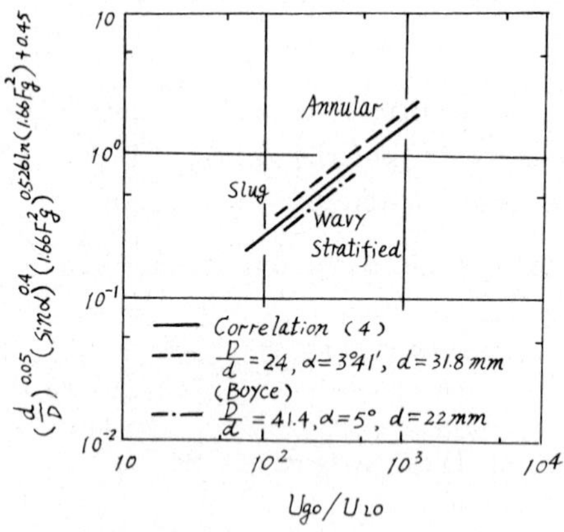

(a) The transition from wavy stratified to slug or annular flow

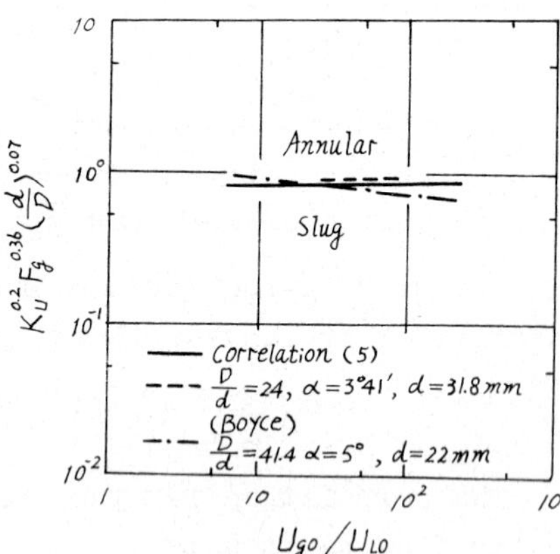

(b) The transition from slug to annular flow

Fig.9 A comparison of correlations (4) and (5) with Boyce's experimental data

$$\left[\frac{|(dp/dx)_{LO}|}{(\rho_L-\rho_g)g}\right]^{0.5}\cdot\left[\frac{\sigma}{g(\rho_L-\rho_g)d^2}\right]^{-0.25}=2.3\left(\frac{d}{D}\right)^{0.2}\left(\frac{u_{go}}{u_{LO}}\right)^{0.87\left(\frac{d}{D}\right)^{0.6}} \qquad (6)$$

In above correlations, $F_g=\sqrt{\frac{\rho_g}{\rho_L-\rho_g}}\cdot\frac{u_{go}}{\sqrt{gd}}$, $K_u=\frac{u_{go}\rho_g^{1/2}}{\{g(\rho_L-\rho_g)\sigma\}^{1/4}}$, $|(dp/dx)_{Lo}|$ is the frictional pressure gradient for liquid flowing alone in helical coils. $|(dp/dx)_{LO}|=\frac{4}{d}C_{LC}\frac{1}{2}\rho_L u_{LO}^2$, C_{LC} can be calculated from relevant correlations.

The calculation results of correlations (4)-(6) have already been shown in Figs. 3-8 by solid curves. The values of the constants which are used in calculation are as follows: $\rho_L=1000 kg m^{-3}$, $\rho_g=1.3 kg m^{-3}$, $\sigma=0.07 N m^{-1}$, $\mu_L=1\times10^{-3} kg m^{-1} s^{-1}$, $g=9.8 m s^{-2}$, C_{LC} is given by Ito's correlation $\left(\sqrt{\frac{D}{d}}C_{LC}=0.079/\{Re(\frac{d}{D})^2\}^{1/5}\right)$. Agreement between correlations and the experimental data is satisfactory.

Correlations (4) and (5) are also agreeable to Boyce's corresponding experimental data for flow pattern transitions in a helical coil. A comparison between both cases is shown in Fig.9.

6. CONCLUSIONS

1. The flow patterns observed for gas-liquid two-phase flow in helical coils are similar to those seen in horizontal two-phase flow. But they also have their own special characteristics.

2. Using the dimensional analysis method and based on the experimental data, correlations of (4), (5), (6), respectively for the transitions of wavy stratified-intermittent or annular flow, intermittent-annular flow and intermittent-dispersed bubble flow, are obtained. The calculation results from above three correlations agree well with the experimental data.

3. The calculation results of correlations (4) and (5) are in agreement with Boyce's corresponding data of flow pattern transitions.

4. As the helix angle increases, the boundaries between plug and slug flow regimes move towards the direction where air flowrates are lower and the region of wavy stratified flow shrinks rapidly,

when $\alpha = 12°$, the wavy stratified flow region will disappear for the observed range of gas and liquid flowrates. α nearly has no effects on the transitions to annular flow and to dispersed bubble flow.

5. For the slug-annular and plug-slug transitions, the boundaries locate at the lower air flowrates with decreasing D, but the region of wavy stratified flow shrinks slightly. At the same air flowrates, the water flowrates during the transition to dispersed bubble flow increase when D decreases and the slope of the transition boundary line in the flow pattern map has the tendency to increase.

NOTATION

C_{LC}: Friction factor of helical coil;

D: Coil diameter, m;

d: Inside diameter of tube, m;

g: Acceleration due to gravity, ms^{-2};

u: Fluid velocity, ms^{-1};

u_{go}: Gas (air) superficial velocity, ms^{-1};

u_{Lo}: Liquid (water) superficial velocity, ms^{-1};

α: Helix angle, deg.;

μ: Viscocity, $kg\ m^{-1}s^{-1}$;

ρ: Density, $kg\ m^{-3}$;

σ: Surface tension, Nm^{-1};

$|(dp/dx)_{Lo}|$: Frictional pressure gradient of liquid flowing alone in coil, Nm^{-3}.

Subscripts

L: Liquid;

g: Gas.

Dimensionless groups

Re: Reynolds number;

Fr: Froude number;

We: Weber number;

Ku: Kutadelaze number, $\frac{u_{go} \rho_g^{1/2}}{[g(\rho_l - \rho_g)\sigma]^{1/4}}$;

F_g: $\sqrt{\frac{\rho_g}{\rho_l - \rho_g}} \frac{u_{go}}{\sqrt{gd}}$.

REFERENCES

1. Chen Xue-jun & Tu Chun-jin 1980 The flow and heat transfer properties of steam generator of helical coil type. Xi'an Jiaotong University, Scientific and technical report 80-317

2. Baker, O. 1954 Simultaneous flow of oil and gas. Oil and Gas J. 53, 185-195

3. Mandhane, J. M. et al. 1974 A flow pattern map for gas liquid flow in horizontal pipes. Int. J. Multiphase Flow 1, 537-553

4. Weisman, J. et al. 1979 Effects of fluid properties and pipe diameter on two-phase flow patterns in horizontal lines. Int. J. Multiphase Flow 5, 437-462

5. Akagawa Koji 1974 Gas-Liquid Two-Phase Flow

6. Yin-Yun Hsu & Robert W, Graham 1976 Transport Processes in Boiling and Two-Phase Systems.

7. Boyce, B. E, et al. 1969 Hold-up and pressure drop measurements in the two-phase flow of air-water mixtures in helical coils. Cocurrent Gas-Liquid Flow

8. Whalley, P. B. 1980 Air-water two-phase flow in a helically coiled tube. Int. J. Multiphase Flow 6, 345-356

9. Taitel, Y & Dukler, A.E. 1976 A model for predicting flow regime transition in horizoutal and near horizontal gas liquid flow. AIChE. J. 22, 47-55

10. Taitel. Y. et al. 1980 Modelling flow pattern transitions for steady upward gas-liquid flow in vertical tubes. AIChE. J. 26, 345-354

11. Butterworth, D & Hewitt, G. F. 1978
Two-Phase Flow and Heat Transfer

12. Barnea, D. et al. 1980
Flow pattern transition for gas-liquid flow in horizontal and inclined pipes. Int. J. Multiphase Flow 6, 217-225

13. Ito, H. 1959 Friction factors for turbulent flow in curved pipes. J. Basic Engineering, Trans. ASME. D81, 123-134

Multi-Phase Flow and Heat Transfer III. Part A: Fundamentals
edited by T.N. Veziroğlu and A.E. Bergles
Elsevier Science Publishers B.V., Amsterdam, 1984 — Printed in The Netherlands

FLOW REGIME TRANSITIONS FOR AIR-WATER FLOW IN HELICALLY COILED TUBES

M. Kaji, K. Mori, S. Nakanishi and S. Ishigai
Department of Mechanical Engineering
Faculty of Engineering, Osaka University
Yamada-oka, Suita, Osaka 565, Japan

ABSTRACT

An experimental study of flow regime transitions for air-water two-phase flow in helically coiled tubes was conducted. Three coiled Perspex tubes of 10 mm I.D. with different coil diameters of 165, 320 and 500 mm, respectively, were used. The helix angle was 8 degrees and the flow direction was upward. Superficial gas velocity was varied from 0.2 to 75 m/s and liquid velocity, from 0.01 to 4 m/s. Test pressure was the atmospheric pressure. After one turn passage from the coil inlet, dispersed bubble, plug, slug, wavy stratified, annular and wavy annular flows were identified by visual observation and by an improved conductance probe technique. Flow regime transition boundaries were compared with those of horizontal or upward inclined flows. Transition boundary between slug and annular flows was close to the map of Mandhane et al and the annular - wavy stratified flow boundary was close to the Baker's map. Annular flow regime area became narrower as the coil diameter decreased. With emphasis on the annular flow, liquid film thickness around the tube periphery was measured. Wavy characteristics of liquid film flow were also measured by the conductance probe technique.

1. INTRODUCTION

Hydrodynamic characteristics of two-phase flow in a helically coiled tube have not been sufficiently clarified, although helically coiled steam generators are used in many industries such as chemical plants, LMFBR (Liquid Metal Fast Breeder Reactor) and so on. A number of reports have been published on CHF (Critical Heat Flux) data [1] and pressure drop data [2] of coiled tubes for practical application.

In our previous paper [3], flow rates of liquid droplets entrained in a coiled tube were measured and the secondary flows caused by centrifugal forces were observed. Dryout mechanism in the coiled tube was discussed comparing the relation between dryout quality and droplet flow rate for the coiled tube with that for the straight tube. To predict the onset of CHF condition and the pressure drop in a coiled tube with accuracy, knowledge of the flow characteristics, such as flow regimes and phase distributions in tube cross section , is required.

Few investigators have dealt with these subjects and carried out experiments on air-water two-phase flows. Void fraction and pressure drop were measured and empirical correlations were presented by Rippel et al [4] and Akagawa et al [5]. Liquid film thicknesses around the tube periphery were measure by Kozeki [6] and Whalley [7] for the annular flow. Their reports indicated that the film thickness was the maximum at the inside (closest to the helical axis) of the tube and the minimum at the outside (remotest from the helical axis) . Liquid film inversion in the coiled tube was analyzed by Banerjee et al [8]. These studies, however, were carried out under limited experimental conditions.

The present paper reported the results obtained in a series of experimental works on adiabatic air-water flows in helically coiled tubes. Three coiled tubes of 10 mm I.D. with different coil diameters of 165, 320 and 500 mm, respectively, were used to investigate the effect of coil diameter. Flow regime map was established for each coiled tube and compared with that of horizontal or upward inclined straight tube flow. With emphasis on the annular flow regime, liquid film flow behaviors around the tube periphery were investigated.

2. EXPERIMENTAL APPARATUS

Fig.1 shows a schematic of the experimental apparatus. Compressed air was fed to a mixing chamber through a pressure regulator, a control valve and a flow meter. Water was fed by a gear pump to the mixing chamber after the flow rate was measured. Two types of mixing chamber as shown in Fig.2 were used. Type (a) had a porous wall, through which water permeated and was mixed with air in the tube. Type (a) was mostly used in the present work. Type (b) had a small diameter tube nozzle for discharging the air in the water flow. This one was suitable to realize a steady bubble flow, since it had less compressible capacity in the chamber.

Three helically coiled tubes were made of 10 mm I.D. transparent acrylic resin tube which enabled visual observation of the flow within. Each coiled tube had a coil diameter D = 165, 320 or 500 mm, respectively, a helix angle of about 8° and a length corresponding at least two times full turn. The cross section of the coiled tube was slightly deformed as a oval, but the difference between the major and the minor diameters was about 0.5 mm at the most for D = 165 mm. The helix axis was vertically located and the flow direction was upward.

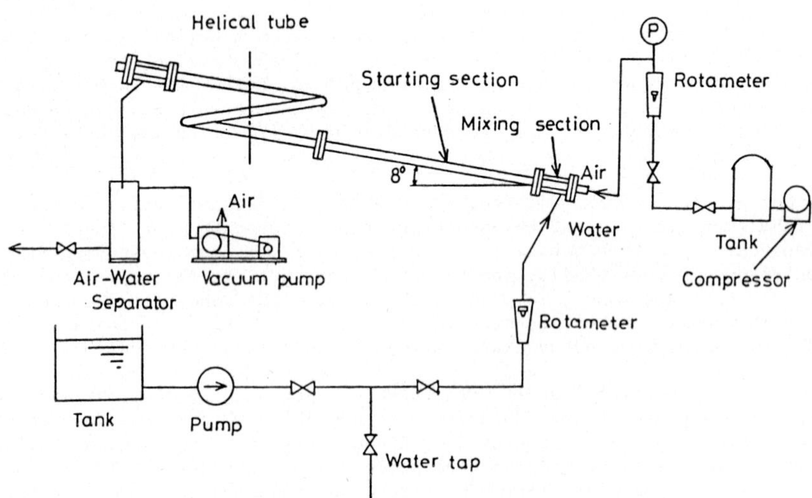

The starting section was made of a 1 m long straight acrylic resin tube of 10 mm I.D. and placed with 8° inclined upward. At the outlet of the coiled test section, liquid film was removed through a porous wall by a vaccum pump. This was useful to observe the cross sectional view of the annular flow.

The flow regime was identified mainly by visual observation at the location of about 540° turn from the coiled tube inlet. Flow regime identification by visual observation, however, was difficult when the velocity was high. Barnea et al [9] proposed to use a conductance probe technique and obtained satisfactory results in horizontal and vertical flows. Their method is useful only for the straight tube. An improved conductance probe technique was developed and applied in the present work.

Fig.3 shows probes installed at different locations to detect the electric coductivity in the flow. Probe A consisted of a couple of brass wire of 1 mm O.D. installed flat with the internal surface of the tube wall. Four couples of probes were located around the tube periphery. It was designated to detect the surface wetness and to measure the liquid film thickness on the tube wall.

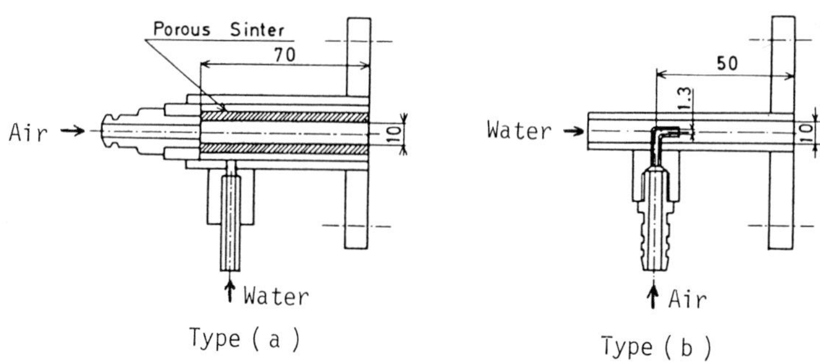

Fig.2. Schematic drawings of mixing chambers

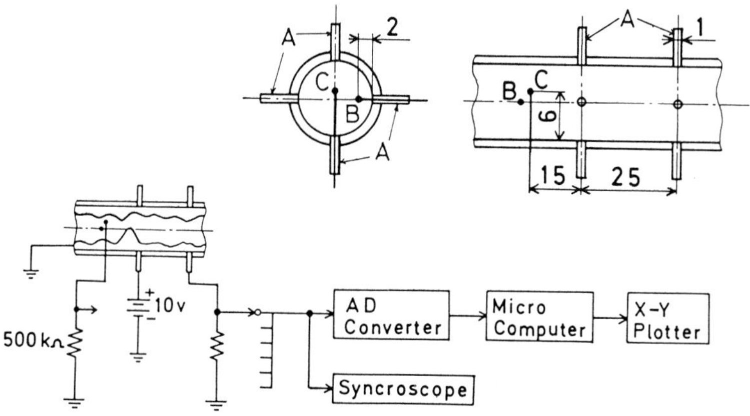

Fig.3. Arrangement of conductance probes and data processing system

Probes B and C were made of steel wire coated with insulation tube of 0.65 mm O.D.. The top ends of these probes were uncovered to measure different electrical conductivity of air and water. B was located in the middle and 2 mm outward extruded from the inside tube wall. It was useful to detect small gas bubbles in bubbly flow. C was located 6 mm above the bottom of the tube. The locations of B and C were appropriately determined from the results of preliminary experiments. When these probes are in water, the electrical circuit between probes A and B, and A and C, respectively, are closed. The electrical diagram and the signal processing system are shown schematically in Fig.3. These probes were sited close to the test section outlet.

The needle contact method was used to measure liquid film thickness around the tube periphery for annular flow. The needle diameter was 0.8 mm and direct current was applied. Time-averaged film thickness was determined by the integral mean value of the probability density of liquid contact versus needle distance. The wavy characteristics of liquid film were measured by conductance probe technique. Four couples of electrodes were located around the tube periphery with every 90°. In this case, the distance of each couple of electrodes was selected as 5 mm according to the required accuracy.

3. RESULTS

Flow regimes and their transition boundaries of two-phase flow in coiled tubes have been observed by several investigators [5-8]. Experimental conditions of these works, however, were limited and few kinds of flow regimes were identified. From their results it is suggested that the flow regimes realized in a coiled tube are similar to those in a horizontal or slightly inclined straight tube. This was also confirmed in our preliminary experiment. In the following, flow regime observations were carried out and identified flow regimes were compared mainly with those of horizontal flow.

3.1. Flow Regime Observations

Fig.4 shows examples of typical photographs taken from the top or side of the tube for various flow regimes. Identified flow regimes in the present experiment were dispersed bubble, plug, slug, annular, stratified wavy and wavy annular flows. Representative output traces of probes A, B and C for each of the flow regimes are shown in Fig.5 (a) - (f) , in which abscissa and ordinate represent time and output voltage, respectively. Output signals were recorded by a data processing system as shown in Fig.3 and were traced by a X - Y plotter. When the tube is fully filled with water, the output signal indicates the maximum voltage due to good conductivity. When the probe is not wetted, the output signal becomes zero. The experimental condition for each flow regime photograph in Fig.4 corresponds to the condition shown in Fig.5. W_{go} and W_{lo} are superficial velocities of gas and liquid phase, respectively.

<u>Dispersed bubble flow</u>. (Figs.4(a) and 5(a)) Small gas bubbles were dispersed in a continuous liquid phase. The bubbles were distributed not uniformly in the tube cross section but were maldistributed in the upper part of the tube like horizontal flow. The top view of the flow (Fig.4(a)) shows that the concentration of bubbles near the inside (closest to the coil axis) wall is high and that the bubble shape is deformed from a sphere. The gas bubble flow near the inside wall was detected as small spike signals by probe A_{in} as well as B. The other probes indicated almost constant maximum voltage.

In the inlet region of the coiled tube, small bubbles uniformly distributed in the straight tube were gathered toward inside and coalesced with each other. As the liquid velocity increased, inhomogeneity of bubble distribution became more significant. The inhomogeneous distribution of bubbles is due to the secondary flow caused by centrifugal forces. This is the reason why the coiled tube flow is different from the straight horizontal flow.

Plug flow. (Figs.4(b) and 5(b)) In this flow regime, gas plugs were elongated as the gas velocity increased. Fig.4(b) shows the front of gas plug with a liquid layer in the bottom part of the tube. When the liquid velocity was high,

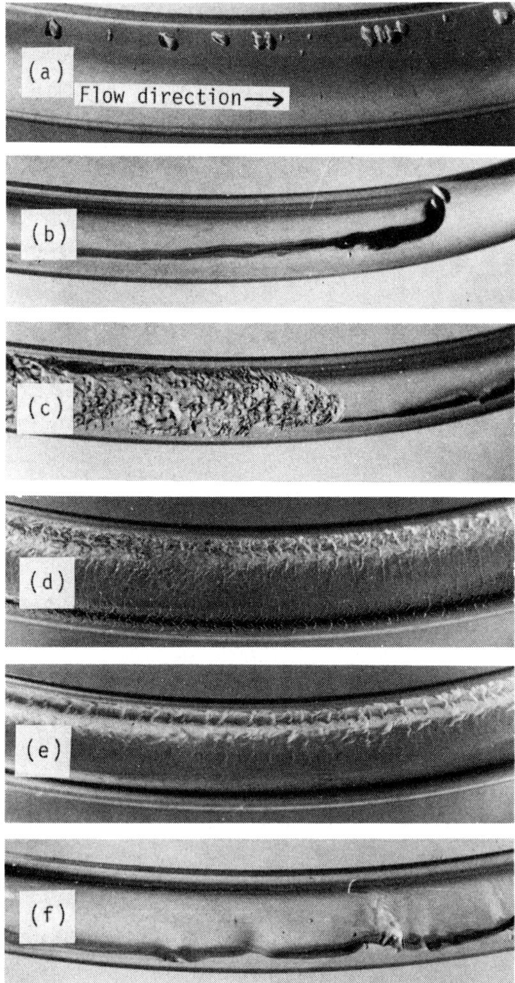

Fig.4. Typical photographs of various flow regimes; a - Dispersed bubble (top view), b - Plug (side view), c - Slug (side view), d - Annular (top view), e - Stratified wavy (top view), f - Wavy annular (side view)

the plug had a cannon shot shape, but was not symmetrical with the tube axis unlike the straight tube flow. The plug flowed in the upper and inside part of the tube as was reported by Akagawa et al [5].

Probe C output in Fig.5(b) clearly indicated passage of plugs. There were no entrained bubbles in the liquid slug. Probe B also detected gas plug passage. A small peak during plug passage indicated that the liquid film drained from the top along the tube wall. This flow regime was classified as elongated bubble flow by Barnea et al [9,10].

Slug flow. (Figs.4(c) and 5(c)) Liquid slugs aerated by gas bubbles move in tube with high velocity and the test section vibrates violently. The photograph shows the front end of liquid slug moving from left to right. The slug flow is often classified as intermittent flow [9] like the plug flow. The distinction between them, however, is difficult to define. In the present work, we identified the flow liquid slugs free of entrained gas bubbles as plug flow and otherwise as slug flow. When the velocity was high, the flow regime was identified based on the probe output traces instead of visual observations.

In Fig.5(c) small pulses of probe C output in the liquid slug region indicated the entrained gas bubbles. Rough interfaces between the gas and liquid phases behind the liquid slug were detected by probes B and C. It was easy to distinguish the probe output traces between plug and slug flows.

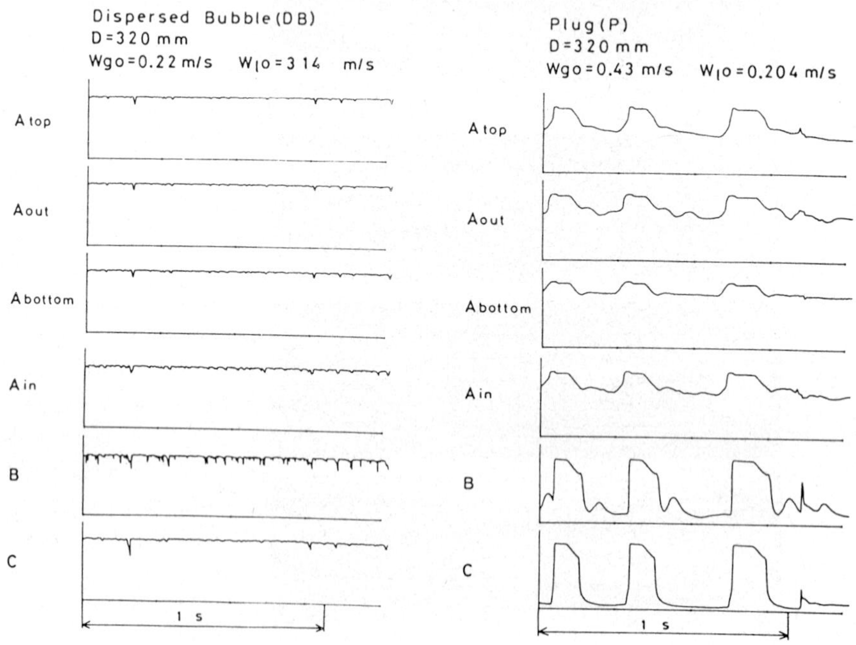

Fig.5 (a). Probe output of dispersed bubble flow

Fig.5 (b). Probe output of plug flow

Annular flow. (Figs.4(d) and 5(d)) Internal tube wall was wetted by water around the whole periphery. Numerous ripple waves were observed on the liquid film surface. The liquid film was most thick at a location between the bottom and inside part of the tube. As the gas velocity increased, the most thick part was moved from bottom toward inside along the tube periphery. Diturbance waves propagated in the most thick part of liquid film were observed. Liquid droplets entrained from the disturbance wave into the gas core were ejected toward outside (remotest from the coil axis) by centrifugal forces and deposited as liquid film on the outside wall. Thereafter, liquid film flowed from outside to inside along the periphery and formed twin helical flows.

In Fig.5(d), probe A traces indicate wavy character of the liquid film. Amplitudes of fluctuation were small at the outside and large at the bottom as well as inside. The output voltage of probe A does not accurately correspond to the local film thickness, because the electrode distance of each pair is so long compared with the internal tube diameter that electric currents between probes interfere with each other. Probe C detected very low output voltage, which corresponded to the low conductivity of liquid droplet flow.

Stratified wavy flow. (Figs.4(e) and 5(e)) When the gas velocity was high and the liquid velocity was low, partial liquid film breakup on the tube periphery was observed. Liquid film flowed continuously on the inside wall and bottom of the tube. At the outside and top of the tube, rivulets formed by deposited liquid droplets were often observed when the liquid velocity did not so much decrease. We identified this flow regime as stratified flow. Only the stratified wavy flow was realized in the present work and the stratified smooth flow was not observed. This is due to upward inclination of the tube. When the flow di-

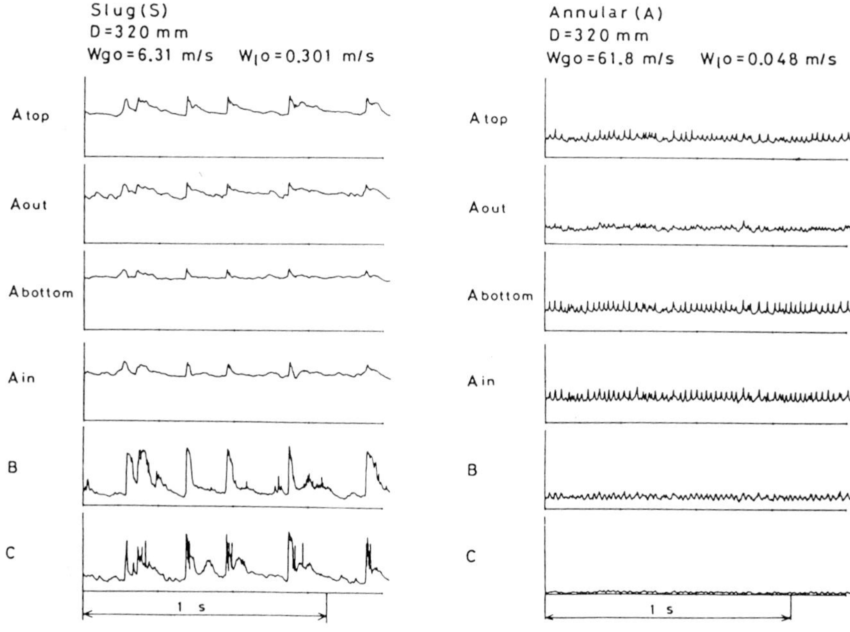

Fig.5 (c). Probe output of slug flow

Fig.5 (d). Probe output of annular flow

rection is inclined upward, high gas velocity which induces waves in the gas-liquid interface is required to convey the liquid phase upward against the gravitational force.

Probes A_{top} and A_{out} were dry and indicated absolutely zero voltage. Probe C also detected zero signal. Output traces of A_{bottom} and A_{in} indicate the wavy character of the continuously flowing liquid film. Although probe B was not exposed to liquid film, it detected the liquid phase. This might be due to water flowing up to the probe top end along the support.

Wavy annular flow. (Figs.4(f) and 5(f)) This flow regime was observed in the transition region between the annular, slug and stratified wavy flows. The liquid film thickness was maximum at the bottom and minimum at the top. No film breakups were observed. Large waves were propagated on the most thick part of liquid film, in which no gas bubbles were involved. When a large wave approached liquid film was thickened on the upper part of the periphery. Liquid phase had hardly axial velocity component at the top of the tube and drained down along the wall after the wave passed by.

The probe output traces are similar to those of slug flow. However, the maximum voltage of probe C is lower than that of slug flow. The slugging of liquid phase was not observed in this regime. The probe outputs B and C indicate that the gas-liquid interface is rather smooth compared with the slug flow. This was also confirmed by visual observations (Fig.4(f)) .

3.2. Flow Regime Map

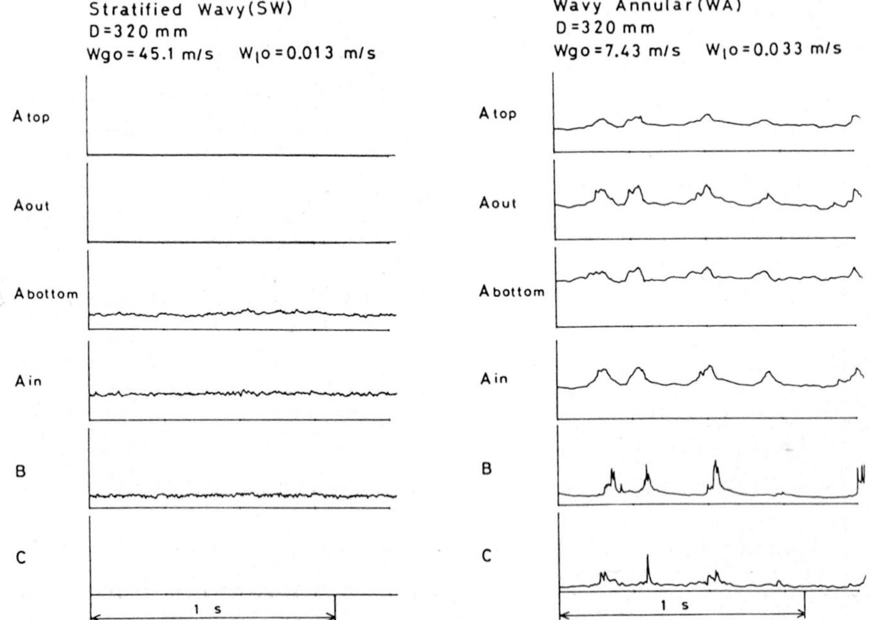

Fig.5 (e). Probe output of stratified wavy flow

Fig.5 (f). Probe output of wavy annular flow

The results obtained in the present experiment are shown in Figs.6 - 8 for various coil diameters, where w_{go} and w_{lo} are superficial velocities of gas and liquid phases, respectively. Transition boundaries between various flow regimes are compared in these figures with those by Baker [10] and Mandhane et al [11] for horizontal flow and those by Barnea et al [12] for 10° upward inclined straight tube flow. In the present experiment, the stratified smooth flow was not observed unlike the horizontal flow. Barnea et al [12] also did not observe the stratified smooth flow for upward inclination except for less than 0.25° from the horizontal. This is due to the effect of gravitational force.

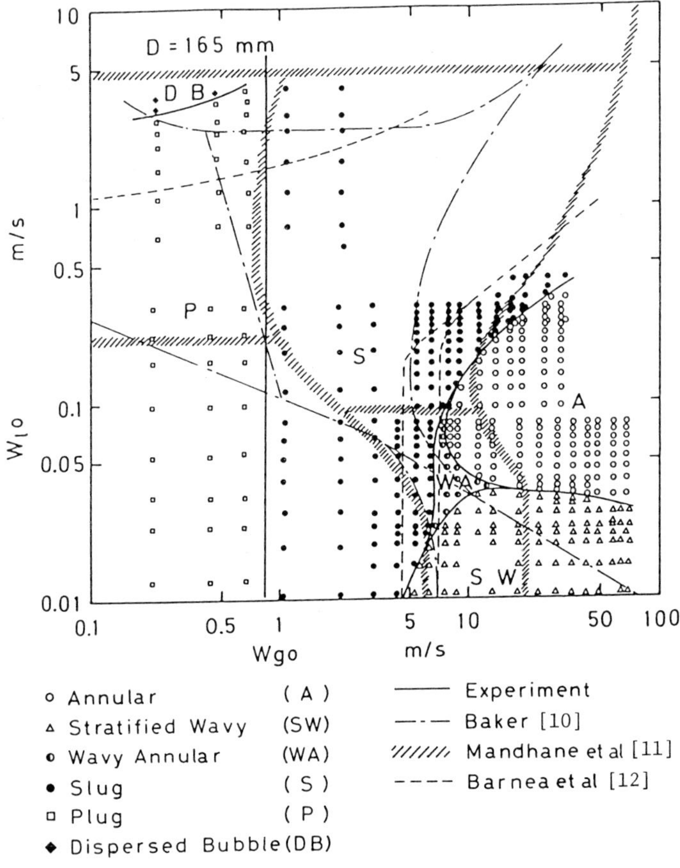

Fig.6. Flow regime map for coiled tube (D = 165 mm)

The boundary between dispersed bubble and plug flows was located at higher liquid velocities compared with the results by Baker and Barnea et al. In the coiled tube, the plug flow was likely to occur because gas bubbles were highly concentrated and coalesced with each other near the inside of the tube due to centrifugal forces. The dispersed bubble - plug flow boundary, however, was not clearly identified in the coiled tube, since gas bubbles were deformed from sphere as shown in Fig.4(a).

The boundary between plug and slug flows agreed well with that of horizontal flow. This boundary was determined based on the probe output signals. The slug - annular flow transition boundary was close to the map of Baker when the liquid velocity was low. As the liquid velocity increased, it became closer to the map of Mandhane et al. This boundary was slightly sensitive to the coil diameter. As the coil diameter became smaller, the slug - annular flow boundary was shifted toward higher gas velocity.

The annular - stratified wavy flow transition boundary was located at different gas and liquid flow rates from the horizontal flow. In the map of Mandhane

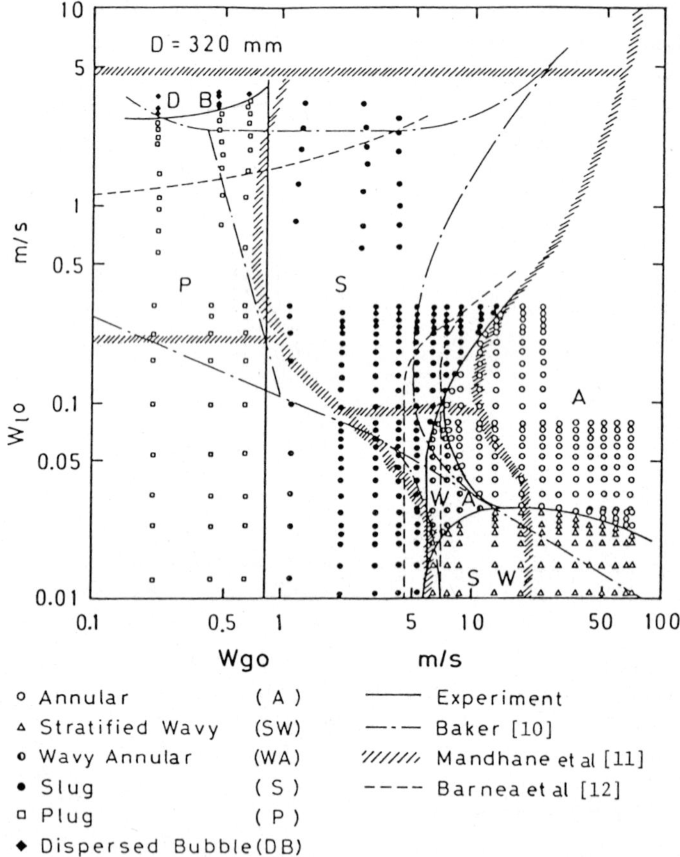

Fig.7. Flow regime map for coiled tube (D = 320 mm)

et al, the annular flow is seen at relatively high gas velocity even if the liquid velocity is extremely low. This boundary was rather close to the map of Baker. The liquid film stratification at the inside part of the coiled tube was likely to occur due to the centrifugal forces. Therefore, this boundary was shifted toward higher liquid velocity as the coil diameter became small. The annular flow regime area of the small coil diameter tube was narrower compared with the large coil diameter one.

The annular - wavy annular flow transition boundary was close to the slug - annular flow boundary of the map of Baker. This coincides with the fact that the probe output traces of the wavy annular flow were similar to those of the slug flow. On the contrary, Barnea et al [12] identified the wavy annular flow for upward inclined tube when the gas velocity was 4 - 7 m/s and the liquid velocity was less than 0.2 m/s. The wavy annular region of the present result partly overlapped with their result. The slug - wavy annular flow boundaries for coiled tubes were located at slightly higher gas velocity than the map of Barnea et al. The wavy annular flow was observed under approximately the same flow rate conditions for different coil diameters.

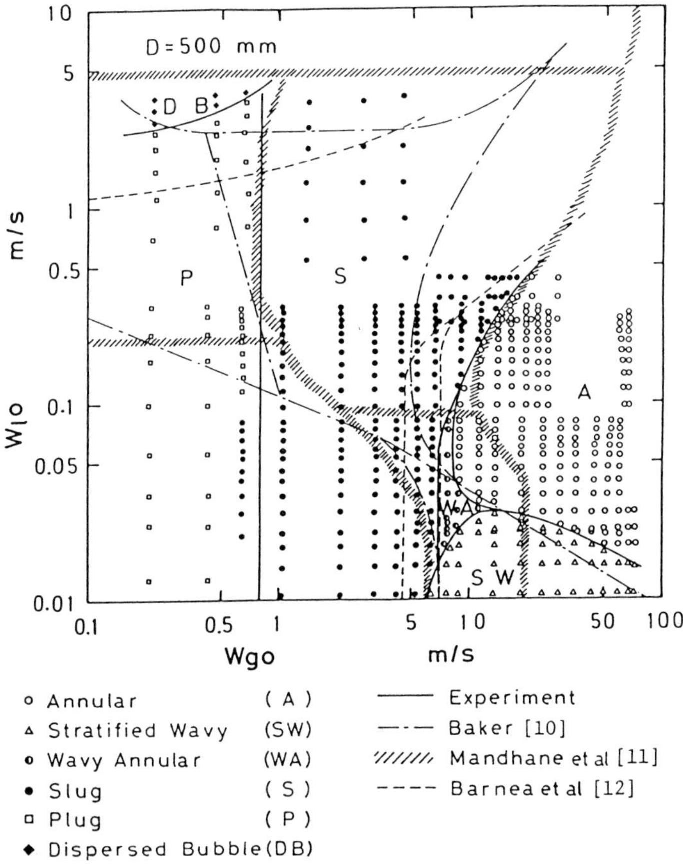

Fig.8. Flow regime map for coiled tube (D = 500 mm)

3.3. Liquid Film Flow Characteristics of Annular Flow

Fig.9 shows liquid film thickness around the tube periphery measured by the needle contact method. Measurement was carried out for the coil diameter D = 320 mm tube and a straight tube. Experimental data are plotted as integral mean values of thickness h_m and data of $D = \infty$ correspond to the results obtained for 8° upward inclined straight tube. The liquid film was most thick at the inside of the tube and thinnest near the top as was pointed out by Whalley [7]. A small peak on the outside wall indicates the deposition of liquid droplets entrained from the liquid film flowing on the inside wall. As the gas velocity increased, liquid film became more thick at the inside. When the liquid velocity was low, partial film breakup was likely to occur at the outside.

Fig.10 shows examples of power spectral densities of conductance probe outputs for various locations around the tube periphery. On the whole, similarities in wavy characteristics of liquid film flow are seen between the top and outside, and the bottom and inside, respectively. When the gas velocity was relatively low, a distinguished peak of wave frequency was detected in Fig.10(c). When the gas velocity was high, on the contrary, the dominant frequency of the wave was not clear. This might be attributed to three dimensional ripple waves induced on the film surface.

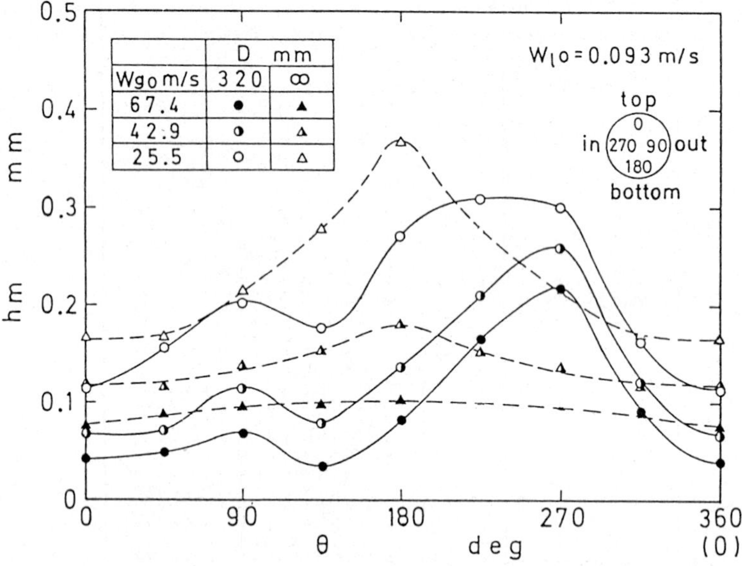

Fig.9. Liquid film thickness around the tube periphery for a coiled tube (D = 320 mm) and a straight tube

4. CONCLUSIONS

Flow regime transitions of air-water two-phase flow in helically coiled tubes were experimentally investigated. Flow regime identification was made by visual observation and an improved conductance probe technique. Flow regime maps for three coiled tubes having different coil diameters were compared with those for horizontal and upward inclined straight tube flows. Results obtained were as follows:

1. Dispersed bubble, plug, slug, annular, stratified wavy and wavy annular flows were observed for each coiled tube. Stratified smooth flow was not realized due to the inclination of the tube. These flow regimes were clearly identified by the conductance probe output traces.

2. The transition boundaries between dispersed bubble, plug and slug flows agreed fairly well with those for the horizontal and upward inclined flows. The annular - stratified wavy flow transition boundary was located at higher liquid velocity than the horizontal flow. The wavy annular flow was observed at approximately same flow rates conditions as those for the upward inclined flow.

3. Liquid film thickness around the tube periphery was measured by needle contact method as well as conductance probe method. As the gas velocity increased, the liquid film became more thick at the inside of the coil. Partial film breakup was likely to occur at the out side. Similarities of wavy characteristics between inside and bottom, and outside and top of the tube, respectively, were observed in the power spectral density functions of the conductance probe outputs.

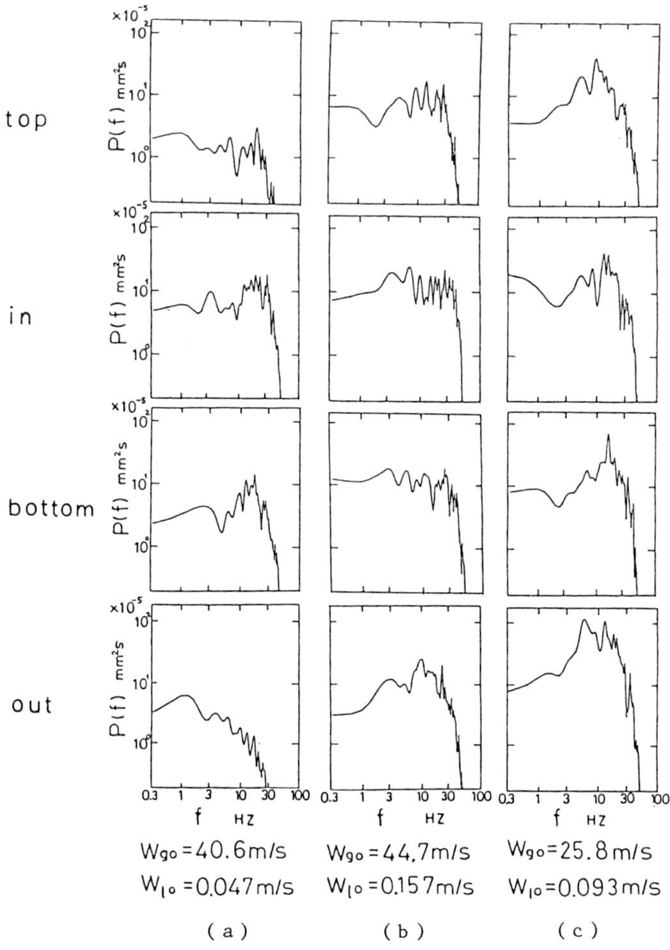

Fig.10. Power spectral density function of waves in liquid film

ACKNOWLEDGEMENT

The authors wish to thank Mr. A. Kishimoto for his assistance in the experiment.

REFERENCES

1. See, for example, Jensen, M.K. and Bergles, A.E., "Critical Heat Flux in Helical Coils with a Circumferential Heat Flux Tilt toward the Outside Surface", Int. J. Heat Mass Transfer, Vol.25, No.9, 1982, pp.1383-1395.

2. See, for example, Anglesea, W.T., Chambers, D.J.B. and Jeffrey, R.C., "Measurement of Water / Steam Pressure Drop in Helical Coils at 179 bars", Symp. Multi-phase Flow Systems, University of Strathclyde, Glasgow, Paper I2, 1974, Published as Instn. Chem. Engrs. Symp. Series, No.38.

3. Kaji, M., Nakanishi, S., Ishigai, S., Kume, M. and Mori, K., "Dryout and Post-Dryout Heat Transfer in Helically Coiled Tubes", Proc., 2nd Int. Topical Meeting on Nuclear Thermalhydraulics, Santa Barbara, California, Vol.1, 1983, pp.271-278.

4. Rippel, G.R., Eidt, Jr.C.M. and Jordan, Jr.H.B., "Two-Phase Flow in a Coiled Tube", Indust. Engng. Chem. Process Design and Development, Vol.5, No.1, 1966, pp.32-39.

5. Akagawa, K., Sakaguchi, T. and Ueda, M., "Study on a Gas-Liquid Two-Phase Flow in Helically Coiled Tubes", Bulletin of the J.S.M.E., Vol.14, No.72, 1971, pp.564-571.

6. Kozeki, M., "Film Thickness and Flow Boiling for Two-Phase Annular Flow in Helically Coiled Tube", Proc. Int. Meeting on Reactor Heat Transfer, Karlsruhe, W.Germanny, Paper No.16, 1973, pp.351-372.

7. Whalley, P.B., "Air-Water Two-Phase Flow in a Helically Coiled Tube", Int. J. Multiphase Flow, Vol.6, No.4, 1980, pp.345-356.

8. Barnerjee, S.,Rhodes, E. and Scott, D.S., "Film Inversion of Coccurent Two-Phase Flow in Helical Coils", AIChE J. Vol.13, No.1, 1967, pp.189-191.

9. Barnea, D., Shoham, O. and Taitel, Y., "Flow Pattern Characterization in Two Phase Flow by Electrical Conductance Probe", Int. J. Multiphase Flow, Vol.6, No.5, 1980, pp.387-397.

10. Baker, O., "Simultaneous Flow of Oil and Gas", Oil and Gas J., Vol.53, 1954, pp.185-190.

11. Mandhane, J.M., Gregory, G.A. and Aziz, K., " A Flow Pattern Map for Gas Liquid Flow in Horizontal Pipes", Int. J. Multiphase Flow, Vol.1, No.4, 1974, pp.537-553.

12. Barnea, D., Shoham, O., Taitel, Y. and Dukler, A.E., "Flow Pattern Transition for Gas-Liquid Flow in Horizontal and Inclined Pipes, Comparison of Experimental Data with Theory", Int. J. Multiphase Flow, Vol.6, No.3, 1980, pp.217-226.

Multi-Phase Flow and Heat Transfer III. Part A: Fundamentals 215
edited by T.N. Veziroğlu and A.E. Bergles
Elsevier Science Publishers B.V., Amsterdam, 1984 — Printed in The Netherlands

FLOW PATTERN TRANSITIONS FOR GAS-LIQUID TWO-PHASE FLOW IN THE BEND OF VERTICAL HAIRPIN TUBES

Ting-Kuan Chen and Zhi-Hua Yang
Department of Power Machinery Engineering
Xian Jiaotong University
Xian, Shaanxi Province, China

ABSTRACT

This paper describes the results of experiments on transitions between two-phase flow patterns during co-current air-water flow in 21.5 mm I.D. vertical hairpin tubes with a 180^{0} bend of radius 694, 500 and 320 mm, respectively, over wide velocity ranges. The transition data are compared to the Mandhane et al.'s and Weisman et al.'s transition correlations for horizontal tubes. An overall flow pattern map based on the present tests is established in terms of superficial gas and liquid velocities V_{SG} and V_{SL}, Correlations for determining flow pattern transitions are suggested.

1. INTRODUCTION

The vertical hairpin tubes have been used in the waste heat boilers in some of our larger synthetic ammonia fertilizer plants with daily output of a thousand tons. Heat is recovered from high temperature technological gas, generating saturated steam up to 180 t/h at 106 bar. This is significant for raising the economic efficiency of the plant. The boiler has advantages of tightly arranged construction, better characteristic of heat expansion of the heating tubes, and good working conditions of the thick tube plate at lower temperature area, etc. However, in operation of the boiler the accident of tube failure in the bend due to overheating occurred sometimes, seriously affecting the safety of production. In addition, hairpin bends are widely used in various heat exchangers.

It is well known that in bends the centrifugal forces have a strong influence on the distribution of the phases, which in turn affect burnout. However, up to now the effect of bends has rarely been investigated, so we intend to do a series of research work to determine the two-phase flow,heat transfer performence. At first, in order to clarify the cause of the tube failure in the waste heat boilers with vertical hairpin tubes, we have carried out a great number of experimental tests about flow pattern transitions for ambient air-water flow in the hairpin bend of vertical tubes, and now we are conducting the investigation of the tube-wall temperature along the bend of tubes under the operating conditions (p=106 bar) in our high pressure electrically heated water loop experimental installation.

When gas-liquid mixtures flow in a tube, the flow phases may distribute in a variety of patterns. Predicting the flow patterns has been an important problem, because a lot of design parameters such as pressure drop and heat and mass transfer are strongly dependent on the flow pattern. Although extensive research on flow patterns has been conducted, most of the research has been concentrated on either horizontal or vertical flow.

The usual approach in research is to visually observe the flow pattern through a transparent test section,and then to map the experimental data by plotting transition boundary lines on a two-dimensional plot. The earliest map for gas-liquid flow was proposed by Baker (1) who plotted superficial gas mass flow against the ratio of the superficial liquid to superficial gas mass flow and introduced correction factors which are functions of the physical properties of the two phases in an attempt to generalize the plot. A recent, widely used flow map was given by Mandhane et al. (2) in terms of superficial velocities of liquid and gas based on a large number of observations of flow patterns. Taitel & Dukler (3) attempted to develop a physical model capable of predicting flow pattern transitions in two-phase flow. In this work the flow pattern boundaries were predicted analytically and every transition boundary can be represented by two dimensionless groups. More recently, Weisman et al. (4) have obtained extensive new data on the effect of fluid properties and tube diameter on flow pattern transitions. They presented revised dimensionless correlations and established an overall flow pattern map in terms of V_{SG} and V_{SL}.

This work reports the results of experiments on flow pattern transitions for air-water flow in the bend of 21.5 mm vertical hairpin tubes. An overall flow pattern map is established in terms of superficial liquid and gas velocities. Correlations for determining flow pattern transitions are suggested.

2. EXPERIMENTAL APPARATUS AND PROCEDURE

The experimental loop is shown schematically in figure 1. Water in tank is pumped by the centrifugal pump through the pressurizer, flow rate adjusting valve and the flow meter to the mixer.

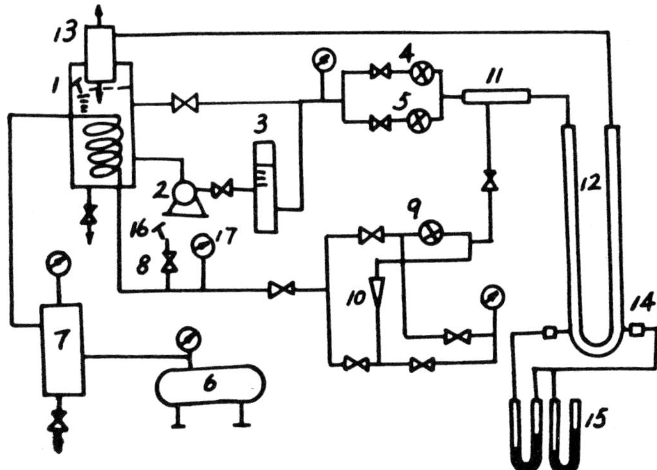

Fig. 1. Schematic Diagram of Experimental Loop
1. water tank 2. water pump 3. pressurizer 4. 10 mm turbine flowmeter 5. 25 mm turbine flowmeter 6. air compressor 7. oil separator 8. flow adjusting valve 9. 15 mm turbine flowmeter 10. 10 mm rotameter 11. air-water mixer 12. test tube 13. air-water separator 14. air-water separation bottle 15. U-tube manometer 16. thermometer 17. manometer

Compressed air from the compressor is delivered through an oil separator, cooled in the water tank, and then passed through the flow adjusting valve and the flow meter into the mixer. The type of mixer is that air is introduced circumferentially into water from a inner tube through uniformly distributed small holes. The mixer can mix air and water uniformly and eliminate pulsations. The air-water mixtures from the mixer then flows through the test section to the cyclone air-water separator. The water returns to the tank.

The test sections were made of plexiglass tubing with inner diameter of 21.5 mm, straight section height of 2 m. The radii of the bends were 694, 500 and 320 mm, respectively. The flow rates in the experiments ranged from V_{SL}= 0.04 m/s to 4.3 m/s, and from V_{SG}= 0.2 m/s to 22 m/s. The water flow rate was measured by 10 mm, and 25 mm turbine flowmeters. The air flow rate was measured by 10 mm rotameter and 15 mm turbine flowmeter. The pressure was measured by standard manometer. The temperature was measured by mercury-in-glass thermometer. The differential pressure of the bend and the static pressure at the exit of the bend were measured by U-tube manometers. Air-water separation bottles were fitted in each tapped line to prevent air entry into the connecting pipe and eliminate pressure pulses.

The flow pattern in the test section was determined by visual observation. Careful preliminary runs were made for horizontal flow in order to define and classify the flow pattern. Experiments were carried out with vertical hairpin tubes for radii of 694, 500 and 320 mm, respectively. In the tests observations were taken at the middle of the bend, i.e. at the lowest bottom of the tube. In order to map out the flow pattern, it was necessary to find the transition flows from one pattern to another. This was done by running at a series of water (or air) flow rates, and at each water (air) flow varying the air (water) flow until the pattern changed. Obviously, a change of pattern generally occured over a range of conditions and the observations were to some extent subject to different interpretation with different operators. Nevertheless the resulting scatter about the transition boundaries was small.

3. THE DEFINITION OF FLOW PATTERNS

The designation of flow patterns has been based largely on individual interpretation of visual observation. Hence, it is important to specify as precisely as possible features of the flow used to characterize the pattern designed. According to observations, the basic flow patterns in the hairpin bend of vertical tubes are similar to those in horizontal straight tubes. Of course, due to the effect of centrifugal forces there are also some differences between them. The following definition was used in reporting observations. Figure 2 shows in diagrammatic form the various flow patterns.

Stratified flow (s): In this case, Due to both gravitational and centrifugal forces acting on the fluid, the liquid flows in the outside of the bend with gas in inner side. The interface between them is distinct.

Wavy flow (W): The liquid and gas are separated as above, but the interface is wavy.

Elongated bubble flow (EB): Characteristic bullet-shaped bubbles located adjacent to the inner surface of the bend move

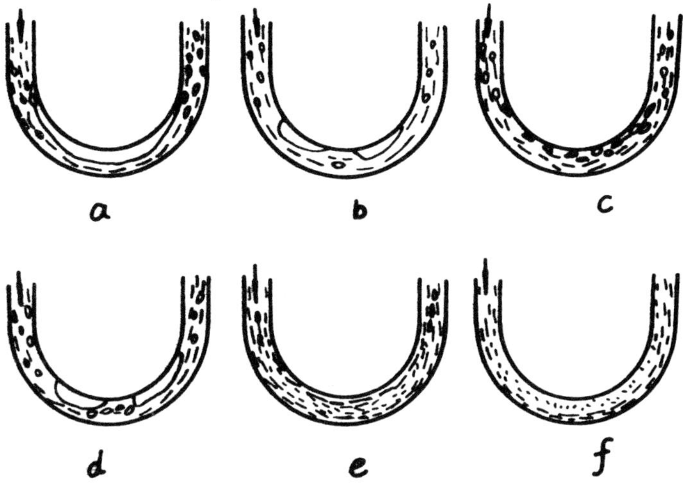

Fig. 2. Flow Patterns in Bend of Vertical Hairpin Tubes
a. stratified flow b. elongated bubble flow c. bubble flow
d. slug flow e. annular flow f. dispersed bubble flow

downstream. They are separated by sections of continuous liquid. Meanwhile, there is a stratified liquid layer flowing along the outside of the bend.

Bubble flow (B): The gas bubbles move through liquid at inner side of the bend.

Slug flow (S): Large frothy slugs of liquid which fill the tube are separated by gas zones which contain a stratified liquid layer flowing along the outside of the bend. The slug may be aerated with distributed small bubbles which are concentrated toward the inner side of the bend.

Annular flow (A): Annular flow occurs at high gas flow rates. The liquid flows as a film around the tube wall. The gas which may contain entrained liquid droplets flows in the centre. The film in the outside of the bend is much thicker than the film in the inner side owing to gravitational effects giving drainage around the periphery.

Dispersed bubble flow (DB): At high liquid flow rates the gas is dispersed in the form of small bubbles within a continuous liquid phase. Because of the centrifugal effects, the bubble density at inner side of the bend will be much higher than that at outer side.

Visual criteria were established for each of the flow patterns. Annular flow was distinguished from wavy or slug flow by the existence of a continuous liquid film around the entire tube circumference. Dispersed bubble flow described the essentially homogeneous froth seen at high liquid flow rates but lower gas flows. The flow pattern may be considered as dispersed once small bubbles are observed in a continuous liquid and the bubbles are distributed more or less uniformly. Slug and bubble flow was distinguished from dispersed or annular flow by the existence of gas volumes which remain at inner side of the bend. Distinctive periodic pressure pulsations may be seen in slug flow. The stratified flow and elongated bubble flow is quite sharp and easy to identify.

4. EXPERIMENTAL RESULTS

The experimental results of the flow pattern transitions for air-water system in three different radiuses of bend 694, 500 and 320 mm of 21.5 mm I.D. vertical hairpin tube are illustrated in figure 3 in terms of superficial liquid and gas velocities V_{SL} and V_{SG} and compared to the Mandhane et al.'s and Weisman et al.'s transitions for horizontal tubes as shown in figure 4. In the maps,the shaded areas indicate the regions of transition between flow patterns. The individual symbols express the observed point at which the flow pattern transition occurred. The broken lines represent the Mandhane et al.'s transition lines and the dot-and-dash lines are Weisman et al.'s correlations for horizontal flow.

The flow patterns which exist in bend of vertical hairpin tubes are similar to that in horizontal straight tube, i.e. the stratified flow , wavy flow, elongated bubble flow, bubble flow, slug flow, annular flow and dispersed bubble flow were observed in tests. The major changes from the horizontal flow are the expansion of stratified flow and restriction of the wavy flow and intermittent flow regions. Both the annular and dispersed transitions are quite similar to those in horizontal tubes. The effects of the bend radius on flow pattern transitions are small in the range of tests.

The transition to dispersed flow occurs at a nearly constant liquid velocity about V_{SL}= 4.2 m/s which is higher than that in Weisman et al.'s value but agree well with Mandhane et al.'s line. Annular flow occurs at high gas flow rates. The required gas flow for the transition increases with decreasing liquid flow. This is in generally good agreement with the Weisman et al.'s correlation. It should be noted that the slug-annular transition presented by Mandhane et al. shows a decidedly different slope.

The major effect of bend on the flow patterns is the considerable expansion of the stratified flow region. In horizontal tubes, the transition to intermittent flow takes place at the liquid velocity of about 0.2 m/s, but in the present tests of bends the liquid velocity for transition to bubble flow is up to more than 1.5 m/s. In stratified region there is a small elongated bubble area, The bullet-shaped bubbles may be seen at inner side of the bend, but it disappear soon with increasing liquid flow. Because bends can suppress waves, the wavy flow pattern shrinks to a small region. The transition between stratified and wavy flow occurs at much

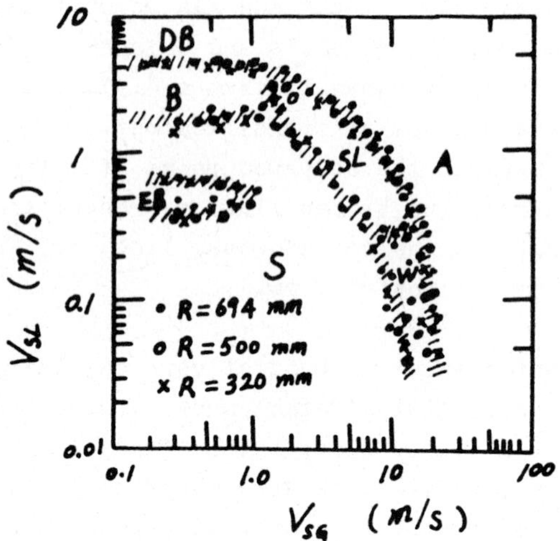

Fig.3. Flow Pattern Map for the Bend of Vertical Hairpin Tubes

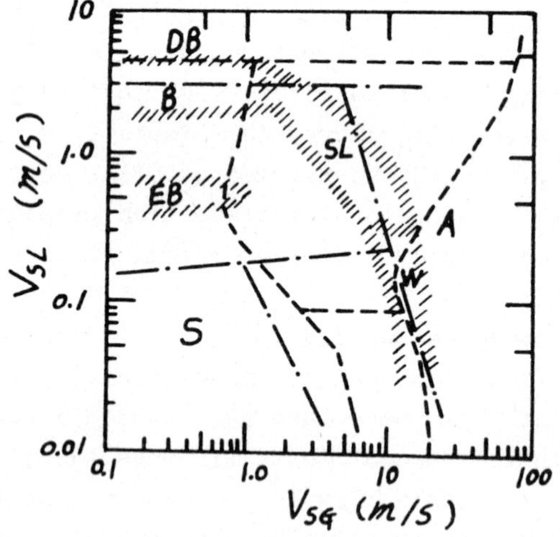

Fig. 4. Comparision of Transitions in Hairpin Bends with Flow Pattern Map for Horizontal Tubes

//////// experimental transition in the present work

------ Mandhane et al.'s transition lines

—·— Weisman et al.'s correlations

higher gas flow rates than that in horizontal tubes. Accordingly, the slug flow also shrinks to a smaller range than that in horizontal flow. At low gas flow rates, the bubble flow was observed. In this region, the bubbles are not dispersed throughout the liquid but are found at the top of the bend.

5. ANALYSIS OF RESULTS

Gas-liquid two-phase flow is an extremely complex phenomenon. The bends make the flow more complex because of adding the effects of centrifugal force and secondary flow. Centrifugal forces acting on the fluid may be expressed by $\rho V^2/R$. As the density of the liquid is usually considerably larger than that of the gas, the liquid would be forced to the outer side of the bend which assists gravity in stratifying the flow. As a result, the stratified flow region is much greater than that in horizontal tubes. However, the centrifugal forces acting on gas increase with increasing gas velocity. In annular flow owing to the presence of greater slip velocity between the gas and liquid phases, the gas velocity is much greater than the liquid. Consequently, it may be considered that in annular flow the centrifugal forces acting on both phases are quite near, and therefore the transition to annular flow is similar to that in horizontal flow. In addition, in a curved tube the centrifugal force of a flowing fluid produces a pressure gradient in a cross section. This pressure gradient yields a pair of secondary flows which forms a couple of vortexes to make the fluid near the wall flow along the wall from outside to inner side of the tube. This is quite favorable in annular flow for maintaining the film at the top of the bend.

According to the present tests, the basic flow patterns are similar to those in horizontal tubes. In order to simplify the flow pattern classification, the flow may be considered to be described by the annular, dispersed, intermittent and separated flow patterns. The intermittent flow includes slug and bubble flow. Separated includes stratified and wavy flow. The elongated flow region, which also belongs to intermittent, is very small and exists in the stratified region, so it is unimportant. Thus, with this simplification, the flow pattern transitions of major interest are: transition to annular flow, transition to dispersed flow and the separated- intermittent transition.

Transition to annular flow. Annular flow can be visualized as developing from slug flow when the aeration in the slug becomes sufficiently high to form a continuous gas phase. Many machanisms have been proposed for maintaining a continuous liquid film along the tube. Although so far there is insufficient evidence to show which of the mechanisms is the most important during a given set of flow conditions, it may be considered that transition to annular flow is dependent essentially on gas inertial forces to overcome the liquid gravity. Hence, the modified Froude number

$$F = \left(\frac{\rho_G}{\rho_L-\rho_G}\right)^{0.5}\frac{V_{SG}}{(gD)^{0.5}} \qquad (1)$$

which characterizes this effect is usually used to correlate the test data for transition to annular flow, or the Kutateladze number

$$Ku = \frac{\rho_G^{0.5}V_{SG}}{[(\rho_L-\rho_G)g\sigma]^{0.25}} \qquad (2)$$

is used to further consider the effect of surface tension.

Weisman et al. obtained a better correlation by correlating the transition in terms of the Froude number, Fr, and the Kutateladze number, Ku, as follows

$$Ku^{0.2}Fr^{0.18} = 1.9(V_{SG}/V_{SL})^{0.125} \qquad (3)$$

where

$$Fr = V_{SG}^2/(gD) \qquad (4)$$

The Weisman et al.'s recent work (5) indicated that the annular flow transition is affected only very slightly by the angle of inclination. The correlation of (3) may be used not only for horizontal flow, but also for vertical and upwardly inclined flow.

For this reason, in the present work we also used the Froude number and Kutateladze number to correlate the data as shown in figure 5. and obtained the correlation as follows

$$Ku^{0.2}Fr^{0.18} = 2(V_{SG}/V_{SL})^{0.2} \qquad (5)$$

or it may be simplified by using the same exponent for both Fr and Ku, as follows

$$KuFr = 40(V_{SG}/V_{SL})^{1.2} \qquad (6)$$

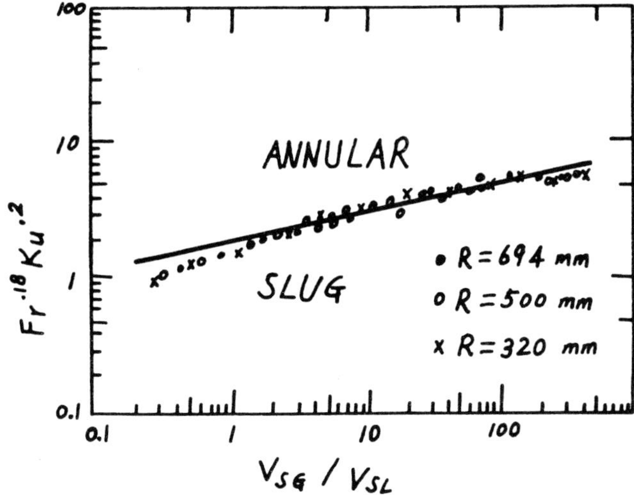

Fig. 5. Correlation to Transition to Annular Flow

Transition to dispersed flow. Taitel & Dukler have suggested that the transition to dispersed flow in horizontal tube takes place when the turbulent fluctuations are strong enough to overcome the buoyant forces tending to keep the gas at the top of the tube. The turbulent forces may be related to the pressure drop and hence they proposed that the data may be correlated in terms of a dimensionless ratio, T, as follows

$$T = \left[\frac{|(dp/dx)_{SL}|}{(\rho_L - \rho_G)g} \right]^{1/2} \qquad (7)$$

Weisman et al. further used the dimensionless group $[\sigma/(\rho_L-\rho_G)gD^2]^{-1/4}$ to correct for surface tension and diameter effect. They obtained the correlation

$$\left[\frac{|(dp/dx)_{SL}|}{(\rho_L - \rho_G)g} \right]^{1/2} \left[\frac{\sigma}{(\rho_L - \rho_G)gD^2} \right]^{-1/4} = \text{const} \qquad (8)$$

At high liquid rates it is to be expected that inertial forces will greatly exceed gravitational forces and hence the orientation of the test section should have little effect. According to the Weisman et al.'s tests in vertical flow, dispersed flow appears to be in approximately the same positions as seen in horizontal flow. So the correlation (8) also may be used for predicting the dispersed transition in vertical and inclined flow.

Thus, it is considered that the Weisman et al.'s dimensionless groups may also be used to correlate the data in the present work. The final correlation obtained is

$$\left[\frac{\left|(dp/dx)_{SL}\right|}{(\rho_L-\rho_G)g}\right]^{1/2}\left[\frac{\sigma}{(\rho_L-\rho_G)gD^2}\right]^{-1/4} = 2.5 \qquad (9)$$

where

$$\left(\frac{dp}{dx}\right)_{SL} = \frac{2f}{D}\rho V_{SL}^2 \qquad (10)$$

with friction factor, f, which can be predicted by the Blasius equation, namely

$$f = C\left(\frac{\rho_L D V_{SL}}{\mu_L}\right)^{-n} \qquad (11)$$

where C and n are taken as 0.046 and 0.2, respectively.

Transition between separated and intermittent flow. In stratified flow, as the liquid or gas rate is increased, waves are formed at the surface of the liquid. When the waves grow big enough to touch the upper surface of the tube, the flow pattern change occurs from separated to intermittent flow. The onset of the transition is due to the inertial gas-phase forces, which tend to create the wave, overcoming the gravitational forces, which tend to collapse it. Taitel & Dukler used modified Froude number, F, to correlate the data and obtained satisfactory results. Weisman et al. have got a better correlation in terms of the gas $(Fr)^{1/2}$ vs. V_{SG}/V_{SL} for horizontal flow. that is

$$\frac{V_{SG}}{(gD)^{1/2}} = 0.25(V_{SG}/V_{SL})^{1.1} \qquad (12)$$

Based on Weisman et al.'s dimensionless groups, the data obtained in present work were correlated as shown in figure 6 for slug flow and figure 7 for bubble flow. The final correlations proposed are

for slug flow

$$\frac{V_{SG}}{(gD)^{1/2}} = 3.8(V_{SG}/V_{SL})^{0.55} \qquad (13)$$

for bubble flow

$$V_{SG}/(gD)^{1/2} = 4.2\,(V_{SG}/V_{SL}) \qquad (14)$$

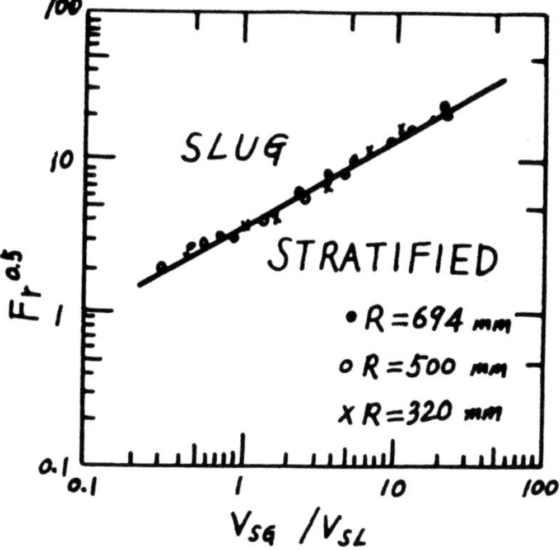

Fig. 6. Correlation for Transition to Slug Flow

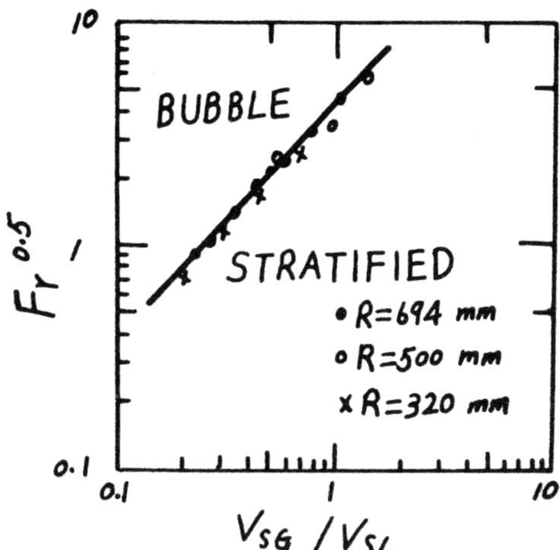

Fig. 7. Correlation for Transition to Bubble Flow

6. OVERALL FLOW MAP

The correlation proposed in (5), (9), (13) and (14) may be readily used to construct an overall flow map for a given fluid and tube diameter. This has been done for the ambient air-water system in a 21.5 mm tube bend in terms of V_{SL} and V_{SG}, as shown in figure 8. For other fluids and diameters, the values of ϕ_1 and ϕ_2, which are derived from the proposed correlations, are obtained in accordance with table 1. For dispersed flow the parameter exponents are corrected according to Weisman et al. (4).

It may be seen from the flow pattern map of figure 3 presented in this paper that transition from one pattern to another generally takes place over a range of conditions. Consequently, the flow pattern map of figure 8 attempts to take this behavior into account by indicating the transitions as shaded regiens of approprite width.

Table 1. Fluid property and tube diameter correlations to overall flow map

Flow pattern	ϕ_1	ϕ_2
Transition to dispersed flow	$(\frac{\rho_L}{\rho_{SL}})^{-0.33}(\frac{D}{D_S})^{0.16}(\frac{\mu_{SL}}{\mu_L})^{0.09}(\frac{\sigma}{\sigma_S})^{0.24}$	1.0
Transition to annular flow	1.0	$(\frac{\rho_{SG}}{\rho_G})^{0.28}(\frac{\Delta\rho}{\Delta\rho_S})^{0.14}(\frac{\sigma}{\sigma_S})^{0.14}(\frac{D}{D_S})^{0.5}$
Separated-slug transition	$(\frac{D}{D_S})^{0.91}$	1.0
Separated-bubble transition	$(\frac{D}{D_S})^{0.5}$	1.0

s denotes standard conditions, D_S= 21.5 mm, ρ_{SG}= 1.3 kg/m^3, ρ_{SL}= 1000 kg/m^3, μ_{SL}= 0.001 Ns/m^2, σ_S=0.07 N/m, $\Delta\rho = \rho_L - \rho_G$.

7. CONCLUSIONS

Experiments have been carried out to determine the flow pattern transitions for the flow of air-water mixtures in the bend of vertical hairpin tubes 21.5 mm internal diameter with radius 694, 500

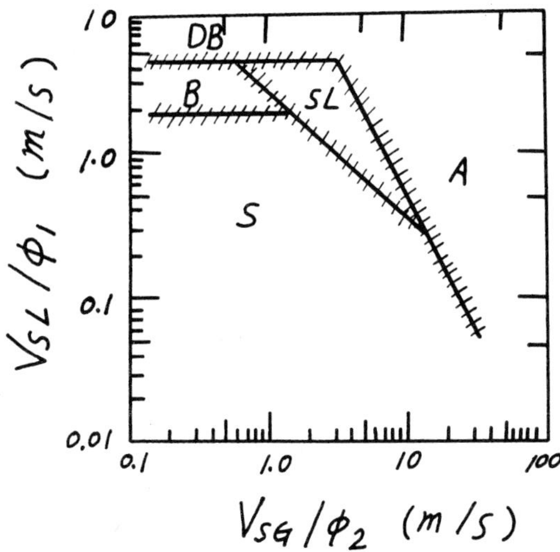

Fig. 8. Overall Flow Pattern Map

and 320 mm, respectively, over wide velocity ranges. The data obtained in the present work have provide a clear picture of flow pattern transitions in the vertical hairpin bends. The flow patterns observed in these bends are similar to those seen in horizontal two-phase flow. The data obtained show that the changes from the horizontal flow pattern map are the expansion of stratified flow and restriction of the wavy and intermittent flow regions. Both annular and dispersed transitions are quite near to those in horizontal tubes. The effects of the bend radius on flow pattern transitions are small in the range of tests.

Four basic flow pattern transitions, i.e. transition to annular, dispersed, slug and bubble flow are considered. Dimensionless correlations which agree with the data in the present work have been suggested in equations (5), (9), (13) and (14). Based on these correlations, an overall flow pattern map has been constructed in figure 8. For air-water at room conditions in 21.5 mm I.D. tube bend, the corrections ϕ_1 and ϕ_2 are equal 1.0. For other fluids and tube diameters, the ϕ_1 and ϕ_2 are obtained in accordance with table 1.

NOMENCLATURE

D	Tube diameter, m
dp/dx	Pressure drop per unit length, N/m^3
F	Modified Froude number
f	Friction factor
Fr	Froude number
g	Acceleration of gravity, m/s^2
Ku	Kutateladze number
p	Pressure, bar
R	Bend radius, m
V	Velocity, m/s
$\Delta\rho$	$(\rho_L - \rho_G)$
μ	Viscosity, Ns/m^2
ρ	Density, kg/m^3
σ	Surface tension, N/m

SUBSCRIPTS

G	Gas
L	Liquid
SG	Superficial for gas flow alone in the tube
SL	Superficial for liquid flow alone in the tube

REFERENCE

1. Baker, O. 1954 Design of pipeline for simultaneous flow of oil and gas. Oil and Gas J. 53(12), 185-195.
2. Mandhane, J. M., Gregory , G. A. and Aziz, K. 1974 A flow pattern map for gas liquid flow in horizontal pipes. Int. J. Multiphase Flow 1, 537-554.
3. Taitel, Y. & Dukler, A. E. 1976 A Model for predicting flow regime transition in horizontal and near horizontal gas liquid flow. AIChE J. 22, 47-55
4. Weisman, J. Duncan, D. Gibson, J. and Crawford, T. 1979 Effects of fluid properties and pipe diameter on two-phase flow patterns in horizontal lines. Int. J. Multiphase Flow 5. 437-462.
5. Weisman, J. & Kang, S. Y. 1981 Flow pattern transitions in vertical and upward inclined lines. Int. J. Multiphase Flow 7. 271-291.

Multi-Phase Flow and Heat Transfer III. Part A: Fundamentals
edited by T.N. Veziroğlu and A.E. Bergles
Elsevier Science Publishers B.V., Amsterdam, 1984 — Printed in The Netherlands

TWO DIMENSIONAL MODELLING OF ADIABATIC ANNULAR TWO PHASE FLOW

R.B.Grover
Reactor Engineering Division
Bhabha Atomic Research Centre
Bombay 400085, INDIA

ABSTRACT

A two dimensional model for the analysis of adiabatic annular two phase flow through tubular geometry based on the solution of continuity and momentum equations is presented here. The numerical method used is an adaptation of Patankar-Spalding method. For the evaluation of Reynold's stresses a mixing length model has been postulated for annular two phase flow. The model has been validated against available experimental data from air-water and steam-water systems.

1. INTRODUCTION

Many attempts have been made in the past to develop mathematical models for the study of annular two-phase flow, because of its importance in equipment involving two-phase flow. Empirical correlations and models which have been developed to predict various features of annular two phase flow are reviewed by Hewitt & Hall Taylor [1], Hewitt [2], Govier & Aziz [3] etc. All the mathematical models developed so far are one-dimensional and are based on a combination of empirical correlations. In order to reduce the dependence on empirical correlations, a model has been developed for adiabatic annular two phase flow, which is based on the solution of two dimensional continuity and momentum equations for tubular geometry. Assumptions used in the development of this model are as follows:

i) The flow is steady and incompressible.
ii) Liquid travels on the pipe wall as a thin film and in the pipe core as a droplet spray.
iii) Gas/vapour travels in the core.
iv) The flow field is circumferentially symmetrical.
v) There is a continual interchange of the liquid between the core fluid and the liquid film.
vi) Core fluid is a homogeneous mixture of liquid droplets and vapour/gas.

Order of magnitude analysis of the basic equations indicated that thin shear layer approximations are valid for this problem [4]. Thin shear layer equations can be solved by a variety of numerical methods. Commonly used and well documented methods are due to Patankar and Spalding [5], Keller and Cebeci [6] and Nelson and Pletcher [7]. All these schemes were examined to judge their suitability for solving the present problem which has the following unique characteristics.

i) Presence of an interface between the two phases, normal to which there may be a mass flux.
ii) Changes in axial mass flux of the two phases and the flow area occupied by the two phases as one proceeds along the axis of the tube in the flow direction.

Of the three methods examined, Patankar-Spalding method was found to be most suitable for the present problem. This is because it uses dimensionless stream-function as the cross-stream variable, numerical value of which does not change with changes in mass flux. To improve its predictions for internal flows through circular cross-sections, a few modificationsto the original scheme were made [8, 9]. Based on the modified numerical scheme, a program ANFLO for the analysis of annular two phase flow has been written. The following sections outline the basic equations and the solution procedure used in this analysis.

2. BASIC EQUATIONS AND SOLUTION PROCEDURE

Fig.1 gives a schematic of annular two phase flow used in the development of this model. For the liquid film and the core

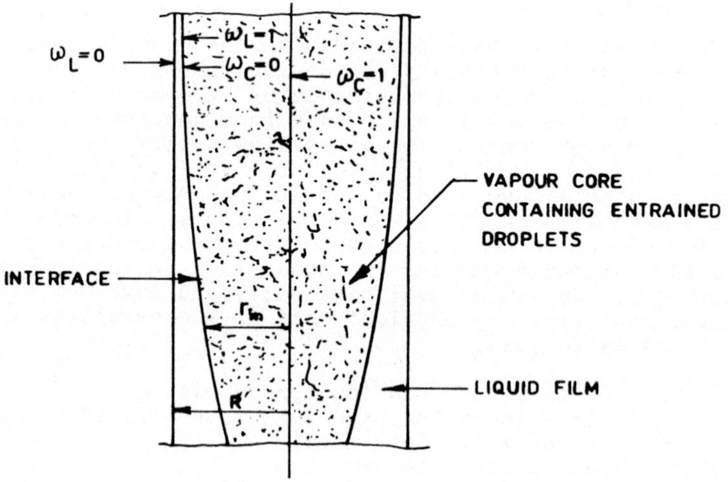

FIG.1. A SCHEMATIC OF ANNULAR TWO PHASE FLOW

fluid, the momentum equations in the von Mises coordinates are,

$$\rho_L U \frac{\partial U}{\partial Z} = \rho_L U \frac{\partial}{\partial \psi} (\tau r) - \frac{dP}{dZ} - \rho_L g , \qquad \ldots(1)$$

$$\rho_C U \frac{\partial U}{\partial Z} = \rho_C U \frac{\partial}{\partial \psi} (\tau r) - \frac{dP}{dZ} - \rho_C g . \qquad \ldots(2)$$

where τ is defined as follows,

$$\tau = \mu_{eff} \frac{\partial U}{\partial r} . \qquad \ldots(3)$$

Density of the core fluid ρ_C is given by the following expression:

$$\rho_C = \rho_L + (1 - \beta) \rho_V , \qquad \ldots(4)$$

where $\beta = C/\rho_L$ . ...(5)

The concentration C is calculated on the assumption that the core fluid is a homogeneous mixture of the vapour and the liquid droplets and there is no slip between the vapour and liquid droplets. This gives,

$$\frac{W_V}{(1 - \beta)\rho_V} = \frac{W_{LE}}{\rho_L} . \qquad \ldots(6)$$

$$\text{Therefore } \beta = \frac{W_{LE}\rho_V}{W_V\rho_L + W_{LE}\rho_V} . \qquad \ldots(7)$$

If m_{LE} is the total liquid mass flux at the interface, then,

$$\Delta W_{LE} = 2\pi m_{LE} r \Delta Z . \qquad \ldots(8)$$

$$\text{Therefore } W_{LE,d} = W_{LE,u} + \Delta W_{LE} . \qquad \ldots(9)$$

Liquid mass flux at the interface, m_{LE} is given by the following relation,

$$m_{LE} = E - D , \qquad \ldots(10)$$

where E is the entrainment flux, while D is the deposition flux. Deposition flux D at any axial location is calculated from the relation,

$$D = kC . \qquad \ldots(11)$$

For calculating E, an improved version of Whalley and Hutchinson correlation was developed, which is given later. Boundary conditions are:

$$U\,(Z,R) = 0\,, \qquad \ldots(12)$$

and $$\frac{\partial U\,(Z,0)}{\partial r} = 0\,. \qquad \ldots(13)$$

At the interface between the liquid film and the core fluid, the following conditions are to be satisfied.

$$\text{i)}\ U_L \Big|_{in} = U_C \Big|_{in}\,, \qquad \ldots(14)$$

$$\text{ii)}\ \mu_{eff} \frac{\partial U}{\partial r}\Big|_{L,in} = \mu_{eff} \frac{\partial U}{\partial r}\Big|_{C,in} \qquad \ldots(15)$$

The momentum and the stagnation enthalpy equations are to be solved using the boundary conditions and the interfacial conditions. To do so first the basic equations are expressed in Patankar-Spalding co-ordinate system using the following transformations:

$$\omega_L = \frac{\psi(Z,r) - \psi(Z,R)}{\psi(Z,r_{in}) - \psi(Z,R)}\,, \qquad \ldots(16)$$

$$\omega_C = \frac{\psi(Z,r) - \psi(Z,r_{in})}{\psi(Z,0) - \psi(Z,r_{in})}\,. \qquad \ldots(17)$$

Transformed equations are converted to finite difference equations. Momentum equations (1) and (2) and interfacial conditions (14) and (15), yield equations of the type

$$U_i = A\, U_{i+1} + B\, U_{i-1} + C + E\,\frac{dP}{dZ}\,. \qquad \ldots(18)$$

Here it may be noted that for the node at the interface coefficients C and E are both equal to zero. Using the definition of the stream function and the transformations given by equations (16) and (17), one gets another relation as follows:

$$\left[\psi(Z,r_{in}) - \psi(Z,R)\right]\int_0^1 (\rho_L U)^{-1} d\omega_L + \left[\psi(Z,0) - \psi(Z,r_{in})\right]$$

$$\int_0^1 (\rho_C U)^{-1} d\omega_C = -\frac{R^2}{2}\,. \qquad \ldots(19)$$

Equation (18) and discretised form of equation (19) are solved together using the boundary conditions (12) and (13) to obtain velocity profiles and the pressure gradient. Solution procedure is an adaptation of Newton-Raphson method [8, 9].

3. TURBULENCE MODEL

3.1 Formulation

In single phase flow in tubes the velocity profile in the region of axis can be correlated for both smooth and rough walls by the velocity deficiency law,

$$\frac{U_{max} - U}{U^*} = \frac{1}{K} \ln\left(\frac{R}{R - r}\right) , \qquad \ldots(20)$$

where U_{max}, U and U* are the maximum, local and friction velocities respectively and K is von Karman universal constant. Gill et al. [10] conducted a series of experiments on adiabatic air-water flows in tubular geometry to determine flow distribution and velocity profiles. The values of K were calculated from this data. It was observed that for a given air flow rate, the value of K falls rapidly with increase in liquid flow rate, although the logarithmic form of the velocity deficiency law (eqn.20) is retained. Data collected at CISE laboratory has been analysed by Alia et al. [11] and they have obtained a value of $K \sim 0.27$ for argon-water mixtures and ~ 0.31 for argon-alcohol mixtures. Change in K indicates a change in the level of turbulence and one can conjecture the following possible reasons for this change.

i) Presence of entrained droplets in the core which influences the velocity profiles
   a) by changing the density of the core fluid and
   b) by suppressing the turbulent eddies, thus reducing the mixing length and hence K.
ii) Presence of the liquid film in the tube influences the velocity profiles
   a) by reducing the area available for the flow of the core fluid and
   b) by presenting a rough surface due to the presence of the disturbance waves at the interface.

In any turbulence model intended for the calculation of annular two phase flow all these factors have to be suitably accounted for. In the present work an attempt has been made in this direction. The liquid film can be compared to ordinary boundary layer flow and turbulence model can be accordingly chosen. The turbulence model chosen is given hereafter.

$$\text{For } y^+ < 5 , \mu_{eff,LF} = \mu_L . \qquad \ldots(21)$$

$$\text{For } y^+ > 5 , \mu_{eff,LF} = \mu_L + \rho_L\, l^2 \left|\frac{\partial U}{\partial y}\right| , \qquad \ldots(22)$$

where l is given by smaller of the following two expressions,

$$l = 0.089\ m , \qquad \ldots(23)$$

and $l = 0.4y \left[1 - \exp(-y^+/26)\right]$ . ...(24)

A few other expressions were also tried for calculating turbulent viscosity. It was found that overall results such as pressure drop, film thickness etc. predicted by the calculations are insensitive to such changes.

For the fluid core turbulence model was chosen based on intuitive reasoning and analysis of available experimental data regarding film thickness and pressure drop. For fully developed pipe flows, mixing length profile is given by Nikuradse equation. For developing flows, equation of the type l = constant X δ, where δ is the boundary layer thickness, is more appropriate. Annular two phase flows analysed here are closer to fully developed flow. Therefore, it was thought appropriate to use Nikuradse mixing length profile as the basic profile and modify it to suit the requirements of the two phase flow. The chosen model is given hereafter.

$$\mu_{eff,C} = \mu_V + \rho_m l^2 \frac{\partial U}{\partial y} , \qquad ...(25)$$

$$\text{where } \rho_m = \rho_V \left[1 - \exp(-\rho_C/\rho_V + 1)\right] + \rho_V , \qquad ...(26)$$

$$l = \eta l_N , \qquad ...(27)$$

$$l_N = R_{eff} \left[0.14 - 0.08 \left(1 - \frac{y}{R_{eff}}\right)^2 - 0.06 \left(1 - \frac{y}{R_{eff}}\right)^4\right], \qquad ...(28)$$

$$\eta = G_C/(G_{LF} + G_C) , \qquad ...(29)$$

$$R_{eff} = r_{in} + m_{eff} . \qquad ...(30)$$

When the liquid film thickness is less than 0.000150 meters,

$$m_{eff} = m(\rho_C/\rho_L) . \qquad ...(31)$$

When it is greater than 0.000150 meters,

$$m_{eff} = m\rho_C/\rho_L + m(1 - \rho_C/\rho_L) \left[1 - \exp\left(-\frac{m + 0.000150}{0.000150}\right)\right]. \qquad ...(32)$$

Transition from low film thickness to high film thickness was fixed at 0.000150m. This was based on analysis of experimental data. However, no optimization was done for this purpose. Further, equations (27) to (32) are so chosen that if $G_{LF} = 0$, and $G_{LE} = 0$, mixing length profile is given by Nikuradse equation.

3.2 Validation

To solve the basic equationsgiven in section 2 using the turbulence model given in section 3.1, a computer program ANFLO was written. With a view to validate the turbulence model this program was used to analyse adiabatic annular two phase flows for cases corresponding to which experimental data is available. Data on air-water and air-Genklene flows generated at Harwell by Whalley et al. [12] was chosen for this purpose. Experimentally determined entrained liquid fraction along with assumed velocity profiles were given as input to the programme ANFLO. As the intention was to verify turbulence model only, amount of liquid entrained in the core was held constant (equal to the experimental value). Computations were terminated when parameters like pressure gradient, film thickness, void fraction etc. showed no change with further forward marching. Computed film thickness and pressure gradient are plotted against corresponding experimentally determined values in fig.2 and 3. It can be seen from these figures that experimental and calculated values are in reasonable agreement.

To further check the validity of the turbulence model predicted velocity profiles were compared with the data of Gill et al. [10]. Fig.4 gives these velocity profiles. Gill et al. integrated the linear velocity profiles to yield the total apparent flow of the gas in the core. They have reported discrepancies between the input and the integrated rates. Discrepancy was defined as follows:

$$\text{Discrepancy } \% = 100 \times \frac{\text{Integrated rate - Input rate}}{\text{Input rate}}$$

Discrepancy values are also reported in the figure. Considering the discrepancy values, the predictions are in reasonably good agreement with the data.

4. ENTRAINMENT CORRELATION

After having developed and validated the turbulence model, next step was to develop a correlation for entrainment. In this regard Hutchinson and Whalley [13] have identified the dimensionless interfacial shear stress S ($= \tau_{in} m/\sigma$) as the important dimensionless group characterising the process of entrainment. They plotted equilibrium concentration, C_E against the dimensionless shear stress S. Later Whalley et al. [12] plotted entrainment rate, E against dimensionless shear stress S. However both these methods do not yield a complete collapse of the available data around a single curve. It was found that it is preferable to plot dimensionless equilibrium concentration, C_{ED} defined by,

$$C_{ED} = C_E/\rho_L \qquad \ldots(33)$$

against S. For this purpose, the following sets of data were analysed.

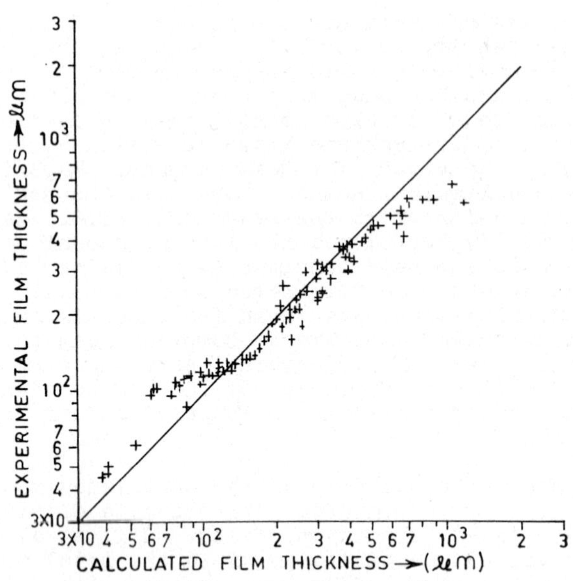

FIG. 2. TEST OF TURBULENCE MODEL

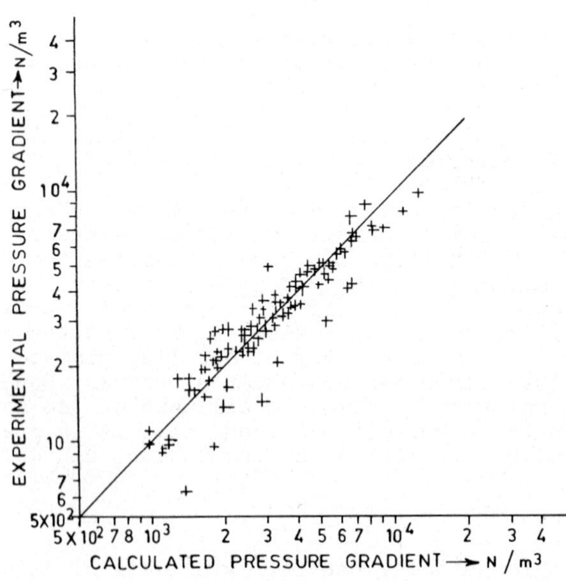

FIG. 3. TEST OF TURBULENCE MODEL

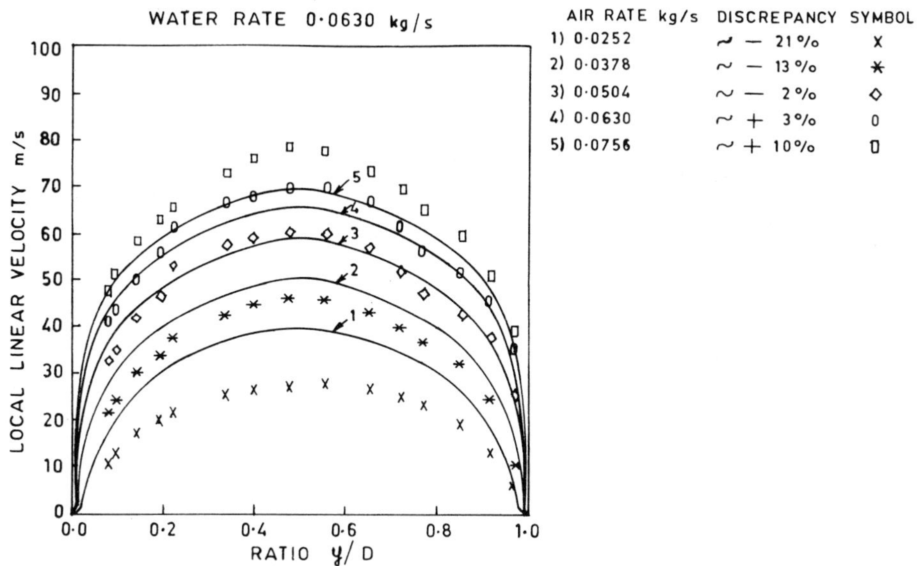

FIG. 4. PROFILES OF LINEAR VELOCITY, COMPARISON OF MODEL PREDICTIONS WITH DATA OF GILL et al. (10)

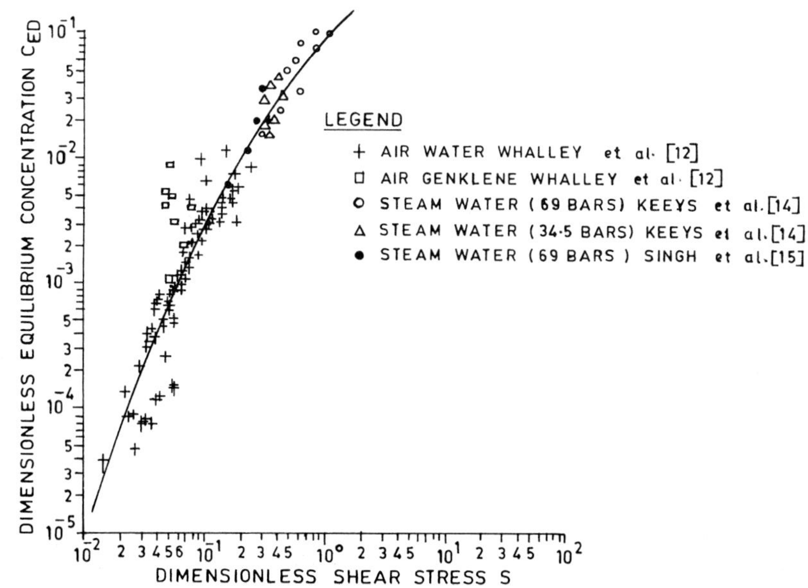

FIG. 5. CORRELATION FOR CONCENTRATION OF ENTRAINED DROPLETS

i) Data on air-water and air-Genklene flows generated by Whalley et al. [12]
ii) Data on steam-water flows generated by Keeys et al. [14]
iii) Data on steam-water flows generated by Singh* et al. [15].

These sets of data were chosen for comparison because these have been generated with the specific objective of understanding the distribution of the two phases in the annular flow and attempt has been made by the investigators to achieve equilibrium conditions.

In fig.5 data is plotted as dimensionless concentration, C_{ED} versus dimensionless interfacial shear stress, S. On this plane, data correlates very well. This correlation is, therefore. recommended for predicting entrained fraction in annular two phase flow. A polynomial was fitted to the data points using weighted least square fit. It was necessary to assign different weights to data as there are too many air-water points and too few steam-water points. Also Air-Genklene data does not seem to follow the general trend. Therefore the following weights were assigned:

S.No.	Fluid Combination	Weights
i)	Air-water	1.0
ii)	Steam-water	4.0
iii)	Air-Genklene	0.5

Using these weights, the following polynomial was obtained for the data in fig.5.

$$C_{ED} = 1.00268652 + 1.06667277\ S - 0.469955173\ S^2 \qquad ...(34)$$

5. PREDICTIONS FOR FULLY DEVELOPED ADIABATIC TWO PHASE FLOWS

The entrainment correlation given in section 4 was incorporated in the program ANFLO and it was used to study various aspects of two phase adiabatic flows. For this purpose, mass flux, tube diameter, steam quality and thermodynamic properties were specified as input to the program. For unknowns like void fraction, entrained fraction of liquid, pressure gradient and velocity profile, suitable values were assumed. For the first forty forward steps, a low value (= 0.01) was used for the deposition coefficient. Later this was increased to 0.1 to expedite convergence. It may be noted that for fully developed adiabatic flows actual value of deposition coefficient has no significance. It is important only for developing adiabatic flows and flows

* Data of Singh et al. was not available in tabular form. Therefore data were read from fig.5 of their paper cited in the reference.

with heat addition. Computations were terminated when liquid concentration in the core became constant. At that time all other parameters viz. pressure gradient, film thickness, void fraction etc. converged to a constant value. *Results of this study are summarised in this section. Wherever possible predictions are compared with experimental results.

5.1 Flow Distribution

Fig.6 gives entrained flow as a fraction of total flow versus steam quality for three different mass fluxes, viz. 544 kg/sm^2, 1360 kg/sm^2 and 2720 kg/sm^2 under equilibrium conditions. Also plotted in the same figure are the data of Keeys et al. [14], Singh et al. [15] and Subbotin et al. [17]. Predictions compare well with data. Maxima exhibited by plots in Fig.6 can be explained as follows. At low qualities vapour velocity in the core is not enough to support large amount of water. As quality increases vapour velocity and hence amount of water in the core increases. However, at the same time total available water continuously decreases due to the increase in quality. Therefore plots indicating entrained water flow fraction pass through a maximum.

Fig.7 gives film thickness as a function of quality and compares the prediction with the data of Janssen et al. [18]. Data does not compare well with the predictions particularly at low qualities, however, overall agreement is good.

5.2 Pressure drop and void fraction

In a review of various correlations for the determination of pressure drop, Weisman and Choe [19] have indicated that the correlation of Baroczy [20] provides reasonable results for two phase flows in a vertical pipe. Predictions of the program ANFLO are therefore compared with Baroczy's correlation. Fig.8 gives two phase friction factor multiplier as a function of quality for a mass flux of 1360 kg/sm^2. Besides present predictions and Baroczy's correlation, a plot based on Thom's [21] correlation is also given. Friction factor multiplier as used here is defined in terms of pressure gradient obtained if the entire mass flow rate was in single phase viz.

$$\phi^2_{lo} = \frac{(dP/dZ)_{tp}}{(dP/dZ)_{lo}}$$

* It may be indicated here that the program ANFLO has been developed primarily for flows with heat addition [16]. It has been used for studying the characteristics of the developed flow using the method described earlier. It is recognised that for studying the characteristics of the developed flow, a direct computation would have been preferable to the asymptotic computation used herein.

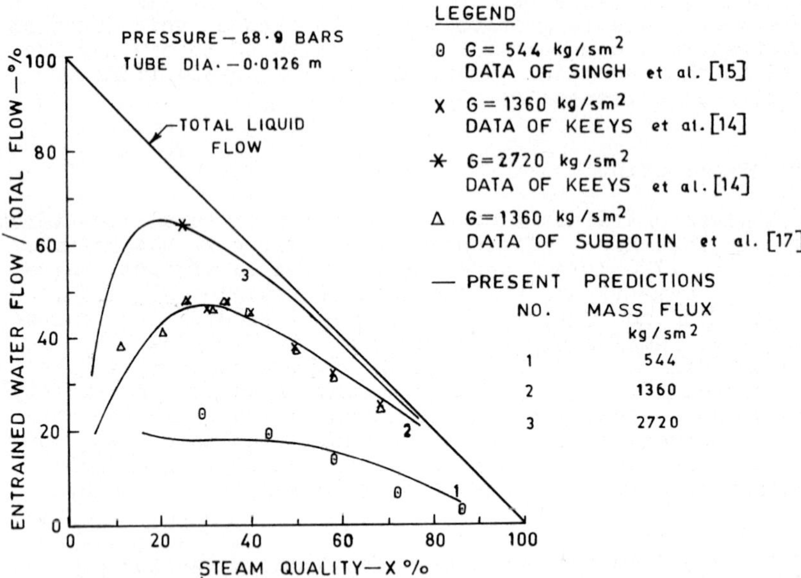

FIG. 6. HYDRODYNAMIC EQUILIBRIUM CURVES AT HIGH PRESSURE

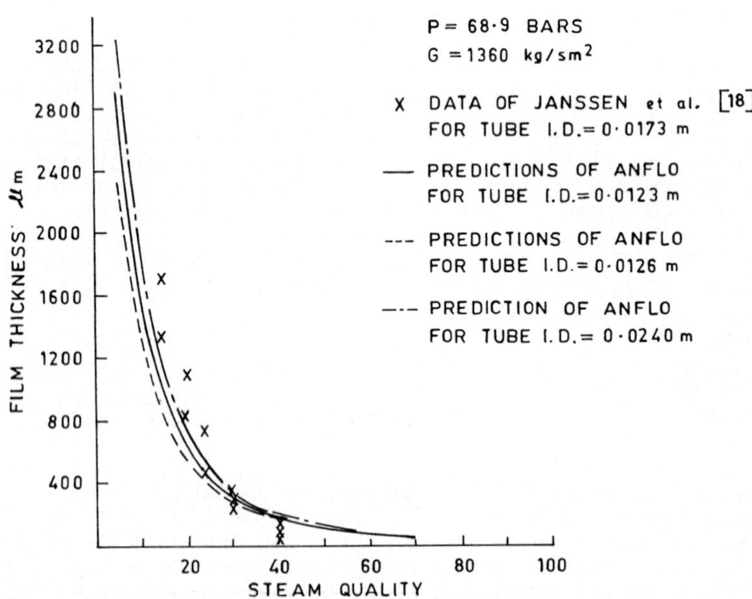

FIG. 7. FILM THICKNESS VERSUS STEAM QUALITY

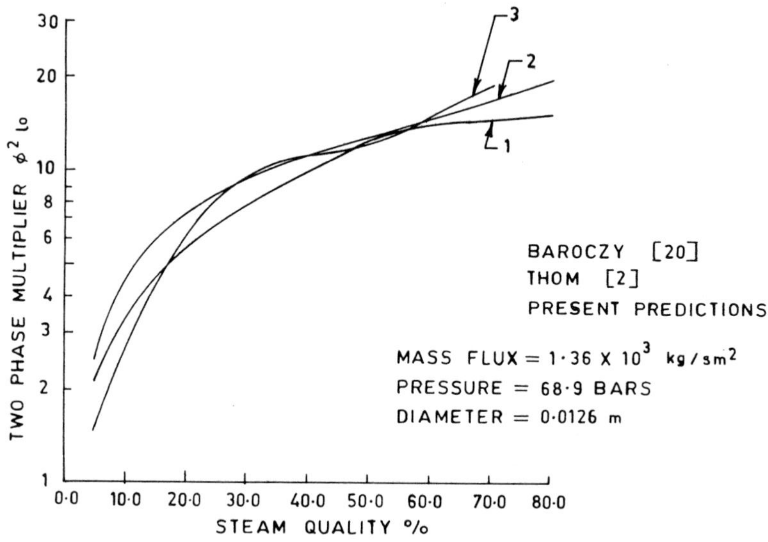

FIG. 8. COMPARISON OF PREDICTIONS OF TWO PHASE MULTIPLIER WITH BAROCZY'S AND THOM'S CORRELATIONS

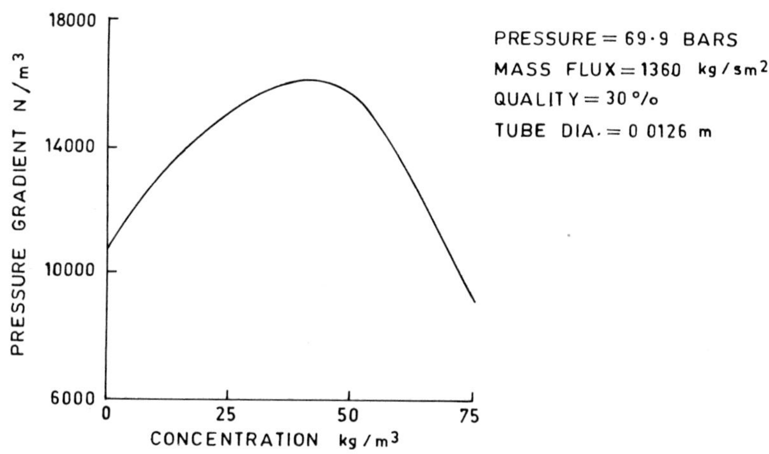

NOTE: CONCENTRATION IF THE FLOW IS ASSUMED HOMOGENEOUS IS 74·9 kg/m³ TO ESTIMATE PRESSURE GRADIENT FOR THIS CASE DATA FROM TABLE 2·1 a OF WALLIS [22] WAS USED.

FIG. 9. EFFECT OF LIQUID CONCENTRATION IN THE CORE ON PRESSURE GRADIENT

where $(dP/dZ)_{lo}$ is the pressure gradient calculated assuming the entire flow to be liquid. Agreement between present predictions and Baroczy's correlation is generally good.

It may be noted that pressure drop is a strong function of level of liquid entrainment in the core and data which is not taken under equilibrium conditions can greatly differ from equilibrium data. This is illustrated in fig.9, where pressure gradient is plotted against concentration of the liquid in the core. Concentration under equilibrium conditions for this case is 52 kg/m^3 as indicated in the figure.

Fig.10 gives void fraction as a function of quality for three different mass fluxes at 68.9 bars. Thom's correlation follows plot No.2 (corresponding to a mass flux of 1360 kg/sm^2) of fig.10. Many of the correlations available for the prediction of void fraction ignore the influence of mass flux on void fraction, which based on present computations look significant. Similar influence of mass flux on void fraction was observed by Zuber et al. [23] in their experiments with Freon-22. Fig.11 gives void fraction as a function of concentration of liquid in the core at fixed quality. As in the case of pressure gradient, void fraction also is a function of the distribution of two phases. Distribution of the two phases in the core and the film depend on the method of creation of the two phase flow. Hence pressure gradient and void fraction depend not only on mass flux, quality, diameter and pressure, but also on the method of creation of the two phase flow. Therefore, it is not possible to predict pressure gradient and void fraction accurately based on simple algebraic correlations as is the normal practice.

6. PREDICTIONS FOR DEVELOPING ADIABATIC TWO PHASE FLOWS

Gill et al. [24, 25] have obtained air-water data in the developing inlet region of a vertical round tube. For introducing water, porous wall section was used. First observation was taken at a distance of 0.15m (6") above the top of the porous injector. Conditions at this point were taken as the initial conditions for the computations. For the computations, only unknown is the deposition coefficient k, Whalley et al. [26] have proposed a graphical correlation for the evaluation of k. As per this correlation value of k should be about 0.25. Computations were done using the program ANFLO for a range of k values lying between 0.01 to 0.25. The results of computations and the experimental data are given in fig.12. Computations corresponding to k = 0.025 agree well with the data. This value of k is not in agreement with the correlation of Whalley et al. This correlation for k was obtained from dryout studies. It does not seem to be applicable to two component , two phase flows such as air-water flows. However, because of limited amount of data available for two component flows, it is not possible to give any different correlation.

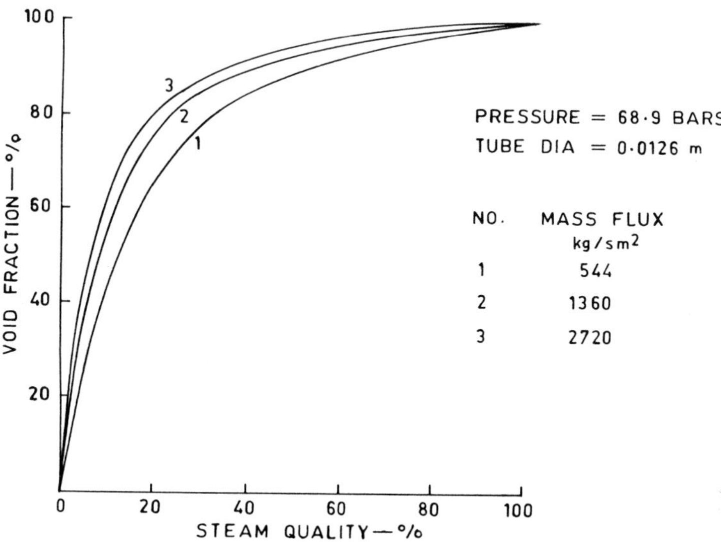

FIG. 10. EFFECT OF MASS FLUX ON VOID FRACTION

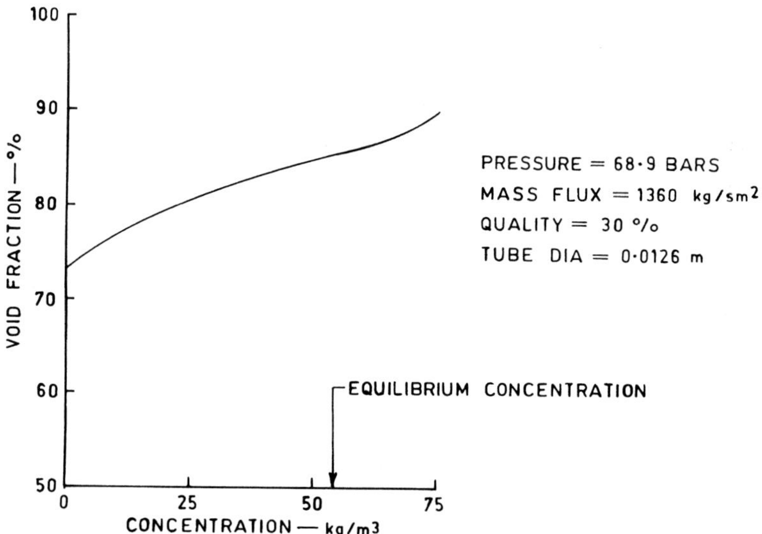

NOTE: CONCENTRATION IF THE FLOW IS ASSUMED HOMOGENEOUS IS 74·9 kg/m3. VOID FRACTION FOR THIS CASE WAS CALCULATED USING HOMOGENEOUS MODEL.

FIG. 11. EFFECT OF LIQUID CONCENTRATION IN THE CORE ON VOID FRACTION.

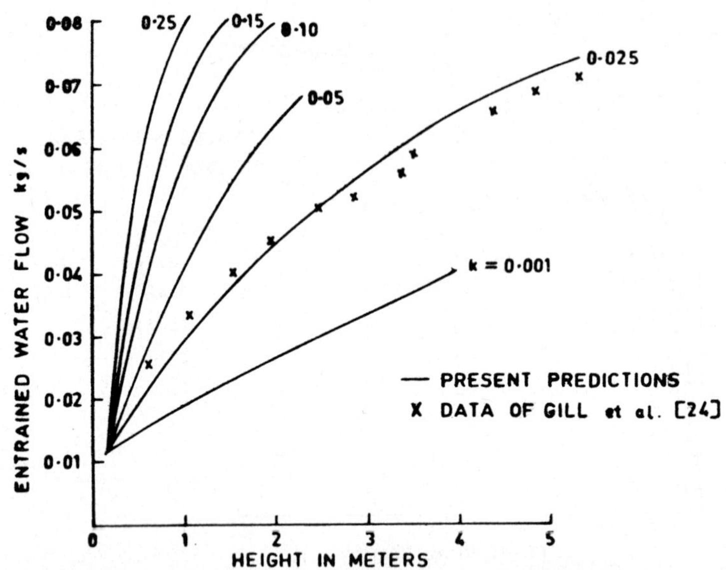

FIG. 12. COMPARISON OF PREDICTIONS OF ENTRAINED LIQUID FLOW VERSUS HEIGHT ABOVE INJECTOR FOR DIFFERENT VALUES OF DEPOSITION COEFFICIENT (k) WITH EXPERIMENTAL DATA.

7. CONCLUDING REMARKS

A very general two dimensional model has been developed for adiabatic two phase flows. The computer program based on this model i.e. ANFLO has been used successfully for the prediction of developed and developing two phase flows. The entrainment correlation of Whalley and Hutchinson has been improved.

The turbulence model used in the program ANFLO is empirical and further work is needed to replace it with a model having sound physical basis. Also the interfacial conditions (14) and (15) employed in this paper are classical and do not account for the wavy nature of the interface. This needs to be improved.

ACKNOWLEDGEMENTS

Author would like to thank Shri S.K. Mehta and Shri V. Venkat Raj, Reactor Engineering Division, B.A.R.C., for going through the manuscript and making useful comments. Thanks are also due to Dr. V.H. Arakeri, Indian Institute of Science, Bangalore, for his comments on the analysis reported.

NOMENCLATURE

C : Concentration, kg/m^3

C_{ED} : Dimensionless equilibrium concentration

D : Rate of deposition, kg/sm^2

E : Rate of entrainment, kg/sm^2

G : Mass flux, kg/sm

h_{fg} : Latent heat, J/kg

k : Mass transfer coefficient, m/s.

K : Von Karman constant

l : Mixing length, m

m : Liquid film thickness, m

P : Pressure, N/m^3

r : Radial co-ordinate, m

R : Radius of pipe, m

S : Dimensionless shear stress

U : Axial velocity, m/s

W : Mass flow rate, kg/s

y : Distance from the wall, m

Z : Axial co-ordinate, m

β : See eqn.5

η : See eqn.29

μ : Viscosity, kg/ms

ρ : Density, kg/m^3

τ : Shear stress, N/m^2

ω : Dimensionless stream function

ψ : Stream function

Subscripts

av : average

C : core

d : downstream

eff : effective

E : entrained

F : film

i : ith node

in : interface

L : liquid

u : upstream

V : vapour

REFERENCES

1. Hewitt, G.F., and Hall Taylor, N.S., Annular Two-phase Flow, 1970, Pergamon Press, Oxford.

2. Hewitt, G.F., 6th International Heat Transfer Conf., 1978, Vol.6, pp. 143-171.

3. Govier, G.W. and Aziz, K., The Flow of Complex Mixtures in Pipes, 1972, Van Nostrand Reinhold Co., New York.

4. Grover, R.B., "Two dimensional modelling of annular two phase flow", Ph.D. Thesis, 1981, I.I.Sc., Bangalore, India.

5. Patankar, S.V., and Spalding D.B., Heat and Mass Transfer in Boundary Layers, 1970, Inter-Text Books, London.

6. Keller, H.B., and Cebeci, T., Second International Conf. on Numerical Methods in Fluid Dynamics, 1971, Springer Verlag, pp. 92-100.

7. Nelson, R.M. and Pletcher, R.H., Heat Transfer and Fluid Mechanics Institute, 1974, pp. 154-170.

8. Grover, R.B., National Heat & Mass Transfer Conf., 1980, Hyderabad, Vol.2, pp. 281-292.

9. Grover, R.B., J. Thermal Engg., 1980, Vol.1, pp. 154-161.

10. Gill, L.E., HEWITT, G.F. and Lacy P.M.C., Report No. AERE-R3955, 1963.

11. Alia, P., Cravarole, L., Hassid, A., Pedrocchi, E., Energia Nucleare, 1968, Vol15, pp. 241-254.

12. Whalley, P.B., Hewitt, G.F., and Hutchinson, P., Report No. AERE-R7521, 1973.

13. Hutchinson, P. and Whalley P.B., Chem.Eng. Sci., 1973, Vol.28, pp. 974-975.

14. Keeys, R.K.F., Ralph, J.C. and Roberts, D.N., Report No. AERE-R6293, 1970.

15. Singh, K., St. Pierre, C.C., Crago, W.A. and Moeck E.O., AIChE.J., 1969, Vol.15, pp. 51-56.

16. Grover, R.B., Paper No. TF17, Seventh International Heat Transfer Conf., 1982, Munich.

17. Subbotin, V.I., Sorokin, D.N., Nigmatulin, B.J., Milashenko, V.I., and Nikolayev, V.E., Paper No. FB 10, Sixth International Heat Transfer Conf., 1978, Toronto.

18. Janssen, E., Kervinan, J.A., and Kim, H.T., Report No. GEAP-10341.

19. Weisman, J. and Choe, W.G., Proc. Two Phase-flow & Heat Transfer Symp. Workshop, Florida, 1976, Vol.1, pp. 193-224.

20. Baroczy, C.J., AIChE. Chem. Eng. Prog. Symp. Series No. 64, 1966, Vol.62, pp. 232-249.

21. Thom, J.R.S., Int. J. Heat & Mass Transfer, 1964, Vol.7, pp. 709-724.

22. Wallis, G.B., One dimensional two phase flow, McGraw Hill Book Co., 1969.

23. Zuber, N., Staub, F.W., Bijwaard, G. and Kroeger, P.G., Report No. GEAP-5417, 1967.

24. Gill, L.E., Hewitt, G.F. and Hitchon J.W., Report No. AERE-R 3954, 1962.

25. Gill, L.E. and Hewitt, G.F., Report No. AERE-M1202, 1966.

26. Whalley, P.B., Hutchinson, P. and Hewitt, G.F., Paper No. B6.11, Fifth International Heat Transfer Conf., 1974, Tokyo.

Multi-Phase Flow and Heat Transfer III. Part A: Fundamentals
edited by T.N. Veziroğlu and A.E. Bergles
Elsevier Science Publishers B.V., Amsterdam, 1984 — Printed in The Netherlands

A STUDY ON LIQUID FILM FLOW CHARACTERISTICS IN ANNULAR TWO-PHASE FLOW

Yoshimichi Hagiwara, Tatsunori Miwada, Kenjiro Suzuki and Takashi Sato
Department of Mechanical Engineering
Faculty of Engineering
Kyoto University
Kyoto 606, JAPAN

ABSTRACT

Structure of interfacial waves is studied in a liquid film of a downward annular two-phase flow, both for ripple wave and disturbance wave flow regimes. Experiments were performed for the thin liquid film flowing down in an acrylic tube. Fine tracer particles are injected into the film to visualize the waves, and two different visualization techniques are applied. Comparison of the obtained photographs with the time trace of the liquid film thickness is also carried out. For the ripple waves, length scale of each wave and distance between two neighbouring waves were examined. It is shown that the scale and the distance decrease with an increase of the gas flow rate. This tendency is a partial reason why the wave passing frequency increases with the gas flow rate. For the disturbance waves, two different patterns can be found: peripherally uniform pattern on its front side, and less uniform one on its back side. Large but peripherally uniform flow fluctuation is also observed inside the liquid film at the instant of the disturbance wave arrival.

1. INTRODUCTION

Annular two-phase flow is encountered in various practical heat or mass transfer systems, in condenser, in steam generator, in heat pipe, in some chemical process and so on. As a result of the interaction between the gas phase flow and the liquid phase flow, various waves are found at their interface. The waves in turn have a influence on the transfer characteristics of momentum, heat and mass in the liquid film or in the gas phase flow near the interface [1,2]. Therefore, it is important to study on the interfacial characteristics of the liquid film in detail.

Statistical studies on the wave characteristics have been carried out actively: the time trace of the liquid film thickness gives some useful informations on statistical quantities, i.e. the wave amplitude, the wave passing frequency, the wave propagating velocity and so on [3,4]. The wave pitch can also be calculated from such imformations. These statistical quantities are important in the estimation of the heat or momentum transfer enhancement by the waves [5,6].

However, the informations obtained from the liquid film thickness trace are not sufficient for the detailed study of the spatial wave structure or the wave effects on the flow behaviour inside the liquid film. Visual study on the liquid film may serve as a means for better comprehension of the spatial wave structure or the flow behaviour inside the liquid film. Woodmansee and Hanratty [7] observed the entrainment of droplets from the roll wave in horizontal two-phase flow using a high speed camera. They found the sequential events of entrainment from their film. Collier and Hewitt [8] reported the impingement of droplets on the liquid film surface in annular two-phase flow. Hall Taylor et al. [9] visualized the wave motion using high speed cine film and conductance method simultaneously. They obtained propagating velocity of each disturbance wave and discussed the interactions between two or three neighbouring disturbance waves. These studies do not give sufficient informations about the spatial structure of the waves.

This experimental study aims at examining the fundamental spatial features of the interfacial waves in both of ripple and disturbance waves of a thin liquid film in a downward annular two-component two-phase flow. Two flow visualization techniques are applied to observe the spatial characteristics of the waves and its effect on the flow behaviour. Relationship between the visualized results and the results from the time trace of the liquid film thickness is also discussed.

2. EXPERIMENTAL PROCEDURES

2.1. Apparatus

The flow system used in this study is shown in Fig. 1. It is the same as that used in reference [10]. Air is supplied as a gas phase fluid by a compressor 1. Its main portion comes into the test section 7 through the duct 5. The remaining portion is issued from the atomizer 11. Both air flow rates are measured separately with orifice flow rate meters 4 and 10. Water is supplied to the atomizer as a liquid phase fluid by a stainless steel pump 12. The water flow rate was measured with a capillary flow rate meter 15.

The atomizer consists of seven small air-atomizers equally spaced and axi-symmetrically allocated. It is mounted upstream of the contraction nozzle 6 in a way seen in Fig. 2. Almost all droplets from the atomizer are deposited to the contraction nozzle and annular liquid film is formed at the nozzle.

The test section is made up of an acrylic pipe of 26.4mm inner diameter and 1850mm in length. It is connected to the contraction nozzle, and the axial distance x is measured from the outlet of the contraction nozzle. (See Fig. 2.)

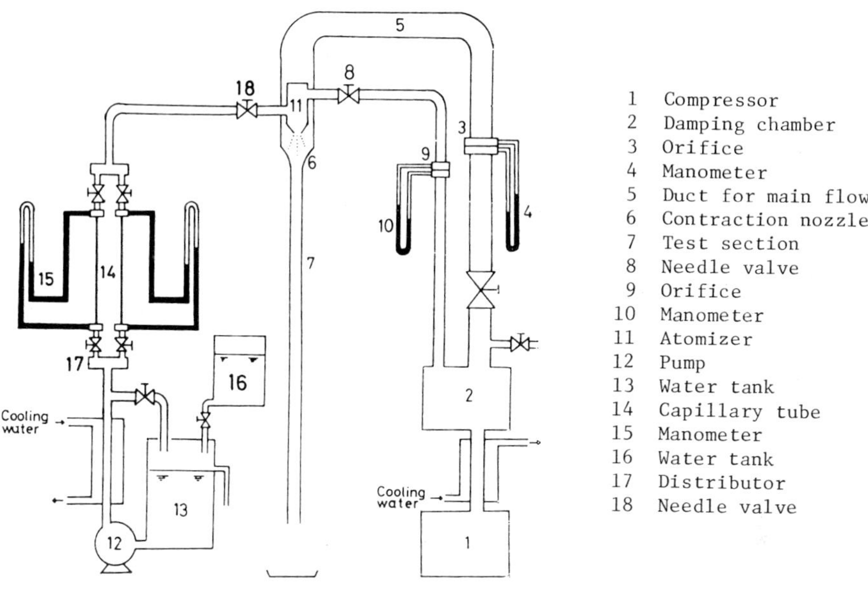

Fig. 1. Schematic Diagram of the Flowing System

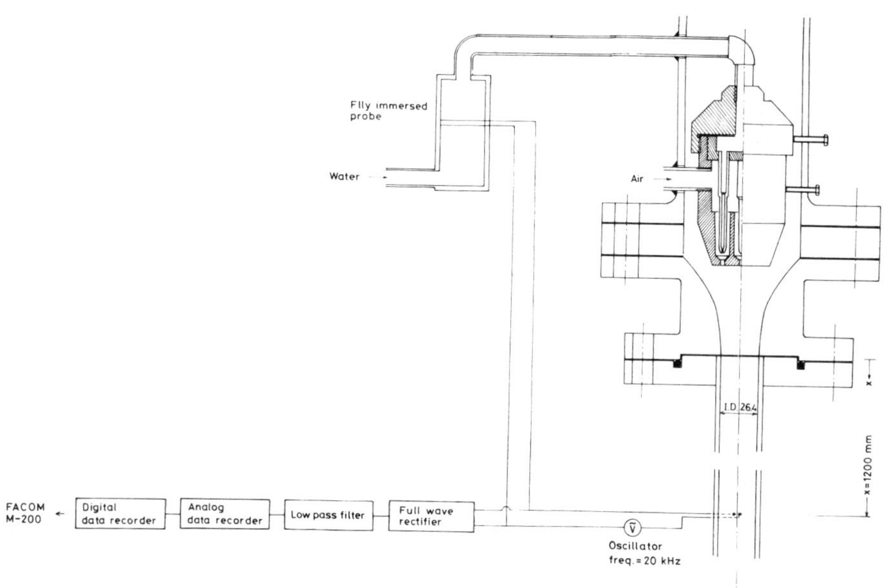

Fig. 2. Atomizer, Test section and Signal Processing Circuit

2.2. Liquid Film Thickness Trace

The time traces of the liquid film thickness were obtained using a two-wire conductance probe in the same way as in reference [10]. (See Fig. 2.) This is composed of two 0.5mm-diameter stainless steel wires spaced 2.5mm circumferentially. It was mounted perpendicularly to the inner surface of the test section tube at x=1200mm (x/D=45.5), where the flow is almost developed at any flow rate condition.

Fig. 2 also shows the signal processing circuit for the output signal from the conductance probe. Its rectified signal was low-pass-filtered, and the finally obtained signal was recorded in a magnetic tape. The recorded signal was digitized and was processed by a computer (FACOM M-200) to get statistical informations of waves. For the purpose of ensuring linear relationship between the final signal and the liquid film thickness, another fully immersed conductance probe was used for correction [10]. This probe is also useful to compensate the water properties effect on the output signal. Static calibration of the linear relationship between the liquid film thickness and the final signal was made in a shallow open channel.

The apparent propagating velocity of each wave was determined as follows: another conductance probe was located at x=1220mm, 20mm apart from the first one in the flow direction. The final output signals from the both probes were recorded simultaneously and digitized. The propagating velocity was estimated from the time difference between the corresponding crests appearing in each signal and the distance between the two sets of the probes.

2.3. Visualization of Wave Structure

A dilute suspension of fine TiO_2 particles was used as test liquid, for the visualization of the interfacial wave structure. Scattered light from the white TiO_2 particles suspended as a pigment was photographed by a camera. Although this method is essentially a kind of tracer method, what is important in the present study is not the streaks of the tracer particles but the brightness of the scattered light from the particles.

The test liquid was obtained by injecting a dilute solution of 1 part TiO_2 particles to 40 parts water by volume into the liquid film at x=100mm with a fine syringe needle. As TiO_2 particles diffuse easily and quickly, so it has been mixed completely before the liquid film arrives at the camera position x=1200mm. The particles themselves do not alter the flow pattern remarkably, because the diameter of the TiO_2 particles is about two orders of magnitude smaller than the mean film thickness. The injection rate of TiO_2 solution was kept less than 0.05g/s, which is two orders of magnitude smaller than the liquid film flow rate. Therefore, the liquid film flow or liquid properties may not be changed by the particle injection so remarkably.

A schematic view of the observing section is shown in Fig. 3. Light source was a photoflood lamp and was set out-of-line with

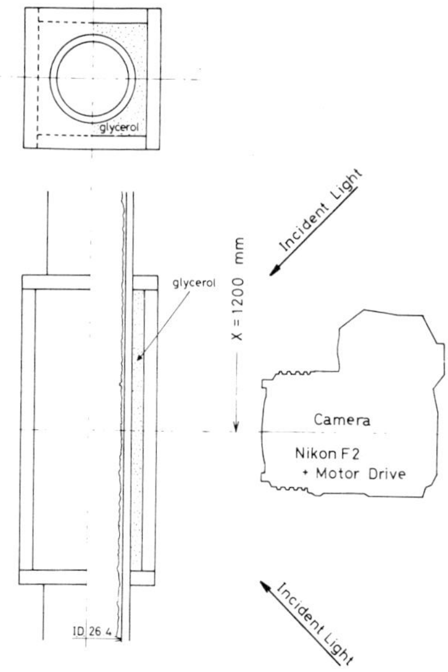

Fig. 3. Schematic View of the Observing Section

the camera mounted perpendicularly to the test section tube. The camera was set so as to focus on the TiO_2 particles in the liquid film on the tube inner surface of the nearer side to the camera. To reduce the light reflection from the outer surface of the test pipe, the test tube was surrounded by another rectangular acrylic duct and the space between the two ducts was filled with glycerol. (See reference [11]) Glycerol has approximately the same refractive index as the transparent acrylic ducts.

The exposure time was 1/2000s and the film used was Kodak Tri-X Pan, black and white negative film.

A simple calibration test of the relationship between the liquid film thickness and the brightness of the scattered light was carried out prior to the observation. The brightness was observed for the same dilute suspension of TiO_2 particles flowing in an annular duct, whose spacing is nearly equal to the liquid film thickness. Qualitative relationship between the brightness and the liquid film thickness was found by changing the spacing of the annular duct.

Incidentally, the same visualization was carried out in the test tube for the conductance probe in Fig. 2, to examine the relationship between the time trace of the liquid film thickness and the obtained photographs. The pulse signals of the motor drive device and the trace of the liquid film thickness were recorded simultaneously. Comparing these signals with each other, the instantaneous liquid film thickness was obtained at the moment when the photograph was taken.

2.4. Visualization of Flow Behaviour

To visualize the flow behaviour inside the liquid film, an ordinary tracer method was adopted. Aluminum powder was used as a tracer particle. The reflected light streaks of the powder were photographed by the same system shown in Fig. 3. The aluminum powder has flat surface, because the shape of the powder is not spherical but scaly. The flat surface of the powder will become parallel to the flow direction. When the powder follows the fluctuation of flow, the direction of the flat surface also changes. Thus, the change of the reflected light from the surface of the powder means that the powder follows some kind of flow fluctuation [12].

Since the aluminum powder can not be mixed easily with water, a very small amount of surface active agent was added to produce a dilute suspension of aluminum powder. Both the injection method and the injection rate of the suspension are the same as those for the TiO_2 particles suspension. The equivalent diameter of the aluminum powder is less than 10 micronmeter, which is smaller than the mean film thickness by more than an order of magnitude.

The exposure time was 1/125s and 1/250s, respectively, for the two lower air flow rate conditions, and for the highest air flow rate to be mentioned in the next.

3. EXPERIMENTAL CONDITIONS

The experiment was made at three air flow rates:

W_g g/s		Re_g		U_{jg} m/s
21.1	:	5.5×10^4	:	32
15.5	:	4.0×10^4	:	23
11.5	:	3.0×10^4	:	17

where U_j means the cross sectional average velocity. The air flow is fully turbulent in every case. The water flow rates chosen are as follows:

W_1 g/s		Re_1		U_{j1} m/s
11.7	:	5.0×10^2	:	2.1×10^{-2}
5.00	:	2.1×10^2	:	9.1×10^{-3}

Only ripple waves were observed in the lower liquid flow rate condition. However, the disturbance waves were observed as well in the higher flow rate condition. In this flow rate condition, the entrainment of droplets from the crests of the disturbance waves may not be significant as described in reference [10]. Incidentally, the observation was also performed for the intermediate flow rate condition of W_1=8.33g/s, which corresponds to the transition from the ripple wave regime to the disturbance wave regime.

4. RESULTS AND DISCUSSIONS

4.1. Interfacial Wave Structure

Typical photographs of the scattered light from the TiO_2 particles in the ripple wave regime are shown in Fig. 4. A lot of small fish-scale-like patterns bounded by dark lines are seen in the figure. It was confirmed from the calibration test that the black lines in the photographs correspond to the part where the liquid film is thin, namely, trough of interfacial wave. On the other hand, white part corresponds to the crest of the wave. Therefore, a lot of ripple waves are found to be formed closely and randomly on the interface.

To study such spatial structure more quantitatively, attention was paid to two different spatial scales for the waves. One is the equivalent diameter of each area bounded by the dark line Λ_R. This is defined as the root of the product of the longest diameter and the shortest one found in each wave pattern. If a pattern is circular, this coincides with its circular diameter. Histograms of the equivalent diameter Λ_R are shown in Fig. 5(a). The arrow in each figure indicates its mean value. Each figure shows that the histogram has neary normal distribution around its mean value, which is supported by the chi-square test. Therefore, it is found that the interface is covered with various ripple waves having random sizes.

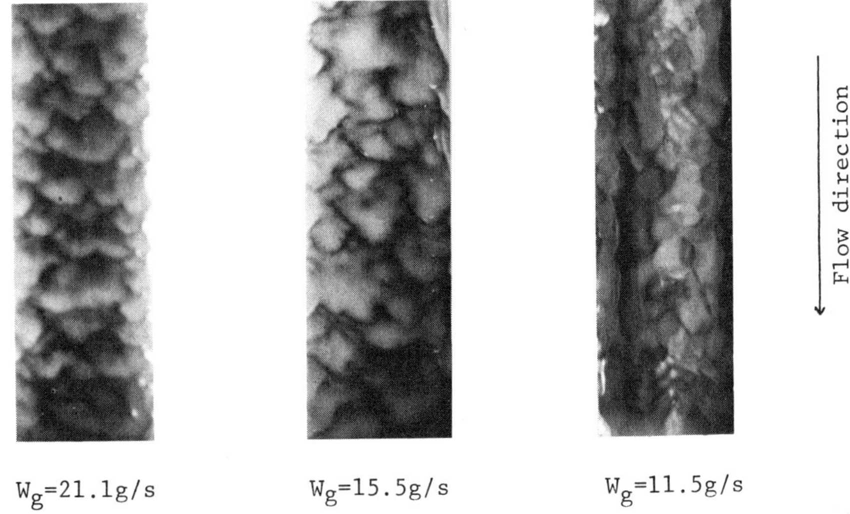

Fig.4. TiO_2 pattern (W_l=5.00g/s : Ripple wave regime)

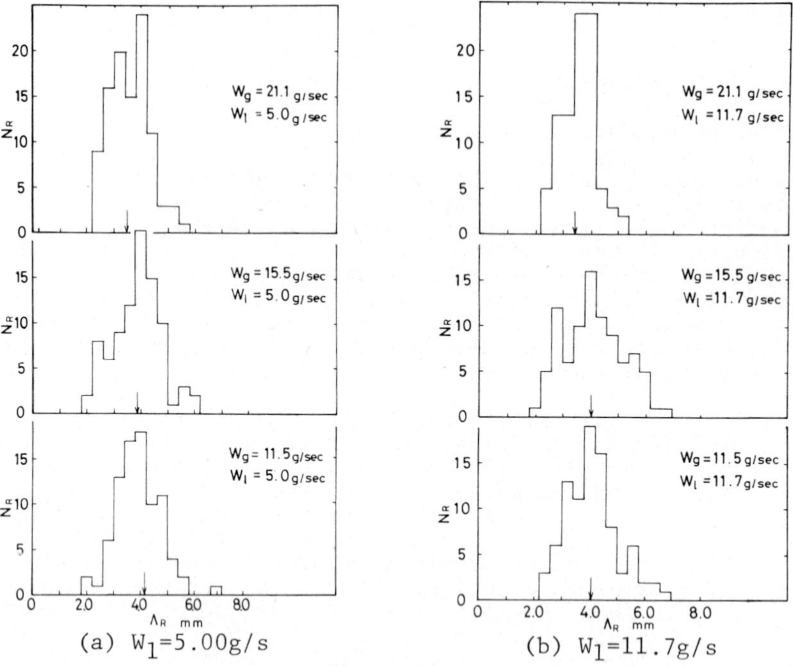

Fig. 5. Histogram of Length Scale of Ripple Waves

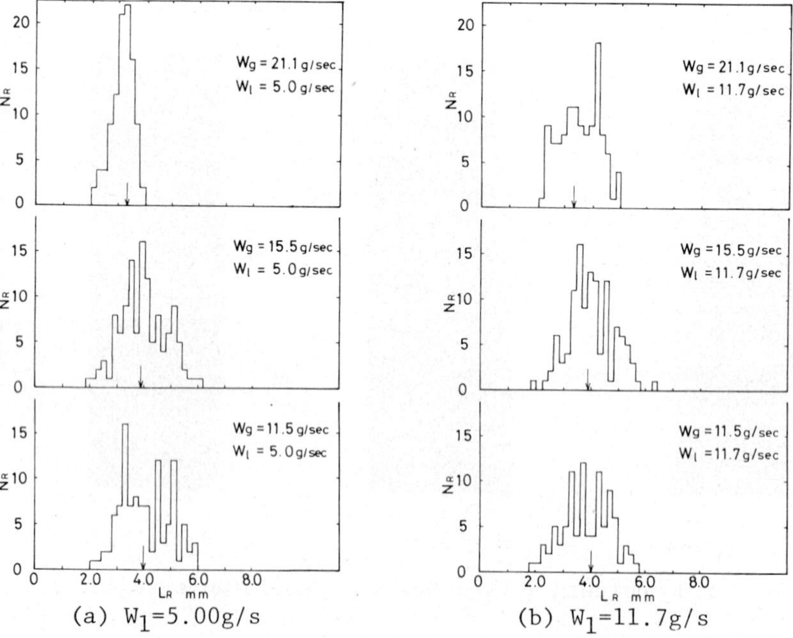

Fig. 6. Histograms of Distance between Two Neighbouring Ripple Waves

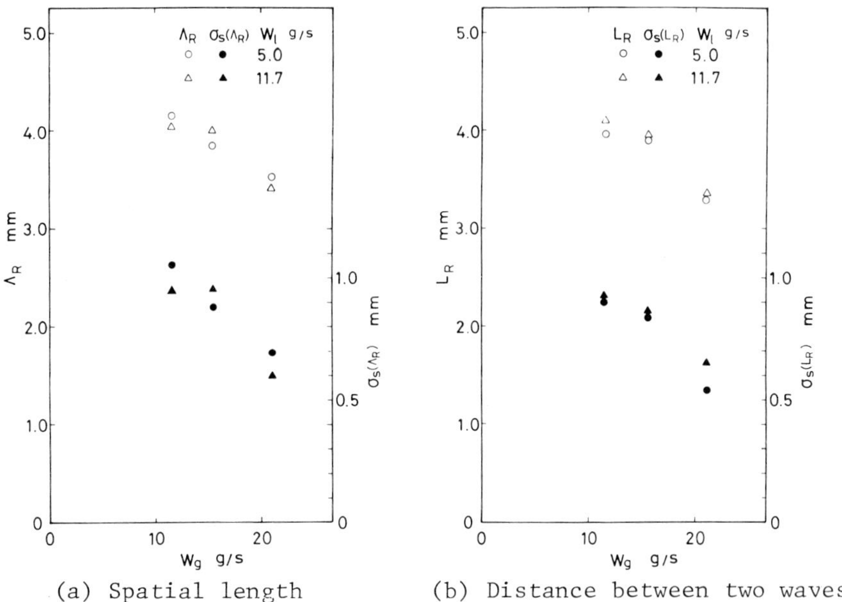

(a) Spatial length (b) Distance between two waves

Fig. 7. Dependence of Spatial Scales on Air Flow Rate

The other scale is the distance L_R between the imaginary centres of the two neighbouring ripple waves. Fig. 6(a) shows the histograms of L_R, where an arrow also indicates the mean value. The distribution of L_R also yields nearly to the normal distribution.

Fig. 7 shows the dependence of such spatial scales on the air flow rate. Not only the average values of such scales themselves but also their standard deviations σ_s decrease with the increase of the air flow rate. It means that smaller ripple waves come to cover the interface more closely, as increasing the air flow rate. The average values of L_R and $\sigma_s(L_R)$ are almost equal to Λ_R and $\sigma_s(\Lambda_R)$ respectively in any air flow rate condition.

The wave passing frequency, obtained from the film thickness trace, has been found to increase with the air flow rate [6]. This has been confirmed from the film thickness spectrum. This is partly because the ripple wave propagating velocity increases with the air flow rate [6], but also because Λ_R decreases with the increase of the air flow rate as found in the present study.

Comparing the time trace of the ripple wave at x=1200mm with that at x=1220mm, the ripple wave does not seem to change in its shape so remarkably during a short period. Thus, it may be assumed that the liquid film flows down, remaining its wave structure unchanged within a short distance. Permitting the assumption that the wave structure shown in the photographs flows down unchanged, the averaged streamwise pitch of wave obtained

Table 1. Streamwise Pitch of Ripple Wave

W_g (g/s)	Streamwise pitch (mm)	
	TiO_2 pattern	Film thickness trace
	p	U_R/f_R
21.1	5.0	6.8
15.5	6.4	6.6
11.5	6.1	7.2

from the distance between the two subsequent bright patterns should be equivalent to the counterpart obtained from the liquid film thickness trace calculated as the averaged propagating velocity U_R divided by the averaged passing frequency f_R. These averaged pitches are compared in Table 1, and are found to agree well with each other.

In the case of the transition flow rate condition from ripple wave to disturbance wave, larger waves are found to cover the interface than ripple waves mentioned above. (See Fig. 8.) It is a reason why the wave passing frequency slightly decreases with the increase of the liquid flow rate from the ripple wave flow regime to the transition flow regime.

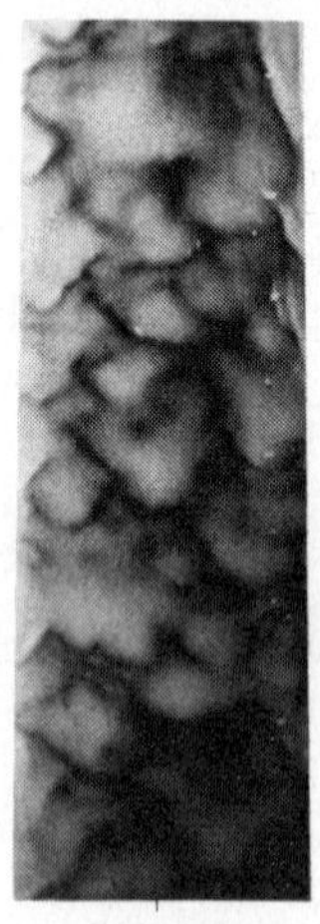

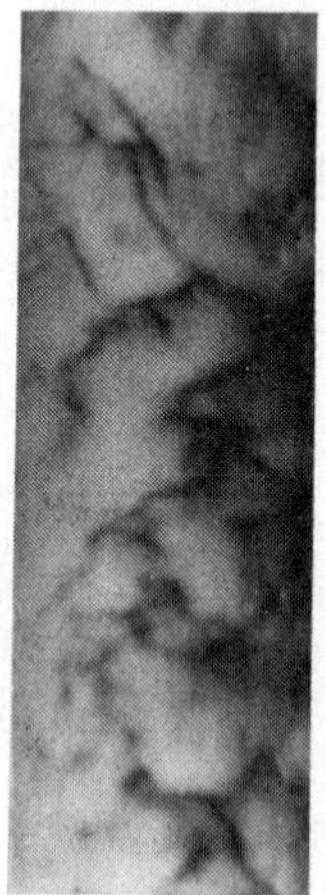

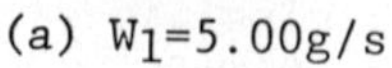

(a) W_l=5.00g/s (b) W_l=8.33g/s

Fig. 8. Comparison of TiO_2 pattern (W_g=15.5g/s)

Two kinds of waves different in size are observed in the disturbance wave regime, as shown in Fig. 9. Small waves like ripple waves in Fig. 4 can also be found. The above mentioned spatial scales are also determined for the small waves. Histograms of Λ_R and L_R are shown in Fig. 5(b) and 6(b), respectively. Similar trends to those found in the ripple wave condition can be found in these figures. Dependence of the average values of Λ_R and L_R and their standard deviations on the air flow rate seems to be the same as those in ripple wave condition (See Fig. 7).

Another pattern much larger than the one discussed in the above can also be observed, which corresponds to the disturbance wave. This was confirmed by the study of the simultaneously obtained photographs ans the time trace of the film thickness. The disturbance wave pattern is found to consists of two different parts:

(1) The peripherally uniform pattern, found on the front side of disturbance wave

(2) The subsequent trail pattern, being less uniform and like elongated ripple waves, found on the back side of disturbance wave

The averaged streamwise length of each pattern measured on the photographs is summarized in Table 2. In the table, their counterparts estimated from the film thickness trace are also shown. The front side length was calculated by the product of the increasing period Δt_F of the film thickness and the disturbance wave propagating velocity U_{DW}. The back side length was determined similarly, but using the decreasing period of the film thickness after the maximum height. These calculated lengths

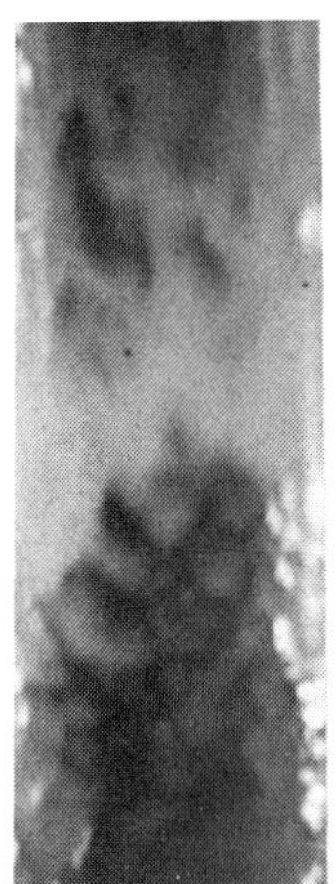

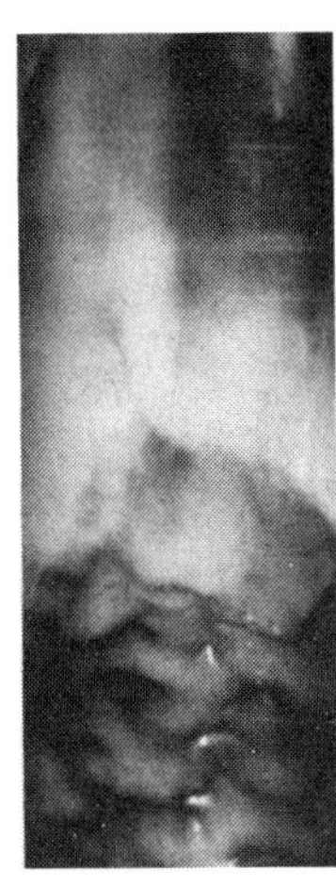

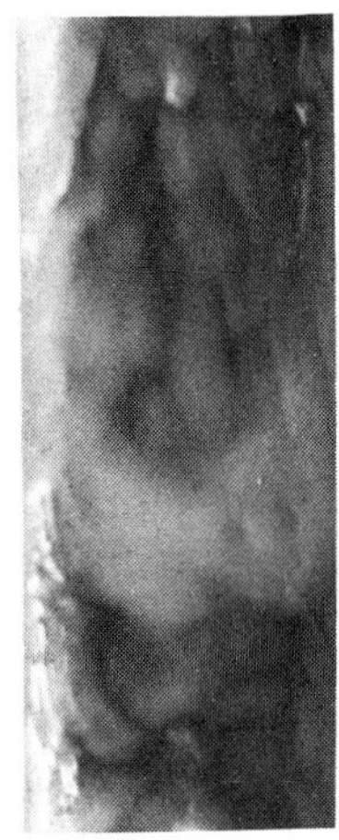

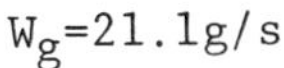

W_g=21.1g/s W_g=15.5g/s W_g=11.5g/s

Fig. 9. TiO_2 pattern (W_l=11.7g/s : Disturbance wave regime)

Table 2. Streamwise Lengths of Disturbance Wave

W_g (g/s)	Length (mm)			
	TiO_2 pattern		Film thickness trace	
	l_{DW}(uniform)	l_{DW}(trail)	Front side ($\Delta t_F \times U_{DW}$)	Back side ($\Delta t_B \times U_{DW}$)
21.1	11	(20)*	8.4	21
15.5	10	(20)*	8.7	20
11.5	11	(20)*	8.1	17

(*: Measuring error may not be negligible.)

agree fairly well with the presently visualized results. Thus, the small fish-scale-like structure of ripple waves disappears when the front side of the disturbance wave is passing. However, transition process from the peripherally uniform structure to the small ripple wave structure starts again just after the passing of the maximum height of the disturbance wave. The less uniform structure on the back side of the disturbance wave may correspond to this transition region.

Incidentally, it is found that the total streamwise length of the disturbance wave does not change so remarkably as increasing the air flow rate as shown in Table 2.

4.2. Flow Behaviour in the Liquid Film

Fig. 10 shows typical photographs of the reflected light streaks from the aluminum powder. In the ripple wave regime, merely small lumps of bright streaks are found. This means that only weak or mild fluctuations occur inside the liquid film in this regime, therefore, that the film may flow down calmly.

In the disturbance wave regime, similar small lumps of bright streaks are observed when only the small waves are passing. On the other hand, a lot of bright streaks are found to distribute uniformly when the disturbance wave passes as seen in Fig. 11. It reveals that the relative large but peripherally uniform flow fluctuation is caused by the propagation of the disturbance wave. Incidentally, the streamwise length for the uniform lump of streaks is shorter than the total length of the disturbance wave pattern obtained by the methods mentioned above. This means perhaps that the length scale of the disturbance caused in the liquid film due to the arrival of the disturbance wave is smaller compared with the length scale of the disturbance wave itself.

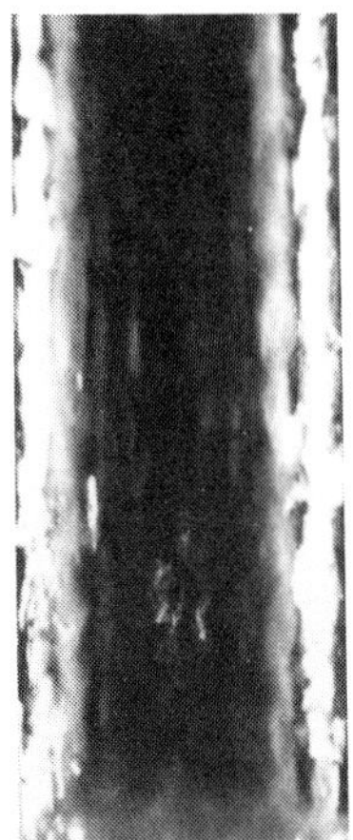

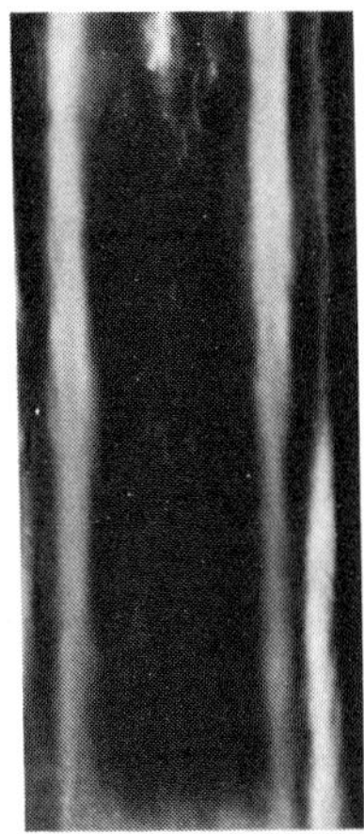

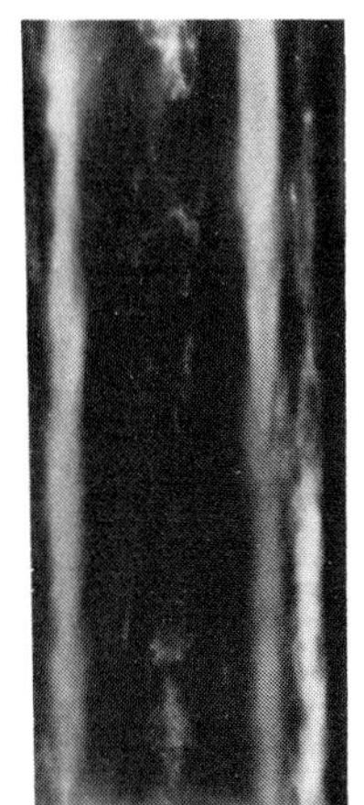

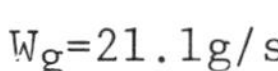

W_g=21.1g/s W_g=15.5g/s W_g=11.5g/s

Fig. 10. Reflected Light Streaks from Aluminum Powder (W_l=5.00g/s : Ripple wave regime)

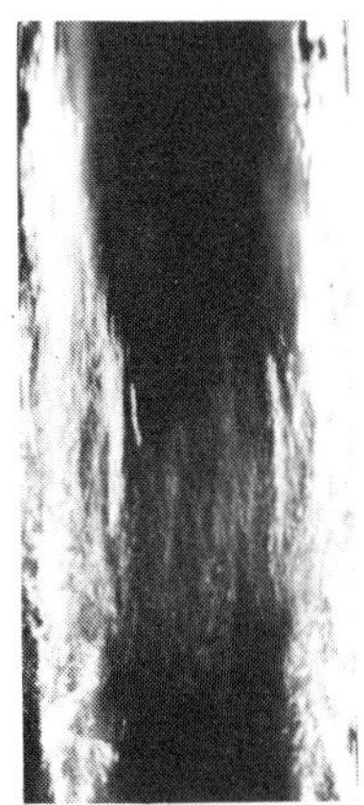

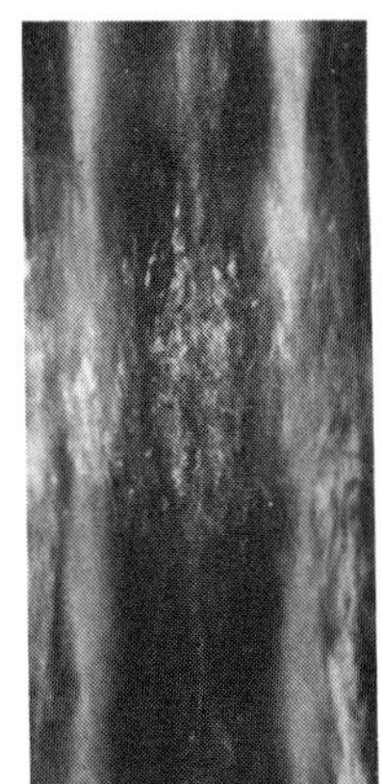

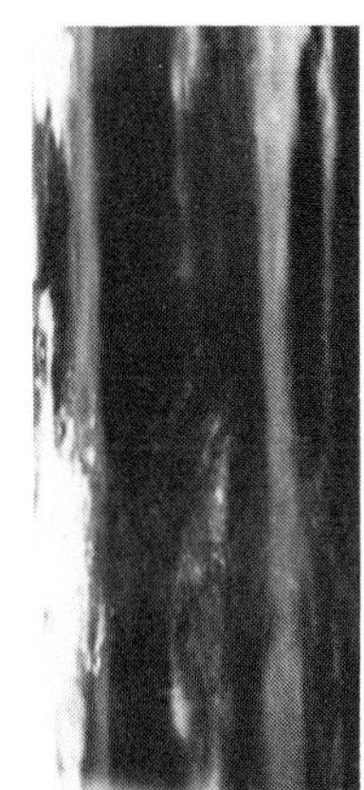

W_g=21.1g/s W_g=15.5g/s W_g=11.5g/s

Fig. 11. Reflected Light Streaks from Aluminum Powder (W_l=11.7g/s : Disturbance wave regime)

5. CONCLUSIONS

Spatial structure of interfacial waves was studied experimentally in a thin liquid film of a downward annular flow. Interfacial waves were observed using flow visualization techniques. Liquid film thickness trace was obtained by the conductance method The main results obtained are as follows:

1. Ripple waves are formed closely and randomly. Smaller ripple wave come to cover the interface more closely, as increasing the air flow rate.

2. The wave passing frequency increases partly because the spatial length scale of ripple wave decreases with the increase of the air flow rate.

3. The streamwise pitch estimated from the photographs agrees well with the counterpart obtained from the trace of the liqud film thickness.

4. Larger waves than ripple waves are found to cover the interface in the case of the transtion flow rate condtion from ripple wave regime to disturbance wave regime. It is a reason why the wave passing frequency slightly decreases with the increase of the liquid flow rate.

5. The front side of the disturbance wave has peripherally uniform pattern. On the other hand, the back side has less uniform pattern like elongated ripple wave. It means that the ripple wave pattern disappears when the front side of disturbance wave is passing but that it starts to appear again just after the crest of disturbance wave was passed.

6. The liquid film flows down calmly in the ripple wave regime, while large but peripherally uniform fluctuation is caused by the propagation of the disturbance wave.

NOMENCLATURE

D pipe diameter
L distance between two neighbouring wave
l streamwise length of disturbance wave
N number in histogram
p streamwise pitch of ripple wave
Re Reynolds number
U axial velocity
W mass flow rate
x axial coordinate
y coordinate perpendicular to wall

Greek

Λ spatial length scale of wave
σ_s standard deviation
Δt period

Subscripts

B back side of disturbance wave
DW disturbance wave
F front side of disturbance wave
g gas phase
j cross sectional averaged value
l liquid phase
R ripple

REFERENCES

1. Brumfield, L. K. and Theofanous, T. G., J. Heat Transfer, Vol.98, 1976, p.496.
2. Suzuki, K., Hagiwara, Y. and Sato, T., Int. J. Heat Mass Transfer, Vol.26, 1983. (to appear)
3. Chu, K. J. and Dukler, A. E., AIChEJ. Vol.21, 1975, p.583.
4. Webb, D. R. and Hewitt, G. F., Int. J. Multiphase Flow, Vol.2, 1975, p.35.
5. Dukler, A. E., Chem. Eng. Educ. Vol.11, 1977, p.108.
6. Hagiwara, Y., Suzuki, K. and Sato, T., Mem. Fac. of Engng, Kyoto Univ., Vol.44-2, 1982, p.309.
7. Woodmansee, D. E. and Hanratty, T. J., Chem. Engng Sci., Vol.24, 1969, p.299.
8. Collier, J. G. and Hewitt, G. F., British Chem. Engng, Vol.11, 1966, p.1526.
9. Hall Taylor, N., Hewitt, G. F. and Lacey, P. M. C., Chem. Engng Sci., Vol.18, 1963, p.537.
10. Hagiwara, Y., Suzuki, K., Sato, T. and Chigusa, N., Proc. ASME-JSME Thermal Engng Joint Conference, Vol.1, 1983. (to appear)
11. Jacowitz, L. A. and Brodkey, R. S., Chem. Engng Sci., Vol.19, 1964, p.261.
12 Handbook of Flow Visualization (ed. Asanuma, T.) Asakura Book Company, 1977. (in Japanese)

Multi-Phase Flow and Heat Transfer III. Part A: Fundamentals
edited by T.N. Veziroğlu and A.E. Bergles
Elsevier Science Publishers B.V., Amsterdam, 1984 — Printed in The Netherlands

DROPLET MOTION IN TWO-PHASE ANNULAR FLOW

Mohamed M. El-Kassaby
Department of Engineering
Purdue University Calumet
Hammond, IN 46323, U.S.A.

E. N. Ganic
Faculty of Mechanical Engineering
University of Sarajevo
Sarajevo 71000, Yugoslavia

ABSTRACT

In two-phase annular flow, a significant fraction of the liquid phase can be entrained as liquid droplets flowing in the gas core. The droplet can leave the liquid film on the wall by various entrainment mechanisms, including also the bubble explosion as a result of boiling of a thin liquid film.

The initial velocity of the droplet leaving the liquid film on the wall is an important parameter for predicting the trajectory of the droplet. If this initial velocity is not high enough, the droplet will just enter the gas stream and will be then subjected to random motion caused by turbulent eddies. The droplet will eventually be deposited on the wall by a diffusion mechanism. However, if the initial velocity of the droplet is high enough, the droplet of a given size will penetrate the turbulent gas stream and will be deposited on the opposite side of the tube.

Equations for droplet motion in r and z directions (r is measured from the wall towards the center line and z is pointed in the flow direction) are developed here to predict various droplet trajectories. A comparison with the available data is given. Overall, the analysis of droplet motion presented in this study is useful to study heat and mass transfer in annular flow.

1. INTRODUCTION

In annular two-phase flow, a significant fraction of the liquid phase can be entrained as liquid droplets flowing in the gas core. The droplet can leave the liquid film on the wall by various entrainment mechanisms as shown in Fig. 1, as surveyed by Ishii & Golmes [1]. Also, by the bubble explosion, as a result of boiling a thin liquid film in the annular flow, the droplets can be entrained.

The initial velocity of the droplet leaving the liquid film on the wall by either mechanism mentioned above is an important parameter for predicting the trajectory of the droplet. If this initial velocity is not high enough, the droplet will just enter the gas stream and will be then subjected to random motion caused by turbulent eddies. The droplet will eventually be deposited on the wall by a diffusion mechanism [2].

However, if the initial velocity of the droplet is high enough, the droplet of a given size will penetrate the turbulent gas stream and will be depos-

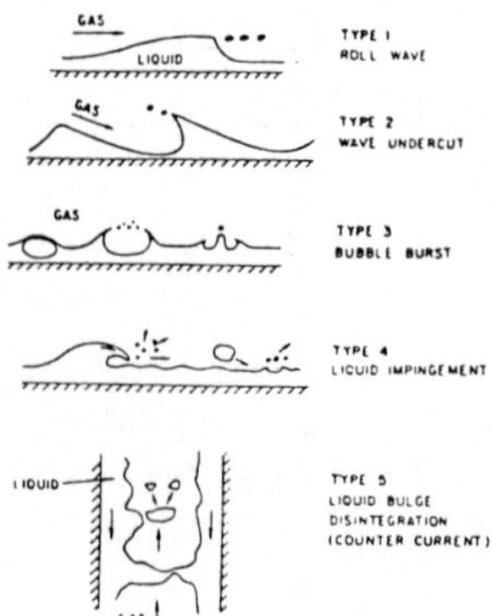

Figure 1: Various entrainment mechanisms (Ishii & Golmes, 1979)

ited on the opposite side of the tube. As seen later in the analysis, it is assumed that there the droplet will be deposited on the opposite side of the wall if its initial velocity is slightly higher than the "critical initial velocity," which is defined here as the initial velocity of the droplet which enables the droplet to just reach the center line of the flow channel.

Equations for droplet motion in r and z directions (r is measured from the wall towards the center line, as shown in Fig. 3, and z is pointed in the flow direction) are developed here to predict the "initial critical velocity" defined above. These analyses can be used to determine how far the particle will move upward, especially in a small test section, when the droplet might travel a greater distance than the length of the test section.

2. THEORETICAL ANALYSIS

2.1. Assumptions Used

1. The gas stream is fully developed turbulent flow.

2. The size of the particle (droplet) is small so that its motion relative to the fluid velocity obeys Stokes' law of resistance.

3. The traveling time is much greater than the time scale of energy containing eddies. This will be true if the velocity of the droplets is several times U^*. For the above reasons we simplify the real motion of the turbulent eddies by assuming that the rms radial velocity of the gas is always radially outwards and equal to U^*, which is the mean value measured by Laufer [3]. Therefore the motion of the particle is resisted by the gas motion in the range for the tube wall up to the center line. If the particle reaches the center line it will be accelerated by the gas motion towards the other side of the tube wall.

4. The particle entrained from the liquid film will move radially towards the center line, i.e., in a direction where there is the highest resistance to

the motion (see fig. 2). With this assumption it was not necessary to include the θ direction, i.e., only the $\theta = 0$ case is studied, since values of rms velocity are best known in the radial direction $0 < r < r_o$ [3]. Also, z-direction is included here since it provides an indication of how far the droplet might move in the axial direction due to its initial ejection velocity. This is important for the case of short test sections where, in some cases, the particle's vertical traveling distance due to its initial ejection velocity might be even greater than the length of the test section.

2.2. Equation of Motion

The equation of motion of a single droplet in the presence of fluid crossing the flow with an average axial velocity $\bar{U}_g$,

in the r-direction:

i) $0.0 \leq r/D \leq 0.5$

$$\frac{\pi}{6} d_p^3 \rho_p \frac{dV_r}{dt} = 3 \mu_g d_p (-U_g - V_r) \qquad (1)$$

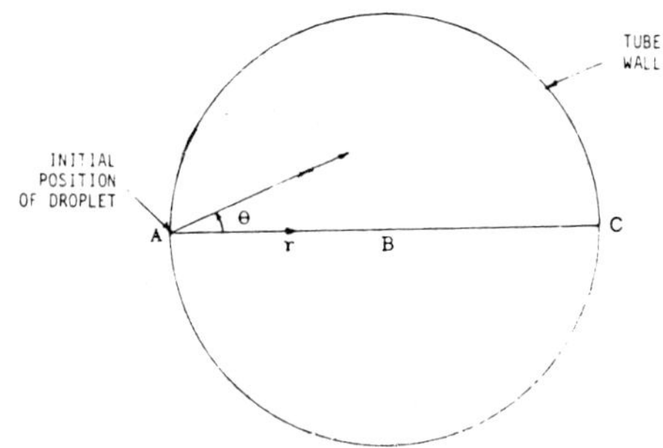

Figure 2: The pass which is considered is A-B-C

ii) $0.5 \leq r/D \leq 1.0$

$$\frac{\pi}{6} d_p^3 \rho_p \frac{dV_r}{dt} = 3 \pi \mu_g d_p (U_g - V_r) \qquad (2)$$

where the term on the left-hand side is the force required to accelerate the particle, and the term on the right-hand side is the viscous resistance force according to Stokes' law.

It is to be noted that the difference between Eq. (1) and Eq. (2) is the sign of U_g, which has been assumed to be positive outwards from the center line. In Eq. (1) the velocity Ug is against the motion of the particle, while in Eq. (2) the velocity is in the same direction of the motion of the particle.

in the z-direction:

$$\frac{\pi}{6} d_p^3 \rho_p \frac{dV_z}{dt} = 3 \pi \mu_g d_p (\bar{U}_g - V_z) - \frac{\pi}{6} d_p^3 \rho_p g \tag{3}$$

where the term on the left-hand side represents the force required to accelerate the particle motion in the axial direction, the first term in the right-hand side is the viscous resistance force according to Stokes' law, and the second term represents the gravitational force.

The boundary conditions for Eqs. (1 - 3) are:

at the time t = 0

1) $r = 0,\ z = 0$

2) $V_r = V_{ro},\ V_z = V_{zo}$

where V_{ro} and V_{zo} are the initial velocities in the r and z directions, respectively. The boundary conditions used to solve Eq. (2) was taken considering the final condition of Eq. (1). Applying the previous boundary conditions to Eqs. (1), (2) and (3), and replacing the term U_g by U* in Eq. (1) and (2) as mentioned in assumption 3., the following solutions are obtained:

in the r-direction:

i) $0 \leq \frac{r}{D} \leq 0.5$

$$r = \frac{U^* + V_{ro}}{a} [1 - e^{-at}] - U^*t \tag{4}$$

$$V_r = (U^* + V_{ro})\, e^{-at} - U^* \tag{5}$$

ii) $0.5 \leq \frac{r}{D} \leq 1$

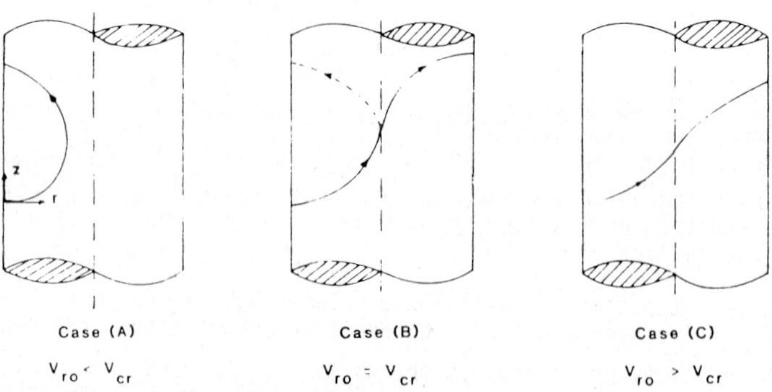

Figure 3: The pass possibilites which can be taken by single droplet ejected from the wall surface

$$r = C_1 + C_2 e^{-at} + U^*t \tag{6}$$

$$V_r = -aC_2 e^{-at} + U^* \tag{7}$$

in the z-direction:

$$z = \left[\frac{-g + a\bar{U}}{a^2} - \frac{V_{zo}}{a}\right]\left(e^{-at} - 1\right) + \left(\frac{-g + a\bar{U}_g}{a}\right)t \tag{8}$$

$$V_z = \left[V_{zo} - \left(\frac{-g + a\bar{U}_g}{a}\right)\right]e^{-at} + \left(\frac{-g + a\bar{U}_g}{a}\right) \tag{9}$$

where

$$a = \frac{18\mu_g}{\rho_p \, d_p^{\,2}}$$

$$C_1 = \frac{1}{a}\left[\frac{Da}{2} + V_{\mathbb{C}} - U^* \left(1 + at_1\right)\right] \tag{10a}$$

$$C_2 = \frac{1}{a}\left(U^* - V_{\mathbb{C}}\right)e^{at_1} \tag{10b}$$

and

$V_{\mathbb{C}}$ = particle velocity at the center line

t_1 = time required for droplet to reach the center

Note that t_1 can be evaluated from Ea. (4), i.e., finding the value of t for which r = D/2. Hence, the resultant velocity of the particle can be easily evaluated as

$$\bar{V}_p = \sqrt{V_r^{\,2} + V_z^{\,2}} \tag{11}$$

2.3. The Initial Critical Velocity

The concept of the so-called "initial critical velocity" is introduced here.

The initial critical velocity is the r-component of the initial ejection velocity which enables the droplet to just reach the center line.

To clarify the concept of critical initial velocity, the possibilties which can be taken by the initial velocity, V_{ro}, according to the set of cases shown in Fig. 3 can be examined. There may exist one of three possibilities:

<u>Case A</u>: $V_{ro} < V_{cr}$

The droplet cannot overcome the fluid resistance, and it can only take the path shown in Fig. 3.

<u>Case B</u>: $V_{ro} = V_{cr}$

The droplet will reach the center line with zero horizontal component and hence, the turbulent eddies will carry it towards either side of the cylinder.

<u>Case C</u>: $V_{ro} > V_{cr}$

The droplet will reach center line with value = $V_{℄}$, and will continue by its own inertia and/or by the turbulent eddies to the other side of the tube.

Next, let's study the important case existing at which $V_{ro} = V_{cr}$ should be examined. According to the boundary condition, at

$t = t_1$ (t_1 = time required to reach center line)

$V_{℄} = 0$

and

$r = D/2$

By substituting these values into Eqs. (4) and (5), we get

$$\frac{D}{2} = \left(\frac{U^* + V_{cr}}{a}\right)\left(1 - e^{-at_1}\right) - U^* t_1 \tag{12a}$$

$$0 = \left(U^* + V_{cr}\right) e^{-at_1} - U^* \tag{12b}$$

From Eq. (12b)

$$t_1 = -\frac{1}{a} \ln\left(\frac{U^*}{U^* + V_{cr}}\right) \tag{13a}$$

Substitute (13a) into (12a) to produce

$$\frac{D}{2} = \left(\frac{U^* + V_{cr}}{a}\right) - \frac{U^*}{a}\left[1 - \ln\left(\frac{U^*}{U^* + V_{cr}}\right)\right] \tag{13b}$$

Solving Eq. (13b) we obtain

$$C\, e^{-V_{cr}/U^*} = \frac{1}{1 + \frac{V_{cr}}{U^*}} \tag{13c}$$

where

$$C = e^{\frac{a}{U^*}\frac{D}{2}}$$

Let

$$w = \frac{V_{cr}}{U^*}$$

then

$$C\ e^{-w} = \frac{1}{1 + w} \tag{14}$$

Using the method of trial and error, we can find the value of w which satisfies Eq. (14). Let the value of w which satisfies Eq. (14) be w*, then

$$V_{cr} = w^* \cdot U^* \tag{15}$$

3. RESULTS AND DISCUSSION

Equation (14) shows that there is one value of w for each value of c. Knowing that c has one value for each D/2U*, therefore there is a unique relation between a* and w*, where

$$a^* = \frac{a\ D}{2U^*} = \frac{g\mu_g\ D}{\rho_p\ d_p^{\ 2}\ U^*} \tag{16}$$

The values calculated for w* at each corresponding value for a* covering a large range of d_p is tabulated in Table 1. and plotted in Fig. 4, which shows that as d_p increases, a* decreases correspondingly with w*. This graph is useful for finding V_{cr} when d_p and flow conditions are given.

Parametric studies have been done to show the effect of initial velocity, drop diameter and the Reynolds number on the trajectory of a single droplet using the present model. The tube diameter has been selected as 0.0127 m (0.5 in.), which has been used in many experimental studies on drop deposition (Ganic and Mastanaiah, [2]). In this particular study, the air was chosen as fluid, and the droplet was taken as a water droplet.

3.1. Effect of the Radial Component of Initial Velocity (V_{ro})

To study the effect of V_{ro}, Re = 23000 and U* = 1.4 m/s were considered, besides the general specification explained before. This effect is shown in Fig. 5.

The droplet diameter has been selected as 20 μm, which is the approximate mean value obtained in many studies on drop deposition [2]. V_{ro} changes here from 5U* to 8U*. For d_p = 20 μm, the initial critical velocity of 5.6 m/s is obtained. The following observations have been noted:

1. For the case V_{ro} = 5U* (i.e., V_{ro} less than the critical velocity) the droplet returns back to the same side. This occurs because the particle

Table 1. The Universal Solution for the Critical Velocity

a*	$w = V_{cr}/U^*$
83.7038	11.205
2.17597	3.73
0.96709	2.098
0.5439	1.434
0.348155	1.082
0.087038	0.478
0.0386839	0.304
0.021759	0.222
0.013924	0.175
9.6709×10^{-3}	0.1455
7.104×10^{-3}	0.1255
5.43×10^{-3}	0.108
4.29×10^{-3}	0.095

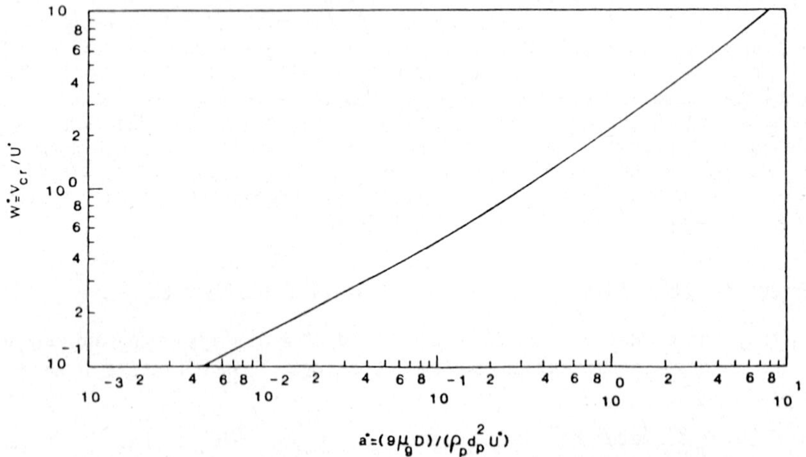

Figure 4: The Universal solution for critical velocity

does not have enough inertia to overcome the resistance of fluid turbulent eddies to be able to reach the center line.

2. If V_{ro} is greater than V_{cr} (here for the cases V_{ro} = 6U*, 7U*, and 8U*), the droplet will hit the opposite side of the tube. In this range, $V_{ro} > V_{cr}$, increasing V_{ro} will decrease the time required to hit the opposite side of the tube, which will not enable the oncoming fluid to carry the droplet a very great distance. Hence, the higher the initial velocity, V_{ro}, the lesser the values for z/D and $\overline{V}_p$.

3.2. Effect of Droplet Diameter

The study has been done with Re = 23000, V_{ro} = 5U*, V_{zo} = 0, and droplet diameter varying from 10 μm up to 40 μm. The results are plotted in Figs. 6-8. In Fig. 6, the relationship between $\overline{V}_p/U^*$ and r/D are plotted. As d_p increases V_{cr} decreases. This is because increasing d_p increases the inertial force that enables the droplet to go further. In Fig. 7, the value of V_r at different locations during the droplet trajectory are plotted. We note that for particles having initial velocity less than the critical velocity, the velocity in r direction (radial direction) decreases with a high rate until it reaches zero value (at location before the center line). It then returns back to the same side of ejection, i.e., V_r now has a negative value.

For those particles having an initial velocity greater than the critical velocity, the particle velocity decreases to a certain value at the center line because of the gas resistance. If this value is still greater than U* after crossing the center line, the particle velocity will also decrease, but at a lower rate. This is because the velocity of the particle is still greater than the gas velocity and therefore it is resisted by the gas. If the velocity of the particle at the center line is less than U*, then it will be accelerated by the gas motion after crossing the center line.

In Fig. 8, the vertical distance, z/D, is plotted versus the radial distance, r/D; and in Fig. 9, the axial velocity, V_z/U^*, is plotted versus the radial distance, r/D. It is noted that from both Figs. 8 and 9, for those droplets having initial velocities less than the critical velocity, an increase in the droplet size will increase the vertical distance z/D and the axial velocity V_z/U^*. In contrast, for droplets having an initial velocity greater than the critical velocity, as the droplet size increases, the vertical distance decreases due to the same reason explained before.

3.3. Effect of Reynolds Number

The study has been done for d_p = 30 μm and initial velocities V_{ro} = 5U* and V_{zo}, with the following values for Re considered: 20000, 30000, 60000, and 80000. The results are plotted in Fig. 10. From the graph, it can be seen that as the Reynolds number increases, the total velocity $\overline{V}_p$, decreases and z/D decreases. The reason for this is that the motion of the droplet from one side to the other depends mainly on the horizontal component of the velocity V_r, which depends on turbulent radial velocity U*. Since V_{ro}/U^* is taken to be constant, increasing Re will increase U*, which correspondingly increases V_r, which in turn decreases the traveling time. Consequently, this effective reduction in traveling time will reduce the vertical distance, z/D.

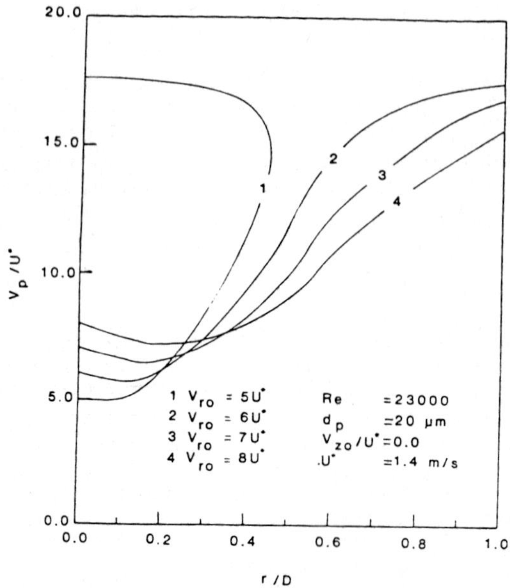

Figure 5: The effect of initial velocity in radial direction

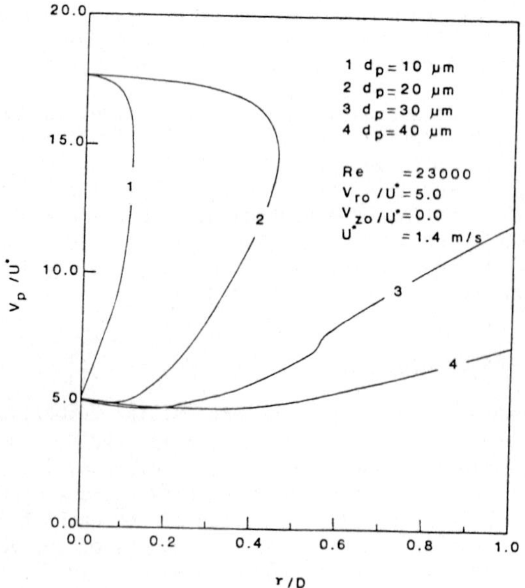

Figure 6: The effect of droplet diameter on the trajectory of the droplet

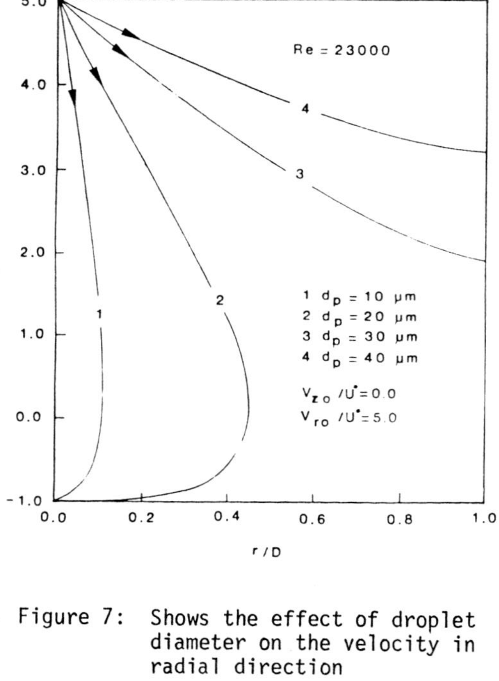

Figure 7: Shows the effect of droplet diameter on the velocity in radial direction

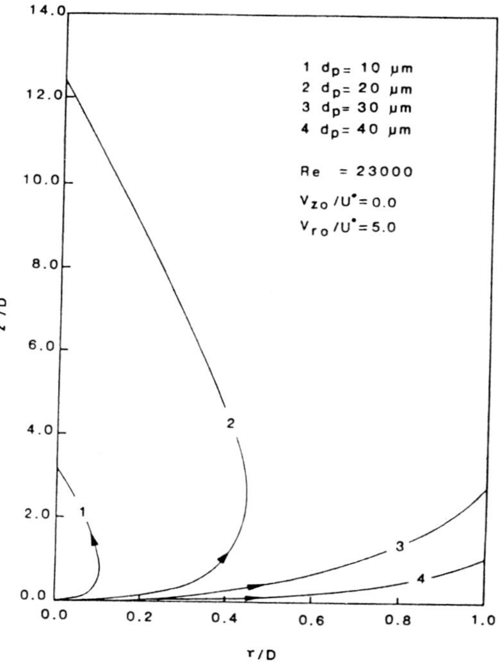

Figure 8: Shows the effect of droplet diameter on the vertical distance

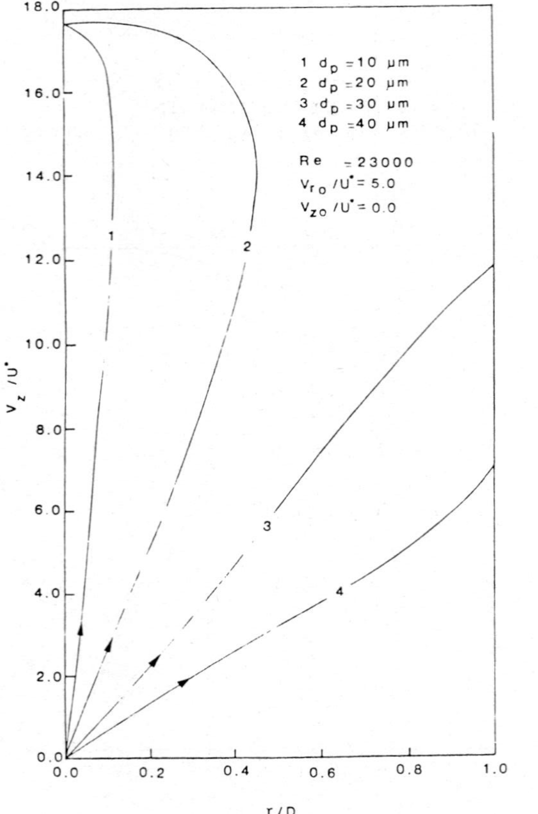

Figure 9: Shows the effect of droplet diameter on the vertical velocity

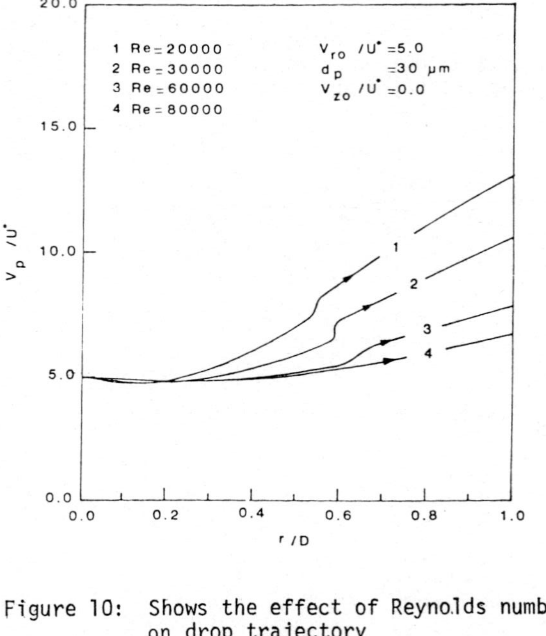

Figure 10: Shows the effect of Reynolds number on drop trajectory

3.4. Effect of the Vertical Component of Initial Velocity (V_{zo})

The study has been done for Re = 23000, V_{ro} and the droplet diameter having been fixed, with only V_{zo} being permitted to change. The results are plotted in Fig. 11 from which we can see that V_{zo} does not have any effect on critical velocity. This is logically correct, since the equation of V_{cr} does not have any term including V_{zo}. It can be seen that the only effect of increasing V_{zo} will be the increase of both the vertical distance, z/D, and the total velocity, V_p.

4. COMPARISON WITH EXPERIMENTAL DATA

In an attempt to visualize the real motion of a droplet in annular flow, Whalley et al. [4], Hewitt and Whalley [5], and James et al. [6] have used a laser axial view technique. More details about the experiment and their techniques are given by [5].

Due to the shortcomings of their laser axial view technique, they could not record the motion of droplet of diameters smaller than about 150 μm. James et al. [6] introduced a new model to predict the position of a droplet in a plane normal to the tube axies as a function of time. At zero time the droplet was assumed to be at the tube wall and traveling towards the axis of the tube at a prescribed velocity. In that model the description of turbulence used by Hutchinson et al. [7] had been used, which indicated that the turbulent structure in the cross section is assumed to be made up of a series of eddies, all of equal length scale, time scale and velocity scale, but with random direction. Droplet trajectory was computed for a range of sizes (1000 μm - 50 μm) and ejection velocities (0.1 m/s - 1.0 m/s) for the following system parameters: $\overline{U}_g$ = 29 m/s, D = 3.2×10^{-2} m, μ_g = 1.76×10^{-5}, Ns/m^2, ρ_p = 1000 kg/m^3 and ρ_g = 1.2 kg/m^3. For each droplet diameter and ejection velocity sixty droplet trajectories were calculated. James et al. analyzed these trajectories to estimate the fraction of the droplets which were deposited in the quadrant opposite to the point of ejection. The obtained results are plotted in Fig. 12. This figure gives an indication of the percentage of droplets which are deposited primarily under the invluence of their own initial ejection velocity. The region to the right of the upper line in this graph represents all the ejected droplets that are deposited in the quadrant opposite the ejection point. Below the lower line, less than one-half of the droplets are deposited in the opposite quadrant. The calculated critical initial velocity for the same condition used by [6] is given in Table 2 at different droplet diameters, and is also plotted in Fig. 12 by the solid line. It can be noticed that the initial critical velocities data, represented by the solid line on Fig. 12 lies almost on the upper line obtained by [6]. This means that the critical initial velocity given by Eq. (14) is a good prediction. Especially at V_{ro} = V_{cr}, the present model shows that the particle is deposited either on the same side of the wall or on the other side. Also, if V_{ro} is greater than V_{cr}, the particle will be deposited on the other side of the wall.

5. CONCLUSION

Comparison of the analysis presented here with data of James et al. (1980) indicates its usefulness. Specific conditions related to droplet trajectory are listed below.

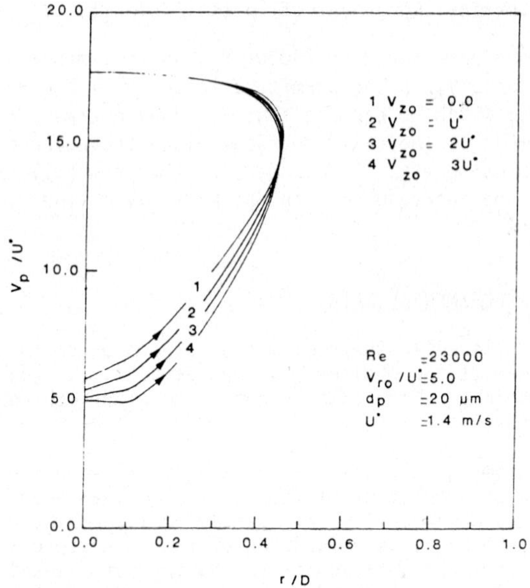

Figure 11: Shows the effect of initial velocity in vertical component on drop trajectory

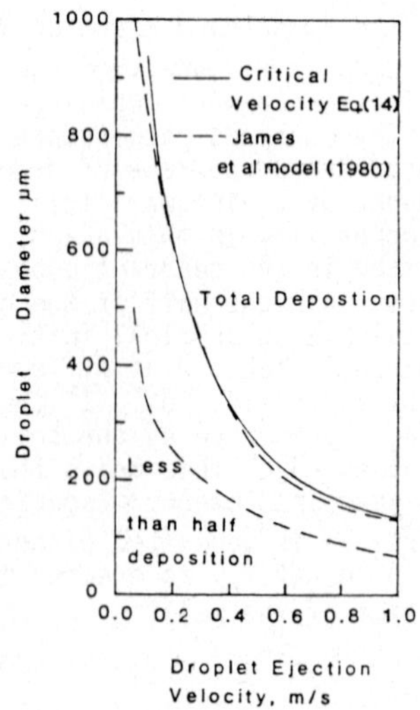

Figure 12: Comparison between the present model with results obtained by James et al. (1980)

Table 2. Critical Velocities Evaluated at Different Droplet Diameters - $Re = 63272$, $\mu_g = 1.76\times10^{-5} N\ s/m^2$ $\rho_p = 1000\ Kg/ms$, $\rho_g = 1.2\ Kg/m^3$

$d_p(\mu m)$	100	200	300	400	500	600	700	800	900
$V_{cr}(m/s)$	1.575	0.6954	0.44	0.3219	0.253	0.21	0.18	0.156	0.1377

- If $V_{ro} < V_{cr}$, the droplet which originates its flight at the wall will never reach the other side of the cylinder.
- If $V_{ro} < V_{cr}$, increasing V_{ro} increases the vertical distance z/D and V_p.
- If $V_{ro} > V_{cr}$, increasing V_{ro} decreases the vertical distance z/D and V_p.
- Increasing d_p decreases V_{cr}.
- For those droplets having $V_{ro} < V_{cr}$, increasing d_p causes the increase of $\overline{V}_p$, a/D.
- For those particles having $V_{ro} > V_{cr}$, increasing d_p causes the decrease of both $\overline{V}_p$ and z/D.
- Increasing Re causes the decrease of $\overline{V}_p$, z/D and V_{cr}.

This simple analysis of droplet motion presented here is useful to the study of mass and heat transfer in annular flow.

NOMENCLATURE

a = constant equal to $\frac{18\mu_g}{\rho_p d_p^2}$, $\frac{1}{s}$

a* = nondimensional constant, $\frac{a\ D}{u^{*2}}$

C_D = drag coefficient

D = tube diameter, m

d_p = droplet diameter, m

g = acceleration due to gravity, m/s^2

m = droplet mass , kg

r = radial distance measured from the wall towards the centerline, m

R = radius of tube, m

Re = tube Reynolds number, $\frac{\rho \bar{U}_g D}{\mu g}$

t = time, s

t_1 = time required for droplet to reach the center line, s

U_g = radial (rms) fluctuation velocity of the gas, m/s

U* = friction velocity, $\bar{U}_g \sqrt{f/2}$, m/s

$\bar{U}_g$ = mean gas axial velocity, m/s

V_d = droplet velocity at center line in x direction, m/s

V_r = droplet velocity in radial direction, m/s

V_{ro} = initial droplet velocity in radial direction, m/s

V_z = droplet velocity in axial direction, m/s

V_{zo} = initial droplet velocity in axial direction, m/s

V_{cr} = critical velocity, m/s

V_p = total velocity of the droplet, m/s

z = axial distance from the location of the droplet, m

Greek Symbols

ρ_g = density of gas, kg/m^3

ρ_p = density of particle, kg/m^3

μ_g = dynamic viscosity for the gas, N.S/m^2

ν_g = kinematic viscosity, m^2/s

REFERENCES

1. Ishii, M. and Grolmes, M.A., "Inception Criteria for Droplet Entrainment in Two-Phase Concurrent Flim Flow," AIChE J., Vol. 21, No. 2, p. 308 (1975).

2. Ganic, E.N. and Mastanaiah, K., "Investigation of Droplet Deposition from a Turbulent Gas Stream," Int. J. Multiphase Flow, Vol. 7, pp. 401-422 (1981).

3. Laufer, J., "The Structure of Turbulence in Fully Developed Pipe Flow," NACA Report 1174 (1953).

4. Whalley, P.B., Hewitt, G.F. and Terry , J.W., "Photographic Studies of Two-Phase Flow Using a Parallel Light Technique," U.K.A.E.A. Report, AERE-R9389 (1979).

5. Hewitt, G.F. and Whalley, P.B., "Advanced Optical Instrumentation Methods," to appear in Int. J. Multiphase Flow.

6. James, P.W., Hewitt, G.F. and Whalley, P.B., "Droplet Motion in Two-Phase Flow," AERE-R9711 (1980).

7. Hutchinson, P., Hewitt, G.F., and Dukler, A.E., "Deposition of Liquid or Solid Dispersions from Turbulent Gas Streams: A Stochastic Model," Chem. Engng. Sci. 26, pp. 419-439 (1971).

Multi-Phase Flow and Heat Transfer III. Part A: Fundamentals
edited by T.N. Veziroğlu and A.E. Bergles
Elsevier Science Publishers B.V., Amsterdam, 1984 — Printed in The Netherlands

A FURTHER STUDY IN DROP DEPOSITION

M. M. El-Kassaby
Department of Engineering
Purdue University Calumet
Hammond, IN 46323, U.S.A.

E. N. Ganic
Faculty of Mechanical Engineering
University of Sarajevo
Sarajevo 71000, Yugoslavia

ABSTRACT

Theoretical and experimental studies have been carried out to further investigate the turbulent deposition of particles from two-phase droplet flow onto the smooth wall of a vertical tube. The work of Ganic & Mastanaiah (1981) for Stoke Regime has been extended by this study to include the Oseen regimes ($Re_p \leq 5$). The proposed theory satisfactorily describes the existing data, as well as the new data taken for air-water droplet flow passing through a 1295mm diameter stainless steel tube at Re = 27000, 36000, 43000, and 55000.

1. INTRODUCTION

The determination of droplet deposition rates is of interest in many applications, such as steam generators and nuclear reactor cooling systems. In two-phase, turbulent gas and particle flow through a vertical tube, having a diameter in excess of about a micron, deposition on the wall occurs primarily by turbulent diffusion, in the absence of electrostatic effects and large temperature gradients.

The experimental deposition data are usually presented in terms of a dimensionless deposition velocity K_d/U^* as a function of the dimensionless particle relaxation time τ^+, based on the Stokes drag force, defined as

$$\tau^+ = \frac{d_p^{\,2}\ \rho_g \rho_p\ U^{*2}}{18 \mu_g^{\,2}} \tag{1}$$

In many practical two-phase flow problems, the deposition of large particles having d_p in excess of about 20-30 μm is of interest. (A review of various theoretical and experimental studies on the deposition of large particles was reported recently by Ganic and Mastanaiah, 1981.) For vertical flow systems, deposition data in this range include those developed by Sehmel (1968), Ilori (1971), Forney & Spielman (1974), Liu & Agarwal (1974), Agarwal (1975) for unifor-sized particles, Cousins & Hewitt (1968) and Farmer et al. (1970) involving the drop size distribution.

The theoretical analysis of the deposition of particles in the above range include those of Friedlander & Johnstone (1957), Forney and Spielman

(1974), Cleaver & Yates (1957), Namie & Ueda (1973), and Hagiward et al. (1979). The various limitations of the above models and the conceptual deficiencies associated with them are described in detail in Ganic & Mastanaiah (1981), who have proposed a physically more acceptable model for the deposition motion. The latter model considers that for large particles with a dimensionless stopping distance (based on Stokes drag) $S^+ > 30$, the particles move from the tube center to the periphery of the buffer layer by turbulent diffusion, then reach the wall by free flight due to the inertia imparted to them by the turbulent eddies at the start of free flight. Their analysis, using Tchen's theory (1947) for particle diffusion was valid within the Stokes regime. The main limitation of their theory is that it cannot be applied for particles outside the Stokes regime, i.e., $Re_p > 1$, where

$$Re_p = \rho_g \frac{|V_p - V_g| d_p}{\mu_g} \tag{2}$$

where V_p and V_g represent rms velocity fluctuation of particle and fluid respectively. The main purpose of the present work is to extend the above theory to include the Oseen regime ($Re_p \leq 5$).

2. THEORETICAL ANALYSIS

The physical model used here for the deposition of large particles is similar to that proposed by Ganic & Mastanaiah (1981). The most important difference between the present work and the above is in the calculation of the particle to fluid diffusivity ratio, δ.

2.1. Deposition Velocity

The following assumptions, originally made by Mastanaiah (1980), are repeated and some revised in the analysis here.

a) The flow is fully developed.

b) The concentration profile is developed so that the entrance effects are not present.

c) Drop concentration is small enough to consider that the fluid turbulence characteristics are unaltered.

d) Particle size is such that the explanation of its motion relative to the fluid includes the Oseen regime ($Re_p < 5$).

e) There is no wall rebound or re-entrainment of drops.

f) The local mass flux of droplets in the turbulent core varies linearly from zero at the tube center to a value of N_o at $y^+ = 30$, thus remaining constant to a value equal to N_o in the region $0 < y^+ < 30$. Therefore, the local mass flux can be given as

$$N = \frac{N_o}{\left[1 - \frac{30}{r_o^+}\right]} \left(1 - \frac{y^+}{r_o^+}\right) \tag{3}$$

g) Both the fluid and the droplet move radially towards the wall with the radial rms fluctuating velocity.

The rate law for the diffusion of droplets due to concentration of gradients is given by

$$N = \varepsilon_p \frac{dc}{dy} \tag{4}$$

where ε_p is the particle diffusivity and equal to $\delta\ \varepsilon_g$. Eq. (4) for the particle flux is based on the assumption that the particles in the gas core interact with the turbulent eddies in such a way that particle dispersion is condidered essentially diffusive. The eddy diffusivity (ε_g) expression proposed by Reichardt (1951) is also considered here and is given by

$$\varepsilon_g = \frac{K\ r_o\ U^*}{6} \left(1 - \left(\frac{r}{r_o}\right)^2\right)\left(1 + 2\left(\frac{r}{r_o}\right)^2\right) \tag{5}$$

Here K is the mixing length constant and is equal to 0.4. Integration of Eq. (4) between $y^+ = 30$ and a point y^+ in the core leads to the concentration distribution as given by

$$C - C_b = \frac{N_o}{\left(1 - \dfrac{30}{r_o^+}\right)U^*\delta K}\ \ell n \left[\frac{1 + 2x}{1 - x}\right]_{X=(1 - y^+/r_o^+)}^{x=(1 - 30/r_o^+)} \tag{6a}$$

At $y^+ = r_o^+$, we have

$$C_c - C_b = \frac{N_o}{\left(1 - \dfrac{30}{r_o^+}\right)U^*\delta K}\ \ell n \left[\frac{1 + 2\left(1 - \dfrac{30}{r_o^+}\right)^2}{1 - \left(1 - \dfrac{30}{r_o^+}\right)^2}\right] \tag{6b}$$

The deposition velocity is defined as

$$K_d = \frac{N_o}{\overline{C}} \tag{7a}$$

with the bulk concentration $\overline{C}$ given by

$$\overline{C} = \frac{\int_0^{r_o} C\ U\ r\ dr}{\int_0^{r_o} U\ r\ dr} \tag{7b}$$

where U(r) is the radial velocity distribution in the tube. In evaluating $\overline{C}$, it is assumed that $C = C_b$ for $0 < y^+ < 30$. This assumption was studied in detail by Ganic & Mastanaiah (1980). By definition of the mass flux, N_o can be expressed as

$$N_o = V_{pb} C_b \tag{7c}$$

From Ganic & Mastanaiah (1981)

$$V_{pb} = 0.75\, U^* \sqrt{\delta} \tag{7d}$$

Equating Eq. (7a) and (7c) using Eq. (7d), we obtain a relation for C_b

$$\frac{C_b}{\bar{C}} = \frac{K_d}{0.75\, U^* \sqrt{\delta}} \tag{7e}$$

The value of $\bar{C}/C_c$ is determined from Eqs. (6a), (6b), (7b) and (7e). In Eq. (7b), the following distribution for $u^+ = (U/U^*)$ is considered (Kays, 1966):

$$U^+ = y^+ , \qquad 0 < y^+ < 5 \tag{8a}$$

$$U^+ = 5\, \ell n\, y^+ - 3.05, \qquad 5 < y^+ < 30 \tag{8b}$$

$$U = 5.5 + 2.5\, \ell n \frac{y^+ \, 1.5 \; 1 + r/r_o}{1 + 2\, r/r_o{}^2} \; , \; 30 < y^+ < r_o{}^+ \tag{8c}$$

Using Eqs. (6c), (7a) and (7e), the dimensionless deposition velocity is given by

$$\frac{K_d}{U^*} = \frac{\delta\,(C_c / \bar{C})}{\dfrac{2.5}{\left(1 - \dfrac{30}{r_o^+}\right)} \ell n \left[\dfrac{1 + \left(1 + 2\left(1 - \dfrac{30}{r_o^+}\right)\right)^2}{1 - \left(1 - \dfrac{30}{r_o^+}\right)^2} \right] + \sqrt{\dfrac{\delta}{0.75}}} \tag{9}$$

2.2. Particle to Fluid Diffusivity Ratio δ

The diffusivity ratio δ is defined as (Hinze, 1975)

$$\delta = \frac{\varepsilon_p}{\varepsilon_g} = \frac{\overline{v_p'^2}}{\overline{v_g'^2}} = \frac{V_p{}^2}{V_g{}^2} \tag{10}$$

The diffusivity ratio, δ, can be determined from the equation of motion of the particles in a turbulent flow field neglecting the gravitational and preseure gradient forces. Taking into consideration assumption number g, the equation of motion can be written as

$$m \frac{dV_p}{dt} = C_D \frac{\pi d_p^2}{8} \rho_g (V_g - V_p)^2 \tag{11a}$$

where the drag coefficient C_D was taken for the Oseen regime (McCormik & Lawrence, 1973), which is valid up to $Re_p = 5$, as

$$C_D = \frac{24}{Re_p} \left[1 + \frac{3}{16} Re_p\right] \tag{11b}$$

Substituting Eq. (2) into (11b) and then into (11a), we get

$$\frac{dV_p}{dt} = a|V_g - V_p| + b(V_g - V_p)^2 \tag{12a}$$

where

$$a = \frac{18\mu_g}{d_p^2 \rho_p} \tag{12b}$$

and

$$b = \frac{27 \rho_g}{8 d_p \rho_p} \tag{12c}$$

If $V_g - V_p = w$, then Eq. (12a) can be rewritten as

$$\frac{dw}{dt} = -b w^2 - aw + \frac{dV_g}{dt} \tag{13}$$

To solve the last differential equation, we have to know dV_g/dt. For the time being, let $dV_g/dt = C_1$ (see Section 2.4) and use the Riccati technique (Korn & Korn, 1968) to convert Eq. (13) from nonlinear form, as follows. Let

$$w = \frac{\xi'}{b \xi} \tag{14a}$$

then

$$\frac{dw}{dt} = \frac{b \xi \xi'' - b \xi'^2}{(b \xi)^2} \tag{14b}$$

Substituting into Eq. (13), we obtain

$$\xi'' + a\,\xi' - b\,\xi\,C_1 = 0 \tag{15}$$

which is an ordinary differential equation, having the general solution

$$\xi = K_2\, e^{C_3 t} + K_3\, e^{C_4 t} \tag{16}$$

provided that $dV_g/dt > 0$ (see Fig. 1) where

$$C_3 = -\frac{a}{2} + \frac{1}{2}\sqrt{a^2 + 4bC_1} \tag{17a}$$

$$C_4 = -\frac{a}{2} - \frac{1}{2}\sqrt{a^2 + 4bC_1} \tag{17b}$$

The constants K_2 and K_3 can be determined from the boundary condition.

Eq. (14a) can be rewritten as

$$w = \frac{1}{b}\left[\frac{K_2\, C_3 e^{C_3 t} + K_3\, C_4\, e^{C_4 t}}{K_2\, e^{C_3 t} + K_3\, e^{C_4 t}}\right] \tag{18}$$

assuming the boundary condition at the center line is as follows:

$$\text{at} \qquad t = 0,\ w = w_o \tag{19a}$$

where w_o = w at the center line. This boundary condition gives

$$K_2 = K_3 C_5 \tag{19b}$$

where

$$C_5 = \frac{C_4 - w_o b}{w_o b - C_3} \tag{19c}$$

Substituting Eqs. (19c) and (19b) into (18), we get

$$V_g - V_p = w = \frac{1}{b}\left[\frac{C_5\, C_3\, e^{C_3 t} + C_4\, e^{C_4 t}}{C_5\, e^{C_3 t} + e^{C_4 t}}\right] \tag{20}$$

The selection of the initial condition w_o is discussed in detail by El-Kassaby (1982), in the range of the present data (6000 < Re < 50000), w_o is selected to be 0.2 U*.

2.3. The Stopping Distance and the Relaxation Time

A stopping distance S is defined as the distance a particle would travel through a stagnant fluid ($V_g = 0$) with a prescribed velocity. Considering Eq. (12a), the stopping distance can be derived as

$$S = \frac{1}{b} \ln\left(\frac{a + b V_{p_o}}{a}\right) \tag{21a}$$

where V_{po} is the free flight velocity at the edge of the buffer layer taken as 0.9 U* (Friedlander & Johnston, 1957). Eq. (21a) can be written in dimensionless form as follows:

$$S^+ = \frac{SU^*}{\nu_g} = \frac{18\, d_p\, \rho_p\, U^*}{27\, \mu_g} \quad \ln\left(1 + \frac{2.7\, \rho_g d_p\, U^*}{16\, \mu_g}\right) \tag{21b}$$

The relaxation time τ can be expressed as follows

$$\tau = \frac{S}{V_{po}} \tag{22a}$$

which can be rewritten in nondimensioned form as follows

$$\tau^+ = \frac{\tau U^{*2}}{\nu_g} = 0.3292\, \frac{d_p U^*}{\nu_g} \quad \ln\left[1 + \frac{2.7\, d_p\, U^*}{16\, \nu_g}\right] \tag{22b}$$

2.4. Verification of the Assumption $dV_g/dt = C_1$

Measurements by Laufer (1953) have shown that the distribution of V_g versus the radial distance is almost the same for different Re, in the range $1 \geq y/r_o \geq 0.1$ (Fig. 1), and for this range v_g varying from 0.72U* at center line up to 1.1U* at y/r_o _ 0.1, which will be used later as a boundary condition. From the previous analysis and referring to Fig. 1, it can be shown that over almost the whole radial distance, the relation of V_g/U^* versus r/r_o can be represented by a parabolic relation which predicts a linear relationship with respect to time. The above mentioned fact was the main idea used in solving Eq. (13) by setting

$$\frac{dV_g}{dt} = C_1 \tag{23}$$

Eq. (23) can be rewritten as follows:

Author	D (mm)	d_p (μm)	Re	τ^+	k_d/U^*				% Error			Re_d
					Exp .	present theory	G & M*	F & J**	present theory	G & M*	F & J**	
Sehmel (1968)	71	28.0	35,000	40	.16	.1256	.123	.054	21	23.1	66.2	.05
	29	8.5	60,000	55	.12	.1123	.110	.050	6.4	8.3	58.8	.04
Ilori (1971)	29.8	9.0	50,000	48	.12	.116	.114	0.051	3.3	5	57.5	0.05

* G & M - theory of E.N. Ganic & K. Mastanaiah (1981)

** F & J - theory of Friedlander & Johnstone (1957)

Table 1: Comparison of predicted deposition velocities with the experimental data of Sehmel (1968) and Ilori (1971)

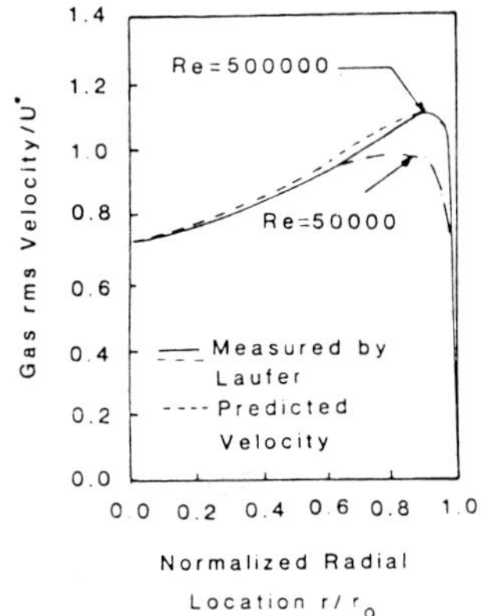

Figure 1: Comparison between the measured value of the rms velocity and the predicted one.

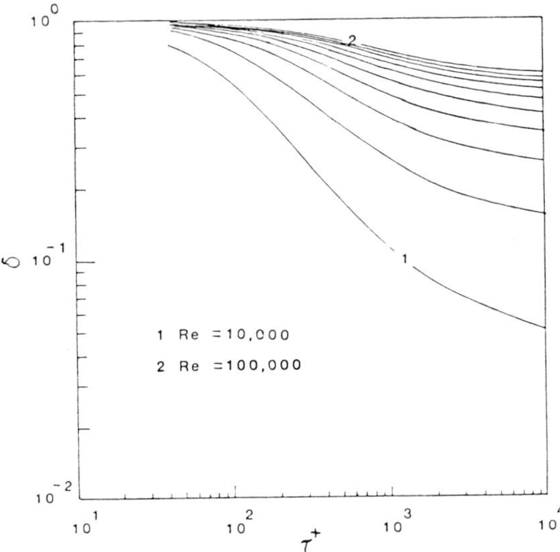

Figure 2: Prediction of the diffusivity ratio δ as a function of τ^+ at different values of Re, starting with Re = 10,000 with increments of 10,000 ending with Re = 100,000

$$V_g \frac{dV_g}{dr} = C_1 \tag{24}$$

By integrating Ea. (24) we get

$$\frac{V_g^{\,2}}{2} = C_1 r + C_6 \tag{25}$$

where C_6 is constant, depending on the boundary condition.

Knowing that $r = r_o - y$, then

$$\frac{V_g^{\,2}}{2} = C_1\, r_o \left(1 - \frac{y}{r_o}\right) + C_6 \tag{26}$$

Multiplying the previous equations by U^*/ν_g, to get it in dimenstionless form, we obtain

$$\frac{V_g^{\,2}}{2} \cdot \frac{U^*}{\nu_g} = C_1\, r_o^{\,+} \left(1 - \frac{y^+}{r_o^{\,+}}\right) + C_6 \frac{U^*}{\nu_g} \tag{27}$$

Using the boundary condition as measured by Laufer (1953) (see Fig. 1), as follows,

$$\left.\begin{array}{lll} V_g = 0.72\ U^* & , & \text{at the center line} \\ \text{and} & & \\ V_g = 1.1\ U^* & , & \text{at } y/r_o \simeq 0.1 \end{array}\right\} \tag{28}$$

Substituting these boundary conditions into Eq. (27) we get

$$C_1 = \frac{0.38422\ U^{*3}}{r_o^{\,+}\ \nu_g} \tag{29a}$$

and

$$C_6 = 0.2592\ U^{*2} \tag{29b}$$

Substituting Eqs. (29a) and (29b) into Eq. (27), we get

$$V_g = U^* \left[0.7684 \left(1 - \frac{y^+}{r_o^{\,+}}\right) + 0.5184\right]^{1/2} \tag{30}$$

V_g can be written as a function of time, using Eqs. (24) and (29a) as

$$V_g = \frac{0.3842\ U^{*3}\ t}{r_o^+\ \nu_g} + 0.72\ U^* \quad (31)$$

The mean rms velocity can be evaluated by integrating Eg. (30) with respect to y^+ to get

$$\overline{V}_g = 0.923213\ U^* \quad (32)$$

The plotting of Eq. (30) onto Fig. 1 (shown by small dashed lines) shows a good agreement with Laufer's measurements (1953).

2.5. Solution Technique

Eq. (20) was used to find the local value of $w = V_g - V_p$. Then using Eq. (31), the local value of V_g was determined. Looking at Eq. (20), one can find the local value of V_p and then from Eq. (10) find δ. Finally, substituting into Eq. (9), one can determine the value of K_d/U^*.

In many practical situations, the suspended droplets may not be of uniform size, but have a size distribution. In such cases, an appropriate mean diameter has to be specified for evaluating τ^+. The arithmetic mean drop size is taken as the effective droplet diameter in the present calculations.

3. PREDICTION OF PARTICLE DIFFUSIVITY RATIO

Using Eqs. (20), (31) and (10), we can evaluate δ at different values of Re and τ^+. The latter parameter is given in Eq. (22). The results are plotted in Fig. 2 , which shows the following observations.

At constant Re, as τ^+ increases, δ decreases.

At constant τ^+, as Re increases, δ increases.

The usefulness of this graph is that one can easily find δ at any given condition, i.e., if we know τ^+ and Re, we can find δ.

4. COMPARISON WITH EXPERIMENTAL DATA

The accuracy of the proposed theory is now tested by comparing the calculated results with the existing experimental data as well as with the present measurements.

4.1. Deposition Data for Monodisperse Particles

Much of the deposition data in the literature are obtained using monodisperse particles. Such data are not associated with uncertainties in particle size and therefore present an excellent basis for comparison with theory. For monodisperse particles, the present theory can be compared with the reliable data of Agarwal (1975) over a wider range of Re and τ^+, with a few data points of Sehmel (1968) and Liu (1971) within a narrow range of τ^+ between 40 and 60, and finally with the data of Ganic & Mastanaiah (1981).

Deposition data of Agarwal (1975). Agarwal (1975) has obtained deposition data for uniform-size uranine-tagged olive aerosol in a vertical down-flow of air in a 3.27 mm I.D. glass tube with L/D = 91.7 at Re = 6000, and in a 13.8 mm I.D. copper tube with L/D = 73.9 at Re = 50000. The test section tubes are smooth. The drops are generated by means of a vibrating orifice monodisperse aerosol generator. The maximum size of the droplets used is 21 μm, and the droplet to fluid density ratio is about 713. The deposition velocity is determined from the amount of aerosol deposited on the deposition section only, and hence no entrance effects are present.

The calculated deposition velocity K_d/U^* versus τ^+ using the present analysis is compared with Agarwal (1975) in Figs. 3 and 4, which show only data for $\tau^+ > 40$. The calculated results from the theories of Friedlander & Johnstone (1957) and Ganic & Mastanaiah (1981) are also displayed. It is clear that the theory of Friedlander & Johnstone (1957) was unable to predict the deposition velocity, and that of Ganic & Mastanaiah (1981) slightly under-predicts the measurements. However, the present model represents the best prediction for the experimental data, as it takes into account the presence of the inertia forces of fluid (Oseen regime).

Deposition data of Sehmel (1968) and Ilori (1971). Sehmel (1968) has obtained deposition data in an upward vertical flow of air and methylene blue aerosol in 71 and 29 mm I.D. tubes. The maximum particle diameters used are 28 and 8.5 μm for the larger and smaller diameter tubes respectively. The data of Ilori (1971) are obtained in an upward vertical flow of air and 80 percent methylene - 20 percent uranine in a 29.8 mm vertical tube at Re = 50000. The droplet to fluid density ratio is about 1317, and the size of the droplets is in the range of 6.9 μm.

The present theory, the theory of Friedlander & Johnstone (1957), and the theory of Ganic and Mastanaiah (1981) are compared in Table 1 with the data of Sehmel (1968) and Ilori (1971). It is evident that the present theory is in satisfactory agreement with the data, while the theory of Friedlander & Johnstone (1957) considerably under-predicts the data. It can also be noticed that the present theory has better agreement than that of Ganic & Mastanaiah (1981).

Deposition data of Ganic & Mastanaiah (1981). Ganic & Mastanaiah (1981) have obtained deposition data in an upward vertical flow of air and water in 12.7 mm I.D. acrylic tubes. The maximum particle diameter used is 45 μm. The present theory, the Mastanaiah theory (1980), and the theory of Friedlander & Johnstone (1957) are compared in Figs. 5 and 6. The present theory shows good agreement, as does that of Mastanaiah (1981) up to $\tau^+ = 800$ (Re = 94600). However, it will be noticed in Figs. 5 and 6 that data by Mastanaiah is underpredicted while that of the present theory is overpredicted. The present theory, however, follows the profile of the plotted experimental data better.

4.2. Present Deposition Measurements

Experimental apparatus and procedure. In order to provide a further check on the proposed deposition model, the theory is also compared with the present deposition data. The authors have recently obtained deposition rate measurements for air-water system in a vertical tube at near atmospheric pressure

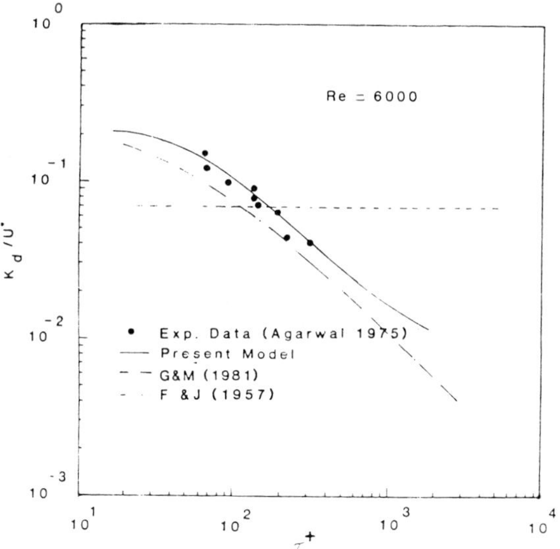

Figure 3: Comparison of predicted deposition velocity with the experimental data of Agarwal (1975)

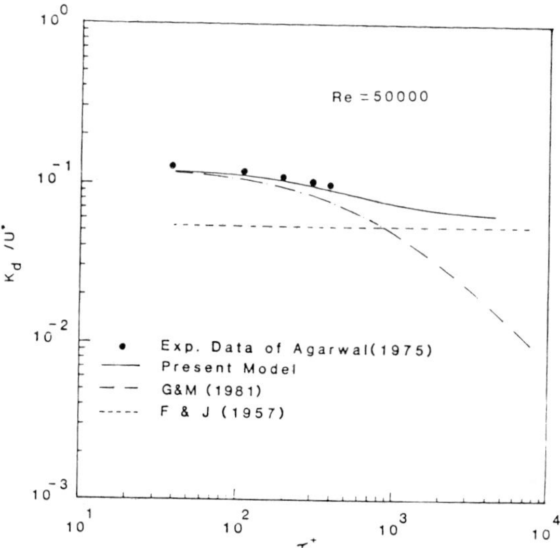

Figure 4: Comparison of predicted deposition with the experimental data of Agarwal (1975)

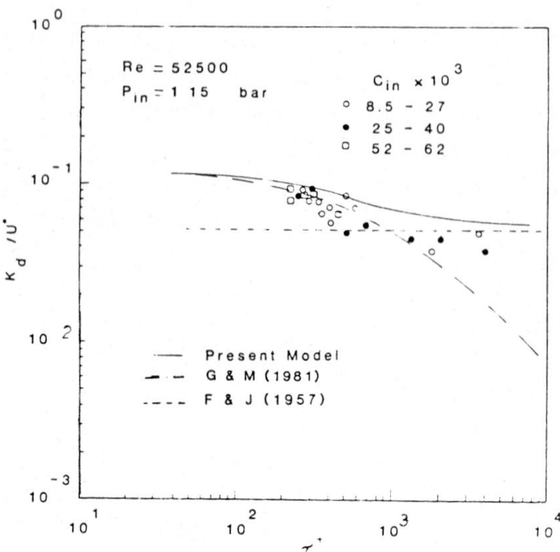

Figure 5: Comparison of predicted deposition velocity and the experimental data of Mastanaiah (1980)

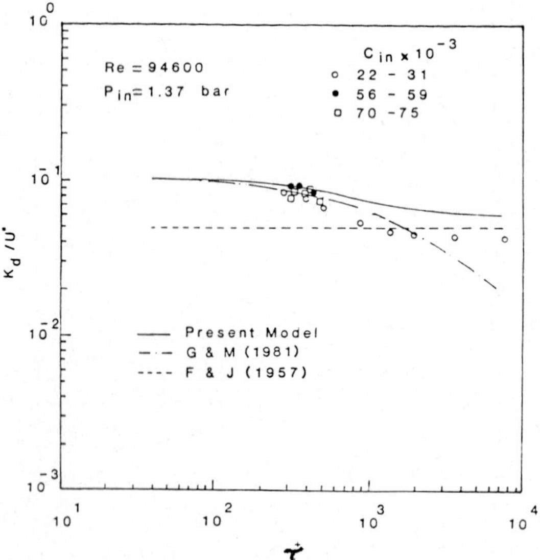

Figure 6: Comparison of predicted deposition velocity with the experimental data of Mastanaiah (1980)

using the same apparatus used by Mastanaiah (1980). Details of experiments are given in Mastanaiah (1980). The Sauter mean diamater of droplet size distribution d_{32} is calculated from the Nukiyama & Tanasawa correlation (see Nukiyama & Tanasawa 1938, 1939), which gives a good agreement with the measured drop size done by Mastanaiah. The values of d_{10} are obtained from d_{32}, using the relations for the drop diameter developed by Tatterson et al. (1977). The technique of measuring the deposition rate is similar to that applied by Cousins & Hewitt (1968). During the experiment, the liquid layer formed at the inlet of the test section is completely removed. The drops then migrate towards the wall and form a thin liquid film that grows continuously along the test section. The liquid film formed at the test section exit is also removed completely. The flow rate of liquid collected at the test section exit gives the amount of liquid deposited over the entire test section length, and thence the fractional deposition F. With a knowledge of F, the deposition velocity k_d is determined from

$$F = 1 - \exp\,[-4(k_d/\bar{U})\ (L/D)] \tag{33}$$

The absence of entrainment for the present experimental conditions is confirmed using the correlations of Troung Quang Minh & Huyghe (1965), which suggests entrainment E of less than 0.5 per cent for Re = 55,200. In view of these factors the entrainment effect is not taken into account in deducing the deposition velocities. The flow rate of air removed through the extractor is, however, measured to be less than about 1 per cent of the total air flow rate through the test section, and is therefore considered to have no significant effect on the measured deposition data. An error analysis done by Mastanaiah (1980) has indicated that the measured deposition velocities are within about 12 per cent accuracy. The reproducibility of the data has been assured by repeating some of the runs. The data in general are consistent without significant scatter and are therefore considered reliable.

Comparison of data obtained from the theory with that of the experimental work. The measured dimensionless deposition velocity for Re = 26700, 36000, 43100 and 55200 are depicted in Figs. 7 through 10. The conditions for a different set of data are given in Table 2. It can be shown in Fig. 7 that K_d/U^* decreases with τ^+ up to some value of τ^+ beyond which it remains nearly independent of τ^+. A similar observation was noticed and reported by Ganic & Mastanaiah (1980). The physical reason for this effect is that as τ^+ increases, the droplet size increases, and also the weight increases, so the droplet becomes too heavy to follow the flow. Up to a certain size, the momentum of the particle will compensate the decrease in velocity, getting almost the same deposition rate without dependency on τ^+.

The prediction of K_d/U^* from the present theory, the theory of Ganic & Mastanaiah (1981), and the theory of Friedlander & Johnstone (1957) is also compared in Figs. 7 through 10. It is observed from Fig. 7 that the present theory is in good agreement with the data, even having the same trend.

In Fig. 8, the present theory is over estimated over the entire range. This is due to the experimental error which was estimated at 12%, while in Figs. 9 and 10 the error becomes relatively greater. The reason for this is that as Re increases, the air flow rate increases, causing some evaporation from the liquid deposited on the wall, resulting in the smaller amount of water deposition measured.

In view of the above comparisons, it is evident that the present theory offers a considerable improvement over previous theories for $Re_p < 5$.

Re	Inlet press. (bar)	d_{10} (μm)	$\overline{C}_{in} \times 10^{-3}$ (kg/m^3)	τ^+ (Oseen regime) eq. (17b)	τ^+ Stokes regime
26790	1.09	22.8-53.13	12.7-56.8	251-1180	308-1632
36040	1.134	17-50	15-36	235-1522	285-2367
43150	1.185	15.5-51	24-31	258-1965	316-3240
55230	1.254	14-53	29-45	279-2817	346-5108

Table 2: The conditions for experimental data obtained

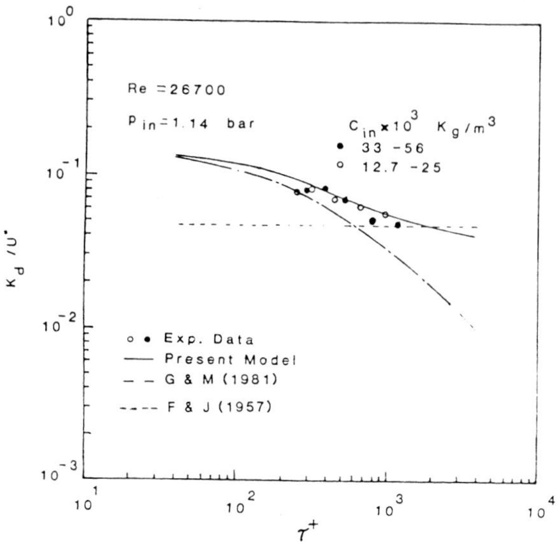

Figure 7: Comparison of predicted deposition velocity with the present experimental data at Re = 26700; gas flow rate w_g = 5.05x10^{-3} Kg/s; liquid loading w_ℓ=2.5x10^{-3} Kg/s and 0.179x10^{-3} Kg/s

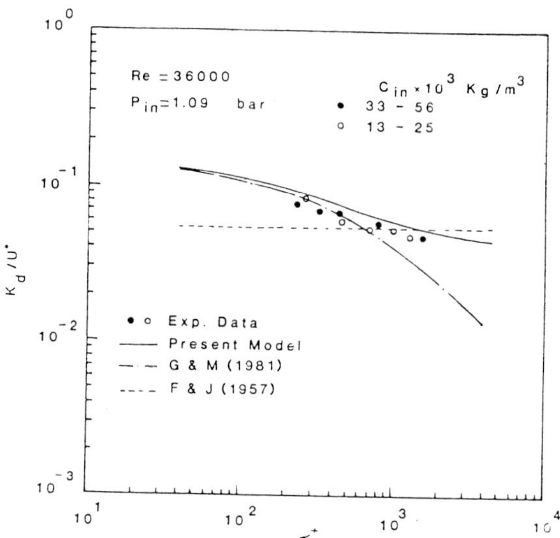

Figure 8: Comparison of predicted deposition velocity with the present experimental data at Re = 36000; gas flow rate w_g = 6.649x10^{-3} Kg/s; liquid loading w_ℓ = 1.35x10^{-3} Kg/s and 0.179x10^{-3} Kg/s

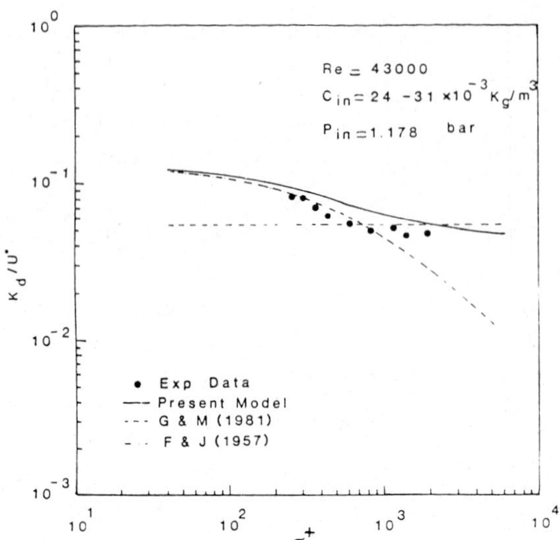

Figure 9: Comparison of predicted deposition velocity with the present experimental data at Re = 43000; gas flow rate w_g = 1.96×10^{-3} Kg/s; liquid loading w_ℓ = 1.79 Kg/s

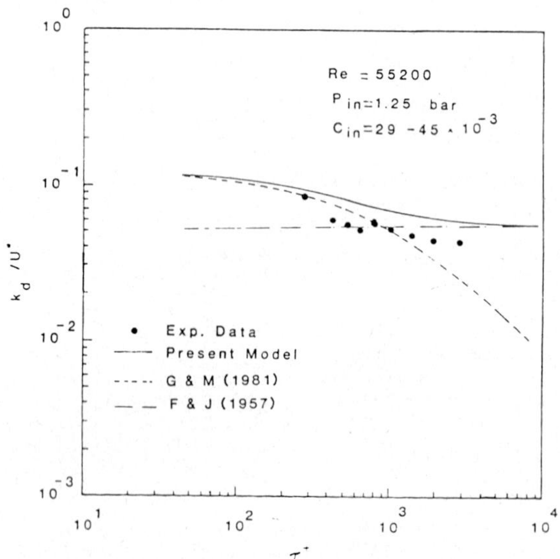

Figure 10: Comparison of predicted deposition velocity with the present experimental data at Re = 55200; gas flow rate w_g = $1o.17\times10^{-3}$ Kg/s; liquid loading w_ℓ = 1.79×10^{-3} Kg/s

5. DISCUSSION AND CONCLUSION

It can be shown that K_d/U^* is a function of Re, τ^+ and ρ_p/ρ_p as follows:

a) From Ganic & Mastanaiah (1981), where it was shown that $C_c/\overline{C}$ is a function of Re and τ^+ only.

b) w (Eq. (20)) is a function of a and b, which means a function of droplet diameter and ρ_p/ρ_g, so that $w = f(\tau^+, \rho_p/\rho_g)$.

c) Since δ is a function of V_g and w, and V_g is a function of Re, then $\delta = f(\tau^+, Re, \rho_p/\rho_g)$.

d) From Eq. (9) K_d/U^* is a function of δ, Re and τ^+. Therefore, $K_d/U^* = f(Re, \tau^+, \rho_p/\rho_g)$.

The proposed expression for deposition velocity given by Eq. (9), using the value of δ by interpolation from Fig. 8, is believed to be of considerable application in predicting droplet deposition rates in two-phase flow. The Oseen drag force produces a better prediction for evaluating the droplet deposition and is valid for a wide range of values for d_p and τ^+. This better predictability should be important in a wide range of applications of two-phase flow.

ACKNOWLEDGEMENT

This work was supported by the National Science Foundation under grant ENG 78-06211.

REFERENCES

1. Agarwal, J.K., Aerosol Sampling and Transport, Ph.D. Thesis, University of Minnesota, Particle Tech. Lab. Publ. 265, 1975.

2. Cleaver, J.W. and Yates, B., "A Sublayer Model for the Deposition of Particles from a Turbulent Flow," Chem. Engng. Sci. 30, pp. 983-92 (1975).

3. Cousins, L.B. and Hewitt, G.F., "Liquid Phase Mass Transfer in Annular Two-Phase Flow: Droplet Deposition and Liquid Entrainment," AERE-R5657 (1968).

4. El-Kassaby, M.M., "Drop Deposition and Heat Transfer in Two-Component Dispersed Flow," Ph.D. Thesis, University of Illinois at Chicago (1982).

5. Farmer, R.A., Griffith, P. and Rohsenow, W.M., "Liquid Droplet Deposition in Two-Phase Flow," Trans. ASME, J. Heat Transfer 92, pp. 587-594 (1970).

6. Forney, L.J. and Spielman, L.A., "Deposition of Coarse Aerosols from Turbulent Flow," J. Aerosol Sci. 5, pp. 257-271 (1974).

7. Friedlander, S.K. and Johnstone, H.F., "Deposition of Suspended Particles from Turbulent Gas Streams," Indust. Engng. Chem. 49, pp. 1151-1156 (1957).

8. Ganic, E.N. and Mastanaiah, K., "Investigation of Droplet from a Turbulent Gas Stream," Int. J. Multiphase Flow 7, pp. 401-422 (1981).

9. Gilliand, E.R., "Diffusion Coefficients in Gaseous Systems," Indust. Engng. Chem. 26, p. 681 (1934).

10. Hagiwara, Y. and Sato, T., An Experimental Investigation on Liquid Droplets Diffusion in Annular-Mist Flow, 2nd Multiphase Flow and Heat Transfer Symposium-Workshop, Miami, Florida, 1979. Also published in Multiphase Transport, T.N. Verziroglu, Ed., Hemisphere, New York.

11. Hinze, J.O., Turbulence, Second Edition, McGraw-Hill, New York, 1975.

12. Ilori, T.A., Turbulent Deposition of Particles Inside Pipes, Ph.D. Thesis, Univ. of Minnesota, 1971.

13. Kays, W.M., Convective Heat and Mass Transfer, McGraw-HIll, New York, 1966.

14. Korn, G.A. and Korn, T.M., Mathematical Handbook, Second Edition, McGraw-Hill, New York, 1968.

15. Laufer, J., "The Structure of Turbulence in Fully Developed Pipe Flow," NACA Report 1174 (1954).

16. Liu, R.Y.H. and Agarwal, J.K., "Experimental Observation of Aerosol Deposition in Turbulent Flow," J. Aerosol Sci. 5, pp. 145-155 (1974).

17. Liu, B.Y.H. and Ilori, T.A., "Aerosol Deposition in Turbulent Pipe Flow," Environ. Sci. and Tech. 8, pp. 351-356 (1974).

18. Lumley, J.L., "Two-Phase and Non-Newtonian Flows," Turbulence, P. Bradshaw, Ed., Topics in Applied Physics 12, pp. 289-324 (1978).

19. Mastanaiah, K., Experimental and Theoretical Investigation of Droplet Deposition and Heat Transfer in Air-Water Dispersed Flow, Ph.D. Thesis, Univ. of Illinois, Chicago Circle, 1980.

20. Mastanaiah, K. and Ganic, E.N., "Investigation of Droplet Deposition from a Turbulent Gas Stream," Int. J. Multiphase Flow 7, pp. 401-422 (1981).

21. McCormack, P.D. and Grave, L., Physical Dynamics, Academic Press, New York, 1973.

22. Namie, S. and Ueda, T., "Droplet Transfer in Two-Phase Annular Mist Flow: II. Prediction of Droplet Transfer Rate," Bull. J. SME 16, pp. 752-764 (1973).

23. Nukiyama, S. and Tanasawa, Y., "An Experiment on the Atomization of Liquid by Means of an Air Stream," Reps. 1 and 4, Trans. JSME 4, p. 86 (1938) and 5, pg. 68 (1939).

24. Reichhartd, H., "Vollstandige Darstellung der trubulenten Geschwindigkeitsverteilung in glatten Leitungen," ZAMM 31, pp. 208-219 (1951).

25. Sehmel, G.A., "Aerosol Deposition from Turbulent Air Stream in Vertical Conduits," Pacific Northwest Laboratory Rep. BNWL-578 (1968).

26. Tatterson, D.F., Dallman, J.C. and Hanratty, T.J., "Drop Sizes in Annular Gas-Liquid Flows," AIChE J. 23, pp. 68-76 (1977).

27. Tchen, C.M., Mean Value and Correlation Problems Connected with the Motion of Small Particles Suspended in a Turbulent Fluid, Martinus Nijhoff, The Hague (1947).

28. Troung, Q.M. and Huyghe, J., "Some Hydrodynamical Aspects of Annular Dispersed Flow: Entrainment and Film Thickness," Symp. Two-Phase Flow, Paper C2, Exeter, England (1965).

NOMENCLATURE

C = drop (particle) concentration, kg/m^3

$\overline{C}$ = bulk concentration, $W_d/(W_g/\rho_g + W_d/\rho_d)$, kg/m^3

C_D = drag coefficient

C_c = center-line concentration, kg/m^3

d_p = droplet diameter, m

E = entrainment

F = fractional deposition

f = smooth tube friction factor, $2\ \tau_w/(\rho_g \overline{U}^2)$

g = acceleration due to gravity, m/s^2

k_d = deposition velocity, m/s

k_d^+ = dimensionless deposition velocity, k_d/U^*

L = tube length, m

m = droplet mass, Kg

N = local rate of radial mass flux of droplets, $kg/(m^2 \cdot s)$

N_o = mass rate of deposition of droplets, $kg(m^2 \cdot s)$

Re_p' = drop Reynolds number based on relative radial fluctuating velocity, $d|\overline{U} - U_d|/\nu_g$

r = radial distance from the tube center, m

r_o = tube inner radius, m

r_o^+ = $u^*\ r_o/\nu_g$

S = stopping distance based on Stokes drag, $d^2\rho_p v_{po}/18\mu_g$

S^+ = dimensionless stopping distance, $S\ u^*/\nu_g$

t = time, s

$\overline{U}$ = superficial mean gas velocity, m/s

U^+ = U/u^*

U^* = friction velocity, $\overline{U}\sqrt{f/2}$, m/s

V_g = radial rms fluctuating velocity, m/s

V'_g = radial fluctuating velocity of gas, m/s

v_o = relative velocity between the secondary gas and the liquid jet in the atomizer, m/s

V_p = radial rms fluctuating velocity of drop, m/s

V_{pb}, V_{po} = value of v_p at start of free flight ($y^+ = 30$), m/s

y = radial distance from wall surface, m

y^+ = dimensionless radial distance, $U^* y/\nu_g$

Greek Symbols:

σ = surface tension, N/m

δ = particle to fluid diffusivity ratio, $\varepsilon_p/\varepsilon_f$

ε_f = fluid diffusivity, m^2/s

ε_p = particle diffusivity, m^2/s

μ = dynamic viscosity, $N.s/m^2$

ν = kinematic viscosity, m^2/s

τ = drop relaxation time based on Stokes drag, $d^2\ \mu_p/18\ \rho_g$, S/v_{po}

τ^+ = dimensionless drop relaxation time, $\tau\ u^{*2}/\nu_g$

Subscripts:

d = droplet

g = gas phase, bulk, air-water vapor mixture

p = drop (particle)

Multi-Phase Flow and Heat Transfer III. Part A: Fundamentals
edited by T.N. Veziroğlu and A.E. Bergles
Elsevier Science Publishers B.V., Amsterdam, 1984 — Printed in The Netherlands

AN ANALYTICAL MODEL FOR TWO-PHASE FLOW FIELD OF LIQUID SPRAYS

S. H. El-Emam and H. Mansour
Department of Mechanical Engineering
Faculty of Engineering
Mansoura University
El-Mansoura, Egypt

ABSTRACT

An analytical model has been developed to describe the two dimensional two-phase flow field resulting from the interaction of a gas stream and liquid sprays. Predictions about the gas flow field are made by adopting a boundary layer model in which the description of drop behaviours is also incorporated. The equation of the motion of the drop is expressed through the Lagrangian description and the underlying momentum exchange is treated by adopting the PSI-Cell model.

The proposed analytical model is based on the boundary layer model, where reverse flow is not permitted to occur. Therefore, it can be applied only to narrow cone spray systems. Within this limitation, the model is capable for predicting gas streamlines, gas flow field, decays of drop velocities and drop flight directions during its propagation in the flow field.

1. INTRODUCTION

The prediction of the two-phase flow field of liquid sprays is one of the most important problems in the design of many industrial applications including atomization, spray cooling and spray drying systems. The momentum exchange between the two phasses in a spray is depending on their relative velocities, drop size distributions and angles of drop trajectories.

A theoretical study on the variation of the axial velocity of the entrained air flow into a spray dryer system has been reported by Gluckert [1]. An attempt to analyse the momentum transfer which underlies air entrainment into a flat spray has been made by Briffa and Dombrowski [2]. Mellor et al. [3] have used standard drag coefficient data of smooth spheres to predict the trajectories and velocities of spray drops moving into a uniform air stream. Experimental studies of the hydrodynamic interaction between hollow cone sprays and gaseous entrainment have been undertaken by Binark and Ranz [4], and Benatt and Eisenklam [5].

Despite all these efforts, the design methods used for spray gas flows have until recently been based on simplified models. This is due to the difficulty to predict the flow field from fundamental principles which neglect the effect of spray drops on its environment. A computational procedure which eliminates this difficulty has been developed by Crowe and co-workers [6]. This procedure has been applied by Boysan et. al. [7] to predict the induced air flows in a hollow cone spray and also by El-Banhawy et al.[8] to predict the flow characteristics of a confined kerosene spray flame.

In this paper an analytical model for describing the two-phase flow field of the spray has been described. This method incorporates the basic concepts of the mixing length hypothesis of the boundary layer, and is solved by the finite difference numerical scheme and marching integration technique. The considered field is divided into small grids, each grid is regarded as a control volume and the gas-drop interaction is incorporated by treating the drop as a source of momentum to the gas phase. The equation of the motion of the drop is expressed through the Lagrangian description and the underlying momentum exchange is treated by adopting the PSI-Cell (Particle-Source-In-Cell) model of Crowe et al. [6]. Initial characteristics of the spray drops were based on the Rosen-Rammler distribution function. Prediction about drop velocities, drop locations, gas flow field and gas entrainment into the liquid spray were obtained.

2. THEORETICAL ANALYSIS

An analytical model of the two phase flow field of the spray is shown in Fig. 1. A co-current gas stream at normal temperature is sucked through a duct of cylindrical shape. Drops of liquid spray are introduced through the injection nozzle symmetrically around the axis of the spray. The system is isothermal, where only the momentum exchange between the two phases (drop-gas) has to be cosidered. Also, because of the small liquid to gas interfacial area at the break-up region in the vicinity of the atomizer, very little momentum transfer to the gas takes place from the liquid film and it will not be considered.

Based on the well known Rosin-Rammler distribution function equation of drop initial characteristics, the spray is divided into five main groups. Each group contains j of drop size ranges. The

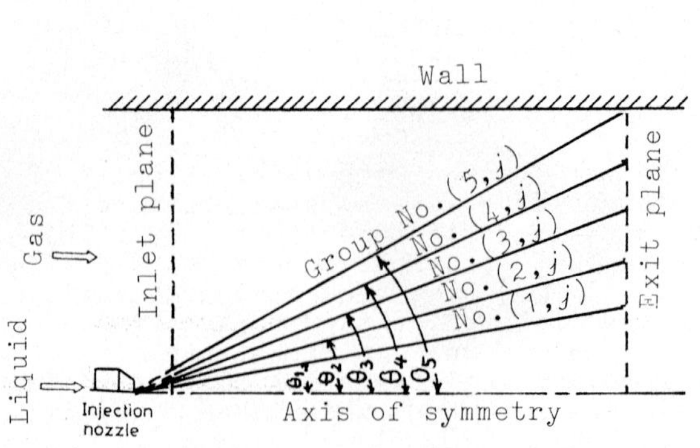

Fig. 1 Model of the two-phase flow field of liquid spray.

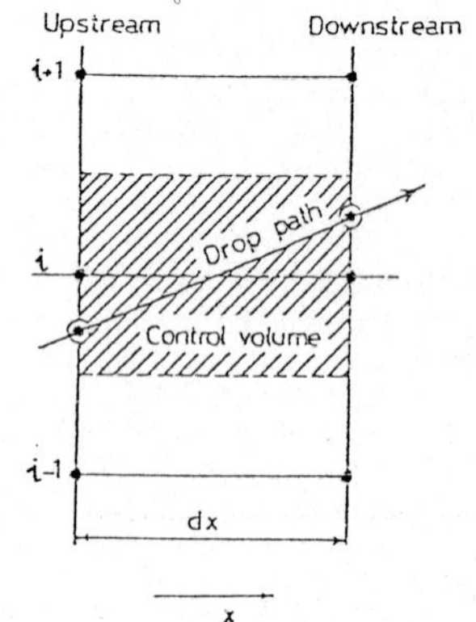

Fig. 2 Control volume.

co-current gas stream and the drop groups moving with their individual initial boundaries enter the flow field at the inlet plane and leave it at the exit plane. The flow field is then subdivided into a series of cells, in which each cell is regarded as a control volume, as shown in Fig. 2. As a drop traverses a given cell of i control volume (shaded area) in the flow field, it will exchange momentum with the gas phase contained in the cell. This momentum exchange is treated with an approach based on that of the PSI-Cell method of Crowe et. al. [6]. Therefore, if a certain size group of drops shows a deceleration while passing the control volume, the loss of its momentum will increase the momentum of the gas phase in the same control volume. This process will also occur for the drops of the other groups which traverse the same cell of the control volume.

The following assumptions have been made:

1. The spray cone angle as well as the relative velocity of the spray and the gas phase have been kept small to insure that there are no counter flow current occurs in the flow field.

2. Only the radial pressure gradient resulting from the momentum augmentation effect of spray drops is considered.

3. The spray drop is assumed to be spherical in shape. Effects of evaporation, condensation and deformation are not considered.

4. The gravitational effect on the gas phase is neglected.

5. The break-up region of the spray in the vicinity of the atomizer is not considered

2.1 Particle Equations

The initial side distribution of the spray drops is based on the well known Rosin-Rammler distribution function equation which can be expressed in the following form [9].

$$R_m(D) = 100 \exp (-D/\bar{D})^n \qquad \ldots\ldots (1)$$

where $R_m(D)$ is the volume percentage oversize, D is the drop diameter, $\bar{D}$ is the diameter above which 36.8% of the total spray volume lies and n is the distribution function between 2 and 4. To reduce the computational effort required, the size distribution can be conveniently divided into an adequate number of discrete intervals, each represented by an average diameter.

For describing drop phase movements, the Lagrangian description is adopted to record and analyse the behaviour of the drops individually. That is, based on the assumptions mentioned before, each drop of the spray is considered as an independent single drop subjected to the forces of gas resistance, gravity and b yancy.

From Newton's second law, the equation of motion for a drop can be written as follows:

$$m_D \frac{d\vec{V}_D}{dt} = C_D \rho_G \frac{A_D}{2} (\vec{V}_G - \vec{V}_D)\cdot|\vec{V}_G - \vec{V}_D| + V_D (\rho_D - \rho_G) + m_D\vec{g} \quad \ldots(2)$$

where m_D is the mass of the drop, A_D is the projected area of the drop, $\vec{V}_G$, $\vec{V}_D$ are gas and drop velocity vector respectively. V_D is the volume of the drop, ρ_D and ρ_G are drop and gas density respectively, C_D is the drag coefficient of the drop and $\vec{g}$ is the gravity vector.

The drag coefficient of a drop moving into the gas stream depends on the relative Reynolds number, R_e . where:

$$R_e = \rho_G D\cdot|\vec{V}_G - \vec{V}_D| / \mu_{G,eff} \quad \ldots\ldots\ldots (3)$$

The relation between the drag coefficient and the relative Reynolds number can be expressed with reasonable accuracy by the following equation [10]:

$$C_D = \frac{24}{R_e} (1 + 0.15\, R_e^{0.687}) \quad \ldots\ldots\ldots(4)$$

Substituting eq. (4) into eq. (2), the equation of motion for a drop can be written in the following form:

$$\frac{d\vec{V}_D}{dt} = 18 \frac{\rho_G}{\rho_D} \frac{1}{D\, R_e} (1 + 0.15\, R_e^{0.687})(\vec{V}_G - \vec{V}_D)\cdot|\vec{V}_G - \vec{V}_D|$$
$$+ (\frac{\rho_D}{\rho_G} - 1 + \vec{g}) \quad \ldots\ldots\ldots (5)$$

Equation (5) is used to predict drop velocity during its propagation in the flow field of the spray.

2.2 Gas Equations

Based on assumptions (1) and (2) mentioned before, the gas phase can be treated as an axi-symmetric, two dimensional boundary layer system. The equations of conservation of mass and momentum with the appropriate source terms resulting from the interface interaction are as follows:

Continuity equation;

$$\frac{\partial u_G}{\partial x} + \frac{1}{r}\frac{\partial}{\partial r}(r\, v_G) = 0 \quad \ldots\ldots\ldots (6)$$

Momentum equation in the x direction;

$$\rho_G u_G \frac{\partial u_G}{\partial x} + \rho_G v_G \frac{\partial u_G}{\partial r} = \frac{1}{r}\frac{\partial}{\partial r}(\tau r) - \frac{\partial p}{\partial x} + F_x \quad \ldots\ldots (7)$$

Momentum equation in the r direction:

$$\frac{\partial p}{\partial r} = F_r \qquad \cdots\cdots\cdots (8)$$

where F_x and F_r are the source terms of the momentum exchange resulting from the interface interaction in the x and r direction respectively. Since the problem under consideration involves an axially symmetric flow situation, the velocity components related to the stream function are as follows;

$$u_G = \frac{1}{\rho_G u_G}\frac{\partial \psi}{\partial r} \quad , \quad v_G = -\frac{1}{\rho_G u_G}\frac{\partial \psi}{\partial x} \qquad \cdots\cdots\cdots (9)$$

Using von Mises transformation equations and substituting the continuity equation (6) and the stream function equation (9) into momentum equations (7) and (8), the differential equations of the momentum in $(x - \psi)$ coordinates in the axial and radial directions respectively are:

$$\frac{\partial u_G}{\partial x} = \frac{\partial}{\partial \psi}(\tau r) - \frac{1}{\rho_G u_G}\frac{\partial p}{\partial x} + \frac{v_G}{\rho_G u_G^2} F_r + \frac{1}{\rho_G u_G} F_x \cdots\cdots (10)$$

and

$$\frac{\partial p}{\partial \psi} = \frac{1}{\rho_G r_G r} F_r \qquad \cdots\cdots\cdots (11)$$

Since the flow in the region enclosed by the spray cone is of turbulent nature, the calculation of the shear stress τ, appearing in the momentum equation (10) depends on the mean gas velocity and effective viscosity, where

$$\tau = \mu_{eff}\frac{\partial u_G}{\partial r} \qquad \cdots\cdots\cdots (12)$$

There are many models of turbulence. Some of which are simple to use, but not generally applicable. The other posses a wider range of applicability, but require the solution of one or more partial differential equations. In this work, the Prandtl's mixing-length hypothesis is employed, in which the effective viscosity is calculated from the following equation:

$$\mu_{eff} = u_G + \rho_G \ell_m^2 \cdot |\partial u_G/\partial r| \qquad \cdots\cdots\cdots (13)$$

where ℓ_m is the mixing length and $|\partial u_G/\partial r|$ is the absolute value of the rate of change of streamwise velocity component with respect to the cross-stream direction. The distribution of the mixing length ℓ_m is defined as:

$$\ell_m = K r_W \qquad \cdots\cdots\cdots (14)$$

where K is a constant = 0.435 and r_W is the distance from the wall to the local position towards the axis of symmetry.

2.3 Equations of Source Terms

The actual spray has a range of drop size, drop velocity and drop flight direction. Therefore it is impossible to chase all drops of the spray on an individual basis. The spatial distributions of the source term force components are evaluated with the aid of the information obtained from the solution of the equation of motion of the spray. If drops of S group of size D are traversing the control volume of a given cell, the source term quantities due to the momentum exchange in the x and r directions can be expressed respectively as follows:

$$F_x = \sum_S \frac{\pi}{6} \rho_D D^3 n_S \cdot \Delta u_{D,S} / dV \quad \ldots\ldots\ldots \quad (15)$$

and

$$Fr = \sum_S \frac{\pi}{6} \rho_D D^3 n_S \cdot \Delta v_{D,S} / dV \quad \ldots\ldots\ldots \quad (16)$$

where n_S is the number of drops belonging to S group and dV is the volume of a given cell of the control volume. $\Delta u_{D,S}$ and $\Delta v_{D,S}$ are the change of drop velocity of group S between the downstream and the upstream boundaries of the control volume in x and r direction respectively, i.e;

$$\Delta u_D = (u_D)_{downstream} - (u_D)_{upstream} \quad \ldots\ldots\ldots \quad (17)$$

and

$$\Delta v_D = (v_D)_{downstream} - (v_D)_{upstream} \quad \ldots\ldots\ldots \quad (18)$$

3. CALCULATION PROCEDURE

Predictions of the f w field of the gas phase have been carried out by adopting the convenient method of Patanker-Spalding [11] for a boundary layer model. A description of the spray characteristics is incorporated in this method. Under specific boundary conditions, the solution of the entire set can be obtained through a series of cyclic calculations starting with a guessed initial flow field. In each cycle, the spray equation of motion is integrated using the previous values of the gas velocity components and the force distribution is updated. This distribution is then used to improve the velocity field. Such cycle are repeated until the scheme converges to the solution.

Consider that the flow parameters of the gas phase at the upstream surface of the control volume shown in Fig. 2 are known. Also, if the parameters of a drop traversing the control volume are known at the upstream surface. These parameters are mainly drop location $(x , r)_{upstream}$ and drop velocity components $(u , v)_{upstream}$. Then drop velocity components at the downstream surface of the control volume $(u , v)_{downstream}$ can be calculated using the equation of the motion of the spray (eq. 5).

The location of the drop when leaving the control volume at

the downstream cross-section $(x\ ,\ r)_{downstream}$ is calculated using the following equations:

$$x_{downstream} = x_{upstream} + \left(\frac{(u_D)_{upstream} + (u_D)_{downstream}}{2} \right) . dt \quad \ldots(19)$$

$$, r_{downstream} = r_{upstream} + \left(\frac{(v_D)_{upstream} + (v_D)_{downstream}}{2} \right) . dt \quad \ldots(20)$$

According to the propagation path direction of the drop while traversing a given cell, there are several probabilities to be considered. Fig. 3 shows the different drop paths during its traversing the control volume. For example, it is possible that a drop may pass into some adjacent cells above or below the given cell before it reaches the downstream section, (case 2 or case 3 in Fig. 3).

In the model of the spray shown in Fig. 1, the properties of drops are computed throughout the flow field. The flow field bounded by the inlet and exit planes, which are located 10 mm and 500 mm from the nozzle plane respectively. The flow field inside a duct of 200 mm diameter is divided into 50 divisions of almost equal radial intervals. The integration step in the downstream direction dx was fixed at 0.5 mm.

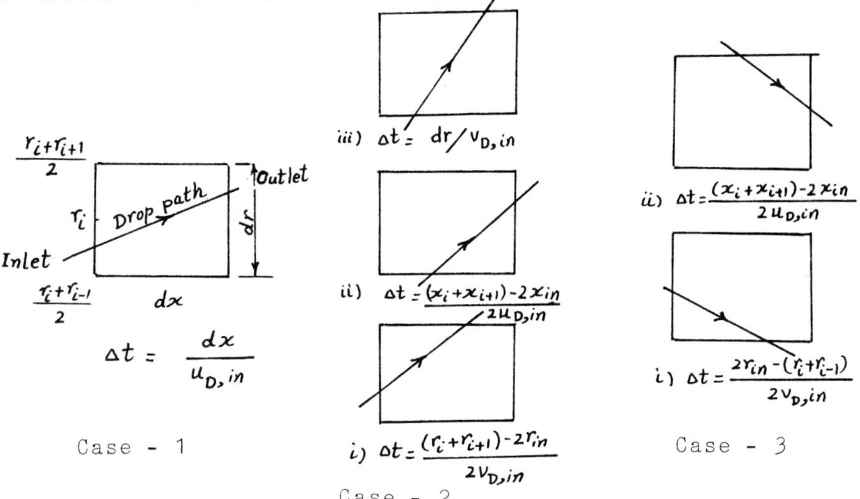

Fig. 3 Cases of drop propagation trajectories.

4. RESULTS

The analytical model was applied to a spray of cone angle $2\theta = 30°$ and the liquid flow rate of 4 ℓ/h injected through a cylinderical duct of 500 mm diameter. A co-current air stream with an initial velocity $u_G = 1.0$ m/s is sucked into the duct around the spray. At first, the spray was divided into five main groups with starting angles of θ_1, θ_2, θ_3, θ_4 and θ_5 equal to 6°, 8°, 10°, 12° and 14°

respectively. Each group was then subdivided into 15 mean size ranges according to their size distribution frequency ratios. The initial drop velocity is fixed at 10 m/s and the spray drop are ranging from 10 μm to 300 μm.

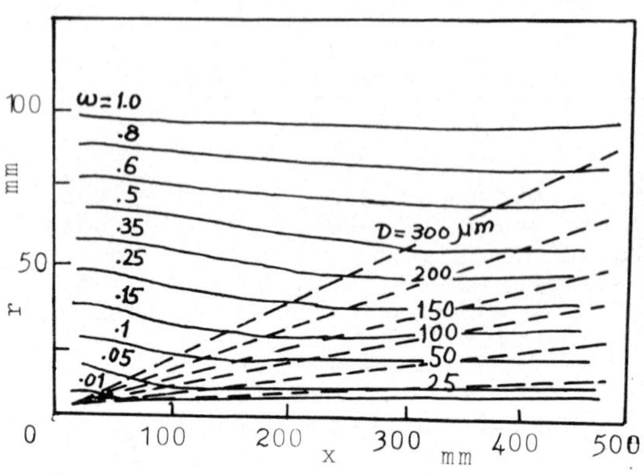

Fig. 4 Predicted air streamlines and drop trajectories.

Fig. 5 Air flow patterns visualized by smoke trace technique, [12].

Figure 4 shows an example of the predicted air streamlines and drop trajectories. It is clear that the degree of inclination of the predicted streamlines is higher at the region near the injection nozzle. The degree of inclination gradually decreases as the distance from the nozzle plane is increased. Inside the spray the streamlines become almost parallel to the axis of the spray.

The appearance of the predicted air streamlines is in well agreement with the observed flow patterns visualized by smoke trace visualization technique [12], as shown in Fig. 5 .

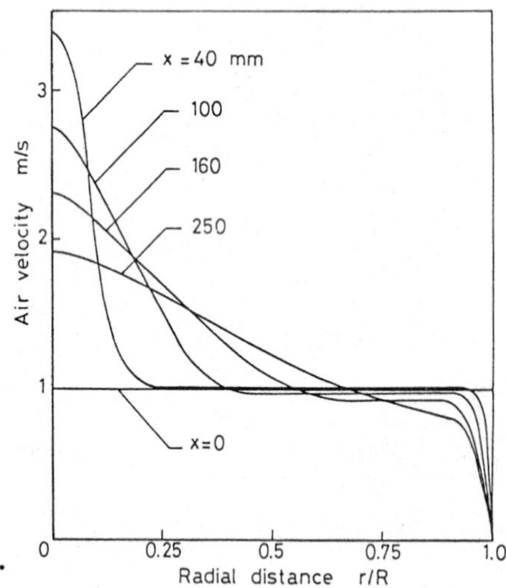

Fig. 6 Predicted air velocity distributions.

Predicted air velocity distributions at various cross-sections along the spray are shown in Fig. 6. It can be seen that air velocities inside the spray cone are higher than those outside the spray cone, and have maximum values on or near the axis of the spray. Also, air velocity inside the spray is decreased as axial distance from the nozzle plane is increased.

Figure 7 shows predicted decays of mean drop velocities with

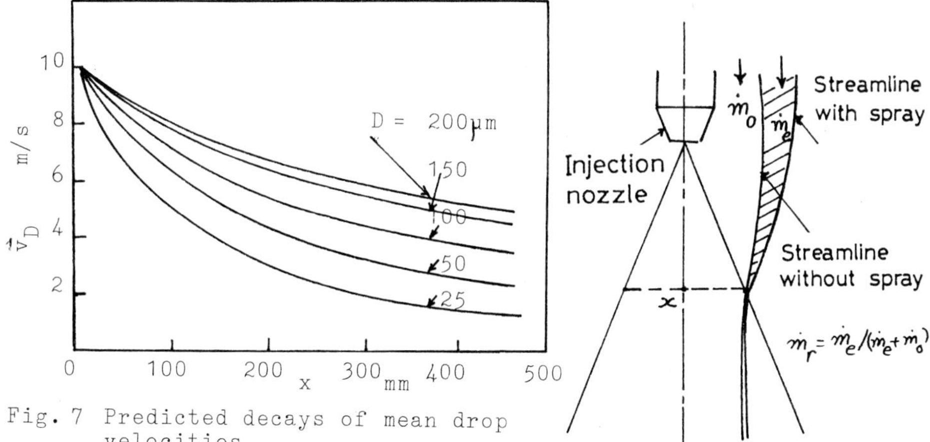

Fig. 7 Predicted decays of mean drop velocities.

Fig. 8 Entrainment mechanism and definition.

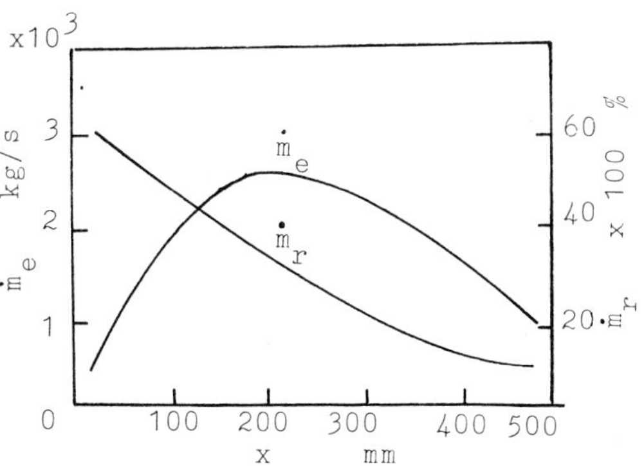

Fig. 9 Predicted entrainment rate and entrainment ratio.

axial distance from the nozzle plane. From this figure, one can notice that, while the initial drop velocities are the same at x = 10 mm the velocity difference is increased as the distance from the injection nozzle plane increases. Also, while large drops move with high velocity in the outer regions of the spray, small drops reach their terminal velocities and their paths become parallel to air streamlines.

Prediction of entrainment rate, $\dot{m}_e$, and entrainment ratio, $\dot{m}_r$, (as defined in Fig. 8), are shown in Fig. 9. In regions near the injection nozzle, the entrainment rate is linearly increased as the axial distance from the nozzle plane increases. However, the entrainment ratio is increased as the axial distance decreases. This is due to the increasing of momentum exchange between dense spray drops with their higher velocities and the surrounding air as the axial distance is decreased.

5. CONCLUSIONS

An analytical model has been developed to describe the two dimensional two-phase flow field of liquid sprays. Predictions of the gas flow field has been carried out by adopting the basic concepts of the mixing length hypothesis of boundary layer models.

The equation of the motion of the drop is expressed through the Lagrangian description while the underlying momentum exchange is treated by adopting the PSI-Cell model.

The following conclusions have been reached:

1. The computational procedure is capable to predict the two-phase flow field of liquid sprays of narrow cone angles.

2. The predicted gas streamlines are in reasonable agreement with the observed flow patterns visualized by smoke trace technique.

3. The model can be adequate for predicting of both qualitative and quantitative aspectives of both gas and drops of the flow field.

4. Measurements of gas entrainment into a liquid spray can be performed and modelled using local states in the flow field under calculation.

NOMENCLATURE

A_D Projected area of the drop.
C_D Drag coefficient.
$\underline{D}$ Drop diameter.
$\bar{D}$ Rosin-Rammler mean diameter.
F_r, F_x Source terms of the total force exerted by the spray per unit volume of the gas in the radial and axial direction respectively.
$\underline{G}$ Gas.
$\vec{g}$ Gravity vector.
i Pertaining to the i th cell of control volume.
j Pertaining to the j th drop size group.
ℓ_m Mixing length.
$\dot{m}_e$ Entrainment rate.
$\dot{m}_r$ Entrainment ratio.
n_S Number of drops belonging to S group of certain drop size.
p Pressure.
r Radial coordinate.
R_e Reynold's number.
$R_m(d)$ Volume percentage of drops oversize, eq. 1.
u Velocity component in x direction.
$\underline{v}$ Velocity component in r direction.
$\bar{V}$ Velocity vector.
x Axial coordinate.

Greek Symbols

ρ Density.
τ Shear stress.
ψ Stream function.
ω Dimensionless stream function.
θ Half angle of spray cone.
μ_{eff} Effective viscosity.

REFERENCES

1. Gluckert, F.A., AIChE J., 8 (1962), 460.

2. Briffa, F.E. and Dombrowski, N., AIChE J., 12 (1966), 708.

3. Mellor, R., Chigier, N.A. and Béer, J.M., Combustion and Heat Transfer in gas Turbine Systems (Ed. Norster, E.), (1971), 291, Pergamon Press.

4. Binark, H. and Ranz, W.E., ASME Paper 58-A-284, (1958).

5. Benatt, F.G. and Eisenklam, P., J. Inst. Fuel, 42-243 (1968-8), 309.

6. Crowe, C.T., Sharma, M.P. and Stock, D.E., Trans. ASME (J. Fluids Engr.), 99 (1977), 325.

7. Boysan, F. and Binark, H., Trans. ASME, 1011 (1979), 312.

8. El-Banhawy, Y. and Whitelaw, J.H., AIAA J., 18-12 (1980-12), 1503.

9. Rosin, P. and Rammler, E., J. Inst. Fuel, 7 (1933), 29.

10. Wallis, G.B., One-Dimensional Two-Phase Flow, (1968), McGraw-Hill.

11. Patanker, S.V. and Spalding, D.B., Heat and Mass Transfer in Boundary Layer, 2nd Ed., (1970), Intertext Book.

12. El-Emam, S.H., A Study on the Flow Field and Combustion of Sprays, Ph.D. thesis, Osaka University, 1981.

Multi-Phase Flow and Heat Transfer III. Part A: Fundamentals
edited by T.N. Veziroğlu and A.E. Bergles
Elsevier Science Publishers B.V., Amsterdam, 1984 — Printed in The Netherlands

HYDRODYNAMICS OF ADIABATIC INVERTED ANNULAR FLOW - AN EXPERIMENTAL STUDY

George De Jarlais and Mamoru Ishii
Reactor Analysis and Safety Division
Argonne National Laboratory
9700 South Cass Avenue
Argonne, Illinois 60439, U.S.A.

ABSTRACT

For low quality film boiling in tubes or rod bundles, the flow pattern may consist of a liquid jet-like core surrounded by a vapor annulus, i.e., inverted annular flow. The stability, shape, and break-up mechanisms of this liquid core must be understood in order to model correctly this regime and to develop appropriate interfacial transfer correlations. This paper reports on a study in which inverted annular flow was simulated in an adiabatic system. Turbulent water jets, issuing downward from long aspect nozzles were enclosed within cocurrent gas annuli. Jet core diameter and velocity, and gas annulus diameter, velocity and species were varied, yielding liquid Reynolds numbers up to 33,000, void fractions from 0.29 to 0.95, and relative velocities from near zero to over 80 m/s. Jet core break-up lengths and, secondarily, core break-up mechanisms, were observed visually, using strobe lighting. Still photographs using a 3 microsecond flash exposure were taken, as the primary source of break-up mechanism information, to observe core surface disturbance wavelengths and dispersed core droplet sizes, and as a secondary source of break-up length data. This experiment is unique in that extensive data for the disintegration of confined coaxial jets (liquid core and gas annulus) have been obtained.

1. INTRODUCTION

Inverted annular flow occurs in confined boiling heat transfer systems, when low quality flow is coupled with post-CHF film boiling. The resulting flow pattern is a liquid core surrounded by a blanketing annulus of vapor. Understanding the hydrodynamics of this flow regime is of practical concern, since inverted annular flow may occur in light water reactor accident situations, in which, after loss of core coolant, core reflood brings coolant into the confined regions between very hot fuel rods [1]. Inverted annular flow may also occur in the transient operation of cryogenic heat transfer or conveyance systems, such as those found in rocket propulsion applications [2].

Given the importance of inverted annular flow specifically, and post-CHF heat transfer phenomena in general, a great deal of work has been done in this area in the last 30 years [1,3]. Yet our understanding of the basic post-CHF hydrodynamics, especially those of inverted annular flow, has not progressed sufficiently, in part because, in typical film boiling experiments, control and measurement of the flow parameters necessary for such an understanding is difficult, if not impossible, to achieve. Confined geometries, and the minimum temperatures and heat transfer rates required for film boiling, often prevent even simple flow visualization. There have been, however, a few inverted annular flow studies in which limited hydrodynamic information was obtained. Simple flow regime observations were made by Laverty and Rosenow [4] and by Kalinin et al. [5] in which the transition of inverted annular flow to inverted slug or dispersed droplet flow was noted. In the analytical study by Jensen [6], break-up lengths of the inverted annular flow liquid core were calculated, along with surface wavelengths. In the transient film boiling experiment of Kurilenko et al. [7] and the steady state film boiling experiment of Ottosen [8], flow regime observations and void fraction measurements were made. Finally, in the FLECHT SEASET series of tests, such as those reported in [9], dispersed droplet sizes and velocities were measured at distances well beyond the CHF point.

As a result of our limited understanding of post-CHF hydrodynamics, including those of inverted annular flow, many film boiling applications are amenable to only limited analysis at present. One example of this can be seen in large-scale LWR safety codes such as TRAC and RELAP, which are essentially constrained by the not-well understood two phase thermo-hydraulics under various accident conditions.

2. SIMULATED INVERTED ANNULAR FLOW EXPERIMENT

In light of the experimental difficulties associated with hydrodynamic studies of inverted annular flow film boiling, an adiabatic simulation of inverted annular flow was performed in this study. This simulation was accomplished by enclosing turbulent water jets, issuing from long aspect nozzles, within cocurrent annular gas jets flowing in transparent tubes. Such a simulation was suggested by Jensen [6], who noted the similarities between the liquid core stability of inverted annular flow and the stability of liquid jets. In this simulation, fluid properties, velocities, and initial geometries were easily controlled and measured. In addition, with no need to maintain film boiling heat transfer, ambient temperatures could be used, steady state flow was assured, and the flow pattern up to the point of core break-up was easily observed.

The experiment was performed in a set of five test series, each with fixed values for initial jet core diameter, D_J, and gas annulus outer diameter, D_O. Within each series, jet core velocity, v_J, and gas annulus velocity, v_G, were varied. In addition, for two of the test series, gas species (and therefore density, ρ_G) was also varied. The series of tests performed are summarized in Table 1.

All tests were performed near atmospheric pressure, with flow directed downward. Upward flow was attempted, but the use of low velocity liquid jets, coupled with easy wetting of the gas annulus outer wall (which would not occur

Table 1. Test Series Geometries, Gas Species

Test Series	Nozzle Diameter (cm)	Nozzle Length (cm)	Outer Diameter (cm)	Initial Void Fraction	Gas Species
A7	0.425	41.	1.66	0.934	Nitrogen
B2	0.763	46.	1.36	0.685	Nitrogen, Helium, R-12
B3	0.902	67.	1.36	0.560	Nitrogen, Helium, R-12
C1	0.604	37.	0.90	0.549	Nitrogen
C2	0.763	46.	0.90	0.281	Nitrogen

in true inverted annular flow film boiling) made this impractical. Even with downward-directed flow, wall wetting by the dispersed liquid core made data acquisition difficult at high gas flowrates.

For each trial run, jet core break-up length, L_B, was determined from strobe-aided visual observation of the flow pattern, with supplementary break-up length information supplied by still photographs. Jet core break-up mechanisms were also determined for each trial run, from still photographs supplemented by visual observations. In addition, core surface disturbance wavelength and dispersed core maximum droplet diameter data were obtained for many trial runs, where photographs provided such information.

The apparatus and procedures for the experiment are described in the following sections.

2.1. Apparatus

The experimental test system is shown schematically in Fig. 1, and a drawing of the basic test section is shown in Fig. 2.

The test section. The test section used to simulate inverted annular flow conditions consisted of three basic components: a lucite base with integral gas plenum, stainless steel tube water nozzles, and Pyrex tubing forming the outer surface of the gas annuli. The water nozzles were coaxially centered within the Pyrex tubing, and gas flow was introduced, from the gas plenum, into the annular gap between the stainless steel and Pyrex tubes. The test section was designed so that various-sized water nozzles and Pyrex tubing could be used to give a range of initial dimensions and void fractions. Test section construction was such that all water nozzles had an aspect ratio (length divided by inner diameter) of at least 50 so that entrance region effects would be minimal at the nozzle exit. Similarly, the length of unobstructed gas flow in the annulus between the stainless steel and Pyrex tubes, upstream of the water nozzle exit, was a minimum of 50 times the steel/Pyrex gap width.

Design and construction of the test section was relatively simple. The base and gas plenum were fabricated from lucite plate and tubing. The gas plenum was filled with coarse stainless steel wool to eliminate large-scale vortices in the gas flow entering the test section, thereby promoting a uniform flow pattern into the gas annulus.

The water nozzles were fabricated from thin-walled (.155 and .255 mm wall thickness) stainless steel tubing. These nozzles were hand-ground, so that tube wall thickness near the nozzle exit decreased smoothly, ending in a knife edge at the exit itself. With this form, no sudden gas flow area change occurred at the nozzle exit, minimizing entrance effects at the start of the simulated inverted annular flow region of the test section. Six spacing pins, made of 1.27 mm diameter copper wire, were soldered onto each steel tube to coaxially center each nozzle tube within an outer tube of Pyrex. To allow for slight variations in D_o along the Pyrex tube length, these pins were designed to act as cantilevered beams, flexing slightly as the steel nozzles were positioned in the Pyrex tubes. Three of these pins were placed near the nozzle exit, equally spaced about the tube circumference, at a distance from the exit at least 50 times the width of the Pyrex/stainless steel annular gap for each test series geometry. This pin/nozzle exit distance was established to allow dissipation of gas flow disturbances caused by the spacing pins, prior to the region of the test section in which the simulated inverted annular flow occurred. The second set of three spacing pins was placed 3 to 7

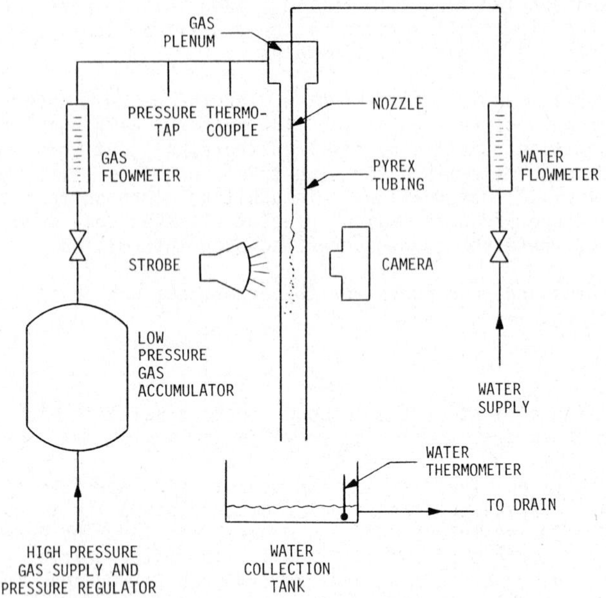

Fig. 1 Schematic Diagram of Test Apparatus

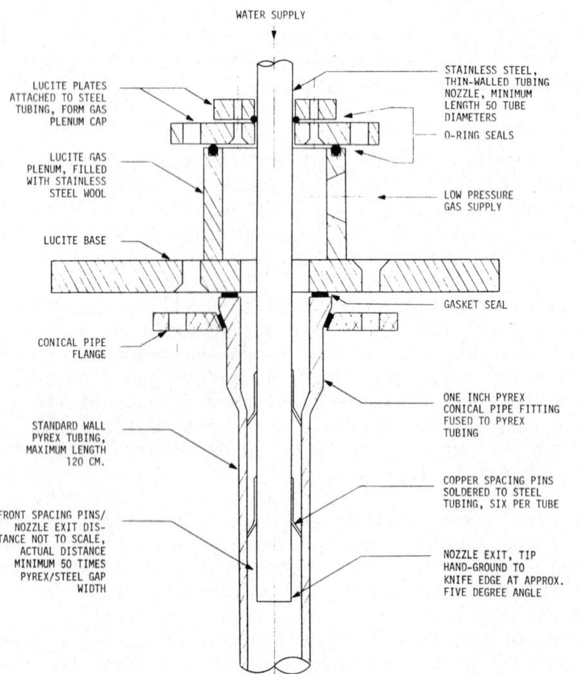

Fig. 2 Drawing of Test Section

cm behind the first set, with the pins spaced about the tube circumference in positions rotated 60° from the first set of pins. Copper spacing pins were not, however, used to align the nozzle of the C2 test series in which the annular gap width was only 0.53 mm. For this narrow gap, the spacing pins became too stiff to accommodate diameter variations (typically ±0.05 mm) in the Pyrex tube. For this test series, coaxial centering was accomplished by placing three narrow ridges, axially oriented, at 120° intervals about the nozzle tube circumference. These ridges consisted of several layers of thin plastic adhesive tape (thickness of 0.05 mm), the number of such layers varied to control alignment, placed over a strip of fabric adhesive tape, which could compress slightly in response to varying Pyrex tube diameter.

The Pyrex tubing of the test section was standard wall stock (1.0 or 1.2 mm wall thickness). At one end of each tube, a 1 inch, Pyrex conical pipe end fitting was fused in place. At the fused tube/pipe joint, a minimum ID equal to the Pyrex tube ID, D_0, was maintained, so that the stainless steel nozzle, with spacing pins (or ridges), could easily be inserted. This conical pipe end fitting was bolted onto the lucite base plate, using an aluminum conical pipe flange. To minimize vibration of the Pyrex tube, it was clamped in place at its free end (the test section exit) and at a point slightly upstream of the water nozzle exit. Standard Pyrex tubing lengths of 120 cm were used for most trial runs. For a limited number of trials in the B2 test series, however, a shortened test section, using only 32 cm of Pyrex tubing, was developed. With this shorter test section, the distance from the core jet break-up point to the test section exit could be minimized, with actual break-up/section exit distances varied by the positioning of the water nozzle within the Pyrex tube. Observations of the dispersed liquid resulting from core break-up could then be made at the test section exit. Such observations, it was found during preliminary experimental trials, could not be made along the test section proper, since liquid of the dispersed core rapidly wetted the inner wall of the Pyrex tube, forming a wavy, opaque, annular film of liquid. The exit of this shortened test section was therefore designed to divert this annular liquid film radially, and allow an unobstructed view of the dispersed slugs or droplets remaining after core jet break-up. This radial diversion of the liquid annular film was accomplished by flaring the free end of the Pyrex tubing outward, and placing a concentric lucite cone, with an aperture equal to D_0, within this Pyrex flare. Water flowing as a film down the Pyrex wall at the test section exit would then flow through the conical annulus of the Pyrex/lucite gap, and away from the test section exit.

Support Systems. Gas and water delivery systems were developed, to supply fluid flows to test section. And, to observe the hydrodynamic behavior of the simulated inverted annular flow within the test section, visual and photographic observation systems were developed as well. These systems, like the test section itself, are shown schematically in Fig. 1.

Nitrogen and helium gas were obtained from high pressure supply tanks. These tanks were connected to a high pressure manifold, from which a pressure reducing valve fed gas into a 0.3 m^3 low pressure gas accumulator maintained at approximately 0.4 MPa. For R-12 (CCl_2F_2) flows, gas was fed directly from supply tanks into the gas accumulator. From this accumulator, gas flowed through a needle valve controlling the flowrate, and then through one of three variable area flowmeters. Use of three flowmeters (cumulative range 0.02-6.3 ℓ/s nitrogen at STP) was necessary to allow for large variations in initial gas flow areas and gas velocities. Gas pressure and temperature were measured at the flowmeter exit. From the flowmeter, gas then flowed into the test section gas plenum, through the annular gap between the water nozzle and the Pyrex tubing, and into the similar annular gap between the Pyrex tubing and turbulent water jet issuing from the water nozzle. Gas was vented directly from the test section exit.

Water, supplied from a potable water system, flowed through a needle valve controlling the flowrate, and then through a variable area flowmeter (range approximately .02-.3 ℓ/s water at 25°C). The water then flowed into the stainless steel long aspect nozzle of the test section. At the test section exit, the water flow was collected in a small plastic tank fitted with a drain line and a fully-submersed thermometer used to measure water temperature.

To serve as a reference length scale for all observations, a transparent ruler marked with 1 mm divisions was clamped along the entire Pyrex tubing length. The Pyrex tubing was then centered in a square tube of lucite, 10 cm wide and 120 cm long. The sides and portions of the back of this square tubing were made opaque, to minimize all lighting of the test section except from directly behind. An externally triggered strobe light was placed behind the test section, on an adjustable mounting bracket which allowed the strobe to traverse the entire Pyrex tubing length. This strobe generated a 3 μs, 0.5 w-s pulse of light, at a maximum rate of 700 pulses/minute. The test section was illuminated by light from this strobe directed onto a white background, mounted on the same bracket as the strobe light and placed approximately 15 cm behind the test section centerline. For visual observations, the strobe was triggered by a manually operated, hand-held switch. Photographic observations of jet core break-up characteristics were made with a Graphlex 4 x 5 camera, using ASA 3000 black and white Land film. The camera was mounted in front of the test section, on an adjustable-height tripod which allowed the camera, like the strobe, to traverse the entire Pyrex tubing length. Lighting for the photographs was provided by the white background illuminated by the strobe light. For the 3 μs exposure times, made possible by the strobe lighting and ASA 3000 film, resolution of small diameter, high velocity droplets was possible (a 0.1 mm drop traveling at 30 m/s will only travel 0.09 mm over a 3 μs time interval). The camera was positioned so that approximately 10 cm of test section could be seen in each photograph.

2.2. Test Procedures

Through the five test series, 741 trial runs were performed using the standard length test section and 17 trial runs were performed using the shortened B2 test section with the 32 cm long, flared Pyrex tubing. Table 2 summarizes the range of gas and liquid velocities, and resultant core break-up lengths, for the various test series. Velocities given are those calculated at the water nozzle exit.

Table 2. Test Series Summary

Test Series	Number of Runs	Gas Species	v_J (m/s)	v_G (m/s)	L_B (cm)
A7	137	Nitrogen	1.4-8.6	0-41	1.7-49
B2	111	Nitrogen	0.7-3.4	1-30	2.2-73
B2	49	R-12	0.7-2.7	4-18	1.6-26
B2	56	Helium	0.7-2.7	24-91	1.6-23
B3	134	Nitrogen	0.9-2.9	1-36	1.8-85
B3	22	R-12	1.4-2.4	5-18	1.2-21
B3	27	Helium	1.4-2.4	25-88	1.4-17
C1	153	Nitrogen	1.1-4.3	2-30	1.0-53
C2	69	Nitrogen	0.7-3.4	2-20	1.2-51

For each test series, vertical alignment of the test section and coaxial alignment of the stainless steel and Pyrex tubing were first established by adjusting the mounting of the Pyrex tubing and the heights of the spacing pins (or ridges) attached to the stainless steel tubing. Such alignment was verified visually and photographically, by observing symmetry within the test section during: 1) the rotation of the water nozzles within the Pyrex, 2) the flight of droplets formed by dripping flow at the nozzle exit, 3) the flight of high velocity liquid jets with minimal gas flow, and 4) the break-up moderate velocity liquid jets with high gas flow.

Once alignment was accomplished, data acquisition could commence. Typically, a series of trial runs were performed at a fixed liquid flow rate and fixed gas species, with monotonically increasing or decreasing gas flow rates. For each trial run, gas temperature, pressure and flowmeter reading, liquid temperature and flowmeter reading, and atmospheric pressure were recorded.

Visual observation of jet core break-up length was then made. The strobe light mounting bracket was gradually moved along the test section length, and the strobe triggered periodically until the region of the test section containing the break-up point was located. The strobe was then triggered in short bursts of 10-20 individual flashes at rates of approximately 150 flashes/minute. At each flash, the point of jet core break-up was noted, using the transparent ruler clamped to the test section as a reference. After 100-200 individual strobe flashes, an average jet core break-up length could normally be ascertained. The difficulties encountered in establishing this break-up length were two-fold. First, break-up length was not limited to a single, unique value for a given set of flow conditions. Rather, break-up occurred within a range of distances downstream of the nozzle exit. This variability in break-up length was similar to that observed by investigators of free-jet break-up. For example, Iciek [10] observed variations in free jet break-up length of up to ±25%, while Chen and Davis [11] found that 150 separate jet length observations had to be made in order to establish an average break-up length, with a 5% probability level and a tolerance of three jet diameters. Secondly, at high gas flowrates, the point of jet break-up became less well defined, as the growth of small volume roll waves on the jet surface, with subsequent droplet entrainment and core surface distortion, created a very complicated flow field. The actual point of jet core break-up became a matter of definition. For low void fractions, the break-up point was defined as the point at which the disintegrating jet continuously wetted the Pyrex, while for high void fractions, the break-up point was defined as the point at which core surface area increased rapidly, due to the growth of large amplitude roll waves, resulting in deformation of the jet core into sheet-like segments. For these flow conditions, flow field visibility was limited by the wetting of the Pyrex wall by roll wave crests and entrained droplets.

When possible, visual observation of jet core break-up mode, or mechanism was also made, using the strobe light. For flow conditions in the transitional regions, where two competing break-up modes were present, such visual break-up mode determinations were difficult to make. When a single mechanism of jet break-up was dominant, however, visual determination of this mechanism could normally be made. When such a visual determination could not be made, one or more photographs were taken of the flow field, near the break-up point. Photographs were also taken at selected high gas velocities, to clarify the definition of the core break-up point. And, finally, one or more photographs were normally taken at every second or third gas velocity value, as gas velocity was monotonically increased or decreased for a given liquid jet velocity. The number of photographs taken at each selected set of flow conditions varied from one up to four or five, depending upon the complexity of the flow field and the usefulness of the information obtained from the photographs.

When photographs taken at the selected flow conditions displayed a relatively representative, uniform surface disturbance wavelength, this wavelength was measured, using as a reference the transparent ruler visible in each photograph, and recorded. In addition, the distance from the liquid nozzle exit to the point of photographic observation was recorded, again using the transparent ruler as a reference.

When photographs taken at the selected flow conditions depicted small droplets being sheared from the jet core surface, the maximum droplet size observed was measured, using the transparent ruler as reference, and recorded, along with the nozzle exit/observation point distance. Typical photographs contained only three to six visible droplets. Size measurements were made by comparing the droplets, under slight magnification (~3X), to a series of templates with widths starting at 0.1 mm and increasing in 0.1 mm increments. For spherical droplets, it was possible to measure droplet diameters within an accuracy of roughly 0.05 mm.

For the 17 trial runs using the B2 test series shortened test section, an additional parameter to be varied was the distance from water nozzle exit to the test section exit. Photographs for these trial runs were normally taken at the test section exit. These photographs usually depicted large sheets and ligaments of liquid resulting from core break-up, and maximum droplet diameters were not recorded. However, at the largest nozzle exit/section exit distances, only dispersed droplets were observed. For one such trial, trial B2-905, all dispersed droplet diameters were measured and recorded. Flowrates of water flowing as dispersed phase in the gas, and flowing as a liquid film on the inside wall of the Pyrex tubing, were measured and recorded for these 17 trial runs.

3. EXPERIMENTAL RESULTS

This adiabatic simulation of inverted annular flow produced a great deal of data on core jet break-up length, break-up mechanisms, interfacial surface characteristics, and dispersed core droplet sizes. Initial results are summarized in the following paragraphs, while a complete tabulation of the experimental results may be found in [12].

3.1. Break-up Length

Typical break-up length data are plotted in Figs. 3 and 4. The data may be divided into two regions, one in which gas flow conditions (v_G, ρ_G, α) had no influence, and a second in which gas flow did influence break-up length. In this first region,

$$L_B \sim v_J^{0.5} \tag{1}$$

This dependence is similar to that observed by Miesse [13] and by Grant and Middleman [14], who found L_B to be proportional to $v_J^{0.4}$ and $v_J^{0.6}$, respectively, for turbulent liquid jets not subjected to significant interfacial drag.

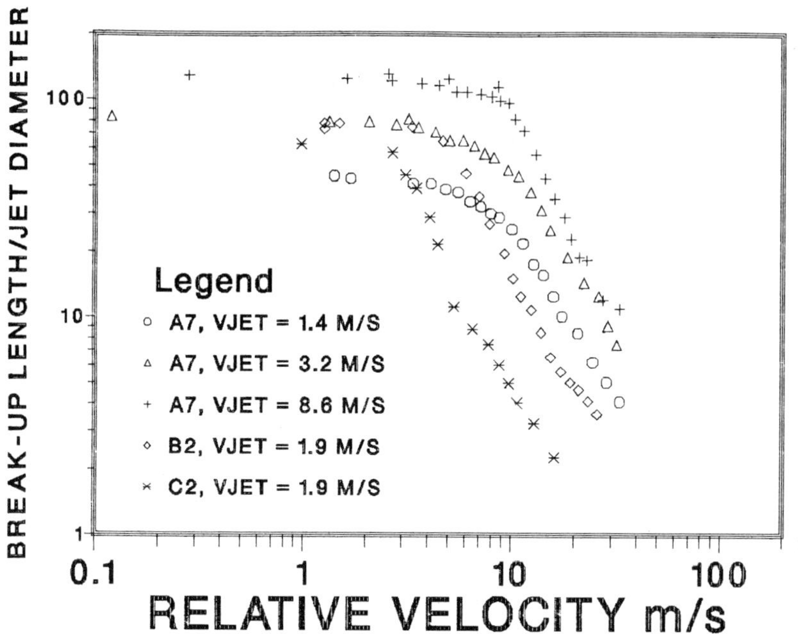

Fig. 3 Break-up Length, Nitrogen Gas, See Table 1 for α, D_0, D_J Values

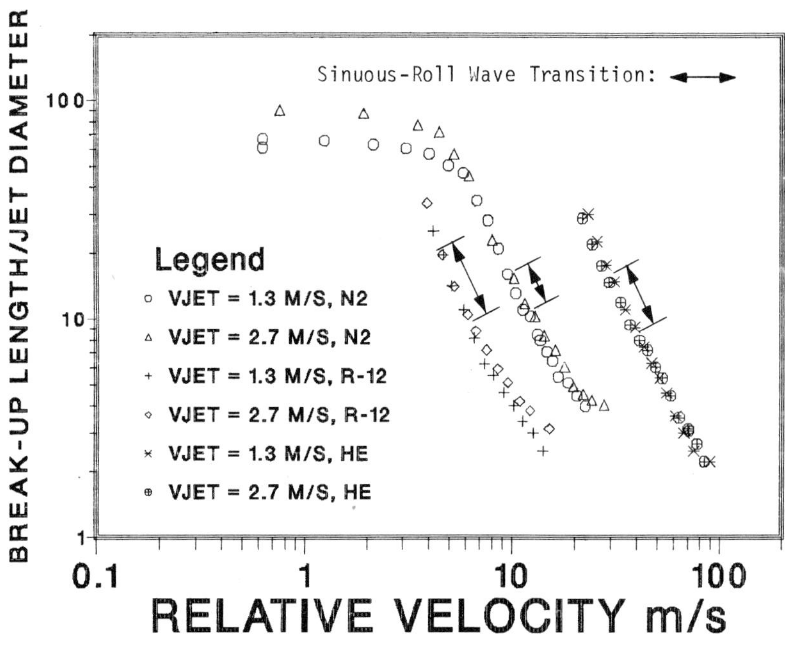

Fig. 4 Break-up Length, Break-up Mechanism, B2 Data, See Table 1

For the region in which gas flow conditions affected break-up length, it appears that

$$L_B \sim v_J^{0.5} \; v_{rel}^{-1.3} \tag{2}$$

where v_{rel} is the absolute value of $(v_G - v_J)$. This is similar to the results of Jensen's [6] calculations, which, in rough approximation, show that

$$L_B \sim v_J^{1.0} \; v_{rel}^{-2.8} \tag{3}$$

for laminar, planer jets, and is in close agreement with the results of Lienhard and Day [15], who found L_B to be proportional to v_J^{-1} in the drag-induced break-up region of free jets (where $v_{rel} = v_J$). The experimental data, like the calculations of Jensen, show that L_B decreases with decreasing void fraction, or with increasing gas density.

3.2. Break-up Mechanisms

Three different break-up mechanisms, varicose deformation of the liquid core, sinuous deformation of the liquid core, and roll wave entrainment dispersion of the liquid core, were observed in this experiment, just as three different free jet break-up mechanisms (varicose deformation, sinuous deformation, and atomization) have been observed [16]. See Fig. 5 for a drawing of these mechanisms. The first mechanism, varicose deformation, was characterized by the growth of axisymmetric waves on the jet core, resulting in alternating points of core expansion and contraction. This break-up mode resulted in transition from inverted annular flow to inverted slug flow, with the liquid slugs having a volume equal to that of a cylinder with diameter equal to initial jet diameter, and length equal to the varicose wavelength with the largest rate of growth. Varicose break-up occurred in the region where gas flow conditions had no effect upon jet core break-up length. See Figs. 3 and 4.

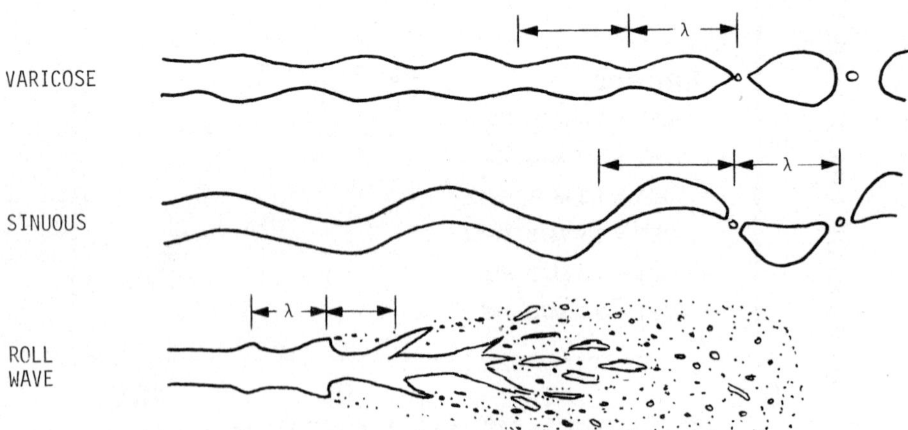

Fig. 5 Jet Core Break-up Mechanisms

The second mechanism of core break-up, sinuous deformation, is characterized by the growth of asymmetric waves which deform the jet core into a snake-like shape. The break-up mode again results in a transition from inverted annular to inverted slug flow. Sinuous break-up first occurs when gas flow conditions began to reduce core break-up lengths.

The third mechanism of jet break-up observed in these experiments is that of roll wave entrainment resulting in a transition from inverted annular flow to dispersed droplet flow. Roll waves first appeared on the crests of sinuous waves, while at higher relative velocities roll waves caused break-up before the core exhibited any sinuous behavior. A single roll wave often developed over a substantial portion of the jet circumference, deforming into a thin, skirt-like sheet. Ligaments or individual droplets could be seen being sheared from the roll wave crests. Except at the highest void fractions, it was not possible to view the entire dispersal of the liquid jet core, due to wall wetting. Because of this, it is possible that after initial core disruption due to large amplitude roll waves, other unobserved mechanisms might also have contributed to droplet formation. The transition from sinuous deformation to roll wave entrainment break-up is indicated in Fig. 4. From Fig. 4 it is apparent that this transition region is roughly determined by the value of ($\rho_G\ v_{rel}^2$). This result is similar to that found by Ishii and Grolmes [17] for inception of roll wave entrainment in annular flow, and that found by Mishima and Ishii [18] for inception of slug formation in horizontal stratified flow. The value of ($\rho_G\ v_{rel}^2$) at the transition from sinuous to roll wave break-up depends upon void fraction, with increasing void fraction requiring larger values of ($\rho_G\ v_{rel}^2$). This result is also similar to that found by Mishima and Ishii [18].

3.3. Jet Core Surface Characteristics

Surface conditions were observed for varicose, sinuous, and roll wave break-up of the jet core. Representative wavelength are plotted versus relative velocity in Fig. 6. For varicose break-up, an average wavelength of

$$\lambda = 5.8\ D_J \tag{4}$$

was observed. This is similar to the value of 4.51 D_J predicted by Rayleigh [19] for varicose break-up of laminar free jets. In the drag-induced break-up regions (sinuous and roll wave break-up), wavelengths steadily decreased with increasing values of ($\rho_G\ v_{rel}^2$).

Regarding the shapes of the surface waves observed, the following generalizations can be made. For low void fractions, roll waves were limited in amplitude by the proximity of the Pyrex wall. At higher void fractions, however, roll waves were able to grow to large dimensions, sometimes deforming the jet into sheets and ligaments with no discernible cylindrical core. This large deformation normally occurred within the last few jet diameters of intact jet core length. Near the inception of roll waves, jets under sinuous break-up no longer had a smooth sinusoidal shape. Wave crests became pointed, and downstream surfaces of each wave were at much steeper slopes than upstream surfaces.

3.4. Dispersed Core Droplet Size

As previously discussed, varicose and sinuous core jet break-up resulted in the formation of large liquid slugs. Subsequent disintegration of these

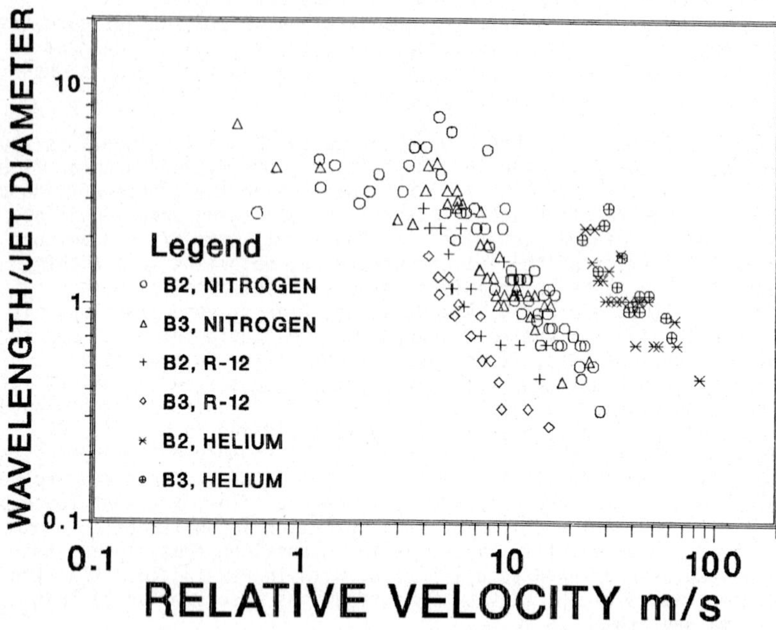

Fig. 6 Jet Core Surface Wavelength, Various v_J Values, See Table 1

slugs was not observed in this experiment, due to wall wetting and limited observation length beyond the point of core break-up.

For roll wave entrainment jet break-up, little droplet size data could be obtained, since roll waves and their entrained droplets wetted the wall after the point of jet break-up, creating an annulus of rough wavy liquid which obscured the dispersed droplet core. However, a limited number of photographs were obtained in which droplets were observed immediately after being sheared from roll wave crests. From these photographs, maximum droplet diameters were measured, within an accuracy of ~0.05 mm. Most droplet diameters were within the range of 0.25-0.55 mm for flows ranging from early roll-wave inception to well established roll wave entrainment break-up of the jet core. These droplet sizes are similar to those observed for high gas velocities in the FLECHT SEASET tests [9], and slightly smaller than those observed by Merrington and Richardson [20] at similar relative velocities, for atomization of free water jets flowing in stagnant air or moving air streams.

In an attempt to observe dispersed core droplet sizes beyond the point of wall wetting and the inception of rough wavy annular film flow, a limited number of trials were run using the shortened B2 test section. Near the point of core break-up, large slugs and ligaments from the distorted jet core, created when roll waves expanded to the Pyrex wall, were observed. These larger dispersed liquid masses continued to break-up beyond the jet break-up point, so that at nozzle/test section exit distances roughly twice as long as the jet core break-up length, only dispersed droplets appeared. Most of these droplets had diameters in the range of 0.25-0.55 mm, the same size as the droplets observed for roll wave crest entrainment. Larger droplets, with diameters of 1-3 mm were also observed, however. These larger droplets are of similar size to those observed for low gas velocities in the FLECHT SEASET tests [9]. Results from this limited number of trials using the shortened test section are not conclusive, since only a small amount of data was obtained. In addition, the effectiveness of the liquid annular film diverter was not established, so that some of the observed droplets may have been formed at the diverter itself. And, finally, the effect of wall wetting makes any observations made beyond the point of jet break-up suspect, when applying them to actual inverted annular flow, where such wetting does not occur.

4. CONCLUSIONS

For this adiabatic simulation of inverted annular flow, the following conclusions can be made:

1. This experimental study was unique in that extensive hydrodynamic information about adiabatic, coaxial jet stability, including liquid core break-up length, break-up mechanisms, core surface characteristics, and dispersed droplet size data was obtained. This data should help in the understanding of film boiling inverted annular flow hydrodynamics. In addition, this data may help in the understanding of the drag-induced break-up of free liquid jets, since jet velocity and relative velocity at the gas/liquid interface were separate variables in this experiment.

2. From preliminary analysis of the data, many similarities can be seen between the phenomenon of coaxial jet disintegration and various aspects of free jet stability, annular and stratified flow, and inverted annular flow film boiling which have been previously investigated [9,13-20]. Results of these previous studies should prove useful in correlating this experimental data.

3. It must be noted that this adiabatic simulation of inverted annular flow, while producing a great deal of useful information, does have certain limitations. First, physical constraints of test section construction limited this study to void fractions no lower than 0.28. Other test section construction and operating constraints limited this study to turbulent liquid core flows, with velocities no lower than 0.7 m/s. Laminar core flow, or low liquid velocities such as those which occur during LWR core reflood, may result in slightly different hydrodynamic behavior. As a third limitation, events beyond the jet core break-up point were not, for the most part, observable, due to wall wetting. Finally, the absence of film boiling conditions (no wall wetting, the presence of an inward-directed radial force at the core surface due to vapor generation) may limit the applicability of these experimental results in the anlaysis of actual inverted annular flow. However, such limitations may be slight, as evidenced by the agreement between droplet size data for this adiabatic simulation and for the FLECHT SEASET heated test section experiments.

ACKNOWLEDGMENTS

This work was performed at Argonne National Laboratory, under the auspices of the U.S. Nuclear Regulatory Commission.

We would like to thank Drs. N. Zuber and M. Young of the NRC for their guidance and assistance in this research, and to thank G. Lambert, E. Sowa and A. Tokuhiro of ANL for help in constructing the experimental apparatus.

NOMENCLATURE

D_J	Jet core diameter
D_o	Gas annulus outer diameter
L_B	Jet core break-up length
v_G	Gas velocity
v_J	Liquid jet core velocity
v_{rel}	Relative velocity ($= \|v_G - v_J\|$)

Greek Symbols

α	Void fraction
λ	Wavelength of maximum growth rate
ρ_G	Gas density

REFERENCES

1. Groeneveld, D.C., "Post-dryout Heat Transfer at Reactor Operating Conditions," presented at the National Meeting on Water Reactor Safety, American Nuclear Society, Salt Lake City, Utah, March 26-28, 1973.

2. Chi, J.W.H., "Slug Flow and Film Boiling of Hydrogen," J. Spacecraft, 1967, Vol. 4, p. 1329.

3. Chen, J.C., Sundaram, R.K., and Ozkaynak, F.T., "A Phenomenological Correlation for Post-CHF Heat Transfer," NUREG-0237, 1977.

4. Laverty, W.F., and Rohsenow, W.M., "Film Boiling of Saturated Nitrogen Flowing in a Vertical Tube," J. Heat Transfer, Trans. ASME, 1967, Vol. 89, p. 90.

5. Kalinin, Ye.K., Berlin, I.I., Kostyuk, V.V., Kochalaev, Yu.S., and Yarkho, S.A., "Heat Transfer in Plug Flow Regime of Film Boiling," Heat Transfer, Soviet Research, 1973, Vol. 5, p. 91.

6. Jensen, R.T., "Inception of Liquid Entrainment During Emergency Cooling of Pressurized Water Reactors," Ph.D. Thesis, Utah State University, 1972.

7. Kurilenko, A.A., Dymenko, S.R., and Kochelaev, Yu.S., "Phase Slip and Heat Transfer to the Liquid in Film Boiling of a Cryogenic Liquid in Piston Flow," J. Eng. Phys., 1980, Vol. 39, p. 961.

8. Ottosen, P., "An Experimental and Theoretical Investigation of Inverse Annular Film Flow and Dispersed Droplet Flow, Important Under LOCA Conditions," Risø National Laboratory, Denmark, Report No. R-424, 1980.

9. Lee, N., Wong, S., Yeh, H.C., and Hochreiter, L.E., "PWR FLECHT SEASET Unblocked Bundle, Forced and Gravity Reflood Task Data Evaluation and Analysis Report, NRC/EPRI/Westinghouse Report No. 10," NUREG/CR-2256, EPRI NP-2013, WCAP-9891, 1981.

10. Iciek, J., "The Hydrodynamics of a Free, Liquid Jet and Their Influence on Direct Contact Heat Transfer - I," Int. J. Multiphase Flow, 1982, Vol. 8, p. 239.

11. Chen, T.-F., and Davis, J.R., "Disintegration of a Turbulent Water Jet," Proc. ASCE, Hyd. Div., 1964, Vol. 90, p. 175.

12. De Jarlais, G., "An Experimental Study of Inverted Annular Flow Hydrodynamics Utilizing an Adiabatic Simulation," M.S. Thesis, Marquette University, Milwaukee, WI, 1983.

13. Miesse, C.C., "Correlation of Experimental Data on the Disintegration of Liquid Jets," Ind. Eng. Chem., 1955, Vol. 47, p. 1690.

14. Grant, R.P., and Middleman, S., "Newtonian Jet Stability," AIChE J., 1966, Vol. 12, p. 669.

15. Lienhard, J.H., and Day, J.B., "The Breakup of Superheated Liquid Jets," J. Basic Eng. Trans. ASME, 1970, Vol. 92, p. 515.

16. Ohnesorge, W., "Die Bildung von Tropfen an Dusen und die Anflosung flussiger Strahlen," Z. Angew Math. Mech., 1936, Vol. 16, p. 355.

17. Ishii, M., and Grolmes, M.A., "Inception Criteria for Droplet Entrainment in Two-Phase Cocurrent Film Flow," AIChE J., 1975, Vol. 21, p. 308.

18. Mishima, K., and Ishii, M., "Theoretical Prediction of Slug Formation in a Horizontal Duct based on Nonlinear Wave Analysis," ANL/RAS/LWR 79-6, 1979.

19. Lord Rayleigh, "On the Instability of Jets," Proc. Lond. Math. Soc., 1878, Vol. 10, p. 4.

20. Merrington, A.C. and Richardson, E.G., "The Break-up of Liquid Jets," Proc. Phys. Soc., 1947, Vol. 59, p. 1.

Multi-Phase Flow and Heat Transfer III. Part A: Fundamentals
edited by T.N. Veziroğlu and A.E. Bergles
Elsevier Science Publishers B.V., Amsterdam, 1984 — Printed in The Netherlands

HOLDUP IN HORIZONTAL GAS-LIQUID FLOW

J.J.J. Chen
Department of Mechanical Engineering
University of Hong Kong
Hong Kong

and

P.L. Spedding
Department of Chemical Engineering
The Queen's University of Belfast
Belfast BT9 5DL
Northern Ireland

ABSTRACT

A review of the more common methods of holdup correlation is presented. The Butterworth form of equation which represents the models of Zivi, Turner & Wallis, Lockhart & Martinelli, Thom, and Baroczy has been justified analytically by Chen & Spedding for certain horizontal situations. In addition, a number of other models including those of Armand, Zuber & Findlay, Nguyen & Spedding, Nicklin et al, and Bankoff, some of which apply to inclined flows, are also found to be special forms of the Butterworth equation. Thus, the Butterworth equation is expected to be a powerful one for holdup presentation for both horizontal and inclined flows. While inclined flows are treated elsewhere, horizontal data are plotted as suggested by the Butterworth equation from which a number of regions may be distinguished: homogeneous flow, annular-mist flow, annular flow, slug and bubbly flow, and stratified flow; correlations are suggested for each region. Thus, a general correlation for the entire range of flow rates is not possible, but specific correlations are necessary for specific flow patterns which cover a limited range of conditions. Further, stratified flow data which are found to exhibit constant holdup at certain flow rates are examined in terms of the Bernoulli equation and the free draining of a liquid-filled tube.

1. INTRODUCTION

In the study of two-phase flow and in the design and operations of process equipments involving two-phase flow, it has now been well recognised that it is important to have a detailed knowledge of the holdup pertaining to the situation. Cases where holdup is required include the calculations of the mean fluid density, the heat contents, the heat transfer coefficients, momentum and hydrostatic components of pressure drops, the prediction of the residence time, the water circulation rates in boilers, etc. However, with the number of calculation methods available in the literature, it is often difficult to choose the appropriate method for the particular situation being considered. Furthermore, even after deciding on a certain calculation method, there is still uncertainty as to the limitations of the method, and the range of conditions under under which it may be extrapolated with reasonable confidence.

2. A REVIEW OF THE RELEVANT LITERATURE

Lockhart and Martinelli (1949) were the first to attempt to formulate a general correlation for holdup in horizontal gas-liquid flow. Their approach was semi-empirical and the correlation was presented as a plot of $\bar{R}$ against X without taking into account the flow pattern. The method was subsequently shown to be quite general (Dukler et al, 1964), and although its degree of accuracy varied with the flow pattern (Chen & Spedding, 1981), it has been recommended as being suitable for the purpose of preliminary design calculations (Turner & Wallis, 1965). The method has also been modified by Baroczy (1965) to apply to the metallic liquid-gas system with the inclusion of a parameter $(\mu_L/\mu_G)^{0.2}/(\rho_L/\rho_G)$.

Expressions for $\bar{R}_G$ for substitution into a formula for estimating the pressure drop were given by Armand (1946). For the case of $Q_G/(Q_G + Q_L) < 0.9$

$$\bar{R}_G = \frac{Q_G}{C(Q_G + Q_L)} \qquad [1]$$

Equation [1] is of course a modification of the homogeneous model to which the term (1/C) was incorporated. Armand also presented expressions for calculating $\bar{R}_G$ for the case of $Q_G/(Q_G + Q_L) > 0.9$, but the equations were rather cumbersome and have not been widely adopted.

A number of attempts have been made to derive analytically the Lockhart-Martinelli holdup correlation by the consideration of specific flow patterns. Levy (1952) was perhaps the first in this respect when he analysed the ideal horizontal annular flow case and obtained a solution for the $\bar{R}$ - X relationship. However, the predicted $\bar{R}_L$ was much higher than that given by the Lockhart-Martinelli correlation.

Turner & Wallis (1966) used a so-called separate-cylinder model to derive an equation relating $\bar{R}$ and X. However, the incorporation of empirical coefficients was necessary in order to achieve agreement with experimental results. The final form was given as

$$\bar{R}_G = \frac{1}{(1 + X^{2/2.5})^{1/2.65}} \qquad [2]$$

and it matches the Lockhart-Martinelli correlation line up to a value of about X = 10.0 (See, for example, Chen & Spedding 1981).

Chisholm (1967) also attempted to justify the Lockhart-Martinelli correlation but without the consideration of any specific flow patterns. The solution, however, could not be obtained analytically and depended on an empirically derived function.

Momentum flux equations were written separately for the gas phase and the liquid phase by Johannessen (1972) for the case of ideal stratified flow. Hydraulic diameters were used by treating the gas as flowing in a conduit and the liquid in an open channel. The holdup $\bar{R}$ was solved in terms of X and the solution presented in a graphical form.

While Johannessen considered only the case of both phases flowing turbulently without accounting for the interfacial shear, Taitel & Dukler (1976) incorporated the effects of interfacial shear and examined the four possible combinations of gas-liquid viscous-turbulent flow regimes.

More recently, Chen & Spedding (1981) proceeded along the formulations of the original Lockhart-Martinelli workings without resorting to momentum flux equations as did Johannessen or Taitel & Dukler and found that there are definite relationships for the $\bar{R}$-X pair for both the ideal annular and the ideal stratified flow situations. In addition, it was shown that the Lockhart-Martinelli correlation may be expressed in the following simple form

$$\bar{R}_L = \frac{X^{2/3}}{k + X^{2/3}} \qquad [3]$$

Chen & Spedding (1981) intuitively suggested that k depends on a number of factors including the pipe diameter. Further, by taking k = 3.5, equation [3] provides an exact match with the Lockhart-Martinelli correlation line over the entire range of X values.

Apart from attempts to derive the Lockhart-Martinelli correlation analytically, a number of models have been proposed to predict the holdup in gas-liquid flow. Zivi (1964) applied the principle of minimum entropy production and by equating the minimum kinetic energy flux obtained, for annular mist flow,

$$\frac{\bar{U}_G}{\bar{U}_L} = \left(\frac{\rho_L}{\rho_G}\right)^{1/3} \left[\frac{1 + \frac{\rho_G}{\rho_L} E_D\left(\frac{1 - x}{x}\right)}{1 + E_D\left(\frac{1 - x}{x}\right)}\right]^{1/3} \qquad [4]$$

For ideal annular flow without entrainment, $E_D = 0$, equation [4] reduces to

$$\bar{R}_G = \frac{1}{1 + \left[\frac{1 - x}{x}\right]\left[\frac{\rho_G}{\rho_L}\right]^{2/3}} \qquad [5]$$

Smith (1971) used the concept of equal velocity heads and obtained a rather complicated form of equation for holdup which reduced to the homogeneous flow model for fully dispersed or misty flow (i.e. $E_D = 1.0$). For $E_D = 0$, the equation becomes

$$\bar{R}_G = \frac{1}{1 + \left[\frac{1 - x}{x}\right]\left[\frac{\rho_G}{\rho_L}\right]^{1/2}} \qquad [6]$$

A number of analysis for flow in vertically upward flow are of interest. Griffith & Wallis (1961) derived the following equation for the bubble velocity in slug flow.

$$U_B = U_{SL} + U_{SG} + U_{BR}. \qquad [7]$$

The term U_{BR} possesses a form similar to that derived by Dumitrescu (1943) and Davies & Taylor (1950). The voidage may be evaluated by

$$\bar{R}_G = (U_{SG}/U_B). \qquad [8]$$

Nicklin et al (1962) suggested that the bubble velocity should be

$$U_B = 1.2(U_{SL} + U_{SG}) + 0.35\sqrt{gD} \quad [9]$$

when the liquid Reynolds number exceeds about 8000. The factor 1.2 accounted for the higher velocity at the pipe axis. The term $0.35\sqrt{gD}$ is simply the rise velocity of a Taylor Bubble in stagnant liquid. It has been argued that the term $0.35\sqrt{gD}$ due to the rise velocity of a Taylor bubble vanishes in the case of horizontal flow when equations [8] and [9] are reduced to the Armand equation [1]. However, it was found that equation [9] applies to both the bubbly and the slug flow patterns. It is to be noted that the form of equation [9] has also been applied to downward flows (Martin 1973, Brown et al 1969, Arosio et al 1976).

Bankoff (1960) assumed that a two-phase mixture may be treated as a single fluid with a variable local density. 'Power law' equations were used to describe the velocity and voidage distribution profiles. The result was

$$\bar{R}_G = \kappa \frac{Q_G}{(Q_G + Q_L)} \quad [10]$$

where κ is a function of the 'power law' exponents and may assume values ranging between 0.6 and 1.0 depending on the values of the exponents. Bankoff further showed that the factor κ was dependent on the system pressure. Zuber (1960) pointed out that the Bankoff model is basically equivalent to the Armand model. Brown, Gomezplata & Price (1969) showed that equation [10] may also be derived if parabolic distribution equations are used.

Spedding & Chen (1982) showed that inconsistencies existed in the limit with both the Bankoff equation [10] and the Zuber & Findlay (1965) equation. Nguyen & Spedding (1977) met these objections by developing a rather complex holdup equation using the field concepts and the concept of fluctuating flow. Under certain conditions, the equation is reduced to the Zuber & Findlay (1965) equation

$$\frac{U_{SG}}{\bar{R}_G} = C_o(U_{SL} + U_{SG}) + B \quad [11]$$

which was derived by considering the effects of local relative velocity between the two phases, the non-uniform flow and the concentration profiles. The term B attained a special meaning in bubbly or slug flow in that it represented the bubble or slug velocity.

From a survey of the literature, Butterworth (1975) intuitively suggested that a number of the more commonly used holdup equations may be represented by a general equation of the form shown in equation [12]

$$\left[\frac{\bar{R}_L}{\bar{R}_G}\right] = A \left[\frac{1 - x}{x}\right]^p \left[\frac{\rho_G}{\rho_L}\right]^q \left[\frac{\mu_L}{\mu_G}\right]^r . \quad [12]$$

The numerical values for A, p, q and r varied with the model considered and these are given in Table 1.

The form of equation [12] for holdup correlation was justified analytically by Chen & Spedding (1983) by writing force balance equations in the liquid phase

Table 1. Values of A, p, q and r in equation [12] and the values of K, a, b and c in equation [13]

Model	A	p	q	r	K	a	b	c
Homogeneous Model	1	1	1	0	1	1	0	0
Zivi (1964)	1	1	0.67	0	1	1	0.33	0
Turner-Wallis (1965)	1	0.72	0.40	0.08	1	0.72	0.32	0.08
Lockhart-Martinelli (1949)	0.28	0.64	0.36	0.07	3.57	0.64	0.28	0.07
Thom (1962)	1	1	0.89	0.18	1	1	0.11	0.18
Baroczy (1965)	1	0.74	0.65	0.13	1	0.74	0.09	0.13

and in the gas phase separately for the cases of ideal stratified and ideal annular flow patterns. In addition, it was shown that the values of A, p, q and r varied with the flow pattern, the flow regime, namely viscous or turbulent, and also with the range of holdup values. Chen & Spedding (1983) also found that in the analysis of data, it is sometimes more convenient to express [12] in terms of the volumetric flow rates:

$$\left[\frac{\bar{R}_G}{\bar{R}_L}\right] = K \left[\frac{Q_G}{Q_L}\right]^a \left[\frac{\rho_G}{\rho_L}\right]^b \left[\frac{\mu_G}{\mu_L}\right]^c \qquad [13]$$

In addition to those models shown by Butterworth to be represented by equation [12] as given in Table 1, it was found that the Armand equation as given in equation [1] may be re-written in the form

$$\left[\frac{\bar{R}_L}{\bar{R}_G}\right] = \frac{1}{C} \left[\frac{1 - x}{x}\right] \left[\frac{\rho_G}{\rho_L}\right] + \left[\frac{1}{C} - 1\right] \qquad [14]$$

The Nguyen & Spedding (1977) or the Zuber & Findlay (1965) form as given in equation [11] may be rewritten as

$$\left[\frac{\bar{R}_L}{\bar{R}_G}\right] = C_o \left[\frac{1 - x}{x}\right] \left[\frac{\rho_G}{\rho_L}\right] + (C_o - 1) + \frac{B}{U_{SG}} \qquad [15]$$

An inspection of equation [12], [14] and [15] shows that, in addition to the models listed by Butterworth as given in Table (1), the models of Armand (1946), Smith (1971), Nguyen & Spedding (1977), Zuber & Findlay (1965), Griffith & Wallis (1961), Nicklin (1962), and Bankoff (1960) are all special cases of the Butterworth's equation given as [12] but with additional factors incorporated. In light of the fact that most derived equations for holdup require empirical determination of constants anyway, and, in addition fall into the Butterworth form, it appears that equation [12] provides a very powerful tool for holdup correlation. Its use had been justified for horizontal flow by Chen & Spedding (1983). By examining the equations of Nicklin et al (1962), Bankoff (1960), etc. for vertically upward flow, and the equations of Martin (1973), Brown et al

(1969), Arosio et al (1976) for vertically downward flow, in a manner similar to the derivation of equation [14] and [15], it may be deduced that the Butterworth form of holdup correlation can also be applied to correlating holdup in other tube inclinations. In fact, Isbin & Biddle (1979) used equation [12] for correlating holdup in vertically upward steam-water flow while Spedding & Chen (1982, 1983) used [13] for correlating holdup data in air-water flow in a 4.55 cm diameter pipe inclined at angles ranging from +90° to -90° to the horizontal.

3. EXPERIMENTAL RESULTS AND DISCUSSIONS

The review of literature shows that the Butterworth form of equation is a very powerful tool for the presentation of holdup data. Accordingly the holdup data of Spedding & Nguyen (1976) and Chen & Spedding (1979) obtained for the atmospheric air-water system flowing in a 4.55 cm pipe over a 2 m test section are plotted in figures 1 and 2 as $\bar{R}_G/\bar{R}_L$ against Q_G/Q_L as suggested by equation [13]. The stratified flow data which are excluded from figure 1 are plotted exclusively in figure 2.

It is possible to distinguish five distinct divisions in figure 1 which are associated with particular flow patterns. For annular flow, with $\bar{R}_G/\bar{R}_L \geq 4.0$, the data fall about a straight line which corresponds very closely to the Butterworth form of equation representing the Lockhart-Martinelli model (refer to equation [13] and Table 1) which overpredicts the value of $\bar{R}_L$. Alternatively, the ratio $\bar{R}_G/\bar{R}_L$ may be derived from equation [3] to give

$$\frac{\bar{R}_G}{\bar{R}_L} = \frac{k}{X^{2/3}} \qquad [16]$$

For the conditions of turbulent-turbulent flow, X may be written as

$$X_{tt} = \left[\frac{1-x}{x}\right]^{0.9} \left[\frac{\rho_G}{\rho_L}\right]^{0.5} \left[\frac{\mu_L}{\mu_G}\right]^{0.1} \qquad [17]$$

Thus, equation [3] derived by Chen & Spedding (1981) may also be expressed in terms of the Butterworth form of equation. The actual data are better described by the empirical equation given as

$$\frac{\bar{R}_G}{\bar{R}_L} = 0.45 \left[\frac{Q_G}{Q_L}\right]^{0.65} \qquad [18]$$

Below $\bar{R}_G/\bar{R}_L = 4$, the data which were predominantly of the slug type, are widely scattered. The bubbly flow data give close agreement with the curved line represented by equation [14], which is derived from the Armand equation. The reason for the wide scatter in the slug flow data is that the test section used was not sufficiently long to give representative results for slug flow. At values of $Q_G/Q_L \geq 15000$, a rather constant value of $\bar{R}_G/\bar{R}_L$ was observed where the flow was observed to be composed of a gas core containing entrained droplets with a thin liquid film on the pipe wall. The experimental facilities employed in obtaining these data were of insufficient capacities to achieve higher values of Q_G/Q_L than shown in figure 1. It is however expected that at higher values

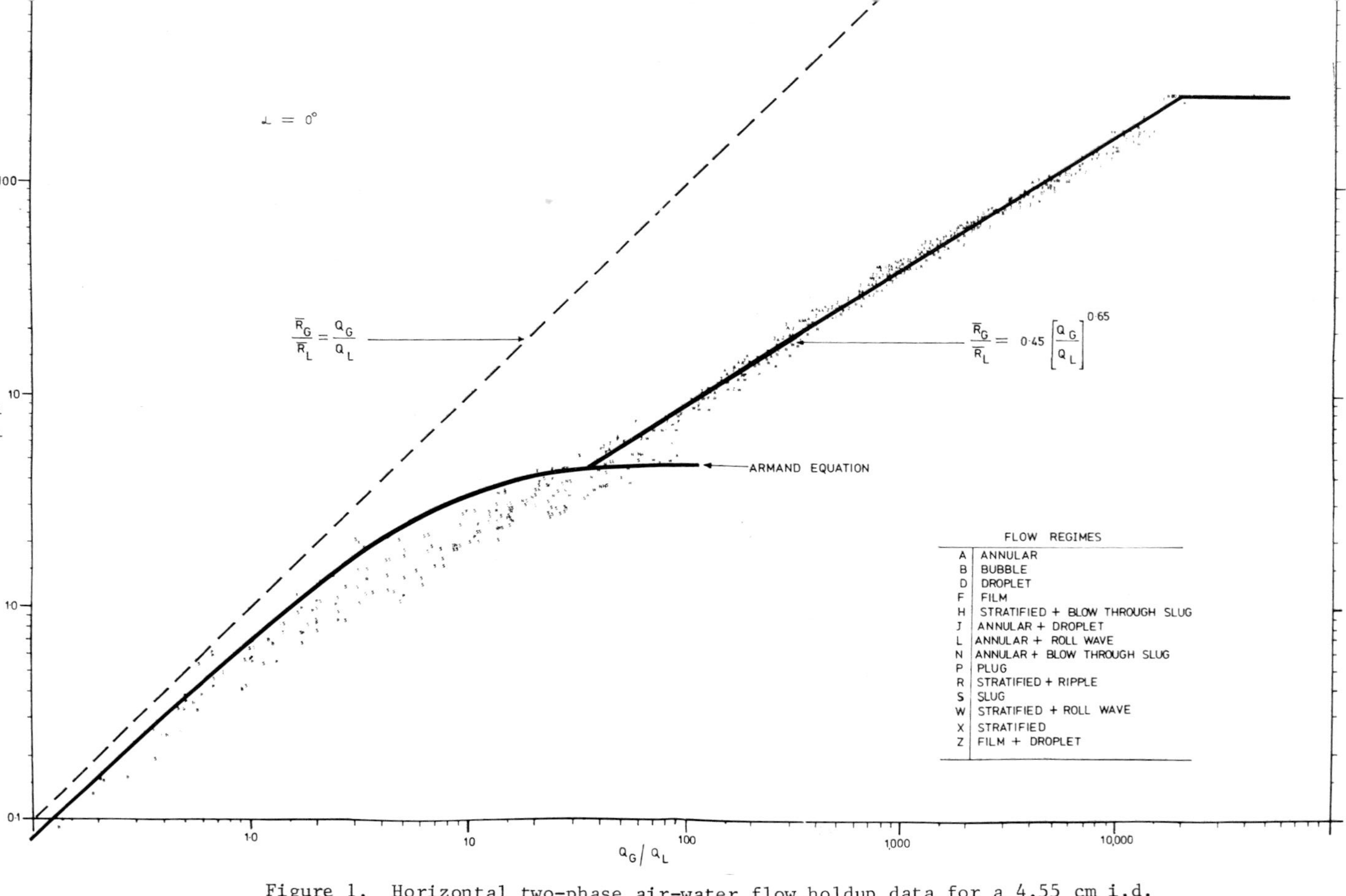

Figure 1. Horizontal two-phase air-water flow holdup data for a 4.55 cm i.d. pipe. Stratified flow data are not included.

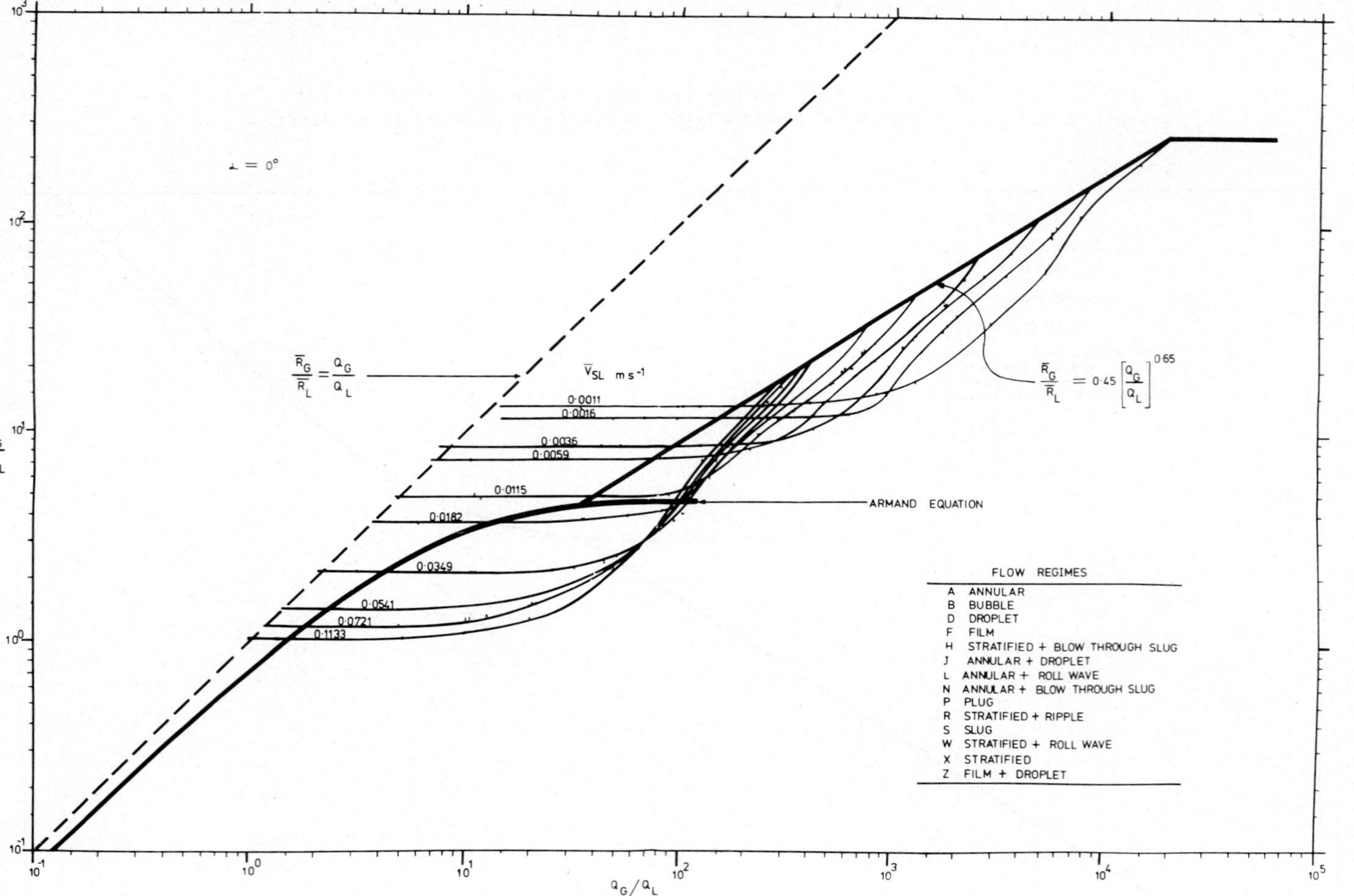

Figure 2. Stratified flow data in horizontal two-phase air-water flow for a 4.55 cm i.d. pipe.

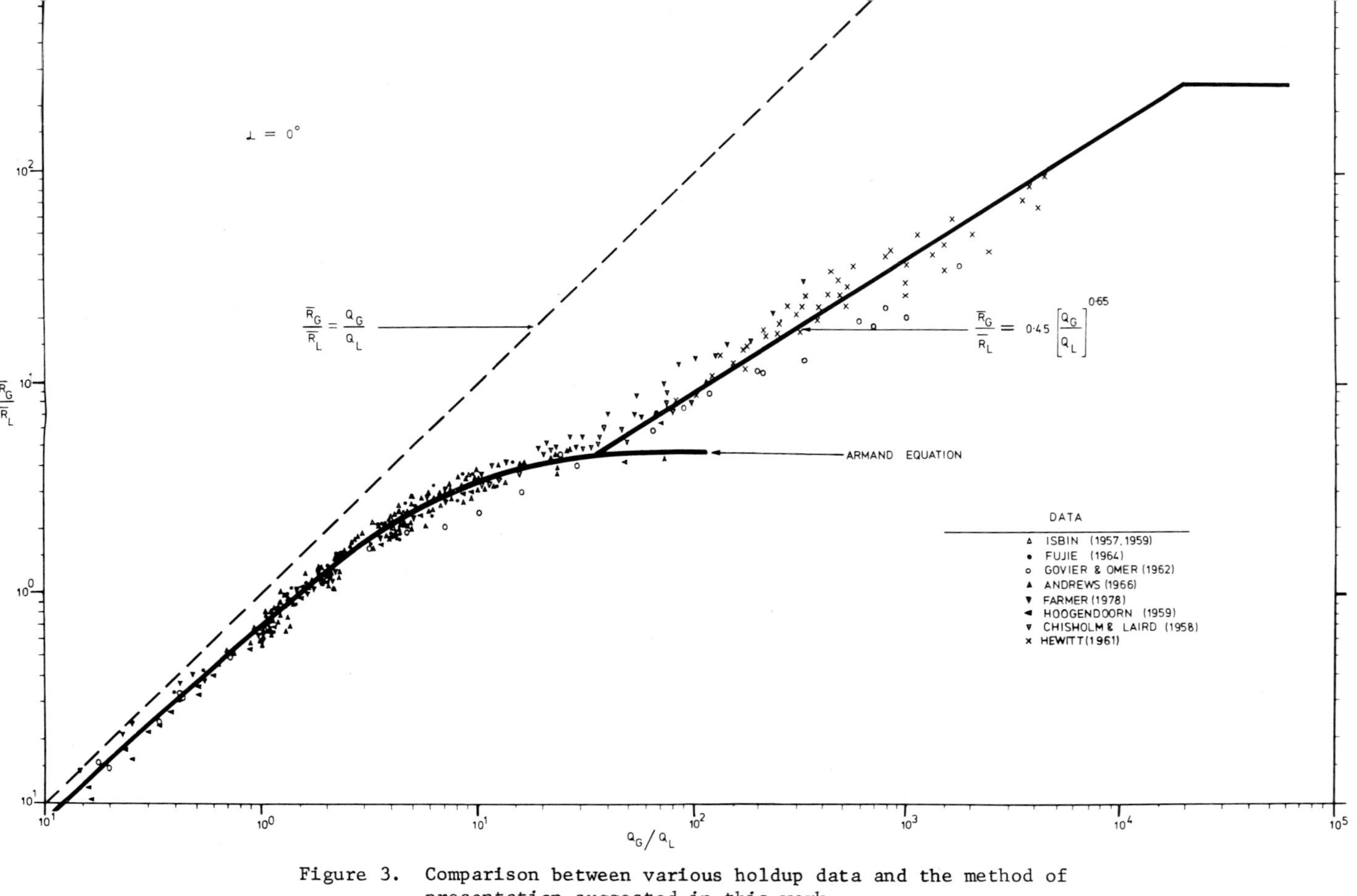

Figure 3. Comparison between various holdup data and the method of presentation suggested in this work.

of Q_G/Q_L, the data would exhibit a steep climb to approach the homogeneous flow regime represented by the $(\bar{R}_G/\bar{R}_L) = (Q_G/Q_L)$ line.

A detailed comparison was made in figure 3 between the form of correlation suggested in figure 1 and a wide range of data for horizontal flow. Some detail need to be presented about these data. The steam water data of Isbin et al (1957, 1959) and of Fujie (1964) for pressure between 7 x 10^5 and 7 x 10^6 Kg m^{-1} s^{-2} and 1.23 cm internal diameter pipe gave good agreement with the correlation. The following air-water data gave good agreement with the correlation; Chisholm & Laird (1958) for a 2.69 cm internal diameter pipe, the Dartmouth correlation of Wallis (1969) and Farmer et al (1978) for a 2.5 cm internal diameter pipe. The annular flow data of Hewitt et al (1961) for a 3.18 cm internal diameter pipe showed reasonable agreement with the correlation. The data of Andrews (1966) obtained on 515 m of 5.25 cm of internal diameter pipe for water-natural gas flow gave a good correlation with the Armand relation. The data of Hoogendoorn (1957) for oil-water flow in a 14 cm internal diameter pipe gave good agreement with the Armand relation for slug type flows but other data were ignored as they were in the stratified regimes. The data used by Lockhart & Martinelli (1949) proved to be very scattered and were not used in this check.

For horizontal two phase flow with the stratified type flow regimes the data in figure 2 initially exhibit a series of horizontal lines at lower gas rates which depend on liquid flow but eventually at higher gas rates join into the holdup correlation lines for the slug and annular type flows. Spedding & Nguyen (1978) suggested that this type of flow regime has parallels with open channel flow. This aspect is examined in detail in the following section using the techniques outlines by Chow (1959) and Henderson (1966).

Applying the Bernoulli equation to the case of horizontal stratified flow for a circular conduit as given in Figure 4,

$$\frac{P}{\rho_L g} + \frac{\bar{V}_L^{\,2}}{2g} = y + \frac{\bar{V}_{L_1}^{\,2}}{2g} = \text{etc.} = E \qquad [19]$$

where P is the hydrostatic pressure ρ_L is the liquid density, y is the liquid depth, g is the gravitational acceleration, $\bar{V}_L$ is the average liquid velocity over a channel cross section and E is the specific energy. For the case of inviscid flow the specific energy must be constant so

$$(E - y)\, A_L^{\,2} = Q_L^{\,2}/2g \qquad [20]$$

where A_L is the cross-sectional area of liquid flow in the conduit which, in the range of interest, may be solved using linear curve-fits, to be a function of y the liquid depth. Since for steady state conditions, the volumetric liquid flow rate must be a constant, then [20] is a cubic with two real roots which are asymptotic to two equations as shown in figure 5. It will be observed that for a given flow rate Q_L there is a minimum specific energy E_c which occurs at the critical depth y_c where the Froude number ($Fr = V_L^2/gy_c$) is equal to unity. Further, at this point of critical flow the conduit will give the maximum flow rate for a given specific energy.

In figure 6 the critical depth of flow is presented for the various experimental conditions employed by Spedding & Nguyen (1976) and Chen & Spedding (1979) when collecting the data used in obtaining figure 2. It will be observed

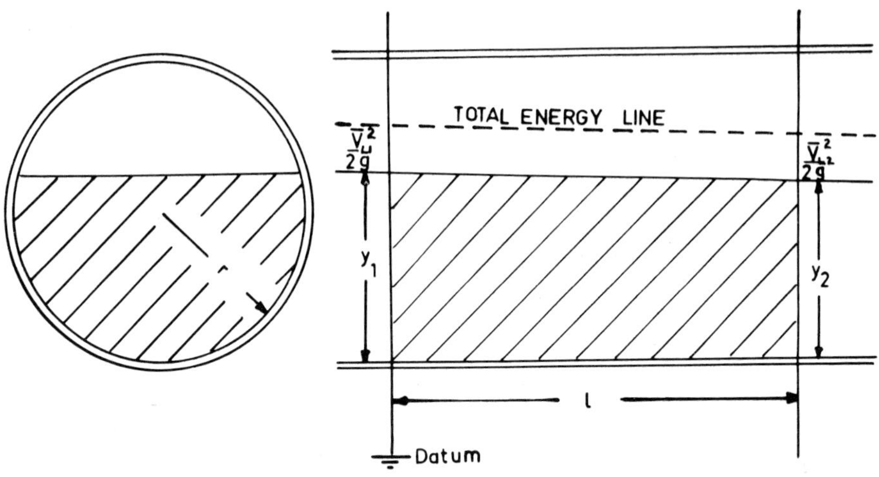

Figure 4. Schematic representation of free surface channel flow in a horizontal circular conduit.

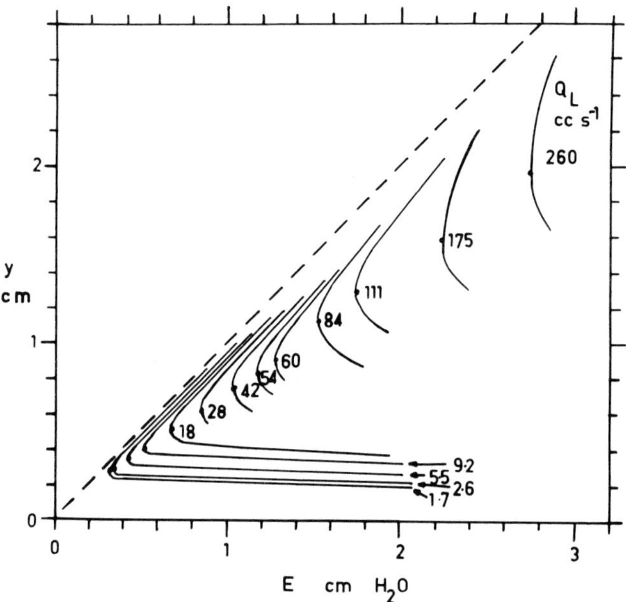

Figure 5. Liquid depth against specific energy for free surface channel flow in a horizontal circular conduit.

that the actual experimental liquid depth obtained is much greater than the critical depth calculated from euqation [3], in fact it corresponds to twice the critical depth in the low liquid flow range. However as the liquid flow rate exceeds about 80 $cm^3\ s^{-1}$, i.e. in the turbulent regime, the experimental data departs from this relation towards the condition of draining of an initially liquid filled horizontal tube. Extending the reasoning of Benjamin (1968) it is possible to calculate out this limiting condition of depth.

Figure 7 illustrates the draining condition in a tube filled with a liquid. The air filled cavity or semi-infinite bubble advances in the upper section of the tube as the liquid drains away in the opposite direction in the lower part of the tube. In order to make the problem more tractable it is assumed that the liquid passes away at such a velocity that the cavity remains stationary with a stagnation point O at its tip and a free boundary above the downstream liquid. Then, A_L the cross-sectional area of liquid flow, is:

$$A_L = (\pi - \alpha + \frac{1}{2} \sin 2\alpha) r^2 = r(1 - \xi) \qquad [21]$$

where

$$\xi = (\alpha - \frac{1}{2} \sin 2\alpha)/\pi \qquad [22]$$

and 2α is the angle subtended at the tube axis by the free surface far downstream, and r is the tube radius.

From the equation of continuity

$$\bar{V}_1/\bar{V}_2 = A_L/\pi r^2 = 1 - \xi \qquad [23]$$

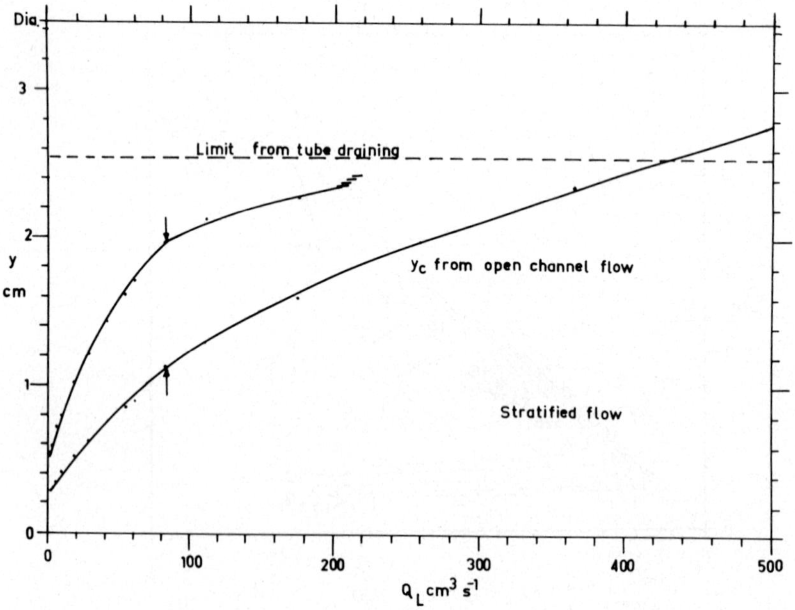

Fig. 6. Depth of liquid flow against liquid rate for liquid flow in a horizontal pipe.

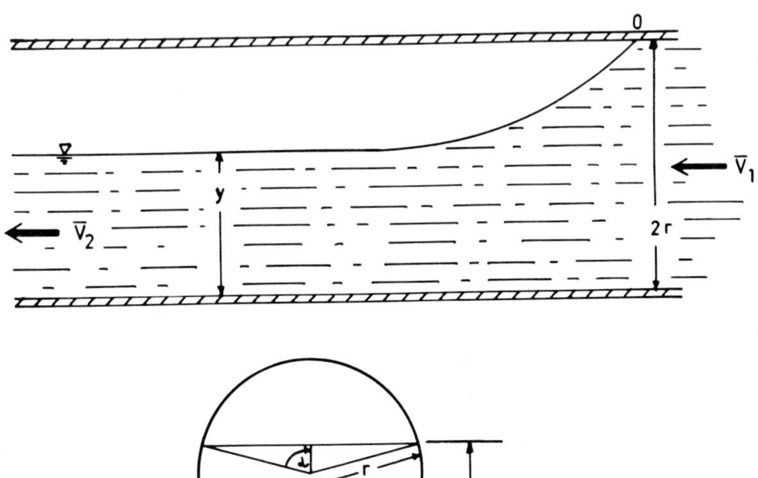

Fig. 7. Schematic representation of the draining condition for a horizontal pipe initially filled with liquid.

Applying Bernoulli's theorem along the free surface between the stagnation point 0 and the asymptotic level of liquid far downstream

$$\bar{V}_2^{\,2} = 2g\ (r - r\cos\alpha - h) \qquad [24]$$

where h is the head loss due to friction. The pressure at the top the cross-section far upstream is

$$P_{T1} = -\frac{1}{2}\,\rho\bar{V}_1^{\,2} \qquad [25]$$

and the pressure in the liquid below has a hydrostatic variation with depth. The total pressure force acting on a cross-section is

$$F_P = (P_{T1} + \rho g r)\pi r^2 \qquad [26]$$

while the total flow force must have the momentum flux added

$$F_{f1} = (P_{T1} + \rho g r + \bar{V}_1^{\,2})\ \pi r^2 \qquad [27]$$

$$= \rho(gr + \frac{1}{2}\,\bar{V}_1^{\,2})\ \pi r^2$$

Downstream, the corresponding total flow force is

$$F_{f2} = 2\rho g r^3 \int_{\alpha}^{\pi} (\cos\alpha - \cos\theta) \sin^2\theta \, d\theta$$

$$= \rho g r \, (A_L \cos\alpha + \frac{2}{3} r^2 \sin^3\alpha) \qquad [28]$$

Since $F_{f1} = F_{f2}$ and using equation [23] and [24] to eliminate the velocity term gives

$$\xi^2 (1 - \cos\alpha) - \frac{2}{3\pi} \sin^3\alpha + \xi \cos\alpha + \frac{h}{r}(1 - \xi^2) = 0 \qquad [29]$$

Figure 8 details graphs of h/d, $\bar{V}_1{}^2/gr$, $\bar{V}_2{}^2/gr$ and $Q_L{}^2/gr^5$ against y/d. At y/d = 0.680 the value of h/d and $\bar{V}_1{}^2/gr$ are at a maximum showing that the speed of motion of the bubble is directly influenced by the applied forces, as reflected by the frictional resistance ratio. Thus when the frictional resistance is positive, y/d > 0.563 that is the flow depth is greater than the minimum value obtained under inviscid conditions because the flow of liquid out of the tube is hindered by the friction. Thus steady flow in which the receding stream fills more than 0.563 d of the pipe is possible if energy loss occurs. To obtain the condition y/d < 0.563, requires the frictional energy loss h to be negative, which would necessitate an external supply of energy to sustain steady flow. As y/d increases from the value of 0.563 at h = 0, $Q_L{}^2/gr^5$ falls steadily from the viscid condition. Thus the rate of liquid flow out of the tube cannot be made larger than the value for free flow without energy loss and the only way in which it can be increased is by pumping the liquid in order to overcome the resistance to flow.

The upsteam Froude number $\bar{V}_1{}^2/gr$ first increases with y/d to a maximum value at y/d = 0.680 and then steadily falls away. The maximum value of the upstream Froude number coincides with the maximum in the h/d curve in a similar manner as reported by Benjamin (1968) for a rectangular channel although the exact value was smaller. The form of the $V_1{}^2/gr$ against y/d graph shows that with a limited range there are two possible values for the downstream depth for each value of upstream velocity. For example, when inviscid conditions pertain the receding liquid stream may be shown to have a Froude number of 1.328 and is supercritical, but it is possible for it to acquire another larger depth in the subcritical range by passing through a hydraulic jump. Therefore, steady flow would be virtually impossible to maintain since any flow instability, for example, induced by wave formation, would precipitate the hydraulic jump which is latent in the particular conditions. The result would be that the receding liquid flow would commence in the supercritical condition but would soon pass into the alternative depth at subcritical conditions. When the hydraulic jump takes place the interface between the liquid and gas phases not only rises but would tend to become blurred due to the onset of gas entrainment in the surface liquid. This would lead readily to a change in flow regime and would explain the blurred region on figure 6 when the data are approximately at the free draining limit condition.

The above working and discussion on the draining condition in a tube initially filled with a liquid, as illustrated in figure 7, obviously is a limiting condition for the current two phase flow situation which is being considered in this work. However the development does have relevance in that it casts some light on the stratified situation under discussion. Firstly, for a pumped or gravity feed liquid condition, the liquid holdup in the tube will be

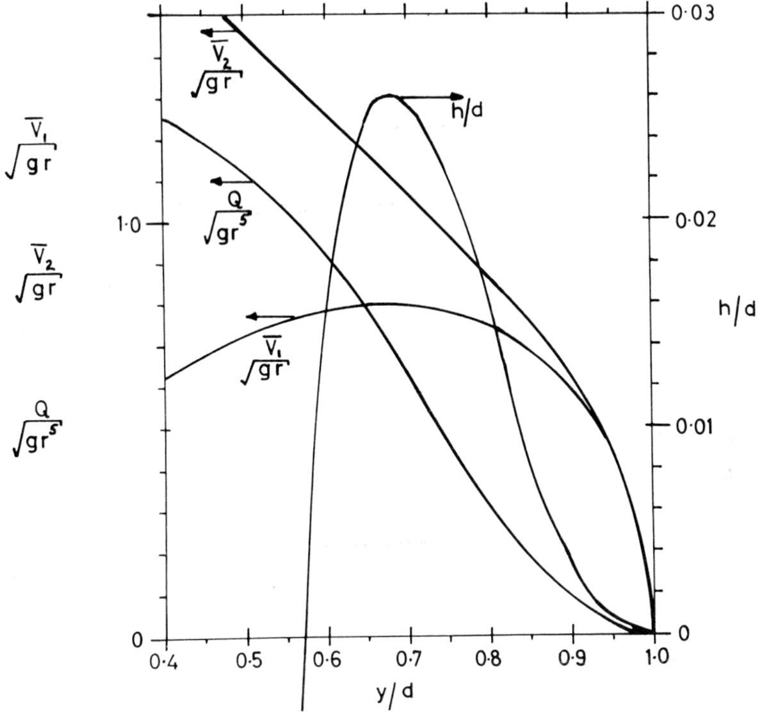

Fig. 8. Graphs of various dimensionless parameters found for the free draining condition of a horizontal pipe considered as a function of (y/d), the dimensionless liquid depth.

in general below that of the free draining condition, as suggested by intuitive reasoning earlier. Secondly, departure from inviscid condition can be expected to increase the liquid holdup. Finally, as the liquid rate is increased for given conditions the onset of surface disturbances will precipitate flow conditions that cause the flow to depart from the stratified regime before the liquid holdup can reach the free draining situation. Chisholm & Laird (1958) have demonstrated that both liquid holdup and two phase pressure drop increase with tube roughness in agreement with the second prediction. The data of figure 6 show that as the liquid rate is increased the liquid holdup does approach the free draining condition, but the flow changes into the mixed type regime before the actual data can coincide with the theoretically derived free draining condition.

4. CONCLUSIONS

As a result of the literature survey, the Butterworth form of equation is shown to be a powerful one for representing holdup data in both horizontal and inclined flows, and no one correlation will allow correct prediction of holdup to be achieved over the entire range of possible flow patterns.

Horizontal flow data are analysed using the Butterworth equation as the basis so that data are presented as plots of (Q_G/Q_L) against $(\bar{R}_G/\bar{R}_L)$. For bubbly and slug flows where $(\bar{R}_G/\bar{R}_L) < 4.0$, the data fall about a curved line represented by the modified Armand equation represented by equation [14]. Annular flow data with $4.0 < (\bar{R}_G/\bar{R}_L) < 275$ fall about a line represented by an empirical equation given by equation [18]. For higher values of (Q_G/Q_L), a somewhat constant value of $(\bar{R}_G/\bar{R}_L)$ was observed which represents approximately annular liquid film-misty flow where the liquid film begins to strip off the tube inner surface. For still higher values of (Q_G/Q_L) for which there is no data, a steep climb of the $(\bar{R}_G/\bar{R}_L)$ is expected so that the homogeneous flow line represented by $(\bar{R}_G/\bar{R}_L) = (Q_G/Q_L)$ is approached.

Horizontal stratified flow data may be better represented using the equations given in Chen & Spedding (1983) which were derived while attempting to justify the Butterworth equation. However, when plotted as $(\bar{R}_G/\bar{R}_L)$ against (Q_G/Q_L) it exhibits constant $(\bar{R}_G/\bar{R}_L)$ at low values of (Q_G/Q_L) but at higher (Q_G/Q_L), they join on to the correlation lines representing the slug and bubbly flow and the annular flow patterns. The low (Q_G/Q_L) data which exhibit constant $(\bar{R}_G/\bar{R}_L)$ were examined in terms of the Bernoulli equation and it was found that the liquid depth correspond to approximately twice the critical depth when the liquid rate is below a certain value, but approaches the condition of the free draining of a pipe initially filled with liquid.

LIST OF SYMBOLS

A	factor in equation [12]
A_L	liquid flow cross sectional area
a	exponent in equation [13]
B	weighted mean drift velocity or initial function
b	exponent in equation [13]
C	Armand coefficient
C_o	Distribution parameter
c	exponent in equation [13]
d	diameter
E	energy
E_D	fraction of liquid entrained as droplets
F	total pressure force
F_r	Froude number
g	acceleration due to gravity
h	head loss due to friction
K	factor in equation [13]
k	a factor
P	hydrostatic pressure
p	exponent in equation [12]
Q	volumetric flow rate
q	exponent in Equation [12]
$\bar{R}$	holdup ($\bar{R}_G$ = void fraction, $\bar{R}_L = 1 - \bar{R}_G$)
r	radius or exponent in equation [12]
V	velocity
$\bar{V}$	true velocity
X	Lockhart-Martinelli parameter
x	dryness fraction
y	liquid depth

Greek Symbols

α	angle
κ	factor in Bankoff's equation
μ	viscosity
ξ	defined in equation [22]
ρ	density

Subscripts

B	bubble
BR	bubble rise
c	critical
f	flow force
G	gas
L	liquid
P	pressure force
S	superficial
T	top
tt	turbulent-turbulent
1	position 1
2	position 2

LITERATURE

Andrews D.E. (1966). The prediction of pressure loss during two phase horizontal flow in two-inch line pipe. M.Sc. Thesis, Univ. Texas.

Armand A.A. (1946). The resistance during the movement of a two phase system in horizontal pipes, IZV., V.T.I. 1, 16-23. AERE Trans. 828.

Arosio S., Bartoloni R., Lorenzi A. & Sotgia G. (1976). Two phase downward flow in large diameter ducts - Preliminary results. Energia Nucleare, 23, (4), 224-232.

Bankoff S.G. (1960). A variable density single fluid model for two phase flow with particular reference to steam-water flow. Trans. ASME, J. Heat Transfer, 265-272.

Baroczy C.J. (1965). Correlation of liquid fraction in two phase flow with application to liquid metals. CEP Symp. Series, 61, (57), 179-191.

Benjamin T.B. (1968). Gravity currents and related phenomena, Journal of Fluid Mechanics, 31, 209-248.

Brown R.W., Gomezplata A. & Price J.D. (1969). A model to predict void fraction in two phase flow. Chem. Engng. Sci., 24, 1483-1489.

Butterworth D. (1975). A comparison of some void fraction relationships for gas-liquid flow. Int. J. Multiphase Flow, 1, 845-850.

Chen J.J.J. & Spedding P.L. (1979). Data on holdup, pressure loss and flow pattern in a horizontal pipe. Univ. Auckland, Engng. Report 214.

Chen J.J.J. & Spedding P.L. (1981). An extension of the Lockhart-Martinelli theory of two phase pressure drop and holdup. Int. J. Multiphase Flow, 7, 659-675.

Chen J.J.J. & Spedding P.L. (1983). An analysis of holdup in horizontal two-phase gas-liquid flow. Int. J. Multiphase Flow, 9, 147-160.

Chisholm D. (1967). A theoretical basis for the Lockhart-Martinelli correlation for two-phase flow, Int. J. Heat Mass Transfer, 10, 1767-1778.

Chisholm D. & Laird A.D.K. (1958). Two phase flow in rough tubes, Trans. ASME, 80, 276-286.

Chow V.T. (1959). Open-channel Hydraulics, McGraw-Hill.

Davies R.M. & Taylor G.I. (1950). The mechanics of large bubble rising through extended liquids and through liquids in tubes. Proc. Roy. Soc., A200, 375-391.

Dukler A.E., Wicks M., Cleveland R.G. (1964). Frictional pressure drop in two-phase flow: A. A comparison of existing correlations for pressure loss and holdup. A.I.Ch.E. J. 10, 38-43.

Dumitrescu V.D.T. (1943). Stromung an einer Luftblase in senkrehten Rohr. Z. Angew. Math. Mech., 23 (3).

Farmer P.R., Horrocks J.K. & Stothart P.H. (1978). Tracer methods for void fraction and dispersion measurement in two phase flow, Measurements in Two Phase Flow. ASME, p. 67-71.

Fujie H. (1964). A relation between steam quality and void fraction in two phase flow. A.I.Ch.E. J., 10, 227-232.

Griffith P. & Wallis G.B. (1961). Two phase slug flow. Trans. ASME J. Heat Transfer, 83C, 307-320.

Henderson F.M. (1966). Open-channel Flow, McMillan.

Hewitt G.F., King I. & Lovegrove P.L. (1961), Holdup and pressure drop measurements in the two phase annular flow of air water mixtures. AERE R3764.

Hoogendoorn C.J. (1957). Gas-liquid flow in horizontal pipes, Chem., Engng. Sci., 9, 205-217.

Isbin H.S. & Biddle D. (1979). Void fraction relationships for upward flow of saturated, steamwater mixtures, Int. J. Multiphase Flow, 5, 293-299.

Isbin H.S., Rodriguez H.A., Larson H.C. & Pattie B.D. (1959). Void fractions in two phase flow. A.I.Ch.E. J. 5, 427-432.

Isbin H.S., Sher N.C. & Eddy K.C. (1957). Void fraction in two phase steam-water flow. A.I.Ch.E. J., 3, 136-142.

Johannessen T. (1972) A theoretical solution of the Lockhart-Martinelli flow model for calculating two phase flow pressure drop and holdup. Int. J. Heat Mass Transfer, 15, 1443-1449.

Levy S. (1952). Theory of pressure drop and heat transfer for annular steady state two phase two component flow in pipes. Proc. 2nd Mid-Western Conf. Fluid Mech. 337-348.

Lockhart M.W. & Martinelli R.C. (1949). Proposed correlation of data for isothermal two phase, two component flows in pipes. C.E.P. 45, 39-48.

Martin C.S. (1976). Vertical downward two phase slug flow. Trans. ASME, J. Fluid Engng., 715-722.

Nguyen V.T. & Spedding P.L. (1977). Holdup in two-phase flow. A. Theoretical aspects. Chem. Engng. Science, 32, 1003-1014.

Nicklin D.J., Wilkes J.O. & Davidson J.F. (1962). Two phase flow in vertical tubes, Trans. I. Chem. E., 40, 61-68.

Smith S.L. (1971). Void fractions in two-phase flow. A correlation based up an equal velocity head model. Heat Fluid Flow, 1, 22-39.

Spedding P.L. & Chen J.J.J. (1982). Holdup in inclined two-phase flow. Pacific Geothermal Conference, Auckland.

Spedding P.L. & Chen J.J.J. (1983). Holdup in Two-phase Flow, Submitted to Int. J. Multiphase Flow.

Spedding P.L. & Nguyen V.T. (1976). Data on holdup, pressure loss, and flow patterns for two phase air-water flow in an inclined pipe. Univ. Auckland Report Engng. 122.

Spedding P.L. & Nguyen V.T. (1978). Bubble rise and liquid content in horizontal and inclined tubes. Chem. Engng. Sci., 33, 987-994.

Taitel Y. & Dukler A.E. (1976). A theoretical approach to the Lockhart-Martinelli Correlation for stratified flow. Int. J. Multiphase Flow, 2, 592-595.

Turner J.M. & Wallis G.B. (1965). The separate cylinder model of two phase flow. Rep. NYO-3114-6, Thayer School of Engineering, Dartmouth College.

Wallis G.B. (1969). One dimensional two-phase flow, McGraw-Hill.

Zivi S.M. (1964). Estimation of steady-state steam void fraction by means of the principle of minimum entropy generation. Trans. ASME, J. Heat Transfer 86C, 247-252.

Zuber N. (1960). On the variable density single fluid model for two phase flow Trans. ASME, J. Heat Transfer, 255-258.

Zuber N. & Findlay J.A. (1965). Average volumetric concentration in two-phase flow systems. Trans. ASME, J. Heat Transfer, 87C, 453-468

Multi-Phase Flow and Heat Transfer III. Part A: Fundamentals
edited by T.N. Veziroğlu and A.E. Bergles
Elsevier Science Publishers B.V., Amsterdam, 1984 — Printed in The Netherlands

BEHAVIOR OF A LARGE BUBBLE IN A HORIZONTAL CHANNEL

T.Sakaguchi, F.Nishiwaki, H.Hamaguchi,
M.Ozawa and K.Shimoyama

Department of Production Engineering,
Kobe University,
Nada, Kobe, Japan

ABSTRACT

This paper describes experimental results of behavior of a large air bubble moving in stationary or flowing water in four kinds of horizontal rectangular channels. The change of the velocity of the large bubble intruding into the channel is expressed by a function of the inflow Froude number, the Eötvös number and a ratio of the distance of the large bubble tip from the channel exit to the channel height. There are five conditions that depend on the large bubble velocity along the channel in the liquid flowing steadily from the inlet. A map is presented to show these regions by three parameters mentioned above. An experimental equation is given for a critical condition that the large bubble rests in the channel. Wave velocities, a large bubble shape, a minimum holdup downstream of the large bubble tip, and liquid velocities around the large bubble have been investigated and it is reported that the former three variables are closely related to the large bubble velocity.

1. INTRODUCTION

This paper describes experimental results of flow behavior of a large air bubble in a horizontal closed channel filled with water. When one end of the horizontal channel is quickly opened, air intrudes as an air cavity or the large bubble from the opened end into the channel which is initially filled completely with stationary water or flowing water.

Gardner and Crow [1] investigated an air cavity movement in the horizontal channel and referred to the air cavity as the "large bubble". They applied the Froude number and the Eötvös number to correlate a velocity of the large bubble intruding into stationary water. This phenomenon is one of the fundamental aspects of the gas-liquid two-phase slug flow in horizontal pipes. It was found that the large bubble velocity was available in a correlation of a gas slug velocity at transient [2] and Weber proposed that the large bubble velocity was useful to correlate the gas slug velocity at steady state [3]. The same fact has been recognized in the gas-liquid two-phase slug flow in vertical pipes [4][5][6]. Dumitrescu [7], Davies and Taylor [8], White and Beardmore [9] and Zukoski [10] investigated behavior of the large bubble in vertical pipes. White and Beardmore proposed a correlation of the bubble rise velocity by the Froude number, the Eötvös number and a physical property group.

Recently, Wallis et al. reported the behavior of the stationary large bubble in the flowing liquid concerning to the problem of Loss-of-Coolant

Accident (LOCA) [11]. Another application of this case was found in the field of dry out phenomena in boiler tubes installed horizontally in a furnace [12], [13].

The above mentioned phenomena are two limiting cases, that is, the moving large bubble with constant velocity in the stationary liquid and the stationary large bubble in the flowing liquid. In between these two limiting cases, more general case exists. It is the dynamic behavior of the large bubble which intrudes into the flowing liquid as it changes its velocity along the channel. The information on this phenomenon can be hardly found in literature. Thus, our purpose in this paper is to report the experimental results on this phenomenon including the two limiting cases mentioned above.

Other applications of this study are found in the gravity current phenomena concerned with the intrusion of salt water into fresh water by a removal of an water gate, a submerged current of muddy water under pure water, and a cold front in air as pointed out by Benjamin [14]. And other several examples are found in the review paper by Simpson [15].

2. MAIN PARAMETERS TO DESCRIBE FLOW BEHAVIOR OF LARGE BUBBLE

Figure 1 shows a sketch of the large bubble intruding into the flowing liquid in the horizontal channel. A left end is a channel inlet and an inlet volumetric liquid flux is denoted by Vin. A right end is a channel exit. The large bubble moves from right to left or reversely according to given conditions. Its velocity, which is referred to as the large bubble velocity, is denoted by Vb. A tip of the large bubble is located under a very small distance from an upper wall. From this tip to the first minimum point of liquid depth which is referred to as a minimum holdup point, a gas-liquid interface curves smoothly. This part of the large bubble is referred to as a bubble nose. Following the bubble nose, there are several waves in some cases. These waves are numbered in sequence. This part with or without waves is referred to as a liquid layer.

The main parameters related to the motion of this large bubble are the large bubble velocity Vb, wave velocities Vw1, Vw2 and others, the liquid velocity V and its distribution around the large bubble, the minimum holdup defined by hmin/H and the bubble tip distance Z from the channel exit end. Their characteristics have been experimentally investigated for various values of the inlet volumetric liquid flux Vin and of the channel height H.

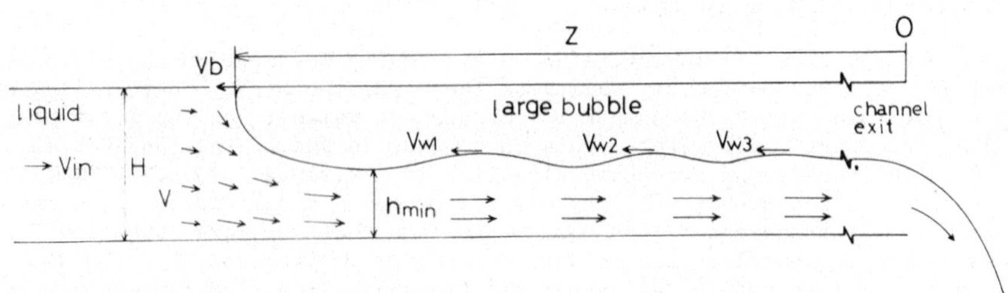

Fig.1 Sketch of large bubble

3. EXPERIMENTAL APPARATUS AND PROCEDURES

Experiments have been carried out with water and air in four rectangular channels of 200 mm width, 2.6 m length and 22.5, 32.5, 42.5 and 53 mm height respectively. Water was fed from an overflow tank located about 3 m above the horizontal channel to the channel inlet through a calming section. In realizing the large bubble in the channel, we took the following procedures. Firstly, a restriction plate was attached on the channel exit end and the channel was filled with water. At this stage, the water flow rate was brought to a required volumetric liquid flux and water flowed out to the atmosphere through the restriction. Then, the restriction plate was quickly removed and air began to intrude into the channel.

Five conductance probes [16] were used for measurements of Vb along the channel, five pairs of conductance probes with two wires for Vw1, Vw2, hmin/H and the large bubble shape measurements, an X-type hot-film anemometry for water velocity distribution and an orifice system for the inlet volumetric liquid flux. The experiments have been conducted for the inlet volumetric liquid flux ranging from 0 m/s to 0.534 m/s and four kinds of the channel height.

4. LARGE BUBBLE VELOCITY

4.1. The Change of the Large Bubble Velocity along the Channel

The position-time diagram of the large bubble tip in the steady counter-current flow of liquid is shown in Fig.2. Data obtained in runs without asterisks stand on each line with approximately constant slope. This means that the large bubble intrudes into the channel with an approximately constant velocity. Those data with asterisks show a tendency that the large bubble intrudes with an approximately constant velocity in the region near the exit but its velocity begins soon to decrease as the large bubble intrudes further into the channel. And it can be presumed for the large bubble to rest at a certain position in a longer channel. The latter data are obtained in the cases that the value of the inflow Froude number $Fri=Vin/\sqrt{gH}$ is larger, the height of the channel is shorter and/or the channel length is quite long. And the former data are obtained in cases of relatively low inflow Froude number

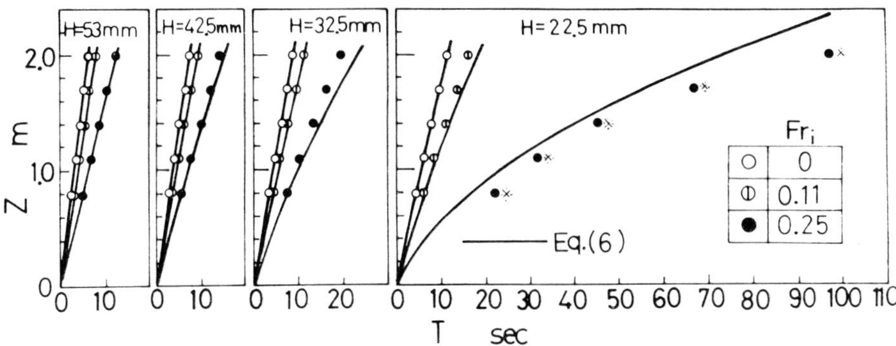

Fig.2 Position-time diagram of large bubble tip

and/or relatively large channel height. But, if the channel length is quite long, the large bubble behavior will become to have the tendency like the latter case. Thus, it can be concluded that the dynamic behavior of the large bubble along the channel is generally expressed by such data that are obtained by the run with the asterisk.

The large bubble velocity in this paper is calculated from a slope of these two neighboring data. In Fig.3, those large bubble velocities in the channel with H=22.5 mm are plotted as one of the experimental results against the tip position Z for various values of Vin. Each solid line is drawn on an arithmetic mean of three experimental data. A deviation of each data from the mean value falls within ±0.002 m/s. And, this confirms a good reproducibility of this experiment.

At lower value of Vin and/or at smaller value of Z, the large bubble velocity is almost constant along the channel. At higher value of Vin and/or at larger value of Z, the large bubble velocity decreases along the channel. And it takes zero at a certain position corresponding to the value of Vin, that is, the large bubble rests in the channel. When the value of Vin is larger than that at the above condition, the large bubble velocity takes a negative value, i.e., the large bubble is pushed back to a direction of the channel exit. Further increase of Vin beyond a certain value results in a total ejection of the large bubble from the channel.

It is found from these figures that there are five conditions classified with reference to the dynamic behavior of the large bubble along the channel moving against or for the liquid flow. The large bubble intrudes with the approximately constant velocity against the liquid inflow. This condition is referred to as the first condition. The condition when the large bubble velocity decreases along the channel is referred to as the second condition. The condition of the stationary large bubble in the liquid flow is referred to as the third condition. We refer to this as a critical condition, too. In the fourth condition, the large bubble is pushed back to the direction of the channel exit for the liquid inflow with the negative velocity. Then, the large bubble is washed out and the channel is full of liquid. This is the fifth condition.

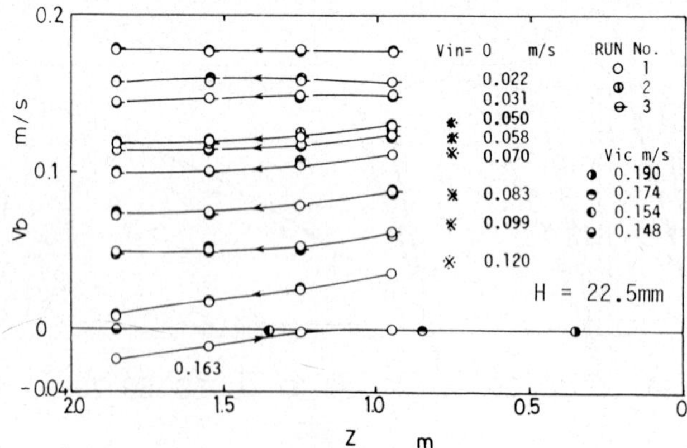

Fig.3 Large bubble velocity along the channel

A threshold condition between the first and the second conditions is given by a derivative of the large bubble velocity with respect to the distance $\delta Vb/\delta z=-0.01$ (1/s) in this paper, thus the second condition begins at the position where the derivative $\delta Vb/\delta Z$ begins to exceed the value of -0.01 after the first condition and ends at the position where the large bubble rests in the channel.

Data in the second condition are marked by asterisk in Fig. 3. In the following paragraph, we discuss on a correlation of the large bubble velocity at the first and the second conditions. Gardner and Crow [1] correlated the large bubble velocity at the first condition in the stationary liquid by the bubble Froude number $Frb=Vb/\sqrt{gH}$ and the Eötvös number $\Sigma=4\sigma/\Delta\rho gH^2$ in the region that the bubble Reynolds number $Reb=HVb/\nu$ is larger than 400. This is followed after results by Zukoski [10] for the large bubble velocity in vertical pipes. Our experimental results are plotted in Fig.4 with Gardner and Crow's and Wilkinson's experimental results [1] [17]. They are correlated well by this coordinate system. Curves by following equations are drawn on the same figure.

Gardner and Crow's equation [1]

$$F_{rb}^{2}\,(2-\frac{h_{min}}{H})/(\frac{h_{min}}{H}) = 1-(\frac{h_{min}}{H})^{2} -2\{k+(1-k)\sqrt{\Sigma}\}\sqrt{\Sigma}+\Sigma \qquad (1)$$

$$0.5F_{rb}^{2}\,/(\frac{h_{min}}{H})^{2} = 1-(\frac{h_{min}}{H})-\{k+(1-k)\sqrt{\Sigma}\}\sqrt{\Sigma} \qquad (2)$$

Wilkinson's equation [17]

$$F_{rb} = 0.5 - 0.68\sqrt{\Sigma} \qquad (3)$$

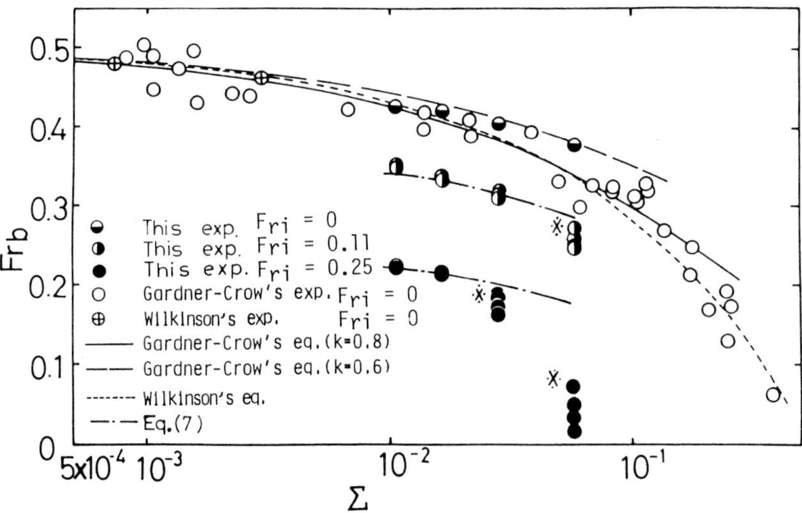

Fig.4 Correlation of large bubble velocity

The curve by Gardner and Crow can be drawn by eliminating hmin in Eqs. (1) and (2). They recomended 0.8 for the value of the experimental constant k.

The experimental results of the large bubble velocity in steady counter-current flow of liquid are plotted on the same figure. The data at the first condition stand on the curve for each inflow Froude number $Fri=Vin/\sqrt{gH}$. But the data at the second condition change from the value at the first condition to zero for the value of the Eötvös number given by the channel height and physical properties of air and water employed in this experiment. Thus, this method is not available to correlate the large bubble velocity at the second condition.

In order to correlate the large bubble velocities in both conditions by a common method, we should take into account the flow resistance by which the large bubble is decelerated along the channel. Liquid viscosity, liquid velocity, liquid holdup and length of liquid layer, i.e., the bubble tip distance are considered as main parameters concerning with this flow resistance. An effect of liquid viscosity is very weak in this Reynolds number range as mentioned above. The liquid velocity, of course, has a large effect on the absolute value of the large bubble velocity as seen in Fig.3, but has relatively weak effect on the decrease of the large bubble velocity, because the derivatives of the large bubble velocity with respect to the distance show similar values among the data. The liquid holdup depends on the height of the channel and the liquid velocity. So, it is tried to correlate the large bubble velocity by using the lengths Z and H in a dimensionless form ; $Z^* = Z/H$ considering that the frictional pressure drop is a function of length/diameter in a circular tube, with other dimensionless numbers, i.e., the bubble Froude number, the inflow Froude number and the Eötvös number.

The correlated results are shown in Fig.5. Experimental results are plotted against a dimensionless quantity $Z^*\Sigma^{1/0.45}$. Though the experimental results from the four channels are involved in data under each inflow Froude number,

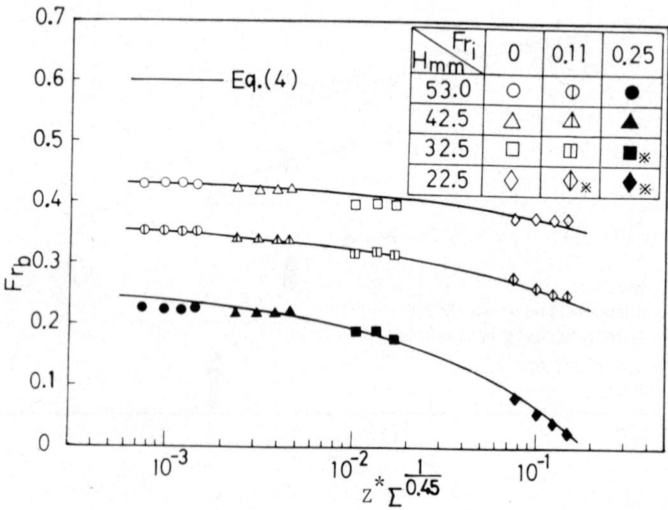

Fig.5 Correlation of large bubble velocity (proposed method)

they fall on each one curve. The bubble Froude number decreases gradually with increasing the value of $Z^*\Sigma^{1/0.45}$ and decreases with increasing the inflow Froude number. The same method is applied to a relation between the large bubble velocity and a time which is a residence time of the large bubble from the beginning of the intrusion to a certain position. The correlated results are shown in Fig.6.

Here, the correlation curves are expressed as follows:

$$F_{rb} = 0.44-0.71F_{ri}-\left(\frac{1}{1.05/F_{ri}{}^{1.08}-2}+0.174\right)\left(Z^*\Sigma^{\frac{1}{0.45}}\right)^{0.45} \tag{4}$$

$$F_{rb} = \frac{\dfrac{1}{0.44-0.77F_{ri}}+0.27\exp(10.1F_{ri})\left(T^*\Sigma^{\frac{1}{0.45}}\right)^{0.55}}{\left\{\dfrac{1}{0.44-0.77F_{ri}}+0.60\exp(10.1F_{ri})\left(T^*\Sigma^{\frac{1}{0.45}}\right)^{0.55}\right\}^2} \tag{5}$$

where $T^*=T/\sqrt{H/g}$.

By integrating Eq.(5), an experimental equation is obtained to express the dynamic behavior of the large bubble tip along the channel. The lines drawn by this equation are shown in Fig.2. And the equation is expressed:

$$Z^*\Sigma^{\frac{1}{0.45}} = \frac{1}{\dfrac{1}{(0.44-0.77F_{ri})T^*\Sigma^{1/0.45}}+\dfrac{0.60\exp(10.1F_{ri})}{(T^*\Sigma^{1/0.45})^{0.45}}} \tag{6}$$

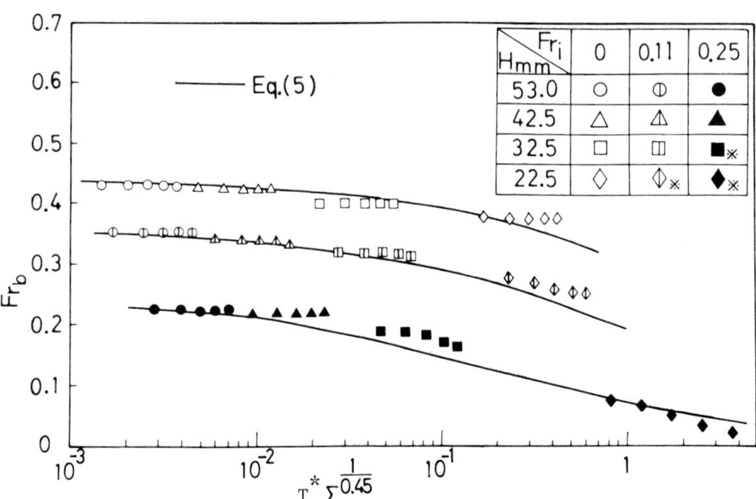

Fig.6 Correlation of large bubble velocity (proposed method)

4.2. Large Bubble Velocity at the First Condition

The large bubble velocities at the first condition in the stationary liquid and the flowing liquid were correlated by the Frb-Fri-$Z^*\Sigma^{1/0.45}$ system as shown in Fig.5. In this section, we discuss that the large bubble velocity at the first condition can be well expressed by a function of only two quantities, that is, the large bubble velocity in the stationary liquid and the inlet volumetric liquid flux. In Fig.7, the drift velocity of the large bubble normalized by that in the stationary liquid: Vd/Vbs=(Vb-Vin)/Vbs, are plotted against the inlet volumetric liquid flux normalized by Vbs: Vin/Vbs. Almost all the data fall on a single straight line independent of the Eötvös number. The normalized drift velocity increases with increasing the normalized inlet volumetric liquid flux. The straight line is expressed by the equation

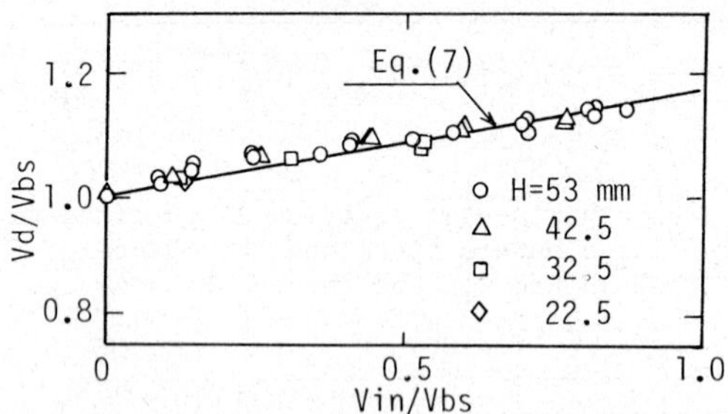

Fig.7 Normalized drift velocity of large bubble

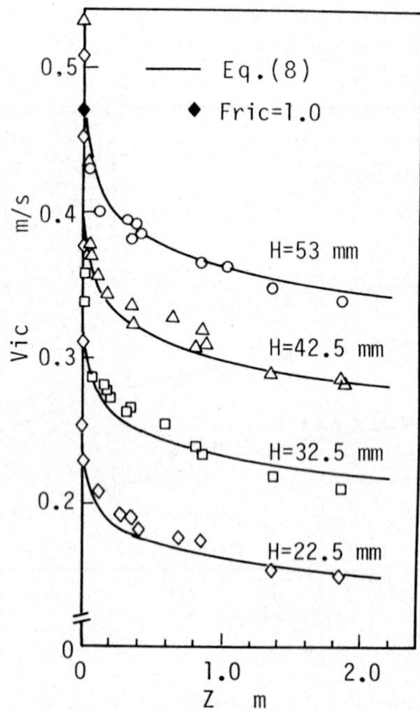

Fig.8 Critical velocity for stationary large bubble

$$\frac{Vd}{Vbs} = 1 + 0.179 \frac{Vin}{Vbs} \tag{7}$$

4.3. Critical Velocity Relating to the Stationary Bubble in the Flowing Liquid

The large bubble rests at a certain position in the channel against the inlet volumetric liquid flux which is referred to as a critical velocity Vic in this study. Experimental results of Vic are plotted against the distance of the large bubble tip from the channel exit for various channel heights in Fig.8. In the perfect fluid, the critical velocity does not change along the channel, but in water, the critical velocity changes along the channel. It decreases significantly near the channel exit and then gradually decreases with increasing distance Z. It increases with increasing channel height. The inlet volumetric liquid flux becomes $\sqrt{gH}=0.47$ m/s for the inlet Froude number Fri=1 in the channel with 22.5 mm height. This value, marked by '◆', approximately agrees with the experimental value of the critical velocity at the channel exit (Z=0). This relation holds also for the cases of other channel heights. This condition, Fri=1, is the condition of no intrusion of the large bubble into the channel.

Providing the critical velocity can be also related with the same dimensionless numbers used in Eq.(4), the critical Froude number $Fric=Vic/\sqrt{gH}$ depends on only the parameter $Z^{*}\Sigma^{1/0.45}$ because of the zero bubble velocity: Vb=0. Most of the experimental results fall on one straight line independent of the channel height and the bubble tip position as shown in Fig.9. Data obtained from the circular tube with 40 mm I.D. are plotted with those by others [11] ∿ [13]. They also fall on the same line. In this case, a tube diameter D is used as the representative length instead of the height H. A solid straight line is expressed by a following equation.

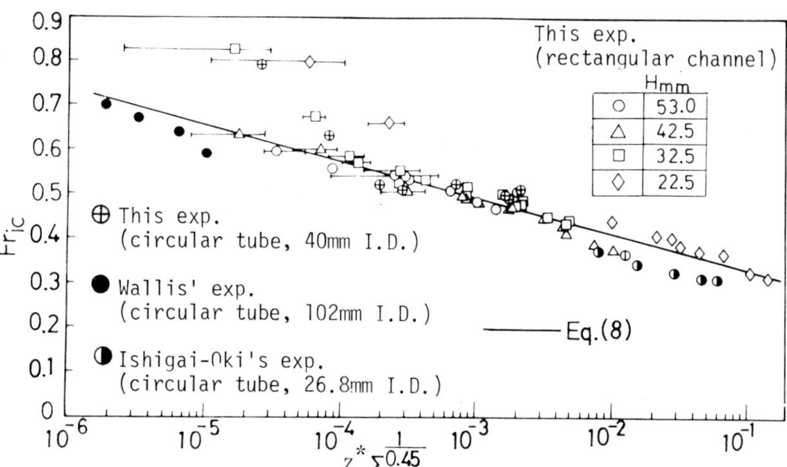

Fig.9 Correlation of critical velocity of large bubble

$$F_{ric} = 0.253 - 0.035\ \ell n(Z^* \Sigma^{\frac{1}{0.45}}) \tag{8}$$

4.4. Flow Regime Map Based on Large Bubble Velocity

As is described above, the motion of the large bubble is classified into five conditions with reference to the characteristics of the large bubble velocity. Gardner and Crow [1] reported that there were three flow regimes with reference to the large bubble velocity and the large bubble shape in the stationary liquid. They classified the flow regimes only by the Eötvös number. In the range of $\Sigma<0.02$, the large bubble velocity is constant and the bubble shape does not change along the channel. This situation corresponds to our first condition. In the range of Σ from 0.02 to 0.105, large waves appear on the liquid layer downstream of the minimum holdup point. For $\Sigma>0.105$, the value of minimum holdup increases and finally the large bubble rests at $\Sigma=0.368$. Recently, Wilkinson [17] carried out an experiment concerning with the large bubble intruding into the stationary liquid in the horizontal channel with a weir at the channel exit and classified the flow situation into three regimes, i.e., the steady flow regime, the unsteady flow regime and the steady

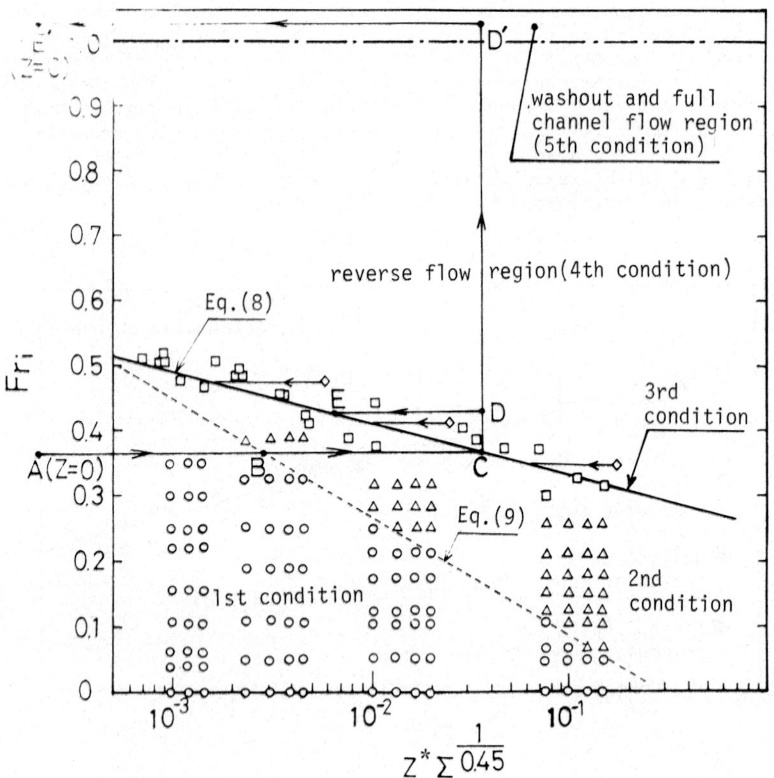

Fig.10 Flow regime map based on large bubble velocity

dissipative regime with reference to the bubble shape, the waves on the liquid layer and the energy dissipation from the bore. We develop it and classify the flow situation of the large bubble with reference to the characteristics of the large bubble velocity related to the inflow conditions of liquid from upstream of the large bubble. Figure 10 is the flow regime map including the large bubble in the stationary liquid. Experimental results in each condition are plotted by different marks on the $Fri\text{-}Z^{*}\Sigma^{1/0.45}$ diagram. The five conditions are expressed as four regions and one line. The dashed line which shows the boundary between the first condition and the second condition which is expressed by the following experimental equation :

$$F_{ri} = -0.1-0.079\ell n(Z^{*}\Sigma^{\frac{1}{0.45}}) \tag{9}$$

The solid line corresponds to Eq.(8), and the dot-dash-line of the boundary between the fourth condition and the fifth condition is the line of the inflow Froude number Fri=1. Of course, each boundary is not a line, but a thin belt. By using this map, we can easily find the behavior of the large bubble velocity and also those of the holdup and the large bubble shape as will be discussed later. Now, we show an example to demonstrate the availability of this map. We suppose the intrusion of the large bubble under a certain condition of the channel height, fluid properties and the inlet volumetric liquid flux. Then we can estimate the inflow Froude number Fri. For example, we take point A, which is assumed to correspond to the channel exit, Z=0. Then, the bubble tip moves from the point A to the right on the straight line of the constant Froude number. This line intersects with the boundary between the first condition and the second condition at the point B. In the region between points A and B, the large bubble moves with constant velocity. When the large bubble intrudes further into the channel, then the bubble velocity decreases and rests at a certain position in the channel corresponding to point C which is a intersect with the critical condition shown by the solid line. Stepwise increase of the inlet volumetric liquid flux brings the state to the point D which belongs to the fourth condition. Then the bubble tip is pushed back to the point E which

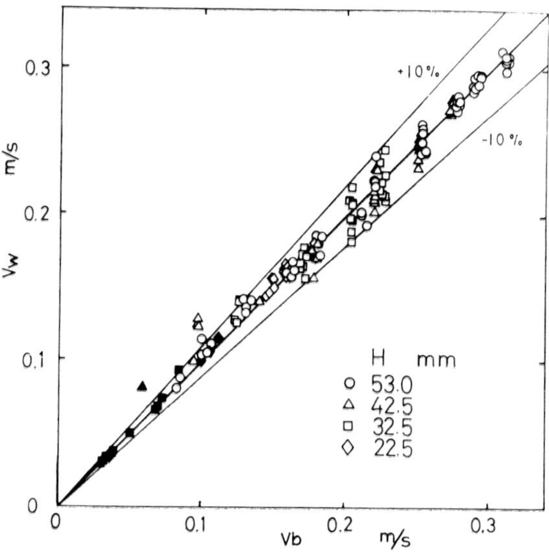

Fig.11 Wave velocity

correspond to the critical condition of the new inflow Froude number equal to that of the point D. When the stepwise increase of the inlet volumetric liquid flux is quite large, the state moves to the point D', which belongs to the fifth condition. Then the bubble is washed out from the channel and the large bubble can be never found in the channel.

5. WAVE VELOCITY

The propagation velocities of the leading two waves on the liquid layer downstream the large bubble nose are plotted against the large bubble velocity in Fig.11. Among the plotted data, the solid marks correspond to those at the second condition and the other correspond to the first condition. Most of the data of the wave velocity coincide with the velocity of each large bubble within the range of ± 10%. That is, the waves on the liquid layer move following after the large bubble tip movement almost with the same speed as its. This tendency is independent of the channel height and the inlet volumetric liquid flux.

6. LARGE BUBBLE SHAPE

When the height of liquid layer and the coordinates are normalized by the channel height [18],[19], then the normalized bubble shapes in the nose part coincide with each other independent of the channel height and of the bubble tip position as shown in Figs.12 and 13. These two figures are those at the first condition. It is clearly concluded from these figures that the shape of

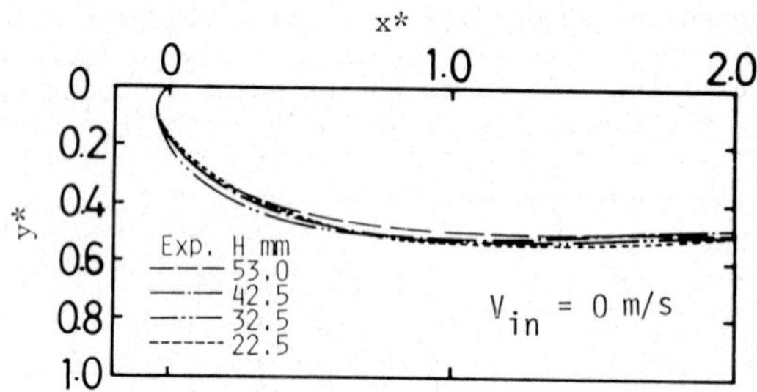

Fig.12 Large bubble shape at the 1st condition

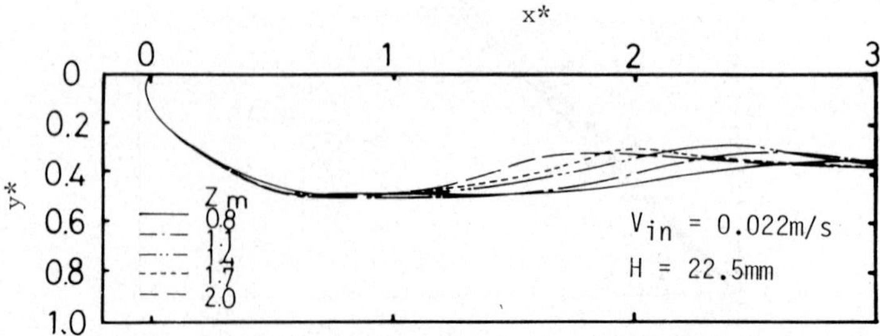

Fig.13 Large bubble shape at the 1st condition

the bubble nose does not change along the channel in the region where the large bubble velocity does not change. The bubble shapes at the second condition where the large bubble velocity decreases with the intrusion of the bubble into the channel are shown in Fig.14. The bubble shape changes significantly at the nose part of the bubble depending on the bubble tip position. From figure 14, it is observed that the minimum holdup increases with increasing the value of Z. The distance between the bubble tip and the wave crest decreases to a certain value nearly equal to the amount of the channel height with increasing the value of Z, then the wave velocity becomes equal to the large bubble velocity. Of cource, the wave velocity is a little higher than the large bubble velocity in this situation, but this deviation of the wave velocity from the bubble velocity is within the range of +10% of the large bubble velocity. These waves on the liquid layer decrease their amplitude with the intrusion of the bubble and at last disappear. The shapes of the stationary bubble, that is, at the critical condition, are shown in Fig.15. As is seen from Fig.8, the position of the stationary bubble tip depends on the inlet volumetric liquid flux. Therefore, the bubble shape for each inlet volumetric liquid flux changes depending on the position of the bubble tip. The minimum holdup increases with increasing the valve of Z or decreasing the valuve of Vic.

From the above discussion, we can conclude that the bubble shape is strongly related to the large bubble velocity both in the stationary liquid and the flowing liquid. The bubble shape does not change along the channel in the region where the large bubble velocity does not change and vice versa. The increase in the minimum holdup which is an important parameter of the bubble shape results in the decrease in the large bubble velocity. This relation

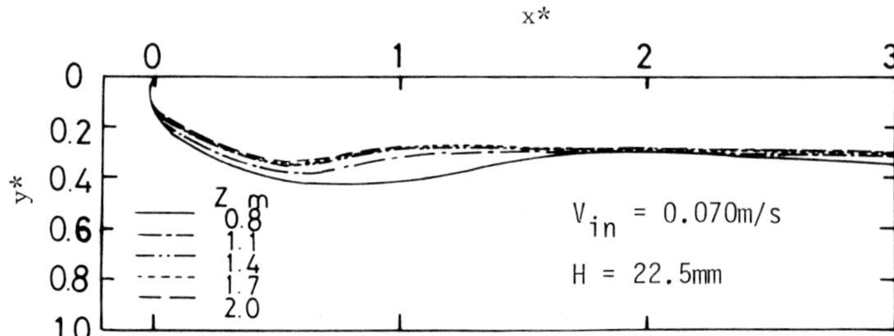

Fig.14 Large bubble shape at the 2nd condition

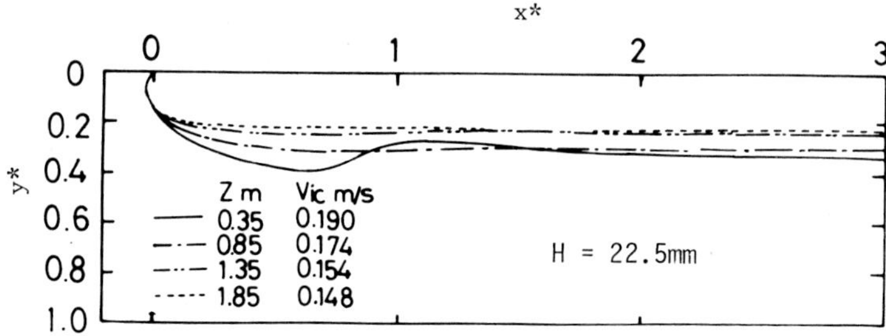

Fig.15 Large bubble shape at critical condition

agrees with the result reported by Wilkinson [17] for the large bubble in the stationary liquid.

7. MINIMUM HOLDUP

As discussed in chapter 6, the minimum holdup changes depending on the large bubble tip distance, the channel height and the inlet volumetric liquid flux just like the large bubble velocity does. So, it is considered that the behavior of the minimum holdup along the channel can be also expressed by a function of Fri and $Z^*\Sigma^{1/0.45}$. The experimental results are plotted against $Z^*\Sigma^{1/0.45}$ in Fig.16. They fall on each curve of the inflow Froude number independent of the channel height and the large bubble velocity. The value of flm increases with increasing the value of $Z^*\Sigma^{1/0.45}$ except the case of Fri=0, and with increasing the inflow Froude number. The data at the first condition exist in the region of flm less than 0.54 and the others are the data at the second condition. Solid lines have been drawn by a following experimental equation.

$$f_{lm} = 5F_{ri}(Z^*\Sigma^{\frac{1}{0.45}})^{0.675} + 0.47 \tag{10}$$

Now, the relationship between the minimum holdup and the bubble Froude number can be obtained by eliminating $Z^*\Sigma^{1/0.45}$ from Eqs.(4) and (10), as follows:

$$f_{lm} = 5F_{ri}\left[\frac{0.44 - 0.71F_{ri} - F_{rb}}{\dfrac{1}{1.05/F_{ri}^{1.08} - 2} + 0.174}\right]^{2/3} + 0.47 \tag{11}$$

Their relation is shown in Fig.17. It is seen from this figure that the value of flm is 0.47 at the first condition in the stationary water : Fri=0 and increases with decreasing the bubble Froude number at the constant inflow Froude number.

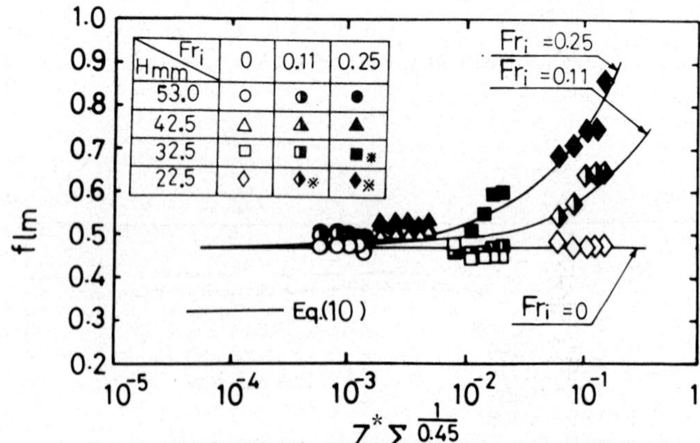

Fig.16 Correlation of minimum holdup

8. LIQUID VELOCITY DISTRIBUTION AROUND THE LARGE BUBBLE

The liquid velocity distribution around the large bubble at the first condition in the stationary liquid is shown in Fig.18, where the coordinates and the liquid velocity are normalized by the channel height and the large bubble velocity respectively. The liquid velocity in the still liquid far upstream of the large bubble tip, of course, is zero. Approaching to the large bubble tip, the absolute value of the velocity increases. The direction of the velocity changes along the bubble surface counterclockwisely from the same direction as the large bubble movement to the downward and then to the reverse direction of the large bubble movement in the liquid layer. The theoretical result by the potential flow model is also shown by dotted lines and agrees well with the experimental result. The detail of the analysis can be found in reference [19].

The velocity distribution around the stationary bubble against the liquid inflow (the critical condition) is shown in Fig.19. Most of the velocity

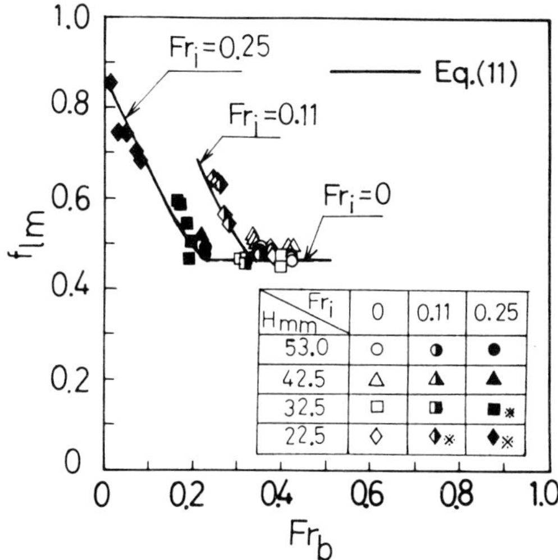

Fig.17 Relation between minimum holdup and bubble Froude number

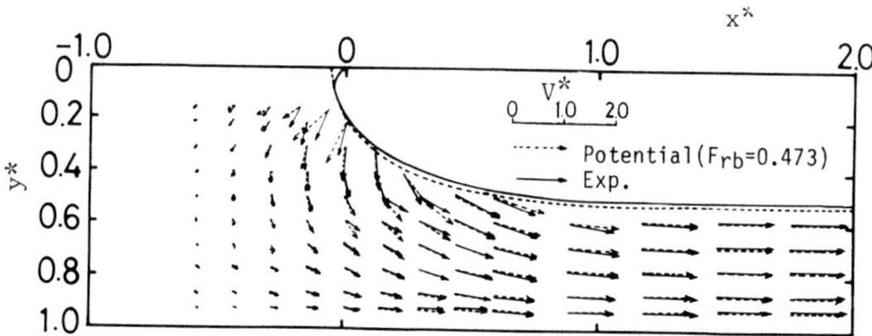

Fig.18 Velocity distribution around large bubble at the 1st condition (Vin = 0 m/s, H = 53 mm)

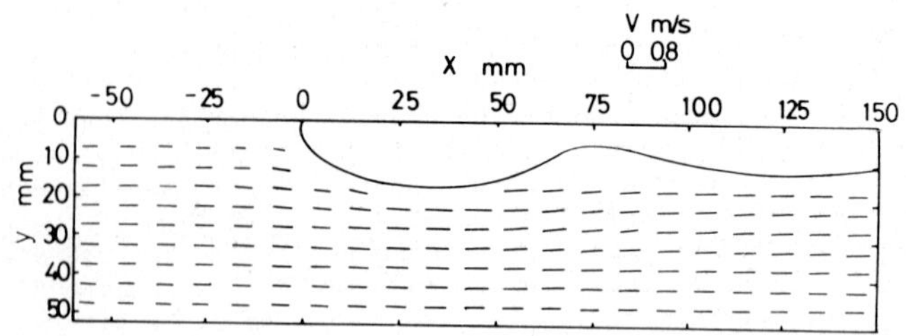

Fig.19 Velocity distribution around stationary large bubble at critical condition (Vin = 0.348 m/s, H = 53 mm)

vectors, except the neighboring region of the large bubble, have the horizontal direction. The velocity vectors tend downward in the region near the bubble nose and they tend upward in the region of the first wave front. The same tendency is found in the liquid velocity distribution of other channel heights. When we observe this velocity distribution in the reference frame of moving with the mean liquid velocity far upstream the large bubble, the velocity of the liquid far upstream become zero and other parts of the velocity distribution have the similar distribution as that in Fig.18. Of course, the absolute values are different from them because of the difference in the drift velocity, the holdup and the large bubble shape.

9. CONCLUSIONS

In concluding this report on the flow characteristics of the large bubble intruding into the stationary liquid or the flowing liquid, we point out the following results.

1) The motion of the large bubble with reference to the large bubble velocity is classified into 5 conditions.

2) The large bubble velocity depends on the inlet volumetric liquid flux, the channel height, the bubble tip distance from the channel exit and the Eötvös number and is well correlated by Fri and $Z^*\Sigma^{1/0.45}$.

3) The normalized drift velocity Vd/Vbs increases with increasing the normalized inlet volumetric liquid flux Vin/Vbs at the first condition.

4) The relation between the inlet volumetric liquid flux and the position of the stationary bubble tip is expressed only by the parameters Fric and $Z^*\Sigma^{1/0.45}$.

5) The wave velocity on the liquid layer is approximately equal to the large bubble velocity.

6) The characteristics of the large bubble velocity are closely related to those of the large bubble shape.

7) The minimum hold-up is well correlated by the parameters Fri and $Z^*\Sigma^{1/0.45}$ as in the case of the large bubble velocity.

8) The characteristics of the large bubble velocity are closely related with that of the minimum hold-up of the bubble nose.

9) The liquid velocity distribution around the large bubble in the reference frame of moving with the mean inlet velocity is similar to that of the large bubble intruding into the stationary liquid.

ACKNOWLEDGEMENT

Authors wish to express their thanks to Messrs. E. Fujii, H. Tanaka, M. Yoshida, Y. Saito and Y. Sakai who supported the experimental part of this study.

NOMENCLATURE

D	: tube diameter	[m, mm]
f_{lm}	: minimum holdup	[-]
g	: gravitational acceleration	[m/s^2]
H	: channel height	[m, mm]
h_{min}	: minimum depth of liquid	[m, mm]
V	: liquid velocity	[m/s]
V_b	: large bubble velocity	[m/s]
V_{bs}	: large bubble velocity in stationary liquid	[m/s]
V_d	: drift velocity of large bubble	[m/s]
V_{ic}	: critical velocity for stationary large bubble	[m/s]
V_{in}	: inlet volumetric liquid flux	[m/s]
V_w	: wave velocity	[m/s]
x,y	: coordinates	[m, mm]
Z	: bubble tip distance from the channel exit	[m, mm]
ρ	: liquid density	[kg/m^3]
$\Delta\rho$	: density difference between liquid and gas	[kg/m^3]
σ	: surface tension	[N/m]
ν	: kinematic viscosity	[m^2/s]
F_{rb}	: bubble Froude number	[-] ($\equiv V_b/\sqrt{gH}$)
F_{ri}	: inflow Froude number	[-] ($\equiv V_{in}/\sqrt{gH}$)
F_{ric}	: inflow Froude number at critical condition	[-] ($\equiv V_{ic}/\sqrt{gH}$)
R_{eb}	: bubble Reynolds number	[-] ($\equiv V_bH/\nu$)
Z^*	: normalized bubble tip distance	[-] ($\equiv Z/H$)
x^*,y^*	: dimensionless coordinates	[-] ($\equiv x/H$, y/H)
Σ	: Eötvös number	[-] ($\equiv 4\sigma/\Delta\rho g H^2$)

REFERENCES

1. Gardner,G.C. and Crow,I.G., "The Motion of Large Bubbles in Horizontal Channels", J. Fluid Mech., 43-2(1970),247.
2. Sakaguchi,T., et al., "Gas-Liquid Two-Phase Transient Slug Flow in a Horizontal Tube", Preprint of the JSME, 814-5(1981),53 (in Japanese).
3. Weber,M.E., "Drift in Intermittent Two-Phase Flow in Horizontal Pipes", The Canadian J. Chem. Eng., 59-3(1981),398.
4. Griffith,P. and Wallis,G.B., "Two-Phase Slug Flow", Trans. ASME., J. Heat Transfer, 83-3(1961),307.

5. Nicklin,D.J., et al "Two-Phase Flow in Vertical Tubes", Trans. Instn. Chem. Engr., 40(1962),61.
6. Zuber,N. and Findlay,J.A., "Average Volumetric Concentration in Two-Phase Flow Systems", Trans. ASME., J. Heat Transfer, 87-4(1965),453.
7. Dumitrescu,D.T., "Strömung an einer Luftblase im Senkrechten Rohr", Z. angew. Math. Mech., 23-3(1943),139.
8. Davis.R.M. and Taylor.G.I., "The Mechanics of Large Bubble Rising Through Extended Liquid and Through Liquid in Tubes", Proc. Roy. Soc. London, Ser. A, 200(1950),375.
9. White,E.T. and Beardmore,R.H., "The Velocity of Rise of Single Cylindrical Air Bubble Through Liquids Contained in Vertical Tubes", Chem. Eng. Sci., 17(1962),351.
10. Zukoski,E.E., "Influence of Viscosity, Surface Tension and Inclination Angle on Motion of Long Bubbles in Closed Tubes", J. Fluid Mech., 25-4 (1966),821.
11. Wallis,G.B., et al., "Conditions for a Pipe to Run Full When Discharging Liquid into a Space Filled with Gas", Trans. ASME, J. Fluid Eng., 99-2 (1977),405.
12. Ishigai,S. and Oki.H., Preprint of the JSME (1960),1 (in Japanese).
13. Ishigai,S., et al., "Dryout at Entrance of Heated Section for the Horizontal Flow of Low-Pressure Steam-Water Mixture", Proc. JSME 1967 Semi-International Symp., Tokyo, Vol.2 (1967),45.
14. Benjamin.T.B., "Gravity Currents and Related Phenomena", J. Fluid Mech., 31-2(1968),209.
15. Simpson, J.E., "Gravity Currents in the Laboratory,Atmosphere and Ocean", Ann. Rev. Fluid Mech., 14(1982),213.
16. Bergles,A.E., "Electrical Probes for Study of Two-Phase Flow", Two-Phase Instrumentation, 11th National ASME/AIChE Heat Transfer Conference (1969),70.
17. Wilkinson,D.L., "Motion of Air Cavities in Long Horizontal Ducts", J. Fluid Mech., 118(1982),109.
18. Sakaguchi,T., et al., "Flow Characteristics of a Large Bubble in Horizontal Channels:Analysis by the Modified SMAC Method", Memoirs of the Faculty of Eng., Kobe University, 29(1982),71.
19. Sakaguchi,T., et al., "Flow Characteristics of a Large Bubble in a Horizontal Channel", Advances in Two-Phase Flow and Heat Transfer, Eds. S. Kakac and M. Ishii, Martinus Nijhoff Publishers, The Hague, The Netherlands, Vol. 1 (1983), 255.

Multi-Phase Flow and Heat Transfer III. Part A: Fundamentals
edited by T.N. Veziroğlu and A.E. Bergles
Elsevier Science Publishers B.V., Amsterdam, 1984 — Printed in The Netherlands

TWO-PHASE FLOW PRESSURE DROP ACROSS THICK RESTRICTIONS OF ANNULAR GEOMETRIES

Saeed Ghandeharioun and E. Ramon Hosler
Department of Mechanical Engineering and Aerospace Sciences
University of Central Florida
Orlando, Florida 32816, U.S.A.

ABSTRACT

This paper presents the rational methods of predicting the steady state two-phase flow (steam and water) pressure drop across the restrictions of annular geometries formed when tubes extend through circular holes in tube support plates. Two approaches are discussed and the major areas of discussion are the orientation of tubes-to-tube support plate holes, and the thickness of tube support plates.

1. INTRODUCTION

In shell and tube type heat exchangers which operate with two-phase (liquid and its vapor) flow such as steam generators in pressurized water reactor in nuclear power plants, support plates are spaced periodically along the tube bundles to maintain the proper geometric arrangements among the tubes. When the shell-side flow is parallel to the tube axis, the flow must pass through the annular shaped clearances between the outside diameter of the tubes and the holes in the support plates (Figure 1), increasing the shell-side pressure drop. Since the support plates contribute a major portion of the pressure drop, predicting shell-side pressure drop for two-phase flow is an important design consideration.

Since there has been little or no work to predict the two-phase flow pressure drop in complicated geometries (such as tube support plates), the goal of this paper is to tie together the previous work on predicting the single-phase pressure drop in annular geometries and two-phase pressure drop in simple geometries (such as tubes and channels) to be able to predict shell-side pressure drop with two-phase flow.

2. PRESSURE DROP OF TWO-PHASE FLOW ACROSS ANNULAR ORIFICES

In general, orifices are used to measure flow rates. However, many flow restrictions, such as clearances between a tube and a tube support plate may be analyzed by treating them as orifices. It is the intent here to evaluate the pressure drop for two-phase flow across annular orifices.

A review of literature indicates that in two-phase flow system, it has been observed experimentally that for a given mass flux, the pressure drop can be much greater than for a corresponding single-phase flow system [1]. The

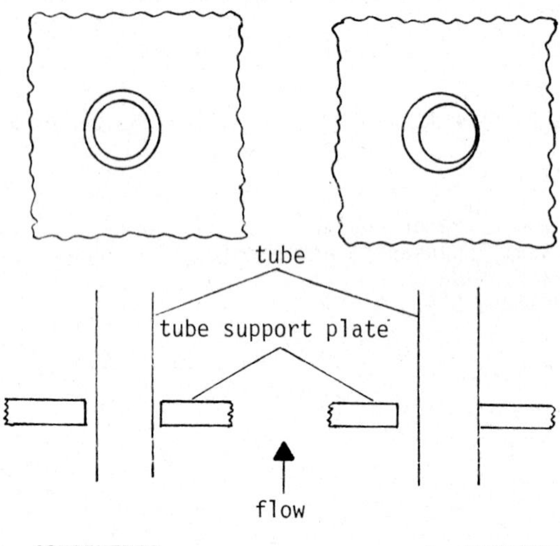

a. CONCENTRIC b. TANGENT

Fig. 1 Schematic of annular gaps between tube and tube support plate

approach, which has been taken to predict two-phase flow pressure drop, is to multiply the equivalent saturated single-phase flow pressure drop by a multiplier, ϕ^2, which is a function of (at least) flow quality and system pressure. Thus

$$\Delta P_{TP} = \Delta P_f \cdot \phi^2 \ (X, P, ...)$$

or

$$\Delta P_{TP} = K^* \frac{G^2}{2g_c \rho_f} \cdot \phi^2 \ (X, P, ...) \tag{1}$$

where G can be the mass flux through an orifice and g_c is the conversion factor, ρ_f is the density of saturated liquid, and K^* is the single-phase loss coefficient, which will be discussed in more detail for various geometries of annular gap between a tube and a tube support plate later in this section.

Since the pressure drop in two-phase flow is closely related to the flow pattern, two principal types of flow models will be discussed in the analysis of two-phase flow pressure drop in this section. They are homogeneous flow model which regards the two-phase to flow as a single-phase possessing mean fluid properties, and the separated flow model which considers the phases to be artifically segregated into two streams; one of liquid and one of vapor.

Before presenting the commonly accepted expressions for two-phase multiplier, ϕ^2, in homogeneous flow model and separated flow model, it is necessary to make the following assumptions;

1. One-dimensional flow
2. Steady-state flow
3. Adiabatic flow across the tube support plate, so that the quality, x, is constant.

4. Pressure drop across support plate is small compared to the total pressure, so that the densities of saturated liquid and saturated vapor, ρ_f and ρ_g, do not change.
5. The parameters affecting voids; i.e.; quality, pressure, and mass velocities are nearly the same upstream and downstream of the tube support plate, so that void fraction, α, is nearly constant, and is given by

$$\alpha = \frac{1}{1 + (\frac{1-x}{x})\,(\frac{V_f}{V_g})\, S} \tag{2}$$

where V_f and V_g are the specific volumes of saturated liquid and saturated vapor, respectively. S is the slip ratio, and it is defined as the ratio of the average velocity of the vapor phase to that of the liquid. For homogenous flow the slip ratio is equal to 1.0, and for nonhomogeneous flow is greater than 1.0 due to the fact that the vapor, because its buoyancy, has a tendency to slip past the liquid.
6. Flow properties are in terms of cross-sectional averages taken across the annular gap flow cross section.

Therefore, in order to find the accepted expressions for ϕ^2, it is suggested [1] that for two-phase flow it can be written

$$\Delta P_{TP} = K^* \frac{G^2}{2g_c\, \rho_{TP}} \tag{3}$$

in which ρ_{TP} is the appropriate two-phase density. For the case of saturated homogenous two-phase flow ($\rho_{TP} = \rho_{hom.}$), Equation (3) can be written

$$(\Delta P_{TP})_{hom.} = K^* \frac{G^2}{2g_c\, \rho_f} \left(\frac{\rho_f}{\rho_{hom.}}\right) \tag{4}$$

where $\rho_{hom.}$ is homogenous density as

$$\rho_{hom.} = \frac{1}{V_{hom.}} = \frac{1}{V_f + V_{fg} \cdot X} \tag{5}$$

Thus, by comparing equations (5), (4), and (1), it is found that the two-phase multiplier for homogenous flow is

$$\phi^2 = (1 + \frac{V_{fg}}{V_f} X) \tag{6}$$

or

$$(\Delta P_{TP})_{hom.} = K^* \frac{G^2}{2g_c\, \rho_f} (1 + \frac{V_{fg}}{V_f} X) \tag{7}$$

which is the expression for prediction of pressure drop in two-phase homogenous flow.

The appropriate two-phase density for separated flow is not as well defined [1]. It has been suggested (Chisholm, 1973) that the momentum density should be used. Thus equation (3) becomes

$$(\Delta P_{TP})_{sep.} = K^* \frac{G^2}{2g_c \rho_f} \left(\frac{\rho_f}{\bar{\rho}}\right) \tag{8}$$

where $\bar{\rho}$ is the momentum density and is given by

$$\bar{\rho} = \frac{1}{\frac{(1-x)^2}{\rho_f(1-\alpha)} + \frac{x^2}{\rho_g \alpha}} \tag{9}$$

Thus, by comparing equations (9), (8), and (1), it is found that the two-phase multiplier for separated flow is

$$\phi^2_{sep.} = \frac{(1-x)^2}{1-\alpha} + \frac{V_g}{V_f} \frac{x^2}{\alpha} \tag{10}$$

or

$$(\Delta P_{TP})_{sep.} = K^* \frac{G^2}{2g_c \rho_f} \left[\frac{(1-x)^2}{1-\alpha} + \frac{V_g}{V_f}\frac{x^2}{\alpha}\right] \tag{11}$$

which is the expression for prediction of pressure drop in two-phase separated flow.

For the clearance between a tube and a tube support plate which may be regarded as an annular orifice, the single-phase loss coefficient can be defined as

$$K^* = \frac{1}{C^2} \tag{12}$$

where C is called the overall annular orifice coefficient.

The experimental study by Bell and Bergelin [2] on wide range of the single-phase flow (water and oil) rates through various single annular orifices indicates that the annular orifice coefficient, C, is a function of the annular orifice dimension, the annular orifice Reynolds number, and the orientation of annular orifice; i.e., concentric or tangent position of tube to tube support plate hole (Figure 1), and the results of the experimental analysis of Bell and Bergelin are presented in Figures 2 and 3 with the equations defining orifice length-to-width ratio, and orifice Reynolds number as

$$Z = \frac{2\ L}{D-d} \tag{13}$$

$$Re = \frac{(D-d)G}{\mu} \tag{14}$$

where L is the thickness of the orifice (or thickness of the support plate), D is the diameter of the hole in the support plate, d is the diameter of the tube, G is mass flux through an annular orifice, and μ is the dynamic viscosity

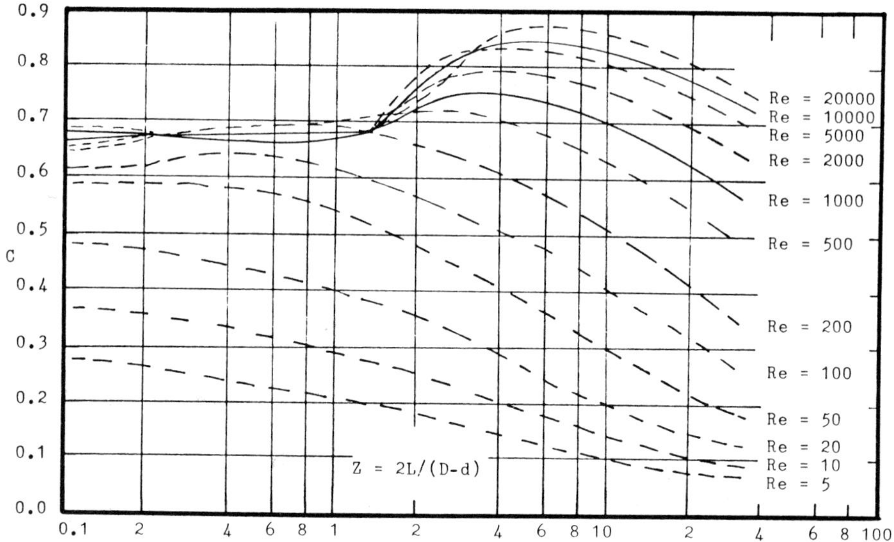

Fig. 2 Summary of concentric-orifice coefficients [2]

of the fluid.

From these curves, it was concluded [2] that for high Reynolds numbers (Re > 5000) in thick orifices with a fixed mass flow rate, the pressure recovery begins at Z - value of about 1.0 up to a Z - value of 6.0. For longer orifice channels the wall friction causes an increase in the pressure drop until the channel length gives Z - numbers in the range of 10 to 100 where the overall orifice coefficients drop to about 0.65, the value for sharp-edge orifice. At higher Z - values the friction resistance lowers the value of overall annular orifice coefficient still more. Also, Bell and Bergelin have developed suitable equations expressing the overall annular orifice coefficient which they are presented in the references [2] and [3].

It should be noted that for two-phase flow through an annular orifice, the orifice coefficient may be estimated at the Reynolds number of the entire flow rate considered in the state of saturated liquid ($Re = Re_f$). Consequently, the results of Bell and Bergelin's analysis for prediction of the overall annular orifice coefficient may be used.

Furthermore, a numerical example is presented in the reference [3], which has been solved based on the foregoing discussion of predicting the two-phase pressure drop across the annular orifice.

3. PRESSURE DROP OF TWO-PHASE FLOW ACROSS SHORT AND LONG LENGTH RESTRICTIONS

This section primarily deals with the Janssen's prediction of steady state two-phase flow pressure drop across short, and long length restrictions, based on a one-dimensional momentum balance.

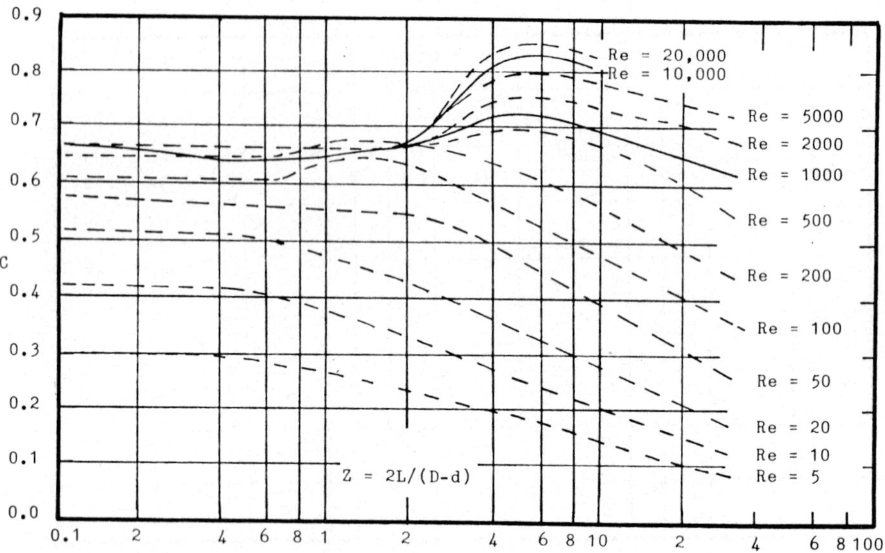

Fig. 3 Summary of tangent-orifice coefficients [2]

A review of literature indicates that Janssen [4] obtained two equations (15 and 16) for prediction of pressure drop across restrictions of circular and rectangular geometries depend on whether the vena contracta occurs inside or outside of the restriction Fig. 4). Thus, for short length restriction where the vena contracta is outside of the restriction, the equation of pressure drop is given by

$$\Delta P_{TP}\left(\begin{matrix}\text{short}\\ \text{restriction}\end{matrix}\right) = \frac{G_1^2}{2g_c\rho_f}\,\frac{1}{\sigma^2\bar{C}^2}\left[\frac{V_g}{V_f}X^2\hat{\alpha}_3\left\{\frac{1}{\alpha_3}-\frac{\sigma^2\bar{C}^2}{\alpha_5^2}\right\}+\right.$$

$$(1-X)^2\,(1-\hat{\alpha}_3)\left\{\frac{1}{(1-\alpha_3)^2}-\frac{\sigma^2\bar{C}^5}{(1-\alpha_5)^2}\right\}-2\sigma\bar{C}\left\{\frac{V_g}{V_f}X^2\left(\frac{1}{\alpha_3}-\frac{\sigma\bar{C}}{\alpha_5}\right)+\right.$$

$$\left.\left.(1-X)^2\left(\frac{1}{1-\alpha_3}-\frac{\sigma\bar{C}}{1-\alpha_5}\right)\right\}\right] \tag{15}$$

where x = flow quality
ρ_f = density of saturated liquid
V_f, V_g = specific volumes of saturated liquid and saturated vapor
G_1 = mass flux in channel
σ = restriction(s) flow area/channel flow area
$\bar{C}$ = area of vena contracta/area of restriction
$\hat{\alpha}_3 = (\alpha_3 + \alpha_5)/2$

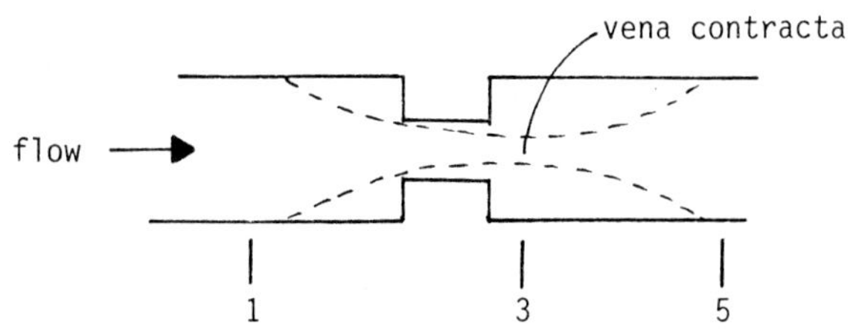

a. Short length restriction

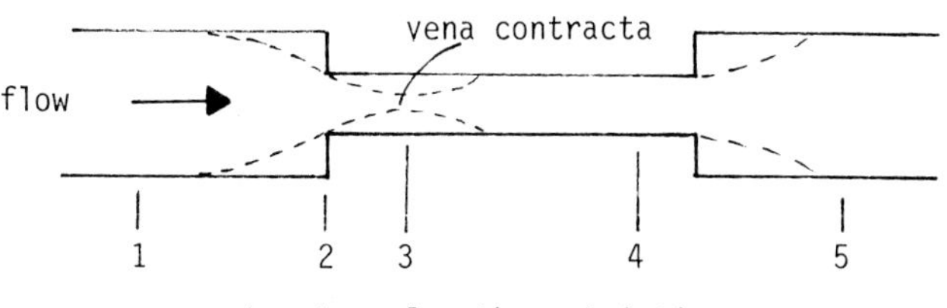

b. Long length restriction

Fig. 4 Short and long length restrictions [4]

and the α's are void fractions at different locations as shown in Figure 4-a with the assumption of slip flow at all locations.

In long length restrictions where the vena contracta occurs within the the restriction, the equation of pressure drop is given by

$$\Delta P_{TP}\left(\begin{matrix}\text{long}\\ \text{restriction}\end{matrix}\right) = \frac{G_1^2}{2g_c\rho_f}\left[\frac{1}{\sigma^2}\left\{\frac{V_g}{V_f}X^2\hat{\alpha}_1\left(\frac{1}{\bar{C}^2\alpha_3^2}-\frac{1}{\alpha_4^2}\right) + (1-X)^2(1-\hat{\alpha}_1)\left(\frac{1}{\bar{C}^2(1-\alpha_3)^2}-\frac{1}{(1-\alpha_4)^2}\right)\right\} - \frac{2}{\sigma^2}\left\{\frac{V_g}{V_f}X^2\left(\frac{1}{\bar{C}\alpha_3}-\frac{1}{\alpha_4}+\frac{\sigma}{\alpha_4}-\frac{\sigma^2}{\alpha_5}\right) + (1-X)^2\left(\frac{1}{\bar{C}(1-\alpha_3)}-\frac{1}{1-\alpha_4}+\frac{\sigma}{1-\alpha_4}-\frac{\sigma^2}{1-\alpha_5}\right)\right\} + \frac{V_g}{V_f}X^2\hat{\alpha}_2\left(\frac{1}{\sigma^2\alpha_4^2}-\frac{1}{\sigma^2}\right) + (1-X)^2(1-\hat{\alpha}_2)\left\{\frac{1}{\sigma^2(1-\alpha_4)^2}-\frac{1}{(1-\alpha_1)^2}\right\}\right] \qquad (16)$$

where $\hat{\alpha}_1 = (\alpha_3 + \alpha_4)\ /\ 2$

$\hat{\alpha}_2 = (\alpha_1 + \alpha_4)\ /\ 2$

and the α's are void fractions at different locations as shown in Figure 4 -b.

On the basis of his own steam-water data, Janssen suggested that the void fractions be estimated by assuming slip flow everywhere except at the vena contracta.

In view of the foregoing equations, it can be seen that the vena contracta ratio is the only major factor to be depended on the geometry of the restriction. Consequently, it would be possible to apply the Janssen's equations to any form of restrictions; i.e., the annular gap between a tube and a tube support plate, providing that the proper value of vena contracta ratio is used.

Since Janssen did not address the question of when a restriction may be considered short or long, the experimental study by Harshe, Hussain, and Weisman [4] with freon and its vapor on both single and multiple hole circular cross-sectional area restrictions suggests that equation (15) may be used for short length restriction, the ratio of restriction length to restriction diameter is less than or equal to 2.0, and equation (16) for long length restrictions when the ratio is greater than 2.0. Moreover, the study recommends that in short length restrictions ($0.5 < L/D \leq 2.0$) and long length restriction (with void fraction greater than 50%), the void fraction at the vena contracta be based on partial mixing at the vena contracta which depends on void fraction and geometry of restriction. It is also suggested that for multiple hole restrictions the ratio of restriction length to restriction diameter should be determined using diameter of a single hole.

However, it should be noted that the validity of foregoing limitations on the ratio of restriction length to restriction diameter remains to be tested for this case study where there is an annular gap, not a circular hole between the tube and the tube support plate. The restriction diameter should be taken as hydraulic diameter of the annular gap.

When there is significant vaporization across the restriction (large difference between the inlet and the outlet void fraction), it is suggested by Harshe, Hussain, and Weisman [4] that the void fraction at the vena contracta be based on the exit quality of restriction. Furthermore, the study recommends the Hugmark correlation [5] for obtaining the relationship between void fraction and quality for slip flow condition.

4. CONCLUSIONS AND RECOMMENDATIONS

The goal of this paper was to present the methods of predicting two-phase flow pressure drop across the tube support plate due to the existence of the small annular gaps between tubes and tube support plate. This was achieved by considering two approaches:

1. The first method considers annular clearance between the tubes and the tube support plate as annular orifices, assuming that the entire two-phase flow is considered in the state of saturated liquid. The approach is found to be easy to use based on the available data and equations for finding the overall annular orifice coefficients, and simple calculation of two-phase multiplier. However, the method has lack of data in the area of the annular orifice coefficient for the cases of multiple annular orifices, and various eccentric orifices other than tangent.

2. The second method deals with the Janssen's prediction of the two-phase flow pressure drop across short and long length restrictions. Since the Janssen's approach deals with the nature of the two-phase flow at every location throughout the restrictions, the method should be more accurate. However, there are many difficulties associated with this method; e.g., accurate prediction of the void fractions at various axial locations and the appropriate values of the vena contracta ratios.

Since the Janssen's equations of two-phase pressure drops require the extensive measurements of void fractions at various axial locations, it would be difficult to apply them for design purposes of large scale heat exchangers. However, the first method is easier to use since it does not require any measurements of flow conditions throughout the shell and should provide adequate accuracy for design purposes.

Based on the findings of this study, it is recommended that future investigations are needed to

1. Examine the effect of geometrical parameters; i.e., number of tubes and pitch-to-diameter ratio of tubes, on loss coefficient.
2. Experimentally confirm Janssen's method on restrictions with annular cross-section area of flow.

REFERENCES

1. Lahey, R. T., Jr., and Moody, F. J., The Thermal-Hydraulics of Boiling Water Nuclear Reactor, La Grange Park, IL: American Nuclear Society, pp. 173-245, (1979).

2. Bell, K. J., and Bergelin, O. P., "Flow Through Annular Orifices," ASME Transactions, 79, pp. 593-601, April, (1957).

3. Ghandeharioun, S., "Two-Phase Flow Pressure Drop Across Thick Restrictions of Annular Geometries," Engineering M.S. Thesis, University of Central Florida, Orlando, Florida, (1982).

4. Harshe, B., Hussian, A., and Weisman, J., "Two-Phase Pressure Drop Across Restrictions and Other Abrupt Area Changes," University of Cincinnati for U.S. Nuclear Regulatory Commission, Report Number NUREG-0062, Springfield, VA: National Technical Information Service, April (1976).

5. Collier, John G., Convective Boiling and Condensation, New York: McGraw-Hill Book Company, 1972, pp. 68.

Multi-Phase Flow and Heat Transfer III. Part A: Fundamentals
edited by T.N. Veziroğlu and A.E. Bergles
Elsevier Science Publishers B.V., Amsterdam, 1984 — Printed in The Netherlands

A METHOD OF PREDICTING PRESSURE PROFILES IN HORIZONTAL 37-ELEMENT CLUSTERS

C.W. Snoek and S.Y. Ahmad
Advance Engineering Branch
Chalk River Nuclear Laboratories
Atomic Energy of Canada Limited
Chalk River, Ontario, Canada K0J 1J0

ABSTRACT

A new two-phase friction multiplier correlation has been developed to predict the pressure profile in two-phase flow with heat addition. The main difference from other pressure drop prediction routines is that in addition to the effect of quality, the effects of system pressure, heat and mass flux on the two-phase multiplier are included. The inception of two-phase flow is defined to occur at the onset of significant void in the subcooled boiling regime instead of the boundary between negative and positive thermodynamic equilibrium quality.

Pressure drop data from an experiment with a fully segmented 6-metre long, 37-element fuel string were used to optimize the constants and exponents of this correlation for the friction multiplier.

Adiabatic flow pressure drop data were correlated to predict the single-phase pressure gradient as a function of temperature and mass flow. Measured diabatic pressure gradients were used in conjunction with these single-phase pressure profiles to define the point of onset of significant void (OSV). Two-phase flow was considered to exist beyond this OSV point. In the two-phase flow regime, the frictional pressure gradient was calculated by subtracting the momentum gradient from the measured total pressure gradient. The two-phase friction multipliers thus calculated were correlated with the vapor weight quality, system pressure, mass and heat flux and expressed as a correction factor to the two-phase homogeneous multiplier. Correct asymptotic trends were built into the form of the (dimensionless) correlation to permit extrapolation to high quality conditions.

This pressure drop prediction technique was shown to successfully predict the measured pressure profiles. The more than one hundred overall pressure drop data were predicted with an RMS error of 5.4%. The pressure drop measured during experiments with geometrically different simulated fuel strings and different operating conditions were also successfully predicted.

The paper presents the derivation of the equations in a systematic manner. The effects of the system parameters on the prediction error are discussed in detail.

1. INTRODUCTION

One of the requirements in the design of a water-cooled, nuclear reactor is the two-phase pressure drop prediction of the coolant in the fuel channels. For CANDU reactors, operating with a few percent outlet quality, it is necessary to maintain an inlet header pressure sufficiently high to overcome the primary coolant pressure drop as well as maintain an adequate outlet header pressure.

It is important to be able to predict the coolant channel pressure profile during anticipated transients in order to estimate the safety margin with respect to the critical heat flux. For a given pump characteristic, the coolant flow itself is also a function of the pressure drop.

Since pump energy requirements are a significant fraction of the plant energy budget, the accurate prediction of coolant channel pressure drop is also an economic consideration.

Over the last few years, much effort has been expended trying to predict the drop in pressure of a liquid-vapor mixture moving through a pipe or channel. Despite this effort, no reliable analytical model of two-phase pressure drop has been developed. Therefore, reliance has to be placed on empirical or semi-empirical prediction methods. The empirical models can be broadly divided into two categories, the "homogeneous" and "separated flow" models, briefly described below.

1.1 The Homogeneous Model

In this model both phases are considered to flow as a single-phase with mean fluid properties. The model has been in use for many years mainly due to its relative ease of application.

The model is based upon the following assumptions:

(a) equal vapor and liquid linear velocities,

(b) thermodynamic equilibrium between liquid and vapor phase, and

(c) the use of a suitably defined single-phase friction factor acting upon the two-phase flow.

The two-phase friction factor (f_{tp}) can be assumed to be equal to an "all liquid" flow friction factor defined by a function of the all liquid Reynolds number (GD/μ_ℓ) and the relative roughness (ε/D).

Another approach is to evaluate f_{tp} using a "two-phase" viscosity μ in the regular friction factor relationships. The expression relating $\bar{\mu}$ and the quality X must be chosen such that

$$\text{(a)}\quad \bar{\mu} = \mu_\ell \text{ when } X = 0 \text{ and}$$
$$\text{(b)}\quad \bar{\mu} = \mu_g \text{ when } X = 1. \qquad \ldots (1)$$

Several suggestions for the form of the relationship are listed in Collier (1980).

The most common definition of $\bar{\mu}$ is

$$\frac{1}{\bar{\mu}} = \frac{X}{\mu_g} + \frac{1 - X}{\mu_\ell} \qquad \ldots \quad (2)$$

where X is the thermodynamic quality.

A two-phase multiplier ($\phi^2_{\ell o}$) has been defined as the ratio of two-phase and single-phase slope, where the single-phase slope is based on the total assumed liquid only.

$$\phi^2_{\ell o} = \frac{\left.\frac{dp}{d\ell}\right|_{tp}}{\left.\frac{dp}{d\ell}\right|_{sp}} \qquad \ldots \quad (3)$$

With the use of equation (2), and the Blasius equation for the two-phase friction factor given by:

$$f_{tp} = 0.079(GD_e/\bar{\mu})^{-\frac{1}{4}} \qquad \ldots \quad (4)$$

the homogeneous model $\phi^2_{H.M.}$ may be expressed as:

$$\phi^2_{H.M.} = \left[1 + X\left(\frac{\nu_{vg}}{\nu_f}\right)\right]\left[1 + X\left(\frac{\mu_{fg}}{\mu_g}\right)\right]^{-\frac{1}{4}} \qquad \ldots \quad (5)$$

where ν is the specific volume.

A table of $\phi^2_{H.M.}$ as a function of pressure and quality is presented in Collier (1980).

1.2 The Separated Flow Model

In the separated flow model the phases are considered to be artificially separated into a liquid stream and a vapor stream.

The model is based on the following assumptions:

(a) the velocity of each phase is constant but the liquid and vapor phase velocities are not necessarily equal,

(b) thermodynamic equilibrium between phases, and

(c) the two-phase friction multiplier ($\phi^2_{\ell o}$) and void fraction (α) are related to the independent variables of the flow by empirical correlations.

Lockhart and Martinelli (1949) studied the isothermal two-phase flow in horizontal tubes and proposed a method for calculating the isothermal two-phase pressure drop. The accelerative component of the two-phase pressure drop was included in studies by Martinelli and Nelson (1948).

In Lockhart and Martinelli's approach, it was assumed that conventional friction pressure drop equations can be applied to the flow of each phase, where each phase is considered to occupy a portion of the flow area.

They argued that the two-phase friction multipliers for both fluid and vapor streams could be correlated as a function of a parameter X, where

$$X = \left[\frac{\left.\frac{dp}{d\ell}\right|_f}{\left.\frac{dp}{d\ell}\right|_g}\right]^{\frac{1}{2}} \qquad \ldots (6)$$

and $$\phi_f^2 = 1 + \frac{C}{X} + \frac{1}{X^2} \qquad \ldots (7)$$

$$\phi_g^2 = 1 + CX + X^2 \qquad \ldots (8)$$

where C is a constant depending on whether the flow of liquid and vapor is turbulent or viscous.

Where both phases are turbulent, the parameter X can be shown to be equal to

$$X_{tt} = \left[\frac{1.0 - X}{X}\right]^{0.9} \left[\frac{\rho_g}{\rho_\ell}\right]^{0.5} \left[\frac{\mu_\ell}{\mu_g}\right]^{0.1} \qquad \ldots (9)$$

When the results obtained with the separated flow model are compared with experimental observations, it is clear that in reality the curves described by equations (7) and (8) are not smooth but show discontinuities which can be associated with changes in flow pattern (Collier (1980)). Also, it has been reported that mass flux has an effect on $\phi_{\ell o}^2$ (Collier (1980)).

Baroczy (1965) introduced a "physical property index (Λ)" in an effort to correct the two-phase friction multiplier for the influence of the mass flux. He plotted the property index

$$\Lambda = (\mu_\ell/\mu_g)^{0.2}/(\nu_\ell/\nu_g) \qquad \ldots (10)$$

versus the two-phase multiplier with the quality as parameter for one mass flow rate. A correction term to the two-phase multiplier could be obtained from a plot of property index versus two-phase multiplier ratio with different mass fluxes as parameter.

In addition to the influence of the mass flow on $\phi_{\ell o}^2$, Tarasova and Leontiev (1965, 1966) found that a heat flux imposed on the flow influenced the void fraction and pressure drop. They presented a correlation to correct $\phi_{\ell o}^2$ for the heat flux effect valid for tubes only.

Shoukri et al. (1981) also investigated the influence of the imposed heat flux on the frictional pressure drop of a low pressure steam-water flow.

They found that at certain qualities the application of a uniform heat flux resulted in a significant rise in pressure drop. It was also observed that increasing the mass flow or decreasing the system pressure caused the maximum pressure drop to shift toward lower qualities. This phenomenon was shown to be the result of liquid-vapor interaction at the slug-annular flow regime transition.

Reddy et al. (1982) derived a two-phase friction multiplier correlation from adiabatic steam-water data obtained in bundles. The correlation, developed as a function of pressure, quality and mass flux was compared with several existing friction multiplier correlations. It was found superior in predicting adiabatic data, and marginally better in predicting diabatic pressure drop data.

2. DEVELOPMENT OF THE TWO-PHASE FRICTION MULTIPLIER CORRELATION

2.1 Data Base

The development of the two-phase multiplier correlation is based on 96 pressure drop scans, obtained from an experiment with a six-metre long, 37-element bundle incorporating end plate segmentation simulation. The test section was mounted horizontally. The fuel strings' axial heat flux profile was uniform while the radial heat flux profile peaked at the outer ring. The test section was equipped with eleven pressure taps along its length to measure the absolute pressure at the inlet and outlet and the pressure difference between pressure taps.

Twelve scans were taken at adiabatic conditions, where the fluid going through the test section was subcooled and no void was present. These runs form the basis of the single-phase friction factor correlation. The remaining 84 scans were used to correlate the two-phase multiplier with the independent variables and optimize the constants and exponents.

The pressure drop data were collected in the following ranges:

(a) pressure $6.56 < P < 11.25$ MPa
(b) mass flux $1952 < G < 4792$ $kg{\cdot}m^{-2}{\cdot}s^{-1}$
(c) heat flux $175 < \phi < 1300$ $kW{\cdot}m^{-2}$
(d) inlet temperature $175 < T_{in} < 290°C$

It is recognized that the pressure gradient in a channel as complex as a 37-element simulated fuel string with its segmentations due to end plate simulations, spacer planes and wear pads cannot be a straight line or smooth curve but must have many discontinuities. It was not possible to accurately estimate these pressure drop discontinuities from the experimental pressure drop data; therefore, it was assumed that the pressure fluctuations due to end plates, spacers and wear pads were uniformly distributed along the channel giving the channel a higher friction factor. The single-phase pressure drop could be accurately estimated from this friction factor and the Reynolds number as is discussed in Section 2.3.

2.2 Pressure Drop Considerations

The pressure drop of a flow in a heated fuel bundle with single-phase inlet and two-phase outlet is the sum of four individual pressure drop components:

$$\Delta p|_{tot} = \Delta p|_{sp} + \Delta p|_{tp} + \Delta p|_{mom} + \Delta p|_{elev} \qquad \ldots \quad (11)$$

where

$\Delta p|_{tot}$ = total pressure drop

$\Delta p|_{sp}$ = single-phase (due to liquid only) pressure drop

$\Delta p|_{tp}$ = two-phase friction pressure drop

$\Delta p|_{mom}$ = pressure drop due to the acceleration (change in momentum) of the flow caused by the change in specific volume

$\Delta p|_{elev}$ = hydrostatic pressure drop due to change in elevation

The orientation of the flow channel was horizontal so that $\Delta p|_{elev} = 0$. In the previous section it was discussed that flow obstruction pressure drop was considered to be uniformly distributed along the flow channel and included in Δp_{sp} and Δp_{tp}.

Equation (11) then becomes:

$$\Delta p|_{tot} = \Delta p|_{sp} + \Delta p|_{tp} + \Delta p|_{mom} \qquad \ldots \quad (12)$$

These individual pressure drop components are discussed in the following sections.

2.3 Single-Phase Pressure Drop

In order to predict the single-phase pressure drop in the diabatic system, it was assumed that the adiabatic pressure drop could be used to predict the diabatic system pressure drop up to the point at which sufficient vapor is generated to affect the pressure gradient. This point, sometimes called the point of onset of significant void (OSV), will be discussed in the next section. D'Arcy's equation,

$$f = 2(D_e/\ell)\Delta p(\rho_\ell/G^2) \qquad \ldots \quad (13)$$

was used to calculate the single-phase pressure drop friction factor. This friction factor was then correlated with the Reynolds number, resulting in the following expression:

$$f = 0.05052\ Re^{(-0.05719)} \qquad \ldots \quad (14)$$

where Re - Reynolds number $(G \cdot D_e/\mu_\ell)$.

In a heated channel, the friction factor will decrease slightly due to a higher film temperature. It can be shown that for even a 10% increase in the Reynolds number, the friction factor will not decrease by more than one-half of one per cent.

2.4 Point of Onset of Significant Void

To predict the dividing point between the single-phase and the two-phase part of each experiment, pressure profiles were plotted for a number of test runs. Also plotted was a straight line originating at P_{in} representing the pressure drop had the flow been single-phase adiabatic. Twenty-one such graphs were produced. A sample graph is shown in Figure 1. On each graph the point at which the two-phase pressure profile started to deviate from the single-phase pressure profile was visually established. The subcooling at that point (S_d) was then correlated with the transverse boiling number (Bo), Reynolds number and the inlet subcooling (S_{in}).

The subcooling at the OSV point was correlated with the inlet conditions since in pressure drop prediction routines the inlet pressure and inlet temperature are usually known. Also, it was shown by Ahmad (1970) that the location of bubble detachment is a function of inlet temperature.

An optimization of the constants resulted in the following correlation:

$$S_d = 0.01887(Re/10^6)^{(-0.5640)} \cdot Bo^{0.4539} \cdot S_{in}^{0.8668} \qquad \ldots (15)$$

where S_d - (dimensionless subcooling at OSV point) $= \dfrac{H_f - H_d}{\lambda}$ $(= -X_d)$

Re - (Reynolds number) $= G \cdot D_e/\mu_\ell$

Bo - (transverse boiling number) $= \dfrac{\phi D_e}{\lambda \mu_g}$

and S_{in} - (dimensionless inlet subcooling) $= \dfrac{H_f - H_{in}}{\lambda}$ $(= - X_{in})$

The thermophysical properties in these dimensionless quantities are based on the pressure at the start of the heated length. The Reynolds and transverse boiling number are based on the hydraulic equivalent diameter.

Equation (15) was used to predict the dimensionless subcooling at the OSV point. It was found that at this predicted subcooling the presence of void started to influence the pressure profile. This was the case for the scans from which the correlation was derived as well as the other runs. From the value of S_d the single-phase channel length could be determined.

Equation (15) is valid for the data from which it was derived. Where a condition of low heat flux (or no heat flux at all) is encountered in a channel, Equation (15) will predict a single-phase channel length greater than the heated length, i.e., the entire channel contains single-phase liquid.

The OSV point predictions of Equation (15) were compared with those of Saha and Zuber's local conditions model (1974). Their suggested correlation applicable to the range of the present data is given by:

$$\frac{\phi}{G\, Cp_f(T_{sat} - T_d)} = 0.0065 \qquad \ldots (16)$$

It was found that Saha and Zuber's model consistently overpredicted the experimental single-phase lengths. Void generation in the high enthalpy subchannels is thought to cause the OSV shift to a higher subcooling in the bundle experiment.

TABLE 1: SUMMARY OF PREDICTION ERRORS

ERROR (%)	SEGMENTED BUNDLE	SEGMENTED BUNDLE (ADDITIONAL DATA)	SMOOTH BUNDLE	REBUILT SMOOTH BUNDLE	3-METRE SEGMENTED BUNDLE
AVERAGE	0.32	4.1	-2.1	-2.9	-6.0
RMS	5.4	7.0	6.6	7.3	10.5
MAXIMUM POSITIVE	15.4	14.4	27.9	26.2	33.2
MAXIMUM NEGATIVE	-7.0	-7.7	-11.7	-14.7	-16.8
NUMBER OF DATA	104	56	216	127	168

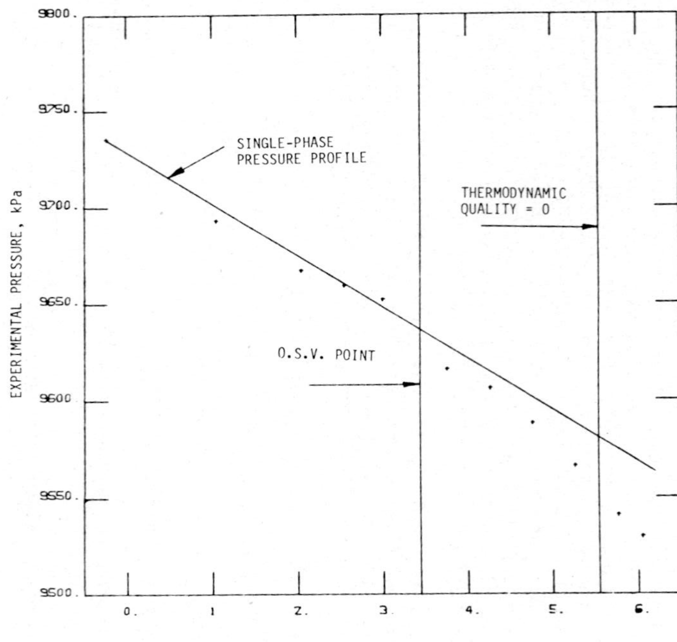

FIGURE 1: COMPARISON OF SINGLE AND TWO-PHASE FLOW PRESSURE PROFILES

2.5 Two-Phase Multiplier

After having obtained the single-phase pressure gradient and the length over which this gradient operated, by subtracting the gradient and length product from the inlet pressure, the pressure at the OSV point could be obtained. The slope at that point was also known, being equal to the single-phase gradient. A parabolic pressure profile was fitted through the remainder, i.e., the two-phase section of the flow channel. The profile was defined by the outlet pressure and the pressure and gradient at the OSV point. It was found that a parabola with these constraints fitted the experimental data very well.

The two-phase multiplier, $\phi^2_{\ell o}$, defined as

$$\phi^2_{\ell o} = \frac{\left.\frac{dp}{d\ell}\right|_{tp}}{\left.\frac{dp}{d\ell}\right|_{sp}} \qquad \ldots \quad (17)$$

was calculated at all pressure tap locations located downstream from the OSV point. The two-phase multiplier at the OSV point was set equal to unity by definition.

The vapor weight quality (Y) defined as

$$Y = \frac{W_g}{W_g + W_\ell} \qquad \ldots \quad (18)$$

was determined from a relationship proposed by Levy (1967):

$$Y = X - X_d e^{(X/X_d - 1)} \qquad \ldots \quad (19)$$

where X - the thermodynamic quality, and

X_d - the thermodynamic quality at the OSV point.

Having obtained the vapor weight quality, a correlation by Armand and Massena (1947, 1960)

$$\alpha = (0.833 + 0.167\ Y) \Big/ \left[1 + \left\{\frac{1 - Y}{Y}\right\} \cdot \frac{\rho_g}{\rho_f} \right] \qquad \ldots \quad (20)$$

was used to determine the void fraction. This enabled the pressure gradient due to the acceleration of the flow to be determined from

$$\left.\frac{dp}{d\ell}\right|_{mom} = G^2 \cdot \left\{ \frac{Y^2}{\rho_g \alpha} + \frac{(1 - Y)^2}{\rho_f (1 - \alpha)} \right\} \qquad \ldots \quad (21)$$

This momentum gradient was subtracted from the gradient obtained by differentiating the parabola fitted through the two-phase data. The remainder is the pressure gradient due to the two-phase friction effect of the flow. The latter gradient was then divided by the single-phase gradient to give the two-phase multiplier $\phi^2_{\ell o}$.

In all, over 950 values of $\phi^2_{\ell o}$ were obtained.

2.6 The $\phi^2_{\ell o}$-Correlation

A multiple regression routine was used to derive a suitable expression for the multiplier correlation. Since other investigators had shown the two-phase pressure drop to be dependent on the mass flux, pressure and heat flux as well as quality (Baroczy (1965), Tarasova (1965,1966) and Shoukri (1981)), $\phi^2_{\ell o}$ was correlated with dimensionless groupings of these parameters. The form of the correlation was made such that correct parametric trends were incorporated. Since at single-phase liquid $\phi^2_{\ell o}$ equals unity by definition and at single-phase vapor equals the value of $\phi^2_{\ell o}$ as given by the homogeneous model, the present $\phi^2_{\ell o}$ correlation was given the form of a "correction factor (CF)" to the homogeneous model $\phi^2_{H.M.}$:

$$\phi^2_{\ell o} = CF*\phi^2_{H.M.} \qquad \ldots \quad (22)$$

The vapor weight quality Y was used to determine $\phi^2_{H.M.}$.

The correction factor must by definition equal unity for both single-phase liquid and vapor. The data indicated that as the void fraction increased from zero, CF increased to a maximum at approximately $\alpha = 0.2$. For values of α greater than 0.2, CF tended again to unity. Therefore, it was decided to correlate these data in two different expressions, one for $0 < \alpha < 0.2$ and the other for values of α beyond 0.2.

For $0 < \alpha < 0.2$

$$\phi^2_{\ell o} = \left[1.0 + 1.0114\ \Lambda^{-0.6698}\left(\frac{Re}{10^5}\right)^{-0.6867}\left(\frac{Bo}{10^3}\right)^{0.6219} Y^{0.3155}\right]\left[1 + Y\frac{\nu_{fg}}{\nu_f}\right]\left[1 + Y\frac{\mu_{fg}}{\mu_g}\right]^{-1/4} \qquad \ldots \quad (23)$$

and for $0.2 \leq \alpha < 1.0$

$$\phi^2_{\ell o} = \left[1.0 + 1.625\ \Lambda^{0.4179}\left(\frac{Re}{10^5}\right)^{-0.9896}\left(\frac{Bo}{10^3}\right)^{-0.0831}\left(\frac{1-Y}{Y}\right)^{-0.2436}\right]\left[1 + Y\frac{\nu_{fg}}{\nu_f}\right]\left[1 + Y\frac{\mu_{fg}}{\mu_g}\right]^{-1/4} \qquad \ldots \quad (24)$$

The properties in the dimensionless groups are based on the local pressure.

Equations (23) and (24) were derived from that portion of the data where

(a) diabatic conditions existed,
(b) the thermodynamic outlet quality was $0.03 < X < 0.42$,
(c) the mass flux was $1952 < G < 4792\ kg{\cdot}m^{-2}{\cdot}s^{-1}$,
(d) the inlet pressure was $6.81 < P_{in} < 11.25$ MPa, and
(e) the outlet pressure was $6.56 < P_{out} < 10.72$ MPa.

The ranges for the Reynolds number, the property index and the transverse boiling number are given below:

$$108{,}000 < Re < 418{,}000$$
$$0.065 < \Lambda < 0.127$$
$$49 < Bo < 365$$

The form of Equation (23) is correct insofar when Y approaches zero, $\phi^2_{\ell o}$ tends to unity and in Equation (24), when Y = 1, $\phi^2_{\ell o} = \phi^2_{H.M.}$.

3. PRESSURE DROP PREDICTION

3.1 Overall Pressure Drop Prediction

The object of the analysis was to derive a two-phase friction multiplier expression that could be used to predict the pressure drop in a horizontal nuclear reactor channel with misaligned 37-element fuel bundles. Of course, the correlation had to be able to predict the pressure drop of the data from which it was derived. This section deals with the approach used to predict this pressure drop. Since the multiplier correlation is based on high flow data, the pressure drop prediction technique outlined below is considered applicable to vertical channels if the gravity component of the total pressure drop is accounted for.

For each applicable run, the single-phase length was determined using Equation (15). Equations (13) and (14) were used to find the single-phase pressure gradient. The product of single-phase pressure gradient and single-phase length was subtracted from the inlet pressure to obtain the pressure at the OSV point.

In order to integrate the two-phase section of the pressure profile, its length was divided into ten sections of equal size. The pressure drop was calculated for each node, the pressure at the end of one node being the pressure at the beginning of the next node.

The quality at the end of a node was initially based on the pressure at the beginning of the node. Since quality is not a highly pressure dependent term, this approximation was justified especially since further iterations over the same node did not result in significant changes in the pressure drop. The nodal average quality was determined and, along with other properties, used in Equations (23) and (24) to obtain $\phi^2_{\ell o}$. The value of $\phi^2_{\ell o}$ was multiplied by the single-phase pressure gradient to find the two-phase pressure gradient. This last value multiplied by the nodal length resulted in the nodal two-phase pressure drop. The quality and thermophysical properties at the nodal boundaries were also used to determine the void fraction using Equation (20). The momentum or acceleration pressure drop was then obtained from Equation (21) and added to the two-phase pressure drop. This resulted in an improved estimate of the pressure at the end of a node. The improved estimate was used to repeat the above calculational procedure. It was found that two iterations per node produced accurate results. Repeating this calculation for all nodes resulted in obtaining the outlet pressure.

4. DISCUSSION OF RESULTS

4.1 Overall Pressure Drop Prediction

A plot of predicted versus measured overall pressure drop values is shown in Figure 2. The figure shows that the pressure drop has been mostly overpredicted. No unusual trends with respect to pressure drop magnitude can be observed from this figure.

The errors in the 104 overall pressure drop predictions can be summarized as follows:

Maximum positive error: 15.4%
Maximum negative error: -7.0%

Average error: 0.32%
RMS error: 5.4%

where the error is defined as

$$\frac{\Delta p(\text{predicted}) - \Delta p(\text{measured})}{\Delta p(\text{measured})} \times 100\%$$

A positive error thus indicates a pressure drop overprediction.

4.2 System Parameters and Errors

To check for unusual error patterns, the following system parameters were plotted against the error in the overall pressure drop prediction:

(a) inlet temperature
(b) mass flow rate
(c) inlet pressure
(d) outlet thermodynamic quality

In Figure 3 the pressure drop error is plotted against the inlet temperature. It appears that at the lower inlet temperatures the correlation overpredicts the pressure drop, whereas at higher inlet temperatures the pressure drop becomes underpredicted. At the higher inlet temperatures the heat flux was generally kept low in order to avoid exceeding the critical heat flux. In a subsequent experiment with higher heat fluxes the same prediction trend was observed. The details can be found in Section 5.5.

The relationship between predictive error and mass flux is shown in Figure 4. This figure shows that the mass flux does not influence the prediction error.

Figure 5 indicates that the pressure drop prediction error is evenly distributed in the range of inlet pressures considered. Figures 3 to 5 indicate that there is no strong trend of pressure drop prediction error with the inlet parameters.

The plot of predicted error versus thermodynamic equilibrium outlet quality is shown in Figure 6. It appears that the error is fairly evenly distributed within the range of outlet qualities considered.

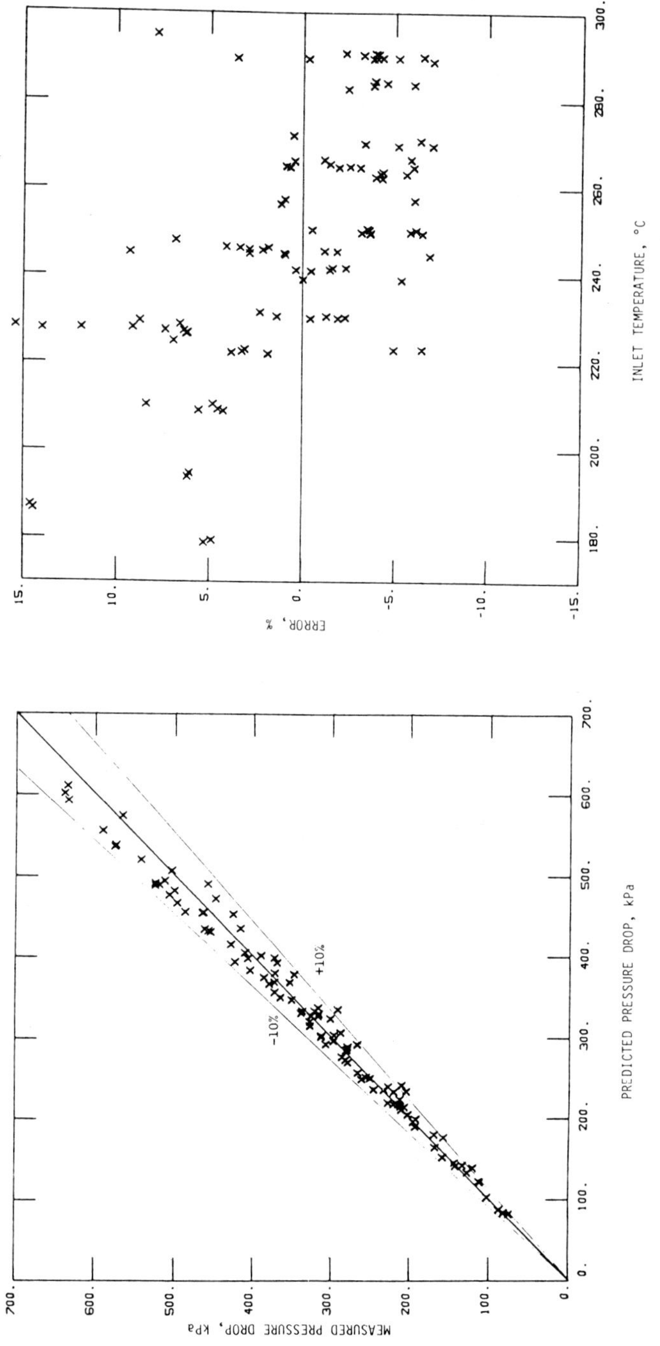

FIGURE 2: COMPARISON OF PREDICTED AND MEASURED TOTAL PRESSURE DROP

FIGURE 3: EFFECT OF INLET TEMPERATURE ON PREDICTION ERROR

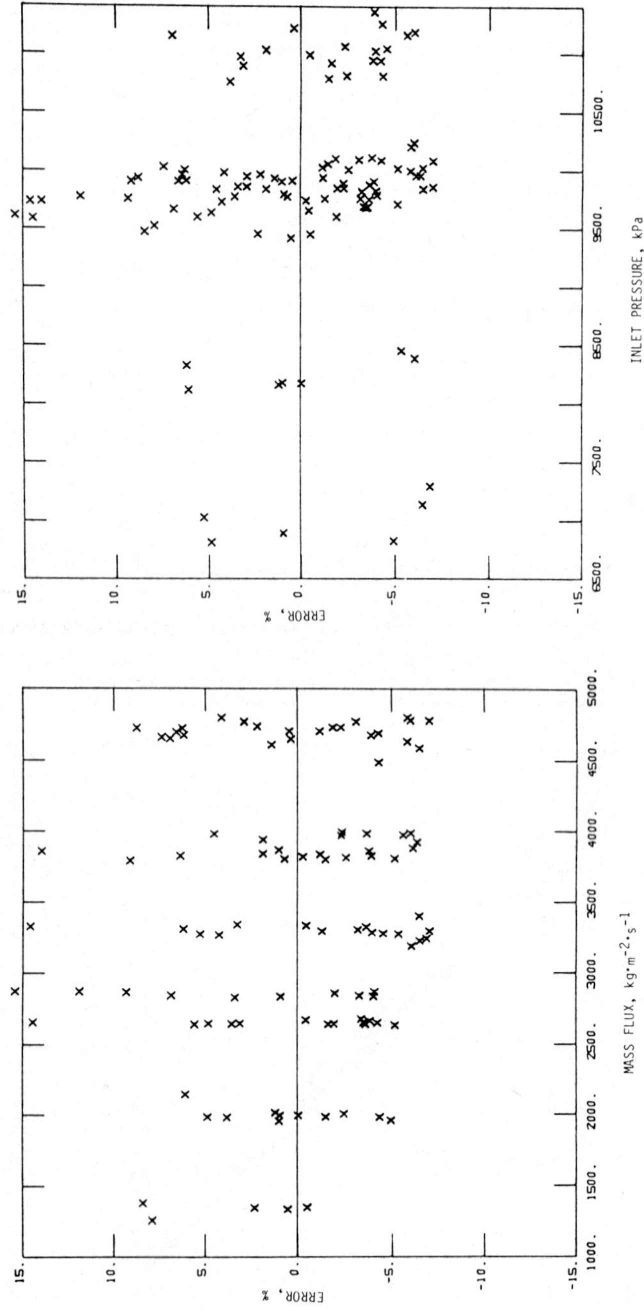

FIGURE 5: EFFECT OF INLET PRESSURE ON PREDICTION ERROR

FIGURE 4: EFFECT OF MASS FLUX ON PREDICTION ERROR

4.3 Pressure Profile Prediction

To test the accuracy of the single-phase adiabatic friction factor correlation and the two-phase friction multiplier correlation in predicting the pressure profile, the predicted pressures at all pressure taps were compared with their measured values for each run. The pressure prediction error for pressure tap (i) was defined as follows:

$$\frac{(P(i)_{pred} - P_{in}) - (P(i)_{meas} - P_{in})}{(P(i)_{meas} - P_{in})} \times 100\% \qquad \ldots \quad (25)$$

which can be simplified to

$$\frac{(P(i)_{pred} - P(i)_{meas})}{P(i)_{meas} - P_{in}} \times 100\% \qquad \ldots \quad (26)$$

The results are shown in Figure 7. The error at the first pressure tap is zero by definition. It is interesting to note that the prediction errors at the second pressure tap are almost all negative.

In adiabatic systems, the temperature and therefore the Reynolds number remain constant as the fluid flows along the bundle. Since the Reynolds number is constant, so must be the friction factor f, and the pressure gradient must be a straight line. On the other hand, in a diabatic system the temperature of the fluid increases down the channel. The density and viscosity are affected by this temperature rise and both decrease in magnitude with an increase in temperature. Since the density decreases at a lower rate than the viscosity, the Reynolds number increases with an increase in temperature. For this reason the friction factor decreases downstream along a channel with diabatic single-phase flow and the pressure loss will be smaller than that experienced with adiabatic flow. However, this trend is not shown by the present data. A careful analysis of the measured adiabatic pressure profiles revealed that pressure tap number two practically always indicated a value of pressure which is below the pressure had the pressure profile been a straight line between the inlet and outlet pressures. This systematic error was also observed in the two-phase pressure profiles. Besides the expected instrument error, the proximity of bundle spacing devices and the physical shape of the pressure tap inlet (burrs) could cause these anomalies in pressure readings.

The prediction error at pressure tap 2 is usually small in absolute terms: from 3.4 kPa up to 9 kPa. The differences between predicted and measured pressure for all pressure taps are shown in Figure 8. Figure 8 is included to show that the per cent errors as shown in Figure 7 do not translate into large absolute values.

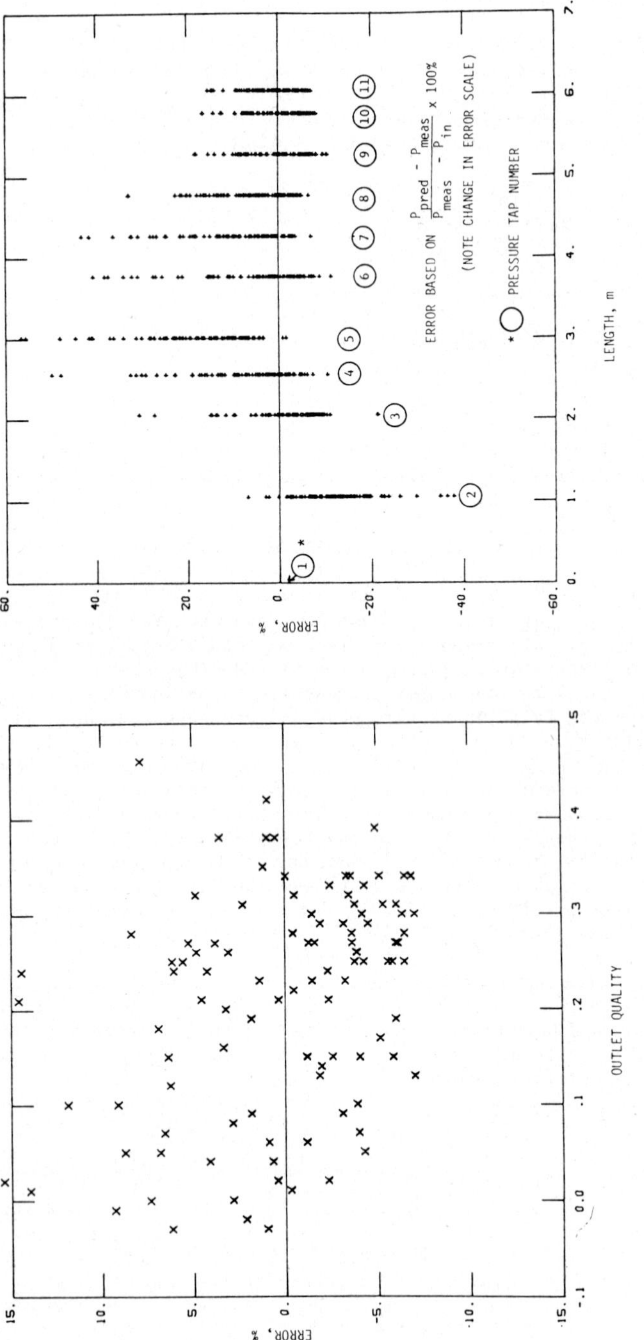

FIGURE 6: EFFECT OF OUTLET QUALITY ON PREDICTION ERROR

FIGURE 7: PRESSURE PROFILE PREDICTION ERROR

5. PRESSURE DROP PREDICTIONS OF OTHER DATA SETS

5.1 Introduction

The applicability of the pressure prediction technique to other data sets with similar or different conditions and geometrically different 37-element clusters, was evaluated by predicting the pressure drop data of these experiments:

(a) additional data obtained with the same segmented bundle,
(b) data obtained with a cluster without end plate simulation (smooth bundle),
(c) data from a rebuilt smooth bundle, and
(d) from the segmented bundle where the upstream half was replaced by a "dummy" unheated length of the same geometry.

5.2 Segmented Bundle

The additional segmented bundle pressure drop measurements and their predictions are compared in Figure 9. The present correlations overpredict the pressure drop in this data set by a few per cent.

5.3 Smooth Bundle

The segmented bundle $\phi_{\ell o}^2$-correlation was also applied to the smooth bundle pressure drop data. The smooth bundle friction factor was derived from single-phase adiabatic data. Figure 10 compares the measured and predicted overall pressure drop. The pressure drop spread is larger than that of the segmented bundle but the RMS error of 6.6% indicates that the derived $\phi_{\ell o}^2$-correlation is also applicable to the smooth bundle geometry.

5.4 Rebuilt Smooth Bundle

Figure 11 shows the results of applying the segmented bundle two-phase friction multiplier to the rebuilt smooth bundle data. The single-phase friction factor used in conjunction with the $\phi_{\ell o}^2$-correlation was optimized from adiabatic data obtained with this bundle. The pressure drop is slightly underpredicted and the error band is larger than the segmented bundle results. The prediction RMS error amounted to 7.3%. However, the correlation is still successful in predicting the pressure drop for this physically different 37-element bundle.

5.5 Three-Metre Segmented Bundle

This bundle was constructed by combining the existing downstream half of the six-metre segmented bundle and a newly fabricated "dummy" upstream section of the same geometry. This upstream section was made such that the electrical resistance was minimal. The heat generation in this section was, therefore, only 7% of the total power. Since this bundle was physically different from the six-metre segmented bundle, a new friction factor correlation was obtained from adiabatic single-phase data. The new friction factor correlation and the segmented bundle $\phi_{\ell o}^2$-correlation were used to predict the

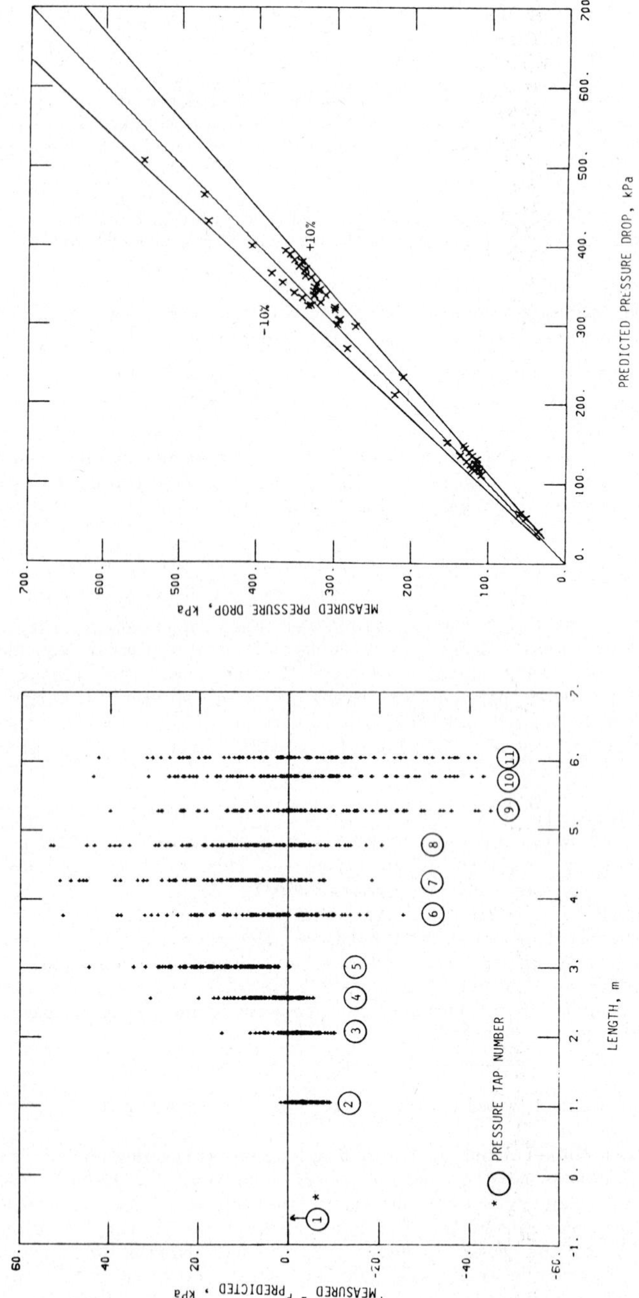

FIGURE 9: COMPARISON OF MEASURED AND PREDICTED TOTAL PRESSURE DROP: ADDITIONAL SEGMENTED BUNDLE DATA

FIGURE 8: DIFFERENCE OF MEASURED AND PREDICTED VALUES IN THE PRESSURE PROFILE

168 three-metre bundle pressure drop data. The pressure drop was underpredicted by 6% and the maximum and minimum errors were 33.2% and -16.8%. The RMS error was 10.5%. The comparison of measured and predicted pressure drop is shown in Figure 12. The effect of the outlet quality on the prediction error is shown in Figure 13 and from this figure it can be seen that as the quality increases, the pressure drop becomes more underpredicted. The other data sets do not display this trend. The heat fluxes encountered in this data set are generally much higher than the heat fluxes on which the subcooled length correlation (Equation (15)) is based. This higher heat flux causes the subcooled two-phase boiling length to increase (due to an increase in the boiling number) possibly beyond the real OSV point. This causes an overprediction of pressure drop in some very low quality runs. The increase in the boiling number also causes a reduction in the $\phi^2_{\ell o}$ values for void fractions above 20%. It is believed that this reduction in $\phi^2_{\ell o}$ is the cause for the pressure drop underpredictions in the majority of the runs.

A summary of prediction errors for the different bundles is shown in Table 1.

6. SUMMARY

The following conclusions can be drawn from the results of this study.

1. The observation that the subcooled diabatic pressure drop always exceeds the pressure drop predicted by the adiabatic pressure gradient was used here by considering the two-phase region of the flow to begin at the onset of significant void (OSV).

2. A correlation was developed which predicted this OSV point with an accuracy acceptable for the present purpose.

3. Over 950 $\phi^2_{\ell o}$ data were derived from the experiment. These data were fitted in a relationship where $\phi^2_{\ell o}$ was a function of vapor weight quality, Reynolds number, transverse boiling number, property index and $\phi^2_{\ell o}$ as predicted by the homogeneous model. The resulting correlation incorporated correct asymptotic trends.

4. The $\phi^2_{\ell o}$-correlation, incorporated into a suitable pressure drop integration technique, predicted the previously obtained 104 dryout pressure drop and pre-dryout pressure drop data with an average error of 0.32% and an RMS error of 5.4%. The maximum positive error was 15.4% (pressure drop overpredicted) and the maximum negative error was -7.0%.

5. The pressure profile was predicted with reasonable accuracy. At the second pressure tap where the fluid in most cases was subcooled, the pressure was mostly underestimated. At tap number 5 the pressure is just about always overpredicted. The cause of the trends in these pressure taps was found to be systematic pressure tap errors. Further downstream the error band narrowed down to +15.4 and -7.0%.

6. The pressure drop prediction technique successfully predicts the experimental pressure drop of 37-element bundles of different design. Since the technique applies to bundles with different end plate geometry etc., it is concluded that it can also be applied to CANDU-nuclear reactor channels.

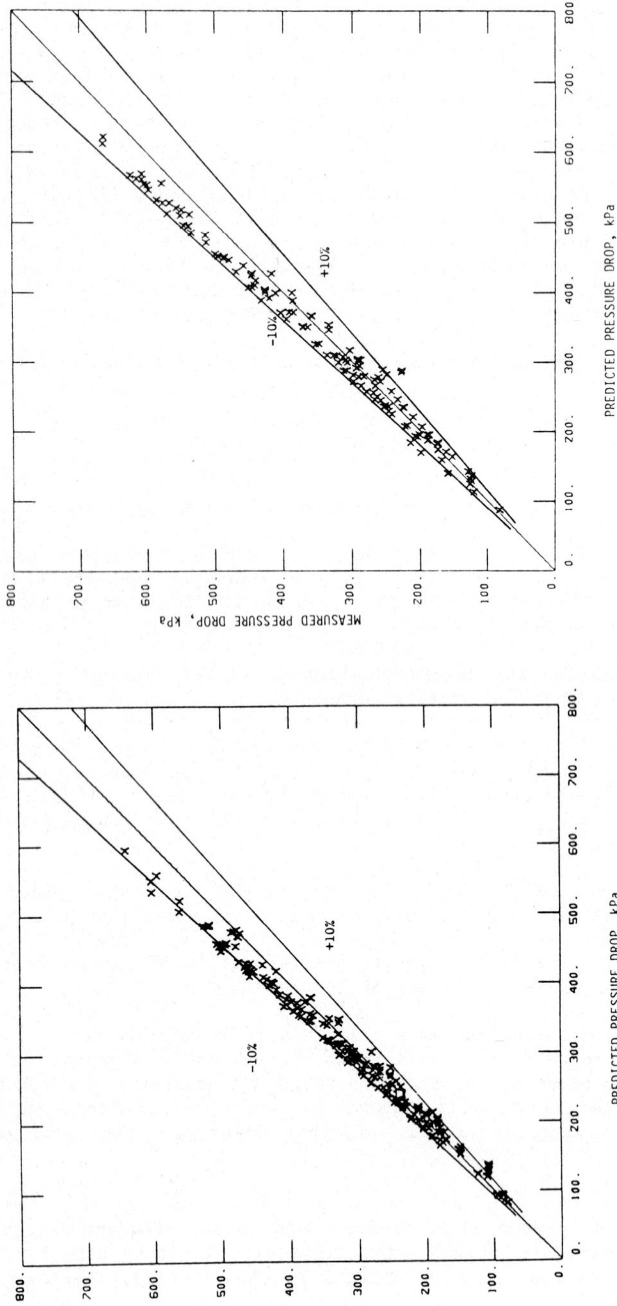

FIGURE 11: COMPARISON OF MEASURED AND PREDICTED TOTAL PRESSURE DROP: REBUILT SMOOTH BUNDLE

FIGURE 10: COMPARISON OF MEASURED AND PREDICTED TOTAL PRESSURE DROP: ORIGINAL SMOOTH BUNDLE

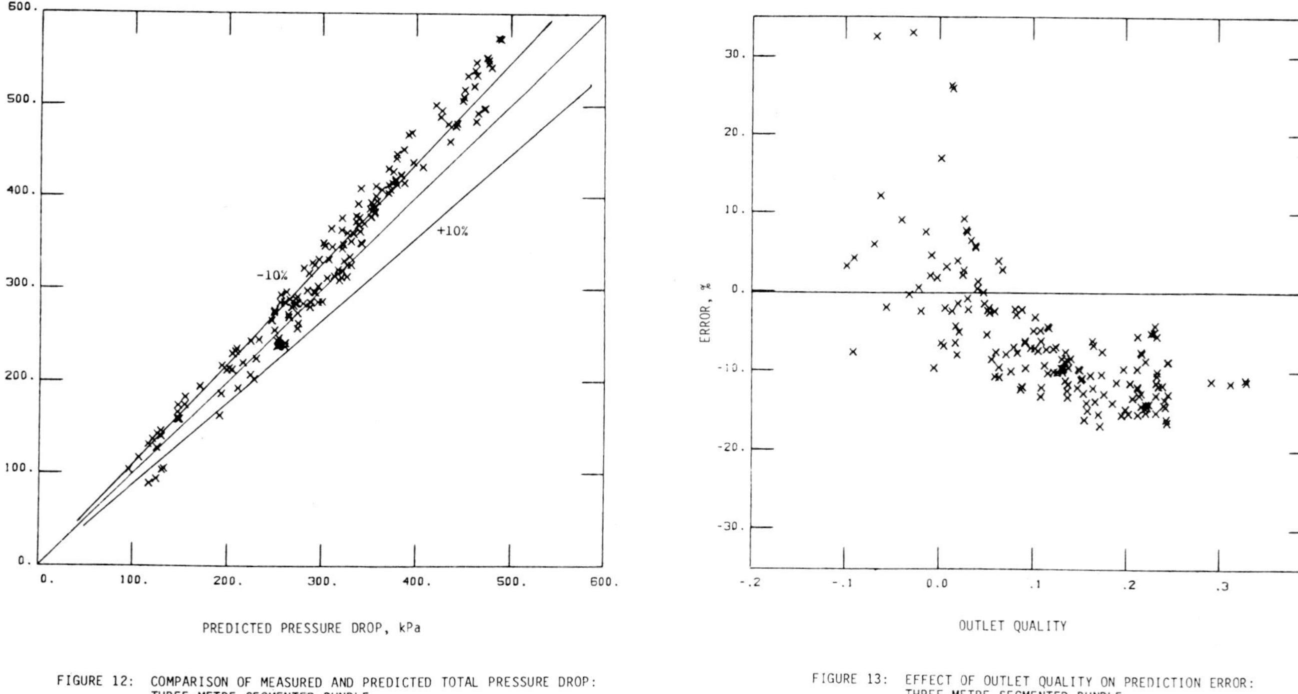

FIGURE 12: COMPARISON OF MEASURED AND PREDICTED TOTAL PRESSURE DROP: THREE-METRE SEGMENTED BUNDLE

FIGURE 13: EFFECT OF OUTLET QUALITY ON PREDICTION ERROR: THREE-METRE SEGMENTED BUNDLE

REFERENCES

Ahmad, S.Y., 1970, "Axial Distribution of Bulk Temperature and Void Fraction in a Heated Channel With Inlet Subcooling," J. of H.T., pp. 595-609.

Armand, A.A. and Treschev, G.G., 1947, "Investigation of the Resistance During the Movement of Steam-Water Mixtures in Heated Boiler Pipes at High Pressures," Izvestiya Veseoyuznogo Teploteknikeskogo Instituta, Vol. 4, pp. 1-5, AERE Lib/Trans 816.

Baroczy, C.J., 1965, "A Systematic Correlation for Two-Phase Pressure Drop," AIChE reprint No. 37, paper presented at 8th Nat. Heat Transfer Conf., Los Angeles, August.

Collier, J.G., 1980, "Convective Boiling and Condensation," 2nd ed., McGraw-Hill, London.

Jones, A.B. and Dight, D.G., 1962, "Hydrodynamic Stability of a Boiling Channel," Part 2, KAPL-2208.

Levy, S., 1967, "Forced Convection Subcooled Boiling - Prediction of Vapor Volumetric Fraction," Int. J. of Heat & Mass Trans., Vol. 10, pp. 951-965.

Lockhart, R.W. and Martinelli, R.C., 1949, "Proposed Correlation of Data for Isothermal Two-Phase Two-Component Flow in Pipes," Chem. Eng. Prog., Vol. 45, No. 39.

Martinelli, R.C. and Nelson, D.B., 1948, "Prediction of Pressure Drop During Forced Circulation Boiling of Water," Trans. ASME, Vol. 70, No. 695.

Massena, W.A., 1960, "Steam-Water Pressure Drop," Hanford Report, H.W. 65706.

Reddy, D.G., Sreepad, S.R. and Nahavandi, A.N., 1982, "Two-Phase Friction Multiplier Correlation for High Pressure Steam-Water Flow," EPRI NP-2522.

Saha, P. and Zuber, N., 1974, "Point of Net Vapour Generation and Vapour Void Fraction in Subcooled Boiling," Paper B4.7, Proceedings 5th Int'l. Heat Transfer Conference, Tokyo.

Shoukri, M., Yanchis, R.J. and Rhodes, E., 1981, "Effect of Heat Flux on Pressure Drop in Low Pressure Flow Boiling in a Horizontal Tube," Canadian Journal of Chemical Engineering, Vol. 59, p. 149.

Tarasova, N.V. and Leont'ev, A.I., 1965, "Hydraulic Resistance with a Steam-Water Mixture Flowing in a Vertical Heated Tube," Teplo. Vyso Temp., Vol. 3, No. 1, pp. 115-123, Jan-Feb.

Tarasova, N.V., Leont'ev, A.I., Hlopuskin, V.I. and Orlov, W.M., 1966, "Pressure Drop of Boiling Subcooled Water and Steam-Water Mixture Flowing in Heated Channels," Paper presented at 3rd Int. Heat Transfer Conf. held in Chicago, Vol. 4, pp. 178-183, Paper 133.

NOMENCLATURE

Cp	Specific heat of the liquid phase	(kJ/kg°C)
D	Diameter	m
f	Friction factor	-
G	Mass flux	$kg{\cdot}s^{-1}{\cdot}m^{-2}$
H	Enthalpy	J/kg
ℓ	Length	m
P	Pressure	kPa
ν	Specific volume	m^3/kg
W	Mass flow	kg/s
X	Martinelli parameter defined by Equation (5)	-
X	Thermodynamic quality	-
Y	Vapor weight quality	-
α	Void fraction	-
Δ	Difference	-
ε	Roughness	m
ρ	Density	kg/m^3
μ	Viscosity	Pa·s
ϕ	Heat flux	kW/m^2
$\phi^2_{\ell o}$	Two-phase friction multiplier	-
λ	Latent heat of vaporization	J/kg

Dimensionless groups

Λ	Property index $(\mu_\ell/\mu_g)^{0.2}\,(\rho_g/\rho_\ell)$
Re	Reynolds number, $G{\cdot}D_e/\mu_\ell$
S_d	Subcooling at OSV point $(H_f - H_d)/\lambda$
S_{in}	Inlet subcooling $(H_f - H_{in})/\lambda$
Bo	Transverse boiling number $(\phi D_e)/(\lambda\mu_g)$

Subscripts

d	Conditions at point of onset of significant void
e	Hydraulic equivalent
f	Saturated liquid
fg	Difference between liquid and vapor properties
g	Vapor
H.M.	Homogeneous model
ℓ	Fluid
ℓo	Total flow assumed liquid
MN	Martinelli-Nelson
sp	Single-phase
tp	Two-phase
tt	Liquid turbulent, gas turbulent

Superscript

-	Mean value

Multi-Phase Flow and Heat Transfer III. Part A: Fundamentals
edited by T.N. Veziroğlu and A.E. Bergles
Elsevier Science Publishers B.V., Amsterdam, 1984 — Printed in The Netherlands

EFFECT OF STREAMLINING ON BUNDLE PRESSURE DROP

L.K.H. Leung and M.E. Salcudean
Department of Mechanical Engineering
University of Ottawa
Ottawa, Ontario
Canada

and

C.W. Snoek and D.C. Groeneveld
Chalk River Nuclear Laboratories
Atomic Energy of Canada Limited
Chalk River, Ontario
Canada

ABSTRACT

Single- and two-phase flow experiments have been performed on two strings of 4-rod bundles: one simulating a string of regular 37-element bundles currently used in CANDU reactors and one simulating a string of streamlined 37-element CANDU bundles. The effects of streamlining and bundle misalignment on the pressure drops are investigated. Pressure variation along each set of bundles is obtained using multi-tube manometers.

It is shown that bundle streamlining improves the pressure recovery at the bundle junction and the spacer planes. It also reduces the dependency of the pressure drop on misalignment of the bundles.

1. INTRODUCTION

Most of the pressure drop in the primary circuit of a nuclear reactor takes place along the fuel bundle. In the case of the horizontal CANDU type fuel bundles, the bundle string pressure drop is composed of skin friction pressure drop, junction pressure drop*, spacer pressure drop, and acceleration pressure drop in two-phase flow. (This is ignored in the present study as the data is obtained under adiabatic conditions).

The skin friction pressure drop for a channel is obtained by using the Darcy-Weisbach equation:

$$\Delta P_F = \frac{fL}{D_e}\frac{\rho_\ell v^2}{2} \qquad (1)$$

where f is the friction factor, L is the length of the bundle, D_e is the equivalent hydraulic diameter, ρ_ℓ is the density of the liquid, and v is the flow velocity. Many authors |1| recommend this equation for the bundle geometry, with the friction factor evaluated using equivalent hydraulic diameter. However,

* In CANDU reactors, the reactor core is made up of several hundred pressure tubes, each containing a string of 12 or 13 50cm-long bundles. The junction pressure drop refers to the pressure drop at the junction of two adjacent 50-cm fuel bundles.

the friction factor has been found experimentally to be also affected by the channel geometry |2,3|. Rehme |4| and Malak et al. |5| have studied this effect and have proposed relations which can be used to estimate the friction factor in bundle geometry.

The pressure drop at the junction of the two bundles can be written as

$$\Delta P_J = K_J \frac{\rho_\ell v^2}{2} \tag{2}$$

where K_J is the junction loss coefficient. It consists of the pressure drops because of (i) obstructions such as an end-plate, and (ii) discontinuity of fuel rods. The pressure drop due to obstruction depends on the size, location, and shape of the blockage. Less pressure is lost for a smaller obstruction located in the slower moving fluid region |6|. For various types of obstructions, the streamlined shape has the best pressure recovery. A blunt type of obstruction causes the highest pressure drop; however, the pressure recovery can be improved by rounding the leading edge of the obstruction |7|. The pressure drop caused by the discontinuity of the elements has been observed in both aligned bundles |3|, and aligned plates |8|. It is simply an effect of flow expansion and contraction. The pressure drop can be reduced, if the areas of the exit and entrance change gradually. In a string of bundles, more pressure drop can occur in the junction area if adjacent bundles are misaligned (see Figure 1) because of the increased blockage area from the misaligned downstream bundle. This additional pressure drop depends on the shape of the end-plug and can be reduced by using a shape that does not promote turbulence.

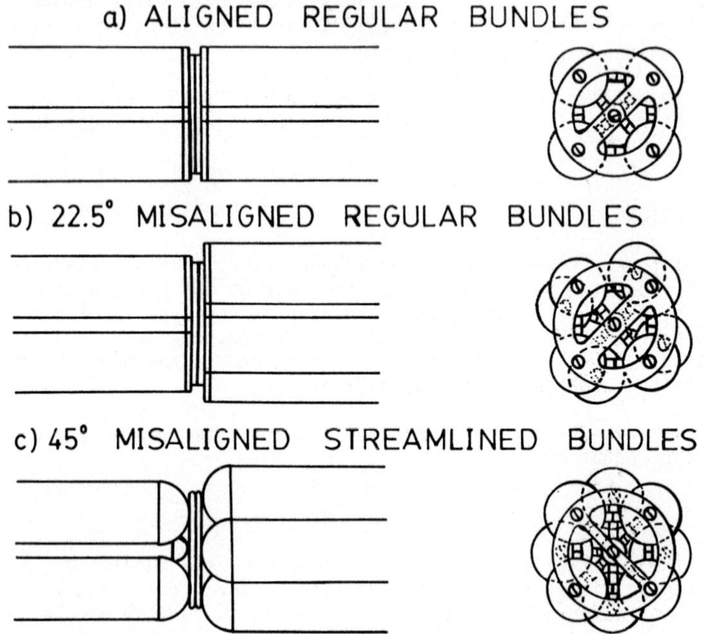

Fig. 1. Change of Flow Blockage Area in Junction Plane due to Bundle Rotation

The spacer which is used to maintain the subchannel space and prevent rod vibration, also interferes with the subchannel flow and increases the overall pressure drop. The spacer pressure drop can be written as

$$\Delta P_S = K_S \frac{\rho_\ell v^2}{2} \tag{3}$$

where K_S is the spacer loss coefficient. It was found that it is not only dependent upon the size and location of the spacer, but that it is also affected by factors such as the shape of the leading edge |9| and that of the trailing edge |10|, the roughness of the rod |11|, and the length of the spacer |12|.

The total pressure drop is the summation of the skin friction pressure drop and the local pressure drops in the junction plane and the spacer plane. It can be written as

$$\begin{aligned}\Delta P_T &= \Delta P_F + \Delta P_J + \Delta P_S \\ &= \left(\frac{fL}{D_e} + K_J + K_S\right) \frac{\rho_\ell v^2}{2} \\ &= K_T \frac{\rho_\ell v^2}{2}\end{aligned} \tag{4}$$

where K_T is the total loss coefficient.

In two-phase flow, the total pressure drop in the horizontal bundles consists of the two-phase frictional pressure drop, the two-phase accelerational pressure drop, the two-phase junction pressure drop, and the two-phase spacer drop. As the channel is horizontal, gravitational components are not present. The frictional pressure drop is usually expressed in terms of the single-phase pressure drop and the two-phase multiplier which is defined as

$$\phi_{LO}^2 = \frac{\Delta P_{TP}}{\Delta P_{LO}} \tag{5}$$

where ΔP_{TP} is the two-phase pressure drop, and ΔP_{LO} is the single-phase pressure drop assuming the total flow to be liquid. Two-phase multipliers as defined in (5) are used also to evaluate the pressure drop through obstructions |13,14,15, 16|.

For the obstruction pressure drop, most of the work was performed on orifices |13,14| and pipe fittings |15,16|. Two-phase multiplier correlations have been proposed for various fixtures. Many authors recommend relations based on homogeneous theory to calculate obstruction two-phase multipliers. Equations are also presented in literature for estimating the spacer pressure drop |17,18|.

The accelerational pressure drop is obtained by evaluating the change of the momentum in the axial direction by using a one-dimensional approach. It is a significant factor in two-phase flow with phase change, but can be neglected when phase change does not occur since the density gradients are small for channel flow.

The total pressure drop is made up of the components mentioned above. It can be written as

$$(\Delta P_{TP})_T = (\Delta P_{TP})_F + (\Delta P_{TP})_J + (\Delta P_{TP})_S$$
$$= (\Delta P_{LO})_F\,(\phi^2_{LO})_F + (\Delta P_{LO})_J\,(\phi^2_{LO})_J + (\Delta P_{LO})_S\,(\phi^2_{LO})_S \quad (6)$$

where $(\Delta P_{TP})_F$, $(\Delta P_{TP})_J$, and $(\Delta P_{TP})_S$ are the two-phase pressure drop for friction, junction, and spacer, respectively; $(\Delta P_{LO})_F$, $(\Delta P_{LO})_J$, and $(\Delta P_{LO})_S$ are the single-phase pressure drop for friction, junction, and spacer, respectively; and $(\phi^2_{LO})_F$, $(\phi^2_{LO})_J$, and $(\phi^2_{LO})_S$ are the two-phase multipliers for the friction, junction, and spacer, respectively.

The objective of the present investigation is to compare the single- and two-phase pressure drops for the simulated regular bundle used currently in the CANDU reactor and those for the modified streamlined bundle.

2. EXPERIMENTAL METHOD

The schematic diagram of the apparatus is shown in Figure 2. Water was circulated by a centrifugal pump and metered by a rotameter. Air was supplied by the central system and metered with two rotameters of different ranges. Both streams were adjusted manually by gate valves and mixed with a spray-type mixer. The mixture was stabilized in a calming section before entering the test-section. The test-section was constructed with a 0.0381-m I.D., 1.83-m long lucite tube. Twenty-three pressure taps were installed along a string of three 0.5-m long bundles. Two panels of Meriam 33KB35 well-type, multi-tube manometers were used to measure the pressure variation along the bundles. The manometric fluid used was water for the single-phase test, and mercury for the two-phase test. Toggle

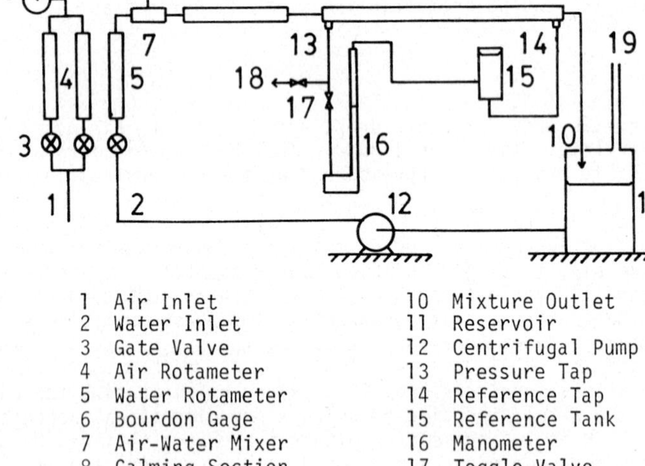

1	Air Inlet	10	Mixture Outlet
2	Water Inlet	11	Reservoir
3	Gate Valve	12	Centrifugal Pump
4	Air Rotameter	13	Pressure Tap
5	Water Rotameter	14	Reference Tap
6	Bourdon Gage	15	Reference Tank
7	Air-Water Mixer	16	Manometer
8	Calming Section	17	Toggle Valve
9	Test-Section	18	To Drain
		19	Air Outlet

Fig. 2. Schematic Diagram of the Loop

valves were installed between the pressure taps and the manometers. The toggle valves could be closed simultaneously to store the pressure readings for recording. By-pass lines connected between the pressure taps and the toggles valves, were controlled by a set of toggle valves. They were used to drain the air bubbles from the pressure lines. After passing through the test-section, the two-phase mixture was led to a reservoir where the phases were separated. The water was recirculated in the loop, and the air was exhausted to the atmosphere.

Three bundles were constructed for both the simulated regular bundles (Figure 3) and the streamlined bundles (Figure 4). Each of the three bundles was composed of four stainless-steel tubes covered with an end-plug in each end, one central rod, two end-plates, sixteen spacers and twelve bearing pads. The end-plug was of the flat-plate type for the regular bundle, whereas it was hemispherical for the streamlined bundle. The central rod was included to obtain the same hydraulic diameter as the 37-element bundle. The shape of the end-plate is illustrated in Figure 5. It was circular with rectangular cross-sectional area for the regular bundle and with circular cross-sectional area for

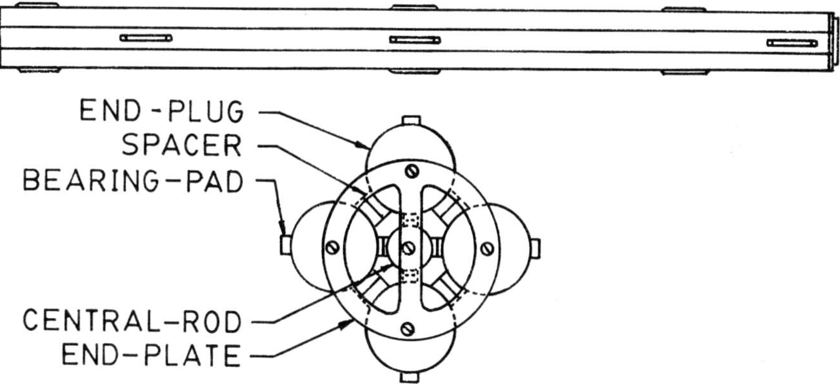

Fig. 3. The Simulated Regular Bundle

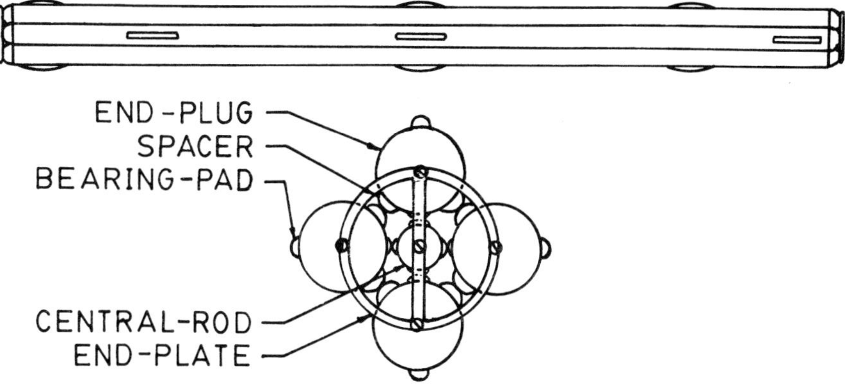

Fig. 4. The Modified Regular Bundle (Streamlined Bundle)

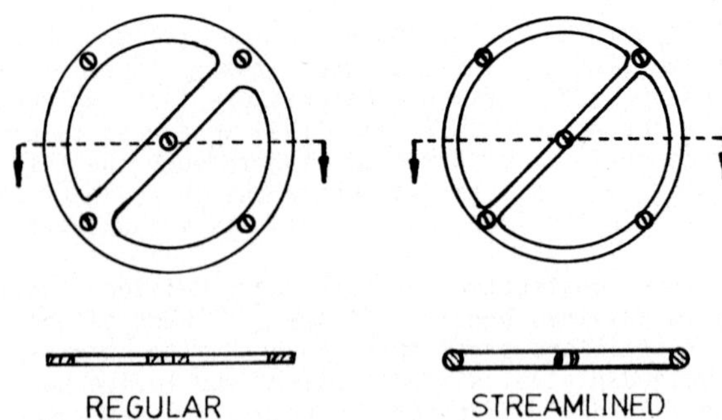

Fig. 5. End-plate Comparison

the streamlined bundle. The sizes of the spacer and bearing pad were similar for both types of bundles, but the edges of these fixtures had been smoothened and rounded for the streamlined bundle (Figure 6).

For single-phase flow, the pressure drop was measured with the Meriam multi-tube manometers using water as manometric fluid. The water flow rate was varied from 0.44 kg/s to 1.88 kg/s.

The pressure drop in two-phase flow was measured using mercury as manometric fluid. The measurements were carried out at water flow rates of 1.57, 1.26, and 0.94 kg/s, with air flow rates varied respectively between 0.00026 to 0.00744 kg/s, 0.00177 to 0.017 kg/s, and 0.00841 to 0.021 kg/s. The water flow rate was adjusted to the desired values, and the air flow rate was adjusted to the largest pre-set value. The toggle valves were opened and then closed simultaneously at the steady manometric fluid level.

The measurements were performed on both types of bundle in various bundle rotations (0°, 22.5°, and 45.0°), for both single- and two-phase flows.

3. RESULTS AND DISCUSSION

3.1 Single-Phase Pressure Drop

The single-phase pressure distribution for the aligned regular bundles is illustrated in Figure 7. The gradual decrease of pressure along the bundles is due to skin friction. The frictional pressure gradient is relatively constant throughout the test-section. The junction pressure drops and the spacer pressure drops are characterized by the sudden decline of pressure in the local region. The location of different planes are indicated by dotted lines. The pressure drops in the junction planes are larger than those in the spacer planes.

The frictional pressure drop can be expressed in terms of a friction factor as shown in Figure 8. Some correlations are illustrated for comparison. The friction factors are the same for both types of bundles and for various

Fig. 6. Spacer Comparison

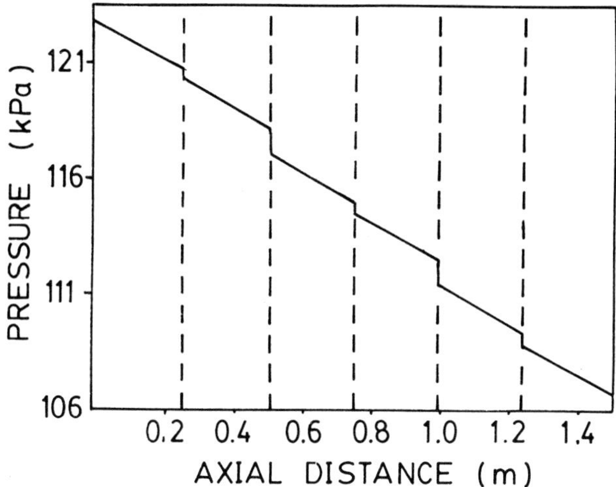

Fig. 7. Single-phase Pressure Variation of the Aligned Regular Bundle

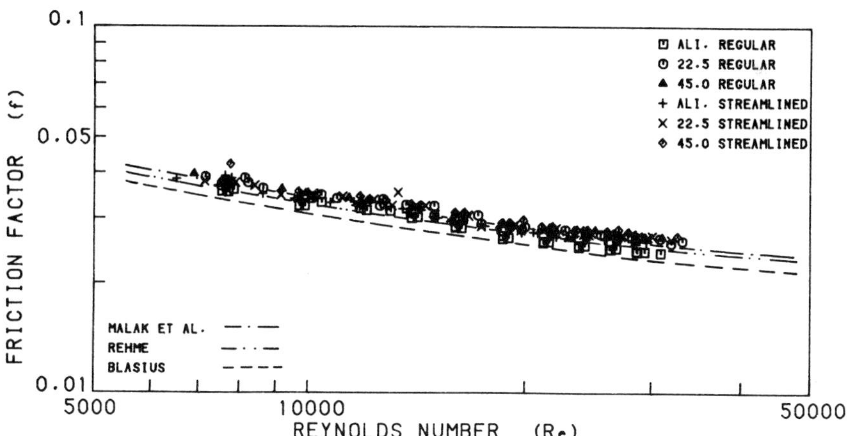

Fig. 8. Friction Factors for the Bundles

misalignments. Both predictions proposed by Rehme |4| and Malak et al. |5| have shown good agreement with the results, whereas the friction factors derived from the relations for smooth tubes are too low. By using the Colebrook equation, the relative roughness is found to be about 0.00076.

The junction loss coefficients for various bundle arrangements are shown in Figure 9. For the regular bundle, the junction loss coefficients are not affected by the change in Reynolds number, but are strongly influenced by the misalignments of the bundles. For a large misalignment angle, the junction loss coefficient is large. This is caused by the additional blockage from the exposed rods. In contrast, the junction loss coefficients for the streamlined bundle vary with Reynolds numbers and are not affected by misalignment. This can be explained by the precedence of friction drag (which depends on Re) over pressure drag which is reduced for the streamlined bundle.

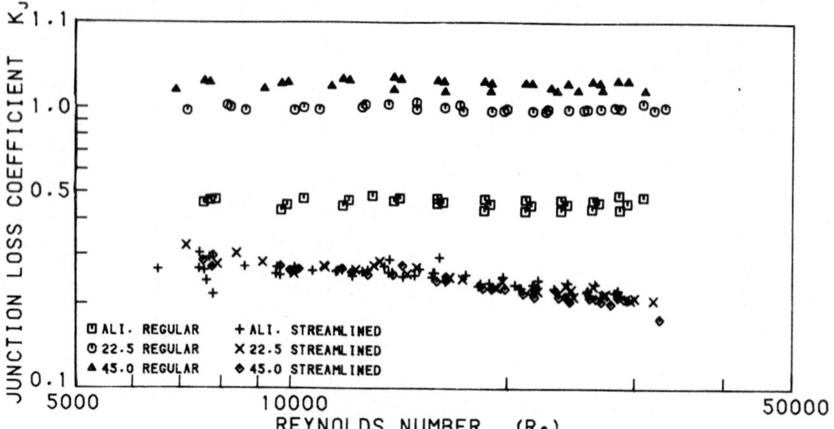

Fig. 9. Junction Loss Coefficients for Various Bundle Arrangements

The spacer loss coefficients, shown in Figure 10, are only slightly affected by the Reynolds number. They are larger for the regular bundle than for the streamlined bundle. This indicates that the pressure recovery can be improved with minor modification to the shape of the obstruction. Even though bundle misalignment does not alter the spacer's blockage area, the spacer loss coefficients appears to be slightly decreased as the misalignment increases. One possible explanation would be that the spacers are located within the separation region created by the junction of misaligned bundles.

The total pressure drop is also expressed in terms of loss coefficient (Figure 11). The loss coefficient decreases as the flow rate increases. Of the total pressure drop, the frictional component contributes approximately 50% for the 45.0° misaligned regular bundle and 75% for the streamlined bundle. The second dominant factor is the junction pressure drop which induces the dependency of the total pressure drop upon the bundle misalignment. Streamlining the bundle junction reduces the junction pressure drop by up to 75% and decreases overall pressure drop up to about 33% (comparison is made between the streamlined bundle and the 45.0° misaligned regular bundle) at larger Reynolds numbers.

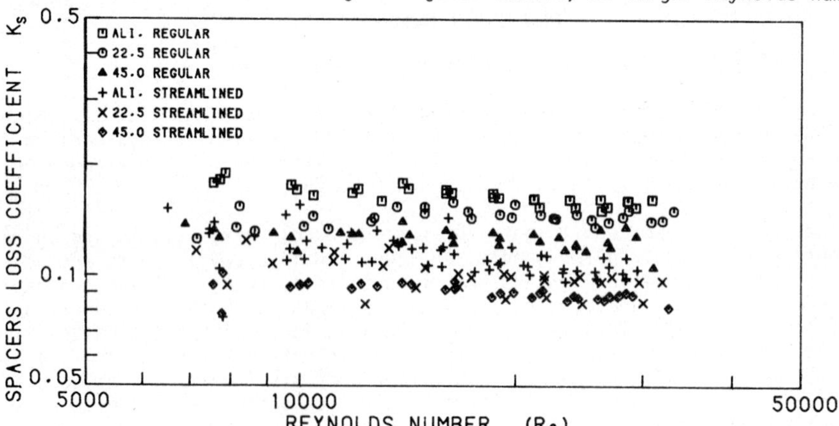

Fig. 10. Spacer Loss Coefficients for Various Bundle Arrangements

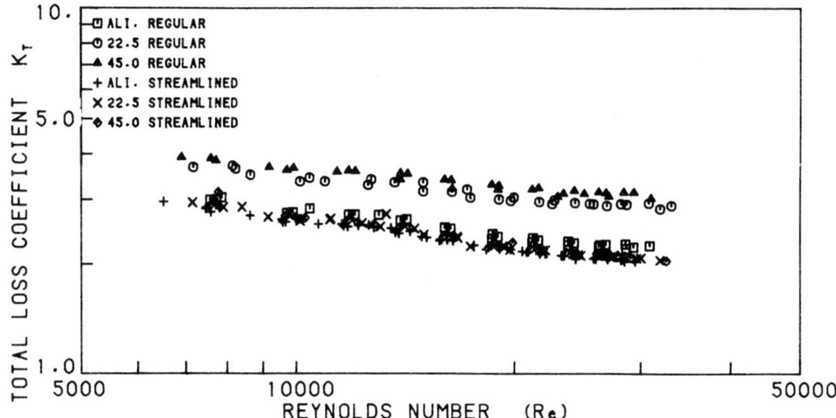

Fig. 11. Total Loss Coefficients for Various Bundle Arrangements

Though improvement has been shown for the spacer pressure drop of the streamlined shape, it is not too significant in the total pressure drop. This is due to the fact that the spacer pressure drop is only about 5% of the total pressure drop.

3.2 Two-Phase Pressure Drop

The two-phase pressure distribution for the aligned regular bundle is illustrated in Figure 12. It follows the same trend as the single phase but the magnitude of the pressure drop is much larger. Since the two-phase mixture is water and air, the accelerational pressure drop is not significant (about 5% of the local pressure). However, it is eliminated from the pressure drop by using an estimated void fraction variation |19|. The pressure decreases because of friction, but the rate of decrease is not uniform along the string of bundles

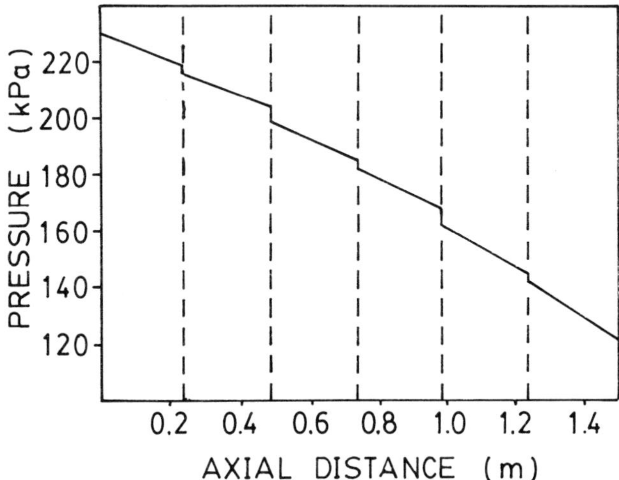

Fig. 12. Two-phase Pressure Variation of the Aligned Regular Bundle

due to the presence of the flow obstructions. The mean frictional pressure drop for the last two bundles was considered in the present calculation. The local pressure drops due to the junctions and the spacers are also shown in Figure 12. As in the single-phase flow, the junction pressure drop is larger than the spacer pressure drop. These local drops are constant along the string of bundles. The small deviation might be caused by the slight changes in density and void distribution as well as experimental inaccurracy. The mean values for each type of local pressure drop were considered in the present calculation. The spacer pressure drop in the first bundle is not considered, however, because it occurs in the entrance region.

The two-phase frictional pressure gradients were measured for different flow rates. As an example, the results for a water flow rate of 1.26 kg/s are illustrated in Figure 13. The two-phase frictional pressure gradient increases with increasing quality. It is not influenced by the misalignment of the bundles. An increase of the friction pressure gradient for increasing water flow rate was also observed. The two-phase frictional multiplier is shown in Figure 14 for the same water flow rate. In the same figure, some correlations are illustrated for comparison. For low quality, all presented correlations predict reasonably well the experimental results. For increased quality, Beattie's correlation gives the best performance.

The two-phase junction pressure drop for 1.26-kg/s water flow rate is shown in Figures 15. The pressure drop increases with increasing water and air flow rates. Similar to the single-phase junction pressure drop, the bundle misalignment affects the pressure drop for the regular bundle but not for the streamlined bundle. The regular bundle with 45.0^{o} misalignment has the largest pressure drop, while the streamlined bundle has the smallest.

The junction two-phase multiplier is shown in Figure 16. It is dependent of the fluid flow rate. The streamlined bundle has the smallest pressure drop, yet it gives the largest two-phase multiplier. This is so because the ratio of the single-phase pressure drop of the regular bundle, and that of the streamlined bundle is larger than the same ratio in two phase flow. That is:

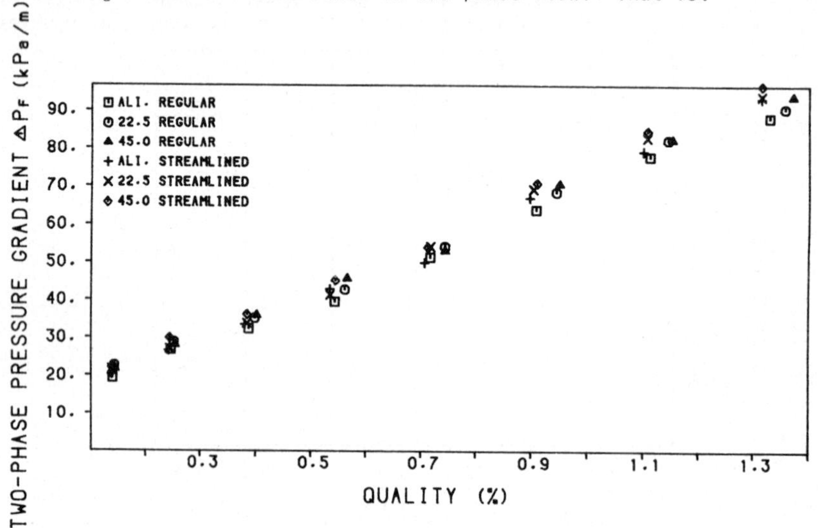

Fig. 13. Two-phase Frictional Pressure Gradient for 1.26 kg/s Liquid Mass Flow Rate

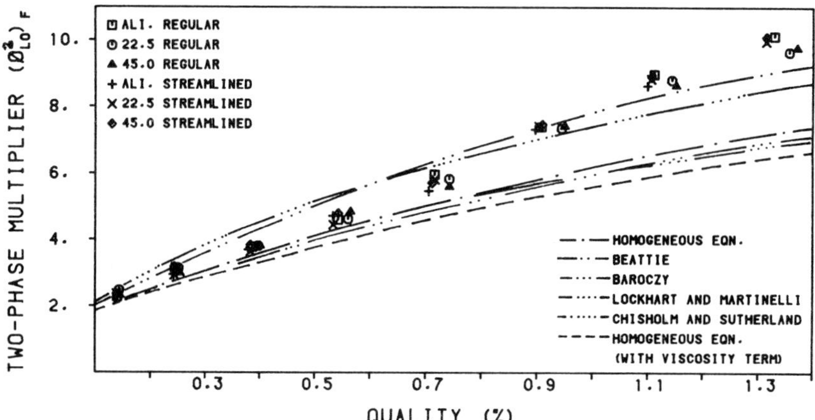

Fig. 14. Two-phase Frictional Multiplier for 1.26 kg/s Liquid Mass Flow Rate

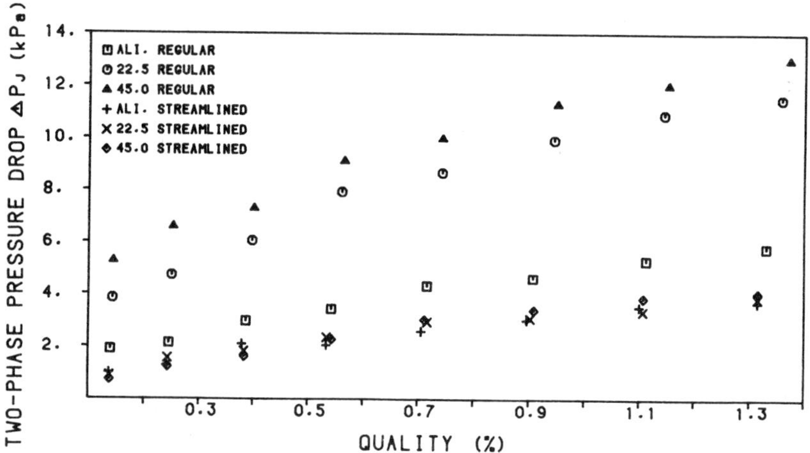

Fig. 15. Two-phase Junction Pressure Drop for 1.26 kg/s Liquid Mass Flow Rate

$$\frac{(\Delta P_{LO})_{Reg.}}{(\Delta P_{LO})_{Str.}} > \frac{(\Delta P_{TP})_{Reg.}}{(\Delta P_{TP})_{Str.}} \tag{7}$$

where the subscripts Reg. and Str. represent the respective pressure drop for the regular bundle and for the streamlined bundle.

Some correlations are shown in the same figure for comparison. Since the junction two-phase multipliers vary for different bundle misalignments and streamlining, none of the correlation is capable of predicting the various multipliers. Overall, the orifice correlation proposed by Beattie is the best for estimating the multiplier for the aligned regular bundle, while the homogeneous two-phase multiplier provides good agreement with the values for the streamlined bundle.

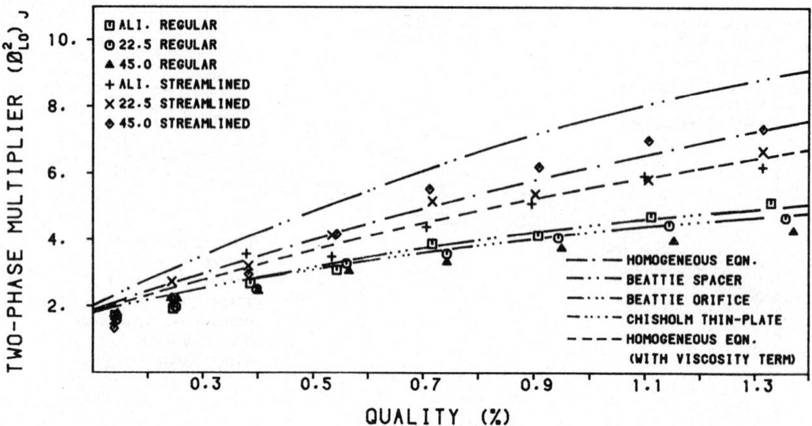

Fig. 16. Two-phase Junction Multiplier for 1.26 kg/s Liquid Mass Flow Rate

The two-phase spacer pressure drop for 1.26-kg/s water flow rate is shown in Figures 17. It is affected by the water and air flow rates, but is not significantly influenced by the bundle misalignments. Though the results are quite scattered, it can be seen that the regular bundle has caused higher pressure drop. The small variation of pressure drop between 0.5 and 1.0% quality is probably caused by the flow transition effect.

The two-phase multiplier for the spacer is shown in Figures 18. It increases with increasing quality and is not affected by the bundle misalignment. For both types of bundles, the two-phase multipliers are close. A comparison between the correlations indicates that the homogeneous theory predicts the experimental results reasonably well, but the Beattie orifice correlation tends to underestimate the results.

The total two-phase pressure drop for 1.26-kg/s water flow rate is shown in Figure 19. The first bundle was not included in order to avoid entrance

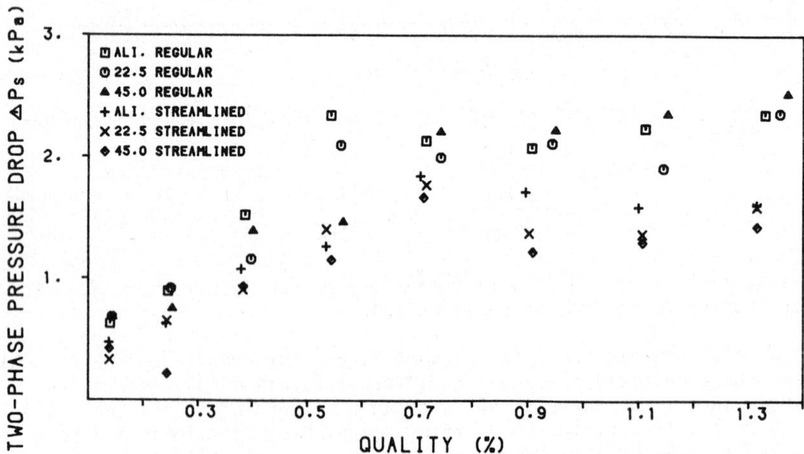

Fig. 17. Two-phase Spacer Pressure Drop for 1.26 kg/s Liquid Mass Flow Rate

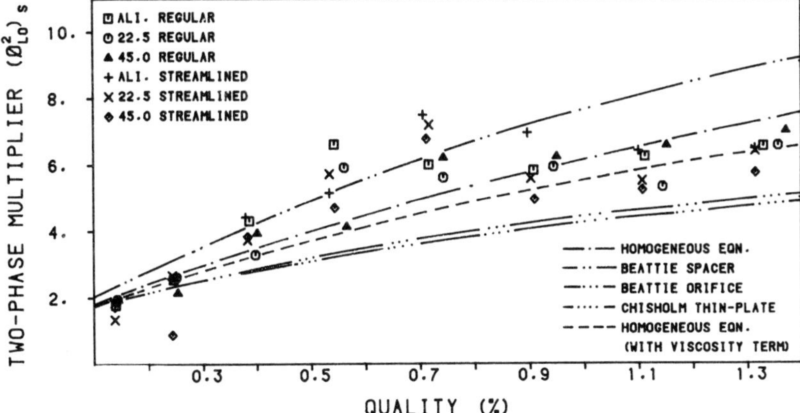

Fig. 18. Two-phase Spacer Multiplier for 1.26 kg/s Liquid Mass Flow Rate

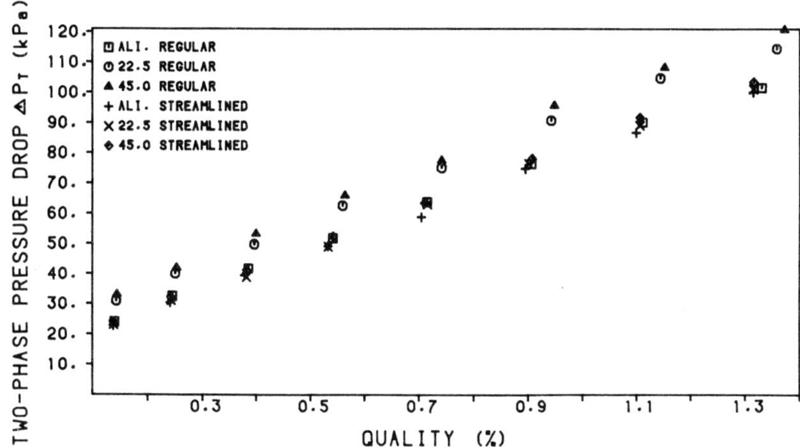

Fig. 19. Total Two-phase Pressure Drop for 1.26 kg/s Liquid Mass Flow Rate

effects. The pressure drop increases with increasing quality. For the regular bundle, it is larger than that for the streamlined bundle, and it is affected by the misalignment of the bundles. The 45.0^o misaligned regular bundle has the largest pressure drop. The pressure drop for the streamlined bundle is not affected by the misalignment, and it is smaller than that for the regular bundle (the larger total pressure drop for the streamlined 45° bundle as compared with the aligned regular bundle is due to the difference in friction pressure loss (see Figure 13)).

The two-phase multiplier for the total pressure drop is shown in Figure 20. It is larger for the streamlined bundles. This is so, because the ratio of the single-phase pressure drop of the regular bundle and that of the streamlined bundle is larger than the same ratio in two-phase flow.

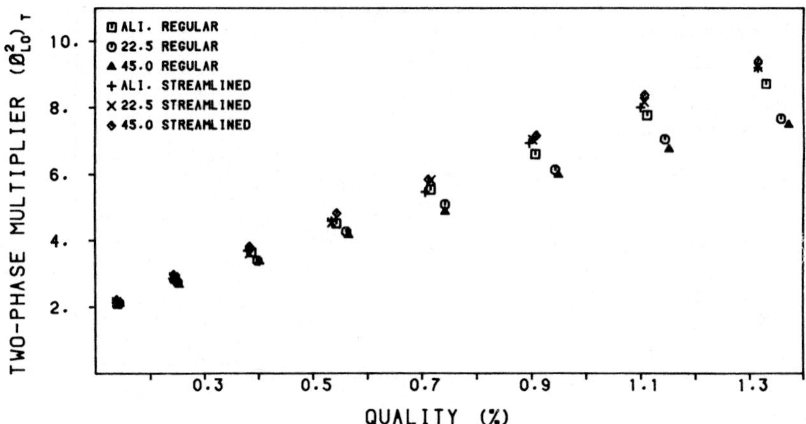

Fig. 20. Total Two-phase Multiplier for 1.26 kg/s Liquid Mass Flow Rate

4. SUMMARY OF RESULTS AND CONCLUSIONS

The pressure drops for the simulated regular bundle and the modified regular bundle (streamlined bundle) have been investigated. The total pressure drop is separated into the frictional pressure drop, the junction pressure drop, and the spacer pressure drop. The frictional pressure drop is the major contributing factor to the total pressure drop, whereas the spacer pressure drop is the least significant one. The results are summarized as follows:

4.1 Single-Phase Flow

The skin friction pressure drops are identical for both types of bundle for various bundle rotations. The proposed equations provided by both Rehme and Malak et al. predict reasonably well the friction factor of the present bundle geometry. The local pressure drops in both junction and spacer planes are larger for the regular bundle. As the misalignment of the bundles increases, the junction pressure drop increases for the regular bundle, but it remains constant for the streamlined bundle.

4.2 Two-Phase Flow

All pressure drops are dependent upon the flow rates of the fluids. More pressure drop is produced for higher fluid velocities. The two-phase frictional pressure drops are identical for both types of bundles in various misalignments. The two-phase junction pressure drop is larger and strongly influenced by the misalignment for the regular bundle. However, it is smaller and constant for various misalignments for the streamlined bundle. Although the streamlined bundle has the smallest junction pressure drop, it has the largest two-phase junction multiplier. This is caused by the difference between the single- and two-phase pressure drop ratio for both sets of bundles. The two-phase spacer pressure drop is smaller for the streamlined bundle. The misalignment of the bundle has no significant effect for either type of bundle. The two-phase total pressure drop, which is the summation of the frictional and local components, is affected by the misalignment of the bundle because of the effect on the junction pressure drop.

The results have been compared with some existing correlations. Generally the homogeneous equation and the correlations suggested by Beattie agree reasonably well with the present experimental data.

In conclusion: the streamlined bundle has shown a better performance than the regular bundle because (i) the streamlined bundle causes the least pressure drops in both junction and spacer area; and (ii) its pressure drops are independent of bundle misalignment.

ACKNOWLEDGEMENTS

The authors would like to thank Mr. J. Woodall and Mr. T. Kennedy for constructing the test-section and the bundles. Also, we greatly appreciate the financial support from Atomic Energy of Canada Limited.

REFERENCES

1. Tong, L.S., Pressure Drop Performance of a Rod Bundle, in Single-Phase Coolant Heat Transfer in Rod Bundles, ed. V.E. Schrock, pp. 57-69, ASME, New York, 1968.

2. Courtaud, M., Ricque, R., & Martinet, B., Etude des Pertes de Charge dans des Conduites Circulaires Contenant un Faisceau de Barreaux, Chem. Eng. Sci., Vol. 21, pp. 881-893, 1966.

3. Le Tourneau, B.W., Grimble, R.E. & Zerbe, J.E., Pressure Drop for Parallel Flow Through Rod Bundles, Trans. ASME, pp. 1751-1758, November, 1957.

4. Rehme, K., Simple Method of Predicting Friction Factors of Turbulent Flow in Non-Circular Channels, Int. J. Heat Mass Transfer, Vol. 16, pp. 933-950, 1973.

5. Malak, J., Hejna, J. & Schmid, J., Pressure Losses and Heat Transfer in Non-Circular Channels with Hydraulically Smooth Walls, Int. J. Heat Mass Transfer, Vol. 18, pp. 139-149, 1975.

6. Salcudean, M., Groeneveld, D.C. & Leung, L., Effect of Flow-obstruction Geometry on Pressure Drops in Horizontal Air-Water Flow, to be published in Int. J. Multiphase Flow.

7. Idel'chik, I.E., Handbook of Hydraulic Resistance, AEC-tr-6630, 1966.

8. Cur, N. & Sparrow, E.M., Experiments on Heat Transfer and Pressure Drop for a Pair of Colinear, Interrupted Plates Aligned with the Flow, Int. J. Heat Mass Transfer, Vol. 21, pp. 1069-1080, 1978.

9. Rehme, K., The Pressure Drop of Spacer Grids in Rod Bundles of 12 Rods with Smooth and Roughened Surfaces, KFK-2697.

10. De Stordeur, A.N. Drag Coefficient for Fuel Element Spacers, Nucleonics, Vol. 19, No. 6, pp. 74-79, 1961.

11. Rehme, K., Pressure Drop of Spacer Grids in Smooth and Roughened Rod Bundles, Nuclear Technology, Vol. 33, pp. 314-317, 1977.

12. Grover, R.B. & Venkat Raj, V., Pressure Drop along Longitudinally-finned Seven-Rod Cluster Nuclear Fuel Ele ıents, Nuclear Engineering and Design, Vol. 58, pp. 79-83, 1980.

13. Murdock, J.W., Two-Phase Measurement with Orifices, Trans. ASME, pp. 419-433, 1962.

14. Chisholm, D., Pressure Gradients During the Flow of Incompressible Two-Phase Mixtures Through Pipes, Venturies and Orifice Plates, British Chemical Engineering, Vol. 12, No. 9, pp. 1368-1371, 1967.

15. Chisholm, D. & Sutherland, L.A., Prediction of Pressure Gradients in Pipeline Systems During Two-Phase Flow, Inst. Mech. Eng. Symp. Two-Phase Flow Systems, Leeds, paper No. 4, 1969.

16. Chisholm, D., Prediction of Pressure Drop at Pipe Fittings During Two-Phase Flow, Proc. 13th International Inst. of Refrig. Congress, Washington, D.C., Aug. 29-Sept. 3, 2, pp. 781-789, 1971.

17. Beattie, D.R.H., A Note on the Calculation of Two-Phase Pressure Losses, Nuclear Engineering and Design, Vol. 25, pp. 395-402, 1973.

18. Lahey, R.T. Jr., Two-Phase Flow in Boiling-Water Nuclear Reactor, NEDO-13388, pp. 59-70, 1974.

19. Messina, W.A., Steam-water Pressure Drop and Critical Discharge Flow-Digital Computer Program, HW-65706, June, 1960.

Multi-Phase Flow and Heat Transfer III. Part A: Fundamentals
edited by T.N. Veziroğlu and A.E. Bergles
Elsevier Science Publishers B.V., Amsterdam, 1984 — Printed in The Netherlands

A MATHEMATICAL MODEL OF A DRUM BOILER

W.J. Turner and T.M. Romberg
CSIRO Division of Mineral Physics
Private Mail Bag 7, Sutherland
New South Wales 2232, Australia

ABSTRACT

A digital computer model of a drum boiler which provides steam for a 500 MW turbine is being developed, based on the network simulation program NAIAD. The economiser, evaporator and superheater are treated as one-dimensional variable area heated pipes; the steam drum and pump by lumped models; and the interaction with the furnace, feedwater system and turbine by time-dependent boundary conditions. The one-dimensional pipe model includes slip between phases, heat diffusion in the pipe wall and a detailed surface heat transfer.

Measurements on an operating boiler of coal to the pulverising mills, feedwater flow, steam flow to the turbine, pressure at the turbine stop valve and drum level have been used to check the model by using the measured coal, feedwater and steam flows as boundary conditions for a calculation of drum level and turbine stop valve pressure. Input-output cross variances of the model and the actual plant are compared.

1. INTRODUCTION

In New South Wales, Australia, a large and increasing fraction of electricity is generated by coal-fired power stations, the remainder coming from hydro stations. In the past, most load following was done with the hydro stations. However, as the fraction of power from hydro stations decreases, this function must also be performed by coal-fired stations. This has stimulated an interest in the control of power station boilers. A mechanistic non-linear model of a 500 MW(e) drum boiler has been developed based on the NAIAD computer program [1-3]. This model and comparisons with measurements on the boiler are presented in this paper.

2. THE PLANT

Unit 8 Wallerawang (Fig. 1) is one of six almost identical 500 MW(e) boilers in the New South Wales Electricity Commission system built to combustion engineering design and installed over the period 1971 to 1980. The major parts of the boiler from a modeller's point of view are the furnace, economiser, steam drum, evaporator, superheater, reheater and feedwater chain. The two-phase mixture from the evaporator tubes enters the upper part of the drum, and is directed into turbo separators by a cylindrical shell mounted inside the drum. Steam leaving

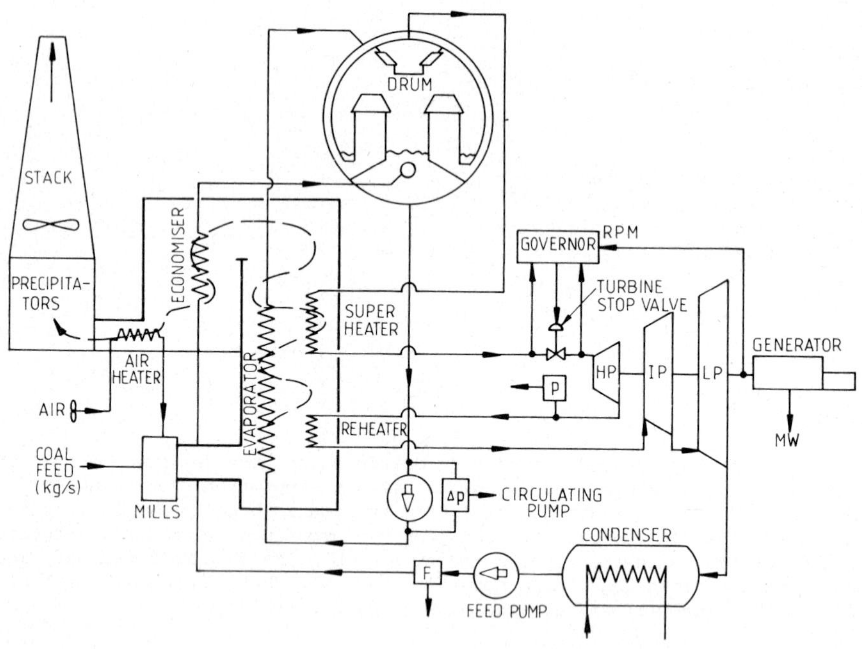

Fig. 1 Wallerawang Power Station

the turbo separators passes through plate dryers on top of the separators into the principal drum space and thence through another set of plate dryers to the superheaters. Economiser water is discharged into the drum through sets of 28 horizontal holes in the feedwater sparge tube above each downcomer. Measurements on Wallerawang 8 of coal to the pulverising mills, feedwater flow, steam flow to the turbine, pressure at the turbine stop valve and drum level were recorded during nominal full power operation in boiler follow mode and later read at 1.6 second intervals. Measurements at 60 second intervals of drum pressure and level, feedwater flow, economiser inlet and outlet temperatures, coal flow, pressure drop across the circulating pumps and generator power were taken from the station log. From this data the mean economiser, evaporator and superheater power were found to be 145, 543 and 325 MW respectively. The mean circulation flow was calculated to be 2243 kg s^{-1} from the pump pressure drop using the pump characteristic. In spite of the economiser outlet temperature being 80°C below saturation, the downcomer temperatures were at saturation. Thus it would appear that the latent heat from 35% of the steam entering the drum from the evaporator passes to the economiser water, heating it to saturation. Presumably at least this percentage of steam leaves the bottom of the turbo separators mixed with the water from the evaporator and is condensed by the feedwater jets in the lower region of the drum.

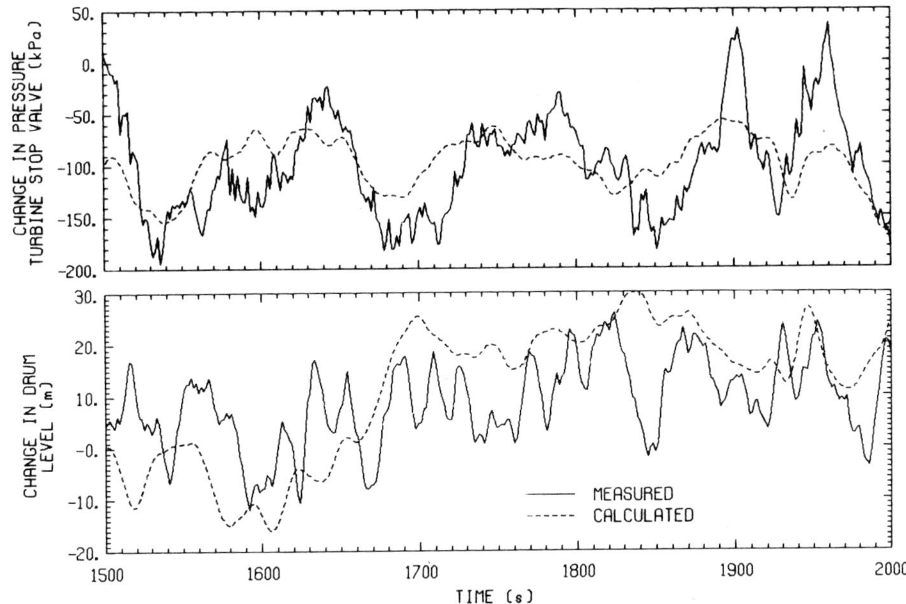

Fig. 2 Comparison of Temporal Variations

A more useful procedure is to compare measured and calculated cross variances of inputs and outputs. This has three advantages: it eliminates any uncorrelated variations such as those due to measurements (provided they do not pass to the plant control system); it summarises the large amount of data; and it shows the lag relationship between signals. The six input-output cross variances of the model and plant were calculated after removing frequencies below 500 μHz and only using the final 4700 seconds of the filtered time series. The plant and model coal-to-drum level and pressure cross variances (Fig. 3) show quite good agreement in the right half plane. The coal-to-power delay of 26 seconds described above was selected to make the peaks in these cross variances coincide. The much lower peak at zero lag in the model coal-to-pressure cross variance is also reflected in the generally smaller fluctuations in pressure shown in Fig. 2. The steam flow cross variances (Fig. 4) are very similar to the coal ones because the action of the control system produces almost identical but opposite sign variations in coal and steam flow. The comparison between the feedwater cross variances (Fig. 5) is interesting, with zones of agreement and others of structural similarity. It appears that the action of the feedwater-to-drum level control is to greatly decrease the cross variance between these parameters.

3. THE MODEL

The NAIAD computer program was written to simulate the thermohydraulic behaviour of water-cooled nuclear power reactors during loss of coolant accidents and has been used for several OECD-NEA-CSNI International Standard Problems [4-8]. However, the code may be used to calculate steady state and transient behaviour in any complex flow networks containing a compressible fluid in single and two-phase states. In the calculation, the flow network is treated as a set of one-dimensional flow paths each of which begins and ends at a connection, e.g. in the boiler model the downcomer begins at the drum and ends at the pump. Thermodynamic equilibrium between the phases is assumed at each flow path cross section. Nodes are placed at the start, finish and at any required intermediate positions in each flow path so that an implicit finite difference method can be used to solve the equations of conservation of mass, momentum and energy for each flow path. All terms are in the conservation equations, including the rate of change and flux of kinetic energy, the rate of change of pressure and gravitational potential energy. There are several types of connections, hence a wide range of network components and boundary conditions can be represented. In the boiler model these are used to represent the drum, pump, economiser inlet flow and steam flow to the turbine.

The walls of flow paths in the network may be heated as in boilers or cooled as in heat exchangers. Radial diffusion of heat in the wall is calculated in up to several cylindrical heat slabs coupled to the hydraulic nodes. In the boiler model, a single heat slab with one node was coupled to each hydraulic node. This represents thermal resistance and storage in the walls. The heat transfer at the wetted surface is calculated by a detailed heat transfer routine which chooses a heat transfer correlation appropriate to the local conditions. In the calculations reported in this paper, non-dryout heat fluxes were calculated from the Dittus Boelter [9], Tong [10] nucleate boiling, and Shrock and Grossman [11] forced convection boiling correlations, and tested for dryout against either the Bernath [12], Biasi [13] or Beattie [14] dryout correlations, depending on the local conditions. A wide range of slip and friction models are available in NAIAD. In this calculation the Smith [15] slip relation and a friction model from Beattie [16], both applicable to annular flow, were used. Thermodynamic equilibrium between phases is assumed at each node and steam properties are read from tables generated by the ASTEM routines [17].

The behaviour at connections is governed by connection equations which may be explicit, linear implicit or non-linear implicit. Initially, a two-region drum model was adopted; however, comparison with the measured data indicated a single region model would be satisfactory and this was used in the calculations reported here. Pressure and enthalpy in the drum are calculated from the drum mass and energy balance assuming that saturated steam flows to the superheater and saturated water to the downcomer. The water level in the drum is determined from the geometry of the drum and the steam quality of the fluid in the drum. The cross-sectional area of the drum at the steam liquid interface is an important parameter in determining the translation of changes in steam volume in the drum to changes in drum level. The area of the turbo separators was included and the area of the outer drum annulus was excluded when calculating this cross-sectional area.

The initial steady state of the boiler was determined by specifying the drum level, drum pressure, feedwater flow, recirculation flow, and economiser, evaporator and superheater power. First the state of the superheater was calculated from the known feedwater flow and steam conditions at the inlet

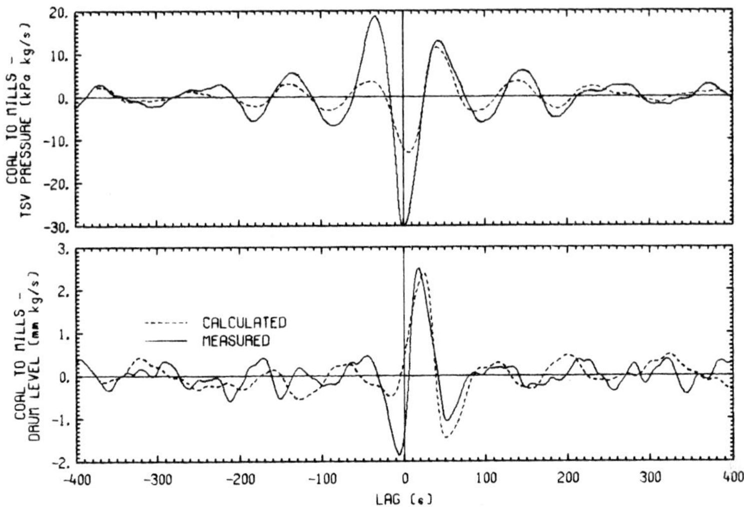

Fig 3. Comparison of Coal Cross Variances

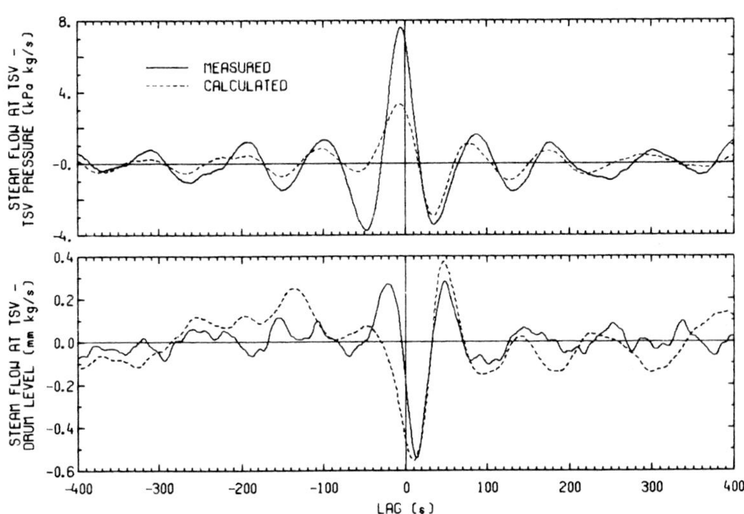

Fig.4 Comparison of Steam Flow Cross Variances

(saturated steam at drum pressure). Good agreement between the calculated and measured pressure drop was obtained. Next, the enthalpy of water leaving the economiser was calculated from a total energy balance of the drum plus evaporator. Thus the state of the economiser could be calculated from the known exit conditions. Similarly, the state of the downcomer was calculated from the known inlet conditions. The pressure drop across the pump was determined from a quadratic fit to the manufacturer's pump characteristic. Since both the flow and the pressure drop are known for the evaporator, a gag coefficient at the inlet was selected to exactly balance the loop at the specified flow.

No attempt was made to model the attemperator system or the reheat cycle. The furnance was represented by a fixed time delay of 26 seconds between coal feed and boiler power. Reasons for this figure are given below.

4. COMPARISON

The data to be used in the comparison were recorded during nominal full power operation in boiler follow mode, i.e. the control system was varying the coal feed to control the turbine stop valve pressure and turbine power, and varying the feedwater flow to control the drum level. The identification of such a system from such data is by no means a simple matter and is much discussed in control literature [18]. Here our aim is much less comprehensive: we wish to show that our model is consistent with the measured data. The difficulties are best illustrated by the fact that when drum level and pressure control systems are added to the model, the calculated initial steady state is stable and no variations occur. In order to compare our model with the measured data, some cause of the variations observed in the plant must be postulated. Some of these variations will be within the section of the plant being modelled, some will be in the control feedback system, and some will be in the measurement process. We have elected to make the most straightforward comparison as a first step: the measured coal, feedwater and steam flow were input to the open loop model of the plant, and the calculated drum level and turbine stop valve pressure compared with the measured data. This is equivalent to an assumption that the source of the variations is outside the portion of the plant being modelled.

This approach presents an immediate difficulty; in open loop, the model, like the plant, is unstable. The enthalpy of saturated steam at high pressure decreases as pressure is raised towards the critical point. Hence an increase in system pressure leads to a decrease in energy flow to the turbine which, in turn, produces a further rise in system pressure. Thus, depending on the round-off error in the initial steady state calculation, system pressure either increases or decreases more or less exponentially. This difficulty was overcome by using an iterative procedure to select constant offsets to be added to the measured coal and feedwater flow such that the open loop model remained finely balanced for the 5700 seconds required. These offsets were of the order of 0.01% and well within the accuracy of the measured data.

A comparison between the calculated and measured pressure and drum level is shown in Fig. 2 for a portion of the simulated period. There is some resemblance between calculated and measured pressures, but the only conclusion one can draw is that, if the model is correct, a major part of variations driving the plant arises within that part being modelled. Not apparent in this figure is a slow fluctuation present in the model outputs but not present in the plant. This is ascribed to slow variations in coal calorific value.

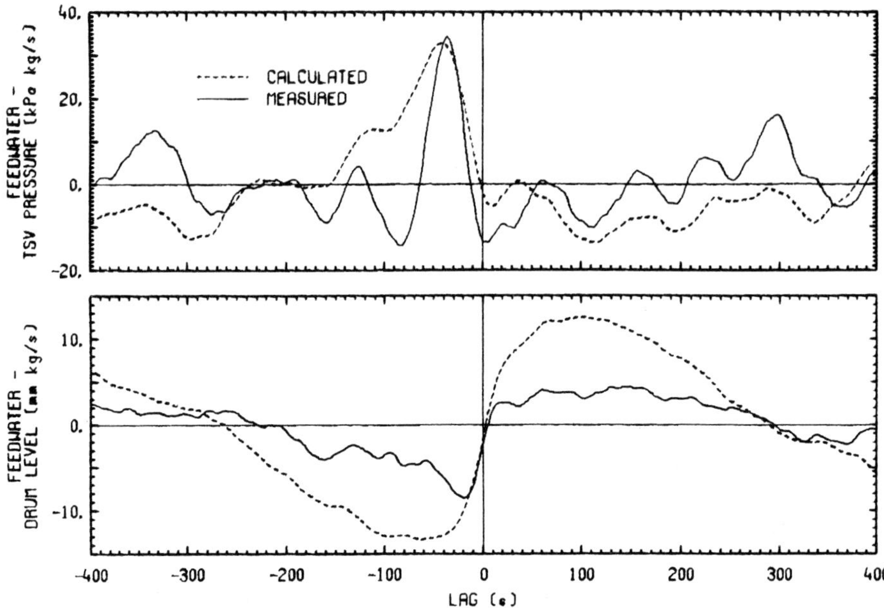

Fig. 5 Comparison of Feedwater Cross Variances

5. CONCLUSION

In general the performance of the model is quite good given that the plant was operated with a feedback loop (boiler follow mode) and the model with no feedback loop. It appears reasonable to ascribe differences to this, hence the results of the comparison are consistent with the model being a good simulation of the open loop plant.

ACKNOWLEDGEMENTS

Access to Wallerawang Power Station and assistance by Mr P.T. Nicholson and Wallerawang Station staff was provided by the Electricity Commission of New South Wales.

REFERENCES

1. Trimble, G.D. and Turner, W.J., "NAIAD - A Computer Program for Calculation of the Steady State and Transient Behaviour (including LOCA) of Compressible Two-Phase Coolant in Networks", Australian Atomic Energy Commission report E378, April 1976.

2. Turner, W.J. and Trimble, G.D., "Calculation of Two-Phase Compressible Flow in a Network", in Computational Methods in Classical and Quantum Physics edited by M.B. Hooper, Advance Publications, Glasgow, 1976.

3. Rainbow, M.T., "An Improved Thermal Model for the Computer Code NAIAD", Australian Atomic Energy Commission report E556, 1983.

4. Turner, W.J. and Trimble, G.D., "Calculation of Transient Two-Phase Flow", OECD-NEA-CSNI Specialist Meeting on Transient Two-Phase Flow, Toronto, 3 August 1976.

5. Turner, W.J. and Trimble, G.D., "Calculation of OECD-CSNI LOCA Standard Problem No. 6 Using the NAIAD Code", CSNI report SINDOC(78) 17, 1978.

6. Turner, W.J. and Trimble, G.D., "Simulation of Depressurisation and Overheating of a Model Nuclear Reactor", Proceedings of SIMSIG-78 Simulation Conference, Australian National University, Canberra, 4-8 September 1978.

7. Turner, W.J. and Trimble, G.D., "Analytical Model for Rewetting in the ERSEC Rig", 18 Congress, International Association for Hydraulic Research, Cagliari, 10-14 September 1979.

8. Turner, W.J., Nucl. Eng. and Design, 60 (1980) 431.

9. Dittus, F.W. and Boelter, L.M.K., University of California Pub. Eng. 2 (1930) 443.

10. Tong, L.S., Boiling Heat Transfer and Two-Phase Flow , John Wiley, New York, p. 118, 1965.

11. Shrock, V.E. and Grossman, L.M., TID 14632, 1959.

12. Bernath, L., Chem. Eng. Progr. Symp. Ser. 56, No. 30, pp 95-116, 1960.

13. Biasi, L. et al., Energy Nucl. (Milan), 14(9) 530, 1967.

14. Beattie, D.R.H., private communication.

15. Smith, S.L., Proc. Inst. Mech. Eng. (London), 184, 1, 1970.

16. Beattie, D.R.H., "Drag Reduction Phenomena in Gas-liquid Systems", International Conference on Drag Reduction, St. John's College, Cambridge, 4-6 September 1974.

17. Moore, K.V., ANCR-1026, 1971.

18. Gustavsson, I., Ljung, L. and Soderstrom, T., Automatica 13, 59-75, 1977.

Multi-Phase Flow and Heat Transfer III. Part A: Fundamentals
edited by T.N. Veziroğlu and A.E. Bergles
Elsevier Science Publishers B.V., Amsterdam, 1984 — Printed in The Netherlands

NUMERICAL SIMULATION OF PHASE SEPARATION IN A BWR STEAM SEPARATOR

Sang S. Wang
Wang Software Service
San Mateo, California, USA

Govinda S. Srikantiah
Electric Power Research Institute
Palo Alto, California 94303, USA

ABSTRACT

The phase separation process in a BWR centrifugal separator under steady operating conditions is studied by applying the EPRI three-dimensional, thermal-hydraulic code ATHOS. The ATHOS code solves the algebraic slip model equations in complex geometries. The slip model employed in this work consists of the mass, momentum and energy equations for the mixture of liquid and vapor in thermal equilibrium, and two algebraic equations for the slips between phases in the axial and radial directions. The axial slip is derived from the drift flux model and the radial slip results from balancing the centrifugal and the drag forces. In the separator configuration, the skimmer, the downcomer and the barrel walls are modeled by use of flow blockages. A variable flow area scheme is implemented to model the axial vanes region. Typical separator operating conditions are used in the calculations and reasonable two-phase flow fields in the entire separator are obtained. The swirling flow motion of the mixture and the radial motion of the liquid phase are observed in the axial vanes region. Furthermore, the calculated void fraction distribution demonstrates a considerable phase separation existing in the separator barrel. The method provides a base for simulating separator performance under varied operating conditions.

1. INTRODUCTION

Steam separators are used in both boiling water reactor (BWR) and pressurized water reactor (PWR) systems to reduce moisture content of wet steam. In a BWR, the separator deck is situated above the reactor core region while in a PWR, it is located near the top of each steam generator. The performance of the steam separators can affect the overall operation and integrity of these systems. For instance, in a BWR, a higher than normal amount of moisture entrained in the steam may result in decreased plant efficiency and premature turbine blade failure. Also, steam carryunder into the reactor downcomer influences the water subcooling at the core inlet, which will in turn affect the thermal-hydraulic performance of the fuel bundles. It is therefore not surprising that reductions of the entrained moisture (carryover), entrained steam in the downcomer (carryunder) and the pressure drop are major concerns in separator design (1).

A steam separator essentially consists of a centrifugal vanes region, liquid removal structures (e.g. skimmers, tangential ports, perforations) and the separator barrel (2). The steam-water mixture entering the separator inlet will undergo a centrifugal separation process to remove most of its water content. The water droplets deposited on the barrel wall will form a swirling liquid film or a water vortex climbing upward on the barrel wall. It is later skimmed off by the skimmers and discharged through downcomers to the water pool surrounding the separators. The dry steam in the core region will exit from the separator outlet into a dryer before being used for turning the turbine. Figure 1 shows a General Electric second-generation steam separator unit along with a detailed axial vanes configuration. The bent vane blades are designed to generate flow rotation which brings forth the centrifugal phase separation. Moreover, the expanding channel also allows the inertial separation effect to remove heavier water droplets. Two groups of skimmers are used to remove the separated liquid phase. The flow restriction ring in the lower downcomer channel can be adjusted to optimize the liquid passage and reduce the steam entrainment in the water pool.

The performance of a separator is a function of a number of operational conditions such as the system pressure, the inlet quality, the total flow rate and the pool water level. The system pressure determines the equilibrium liquid and steam densities as well as their enthalpies. The inlet quality measures the liquid content in the two-phase mixture entering the system. In addition, the geometric configuration actually plays a central role in the physical separation of the liquid and the steam phases. A change of geometry, for instance, from a two-stage skimmers arrangement to a three-stage one, can notably improve separator performance (3).

To adequately predict the behavior of a steam separator, the detailed flow field and phase separation process needs to be calculated. This

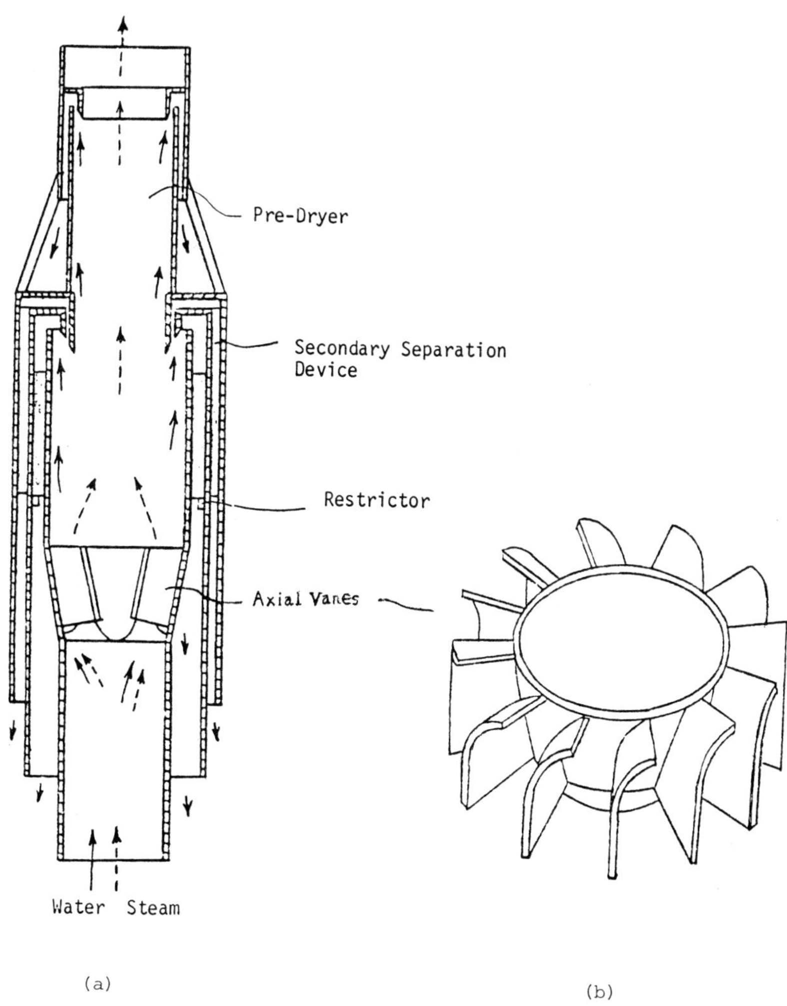

Figure 1. (a) A General Electric second-generation steam separator configuration, (b) A enlarged axial vanes unit.

requires that the separator geometry be modeled and an effective numerical method be employed to compute the two-phase flow evolution in the separator. The recent work of VanderVorst and Steininger (4) showed some genuine progress in calculating the separator flow fields. However, in their axisymmetric calculations, the flow in the centrifugal vanes region and the tangential ports were approximated by simple equations averaged over the circumference.

In this work, the three-dimensional, thermal-hydraulic code ATHOS (5) is applied to model a BWR separator configuration and to compute the two-phase flow field. Flow channels, skimmers and barrel walls are all modeled by flow blockages at the computational cell boundaries. A variable flow area scheme is developed to model the axial vanes structure for redirecting the flow motion. The ATHOS code has been used to simulate the PWR steam generator thermal-hydraulics for several years (5,6). The ATHOS model equations and the numerical method are briefly described in Sections 2 and 4. The newly added algebraic slip model for calculating the phase separation is described in Section 3. This model encompasses a wide range of mixture conditions investigated by Ishii (7). The calculation results presented in Section 5 indicate that our approach can produce realistic flow fields and phase separation.

2. MODEL EQUATIONS

The ATHOS two-phase flow model equations are derived from the fundamental phasic conservation equations (8). In this study, the five differential equations to be solved are the conservations of mixture mass, momentum and energy at steady-state:

a) Mixture mass conservation

$$\nabla \cdot (\rho \vec{V}) = 0 \tag{1}$$

b) Mixture circumferential momentum conservation

$$\nabla \cdot (\rho \vec{V} u r) = -\frac{\partial P}{\partial \theta} \tag{2}$$

c) Mixture radial momentum conservation

$$\nabla \cdot (\rho \vec{V} v) = -\frac{\partial P}{\partial r} - \rho \frac{(ur)^2}{r^3} + \frac{1}{r} \frac{\partial}{\partial r}\left(r \frac{\rho_\ell \rho_g \alpha V_{\ell g}^2}{\rho (1-\alpha)} \right) \tag{3}$$

d) Mixture axial momentum conservation

$$\nabla\cdot\left(\rho\vec{V}w\right) = -\frac{\partial P}{\partial z} - \rho g - f_z - \frac{\partial}{\partial z}\left[\frac{\rho\rho_g\alpha w_{gm}^2}{\rho_l(1-\alpha)}\right] \tag{4}$$

e) Mixture energy conservation

$$\nabla\cdot\left(\rho\vec{V}h\right) = -\frac{\partial}{\partial z}\left[\rho_g\alpha h_{lg} w_{gm}\right] \tag{5}$$

where ρ is the mixture density, h is the mixture enthalpy, P is the local pressure, $\vec{V}$ is the velocity vector with components (u,v,w) in the (θ,r,z) directions, g the gravitational acceleration, void fraction, h_{lg} latent heat, f_z is the axial flow friction, v_{lg} is the radial relative velocity between the liquid and the gas phases and w_{gm} is the axial relative velocity between the gas and the mixture. In Equations (1) to (5), we assume that there is no heat and mass transfer between phases. The liquid and the steam are maintained in thermal equilibrium at the system pressure. The physical properties, such as viscosity, thermal conductivity, etc. in both phases are considered constant in the calculation. The diffusion and the turbulence effects are neglected. The slip velocities v_{lg} and w_{gm} will be derived in the next section. To close these equations, two auxiliary relations are needed, namely,

f) Mixture density

$$\rho = \frac{\rho_l\rho_g}{\rho_g + (\rho_l-\rho_g)(h-h_l)/h_{lg}} \tag{6}$$

g) Mixture void fraction

$$\alpha = \frac{\rho h - \rho_l h_l}{\rho_g h_g - \rho_l h_l} \tag{7}$$

where the densities of the liquid and the steam (ρ_l, ρ_g) and the enthalpies of the liquid and the steam (h_l, h_g) are taken to be determined by the pressure.

3. CORRELATIONS

The important correlations used in this study include the axial slip correlation, the drag coefficient correlations related to the radial slip equation and the axial friction correlation. The Lellouche and Zolotar (9) slip correlation applicable to both the bubbly and the dispersed flows is used to calculate the vapor drift flux velocity w_{gj}. The axial relative velocity between the steam and the mixture can be determined from w_{gj} as follows,

$$w_{gm} = \frac{\rho_l w_{gj} + (C_o - 1)\rho_l w}{C_o \alpha \rho_g + (1 - C_o \alpha)\rho_l} \tag{8}$$

where C_o is the distribution parameter. Detailed derivation of Equation (12) can be found in Reference (5).

The radial slip velocity between the liquid and the gas phases is derived from the two-fluid momentum equation (8) in the radial direction

$$\alpha_k \frac{dP_m}{dr} + \alpha_k \rho_k \frac{u_k^2}{r} - M_{ik} = 0 \tag{9}$$

where k indicates phase, P_m is the mixture pressure, u_k is the circumferential velocity for the k phase and M_{ik} is the interfacial drag force. We assume no slip exists in the circumferential direction, u_k is thus replaced by u, the mixture circumferential velocity. By adding the phase momentum equation (9) and using $\Sigma_k M_{ik} = 0$, we obtain

$$\frac{dP_m}{dr} = -\rho \frac{u^2}{r} . \tag{10}$$

The drag force acting on a spherical particle at steady-state is given by

$$F_d = -\frac{1}{2} C_D \rho_g v_{lg} |v_{lg}| A_d , \tag{11}$$

where C_D is the drag coefficient and A_d is the projected area of the particle. Integrating M_{ik} over the volume associated with a particle, we can link M_{ik} to F_d by,

$$F_d = M_{il} V_d / (1 - \alpha) , \tag{12}$$

where V_d is the particle volume. Substituting Equations (10) - (12) into (9) results in the radial relative velocity between the liquid and the gas phases:

$$v_{lg} = \left[\frac{8\alpha(\rho_l - \rho_g) r_d}{3 C_D \rho_g r} \right]^{1/2} u \tag{13}$$

where r_d is the particle radius and r is the radial distance. If ρ_g is dropped, Equation (13) is reduced to the terminal velocity of a particle moving in a constant centrifugal force field u^2/r, which is balanced by an interfacial drag force F_d. For the flow at low Reynolds numbers where the Stokes drag law is applicable, Equation (13) can be reduced to the radial relative velocity used by Bennett and Kondic (10) in their momentum flux model.

The drag coefficient in Equation (13) depends on the droplet size as well as the mixture flow conditions. For a multiple droplet system such as dispersed flow, Ishii and Chawla's idea of mixture viscosity (7) is used to determine C_D. That is, the deformation of the local flow field due to the existence of other particles can affect the present particle motion similar to an increase of flow viscosity. Ishii and Chawla, after comparing with a substantial amount of experimental data, reached the following set of formulae for the drag coefficient in a dispersed droplet flow:

$$= \begin{cases} \dfrac{24}{R_e}\left(1 + 0.1 R_e^{0.75}\right) \alpha^{-2.5} & \text{for } R_e \leq 1000, \text{ undistorted particles} \\ 0.45 \left\{ \dfrac{1 + 17.67\left[f(1-\alpha)\right]^{6/7}}{18.67 f(1-\alpha)} \right\}^2 & \text{for } R_e > 1000, \text{ undistorted particles} \\ \dfrac{4}{3} r_d \sqrt{\dfrac{u^2(\rho_l - \rho_g)}{\sigma r}} \left[\dfrac{1 + 17.67\alpha^{2.6}}{18.67\alpha^3} \right]^2 & \text{for distorted particles} \end{cases} \tag{14}$$

where σ is the surface tension, function $f(x) = (1-x)^{0.5}(1-x/0.62)^{1.55}$, Reynolds number is defined to be $Re = 2r_d \cdot \rho_g \, |v| / \mu_g$ where μ_g is the steam viscosity. Transition from undistorted small droplets to large distorted droplets occurs when $N_\mu \geq N_t$, where

$$N_\mu = \mu_g / \{ \rho_g \sigma [\sigma r / u^2 (\rho_l - \rho_g)]^{0.5} \}^{0.5},$$

$$N_t = 36\sqrt{2} (1 + 0.1 R_e^{0.75}) / R_e^2 . \qquad (15)$$

Notice that the drag coefficient is independent of the viscosity for the large distorted droplet system. Equations (13) and (15) are used to solve for v_{lg}, which will then be used in Equations (1) - (8) to solve for the dependent variables in the system.

For the axial flow friction, ATHOS uses Thom's (11) correlation with the friction force defined to be $f_z = f A w | w | \phi$, where ϕ is a two-phase multiplier and f is the standard single phase friction factor. Since there is no internal structure in the separator barrel, this axial flow friction is essentially used for modelling the restriction ring in the downcomer channel.

4. NUMERICAL METHOD

The ATHOS numerical method for solving differential equations (1) - (5) is an implicit under-relaxation technique developed by Singhal and Spalding (12). In this method, the pressure and the enthalpy are defined at the center of each computational cell while the velocity components are at the centers of cell surfaces. The differential equations are integrated over the control cells to generate the finite-difference equations. A set of pressure correction equations are derived by combining the cell mass and momentum balances. These equations are solved iteratively as follows: Starting from the initial flow field predictions, the new pressures are obtained by solving the pressure correction equations for the whole domain. The rest of the variables are then solved for by back substitution. These steps are repeated until a fully convergent solution is obtained under the constraints of boundary conditions. The details of the method including the fully implicit time differencing scheme for transient calculations are given in Reference (5).

5. SIMULATION RESULTS

A non-uniform 8x8x22 computational grid in the (θ,r,z) cylindrical coordinate system is used in the present calculations. The BWR second-generation steam separator configuration is set up on this grid as shown in Figure 2. A number of cell boundaries are purposely aligned with the solid walls in the separator to achieve a better resolution of the flow field. The vertical and horizontal solid baffles that form the barrel wall, the skimmers and the downcomer channel walls are modeled using large concentrated flow resistances as the flow blockages. That is, with zero porosity on the cell boundary, flow will not be able to penetrate but only to move in parallel to the wall surfaces. These flow blockages are set during the geometry preparation step prior to the flow calculation step.

A variable flow area scheme has been devised to control the flow direction in the vanes region. By blocking off the horizontal flow area where vane blades exist and adjusting the vertical flow areas according to the relative position of the blade surface in the cells, the flow can be redirected effectively. This is illustrated in Figure 3 where the horizontal flow areas Ah1, Ah2 and Ah3 are blocked while the vertical flow areas Av1, Av2 and Av3 are decreased to represent a concave blade and Avx1, Avx2 and Avx3 are increased to represent a convex blade. The flow in this vane channel is forced towards the positive θ direction. The degree of the redirection is determined by the flow area variations. The vertical flow areas in the calculation are taken from the actual vane configuration.

The boundary conditions used in the calculations are specified in the following manner: At the separator inlet, the mixture velocity W_i, the quality x_i and the mixture enthalpy h_i are specified. At the upper separator outlet the drum pressure P_d is specified and at the downcomer exit the pressure specified is the drum pressure plus the hydrostatic head due to the pool water level.

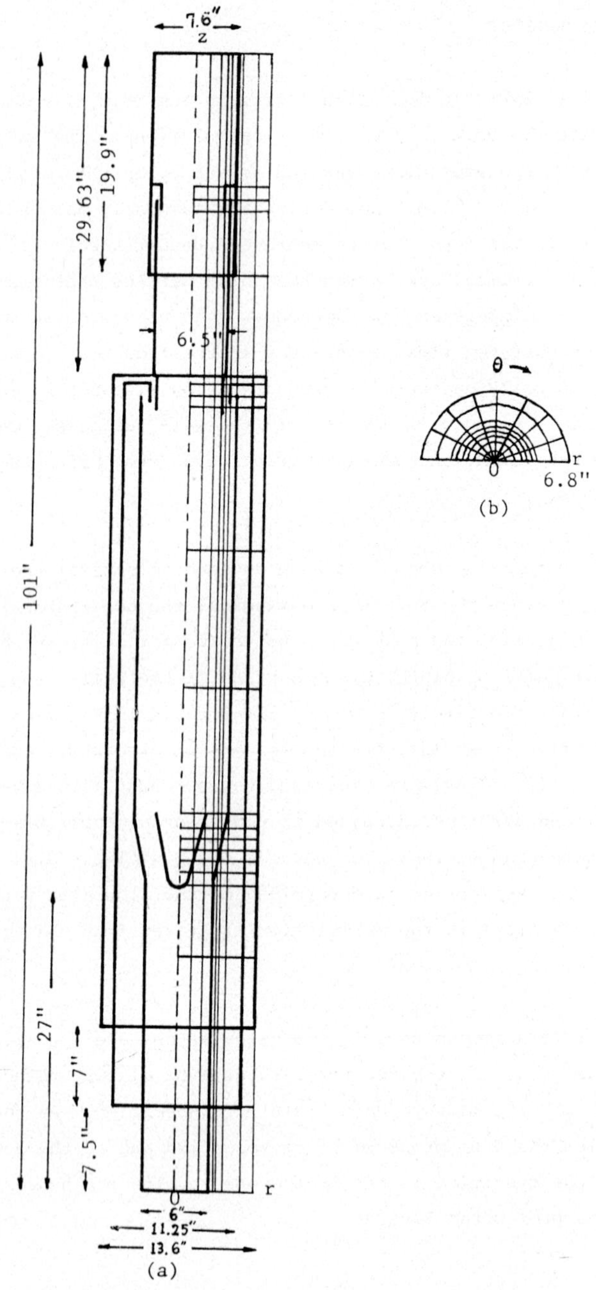

Figure 2. (a) The BWR steam seperator configuration under ATHOS simulation, heavy lines represent the seperator geometry and light lines indicate the computational grid. (b) The cross section grid.

The separator operating conditions incorporated are:

System pressure $P_d = 6.9 \times 10^6\ N/m^2$

Inlet quality $X_i = 8\%$

Total flow rate $G = 53\ Kg/sec$

The mixture inlet velocity is calculated from:

$$w_i = \frac{G\left[\rho_g + X_i(\rho_\ell - \rho_g)\right]}{A_i\,\rho_\ell\,\rho_g}$$

where A_i is thecross-sectional flow area at the inlet.

The results of the flow field calculation are schematically presented in Figure 4. A close-up view of the flow in the neighborhood of the axial

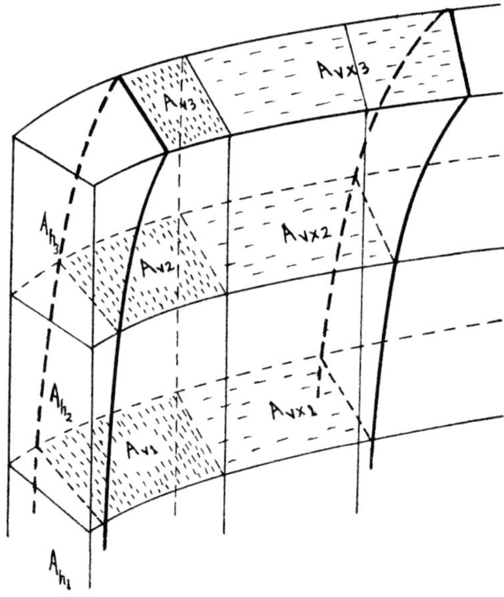

Figure 3. Flow area variation generated by vane blades intersecting the computational cells.

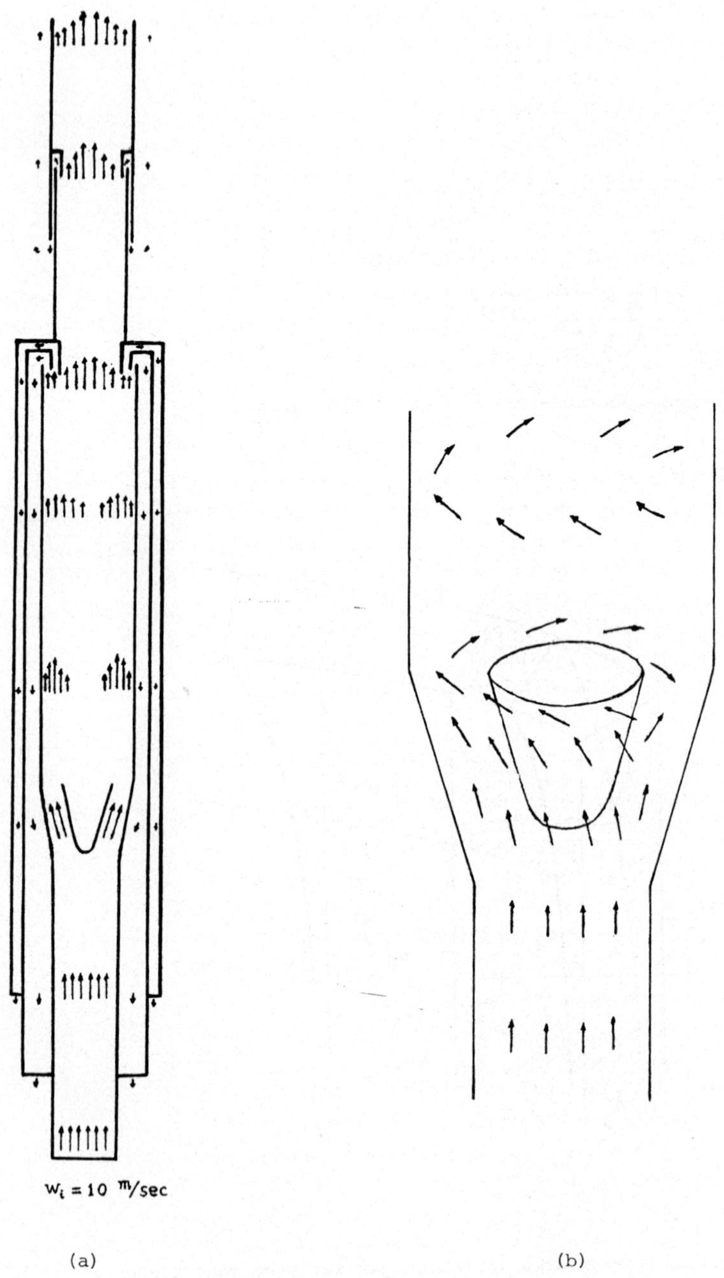

Figure 4. (a) A schematic flow field calculated in the BWR separator by the ATHOS code, (b) Enlarged view of flow field near vane region.

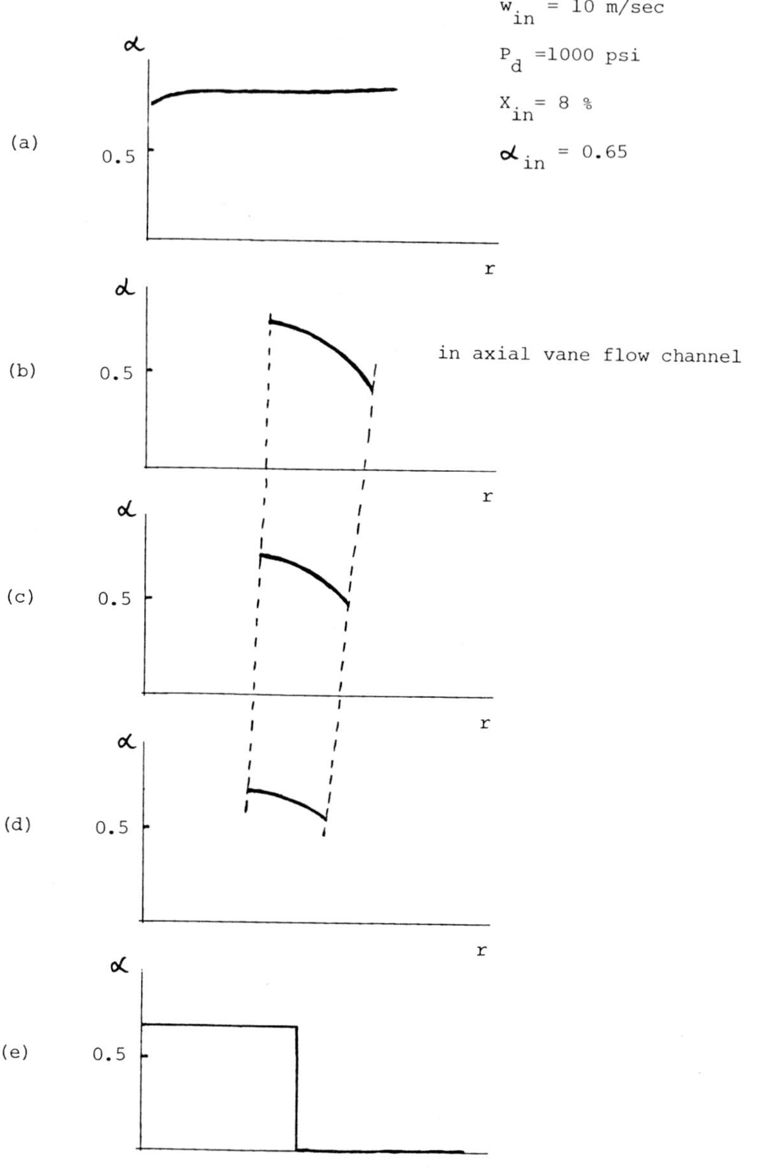

Figure 5. The void fraction distributions at (a) z= 2.42 m, (b) z= 0.83 m, (c) z=0.80 m, (d) z= 0.78m and (e) z=0.05m.

vanes region is also shown. For clarity, the velocity vectors are only plotted at a dozen selected locations. As shown in the figure, the inlet flow which is parallel is transformed into an increasingly swirling flow by the axial vanes. The upward moving flow is seen to become collimated when it passes the lower skimmers region. Two streams of flow also begin to enter the skimmer channels. Thus the calculation has produced a reasonable flow resolution. The pressure drop across the separator which is an important design variable is calculated to be 7.5 psi for this design. This compares very well with the experimental data of Wolf and Moen which is 8 psi (3).

The phase separation result is best demonstrated by void fraction distributions at various axial locations below, within and above the axial vanes region. Figure 5(a) and (e) are the void fraction distributions calculated at the outlet and the inlet locations. They are flat when compared against the void distributions in the axial vanes region. The latter show a strong radial dependence in Figure 5(b)-(d). For instance, at $z = 0.83$m, the void fraction varies from 0.8 on the inner side of the channel to 0.45 at the outer side of the channel, which means more liquid phase already exists at the outer rim of the channel due to the centrifugal separation. Consistently, the radial liquid velocity is calculated to be increasing from 0.5 m/sec at $z = 0.78$m to 1.6 m/sec at $z = 0.83$m. This demonstrates that the radial momentum and slip equations (3) and (13) have resulted in the generation of phase separation.

Upon varying the pool free surface level from 2.25m to 0.5m, we found that the average downcomer discharge velocity increases from 0.32 m/sec to 0.43 m/sec. This trend is again consistent with our expectation.

6. CONCLUSION

In this paper a method is presented for the calculation of the internal two-phase flow field and the phase separation process in a BWR steam separator, by the application of the ATHOS code. This method is based on the development of a new radial slip model applied locally and a detailed representation of the separator internal geometry including the axial vanes. The results of the calculation demonstrate that the internal flow field is adequately resolved and good phase separation is obtained. The method is general enough for application to simulate separators of other geometries.

ACKNOWLEDGEMENTS

Useful discussions with D. Steininger and J. Naser of EPRI about the separator configuration and operation and consultations with A. Singhal and L. Keeton of CHAM of North America on the ATHOS program are highly appreciated. The first author is supported by EPRI RP2353-1.

REFERENCES

1. Carson, W. R. and Williams, H. K., "Methods of Reducing Carry-Over and Reducing Pressure Drop Through Steam Separators," EPRI report NP-1067, November, 1980.

2. Moen, R. H., Kudirka, A. A., Fitzsimmons, G. W. and Burley, E. L., "Advances in Boiling Water Reactor Steam Separation Systems," ASME Paper No. 69-WA/NE-5, 1970.

3. Wolf, S. and Moen, R. H., "Advances in Steam-Water Separators for Boiling Water Reactors," ASME Paper No. 73-WA/PWR-4, 1973.

4. VanderVorst, M. J. and Steininger, D. A., "Calculation of Phase Separation by Centrifugal Force," ASME Paper No. 82-FE-5, 1982.

5. Singhal, A. K., Keeton, L. W., Spalding, D. B. and Srikantiah, G. S., "ATHOS-A Computer Program for Thermal-Hydraulic Analysis of Steam Generators," Vols. 1-3, EPRI report NP-2698-CCM, October 1982.

6. Srikantiah, G. S. and Singhal, A. K., "Modeling and Simulation of Recirculating U-Tube Nuclear Steam Generators," <u>10th IMACS World Congress on System Simulation and Scientific Computation Proceedings</u> Vol. 4, pp 133-135, 1982.

7. Ishii, M. and Chawla, T. C., "Local Drag Laws in Dispersed Two-Phase Flow," ANL report, NUREG/CR-1230, ANL-79-105, December, 1979.

8. Ishii, M., "One-Dimensional Drift Flux Model and Constitutive Equations for Relative Motion Between Phases in Various Two-Phase Flow Regimes," ANL-77-47, October, 1977.

9. Lellouche, G. S. and Zolotar, B. A., "A Mechanistic Model for Predicting Two-Phase Void Fraction in Vertical Tubes, Channels and Rod Bundles," EPRI NP-2246-SR, 1982.

10. Vennett, R. R. and Kondic, N. N., "Momentum Flux Model for Liquid-Vapor Separation," National Heat Transfer Conference, 1977.

11. Thom, J. R. S., Int. J. Heat and Mass Transfer, 7, 709, 1965.

12. Singhal, A. K. and Spalding, D. B., "Multiphase Transport Fundamentals, Reactor Safety Applications, Vol. 1, 373, Hemisphere, Washington, D.C., 1980.

Multi-Phase Flow and Heat Transfer III. Part A: Fundamentals
edited by T.N. Veziroğlu and A.E. Bergles
Elsevier Science Publishers B.V., Amsterdam, 1984 — Printed in The Netherlands

THE HIGH VELOCITY MOISTURE SEPARATORS

P. Talleu and J.P. Cerdan
Research and Development Division
Electricité de France
78400 Chatou, FRANCE

ABSTRACT

One or several steam drying stages are necessary during steam expansion in the turbines of LWR nuclear power stations. The Research and Development Division of ELECTRICITE DE FRANCE has undertaken to design a new type of dryer, the High Velocity Separators. Several aerodynamic calculations and preliminary tests on a steam-water separator test rig, were carried out successfully. The construction of a large demonstration H.V.S. unit was decided and two industrial prototypes have been installed on a 900 MW nuclear power plant.

Many promising test results were obtained on these prototypes. They concern both H.V.S. pressure drop and separation efficiency. In addition, the reliability of H.V.S. technology is now demonstrated.

Owing to the H.V.S. size, the possibility of drying steam directly at the H.P. turbine exhaust will remove any erosion-corrosion problem in the connecting pipes and carbon steel will be available for their construction. Lastly the pipe layout will be simplified, and their length reduced.

All these performances make the High Velocity Separators very attractive and suggest they will be developped on an extensive scale.

1. INTRODUCTION

Electric energy production by the way of nuclear reactors leads generally to poor thermodynamic characteristics at the steam turbines inlet. Since moistures rates in the turbine last stages are to remain within acceptable limits (and this as well for efficiency than for resistance to erosion) one or several steam dryings during the expansion have to be considered. The separation of the liquid phase is generally carried out in large external separators often associated with superheating devices.

The performance of the steam-water separators is decisive for the power plants output and their disponibility (risks of erosion-corrosion of superheaters bundles). On another hand a reduction of the separators size would lead to important profits not only on this equipment but also on the piping and machine room itself.

The Research and Development Division of "ELECTRICITE DE FRANCE" has therefore undertaken to design, in GENNEVILLIERS power station, a test rig reproducing the actual conditions of separators operation. The types of separators corresponding to the most widely used models have been tested under joint programs with turbine manufacturers. An observation of the results obtained with classical separators shows that separation efficiencies remain acceptable with low steam velocities (a few meters per second in the range of usual operating pressures) but strongly decrease when steam velocities are increasing (references 1, 2 and 3).

In large nuclear power plants steam drying at the exhaust of the High Pressure turbine needs then bulky and expensive vessels, due to the considerable steam flows (200 m^3/s for a 900 MWe power station).

Moreover, it is difficult in this case to reach sufficiently good velocities distributions inside the separators.

So the Research and Development Division of E.D.F. attempted to design a new type of separator, the "High Velocity Separator" with very small dimensions and able to operate with velocities more than ten times higher than those for classical separators (reference 4).

Preliminary tests were conducted on the GENNEVILLIERS wet steam test rig and promising results were achieved. Then calculations were worked out, concerning droplets trajectories and steam-flows inside the H.V.S., in the aim to determine an optimum apparatus geometry giving good droplets separation efficiencies and low pressure drops. Aerodynamic part of calculations and design has been developped with collaboration of "Ecole Centrale de LYON", under Professeur K.D. PAPAILIOU'S authority. Theoretical results have been confirmed by air experiments on scale three models. Lastly, performances of the selected geometry were controlled by wet steam tests in GENNEVILLIERS.

Then it was decided to design an industrial prototype in cooperation with ALSTHOM-ATLANTIQUE and STEIN-INDUSTRIE. Two H.V.S. units have been installed during July 1980 in BUGEY power plant at the high pressure exhaust of a 900 MW turbine. Complete tests were planned, and the results have confirmed the very good performances of the H.V.S. compared to conventional separators. It is now possible to look to the H.V.S. integration in projected power stations.

In addition, H.V.S. technology will turn two drying cycles competitive for nuclear power stations (references 5, 23).

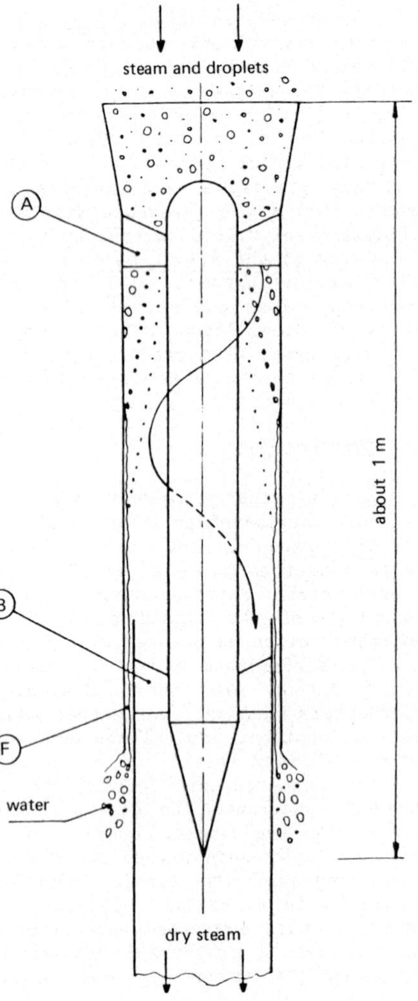

2. MAIN PRINCIPLE OF H.V.S. OPERATION

High Velocity Separators are a special type of centrifugal separator. However, the design of conventional separators often leads to serious water recarry over and, on the other hand, their size and geometry do not make it possible for small drops to be always caught by the walls.

In High Velocity Separators technique, the water elimination is improved by extraction of steam. The amount of extracted steam is generally well adapted to the needs of one of the feeding water reheaters. On the other

FIGURE 1

HIGH VELOCITY SEPARATOR

hand, an improvement in water drop centrifugation can be obtained through a reduction in separator diameter. H.V.S. are therefore made of separating cells fitted in parallel within the connecting pipes of the turbines. A cell operating principle is shown in Figure 1.

At the dryer inlet, a cascade (A) imputes a swirling movement to the water/steam mixture ; the water is thrown into the walls and discharged into an annular space (F), from which a small part of steam is also drawn off. Pressure drop, which is one of the drawbacks of centrifugal separators, is reduced by a second cascade (B) fitted at the cell outlet, in the aim to straighten the flow and to recover the steam kinetic rotation energy. The pressure drop inside the cell itself and inside the down-stream pipes is thus considerably reduced (uniform velocity distribution).

Selection of separation cell dimensions is the result of a compromise. Indeed if the cell diameter is increased, it is necessary, in order to maintain correct centrifugation, to increase either the steam velocity, and hence the pressure drop, or the length of the unit, hence its bulk and pressure drop owing to the comparatively large energy losses in the space between the two cascades. If the cell diameter is too much decreased, the number of cells necessary will be prohibitive. The best solution consists in using cells of a reduced diameter and short length, and assembling these inside the connecting pipes, just at the HP turbine exhaust.

3. THEORETICAL PREDICTION OF H.V.S. PERFORMANCES

For designing High Velocity Separators, several calculations have to be worked out in order to determine the optimum cells geometry (corresponding to the best compromise between droplets separation and pressure drop) and to predict the performance of a given design. Such calculations concern droplets trajectories predictions, blades profiles determinations, blades loading analysis, and losses computations.

3.1. Drop Centrifugation

A computer calculation code was perfected to determine the droplets trajectories in a steam flow. It was thus possible to study a large number of cells of various dimensions. The parameters to be optimized were, on the one hand, the radii r_1 and r_2, and on the other, variation of the deflection angle at the outlet of the distributing cascade (figure 2).

For a given variation of θ, as a function of the radius, the axial and tangential velocities variation laws are fixed. These two absolute speed components are not independent, and should comply with the radial momentum equation.

By stating the following hypotheses :

- friction and gravity forces not taken into consideration,
- stationary and axi-symmetrical flow, at the outlet of each cascade,
- radial velocity components negligible,

the following simplified radial momentum equation is obtained :

$$C_a \cdot \frac{\partial C_a}{\partial r} + \frac{C_u}{r} \cdot \frac{\partial (r \cdot C_u)}{\partial r} = 0$$

C_a : axial velocity.

C_u : tangential velocity.

FIGURE 2

INLET CASCADE BLADE PROFILE

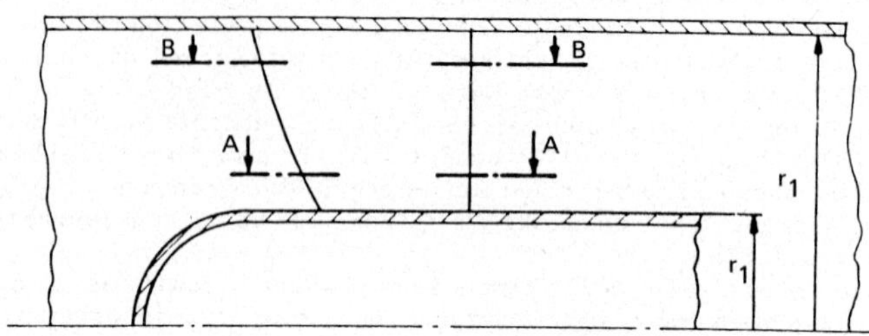

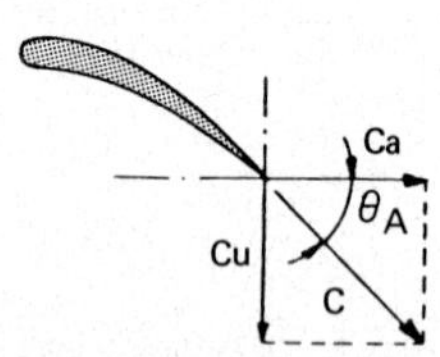

Cross section plane A - A

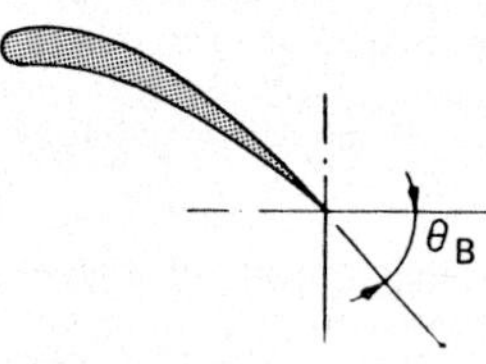

Cross section plane B - B

C : absolute velocity

C_a : axial velocity

C_u : tangential velocity

θ : outlet angle or deflection angle

Moreover, considering the mass flow continuity equation, for a given variation in one of the velocity components, it is possible to calculate the variation of the other, and to determine the deflection angle $\theta = \text{Arctg}\ \frac{C_u}{C_a}$.

The types of flow studied were as follows :

- Free vortex : C_a = constant (type n° 1) :

  In this case :
  $r.C_u$ = constant, $\theta = \text{Arctg}\ \frac{K}{r}$

- Tangential velocity linear variation : $C_u = A - k.r$ (type n° 2) :
  C_a becomes :

  $$C_a = A\ (\frac{B}{A^2} + 6\ \frac{k}{A}\ .\ r - 2\ (\frac{Ak}{r})^2 - 2\ \ln r)^{1/2}$$

- Tangential velocity parabolic variation : $C_u = A - k.r^2$ (type n° 3) :

  An equation similar to the above in r^2, r^4, lnr, is obtained for C_a.

- Constant deflection : θ = constant (type n° 4) :

For this kind of flow, C_a and C_u become :

$$C_a = \frac{D}{r^{\alpha}} , \quad C_u = \frac{k.D}{r^{\alpha}}$$

with :

$$\alpha = \frac{k^2}{k^2 + 1}$$

and :

$$k = \tan \theta$$

In each case, droplet trajectories at the distributing cascade outlet were calculated by using a drag coefficient C_x as a function of REYNOLDS' number Re. Several values of C_x available for undeformable spherical particles (references 6 and 7) were considered :

$$C_x = 0.44 \qquad \text{if Re} > 1000$$

$$C_x = \frac{24}{Re} + \frac{4}{Re^{1/3}} \qquad \text{if } 3 < Re < 1000$$

$$C_x = \frac{24}{Re} + 4.31 - 0.719 \; . \; Re + 0.068 \; Re^2 \qquad \text{if Re} < 3$$

Thus, it was possible, for several droplet sizes to determine the separation efficiencies of many types of distributing cascades (reference 8). Some typical results are shown on tables 1 and 2.

Table 1

Separation performance for an axial distance of 1 m

r_1 = 100 mm
r_2 = 25 mm constant inlet velocity : 50 m/s

Séparation efficiency for various drop sizes Rg*

Type	r_1 (°)	r_2 (°)	Absolute C (m/s)	Rg* = 1µ (%)	Rg* = 2.5 µ (%)	Rg* = 5µ (%)
1	53	17	83 to 52	3	9	28
2	36	34	100 to 57	6	38	100
3	34	41	109 to 58	12	58	100
4	40	40	105 to 58	8	43	100

Table 2

Increase in performance obtained by reducing cell dimensions

$(\frac{r1}{r_2} = 4)$

r_1 (mm)	Rg = 5 µ		Rg = 10 µ	
	d* = 0.5 m (%)	d* = 1 m (%)	d* = 0.5 m (%)	d* = 1 m (%)
150	45	80	74	100
125	58	95	88	100
87	90	100	100	100

3.2. Blading Load

For each studied type of flow, and each cell dimension, it is necessary to examine the aerodynamic load conditions for the cascades, and especially at the blading roots of the outlet cascade.

3.3. Blading Profile Design

Due to the above considerations concerning separation efficiency and blading loads, the n° 3 type of flow ($C_u = A - kr^2$) was retained.

The numerical values of deflection angles, steam velocities and cascade dimensions were fixed by several conditions : separation efficiency, pressure drop, sizing optimisation, erosion considerations etc...

Then the blading profiles of distributing and straightening cascades were chosen in order to get good aerodynamic performance, in a large range of operating conditions.

3.4. Loss Determination

The profile loss and secondary losses of the distributing and straightening cascades were estimated in using existing correlations (references 9 and 10).

Three-dimensional boundary layer calculations were carried out, on the hub and casing, giving losses for the pipe located between both cascades (references 11 and 12). In particular, allowing for the thickness of the boundary layer obtained by the above calculations, it is possible to dimension the annular space dividing the flow into main flow and steam bleeding.

Through-flow calculations, using complete radial momentum equations were performed. On the one hand, this checked out that the simplified radial balance calculation hypothesis was sufficiently accurate, and, on the other hand, enabled study of various cases of distributor feeding (heterogeneity of the flow pattern in front of the cells. All results concerning these cases give practically the same velocity distributions between both cascades, and confirmed the correct operation of the equipment.

3.5. Precise Determination of the Distributor Outlet Angle

A blade to blade calculation code perfected at the "Ecole Centrale de LYON" was used in order to determine flow conditions all around the cascades blades and then to calculate the corresponding outlet angles.

Resolution was obtained by a singularity method in the case of subsonic, adiabatic and friction-free flow (reference 15).

In this calculation code the variation in axial velocity is taken into account by introducing a correlative variation of the flow tubes geometry in the interblade channel.

4. PRELIMINARY AIR TESTS

The purpose of these tests was, on the one hand, to check the theoretical values of the pressure drop coefficients and velocity distribution at the outlet of the cascades, and, on the other hand, to determine optimum designs at the inlet and outlet of the separation cell. In order to respect REYNOLDS number similarity, the model was a 3/1 scale model. It consisted of several assembled elements permitting to test various configurations (figure 3). The variable elements were the distance L between both cascades, the position of the inlet nacelle (O) and the shape of the outlet diffuser (D).

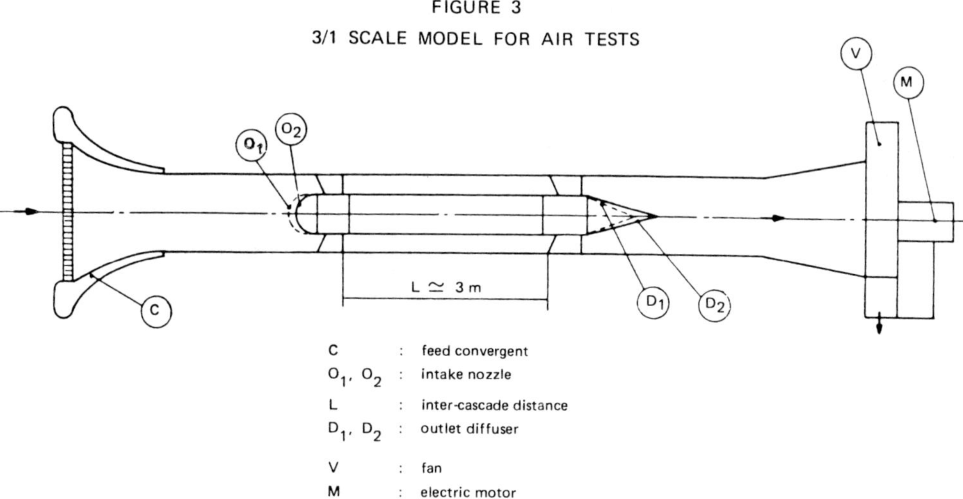

FIGURE 3
3/1 SCALE MODEL FOR AIR TESTS

A test without hub was also performed. This showed an increase of 70 % in pressure loss for a pipe length of 2.5 m.

The pressure loss results were slightly higher than theoretical values. For the final design the pressure drop reaches about 0.5 times the dynamic pressure at the cell inlet (references 13 and 14).

Velocity distribution measurements, although only slightly exploitable due to the lack of experimental means, confirmed the mean theoretical value of the distribution outlet angle, and showed adequate conservation of the velocity distribution within the inter-cascade space (reference 16).

Subsequent to certain differences between the theoretical predictions and the results of these air tests, a study concerning dihedral effects (reference 17) on the flow angle at the outlet of the first cascade was conducted. This showed that dihedral effects could be important, and considerably modify the desired outlet angle of the distributor.

The construction of a new distributor, with blade profiles stacked up on the trailing edge, was therefore undertaken.

Tests concerning this new configuration demonstrated, for the distributor outlet angle, a clear improvement between theory and practice (reference 18).

It was also attempted to visualize some parts of the flow. It was confirmed that the distributor was adequately supplied. Interesting results concerning flow in the cascade were obtained. These results showed that flow within the inter-blade space was practically cylindrical, and that only the zone at the root of the trailing edge blade showed a slight separation of the boundary layer. At the outlet of the straightening cascade the flow was perfectly axial and free of any sensible separation.

5. WET STEAM TESTS

The tests were conducted at the GENNEVILLIERS wet steam test rig on an H.V.S. cell absolutely identical to those used for the industrial prototypes described in the next chapter.

The aim of these tests was first to determine the cell separation efficiency for various operating conditions, and secondly to measure pressure drop and velocities distributions (reference 19).

5.1. Definition of Test Cell

The cell was placed horizontally. Many dimensional details concerning connection of the cell to the test installation are shown in figures 4 and 5.

The cell was located inside a pipe, and several taps (F) enabled velocity and pressure measurements in planes P_1 to P_7.

Static pressures were measured at the wall at point (T). Velocity distribution measurements were carried out by translation of a 5-holes probe.

Steam is bled out through the annular space (L) towards chamber B_2.

5.2. Principle of Operation of Installation

The wet steam mixture is obtained in a homogenization chamber B_1 fed with an injected water flow Q_e, and a steam flow Q_v, produced by a 45 MW boiler.

The steam flow q, drawn from the H.V.S. cell, is directed to an intermediate tank B_3, where gravity preseparation is performed between the liquid phase and the gas phase. The steam is then vented to a condenser.

The drawing rate adjustment is provided by regulating valve V_R.

5.3. Tests Conditions

The general tests conditions at the inlet of the cell were :

- pressure	: 12 bars,
- steam velocity	: 40 m/s to 70 m/s,
- moisture	: 8 % to 15 %,
- steam drawing flow-rate	: 0 % to 15 % of main flow.

This conditions include the nominal operating ones : 50 m/s for steam velocity, 10 % for inlet moisture, 10 % for drawing flow-rate.

Since liquid phase granulometry is an important parameter in the separation process, droplet size measurements were carried out. The used method is grounded on the determination of a light beam absorption through a steam flow (reference 2). The droplet sizes produced by the test rig water injection device are very close to that measured on 900 MW PWR units at the HP turbine exhausts.

5.4. Separation Efficiency of the H.V.S. Cell

Wetnesses measurements. The cell separation efficiency determination requires steam wetness measurements. A general water flow balance, taking into account enthalpy exchanges with steam, gives the wetness at the inlet of the cell. At the outlet, wetness is measured by a tracing method consisting in doping the water phase before injection. The outlet wetness is obtained by a concentration determination of the total flow leaving the cell, after it is condensed. This method, easy to use, is a very accurate one, since wetnesses lower than 0,01 % could be thus measured.

Test configurations. Efficiency tests were conducted with several configurations.

First the distance between the two cascades was adjustable. Three lengths were tested : 600, 800 and 1000 mm.

It was also possible to change the annular space dimensions.

Lastly several hub-distributor connection designs were installed in the aim to determine the effect of a bead welding. This was simulated by different shapes of rings located at (J) (figure 5). Tests without hub were also carried out.

FIGURE 4

GENNEVILLIERS WET STEAM TEST INSTALLATION DIAGRAM

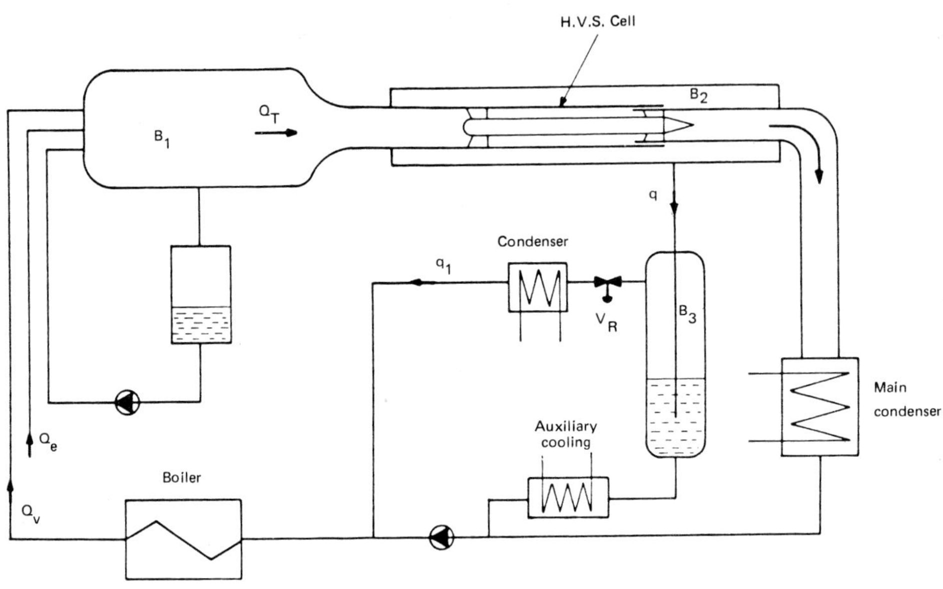

FIGURE 5

H.V.S. CELL DETAILS

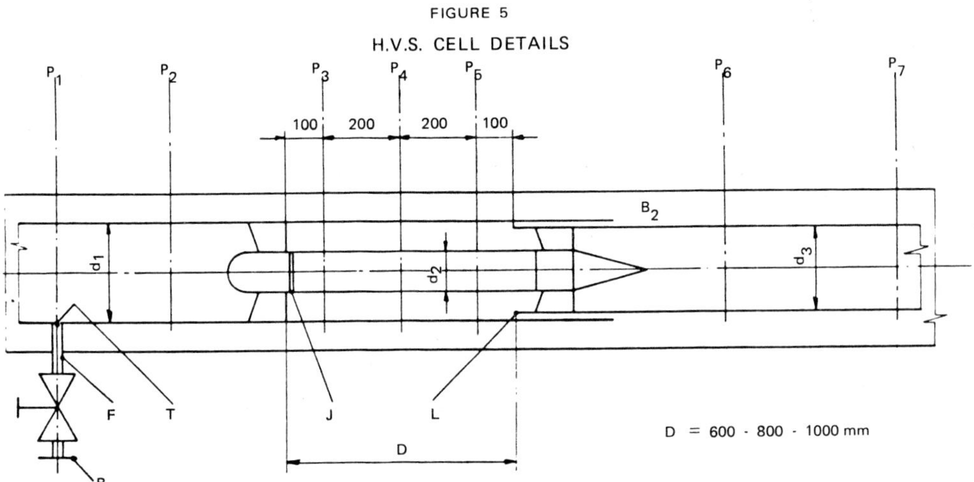

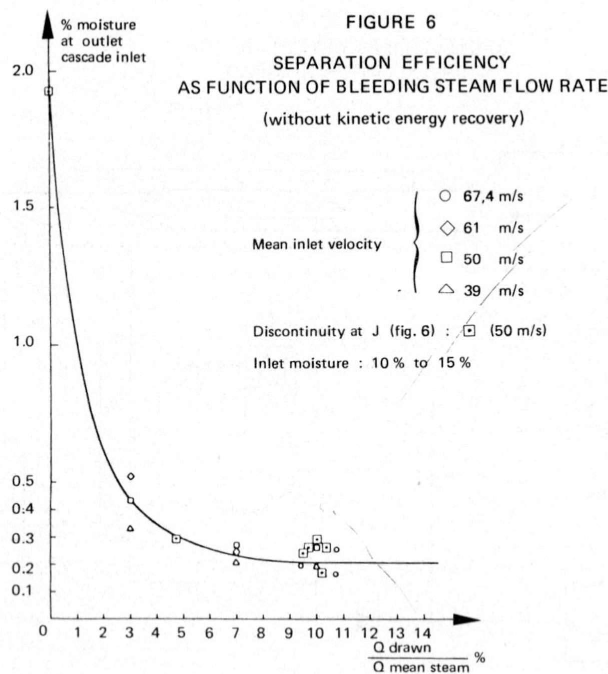

Main results. Typical tests results are showed on figure 6. Kinetic energy recovery through the straightening cascade was not taken into account due to the wetness measurement technique. This recovery results in a steam static enthalpy increasing and the wetness values indicated on figure 6 have then to be reduced of 0.1 % to 0.2 %, depending on the steam flow-rate.

Effect of steam drawing flow-rate. One can observe a rapid decrease in outlet moisture when steam-drawing increases. Thus, with a drawing rate of 3 %, the separation efficiency is improved by a factor of five with respect to the no-drawing operation point.

Beyond a drawing rate of 8 % the outlet moisture remains practically constant.

Improvement in efficiency with steam drawing is due to the influence of the drawing on the flow pattern in front of the annular space. Insufficient drawing indeed will cause stream lines separation and consequently water reentrainment.

Effect of steam inlet velocity. At low drawing rate the cell separation efficiency decreases when steam velocity is increasing. This is due to blocking phenomena which take place at the inlet of the drawing annular space. They result in aerodynamic turbulences and water reentrainment, and become more important when the velocities increase.

Effect of inter-cascade length. Several tests results indicate that the cell separation efficiency is slightly decreasing when the distance between the two cascades is enlarged. In this case, while small droplets separation is improved, the boundary layer development will cause higher turbulences and water reentrainment in front of the drawing annular space.

Effect of drawing annular space dimensions. An increasing of the size of the drawing annular space results in a lower cell separation efficiency and is in fact comparable to a reduction of the steam drawing flow-rate.

Conclusion of the wet steam efficiency tests. With an adequate design of the H.V.S. cell, it is to say a correct sizing of both inter-cascade length and drawing annular space, and for a sufficient steam drawing flow-rate, the H.V.S. separation efficiency is very high. The wetness at the outlet of the straightening cascade at nominal operating conditions, is of about 0.1 %, and the steam leaving H.V.S. would be slightly superheated for increasing flow-rates.

5.5. Pressure Drop and Velocity Measurements

Instrumentation. Pressure drop measurements are achieved by a combination of static parietal pressure measurements and velocity distribution determinations.

Velocity measurements in wet steam flows require special instrumentation and procedure. The probe used is a five-hole probe. After a preliminary calibration in a wet steam test channel, differential pressure measurements between the different holes give absolute steam velocity modulus and direction.

For that, the probe is required to be movable around OY axis while a translation along this allows one to explore the steam flow spatial distribution.

A special care has to be taken for wetness problems and it is necessary to provide for water filling, steam bleeding and air blowing devices concerning probe, pressure gauges and convecting pipes.

However, in the present case, a velocity measurement data analysis involves several problems (reference 20). These are related first to the noticeable drag effects due to the distributor, and secondly to the disturbances, specially near the walls, due to the relative dimensions of the velocity probe head compared with the H.V.S. cell itself. Thus it was not possible to obtain a sufficient number of representative values in sections P_3, P_4 and P_5 (figure 5), where only flow direction indications were available with good accuracy.

Distributor deflection angles. The results of measurements at the outlet of the distributing cascade are shown on figure 7.

It should be noted that there is a good agreement between the predictions based on two-dimensional correlation and experimental results. An appreciable dispersion is however observed near the hub where the mean experimental values are higher that the theoretical ones. This effect can be probably attributed to interactions between hub, probe and boundary layer.

Another cause of dispersion is due to distributing blades drag effect, and it is possible for each measurement plane to determine the area concerned by this effect (since blades position and deflection angles are known) and to eliminate the corresponding test results.

Pressure drop coefficients. Because of the various problems mentioned above, it was only possible to determine the global coefficient of the system with sufficient accuracy. This was estimated as 0.65 times the mean dynamic pressure at the inlet to the H.V.S. cell, for an inter-cascades a distance of 1 m, an inlet wetness near 0 % and a drawing flow-rate of 10 %. It should be noted that agreement with the theoretical values is very good.

For an inlet moisture of about 10 % the pressure drop coefficient is clearly increased by a factor of 2 to 3. A study is at present being conducted to determine the energy required for centrifugation of liquid droplets. This is excepted to provide an explanation for the corresponding increase in pressure drop.

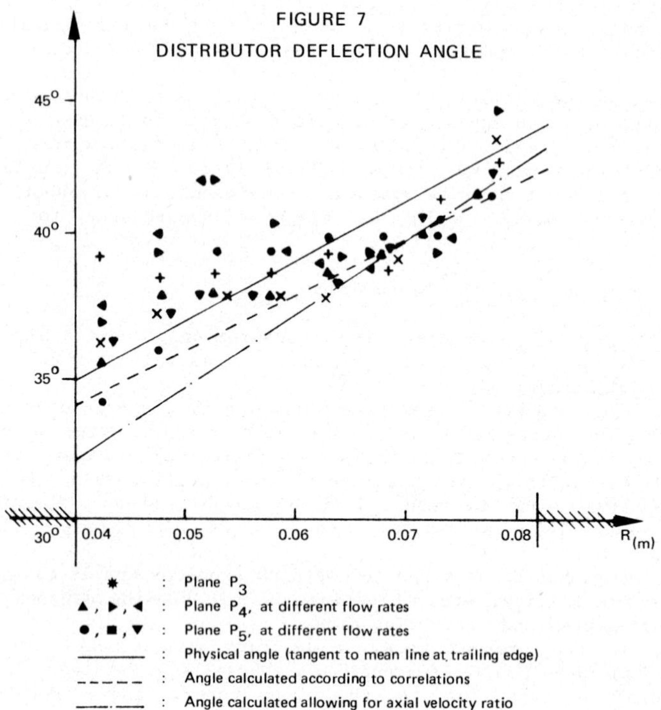

A pressure drop coefficient increasing is observed with a steam drawing rate reduction and a maximum value of 2.6 is reached for operation without drawing. This important increase in pressure drop is understandable because of the high "step" effect due to drawing annular area, at the inlet of the straightening cascade.

6. H.V.S. INDUSTRIAL PROTOTYPE

The theoretical and experimental studies described above resulted in the construction of an H.V.S. industrial prototype. The design of this prototype, together with the installation of two units during July of 1980 on the 900 MW BUGEY PWR power station, was conducted in collaboration with ALSTHOM-ATLANTIQUE and STEIN-INDUSTRIE. The results obtained during 2 years of operation were excellent, concerning as well drying efficiency and pressure drop as reliability.

6.1. Prototype Installation

As indicated above, the best solution for drying large wet steam flows consists in fitting several parallel small separating cells.

This solution leads indeed to low pressure drop and reduced length. The cells dimensions for the prototypes resulted from a compromise between separation efficiency and pressure drop. For each prototype unit a number of 24 cells, of about 0.2 m in diameter, was decided (figure 8). The cells internal arrangement (cascades geometry, steam drawing annular space, etc...) was a result of the calculations and tests described above.

The H.V.S. units installation at the exhaust of BUGEY HP turbine is indicated on figure 9. Their location is not an optimum since the plant was yet in operation from one year before the H.V.S. were installed. In addition, the

FIGURE 8
HVS SYSTEM DIAGRAM

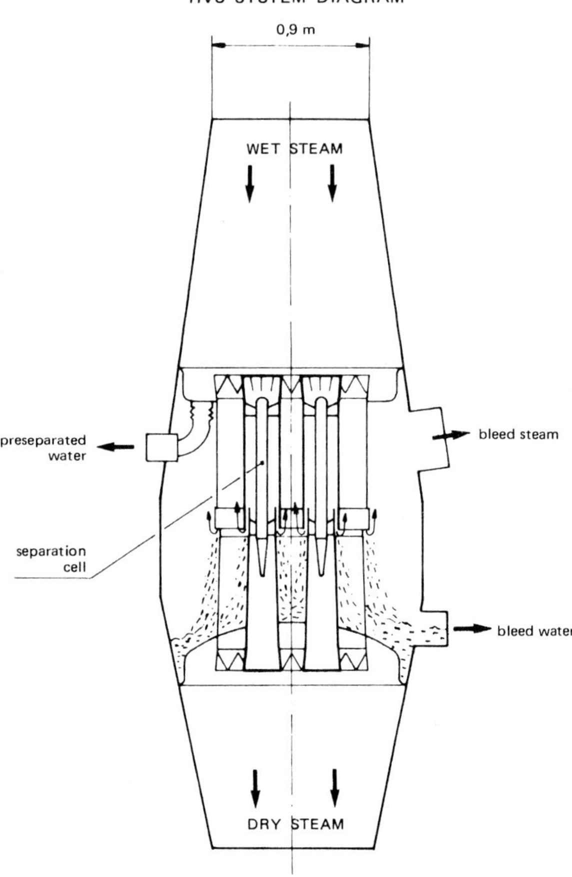

aim of the prototypes was to demonstrate the good performances of H.V.S. even under unfavourable operating conditions.

The steam drawing flow-rate, required for a better H.V.S. efficiency, was a fraction of the existing bleeding at the HP turbine exhaust.

The mean operating conditions for the two prototypes are :

- pressure : 10,4 bars,
- wetness : 0,114,
- mean flow-rate per prototype : 530 tons/hour.

6.2. Bugey Prototypes Tests Results.

Pressure drop measurements. Static pressure drop coefficient (K) for each device was calculated using the feeding dry steam flow rate (Q) and the static pressure drop (ΔP) between the inlet and the outlet of the H.V.S. (reference 22 and 23).

FIGURE 9

INSTALLATION OF H.V.S. PROTOTYPES AT THE 900 MW BUGEY II POWER STATION

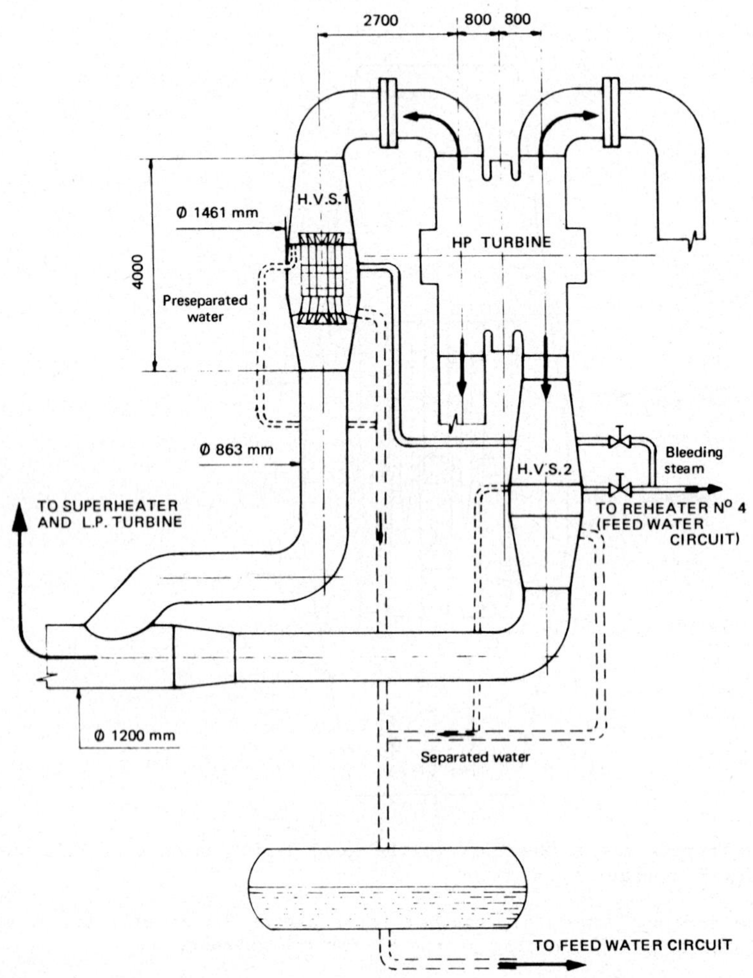

	Pressure (bars)	Q (kg/s)	dynamic pressure (mb)	ΔP (mb)	K
Upper HVS	10.8	115	39,5	108	2,73
Lower HVS	10.8	147	64,6	118	1,83

The difference between both values of K can be explained by an highly heterogeneous feeding of the upper prototype, located at the outlet of a 180° bend. In this region the steam flow is highly disturbed and important velocity differences were observed. Local velocities more than twice the mean velocity were measured.
In actual fact the H.V.S. operation resulted in a very low increase in pressure drop between the HP and LP turbines. This point was demonstrated by comparative measurements performed before and after the H.V.S. installation. Such an effect is due to the steam dryness and to the good velocities distribution at the H.V.S. outlet. Several tests carryed out on FESSENHEIM and BUGEY power plants connecting pipes at the HP turbines exhausts have shown that a moisture level of about 10 % resulted in more than 80 % pressure drop increasing (reference 24). The reduction of wetness due to H.V.S. leads to a pressure drop reduction for the pipes down-stream.
In addition, for projected plants, the small size of H.V.S. will allow a length reduction and a laying out simplification for the connecting pipes. This will result in a pressure drop reduction between HP and LP turbines.

H.V.S. prototypes separation efficiency. A tracing method was used for steam wetness measurements. It consists in injecting a soluble, non-volatile salt inside of a wet steam flow and carrying out a parietal liquidfilm sampling down-stream. Some auxiliary tests are necessary in order to be sure of a good tracer dilution with all the liquid phase.

Due to the droplets size and turbulences at the HP turbine outlet, a few meters are generally sufficient between injection and sampling points. The water flow-rate is directly given by the dilution ratio. Concerning the wetness value itself, a dry steam flow-rate determination is necessary and can be obtained by velocity measurements inside the pipe or by a general flows balance at the turbine exhaust.

Lithium hydroxide L_iOH was selected as a tracer, owing to the possibility it offered of very sensitive concentration measurements.

The injection point was located just after the jonction of the H.V.S. exhaust pipes. Many sampling points were installed down-stream. Several tests were carried out. They all confirmed the excellent H.V.S. separation efficiency. Under normal operating conditions the mean wetness measured at the outlet of the H.V.S. was 0,28 %. (reference 25).

This value is slightly higher than that obtained on one cell during the first wet steam tests (see 5.4.). This difference can be explained as follows :

- Allowing to velocities heterogeneity in front of the prototypes it is likely that some of the cells are misadapted.

- Several observations at the exhaust of HP turbines, showed that water concentrations occured inside the connecting pipes, due to reentrainment from the walls. They may cause local water over loading for some H.V.S. cells.

Various velocity distribution measurements were carryed out in order to determine the drawn steam percentage in the BUGEY prototypes. Due to the limitations of valves on the drawing circuit, a maximum drawn flow-rate of 7 % of the inlet prototypes steam flow was obtained, which is less than the design one (10 %). Moreover, this is a mean value and since the cells are not all supplied in the same manner, some must operate with a much lower drawing rate than nominal. Under these conditions, and as can be seen from figure 6, it is normal to observe a slight deterioration in outlet moisture. An operating test, conducted without drawing, give a wetness value of 1,4 % at the H.V.S. outlet. This value is fully compatible with that obtained before on one cell (figure 6).

H.V.S. Operation at low bleed steam flow-rate. It is very useful and interesting to know the efficiency of both apparatus when the drawn steam percentage decreases from 7 % to 0 %. This test has been carried out for one cell on our test loop, but the results obtained may not be extrapolated for 24 cells. So we decided to perform the same test at Bugey (Reference 26). Great care was taken for the prototypes to determined the efficiency corresponding to very low drawn steam pourcentages with accuracy (about 1 % of the feeding flow). The tests results are shown on figure 10. The evolution of the outlet moisture is quite simular to that for one cell : without bleed steam, an outlet moisture of about 1,8 % is observed ; with few bleed steam (1 %), a sharp decrease of the outlet moisture is observed (about 0,45 %). These tests have mode the demonstration that the H.V.S. prototypes are not sensitive to the drawn steam percentage, and only 2 or 3 % are enough to get a good separation efficiency.

7. IMPROVEMENTS IN WATER AND STEAM EXTRACTION

For some particular applications we suggest to dry the bleed steam at the exhaust of the main H.V.S with one or several H.V.S. cells operating without bleeding, or to bleed off water and steam by the same drawing-out pipe at the bottom of the device and to dry this wet steam in the pipe with one or several H.V.S. cells (without bleeding). Both kinds of secondary H.V.S. have been installed on Bugey H.V.S. prototypes (reference 27).

FIGURE 10

SEPARATION EFFICIENCY OF THE PROTOTYPE AS A FUNCTION OF BLEEDING STEAM FLOW RATE

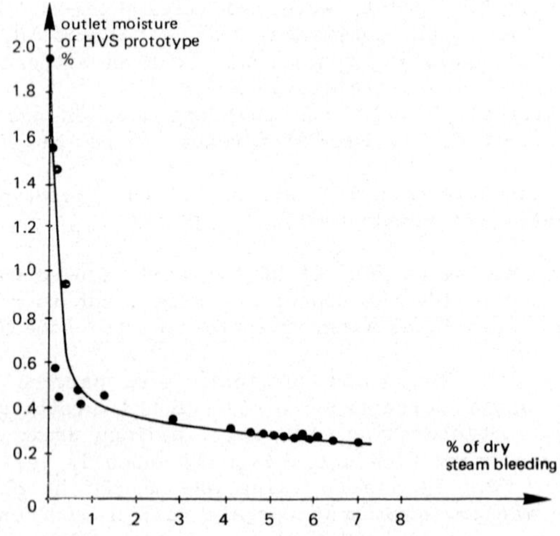

7.1. Secondary H.V.S. on Bleed Pipes.

A diagram is shown on figure 11 (left part). This H.V.S. operates without bleed steam, inlet moisture was included between 30 and 50 % and steam velocities around 30 and 50 m/s depending on the disposable pressure drop for the bleed circuit were employed.

The first generation of secondary H.V.S. yields an outlet moisture of about 3 %. This low wetness in front of the inlet one decreases the pressure drop of the complex bleed circuit because the increasing pressure drop coefficient due to a 3 % moisture is only 1.2. Some improvements in the drawing annular space designing would easily result in an outlet moisture lower than 1 %.

7.2. One Exhaust For Drawing Water and Steam Off

This configuration is shown on figure 11 (right part). This operation system, including a secondary H.V.S. drying the bleed steam with all the separated water, has several advantages : lower pipes around the main H.V.S., bleed steam with a low moisture, better operation especially for horizontal H.V.S.

FIGURE 11

SOME IMPROVEMENT IN WATER AND STEAM EXTRACTION

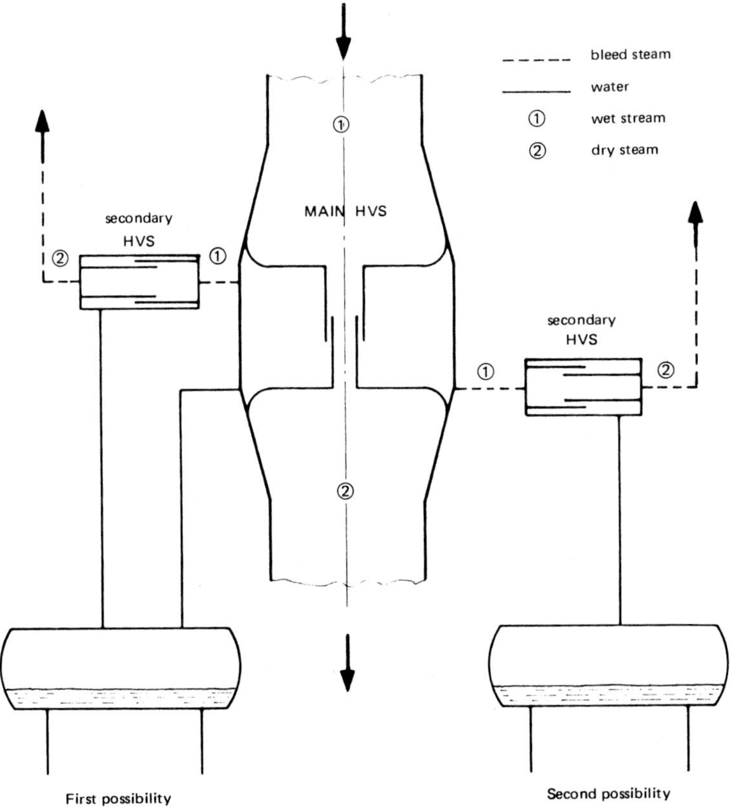

8. NEXT TESTS ON THE WET STEAM TEST RIG OF GENNEVILLIERS.

A six horizontal alls model has been designed, first, to check the good operation with horizontal location when drawing water and steam off at the bottom and then to prove the absence of effect of the flow of water around the cells often it left the annular space (figure 12). These tests will be carried out for several pressures (3 to 12 bars) and different bleed steam flow-rates.

9. CONCLUSION - H.V.S. ADVANTAGES.

Very promising test results were obtained on the H.V.S. industrial prototypes in operation at the BUGEY 900 MW power station. They concern both H.V.S. pressure drop and separation efficiency. A good agreement was observed with the first wet steam test carried out in the GENNEVILLIERS test rigs and with theoretical predictions. In particular, the calculation codes used for H.V.S. design were well validated.

In addition, the reliability of H.V.S. technology is now demonstrated. A complete inspection after more than 11000 hours of operation showed the perfect resistance of the BUGEY prototypes to erosion attacks and to mechanical or thermal operating strengths. This confirms that the design and sizing of the H.V.S. prototypes are well convenient.

FIGURE 12

DESIGN FOR WATER-STEAM TESTS ON 6 HORIZONTAL CELLS

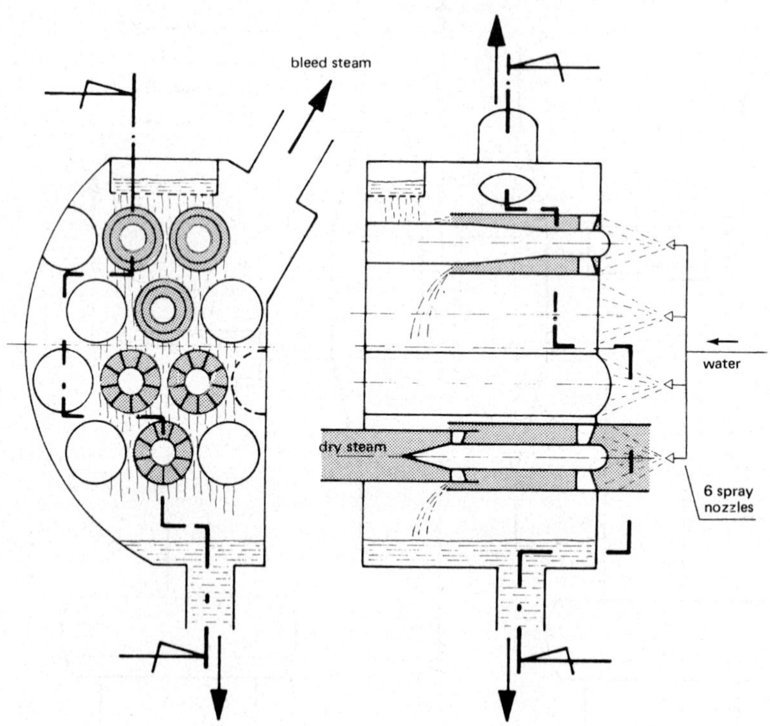

High Velocity Separators will result in sensible cost reductions for nuclear power stations, with respect to investments and efficiency.

Investment charges decrease will be due to a reduction of the separators-reheators size itself and of the machine room dimensions. In addition, owing to the H.V.S. size, the possibility of drying steam directly at the HP turbine exhaust will remove any erosion-corrosion problem in the connecting pipes and carbon steel will be available for their construction. Lastly the pipes layout will be simplified, and their length reduced.

Efficiency increases will be the consequence of a better moisture separation and a lower pressure drop between High Pressure and Low Pressure Turbines. This last point is due to the connecting pipes lenght reduction and to the wetness removal just at the HP turbine exhaust (see 6.2).

Last, but not the least, H.V.S. make it possible to consider a modification of the expansion cycle proper, and, in particular, to look to two-drying cycles with no intermediate superheating. Such cycles offer many advantages (better efficiency and higher availability due to superheater suppression) but they were seldom used owing to the size of the conventional separators. A comparison between a two-drying cycle combined with a PWR nuclear boiler and a one-drying-superheating cycle, shows important gains resulting in a sensible energy cost reduction (reference 5).

REFERENCES

Reference 1 : J.P. CERDAN
"Performances des séparateurs eau-vapeur. Etude comparative".
EDF-DER P32/070/2.14.

Reference 2 : J.P. CERDAN
"Analyse du fonctionnement des séparateurs eau-vapeur".
Revue ENTROPIE n° 83, 1978.

Reference 3 : J.P. CERDAN
"Progrès récents dans l'analyse du fonctionnement des séparateurs eau-vapeur".
A.I.M. LIEGE, Centrales électriques modernes, Journées Internationales d'Etude, mai 1978.

Reference 4 : J.P. CERDAN - P. TALLEU
"Sécheurs à grande vitesse. Description".
EDF-DER HP32/071/1.6.

Reference 5 : J.P. CERDAN
"Une application des SGV : les cycles à 2 séchages. Bilan technico-économique pour le palier 1300 MW".
EDF-DER P32/040/15.

Reference 6 : J.P. CERDAN
"Les sécheurs à Grande Vitesse : performances et perspectives".
EDF-DER - Note interne P32/04/82/1.23, 1982.

Reference 7 : "Coefficient de traînée d'une sphère. Synthèse bibliographique".
EDF-DER, Rapport interne HC062/111/1.1, 1968.

Reference 8 : N.A. FUCHS
The mechanics of aerosols .
PERGAMON-PRESS, 1964.

Reference 9 : P. TALLEU
"Sécheurs à Grande Vitesse. Calcul de trajectoires de gouttes entraînées par un écoulement centrifuge".
EDF-DER P32/071/1.2, mai 1977.

Reference 10 : K.D. PAPAILIOU
"Secondary flows in axial compressors".
V.K.I. L.S. 72, 1975.

Reference 11 : K.D. PAPAILIOU
"Correlations concerning the process of flow deceleration".
Journal of Engineering for Power, April 1975.

Reference 12 : J. HORLOCK - H. PERKINS
"Annulus wall boundary layers in turbomachines".
AGARDOGRAPH n° 185, 1974.

Reference 13 : A. SATTA
"Un metodo di calcolo della strato limite tridimensionale nei diffusori delle turbine a gas".
PISA, 1976.

Reference 14 : C. HENRY - P. SPETTEL
"Description et utilisation du programme singularités versions I et II, FORTRAN IV, de calcul de l'écoulement subsonique à travers une grille d'aubes".
METRAFLU, 1977.

Reference 15 : K.D. PAPAILIOU
"Dispositifs de mise en rotation d'écoulements de vapeur humide - Première partie".
EDF-DER, Note interne, 1978.

Reference 16 : K.D. PAPAILIOU
"Dispositifs de mise en rotation d'écoulements de vapeur humide - Deuxième partie".
EDF-DER, Note interne, 1978.

Reference 17 : "Séparateur E.D.F.".
Rapport n° 2055 A - Institut Aérodynamique de SAINT-CYR, 1978.

Reference 18 : M.H. VAVRA
"Aero-thermodynamics and flow in turbo machines".

Reference 19 : "Séparateur EDF avec nouveau distributeur".
Rapport n° 2069 A, Institut aérodynamique de SAINT-CYR, 1979.

Reference 20 : P. TALLEU
"Sécheurs à Grande Vitesse. Mesure sur une cellule de l'efficacité de séparation en vapeur humide".
EDF-DER - Note interne, en cours de rédaction, 1981.

Reference 21 : K.D. PAPAILIOU
"Dispositifs de mise en rotation d'écoulements de vapeur humide - Troisième partie".
EDF-DER - Note interne, 1980.

Reference 22 : E. DUEYMES
"Sécheurs à Grande Vitesse. Résultats d'une première campagne de mesure de pertes de charge menée à BUGEY II, après mise en place des prototypes".
EDF-DER - Note interne P32/04/81/1.9, 1981.

Reference 23 : E. DUEYMES
"Coefficient moyen de perte de charge d'un SGV - Coefficient de perte de charge tenant compte de l'humidité".
EDF-DER - Note interne P32/04/82/1.20, 1982.

Reference 24 : E. DUEYMES
"Pertes de charge en vapeur humide - Détermination de coefficients correcteurs tenant compte de l'humidité".
EDF-DER - Note interne P32/04/81/1.17 - 1981.

Reference 25 : P. TALLEU
"Sécheurs à Grande Vitesse. Premières mesures d'humidité à l'échappement des prototypes de Bugey".
EDF-DER - Note interne P32/04/81/1.10 - 1981.

Reference 26 : E. DUEYMES
"Séparateurs à Grande Vitesse - Comportement des prototypes de Bugey II à soutirage réduit".
EDF-DER - Note interne - P32/04/82/1.2.3. - 1982.

Référence 27 : E. DUEYMES
"Prototypes SGV - Mise en place de 6 nouvelles cellules S.G.V. réalisée à Bugey II en juin et juillet 1982".
EDF-DER - Note interne - P32/04/82/1.24 - 1982.

Multi-Phase Flow and Heat Transfer III. Part A: Fundamentals
edited by T.N. Veziroğlu and A.E. Bergles
Elsevier Science Publishers B.V., Amsterdam, 1984 — Printed in The Netherlands

SIMULATION OF WET STEAM DISTRIBUTION NETWORKS

R. S. Peterson, G. R. Sullivan, and E. Rhodes
Department of Chemical Engineering
University of Waterloo
Waterloo, Ontario, Canada N2L 3G1

ABSTRACT

A major method for the recovery of heavy oil deposits is wet steam injection. In order to best utilize the energy resources available it is necessary to understand the changes that occur as the wet steam is piped through the network to the injection sites, which are often several miles away.

In order to investigate wet steam distribution networks, a computer simulation has been developed. Homogeneous, steady state models are used to predict the pressure and quality changes, and flow restrictions that would occur in the pipes, valves, and splitters that comprise a network. The simulation of several wet steam distribution networks has been performed, including a portion of a commercial plant in order to verify the models. The results agreed quite well with the data supplied by the facility.

Several simulations with different constraints on outlet conditions were done for each network. Outlet pressures ranged from 6.0 to 18.0 MPa, and outlet qualities from 0.0 to 1.0. The sensitivity of the system overall operating cost to changes in adjustable variables (valve positions, exit pressures, flow rates) was demonstrated to be extremely high. Network constraints on pressure, quality, and energy flow were demonstrated to severely limit operating flexibility in some instances.

The work is being extended to develop general operating strategies, and optimal control strategies for wet steam distribution networks.

1. INTRODUCTION

Wet steam injection can be done with small boilers positioned at each injection site, or alternatively with a few large scale boilers and a steam distribution network. This paper reports a study of steam distribution from a large boiler complex through a network of pipes, valves, and splitters to the injection sites. Since many possible network designs and operating strategies can yield the same energy flow and steam quality at the injection sites, it is worthwhile to examine the economic impact of design and operational decisions.

In order to investigate wet steam distribution networks, a computer simulation has been developed. This is a synthesis of several models; pipes, valves, splitters, mixers, and boilers, which are linked together by a network algorithm. The simulation predicts the changes that occur as the steam passes through the network to the injection sites.

Two approaches to problem solving have been used, namely:

1. The Design Approach; where there is a great deal of flexibility in network layout, pipe sizes, and pressure constraints.

2. The Operations Approach; where the network already exists, and the objective is to use it efficiently. Typically the operating constraints, such as injection site pressures and qualities, are quite stringent.

Two-phase flow in small networks typical of the power industry has been investigated [1], but there is little information available about large scale steam distribution, typical of commercial enhanced oil recovery. This study makes use of a relatively simple, modular, two-phase flow simulation, which allows formulation, and simulation of a network in just a few hours. General trends for pressure, quality and energy distribution are investigated, and cost sensitivity to these variables is evaluated.

2. MODELS

Wet steam distribution is by definition a two-phase flow problem; in this case, liquid water and steam. Since two-phase flow calculations can become quite complex, a number of assumptions have been incorporated into the models in order to simplify the calculations. The system is evaluated at steady state, and homogeneous two-phase flow is assumed. The physical properties are evaluated assuming a saturated system; i.e., liquid and vapour in equilibrium. There are also several minor assumptions specific to the individual models.

Models have been developed for a number of units; a boiler, pipe, splitter, mixer, and valve. The models have been designed as independent modules, which allows one model to be revised without affecting the operation of the others. Each model requires either the inlet or outlet conditions (pressure, quality and flow rate), and a set of physical parameters (pipe diameters, valve coefficients, etc.) specific to the model.

A brief description of the major functions and highlights of each model, including some of the major equations, are given below. Nomenclature for all the equations that follow is provided at the end of the paper.

2.1. Pipe Model

The objective of the pipe model is to predict pressure drop and quality change along a length of pipe. The pressure drop is estimated using a simplified form of the equation reported by Collier [2] for homogeneous flow.

$$P_1 - P_2 = \frac{2f_{tp}LG^2V_f}{D}\left(1 + \frac{x_1 + x_2}{2}\left(\frac{V_{fg}}{V_f}\right)\right) + G^2V_{fg}(x_2 - x_1) \tag{1}$$

The energy balance takes into account the internal energy (enthalpy), the kinetic energy (velocity), and the heat loss to the environment, to estimate the quality change. The overall energy balance is:

$$H_2 + 1/2(V_2)^2 = H_1 + 1/2(V_1)^2 + Q/W \tag{2}$$

2.2. Valve Model

The objective of the valve model is to allow flow control. The valve position is specified from 0.0 (closed) to 1.0 (open). The valve coefficients are used to determine the maximum allowable flow, based on equations reported in the Control Valve Handbook [3] for liquid-gas mixtures. The actual flow is just a fraction of this maximum value.

2.3. Mixer/Splitter Model

The objective of the mixer and splitter model is to predict changes in the flow conditions as a result of branching. Because mixing and splitting are similar operations, they are handled by one model. For example, a splitter with known outlets is analagous to a mixer with known inlets.

In order to estimate pressure changes, branching is treated as either a sudden expansion or a sudden contraction, depending on the ratio of upstream cross sectional area to downstream cross sectional area.

For a contraction the following pressure drop equation reported by Bergles et al. [4] is used.

$$P_1 - P_2 = 1/2(1-1/S^2 + (1/C_c - 1)^2)G_2^{\,2}(xV_g + (1-x)V_f) \tag{3}$$

For an expansion another relationship presented by Bergles et al. [4] is used.

$$P_2 - P_1 = S(1-S)G_1^{\,2}(xV_g + (1-x)V_f) \tag{4}$$

Once the pressure change has been determined, the quality change is evaluated by an energy balance.

2.4. Boiler Model

The boiler model relates inlet and outlet conditions by an energy balance, namely:

$$H_2 = H_{fw} + Q \tag{5}$$

The feedwater enthalpy, H_{fw}, can be estimated by a simple two piece linear function relating enthalpy to subcooled feedwater temperature, T_{fw}, as reported by Lin et al. [5].

$$H_{fw} = aT_{fw} + b \tag{6}$$

and $a = 4.187 \times 10^3$, $b = 1.254 \times 10^4$ $\quad -17.8 < T_{fw} < 154.7$
$a = 4.438 \times 10^3$, $b = -2.633 \times 10^4$ $\quad 154.7 < T_{fw} < 315.6$

3. NETWORK ALGORITHM

Many networking problems can be solved using a form of Hardy Cross analysis, which is based on Kirchoff's laws [6]:

1. The algebraic sum of the flows at any pipe junction is zero.

2. The algebraic sum of the pressure drops around any closed loop of the network is zero.

However, for many applications of wet steam distribution the second law is irrelevant because there are no closed loops in the network. Also, most network algorithms are designed for incompressible flow. However, wet steam distribution is a two-phase problem, and hence a compressible system. Condensation and flashing occur throughout the network, which increases the complexity of the system.

These considerations led to the development of a network algorithm specifically for wet steam distribution in an open loop system. Since, in general, there are no recycle streams in steam distribution, a sequential, modular approach, rather than an equation oriented approach, is used to solve the network.

The network is divided into units; pipes, valves, splitters, mixers, or boilers, which have different models. The network algorithm links all the individual units together and evaluates the overall system taking into account specified conditions and restrictions.

The entire network is described by two data matrices. One contains all the unit specifications(pipe diameters, valve coefficients, etc.,), the other contains the nodal conditions (flow rate, pressure, quality). The network algorithm works through the network, unit by unit, according to the placement of the units in the unit matrix, making calls to the model subroutines as required. The models use the unit specifications, and the known nodal conditions (inlet or outlet) to update the nodal matrix. If one set of conditions (inlet or outlet) is known for a unit then the other set can be calculated. The network is evaluated iteratively and the closure criterion is an overall mass balance on the inlet and outlet nodes.

The network algorithm checks for inconsistencies and overrides initial nodal specifications in some instances. For example, flow (a nodal specification) through a closed valve (a unit specification) is not allowed. In this case the unit specification takes precedence, and the flow would be set to zero at the valve inlet node, and corrections would be made to the nodal flow values upstream.

In some instances a subcooled liquid or a superheated vapour is predicted at one or more nodes. In these cases the quality is adjusted to maintain saturation; i.e., a quality of 0.0 (subcooled case) or 1.0 (superheated case) and the pressure is adjusted to maintain the energy balance. This could lead to inaccuracies in subsequent calculations.

4. NETWORK ANALYSIS

There are a number of factors that contribute to the design and operation of a wet steam distribution network. For example, operating constraints; cost sensitivity to adjustable variables; and distribution of energy, pressure, and quality throughout the network are all important. These factors are discussed in the sections to follow.

4.1. Constraints

For any network there are a number of constraints that could be specified, namely:

1. The minimum pressure at the injection site; which is a function of the reservoir characteristics (capillary pressure, permeability, porosity).

2. The maximum pressure throughout the network; which is fixed by design considerations. In order to operate at higher pressures, stronger (and hence more expensive) pipe, fittings, etc., are required.

3. The maximum quality of the steam; which is usually less than 80% to reduce fouling inside the boiler tubes [7].

4. The capacity of the boilers;

5. The amount of energy delivered to each injection site; which is a function of the mass flow, pressure, and quality.

6. The quality of the steam at the injection site. In general, the effect of the quality is dependent on the reservoir characteristics [8].

7. The maximum mass flow rate; which is set by design specifications on pumps, boilers, pipes, and valves.

These constraints are highly interactive. For example, quality and flow rate constraints restrict the maximum energy flow rate available to individual injection sites, and boiler capacity restricts the total energy available throughout the network. Some examples of the limitations imposed by constraints are presented in the results section.

4.2. Cost Sensitivity

Ideally, the cost (or conversely the revenue) of operating the network could be expressed as a detailed function of the adjustable operating variables (pressure, quality and flow rate) and the physical parameters (pipe diameters, pump sizes, etc.,).

There are several possible costs to consider.

1. The cost of the fuel used by the boilers, which could be expressed as a function of the fuel rate, F(fuel rate).

2. The costs of water treatment, pumping, pump deterioration, and related water costs, which could be expressed as a function of the water flow rate, G(water rate).

3. Capital costs, which could be expressed as a function of the operating pressures, network size, number of boilers, etc., H(P,size,etc.,)

4. Revenue, which is a negative cost related to the amount of oil recovered. This is expected to be a function of the quality of the steam injected, its pressure, and its energy. A number of other parameters could also be included;

such as the timing of the injection cycle, geometry of the injection wells, and reservoir characteristics. The revenue function could be expressed as, R(x,P,E,t,etc.,)

These terms combine to yield a cost function of the form:

$$\text{Cost} = F + G + H - R \tag{7}$$

Since the capital cost and revenue are extremely variable and highly unpredictable, in this study just the first two costs are examined, which are dependent on the fuel rate, and the water rate. The type of boilers, pumps, size of pipes, etc., vary depending on the throughput required, these capital costs could be amortized and included in the water cost for the design case. In the simplest case, the overall cost is expressed as a linear function of the flow rates of the water and fuel, namely:

$$\text{Cost} = K_1(\text{fuel rate}) + K_2(\text{water rate}) \tag{8}$$

The total cost will vary for different conditions depending on the relative costs of fuel and water, which can be represented by the ratio of K_1 to K_2. This ratio will be referred to as the cost ratio in subsequent discussions. Comparing the total cost for several different cost ratios, the cost sensitivity to changes in the adjustable variables(pressure, quality) can be observed.

In order to establish the range of expected cost ratios, the costs of fuel and water treatment have been estimated for a commercial network in Cold Lake, Alberta.

Water treatment is necessary to prevent scaling in the boiler tubes and to prevent corrosion in the distribution system; it is costly, even when fresh water is primarily used. In many instances, the produced water must be recycled, due to fresh water limitations [9]. The produced water is especially dirty, and requires additional treatment, which greatly increases the cost.

Robertson [10] estimates a 5 MW boiler could cost $10,000 - $20,000 per year in chemicals for dosing and for operation of the water softener. For a portion of the Cold Lake commercial network, an energy flow of close to 300 MW is required [7]. This corresponds to approximately $1,200,000 (in the higher case) per year for water treatment. The cost per kg of water used based on operation 365 days per year, 24 hours per day, is $0.00038. The additional cost of treating produced water could substantially elevate this estimate.

Using $34.00 per barrel as the cost of fuel oil, the cost per kg of fuel is approximately $0.25 (assuming a specific gravity of 0.85 for the fuel oil).

These two simplified cost estimates yield a cost ratio of close to 660. The true cost ratio would be expected to range anywhere from 50 to 5000 for the operations case, due to the the variation of individual costs and requirements from one facility to another. For the design case the cost ratio would be substantially lower if capital costs were included in the water cost.

4.3. Energy Distribution

In a large network one of the major difficulties is to control the energy, pressure, and quality distribution. Typically the outlet conditions are specified, and then related back through the network to determine a boiler load. However due to the high degree of interaction inherent in the distribution network some conflicts can arise. For example, in order to achieve a desired pressure and quality at one injection site, restrictions are imposed at other sites.

In order to achieve a specific pressure at the farthest injection site downstream, higher pressures will be necessary at sites further upstream. The quality will vary also; depending on the operating pressure and quality, flashing or condensation can occur due to pressure drops. This is readily apparent from figure 1, which is an enthalpy versus pressure diagram for saturated steam. At qualities below 0.5 only condensation occurs due to an

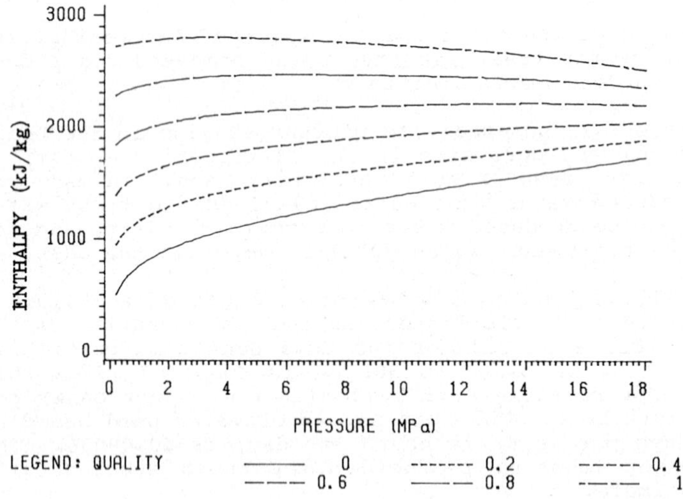

Figure 1: Enthalpy of Saturated Steam

adiabatic pressure drop. At qualities above 0.8, if the pressure is above 6.0 MPa flashing will occur due to an adiabatic pressure drop. Within the intermediate region, qualities 0.5 to 0.8, either flashing or condensation can occur depending on the operating pressure.

Even though the pressures and qualities may vary from one injection site to another, it is still possible to achieve an even energy distribution. In many cases, some variation in injection pressures and qualities is tolerable. In general, an equable energy distribution can be achieved by adjusting the flow rates, and accepting the variations in pressures and qualities that result. However, there are situations where the pressure and quality are tightly constrainted. In these instances the network may be unable to simultaneously meet the quality, pressure, and energy requirements. Some examples of typical network distributions are presented in the results section.

5. RESULTS AND DISCUSSION

Several simulations have been performed in order to investigate general trends in cost sensitivity, pressure and quality changes, and energy distribution. The following representative examples illustrate these trends.

5.1. Commercial Plant

In order to verify the simulation, a portion of a commercial plant in Cold Lake, Alberta [7] has been simulated. Figure 2 is a schematic representation of the commercial network; the injection sites are numbered (1-12). The only control provided is flow control with valves after the boiler and at the injection sites. The network is comprised of 17 pipes, 5 splitters, 12 valves and 1 boiler; the distance from the boiler to the most distant injection site, 12, is 5100m; pipe diameters range from 0.359m to 0.146m.

Figures 3, 4 and 5 compare the simulated pressures, qualities and energy distribution with the data supplied. The results agree quite well with the data provided by the facility. Simulated pressures are within 2%, qualities are within 1%, and energy flows are within 0.5% of the commercial data. This is an example of equal flow distribution, since the flow rate is set at 11.0 kg/s for each injection site in both cases. These curves represent typical pressure, quality and energy distributions. The pressure decreases as the steam moves farther from the boilers. The quality decreases at some points (condensation), and increases at others (flashing).

The magnitudes of the pressure and quality changes are dependent on the operating conditions along the network. Even with

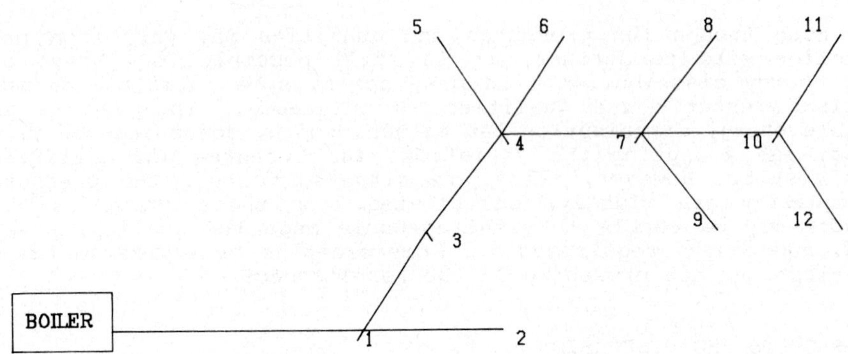

Figure 2: Cold Lake Commercial Network

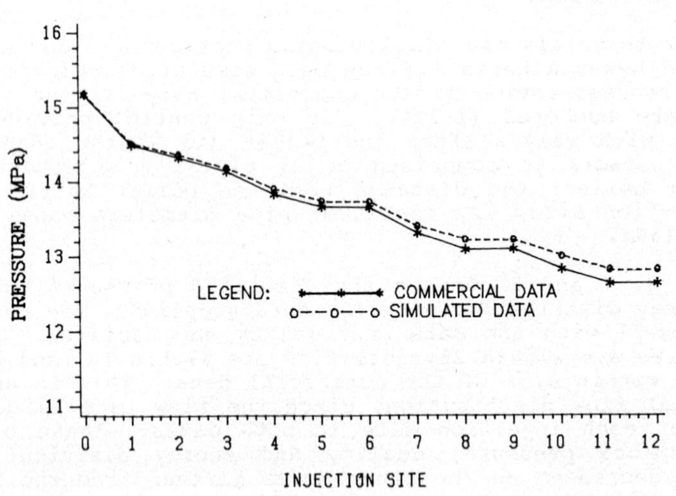

Figure 3: Cold Lake Network, Pressure Distribution

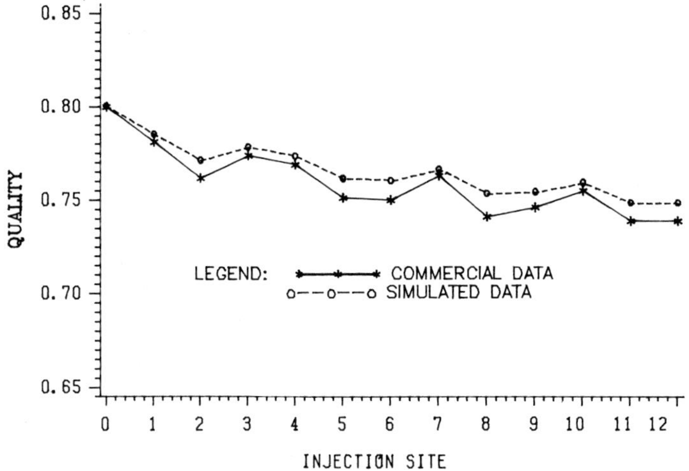

Figure 4: Cold Lake Network, Quality Distribution

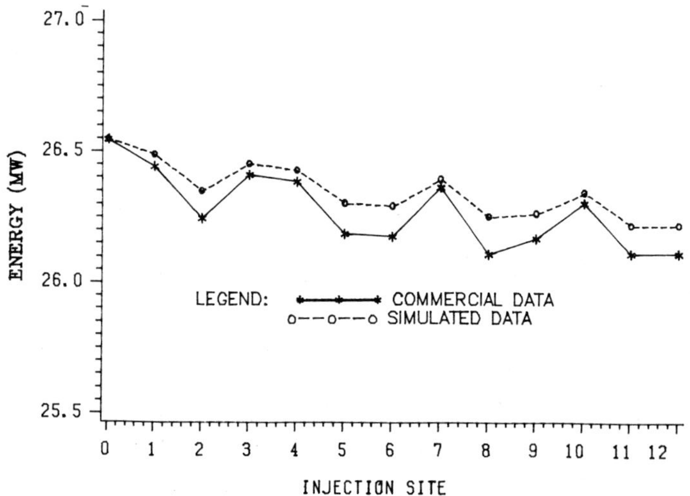

Figure 5: Cold Lake Network, Energy Distribution

pressure and quality variation, the energy flow is fairly even as is illustrated by figure 5.

5.2. Simple Pipe Example

Many operating concerns can be readily identified by examining a simple system, consisting of only a boiler and one long pipe, in this case a pipe 1000m long with an internal diameter of 0.30m. A fairly wide operating range is investigated; pressures in the range 6.0 MPa to 18.0 MPa, and qualities from 5% to 95%, all at a constant energy flow.

First, the relationship between the total operating cost and injection site quality for several different injection site pressures, and different cost ratios is examined. In order to achieve the same energy flow at an injection well, different injection pressure and quality combinations are possible. These conditions require different water and fuel rates at the boiler.

A high water flow rate requires less energy per unit mass to achieve the same energy flow rate as a low water flow. A high water flow at a lower quality can provide the same energy flow as a low water flow at high quality. At a high water flow rate the sensible heat of the incoming water represents a larger portion of the total energy. Therefore less fuel is required to achieve the same energy flow. Figures 6, 7, and 8 illustrate the variation in cost possible for a constant energy flow rate. The cost is plotted against the outlet quality for several pressures on each figure and each figure is for a constant cost ratio. The costs are all normalized, with the lowest cost set to 1.00.

Although these curves were obtained for a single pipe of a specific size, the general trends apply to networks comprised of pipes of many different sizes. This will be illustrated in the network examples.

Figure 6 with a cost ratio of 50 represents a relatively high water cost. In this case it becomes advantageous to reduce the water flow rate. This can be accomplished by increasing the quality, and by operating at lower pressures; both these changes increase the energy per unit mass.

However, if it is necessary or desirable for other reasons to operate at low quality, then it is better to operate at high pressure. This is because the enthalpy of steam for qualities less than about 0.6 increases with increasing pressure. In the region of qualities between 0.6 and 0.7, there is a crossover from high to low operating pressure to maintain the lowest cost. In this region the cost is not as sensitive to the operating pressure. The enthalpy of steam in this region of qualities is essentially independent of pressure for the pressures under consideration (see figure 1). If a facility were to operate at a

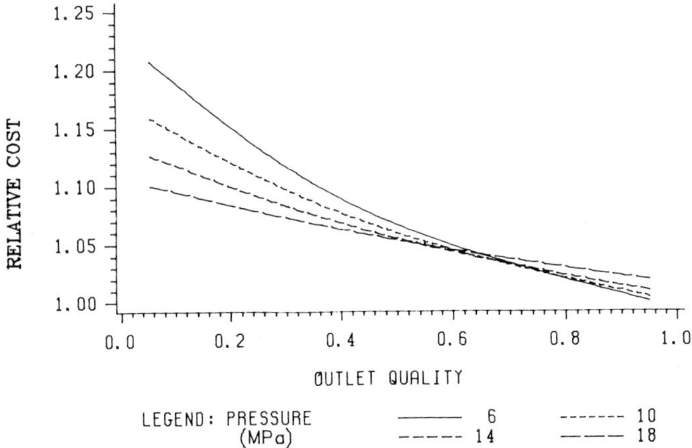

Figure 6: Relative Cost, Cost Ratio of 50

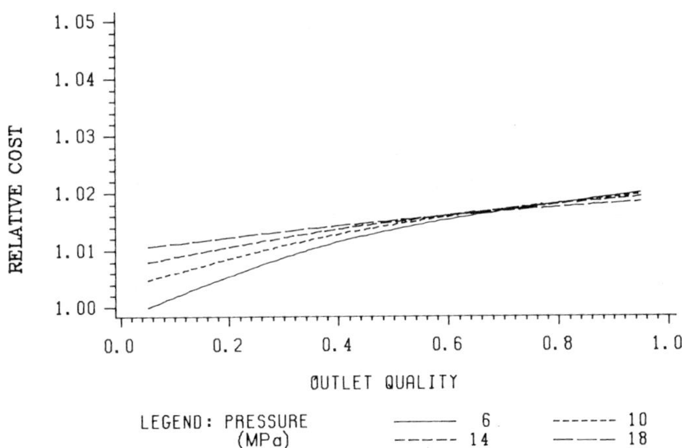

Figure 7: Relative Cost, Cost Ratio of 5000

quality in the region of crossover, the operating cost would not vary substantially with fluctuations in pressure.

The cost is most sensitive to quality changes at the lowest pressure (6 MPa). This is due to the higher heat of vaporization

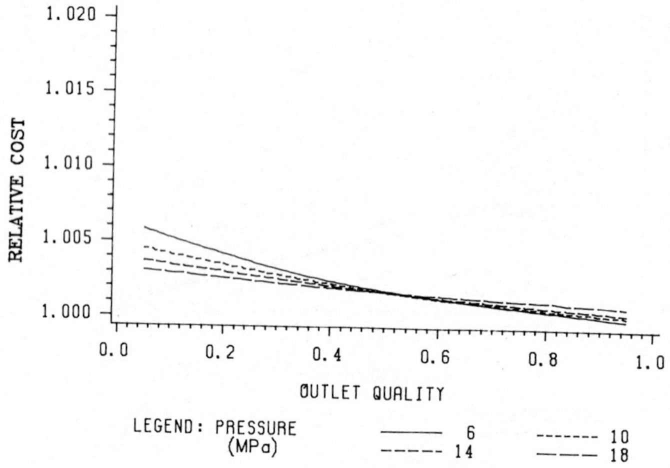

Figure 8: Relative Cost, Cost Ratio of 500

at low pressures. Also the cost is most sensitive to pressure at the lowest quality (5%). The rate of change of enthalpy with pressure is greatest for low quality. The same operating pressure gives the highest and lowest cost, depending on the quality.

Figure 7 covers the same operating conditions but for a cost ratio of 5000, a relatively high fuel cost. It now becomes advantageous to operate at low quality and low pressure, in order to use the lowest fuel rate and highest water rate possible.

Once again the lowest pressure shows the greatest sensitivity of cost to quality. The crossover region in this case is between qualities of 0.6 and 0.8. The overall cost variation is substantially less than in the first case, 2% versus 20%

Figure 8 represents a cost ratio of 500. In this case there is considerably less variation, a maximum of only 0.6%. At this cost ratio, a savings made by reducing the water flow is offset by the extra cost of the additional fuel required, and vica versa. However, even in this situation by judicious selection of operating conditions the operating costs could be reduced by 0.1% to 0.5% depending on other constraints. This could still represent a significant savings.

Another related concern is the effect of quality on pressure drop for a constant energy flow. The mass flow is lower for higher qualities in order to achieve constant energy flow. Figure 9 illustrates the marked increase in pressure drop for higher quality flow. These large pressure drops at high quality could severely restrict operating flexibility.

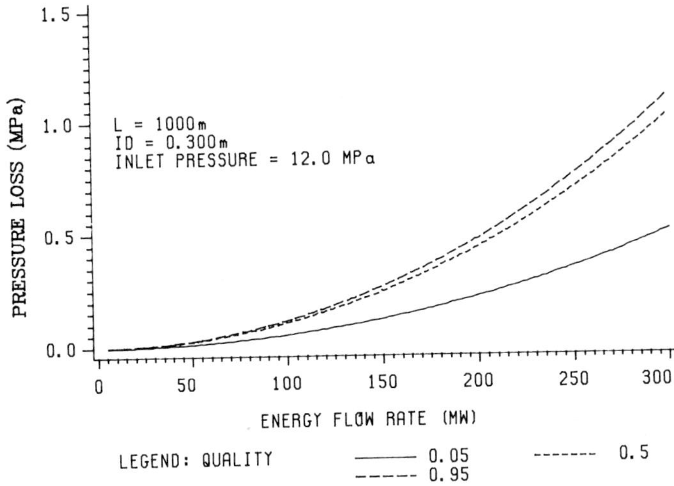

Figure 9: Two-phase Pressure Drop

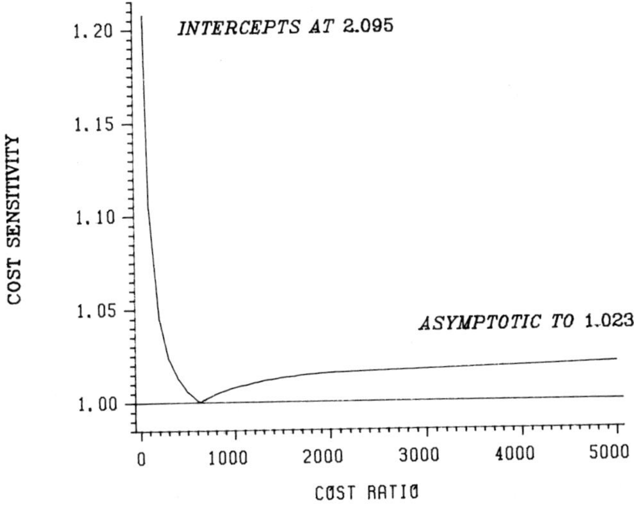

Figure 10: Cost Sensitivity as a Function of Cost Ratio

The variation in cost for a given set of constraints is also important. Figures 6, 7, and 8 have quite different cost variations. The cost sensitivity is defined as the ratio of the highest to the lowest cost for a given cost ratio in the operating region allowed. For example, figure 6 has a high cost of 1.21, and a low cost of 1.00, hence the cost sensitivity is 1.21. Figure 10 is a plot of cost sensitivity versus cost ratio for the simple pipe example. As the cost ratio approaches zero, the cost sensitivity approaches the ratio of the two extremes of water flow rate within the operating constraints. As the cost ratio becomes very large, the cost sensitivity approaches the ratio of the highest to the lowest fuel flow rate. There is a point where the cost sensitivity is actually 1.00. This occurs where the cost ratio corresponds to the ratio of water variation to fuel variation. At this point the cost is insensitive to changes in the operating conditions. Thus, operating near this point would be desirable, although normally, the cost ratio cannot be controlled by the operator. Figure 8 represents an example of a cost ratio in this region. Depending on the relative costs of fuel and water the overall cost can be quite sensitive to the operating conditions.

5.3. Network Examples

The commercial network described previously will be used to illustrate both design and operating considerations. In both cases the energy supplied will be the energy equivalent of 15.0 MPa at 0.80 quality and 10.0 kg/s (24.16 MW) to all injection sites, which is representative ot actual operating conditions.

1. Operations Case;

   For this example a cost ratio of 50 is used for cost analysis. This reflects the higher cost of recycled water and the lower cost of fuels available on site. The following constraints, which were specified by the commercial facility [7] are used.

   a) Maximum operating pressure is 15.5 MPa;

   b) Minimum injection site pressure is 12.4 MPa;

   c) Maximum boiler quality is 0.80;

   d) Maximum steam rate at each injection site is 11.0 kg/s.

The constraints apply to all the injection sites. The energy requirement and the flow rate restriction combine with the pressure constraints to form a quality restriction. For injection site pressures between 12.4 MPa and 15.5 MPa and flow rates at the maximum, in order to achieve the energy required the quality must be greater than 0.55 (see figure 1). Since the quality tends to decrease slightly due to condensation as the steam passes through the network at these flow rates and pressures, the maximum injection quality obtainable is less than 0.80. The maximum quality obtainable at the farthest injection site is approximately 0.75.

The pressure constraints are quite restrictive. Given the maximum boiler pressure (15.5 MPa), the farthest injection site can operate at a maximum pressure of only 13.6 MPa.

The interaction of constraints restricts the outlet conditions at the most distant site as follows: pressure in the range 12.4 to 13.6 MPa. quality in the range 0.55 to 0.75, flow rates in the range 10.0 to 11.0 kg/s. The challenge is to minimize the cost while operating in this rather narrow pressure and quality window.

If the energy flow requirement were reduced the allowable operating ranges would expand. Lower flow rates would be able to deliver the required energy at a lower pressure drop, which would permit operation at higher injection site pressures. Alternatively, lower qualities could be used with the same flow rates and still maintain the required energy flow. The initial energy specification is near the maximum for this network. Given these pressure, quality and flow restrictions, the network cannot deliver more than 10% more energy.

The operating conditions will of course vary throughout the network. For example, in order to operate the most distant injection site at the minimum pressure (12.4 MPa), and approximately 0.75 quality, the boiler must produce steam at 14.5 MPa, and 78% quality. The conditions at the intermediate injection sites lie between these values. The pressure and quality distributions are similar to those described by figures 3 and 4. The overall energy distribution is quite good, all injection sites are within 1.0% of the 24.16 MW specification. The energy distribution is similar to that described by figure 5. It is impossible to operate all the wells at the same pressure, quality and energy flow rate. In order to achieve the same outlet pressures, different flow rates would be required. In this example the injection site flow rates are assumed equal; the pressure and quality vary somewhat to deliver approximately the same energy flow.

Even with the fairly narrow operating ranges imposed on the network, there is still enough flexibility to reduce operating costs. For example, figure 11 shows the cost sensitivity in the operating range defined for this network. There is a 2% difference in cost between the best and worst operating conditions, even for this constrained situation. With total operating expenses expected to be $30 - 50 million per year for a full scale commercial plant, a 2% savings is quite significant.

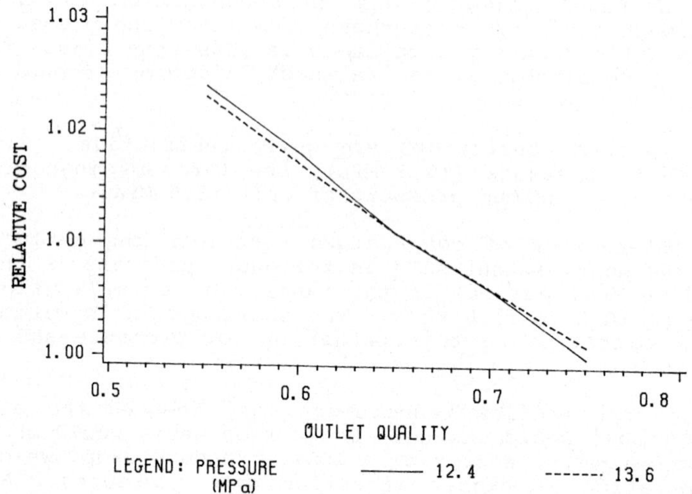

Figure 11: Cost Sensitivity, Operations Case, Cost Ratio of 50

2. Design case;

For an example of the design case the same network is used. The cost ratio is lowered to 20 to account for the effect of water flow rate on capital costs, and the constraints are relaxed as follows:

a) Operating pressure 6.0 - 18.5 MPa,

b) Maximum boiler quality 0.90,

c) Maximum steam rate subject only to physical limitations.

These relaxed constraints allow investigation of the cost sensitivity in a wider region. Other factors to consider are; pressure distribution, quality distribution, and energy distribution.

Once again the energy distribution is quite even, similar to figure 5. The deviation from the specified energy flow is at most 2%. (23.9 - 24.5 MW) for all the operating conditions within the constraints. However there is restriction to flow at low qualities (10% or less). It would be necessary to use larger pipes, and larger valves to achieve the required energy flow at low quality, because the mass flow required is too large for the pipe sizes used in this example.

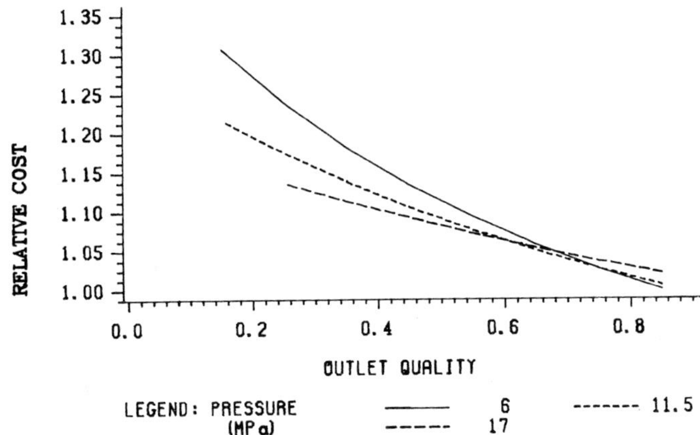

Figure 12: Cost Sensitivity, Design Case, Cost Ratio of 20

The maximum pressure available at the most distant injection site is approximately 17.0 MPa, with the boiler operating at the maximum pressure (18.5 MPa). The maximum outlet quality is approximately 0.80 for a boiler outlet quality of 0.90. The pressure variation throughout the network is most pronounced at lower operating pressures and higher operating qualities.

There is a considerable cost variation over the operating range. From figure 12, for a pressure of 6.0 MPa there is a 31% variation in cost between the operating qualities of 0.15 and 0.85, with the higher cost occurring at low quality. The lowest and highest cost both occur at low pressure (6.0 MPa).

5.4. Network Summary

There are two major factors that restrict the potential operational savings, namely: the cost sensitivity, which is dependent on the cost ratio and the boiler efficiency; and the size of the operational window, which is set by the network constraints established during the design phase.

Within the operating window defined by the network constraints, it is possible to identify the optimum operating conditions based on the operating cost. In general the optimum occurs at the lowest allowable pressure, and at either the highest or lowest quality depending on the cost ratio. Therefore it is important to accurately estimate the cost ratio.

A savings of even 1% per year can be extremely significant when applied to the large scale projects typical of the tar sands.

6. CONCLUSIONS

1. A steady state simulation of wet steam distribution networks has been developed, and verified by commercial data.
2. The simulation enabled the analysis of cost sensitivity to be done, which demonstrated the importance of selecting appropriate operating conditions.
3. Operating constraints can combine to severely limit flexibility, and thereby reduce the potential for savings in the event of cost fluctuations.
4. It is impossible to provide equal injection site pressures and qualities, and an even energy distribution, but an acceptable energy distribution can be achieved by adjusting the mass flow distribution.

ACKNOWLEDGEMENTS

This work has been supported primarily by the Natural Sciences and Engineering Research Council of Canada. We would also like to acknowledge the assistance of Esso Resources Canada Ltd. for provision of plant data.

NOMENCLATURE

A = cross sectional area of pipe (m^2)
C_c = contraction coefficient
D = pipe internal diameter (m)
f_{tp} = the two-phase flow friction factor
G = the mass flux (kg/m^2s)
H = total enthalpy (J/kg)
H_{fw} = enthalpy of the feed water (J/kg)
L = length (m)
P = pressure (Pa)
Q = heat addition or loss (W)
S = ratio of cross sectional areas, A_1/A_2
T = temperature
T_{fw} = temperature of the feedwater (degrees C)
V = velocity (m/s)
V_f = specific volume of the liquid phase (m^3/kg)
V_g = specific volume of the gaseous phase (m^3/kg)
V_{fg} = specific volume of vaporization (m^3/kg)
W = mass flow rate (kg/s)
x = quality

The subscript 1 denotes an inlet condition, and the subscript 2 denotes an outlet condition. For example x_1 is inlet quality and x_2 is outlet quality.

REFERENCES

1. Fouda, A. E., "Two-Phase Flow Behaviour in Networks and Manifolds", PhD Thesis, University of Waterloo, 1975.

2. Collier, John G., "Convective Boiling and Condensation", McGraw-Hill, London, 1972.

3. "Control Valve Handbook", 2nd Edition, Fisher Controls Company, Marshalltown, Iowa, 1977.

4. Bergles, A. E., Collier, J. G., Delhaye, J. M., Hewitt, G. F. and Mayinger, F., Two-Phase Flow and Heat Transfer in the Power and Process Industries . Hemisphere Publishing Corporation, Washington, 1981.

5. Lin, Yu-Hwan, Nielsen, R. S. and Ray, A., "Fuel Controller Design in a Once-Through Subcritical Steam Generator System", Journal of Engineering for Power, vol. 100, pp. 189-196, January, 1978.

6. Gay, B. and Middleton P., "The Solution of Pipe Network Problems", Chemical Engineering Science, vol. 26, pp109-123, 1971.

7. Telford, A. S., Moyensen, A. C. and Chambers, N. A., "Cold Lake High Pressure Steam Distribution System", Paper presented at Welding Institute of Canada International Conference, Calgary, Alberta, Canada, November 10, 1980.

8. Gomaa, Ezzat E., "Correlations for Predicting Oil Recovery by Steamflood", Journal of Petroleum Technology, vol. 32, pp. 325-332, February 1980.

9. Whalley, M. J., Wilson, T.M., "Water Conservation in a Steam Stimulation Project", The Future of Heavy Crude Oils and Tar Sands, First International Conference, Edmonton, Alberta, Canada, June 4-12, 1979, pp. 734-738.

10. Robertson, W. S., Boiler Efficiency and Safety , Macmillan Press Ltd, London, 1981.

Multi-Phase Flow and Heat Transfer III. Part A: Fundamentals
edited by T.N. Veziroğlu and A.E. Bergles
Elsevier Science Publishers B.V., Amsterdam, 1984 — Printed in The Netherlands

STRONG PRESSURE DISTURBANCES PASSING THROUGH A BUBBLY MIXTURE

C. C. Chang[1] and S. G. Bankoff
Chemical Engineering Department
Northwestern University
Evanston, Illinois 60201, U.S.A.

ABSTRACT

Strong steady shock waves propagating through a liquid containing small gas bubbles are investigated numerically. The complete bubble dynamics equation, together with the equations of continuity and motion, are employed under the assumptions of zero relative velocity between bubbles and liquid, and isothermal compression. The resulting pressure and velocity profiles show that there is an exponential increase at the shock front and an oscillatory wave attenuating at the back side of the shock wave. In the weak shock limit the results agree with theory.

INTRODUCTION

The problem of the nature of a pressure disturbance passing through a liquid containing small gas bubbles has received considerable attention, both for its practical and theoretical interest. Such problems occur in practice in connection with underwater signalling, transient operation of process equipment, hypothetical nuclear reactor accidents, and possibly also in connection with vapor explosions. From a theoretical point of view the problem has unique features, such as the possibility of finite-amplitude signal propagation in the form of solitons and oscillatory wave packets. The acoustic theory has been worked out by a number of authors [1-5]. Considerable progress has also been made on the one-dimensional propagation of finite, but weak, perturbations through a bubbly mixture with small void fraction. Defining a small parameter ε, as the ratio of the pressure amplitude to the undisturbed pressure, one obtains the Korteweg-deVries (KdV) or Burgers-Korteweg-deVries (BKdV) equations by retaining terms including $O(\varepsilon^2)$. The BKdV equation differs from the KdV in retaining viscous dissipation effects. A number of authors have contributed to this effort [6-21], among whom van Wijngaarden and Nakoryakov, et al. have been prominent. The properties of the KdV and BKdV equations have been the subject of intense activity in recent years, in view of the applications which have been found for nonlinear phenomena in a variety of fields, including nonlinear dispersive transmission lines, deep-water waves, plasmas, Josephson junctions, shock waves propagating through lattices, quantum physics, etc. A key feature is the possible appearance of solitons (solitary waves), oscillatory wave packets and/or shock waves. Related equations, such as sine-Gordon equation, nonlinear Schrodinger, etc.) which have soliton solutions, are also found in a number of applications. Some interesting properties are the following:

[1]On leave from Institute for Nuclear Energy Research, Republic of China, whose support is gratefully acknowledged.

1. The KdV equation has solitary wave solutions, and also periodic cnoidal wave solutions which can be written in terms of Jacobi elliptic functions. The solitary waves form a one-parameter family, with the taller solitary waves traveling faster, and being narrower, than shorter ones.

2. If a solitary wave is initially placed on the real line to the left of a shorter one, both traveling to the right, it will catch up and pass the shorter one, with no change in wave form or amplitude, despite the nonlinear interaction. Such solitary waves are termed solitons. Any localized initial condition for the KdV equation will evolve into two parts, solitons propagating off to the right and an oscillatory wave with time-decreasing amplitude to the left. The existence of solitons establishes that, if friction is neglected, long water waves do not necessarily continue to steepen in front and become less steep behind. By a nonlinear transformation of the KdV equation it can be shown that there is an infinite sequence of polynomial conservation laws. One such transformation leads an exact solution of the initial-value problem by an inverse scattering method [22]. Another sequence of transformations are the Bäcklund transformations, consisting of a pair of first-order partial differential equations relating the solution of the given equation to another solution of the same equation as to a solution of another second-order equation. These are developed formally through the theory of Lie groups and the differential geometry of manifolds [23].

3. The theory of weakly nonlinear deep-water gravity wave trains leads to a nonlinear Schrödinger equation, which governs the evolution of the envelope of the carrier wave. This equation can be related to the KdV equation by a nonlinear transformation. Three important features exist which are not found in linear wave theories: (i) the existence of envelope solitons (ii) the instability of a uniform wave train to modulational perturbations (Benjamin-Feir instability) [25] (iii) the time-periodic return of the unstable wave train to its initial state (Fermi-Pasta-Ulam (FPU) recurrence) [26]. FPU recurrence for the KdV equation (27) suggests that a wind-wave system on deep water can be characterized to first order by a single nonlinear wave train having a carrier frequency equal to the dominant frequency in the wind-wave model (nonlinear coherent model) [28].

With such a wealth of fascinating features established for weakly nonlinear systems, and in particular the bubbly liquid system, it is natural to inquire whether these same properties can be found for strong disturbances. Analytical solutions are no longer possible, although van Wijngaarden has expanded the equations near the front and back of a strong shock and deduced some valuable features. A direct numerical attack is therefore required, but the results are difficult to interpret. We therefore look for the class of solutions consisting of a steady wave train relative to an observer moving with a prescribed velocity, U. Since the wave train moves steadily into an undisturbed region, the initial conditions are homogeneous, and the prescribed velocity U is an eigenvalue for the nonlinear problem. This problem has been previously considered by Crespo [30] and Hijikata, et al. [31], and is further analyzed by Tan [32].

STATEMENT OF THE PROBLEM

We consider a mixture of gas bubbles, initially all of radius R_o, in a fluid of density ρ_f. In the undisturbed region, the pressure in both the fluid and bubbles is P_o. Because the density of the gas can be neglected, one has

$$\rho = \rho_f(1 - \alpha) \tag{1}$$

where ρ is the density of the mixture. If we assume that the relative velocity between the bubbles and the fluid is negligible compared to the disturbance velocity, U, then in a unit mass of the mixture the mass of gas can be considered to be constant. Therefore

$$\frac{\rho_g \alpha}{1 - \alpha} = \frac{\rho_{go} \alpha_o}{1 - \alpha_o} = \text{constant} \tag{2}$$

The behavior of the bubbles is taken to be isothermal over a large range of frequencies [4]:

$$\frac{P_g}{\rho_g} = \frac{P_{go}}{\rho_{go}} = \text{constant} \tag{3}$$

The conservation of mass requires

$$\frac{\partial \rho}{\partial t} + \frac{\partial(\rho u)}{\partial x} = 0 \tag{4}$$

The momentum equation in which the viscosity effect is neglected is

$$\frac{\partial u}{\partial t} + u\frac{\partial u}{\partial x} = -\frac{1}{\rho}\frac{\partial P}{\partial x} \tag{5}$$

The bubble dynamical equation gives the relation between the pressure in an isolated bubble in an infinite fluid and the far-field pressure, P_∞. It is now assumed that P_∞ is the average pressure P at (x,t), whence

$$P_g - P = \rho_f\left[R\frac{d^2R}{dt^2} + \frac{3}{2}\left(\frac{dR}{dt}\right)^2 + \frac{4\nu_f}{R}\frac{dR}{dt} + \frac{2\sigma}{\rho_f R}\right] \tag{6}$$

Following van Wijngaarden [8], eqs. (1)-(3) can be substituted in eqs. (4)-(6) to give:

$$\frac{\partial P_g}{\partial t} + \frac{P_g\left[P_g(1-\alpha_o) + P_{go}\alpha_o\right]}{P_{go}\alpha_o}\frac{\partial u}{\partial x} + u\frac{\partial P_g}{\partial x} = 0 \tag{7}$$

$$\frac{\partial u}{\partial t} + u\frac{\partial u}{\partial x} + \frac{P_g(1-\alpha_o) + P_{go}\alpha_o}{\rho_f P_g(1-\alpha_o)}\frac{\partial P}{\partial x} = 0 \tag{8}$$

$$\frac{P_g - P}{\rho_f} = -\frac{R_o^2 P_{go}^{2/3} P_g^{-5/3}}{3}\left(\frac{d^2P_g}{dt^2}\right) + \frac{11R_o^2 P_{go}^{2/3} P_g^{-8/3}}{18}\left(\frac{dP_g}{dt}\right)^2 - \left(\frac{4\nu_f}{3P_g}\right)\frac{dP_g}{dt} + \frac{2\sigma P_g^{1/3}}{\rho_f R_o P_{go}^{1/3}} \tag{9}$$

We seek steady solutions for an observer moving with the disturbance velocity U, so that $\frac{\partial}{\partial t}$ may be replaced by $-U\frac{d}{dx}$. Eqs. (7)-(9) can now be written as follows

$$(u-U)\frac{dP_g}{dx} + \frac{P_g\left[P_g(1-\alpha_o) + P_{go}\alpha_o\right]}{P_{go}\alpha_o}\frac{du}{dx} = 0 \tag{10}$$

$$(u-U)\frac{du}{dx} + \frac{P_g(1-\alpha_o) + P_{go}\alpha_o}{\rho_f P_g(1-\alpha_o)}\frac{dP}{dx} = 0 \tag{11}$$

$$\frac{R_o^2 P_{go}^{2/3} P_g^{-5/3} U^2}{3}\frac{d^2P_g}{dx^2} - \frac{11R_o^2 P_{go}^{2/3} P_g^{-8/3} U^2}{18}\left(\frac{dP_g}{dx}\right)^2 + \frac{4\nu_f U}{3P_g}\frac{dP_g}{dx}$$

$$- \frac{2\sigma P_g^{1/3}}{\rho_f R_o \rho_{go}^{1/3}} + \frac{P_g - P}{\rho_f} = 0 \tag{12}$$

Given the initial conditions α_o, P_{go}, and R_o, this set of homogeneous differential equations can be solved numerically for P_g, P and u. The disturbance velocity U is a free parameter which determines the strength of the disturbance.

NUMERICAL RESULTS

Eqs. (10)-(12) were integrated by a fourth-order Runge-Kutta algorithm for suspension of air bubbles in water at 0-1 MPa and 20°C, with initial radius 1 mm, and initial void fractions α_o = 0.005, 0.01, 0.02, 0.03, 0.04, and 0.05. The disturbance velocity U was obtained by modeling the disturbance front as a shock wave, and employing the jump balance equations:

$$\rho_f U(1-\alpha_o) = \rho_f\left(1 - \frac{P_o\alpha_o}{P_1}\right)(u_1+U) \tag{13}$$

$$P_o + \rho_f U^2(1-\alpha_o) = P_1 + \rho_f\left(1 - \frac{P_o\alpha_o}{P_1}\right)(u_1+U)^2 \tag{14}$$

where u_1 and P_1 are the mixture velocity and pressure behind the shock front. By choosing the shock pressure P_1, these equations can be solved for u_1 and U. The shock pressures which were investigated were 0.5, 1, 1.5, 2, 2.5 and 3 MPa. In addition to the water-air system, a mixture of equal volumes of glycerin and water containing helium bubbles was investigated because of its interest for experimental work. It is easier to produce uniform small bubbles, with minimum coalescence, in a viscous liquid, while the low heat capacity of helium makes the isothermal assumption more realistic. The parameters ranges were the same above, except:

$$R_o = 0.5 \text{ mm};\ \nu_f = 7.31 \times 10^{-6}\ \text{m}^2/\text{s};\ \rho_f = 1.145 \times 10^3\ \text{kg/m}^3.$$

Typical results are shown in Figs. 1-3. Many other results can be found in Ref.28. Fig. 1 shows P(t) and $P_g(t)$ for P_1 = 1.5 MPa and α_o = 0.03. One sees a nonlinear oscillatory wave train, where the gas pressure shows very sharp peaks, corresponding to rapid overcompression of the collapsing bubbles. In a real system, where many bubble sizes are present, one would expect considerable incoherence, which would smooth out these peaks. This is, in fact, observed experimentally [32]. The viscous damping of the liquid pressure oscillations is also apparent. While the envelopes of the liquid pressure and velocity oscillations smoothly contract due to viscous damping, the envelope of the gas pressure oscillations shows marked instabilities.

However, these are numerical instabilities associated with the sharp changes in slope, rather than instabilities analogous to Benjamin-Feir instabilities. This is shown by Fig. 2 in which P_1 has been reduced to 0.5 MPa, resulting in apparently regular behavior, and also in later calculations [32] using an integration routine for stiff differential equations. As expected, the envelope of the liquid pressure oscillations contracts more rapidly in the more viscous mixture, as shown in Fig. 3.

DISCUSSION

From Figs. (1)-(3) one can observe an exponential pressure rise at the shock front, in accordance with the weak-shock analysis of van Wijngaarden [19]. An estimate of the shock thickness Δ, in terms of the pressure ratio, P_1/P_o, initial bubble radius, R_o, and initial void fraction, α_o, is given by

$$\Delta = \frac{C\, R_o}{\alpha_o^{1/2}\left(1-\dfrac{P_o}{P_1}\right)^{1/2}} \tag{15}$$

where C is a constant of proportionality.
Table 1 shows that for pressure ratios ranging from 1.5 to 5, the shock thickness decreases with pressure ratio, although less rapidly than predicted by Eq. (15). In fact, as the pressure ratio is further increased, Δ begins actually to increase. This is related to the increase in disturbance velocity and in bubble inertial effects as P_1 increases. On the other hand Table 1, (together with Figs. 18-23 of [29]) shows that the shock thickness is roughly proportional to the bubble size and inversely proportional to the square root of the void fraction for weak shocks (P_1 = 0.15 MPa) in accordance with Eq. (15), but deviate for strong shocks. The constant of proportionality for Eq. (15), based on numerical results, is shown in Table 2 for the weaker shocks. For pressure ratios of 1.5 and 2, the constants have an average value of 3.7 and 4.24, respectively, whereas Noordzij made a theoretical estimate, based on the Korteweg-de Vries equation, of 3.9 for isothermal conditions [19].

Again for weak shocks, the waves at the back side of the shock wave are predicted by Wijngaarden to have the wavelength [30].

$$\lambda = \frac{2\pi R_o\left(\dfrac{P_o}{P_1}\right)^{1/3}}{\left[3\left(\alpha_o - \dfrac{\alpha_o P_o}{P_1}\right)\right]^{1/2}} \tag{16}$$

This equation does not agree with the results of strong shocks (Table 3), although the general trend that the wavelength decreases as both the void fraction and pressure disturbance increase is in agreement with Eq. (16). A comparison of the numerical results with Eq. (16) for pressure ratios of 1.1 and

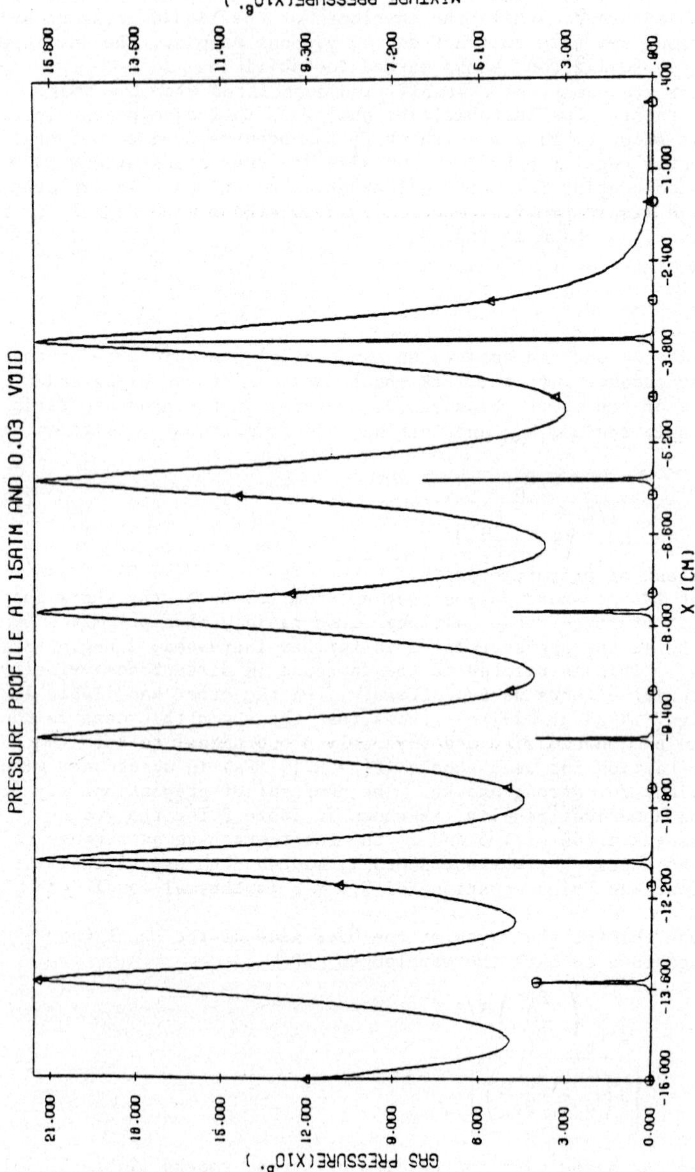

Figure 1. Pressure profiles at 15 atm. and 3 % void fraction for water-air.

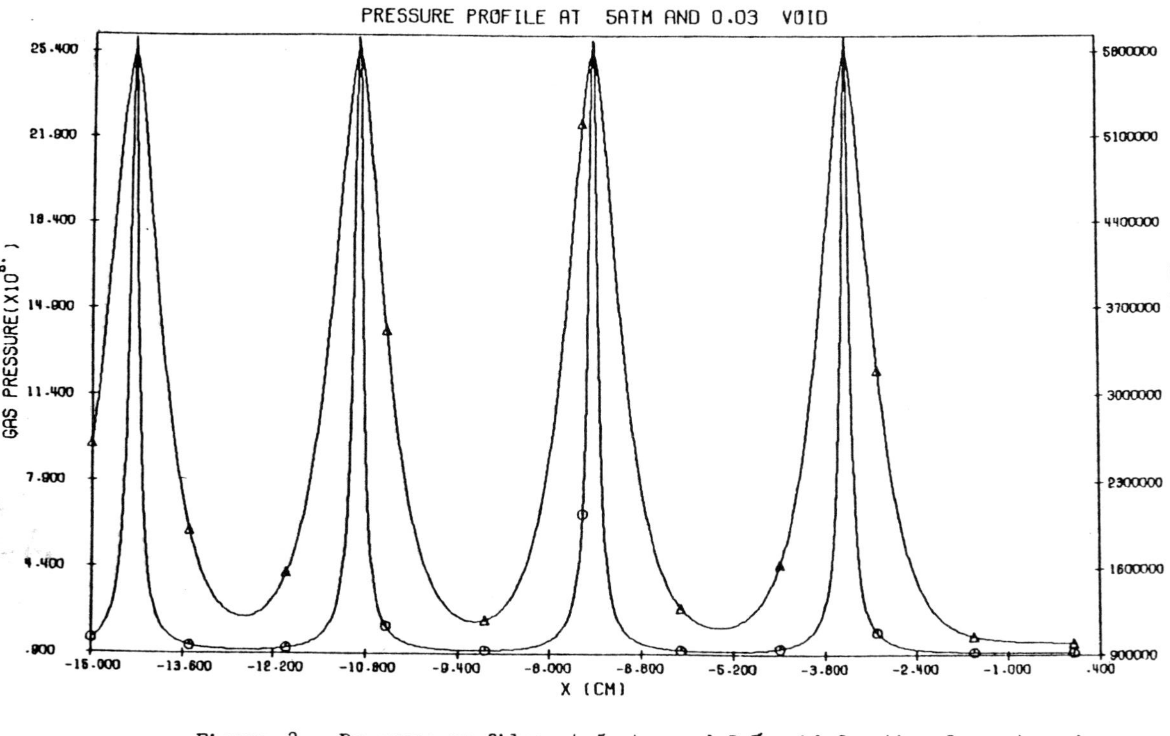

Figure 2. Pressure profiles at 5 atm. and 3 % void fraction for water-air.

LEGEND
- GAS PRESSURE
- MIXTURE PRESSURE

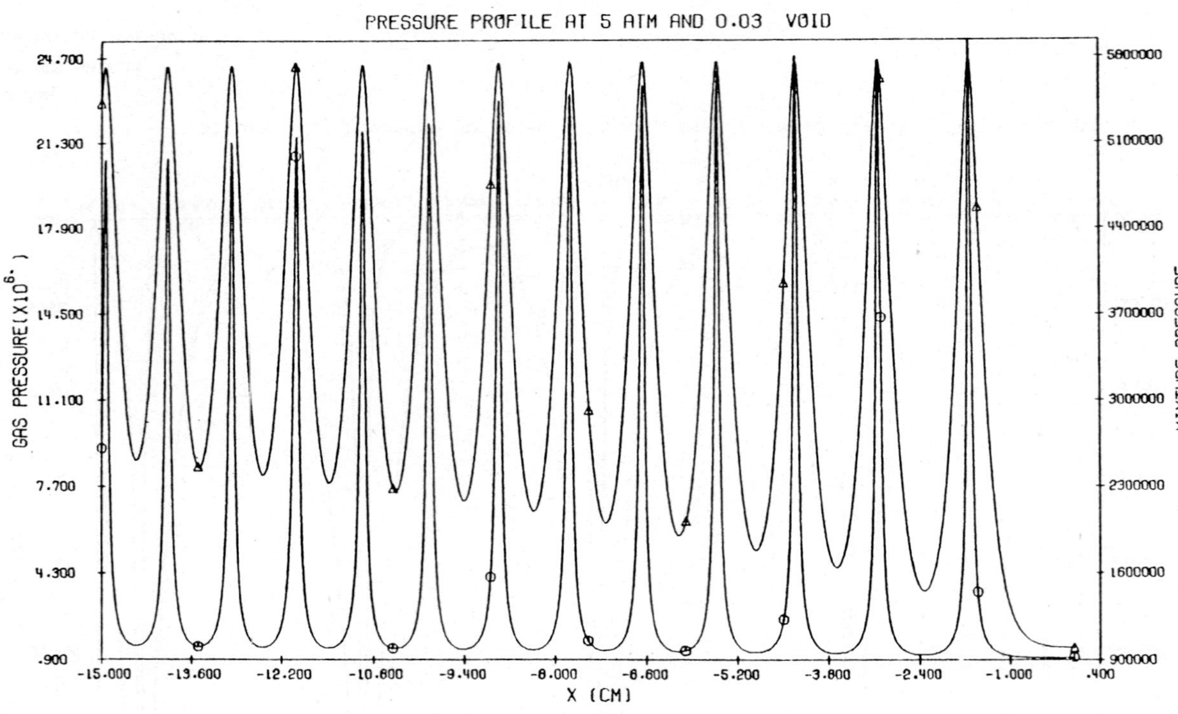

Figure 3. Pressure profiles at 5 atm. and 3 % void fraction for glycerine-water-helium.

Void Fraction \ Pressure, $10^5 n/m^2$	1.5	2	5	10
0.005	4.455	4.185	3.945	4.005
0.01	3.165	2.97	2.79	2.835
0.02	2.25	2.115	1.98	2.01
0.03	1.86	1.74	1.635	1.65
0.04	1.62	1.515	1.425	1.44
0.05	1.455	1.365	1.275	1.29

Table 1. Shock thickness at various conditions for glycerine-water-helium, m x 10^{-2}.

Void Fraction \ Pressure 10^5 n/m^2	1.5	2	5
0.005	3.65	4.18	4.99
0.01	3.64	4.18	4.98
0.02	3.69	4.23	4.95
0.03	3.72	4.24	5.1
0.04	3.77	4.33	5.1
0.05	3.73	4.26	5.1

Table 2. Constant of proportionality for Wijngaarden's shock thickness equation.

10 is shown in Figure 4. One sees that the deviation is quite large for the strong shock. Even for the weaker shock, there is still about a 10% deviation from the linearized prediction, which is in agreement with the magnitude of the dimensionless pressure disturbance. Another factor is that surface tension effects were neglected in the Wijngaarden analysis.

CONCLUSIONS

A class of steady solutions to the problem of a strong shock passing through a bubbly gas-liquid mixture is demonstrated numerically. The assumptions of zero relative velocity and isothermal compression were made, but these are not crucial to the development, and drag forces acting on the bubbles and

Pressure (atm.) / Void Fraction	1.5	2	5	10
0.005	5.98	4.8	3.39	2.99
0.01	4.32	3.4	2.42	2.13
0.02	3.04	2.43	1.72	1.415
0.03	2.5	2.0	1.4	1.24
0.04	2.2	1.74	1.22	1.08
0.04	2.0	1.58	1.11	0.975

Table 3. Wavelength at back side of shock wave for glycerine-water-helium, m x 10^{-2}.

polytropic compression can readily be accommodated. The solutions all exhibited a steep front followed by an oscillatory wave train. No solitary waves were found in the parameter ranges studied. Because of the extremely sharp bubble pressure peaks, the question at the existence of modulational instabilities remains unanswered, but seems unlikely. The wavelength agrees with the theoretical result of Wijngaarden for weak shocks, but diverges sharply for strong shocks ($P_1/P_2 > 2$).

REFERENCES

1. Karplus, H.B., "Propagation of Pressure Waves in a Mixture of Water and Steam". USAEC Report ARF-4123-12, Armour Research Foundation (1961).

2. Grolmes, M.A. & Fauske, H.K., "Propagation Characteristics of Compression and Rarefaction Pressure Pulses in One-Component Vapor-Liquid Mixtures". Nucl. Eng. Des., II(I), 137 (1969).

3. Miyazaki, K. & Fujii-E Y. J. Nucl. Sci. & Tech., 14(5), 327 (1977).

4. Plesset, M.S. and Hsieh, D.Y., Phys. Fluids, 4, 970 (1961).

5. Semenov, N.I. and Kosterin, S.I., "Results of Sound Speed Studies in Flowing Gas-Liquid Systems", Teploenergetika II(6), (1964).

6. Benjamin, T.B., Proc. 6th Symp. Naval Hydrodyn. (ed. Cooper, R.D. and Dodoff, S.W.), 121, Office Naval Res., Washington (1966).

7. Benjamin, T.B., "Wave Formation in Laminar Flow Down an Inclined Plane", J. Fluid Mech., 2, 554 (1957).

8. Wijngaarden, L.v., J. Fluid Mech., 33, 465 (1968).

9. Batchelor, G.K., "Compression Waves in a Suspension of Gas Bubbles in a Liquid", Proc. 8th Symp. on Advanced Problems and Methods in Fluid Dynamics, Tarda (Poland), 1967.

10. Wijngaarden, L.v., "On the Structure of Shock Waves in Liquid-Bubble Mixture", Appl. Sci. Res. 22, 366 (1970).

11. Wijngaarden, L.v., "Sound and Shock Waves in Bubbly Liquids", Congress Göttingen, 1979.

12. Kutateladze, S.S., Nakoryakov, V.E. and Pokusaev, B.G., "Experimental Investigation of Wave Processes in Gas and Vapor-Liquid Media", International Centre for Heat and Mass Transfer Seminar, Dubrovnik, Yugoslavia (1978).

13. Nakoryakov, V.E., Sobolev, V.V. and Shreiber, I.R., "Long Wave Perturbations in Gas-Liquid Mixture", Izv. Akad. Nauk SSSR, Nekh. Zhidk. i Gaza, 5, 71 (1972).

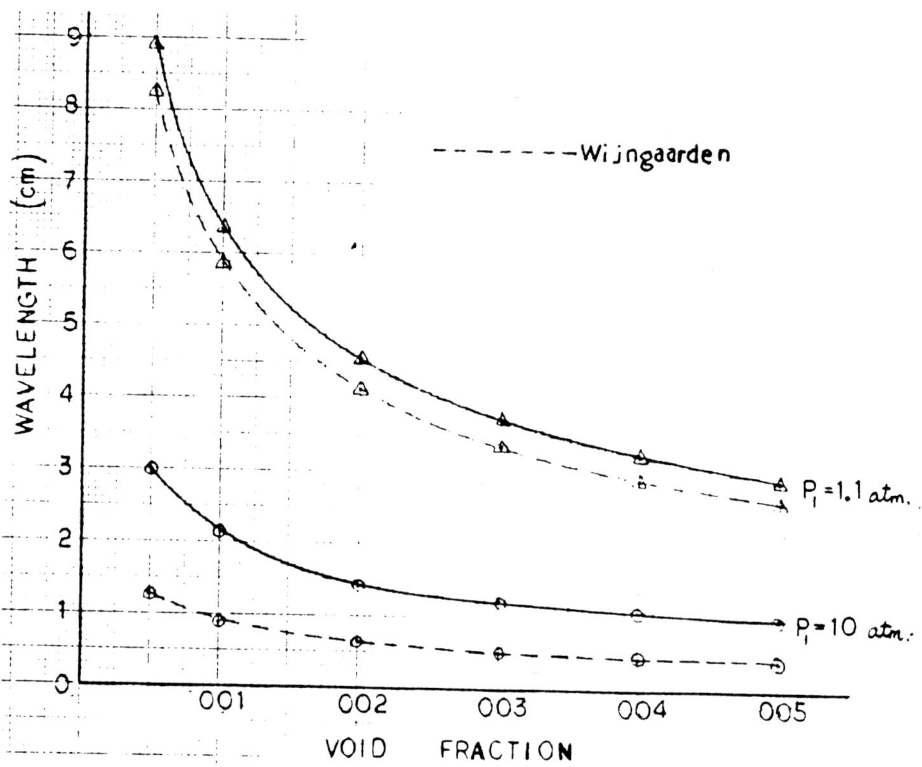

Figure 4. Comparison of wavelength with Wijngaarden's results.

14. Noordzij, L. and Wijngaarden, L.v., "Relaxation Effects Caused by Relative Motion on Shock Waves in Gas-Bubble Liquid Mixtures", J. Fluid Mech., 66, 116 (1974).

15. Nigmatulin, R.I. and Shagapov, V. Sh., "Structure of Shock Waves in Liquid with Gas Bubbles", Izv. Akad. Nauk SSSR, Mekh. Zhidk. i Gaza, 6, 30 (1974).

16. Kuznetsov, V.V., Nakoryakov, V.E., Pokusaev, B.G. and Shreiber, I. R., "Experimental Investigation of Perturbation Propagation in Liquid Gas Bubbles." In: Nonlinear Wave Processes in Gas-Liquid Flows, ed. by Kutateladze, S.S., Novosibirsk, 32 (1977).

17. Kuznetsov, V.V., Nakoryakov, V.E., Pokusaev, B.G. and Shreiber, I.R., "Propagation of Perturbations in a Gas-Liquid Mixture", J. Fluid Mech., 85, 1 (1978).

18. Borisov, A.A., Gelfand, B.E., Gubaidullin, A.A. et al., "Shock Waves Intensification in Liquid with Gas Bubbles". In: Nonlinear wave processes in gas-liquid flows, ed. by Kutateladze, S.S., Novosibirsk, 67 (1977).

19. Wijngaarden, L.v., Ann. Rev. Fluid Mech., 4, 369 (1972.

20. Nakoryakov, V.E., Sobolev, V.V. and Shreiber, I.R., In Wave Processes in Two-Phase Media"(ed. by Kutateladze, S.S.), 5, Novosibirsk: Inst. Thermophys., Siberian Branch USSR Acad. Sci. (1975).

21. Noordzij, L., Proc. IUTAM Symp., Non-Steady Flow of Water at High Speeds , (ed. Sedov, L.I. and Stepanov, G.Yu.), 369, Moscow: Nauka (1973).

22. Miura, R.M., in "Solitons in Action", ed. by Lonngren, K. & Scott, A., Academic Press, N.Y., 1-17 (1978).

23. Hermann, R. ibid, p. 33-59.

24. Yuen, H.C. and Lake, B.M., ibid, p. 39-126.

25. Benjamin, T.B. and Feir, J.E., "The Disintegration of Wave Trains in Deep Water", Part I, J. Fluid Mech., 23, 417 (1967).

26. Fermi, E., Pasta, J. and Ulam, S., "Studies of Nonlinear Problems", in Collected Papers of Enrico Fermi , Univ. Chicago Press, 2, p. 978-988 (1962).

27. Zabusky, N.J. and Kruskal, M.D., Phys. Rev. Letters, 15, 240 (1965).

28. Chang, C.C., M.S. Thesis, Dept. of Chemical Engineering, Northwestern University, Evanston, Illinois (1981).

29. Wijngaarden, L.v., in Progress in Heat and Mass Transfer , Pergamon Press, Oxford, 6, 637 (1971).

30. Crespo, A., Phys. Fluids, 12, 970 (1961).

31. Hijikata, K., Mori, Y., Nagasaki, T. and Nagakawa, M., in Two-Phase Flow Dynamics , A.E. Bergles and S. Ishigai; eds., Hemisphere Publ. Co., N.Y., 239-254 (1981).

32. Tan, M.J., Ph.D. Thesis, Department of Chemical Engineering, Northwestern University, Evanston, Illinois (1982).

Multi-Phase Flow and Heat Transfer III. Part A: Fundamentals
edited by T.N. Veziroğlu and A.E. Bergles
Elsevier Science Publishers B.V., Amsterdam, 1984 — Printed in The Netherlands

ULTRASONIC WAVE PROPAGATIONS IN A BUBBLY GAS-LIQUID TWO-PHASE FLOW

E.C. Morala[1], D. Cheong[1], P.T. Wan[1], G.A. Irons[2] and J.S. Chang[1]

1. Department of Engineering Physics, McMaster University, Hamilton, Ontario, Canada, L8S 4M1

2. Department of Metallurgy and Materials Science, McMaster University, Hamilton, Ontario, Canada

ABSTRACT

The study of ultrasonic wave propagations in a bubbly gas-liquid two-phase flow has been conducted both experimentally and theoretically. Ultrasonic wave interactions with a single bubble are examined for a wide range of bubble sizes, the distance from sound sources, the intersectional areas between waves and bubbles, and the rising velocities in a silicone oil to obtain fundamental waveforms for both transmission and reflection modes. The ultrasound wave propagation waveforms in a bubbly gas-liquid two-phase flow are analyzed based on the waveforms thus obtained from the single bubbles and compared with nearly uniform distributed and sized bubbly air-water two-phase flows for various range of bubble sizes, void fractions and flow rates.

1. INTRODUCTION

The study of ultrasonic wave propagations in a bubbly gas-liquid two-phase flow has been conducted both experimentally and theoretically. The study is potentially important in the area of two phase flow diagnostics [1,2] and nuclear reactor safety [3,4].

Since ultrasonic waves propagate at a relatively faster speed compared to fluid flow, its attenuation is not severely affected by the fluid flow. The study of ultrasonic wave propagations on gas-liquid two phase system has been conducted by various authors and reviewed by Akagawa [5]. However, most of these studies were conducted on the lower range of frequency (less than 0.1 MHz).

In this paper, the 0.5 to 5 MHz frequency range of ultrasonic wave interaction with gas-liquid bubbly flow was studied based on the analysis of a single bubble interaction with ultrasound, for various bubble sizes and rising velocities. Also, the effect sound velocity and time-averaged attenuation of ultrasonic wave as a function of air flow rates or void fraction was measured and some analysis is presented.

2. EXPERIMENTAL APPARATUS

2.1 Single Bubble Experiment

Experimental study on the single bubble interaction with an ultrasonic beam has been conducted with a cylindrical tank as shown in Figure 1. The

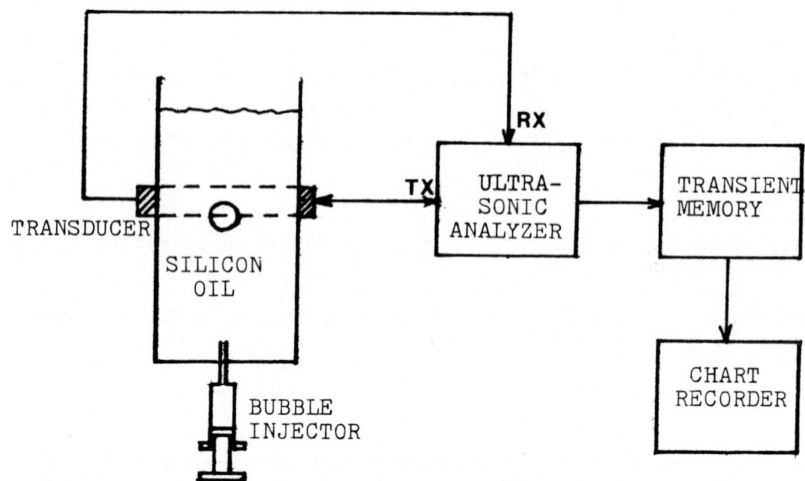

Fig.1: Experimental set-up for a single bubble rising in silicon oil.

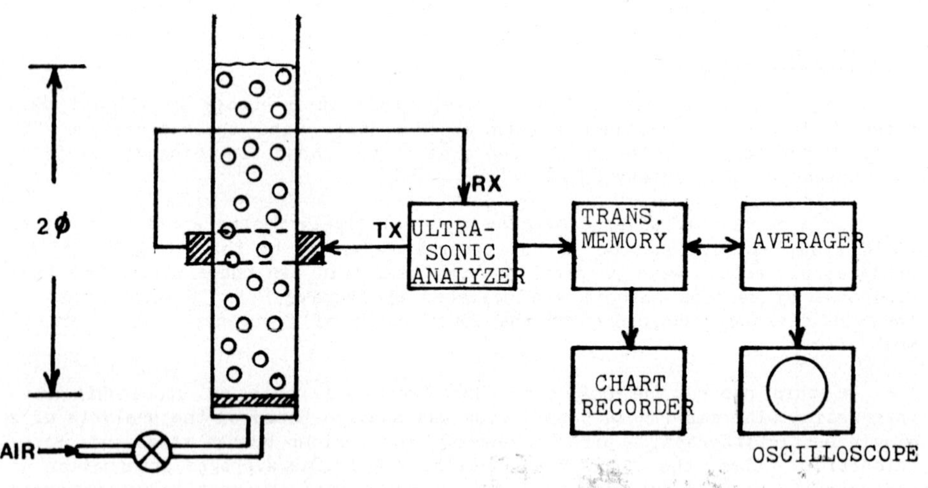

Fig.2: Experimental set-up for a multi-bubbly air water flow.

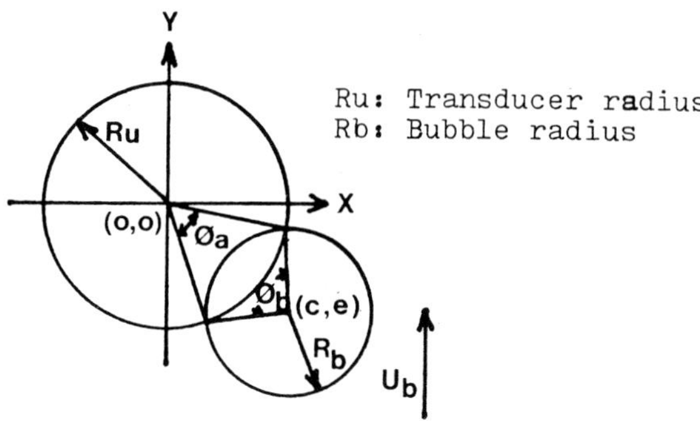

Fig.3: Typical geometric configuration between ultrasonic transducer and rising bubble.

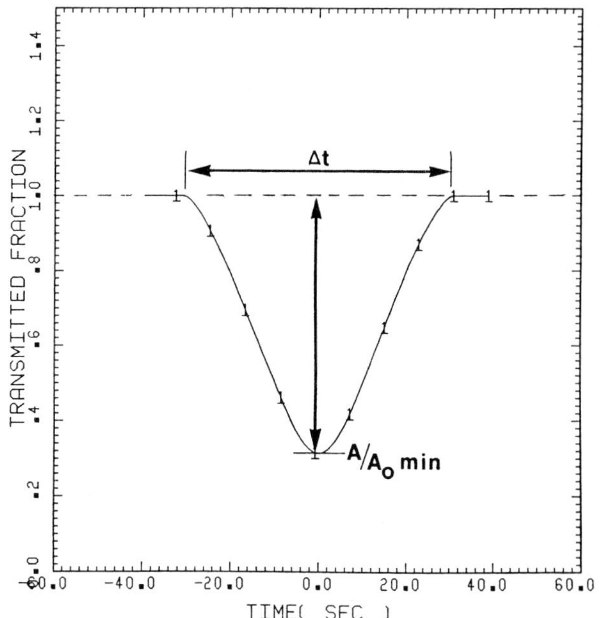

Fig.3B: Typical single bubble ultrasonic waveform Ru=0.3175cm, Rb/Ru =1.0, Ub=0.02cm/s, C/Ru =0.5

single bubble was generated by injecting a known volume of air from a syringe at the bottom of a 10 cm diameter, 20 cm long cylindrical tank filled with silicone oil. The bubble (ranging from 1.2 to 8 mm diameter) slowly rises to intercept the fixed 6.35 mm diameter ultrasonic beam. The bubble sizes, location, and rising velocities are also independently measured by photographic processes. The ultrasonic waves are generated by a Panametrics 5052 UA ultrasonic analyzer, and discharged from a contact probe type transducer. The transducers used were:

(a) 6.35 mm diameter @ 2.25 MHz frequency
(b) 12.7 mm diameter @ 5 MHz
(c) 25.4 mm diameter @ 0.5 MHz

The output of the ultrasonic analyzer is connected to transient memory (Kawasaki MR-50E) and displayed onto an oscilloscope or to a conventional chart recorder.

2.2 Multi-Bubble Experiment Apparatus

Multi-bubbly flow experiments are conducted in a vertical 10 cm wide rectangular duct with a sieve plate at the bottom, as shown in Figure 2. Air is injected from the bottom and there is no net flow of liquid. The gas phase is dispersed uniformly by the plate into bubbles of almost equal sizes along the channel. The void fraction was also determined independently from the liquid level measurements. The output of the ultrasonic analyzer was stored in the Kawasaki MR-50 E transient memory at a digitizing rate of 50 nsec. The time-averaged ultrasonic waveform were obtained in conjunction with a Kawasaki TMC-300 signal averager. In addition, accurate measurements of ultrasonic wave propagation velocity and intensity were obtained from a Tektronix 468 digital storage oscilloscope.

3. INTERACTION OF ULTRASONIC WAVES AND SINGLE BUBBLES

3.1 Theoretical Treatment

The interaction of a single bubble with an ultrasonic beam can be analyzed in a similar manner as with an optical beam [6]. Therefore, the transmitted signal, A, of an ultrasonic beam centered at (0,0) with a bubble centered at (c,e) is proportional to the unblocked projected area of the beam. That is,

$$A \propto \pi R_u^2 - f(\theta_a, \theta_b, c, e) \quad (1)$$

where θ_a, θ_b are defined in Fig. 3a, R_u and R_b are the ultrasonic beam and bubble radii, respectively. Here, it must be noted that the ultrasonic beam cannot penetrate through most of the combination of gas-liquid interphase in the present experimental frequency range of 0.5 to 5 MHz due to the acoustic impedance mismatch of this combination.

The reduction of the transmitted ultrasonic wave signal A/A_o can be calculated as a function of the sound beam radius R_u, the bubble radius R_b, the bubble rising velocity U_b and the relative position between the centers of the sound beam (0,0) and the bubble (c,e) as follows

$$\frac{A}{A_o} = \{R_u^2 [\pi - \cos^{-1} (\frac{R_u^2 - R_b^2 + d^2}{2R_u d}) - R_b^2 \cos^{-1} [\frac{R_b^2 - R_u^2 + d^2}{2R_u d}]$$

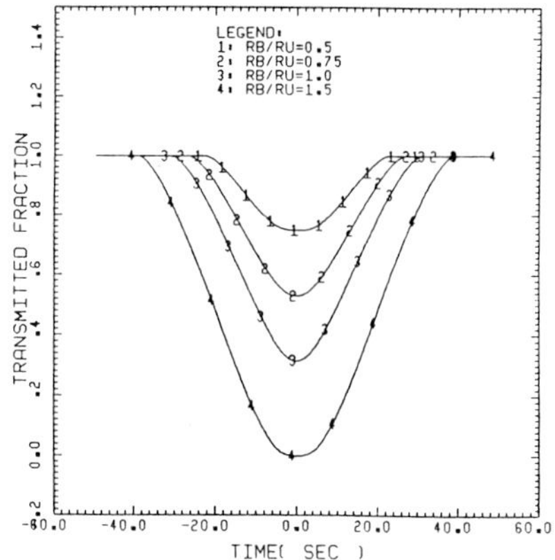

Fig.4: Single bubble ultrasonic waveform as a function of bubble/transducer radius ratio, Rb/Ru, with Ru=0.3175cm, Ub=0.02cm/s, C/Ru =0.5.

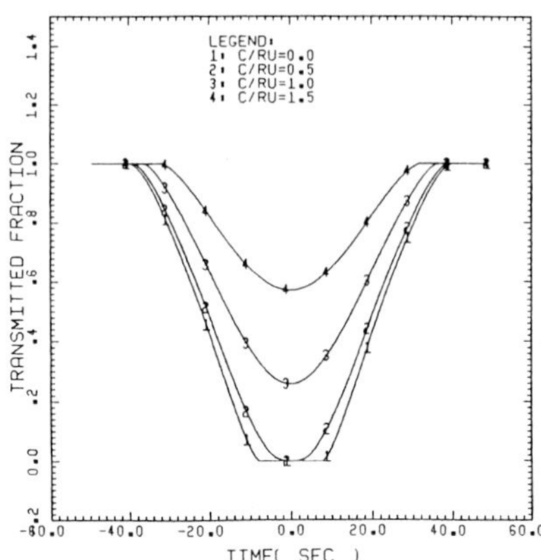

Fig.5: Single bubble ultrasonic waveform as function of distance between centres C/Ru, with Ru=0.3175cm, Rb/Ru =1.5, Ub=0.02cm/s.

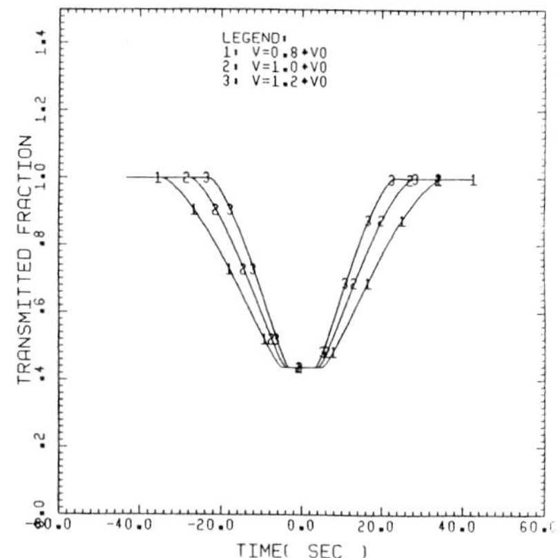

Fig.6: Single bubble ultrasonic waveform as function of bubble rising velocity; Ru=0.3175cm, Rb/Ru =0.75, Ub=0.02cm/s, C/Ru =0.0

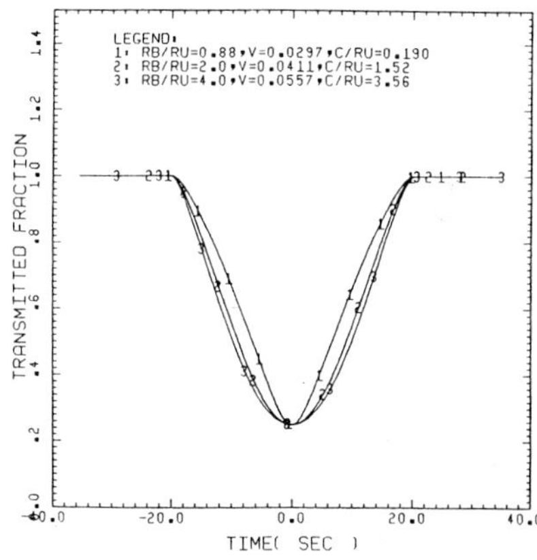

Fig.7: Single bubble ultrasonic waveform; Dt=40sec, Ru=0.3175cm, Min(A/Ao)=0.25 Bubble radius, position and rising velocity adjusted to above constrain

$$+ R_u d \left[1 - \left(\frac{R_u^2 - R_b^2 + d^2}{2R_u d}\right)^{2\ 1/2}\right] \quad \} / \pi R_u^2 \tag{2}$$

where A_o is the transmitted signal without bubbles (πR_u^2), $d = (e^2 + c^2)^{1/2}$ and $e = U_b t$. By best-fitting the theoretical curves from Eq. (2) to the experimental waveform, the bubble rising velocity, bubble size and location c can be estimated. For the limit of c = 0, the minimum transmitted signal $(A/A_o)_{min}$ becomes

$$(A/A_o)_{min} = (1 - R_b^2/R_u^2), \tag{3}$$

and the bubble rising velocity becomes

$$U_b = 2\ (R_b + R_u)/\Delta t \tag{4}$$

where Δt is the time duration of interaction between the bubble and the ultrasonic beam as shown in Fig. 3b.

Typical numerical results of Eq. (2) as a function of the transit time, where t = 0 and e = 0, for various bubble to beam radius ratios, bubble locations and bubble rising velocities as shown in Fig. 4, 5 and 6, respectively. Judging from these figures, it might be thought that the ultrasonic waveforms corresponding to different bubble sizes and rising velocities cannot be distinguished from one another when $c \neq 0$. However, Fig. 7 shows that for the same (A/A_o) and Δt, the waveforms for different R_b/R_u, U_b and c/R_u can be differentiated.

3.2 Experimental Confirmation

A typical experimental waveform and theoretical results from Eq. (2) are compared in Fig. 8. If we ignore the two small peaks ($A/A_o > 1$, shown by arrows in Fig. 8), the numerical results from Eq. (2) can be adjusted to match the experimental waveforms very closely. The two small peaks may be due to divergence effects of the small sound beam, such that the diverged ultrasonic beam may be reflected at the air-water interface as the bubble approaches the beam interaction zone. Figure 9 shows the bubble to transducer radius ratio, determined by different methods, as a function of the minimum transmitted fraction. For small bubble sizes, the radius is determined from photography. For larger bubble sizes, the radius is determined from the known volume of injected air, assuming spherical bubbles. These bubble sizes are compared to those obtained from matching waveforms, assuming c = 0. Figure 9 shows that the ultrasonic waveform method agrees well with the other methods, especially for small size bubbles, $R_b/R_u \lesssim 1$, where optical studies show that discrepancies in large bubble cases are mainly due to elongation of bubbles. A comparison of bubble rising velocity, as determined by the ultrasonic waveform matching and by an optical method is shown in Fig. 10. Fig. 10 shows fairly good agreement between the two methods.

4. INTERACTION OF ULTRASONIC WAVES AND MULTI-BUBBLES

4.1 Ultrasonic Wave Propagation Measurements

The experimental and theoretical studies of sound wave propagation velocity in a bubbly two-phase flow has been done by numerous authors; a review has been given by Akagawa [5]. The most commonly used relationships considering slip velocity of two-phases was given by Henry [7] semi-

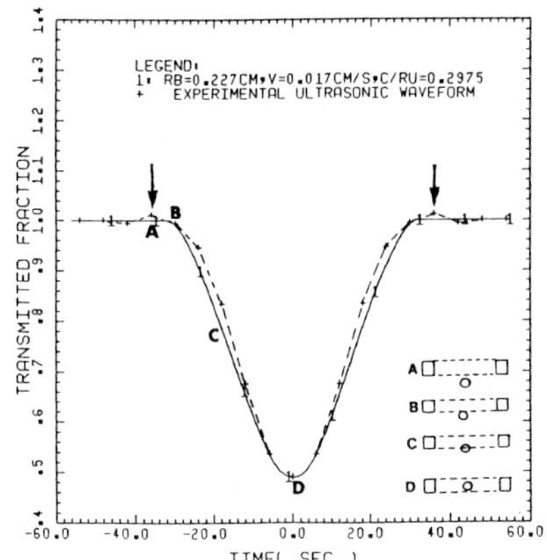

Fig.8: Comparison of experimental and theoretical waveforms: Ru=0.3175cm, min(A/AO)=0.49, interaction time=63sec.

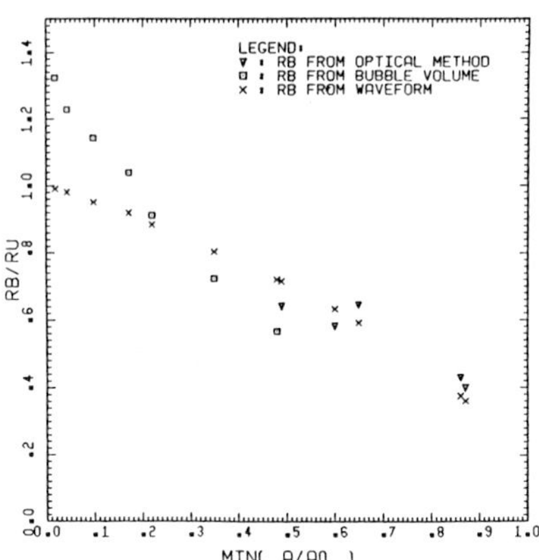

Fig.9: Comparison of normalized bubble radius (RB/RU) from diff. methods vs min. transmitted fraction. RU= 0.3175cm. Spherical bubble assumed from volume.

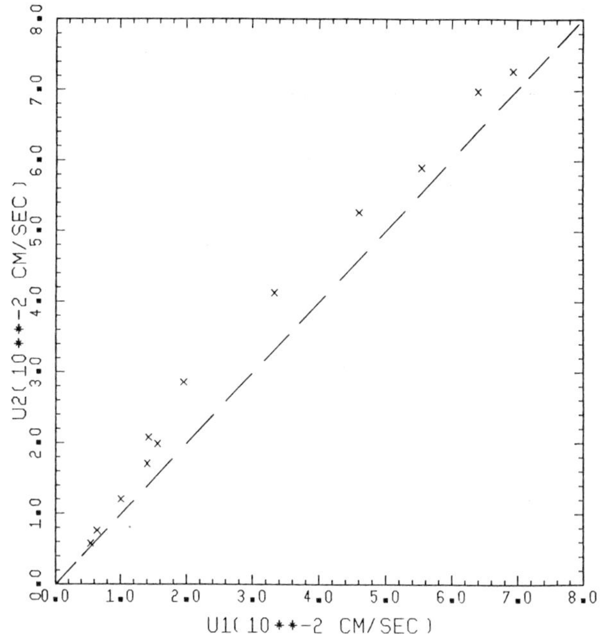

Fig.10: Comparison of bubble rising velocity: U1 from optical methods, U2 from matching ultrasonic waveforms.

empirically as follows

$$\frac{a}{a_g} = \frac{1.035 + 1.67\ \varepsilon_g}{[k\ \{\varepsilon_g^2 + \varepsilon_g(1 - \varepsilon_g)\ \rho_\ell/\rho_g\}]^{1/2}} \tag{5}$$

where a is the two-phase sonic velocity, a_g is the gas single-phase sonic velocity, ε_g is the void fraction, ρ_ℓ and ρ_g are the densities of the liquid and gas phase, respectively, and k = 1.4 for air. Karplus [7] conducted experiment in the ultrasonic region up to 1 KHz, and observed that sound wave propagation velocity increases with increasing frequency of sound. Saneyoshi [9] also considered the effect of bubble size and void fraction on the ultrasonic wave propagation velocity, and he showed theoretically that propagation velocity have a resonance point around 10 to 200 kHz depending on void fraction or bubble radius. However he also showed that for the frequency above 1 MHz, the wave propagation velocity is not significantly influenced by bubble size and void fraction when both values are relatively small.

Due to the high acoustic impedance mismatch that exist between gas and liquid, the ultrasonic wave is significantly attenuated in the presence of a gas-liquid interface along its propagation path. The frequency of the transducer used, also affects the degree of attenuation.

In our experimental range of frequency (.5 to 5 MHz), the ultrasonic wave velocity is not expected to change with void fraction. Typical time-averaged transmitted ultrasonic waveforms averaged for a period of 1.5 sec are shown in Figure 11 for various gas flow rates or void fraction. By measuring the flight time of the ultrasonic wave propagated along a fixed distance we can determine its velocity. Figure 12 shows that the ultrasonic propagation velocity in an air-water bubbly flow is 1.34 km/sec; and is not influenced by a 0 to 25% time-averaged void fraction. Typical time averaged reflected ultrasonic signal (pulse-echo mode) is shown in Figure 13. In this case, the bubbles are causing the ultrasonic pulses to be reflected back to the source.

4.2 Amplitude of Transmitted Ultrasonic Signals

Typical experimental transmitted sound waveforms observed for various gas flow rates are shown in Fig. 14. The waveforms of single bubbles interacting with the sound beam were observed to be similar to that of Eq. (2) and those from the single bubble experiment in silicone oil. When the gas flow rate is relatively small, the single bubble analysis discussed in section 3 can be applied to obtain bubble radius and rising velocity. In multi-bubbly flow, if there are n bubbles in the interaction zone, the transmitted signal becomes

$$A = R_u^2\ [\pi - \sum_{i=1}^{n} \cos^{-1}\left(\frac{R_u^2 - R_{bi}^2 + d_i}{2\ R_u d_i}\right)] - \sum_{i=1}^{n} R_{bi}^2 \cos^{-1}\left(\frac{R_{bi}^2 - R_u^2 + d_i^2}{2\ R_{bi}\ d_i}\right)$$

$$+ R_u \sum_{i=1}^{n} d_i\ [1 - \left(\frac{R_u^2 - R_{bi}^2 + d_i^2}{2\ R_u d_i}\right)^2]^{1/2} \tag{6}$$

if $[(y_i - y_j)^2 + (x_i - x_j)^2]^{1/2} > R_{bi} + R_{bj}$ where (x_i, y_i) and (x_j, y_j) are the centers of any two bubbles and $d_i = (x_i^2 + y_i^2)^{1/2}$. Typical numerical results of Eq. (6) are shown in Fig. 15 and 16 for R_b/R_u = 0.75 for two equal size bubbles. The figures show that the method discussed in section 3 can be applied when $[(y_1 - y_2)^2] \gg (R_b + R_u)^2$.

For relatively high gas flow rates or void fractions, we can obtain approximate bubble sizes and rising velocities by using the time-averaged (A/A_o) signal as a constant basis. Figures 17a and b show the bubble radii and rising velocities as a function of the air flow rate. Fig. 17 and 18 show that the bubble radius and rising velocity can be obtained accurately by the current method for air flow rates less than 0.3 ℓpm and within ± 50% for the present range of experiment. Here the time-averaged (A/A_o) as a function of the air flow rate used in the above analysis is shown in Fig. 18.

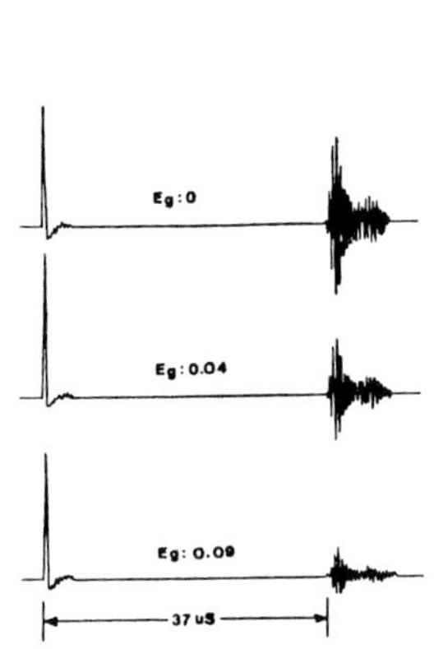

Fig.11: Typical time-averaged waveform as function of void fraction, operated in the pitch-and-catch mode.

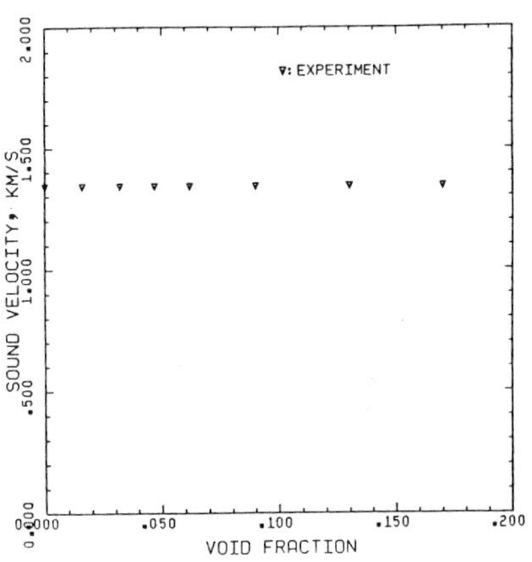

Fig.12: Ultrasonic velocity vs void fraction in a bubbly air water flow.

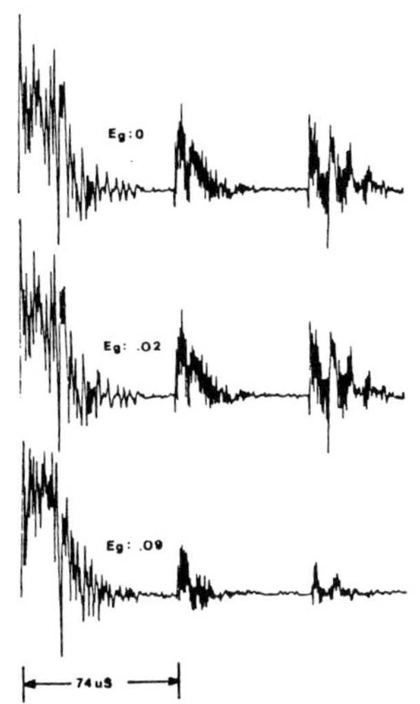

Fig.13: Typical time-averaged waveform as a function of void fraction operated in the pulse-echo mode.

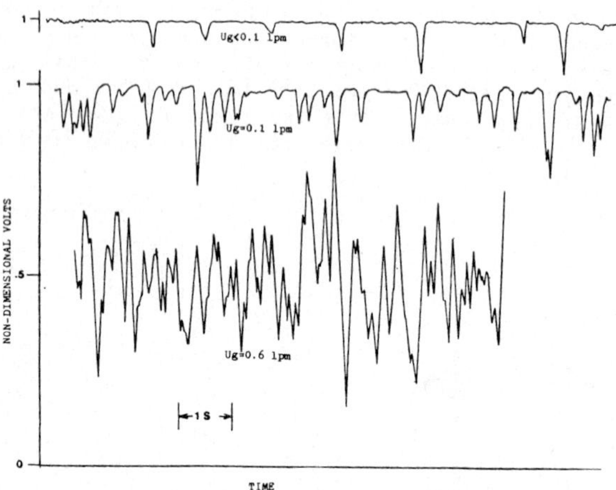

Fig.14: Typical ultrasonic waveform attenuation received at a fixed location as a function of time for different flow rates.

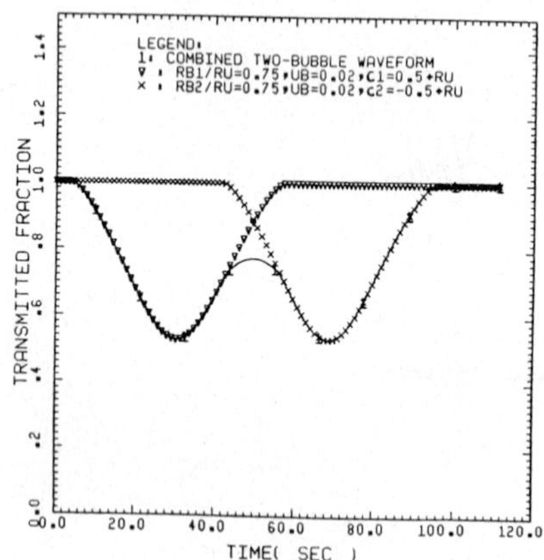

Fig.15: Two-bubble ultrasonic waveform: Ru=0.3175cm, distance between bubble centers= 2.4*Ru.

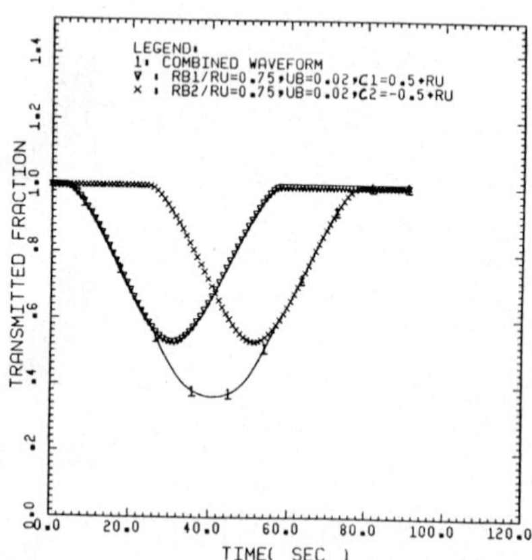

Fig.16: Two-bubble ultrasonic waveforms: Ru=0.3175 cm., distance between bubble centers=1.3*Ru.

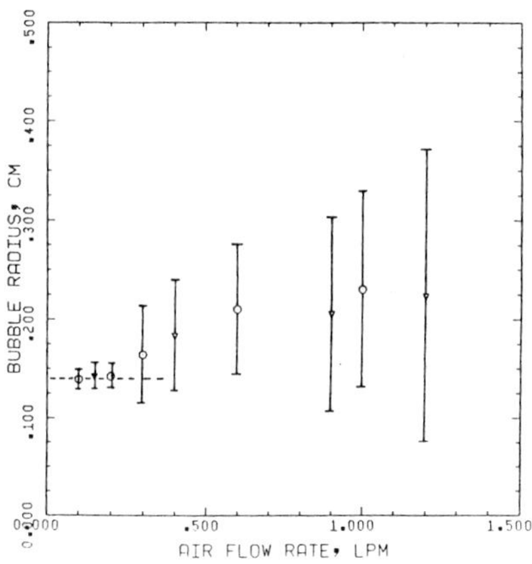

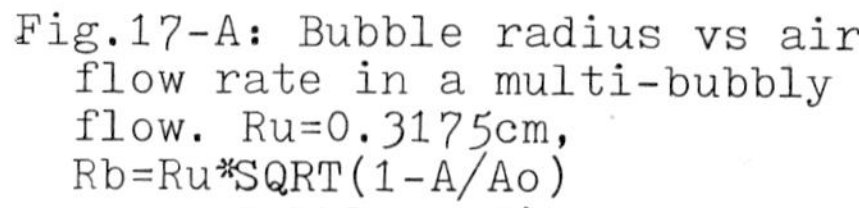

Fig.17-A: Bubble radius vs air flow rate in a multi-bubbly flow. Ru=0.3175cm, Rb=Ru*SQRT(1-A/Ao)

----- Bubble radius as seen from photograph.

o sieve plate with 0.26mm holes.

▽ sieve plate with 0.76mm holes.

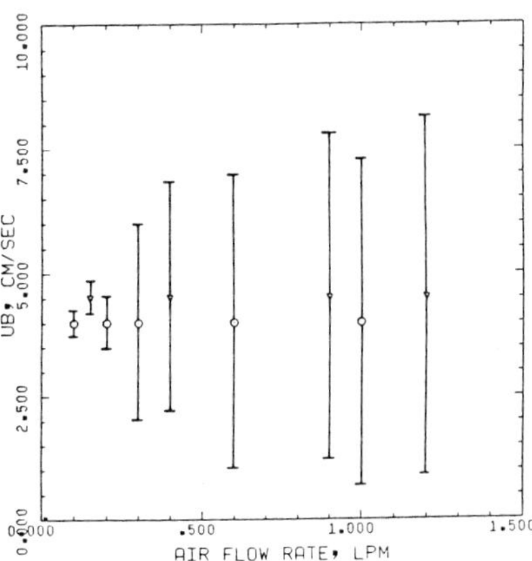

Fig.17-B: Bubble rising velocity, Ub vs air flow rate in multi-bubbly flow. Ub=2*(Rb+Ru)/DT

o, ▽ same as above

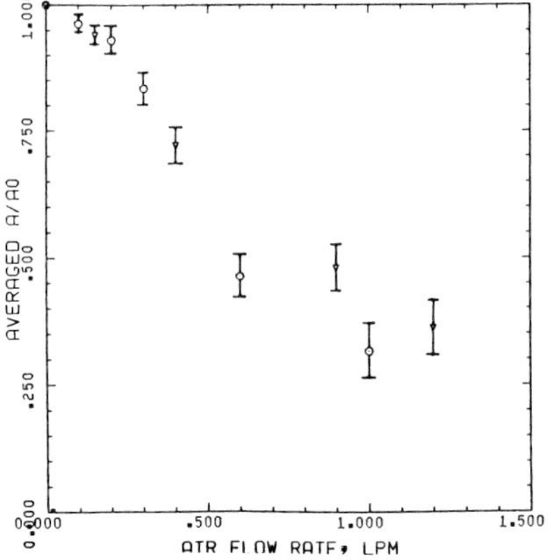

Fig.18: Ultrasonic wave attenuation, Averaged A/Ao vs air flow rate in a multi-bubbly flow.

o, ▽ same as above

5. CONCLUDING REMARKS

Experimental and theoretical studies have been conducted for the ultrasonic wave propagation in a bubbly gas-liquid two-phase flow, and the following concluding remarks have been obtained.

(1) Ultrasonic transmitted waveforms obtained from experiment agree well with theoretical results when only single bubbles are present in the cross section of the ultrasonic beam, except for two small peaks which may be due to small beam divergence effects for certain bubble locations.

(2) Bubble radius and rising velocity can be obtained from transmitted ultrasonic signal from best-fitting the experimental waveforms.

(3) Time averaged ultrasonic propagation velocity is not significantly influenced by the void fraction for the present range of sound frequencies, $0.5 < f < 5$ MHz.

(4) For small gas flow rates or void fractions, an estimate of the bubble size and rising velocity can be obtained by using the present analysis.

ACKNOWLEDGEMENT

The authors wish to express their appreciation to J. Chan and Y. Ichikawa for valuable discussions and comments. This work was supported by the Natural Science and Engineering Council of Canada under account No. 3434 and No. 1311.

REFERENCES

1. Chang, J.S., Ichikawa, Y. and Irons, G.A., "Measurements in Polyphase Flows - 1982", Ed. T.R. Heidrick and B.R. Patel, ASME Publ., New York, 1982, pp. 7-12.

2. Brimley, W. and Chang, J.S., "Thermalhydraulics of CANDU Reactor", R. Bonalumi, et al. Ed. MIES Press, Hamilton, 1982, Ch. 17.

3. E.P. Ying, J. Acoustic Soc. Amer., Vol. 53, 1627, 1973.

4. D.L. Parry, Exxon Nuclear Co., Report XN-310, 1975.

5. Akagawa, K., "Gas-Liquid Two-phase Flow", Corona Pub. Co. Ltd., Tokyo, 1974.

6. Krautkramer, J. and Krautkramer, H., "Ultrasonic Testing of Materials", Springer-Verlag, New York, 1969.

7. Henry, R.E., Chem. Eng. Prog. Symp. Ser., 1969, vol. 11, pp. 1.

8. Karplus, H.B., ARF-4132, 1961 or (see Ref. 7).

9. Saneyoshi, J., Bul. Tokyo Inst. Tech. Ser. D, 1951, vol. 3, pp. 127.

Multi-Phase Flow and Heat Transfer III. Part A: Fundamentals
edited by T.N. Veziroğlu and A.E. Bergles
Elsevier Science Publishers B.V., Amsterdam, 1984 — Printed in The Netherlands

VAPORIZATION OF SUPERHEATED WATER UNDER SUDDEN DEPRESSURIZATION

Daniel Favrat and J.-P. Denisart
Institut CERAC SA, 1024 Ecublens, Switzerland
and I. Ryhming
Prof. Fluid Mech. Lab., Federal Institute of Technology
1015 Lausanne, Switzerland

ABSTRACT

The sudden one-dimensional expansion of a cylindrical volume of superheated water is studied both experimentally and theoretically. In the experimental set-up, the water is initially pressurized by a piston at the closed end of a cylinder and heated to the onset of vaporization which initiates the expansion by releasing the piston counterpressure. Pressure measurements along the path of the piston indicate a significant pressure drop compared to the numerical predictions based on a one-dimensional, homogeneous equilibrium flow. This discrepancy caused by non-equilibrium vaporization delays, affects the whole expansion range tested (volume ratio up to 50). Pictures from a high speed movie illustrate the vaporization phenomena with a sharp separation front between the steam and water phases, as high quality steam concentrates behind the moving piston.

1. INTRODUCTION

After having been exploited at a slow rate for more than half a century, geothermal energy resources have gained in interest during the 1970th and are now part of the alternative energy sources for oil replacement. The vast majority of the pressurized geothermal sources are liquid dominated and consist of reservoirs of superheated water. Under the effect of gravity the geothermal brine is partly vaporized in the well bore and reaches the surface as a multiphase mixture of low steam quality. Submersible pumps are also being thought of in order to avoid vaporization in the well and therefore to increase the mass flow and to obtain superheated brine at the wellhead. Steam is normally separated and expanded through condensing turbines, but new expander concepts able to accept the total flow are being proposed [1,2]. The aim of most total flow expanders is not only to use the pressure energy of the steam but also the pressure contribution of the vaporizing water.

The use of superheated liquid is also known to be a volume efficient way to store energy and has been envisaged in electric grid peak shaving or solar energy applications. Superheated water is also used in auxiliary rockets to boost aircraft power during take-off.

Although there exists an extensive literature covering the field of superheated fluid flow in nozzles, safety valves or resulting from pipe rupture [3], little is published on actual work recovery from the expansion of pressurized hot liquids in a growing cavity. Most of the published experimental data deal with high pressure drops relative to the pressure differentials caused by non-equilibrium phenomena. In the particular case of positive displacement machines, working with relatively low pressure differences, the vaporization delays resulting from thermodynamic non-equilibrium phenomena in the expanding chambers will play a significant role. However, non-equilibrium models, as reviewed by Wallis [3], still heavily rely on empirical coefficients often derived from critical flows in nozzles. The Simpson and Silver model [4] will be discussed here, but the possibility to predict correctly the non-equilibrium flow phenomena in our case was judged too low and justified a new experimental investigation.

2. EXPERIMENTS

The base case chosen for our study deals with an entrapped mass of superheated water, initially at rest, which is suddenly allowed to expand while pushing a piston in a cylinder. Such a configuration allows a clear definition of the initial conditions both theoretically and experimentally. Figure 1 is a schematic drawing of the experimental set-up, showing the piston in its initial position in a cylinder equipped at one end with a heating wall and with a gas damper at the other end. The extremity of the damper can be opened to allow an accurate positioning of the piston during the filling up and the deaeration of the water cavity. The water is then pressurized by applying a counterpressure of air on the back face of the piston, and by slowly heating up the water to the onset of vaporization. The water temperature is indicated by a submerged thermocouple embedded in the front face of the cylinder which triggers the release of the air counterpressure through the rupture of a diaphragm.

Two types of cylinders of 40 mm calibrated diameter have been used in order to allow the visualisation of the flow phenomena and the recording of the pressure by means of pressure transducers along the cylinder walls. The 60 mm long pistons are made of either PTFE or aluminium and sealed with O-rings.

All tests have been run with distilled water and a counterpressure of air.

2.1. Pressure Measurements

The cylinder is made of two concentric tubes. The inner part is made of PTFE in order to prevent excessive heat transfer losses during the heating phase and guarantee a uniform temperature distribution in the water and an outer shell in steel to avoid the deformation of the inner tube. Several pressure taps have been drilled along the cylinder axis for the pressure transducers. Two DISA capacity transducers have been used during the tests and placed successively at three different locations along the cylinder. The pressure signals are recorded on a BIOMATION digital recorder and transferred afterwards on a TEKTRONIX oscilloscope.

Figure 2 shows typical pressure signals obtained with an initial pressure of 600 kPa and a 0.113 kg piston mass. Triggered at the arrival of the air expansion waves, the signals represent successively the residual air pressure in front of the piston and the liquid pressure behind it. These results indicate that, in spite of the rather slow speed of the piston, the front air pressure cannot be neglected and influences the acceleration of the piston. Each trace

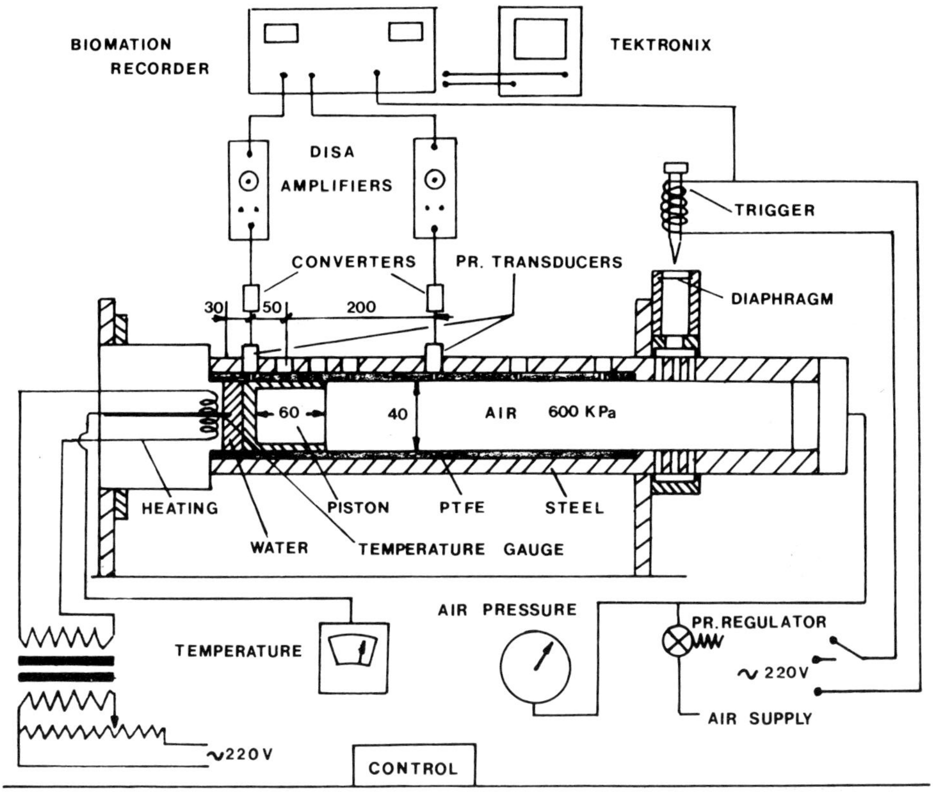

Figure 1 Schematic drawing of the experimental system used for pressure measurements.

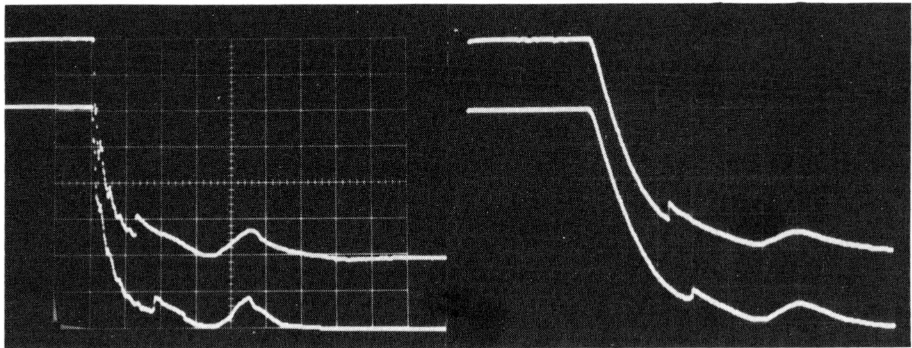

Figure 2 Pressure traces from DISA capacity transducers. (20 msec/div. and 86 kPa/div.)
a) Rupture of a diaphragm b) Electric valve

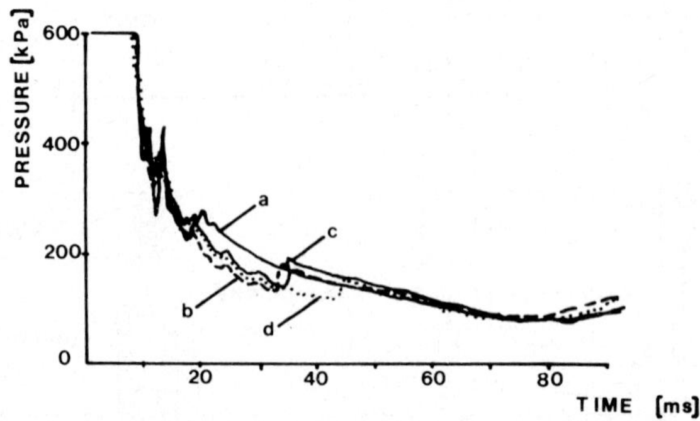

Figure 3 Superimposed transducer signals.
Initial pressure = 600 kPa
Initial temperature = 158.8 °C
Piston mass = 0.113 kg

a) Position = 30 mm
b) and c) Position = 80 mm
d) Position = 180 mm

of Figure 2 shows a pressure jump which corresponds to the passing over the O-ring seal of the piston which is located at 10 mm of the wetted face of the piston. Measuring with two transducers only required a good repeatability between the successive tests, which was generally obtained. However, care was taken to maintain one reference transducer at the same location (position 80 mm) for the sake of comparison. In Figure 3 four pressure traces are combined, giving an order of magnitude of the repetivity errors by comparing the two intermediate signals at position 80. Direct reading of the traces on oscilloscope photographic recordings also result in inherent inaccuracies.

Non-equilibrium phenomena are directly influenced by the expansion rate which in our case depends on the piston mass, the friction forces and the gas counterpressure function. The latter has been modified in the course of the tests by the substitution of the air diaphragm by a fast opening valve. The slower expansion rates achieved (Figure 2b) results in an increased biphase pressure at equivalent expansion ratios. The pressure distribution at the regular piston positions is shown in Figure 4 as a result of the combination of the transducer signals.

It is worth noting that, even for the case of the rupture of a diaphragm, the pressure differential driving the piston is rather low, decreasing quickly to 1/15 of the overall pressure drop. The pressure distributions are uniform through the expanding zone with a slight drop of pressure towards the end wall of the tube where the steam quality remains low, therefore favoring non-equilibrium effects.

2.2 Visualisation

The same set-up, but with a calibrated glass cylinder, was used during the visualisation tests (Figure 5). A rotating prism John Hadland (P.I.) Hyspeed Camera was synchronized with the diaphragm braker to film along the 600 mm path of the piston. A time delay was allowed to accelerate the camera to the selec-

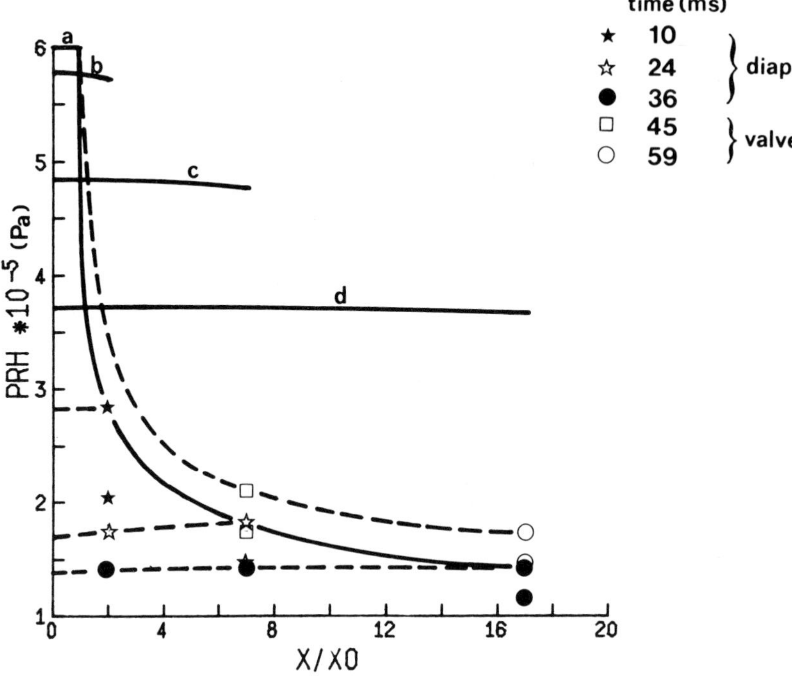

Figure 4 Pressure distribution in the two-phase fluid at different time intervals. Curves a, b, c and d are calculated equilibrium pressure distributions corresponding respectively to the expansion ratio 1, 2, 7, 17.
T_o = 158.8 °C, x_o = 10 mm, piston mass M_p = 0.113 kg,
PRH = mixture pressure, x/x_o = expansion ratio

Figure 5 Experimental set-up for high speed visualisation.

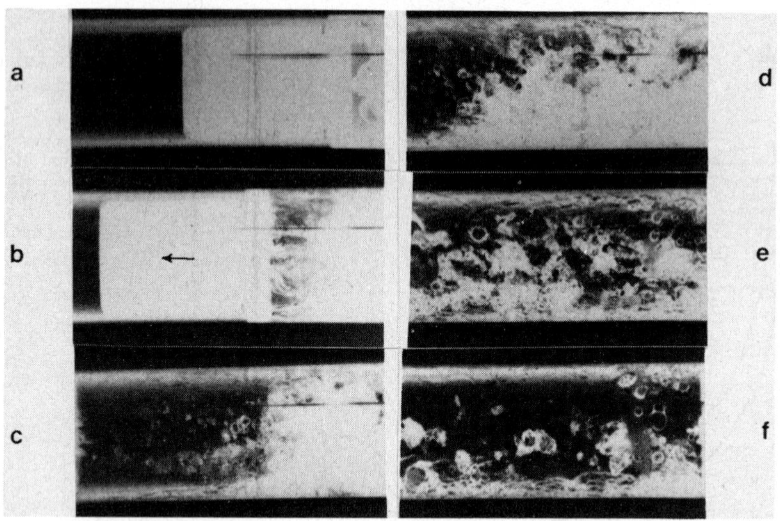

Figure 6 Pictures from a high speed 16 mm movie (3000 frames/sec).
Pin = 600 kPa, Tin = 158.8 C, Xo = 10 mm, Mp = 0.160 kg
Rupture of a diaphragm
t_a = 0., t_b = 26 msec, t_c = 4.6 msec, t_d = 150 msec, t_e = 113 msec, t_f = 150 msec

ted speed of 3000 frames/sec. The lighting was made by a TRISTAR lamp located at a distance of two meters from the tested tube.

The pictures of Figure 6 are taken out of a high speed 16 mm movie and illustrate the vaporization phenomena at six time steps:

a) The piston is still quasi-immobile and one sees the occurence of vaporization pockets around the thermocouple and the heating elements.
b) The piston is in the acceleration phase towards the left and has moved 24 mm which corresponds to a volume ratio of 3.4. Steam is mainly generated or migrates towards the piston side of the cavity. Threads of bubbling water are being pulled out of the face of the piston.
c) The piston is just visible as it leaves the visual field. An abrupt gradient of quality separates the volume into two main zones, with high quality steam following the piston, and the boiling water with a high inertia being left behind.
d) The boiling zone is expanding with a distinct effect of gravity which orients the separation front obliquely.
e) The volume ratio is of the order of 50 and the steam quality is still low in the front part of the cylinder with a strong boiling of the remaining water mass.
f) The pressure behind the piston has fallen under atmospheric, condensation occurs on the walls and the piston is slowly sucked back as shown by the backwards floating coalescent drops.

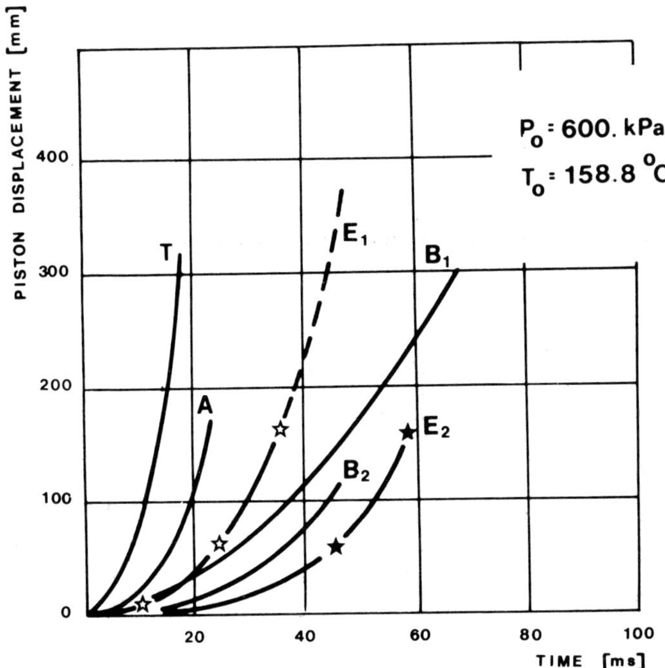

Figure 7 Piston displacement versus time.
T = equilibrium calculation, x_o = 10 mm, M_p = 0.113 kg
A = film measurements, x_o = 25 mm, M_p = 0.113 kg
B_1 = film measurements, x_o = 16 mm, M_p = 0.160 kg
B_2 = film measurements, x_o = 10 mm, M_p = 0.160 kg
E_1 = pressure measurements (diaphragm), x_o = 10 mm, M_p = 0.113 kg
E_2 = pressure measurements (valve), x_o = 10 mm, M_p = 0.113 kg

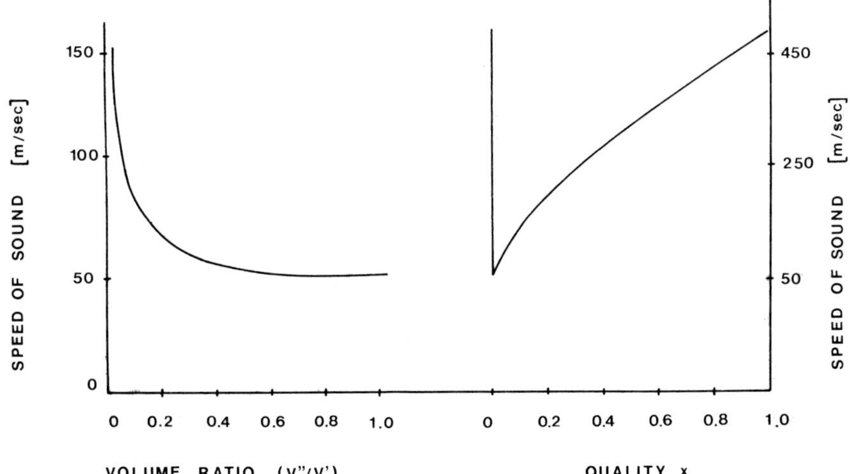

Figure 8 Speed of sound in water-steam mixtures

The piston displacement as a function of time is obtained by direct measurements from the film for various initial water amounts and piston masses (Figure 7). The three points corresponding to the pressure measurements are also reported and suggest a piston displacement which is consistent with the displacement obtained from film measurements. These data show some similarities with uniformly accelerated motions as a result of the small and quasi-uniform driving pressure differential noticed from pressure measurements.

3. CALCULATIONS

As for most fluid machinery performance evaluations, the best reference values are corresponding to an ideal isentropic transformation. When the piston motion is slow relative to the speed of sound of the expanding fluid, the state functions can be considered uniform through the expanding zone and the solution based on the first principle of thermodynamics is straigthforward. However, the speed of sound of two-phase mixtures is strongly affected by the quality ratio as shown in Figure 8. This dependance is particularly pronounced in the range of very small quality values corresponding to the acceleration phase of the piston. At high depressurization rates, the calculation method was therefore due to allow for the presence of gradients of the thermodynamic functions through the expanding zone.

Based on the above considerations and with the prospect of introducing, at a late stage, a non-equilibrium model, calculations have been made numerically using STEALTH Lagrangian finite difference code [5]. The fluid area is divided into discrete zones separated by mesh points. The explicit one-dimensional finite-difference equations are solved using two overlapping meshes, with equations centered at the mesh points for the momentum related variables and centered at zone interiors for the energy related variables. The calculational sequence is shown in Figure 9 where the subscript i is the spatial index and superscript n the time index. The momentum equation is solved first, followed by the conservation of mass equation expressed in terms of relative volume V (ratio between the reference density and the zone density). The energy and the constitutive equations are then solved iteratively and the local speed of sound is recalculated according to:

$$a^2 = \frac{x\,c''_p + (1-x)c'}{x\,c''_v + (1-x)c'} \left[\left(\frac{\rho}{\rho'}\right)^2 \frac{1-x}{a'^2} + \frac{\rho^2}{P}\frac{x}{\rho''} \right]^{-1} \qquad (1)$$

where
- a = two-phase speed of sound
- c = specific heat
- x = steam mass quality
- ρ = density
- P = pressure
- superscript ()' = saturated water
- ()" = saturated steam
- subscript ()p = constant pressure
- ()v = constant volume

Although the formulation is different, this equation is similar to the expression given in Tangren's paper [6], where a' is assumed to be infinite.

The Courant stability criterion determines the next time step on the basis of the new speed of sound. The acceleration of the piston is calculated at each time step according to:

$$\ddot{x}^{n+1}_{i_{max}} = \frac{P^{n+1}_{i_{max}} - P_{air}(t)}{M_p} \frac{\pi d^2}{4} - F/M_p \tag{2}$$

where $P^{n+1}_{i_{max}}$ = Pressure of the closest zone to the piston

P_{air} = Pressure function in the front of the piston

M_p = Piston mass

d = Cylinder diameter

F = Friction force of the piston seals in the cylinder.

The numerical procedure is first checked by repeating Kratz gas gun calculations [7] and the by introducing the state and speed of sound functions for a one-dimensional, homogeneous equilibrium flow of steam and water. The two-phase equation-of-state for water is given by discrete values along the saturation line from which the main thermodynamic functions (pressure, temperature, quality, entropy) are extrapolated.

Eight to fifteen lagrangian zones have been used during the calculation and the results expressed in function of the expansion ratio are shown in Figure 10 for two different acceleration rates. Figure 10a corresponds to the acceleration of a 0.113 kg piston without any air residual counterpressure or friction forces and illustrates the initial pressure and quality gradients occurring at relatively high expansion rates. Figure 10b indicates the more uniform distributions corresponding to an acceleration hindered by the following air counterpressure function:

$$P_{air} = (P_0 - P_{atm})\, e^{-t^n/\tau} + P_{atm}$$

with $\tau = 2.2\ 10^{-3}$ sec, $n = 0.79$.

Steep initial gradients are characteristics of low piston inertia and friction factors relative to the fluid internal energy drop per expansion step. Also of theoretical interest is the capability of the numerical method to calculate the back and fourth motions of the piston until friction damping stops it definitively.

For the rates of expansion considered in this study the predicted pressure distributions quickly flaten when the piston moves away as observed during the tests, although a small negative pressure gradient remains as opposed to the slightly positive gradients measured. This result is a typical influence of the non-equilibrium vaporization delays which depend on the nucleation and bubble site density and therefore more significantly affect the low quality zones. Moreover, the homogeneous flow model fails, of course, to predict the sharp quality gradient observed in Figure 6, which is the combined result of the inhomogeneous vaporization delays and of the faster migration of the vapor phase.

Conservation of Momentum
$$\ddot{x}_i^n = \left(\frac{2}{\alpha^n_{i-1/2}+\alpha^n_{i+1/2}}\right)\left(\sigma^n_{xx_{i+1/2}} - \sigma^n_{xx_{i-1/2}}\right)$$

Explicit Integration
$$\dot{x}_i^{n+1/2} = \dot{x}_i^{n-1/2} + \ddot{x}_i^n \cdot \Delta t^n \qquad x_i^{n+1} = x_i^n + \dot{x}_i^{n+1/2} \cdot \Delta t^{n+1/2}$$

Conservation of Mass
$$V^{n+1}_{i-1/2} = V^n_{i-1/2} + \underbrace{\frac{\overset{\circ}{\rho}_{i-1/2}}{\overset{\circ}{m}_{i-1/2}}\left[\dot{x}_i^{n+1/2} - \dot{x}_{i-1}^{n+1/2}\right]\Delta t^{n+1/2}}_{\Delta V^{n+1/2}_{i-1/1}}$$

Conservation of Energy
$$\tilde{U}^{n+1}_{i-1/2} = U^n_{i-1/2} - P^n_{i-1/2} \cdot \Delta V^{n+1/2}_{i-1/2}$$

$$U^{n+1}_{i-1/2} = U^n_{i-1/2} - \tilde{P}^{n+1}_{i-1/2} \cdot \Delta V^{n+1/2}_{i-1/2}$$

Equation-of-State

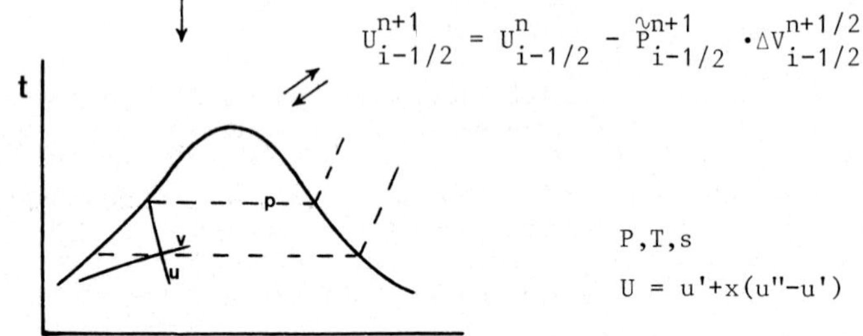

P,T,s

$U = u'+x(u''-u')$

$$\sigma^{n+1}_{xx_{i-1/2}} = P^{n+1}_{i-1/2}$$

Speed of Sound — see equation 1

$$\alpha_i^{n+1} = \frac{\overset{\circ}{\rho}_i}{V_i^{n+1}}\,\Delta x_i^{n+1}$$

New Time Step
$$\Delta t^{n+1} = f(a, \Delta x)$$

Figure 9 Calculational sequence of "STEALTH" Lagrangian code.

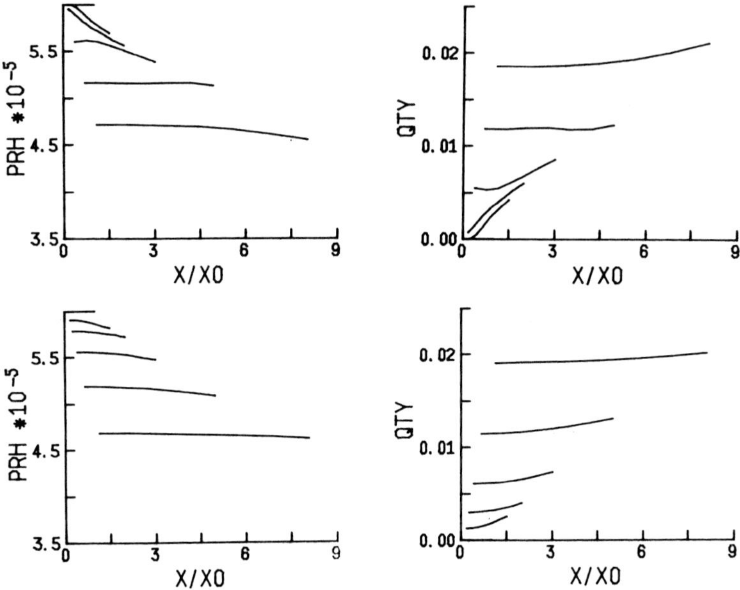

Figure 10 Calculated quality (QTY) and pressure (PRH) distributions versus expansion ratio (x/x_o).
a) without air counterpressure b) with air counterpressure

4. DISCUSSION

These experiments have confirmed the key role played by non-equilibrium phenomena which considerably offset the working pressure of expanding superheated liquids, at least during time intervals compatible with fluid machinery. The transient pressure deficit between the calculated equilibrium pressure and the measured pressure cumulates at 290 kPa for the expansion commanded by the rupture of a diaphragm and at 266 kPa in the case of the opening of a fast electric valve (Figure 11).

Vaporization delays are dependent on the nucleation characteristics and on the interphase heat, mass and momentum transfer. The early formation and growth of the vapor embryos require a minimum activation energy to provide the surface energy and to work against the surrounding fluid or to fill the cavities with vapor. This time consuming process depends on the initial density of embryos, the surface tension and the presence of foreign bodies or gases which encourage a so-called heterogeneous nucleation by reducing the required energy of activation. Once the embryos have reached a critical radius, then the further growth of the new born bubble is mainly influenced by heat conduction in the liquid phase.

For comparison purposes, let us consider the thermal non-equilibrium model established by Simpson and Silver [4] which is summarized by the following equation valid for very low quality flows:

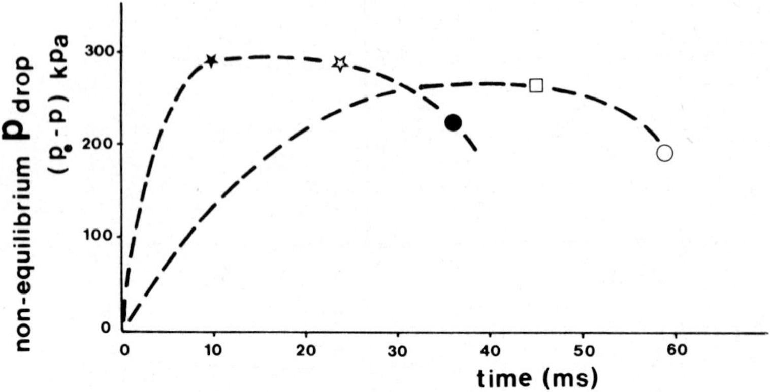

Figure 11 Non-equilibrium pressure drop evolution.

$$\frac{dv}{v} \simeq \frac{32}{81} \frac{\pi A}{\beta^3} (P_V - P_\ell)^3 e^{-B/(P_V-P_\ell)^2} t^{3/2} dt \tag{3}$$

with:

$$A = N_n \left(\frac{2\sigma}{\pi m}\right)^{1/2}, \quad B = \frac{16}{3} \frac{\pi\sigma^3 f(\alpha)}{k T_\ell}; \quad \beta = \frac{\left(\frac{3}{\pi} K_\ell\right)^{1/2} P_o^2 (\Delta u_{vap})^2}{\lambda_\ell R^2 T_\ell^3}$$

N_n = Number of sites for heterogeneous nucleation per unit volume
σ = Surface tension of the liquid
m = Mass of one vapour molecule
α = Contact angle of a liquid on a solid surface
k = Boltzmann's constant
K_ℓ = Thermal diffusivity of the liquid
λ_ℓ = Thermal conductivity of the liquid
Δu_{vap} = Latent heat of vaporization

The particular merit of this model is to illustrate the relative importance of most of the physical factors governing transient vaporization phenomena. However, when the quality increases, the model suffers from not introducing the heat convection which inevitably accompanies the slip velocity associated with the migration of the vapour phase. As pointed out by Simpson and Silver, the parameters N and $f(\alpha)$ are sources of uncertainties so that, in practice, the coefficients A and B are fixed empirically. In our case, if one integrates the equation 3 over the time lapse corresponding to the flat range of non-equilibrium pressure drop (Figure 11), while keeping Simpson and Silver's empirical coefficients, one obtains a predicted non-equilibrium pressure difference of:

- 250 kPa instead of 290 kPa for the case of the diaphragm rupture
- 220 kPa instead of 266 kPa for the case of the electric valve.

The predicted numbers represent only rough evaluations which are to be taken with care. A more suitable approach would be to determine the quality rate by a differential equation combining equation 3 and the void generation term related to the action of the moving piston.

The departure from equilibrium observed during our tests is also signficantly greater than the values extrapolated from the experimental work of Miyazaka et al [8]. These authors measured the transient depression below saturation in a vertical pipe filled with subcooled water and suddenly depressurized at the rupture of a diaphragm. The extrapolation from Miyazaki's curves would give a non-equilibrium pressure drop of 186 kPa.

5. CONCLUSION

The piston in a cylinder set-up used to study the expansion of saturated water has proven to be adequate for the evaluation of the non-equilibrium phenomena affecting the vaporization of very low quality fluids. Steady initial conditions, well defined fluid boundaries and the possibility to adjust the counter-pressure history are among the advantages of the method which makes it suitable for analytical comparison.

The results presented in this paper are based on tests with distilled water and the absence of initial turbulence effects or of forced mixing between the phases, all factors which tend to increase the non-equilibrium pressure drop. The present study is therefore giving a lower bound of the work potentially recoverable out of the expansion of superheated liquids.

In the range of expansion rates considered (16 to 50 kPa/msec), the maximum non-equilibrium pressure drop is only varying within 10%, (266 to 290 kPa).

Further work with detailed water analysis, different pressure ratios and piston masses would be required to better assess the empirical factors to be used in a non-equilibrium model derived from the Simpson-Silver model, which is mentioned in the paper.

ACKNOWLEDGEMENTS

The authors are particularly grateful to Ms. Anne Vallotton for assistance with STEALTH computer analysis.

REFERENCES

1. Austin, A.L., Higgins, G.H., and Howard, J.H., "The Total Flow Concept for Recovery of Energy From Geothermal Hot Brine Deposits", UCRS Report No. 51366, Lawrence Livermore Laboratory, April, 1973.

2. McKay, R., "Helical Screw Expander Evaluation Project", Jet Propulsion Laboratory Publication No. 82-5, 1982.

3. Wallis, G.B., "Critical Two-phase Flow", Int. J. Multiphase Flow, 1980, Vol. 6, pp. 97-112.

4. Simpson, H.C., and Silver, R.S., "Theory of one-dimensional, Two-phase Homogeneous Non-equilibrium Flow", Proc. Instn. Mech. Engr., Symp. on Two-phase Flow, pp. 45-56, 1962.

5. Hofmann, R., "STEALTH, A Lagrange Explicit Finite-Difference Code for Solids Structural and Thermohydraulical Analysis", Science Applications, Inc., California, 1976.

6. Tangren, R.F., Dodge, C.H., and Seifert, A.H., "Compressibility Effects in Two-Phase Flow", J. Applied Physics, 1949, Vol. 20, pp. 637-645.

7. Kratz, H., "Vereinfachte Methode zur Berechnung der Ausstromvorgänge aus Rohren, wenn die Strömung durch die Beschleunigung eines festen Körpers eingeleitet wird", Z. Flugwiss., 1968, Vol 16, No. 4.

8. Miyazaki, K., et al, "Experimental Studies on Characteristics of Pressure Attenuation and Recovery to Saturation for Decompressive Disturbances in Low Pressurized Water", J. Nuclear Science and Technology, 1974, Vol. 11, pp. 284-292.

Multi-Phase Flow and Heat Transfer III. Part A: Fundamentals
edited by T.N. Veziroğlu and A.E. Bergles
Elsevier Science Publishers B.V., Amsterdam, 1984 — Printed in The Netherlands

A SIMPLE APPROACH TO TRANSIENT TWO-PHASE LEVEL SWELL

Michael A. Grolmes
Fauske & Associates, Inc.
16W070 West 83rd Street
Burr Ridge, Illinois 60521, U.S.A.

ABSTRACT

A two-phase level swell occurs when an otherwise static system is subjected to a sudden increase in vapor source. Depressurization of a container of saturated liquid (such as a steam generator or a beverage bottle), the sudden injection of gas, or the onset of volume heating in a saturated pool are typical examples. This paper provides a simple, yet practical evaluation of two-phase level swell.

1. INTRODUCTION

The literature contains several significant evaluations of the problem of pool swell dynamics [1-4]. The purpose here is not to add to or contradict these efforts, of which [1,2] represent exact analytic solutions of a formidable mathematical problem. The purpose here is to illustrate that much useful information can be obtained from a first order lumped system approach which avoids the numerical effort involved in a method of characteristics approach.

The physical condition to be modeled is depicted in Fig. 1. A vessel is shown with a liquid inventory defined by a rest height H_o. At some instant in time vapor production commences, described by a volumetric source rate $\dot{V}$. The pool surface, denoted by H, expands at a rate that is determined by the difference between the volume source rate $\dot{V}$, and the rate at which vapor can escape from the expanding pool. The latter is determined by the vapor volume flux term $j_g(\alpha)$, where the α dependency refers to the vapor volume fraction under the swelled height H.

Two conditions may result. The swell transient may terminate at an asymptotic height H_∞ that is located inside the container, that is $H_\infty < H_x$, where H_x is the height of the container. At this condition $\dot{V}$ is equal to $j_g(\alpha)\, A_x$. The discharge from the vessel through the opening A, will be all vapor. This is referred to as complete vapor separation or disengagement within the container. Depending upon the magnitude of $\dot{V}$ relative to H_o and the dimensions of the container, the asymptotic limit H_∞ may exceed the vessel height H_x. This may give rise to significant impact forces on the vessel head and will certainly lead to a two-phase discharge from the opening A. Methods for treating the two-phase discharge are discussed in Ref. [5]. The level swell treatment follows from the above description.

2. ONE-NODE MODEL

A basic model can be written as a volume balance with the aid of Fig. 1. A level swell statement is:

$$A_x \; \Delta H = [\dot{V} - A_x \; j_g(\alpha) - \dot{V}_f]\Delta\theta \tag{1}$$

The LHS of (1) is the volume displaced by the level rise in time element $\Delta\theta$. On the RHS of (1), the three terms in brackets are the volumetric rate of vapor production, the volumetric rate of vapor release from the pool at H, and the volumetric rate of liquid loss. The last term accounts for liquid to vapor conversion under vaporizing conditions.

Equation (1) can be simplified to:

$$H^{\bullet} = S - j_g(\alpha) \tag{2}$$

where,

$$H^{\bullet} = \Delta H/\Delta\theta$$

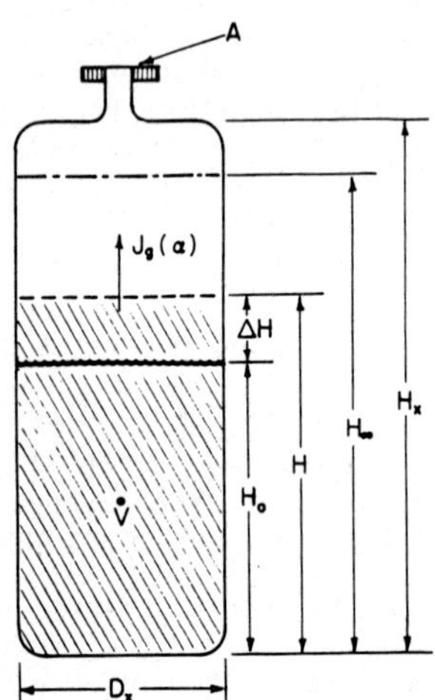

A - Discharge Area

A_x - Vessel Area

D_x - Vessel Diameter

H - Instantaneous Swelled Height

H_o - Rest or Unswelled Height

H_∞ - Maximum Swelled Height

H_x - Height of Vessel

ΔH - Swelled Height Increment

j_g - Vapor Volume Flux

α - $(H - H_o)/H$

$\dot{V}$ - Volume Source Rate

Fig. 1. Level Swell Analysis Diagram

and,

$$S = (\dot{V} - \dot{V}_f)/A_x$$

In (2) S is the net vapor volumetric source term and $j_g(\alpha)$ represents the vapor escape or disengagement term.

3. VAPOR DISENGAGEMENT

For water-like systems in the absence of foaming, the churn turbulent flow regime has been often used as a basis for development of vapor holdup correlations. It is assumed that a correlation of steady-state holdup data can be used in (2). One of the obstacles to extreme accuracy in physical model comparison is the variability in vapor-liquid holdup data. Figure 2 illustrates this variability based on typical literature data, Ref. [6].

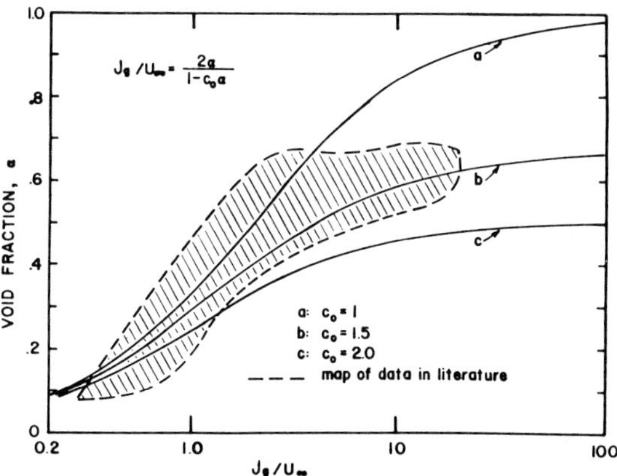

Fig. 2. Comparison of Vapor Holdup Correlation with Literature Data Ref. [6] for Volume Heated Saturated Water

For volume heated systems a correlation of the form:

$$j_g(\alpha) = \frac{2\ \alpha\ U_\infty}{1 - C_o\alpha} \tag{3}$$

provides an adequate representation of the trends in most holdup data for this flow regime. In (3) U_∞ is a rise velocity characteristic of discrete bubbles defined by (4), see Ref. [7].

$$U_\infty = 1.53 \left[\frac{\sigma g(\rho_f - \rho_g)}{\rho_f^2}\right]^{1/4} \tag{4}$$

For water U_∞ has a value of ≃ 0.2 m/s. The parameter C_o was introduced by Zuber and Findlay, Ref. [8] to account for radial void distribution effects. For pipe flow, they found C_o ≃ 1.2. This parameter can obviously be used to "adjust" (3) to fit holdup data in a volume heated system. A value of C_o between 1.5 and 2.0 is often found to be a good estimate. However, in such instances the actual vapor holdup is influenced by axial as well as radial distribution effects brought on by internal circulation in the vessel. It has been found by this author that $C_o = 1.5$ is most often a good choice of the parameter.

4. MODEL REDUCTION

For the churn turbulent flow regime, (3) and (2) can be combined and reduced to a convenient dimensionless form. If one defines a dimensionless superficial velocity ψ as:

$$\psi = j_g(\alpha)/U_\infty \tag{5}$$

(3) becomes,

$$\psi = 2\alpha/(1 - C_o\alpha) \text{ or } \alpha = \psi/(2 + C_o\psi) \tag{6}$$

Since α refers to the average void fraction under the swelled height H,

$$\frac{H}{H_o} = (1 - \alpha)^{-1} \tag{7}$$

When $j_g(\alpha) = S$ the pool has swelled to its limiting height H_∞. At this condition

$$\alpha_{max} = \psi_s/(2 + C_o\psi_s) \tag{8}$$

where $\psi_s = S/U_\infty$ and

$$\frac{H_\infty}{H_o} = (1 - \alpha_{max})^{-1} \tag{9}$$

If one now introduces a dimensionless height,

$$H^* = \frac{H - H_o}{H_\infty - H_o} \tag{10}$$

where $0 < H^* < 1$, and a dimensionless time,

$$\theta^* = S\theta/(H_\infty - H_o) \tag{11}$$

then (2) becomes

$$\dot{H}^* = 1 - H^* \left[\frac{H_\infty (C_o - 1) - C_o H_o}{H (C_o - 1) - C_o H_o}\right] \tag{12}$$

For $C_o = 1$, (12) reduces to

$$\dot{H}^* = 1 - H^* \tag{13}$$

which has the solution

$$H^* = 1 - e^{-\theta^*} \tag{14}$$

For $C_o \neq 1$, (12) can be reduced to

$$\dot{H}^* = \frac{1 - H^*}{1 - \phi H^*} \tag{15}$$

where,

$$\phi = \frac{(C_o - 1)\ \psi_s}{2 + (C_o - 1)\ \psi_s} \tag{16}$$

For the churn turbulent flow regime all possible solutions of (15) for various combinations of C_o and source strength ψ_s are bounded by the $\phi = 0$ and $\phi = 1$ curves are shown in Fig. 3.

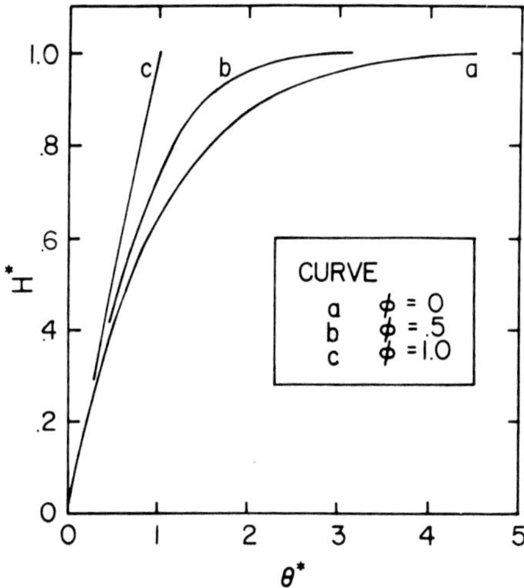

Fig. 3. Dimensionless Pool Swell Height H* vs Dimensionless Time θ* Indicating the Bounds of all Possible Solutions to Eq. (15) for the Churn Turbulent Flow Regime

Curve C in Fig. 3 ($\phi = 1$) represents a limiting case of large source strength S. In this case, the vapor escape is negligible and the swell proceeds to the limit at the maximum rate defined by,

$$\dot{H}^* = 1 \tag{17a}$$

or in dimensional form

$$\dot{H}_{max} = S \tag{17b}$$

Curve a in Fig. 3 ($\phi = 0$) represents the limiting case of ideal churn flow behavior defined by $C_o = 1$ which is actually (14). These results indicate that all possible solutions for the churn turbulent flow regime reach their asymptotic limit within dimensionless time $1 < \theta^* < 5$. (See footnote).

Often a case of interest is a quick estimate of the time and velocity of a level swell reaching a height H that is less than H_∞ such as would occur in an impact condition on a vessel head. Examining Fig. 3 would indicate that if $H < 0.5\ H_\infty$, the impact velocity is not significantly less than the maximum velocity defined by (17).

5. COMPARISON WITH EXACT SOLUTION

Epstein [1] has provided an exact formulation of the level swell problem for several limiting cases of volume heating and ideal flow regimes. Epstein uses a different choice of dimensionless variables. For pool height he defines,

$$h = \psi\ H/H_o \tag{18}$$

and for dimensionless time

$$\tau = \psi_s\ U_\infty\ \theta/H_o \tag{19}$$

The exact solution for a volume heated pool swell with an ideal churn turbulent vapor disengagement model is given by Epstein as:

$$h(\tau) = (1 + \tau)\ (1 + \psi_s) - \frac{1}{2}\,[(1 + \tau)^2 + 1] \tag{20}$$

The corresponding one node model, Eq. (15), cast in the same dimensionless form as (18) and (19) above, is

$$\dot{h} = \psi_s - 2(h - \psi_s)/[h - C_o(h - \psi_s)] \tag{21}$$

The comparison of (20) and (21) are shown in Fig. 4 for the special case $\psi_s = 1$ and $C_o = 1$. The two solutions compare favorably over most of the expansion range. The major difference is the fact that the one node model reaches the correct limit in twice the time as the exact solution. However, because of the exponential character of the approach to the swell limit H_∞, the difference in level swell during the earlier stages is not significant.

6. EVALUATION OF VAPOR SOURCE TERM

The volumetric source term S may be constant or varying with time. There are however, several cases for which the source term can be easily evaluated.

This analysis applies to the churn flow regime and does not characterize foaming fluids which introduce additional analytical complexities discussed in Ref. [9]. However, the level rise in a foaming system can be closely approximated by (17b) for a significant interval.

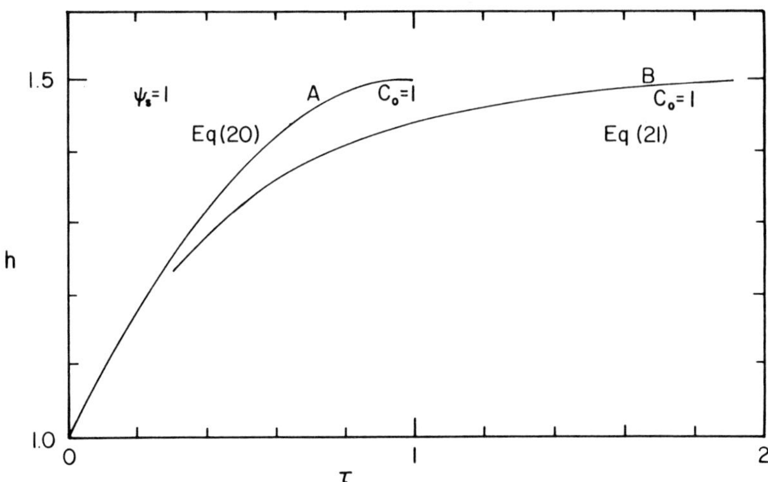

Fig. 4. Comparison of One Node Model Eq. (15) - Curve B, with Exact Method of Characteristics Solution - Curve A

Volume heated system at constant pressure - Under constant pressure conditions for a saturated liquid with uniform volume heating, the source S is

$$S = q\ H_0(R - 1)/\lambda \tag{22}$$

where q is the specific power density (W/kg), H_0 is the liquid rest height, R is the vapor to liquid specific volume ratio, and λ is the latent heat of vaporization.

System pressure above ambient - for a saturated liquid in a vessel at a pressure above ambient, the source term is determined by vapor flow through the vessel opening A. This is the case with or without volume heating. Volume heating, along with A, merely determines the rate of pressure change. In this case,

$$S = v_{fg}\ GA/A_x \tag{23}$$

where,

$$G = \hat{F}\ P/\sqrt{R_g T} \tag{24}$$

for sonic vapor flow* and $\hat{F}$ is a function of the vapor specific heat ratio γ

$$\hat{F} = \left[\frac{2}{\gamma + 1}\right]^{\frac{1}{\gamma - 1}} \left[\frac{2\gamma}{\gamma + 1}\right]^{\frac{1}{2}} \tag{25}$$

In (23) and (24) all properties are evaluated at vessel or stagnation conditions.

Level swell is often encountered in the depressurization of a vessel containing a saturated liquid. In such cases the value of S can be evaluated at stagnation conditions, and can be assumed to be constant over the period of level swell. This is a reasonable approximation since the time for the level swell to reach its asymptotic limit, or the limit of the vessel height is usually very short compared to the overall time scale of the depressurization event. Also, S varies as the $\sqrt{T}$ over the depressurization event and this variation over the time scale of the level swell event is negligible.

The application of these concepts are valid so long as inertial effects are small relative to hydrostatic head considerations. Epstein [1] has shown that inertia effects are not important so long as

$$\psi_s << g\ H_o/U_\infty^2 \tag{26}$$

7. DATA COMPARISONS

The advantage of this short formulation of the level swell analysis over more exact methods are calculational ease, and most important the ease with which reasonable accuracy can be obtained. Figure 5 illustrates a comparison of (15) with data of Ref. [10]. It is seen that most of the level swell follows the maximum rate as defined by $\dot{H}^* = 1$.

Figure 6 illustrates an integral calculation of data from a large vessel test described in Ref. [11]. The key features of the test are:

Vessel volume	4.3 m^3	Vent area	11.2 x 10^{-4} m^2
Diameter	1.06 m	Liquid inventory	1.85 x 10^3 kg
Height	4.54 m	Saturation pressure	6.69 MPa

Figure 6 shows the pressure and swelled height as a function of time. Also indicated is the calculated rest height which is decreasing because of a loss of mass due to vapor flow from the vessel opening. Similar results can be obtained for the data cited in Ref. [4] as indicated in Fig. 7. In all of these comparisons the source term S was determined from (23) and (24), H_∞ from (8) and (9), and the level swell from (15). For quick estimates, Fig. 3 can be used in place of direct integration of (15).

8. SUMMARY

A one node level swell model is often adequate for quick assessments provided one is aware of the limitations. This paper has indicated that a one node model is in good agreement with exact solutions over most of the time scale of the level swell event and, of course, proceeds to the same asymptotic limit. The one node model is also shown to be capable of reproducing the key

---

*For subsonic flow use

$$G = \frac{P}{\sqrt{R_g T}} \left[\frac{2\gamma}{\gamma - 1} (1 - P/Pa)^e) \ (P/Pa)^{2/\gamma} \right]^{1/2}$$

where $e = (\gamma - 1)/\gamma$ and Pa is the ambient pressure.

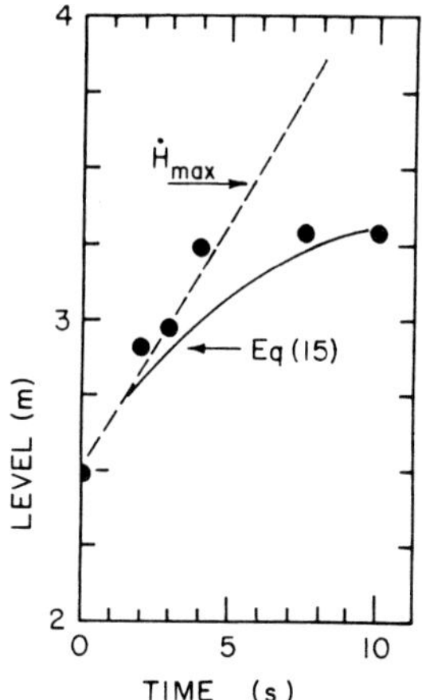

Fig. 5. Comparison of this Analysis with Saturated Water Blowdown Data of Ref. [10], P_{sat} = 7.17 MPa, A = 48 x 10^{-6} m^2, A_x = 7.3 x 10^{-2} m^2, H_o = 2.48 m, $\dot{H}_{max}$ = S = 0.17 m/s from Eq. (23) and (24)

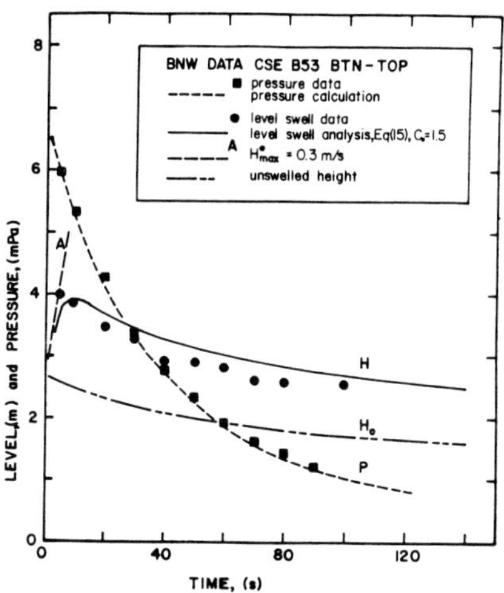

Fig. 6. Comparison of Level Swell Analysis of this Paper with Integral Saturated Water Blowdown Data of Ref. [11]

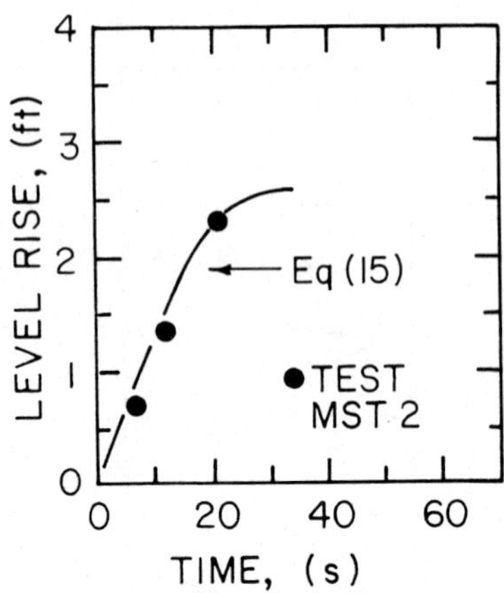

Fig. 7. Comparison of Level Swell Analysis of this Paper with Data Reported in Ref. [4]

features of experimental data even at large scale. For the churn flow regime, the solution to (15) as indicated in Fig. 3 can be used in a universal sense.

NOMENCLATURE

A	Vessel discharge area
A_x	Vessel area
C_o	Distribution parameter in (3)
D	Vessel Diameter
G	Mass flux
g	Gravity 9.8 m/s^2
H	Height
h	Dimensionless height used in (18)
$\dot{H}$	$\Delta H/\Delta\theta$
$\hat{F}$	Defined by (25)
j_g	Vapor volume flux
P	Pressure
q	Heat source
$\dot{V}$	Volume source rate
α	Vapor volume fraction below swelled height
Δ	Change
θ	Time
σ	Surface tension of liquid phase
λ	Latent heat
ρ	Density
ψ	Dimensionless volume flux
ϕ	Defined by (16)
τ	Dimensionless time used in (19)
γ	Specific heat ratio for gas
R	Specific volume ratio v_g/v_f
R_g	Gas constant
S	Source strength expressed as a volume flux

T Temperature

v Specific volume

U_∞ Bubble rise velocity Eq. (4)

SUBSCRIPTS

x Refers to vessel

o Rest

∞ Asymptotic

s Value of ψ when $j_g(\alpha) = S$

g Vapor

f Fluid

fg Difference between vapor and fluid properties

max Maximum

SUPERSCRIPTS

* Dimensionless parameter

· Time derivative

REFERENCES

1. Esptein, M., "Transient Behavior of a Volume Heated Boiling Pool," ASME Paper 75-WA/HT-31, 1975.

2. Vea, H. W. and Lahey, Jr., R. T., "One Exact Analytical Solution of Pool Swell Dynamics During Depressurization by the Method of Characteristics," Nucl. Eng. and Design, Vol. 45, pp. 101-116, 1978.

3. Condiff, D. W., "Note on Pool Swell Dynamics During Depressurization," Nuclear Engineering and Design, Vol. 52, No. 1, pp. 57-58, 1969.

4. Moody, F. J., "Liquid/Vapor Action in a Vessel During Blowdown," Trans. ASME, Journal of Eng. for Power, Seven A, No. 1, Vol. 91, pp. 53-61, 1969.

5. Grolmes, M. A. and Fauske, H. K., "An Evaluation of Incomplete Vapor Phase Separation in Freon-12 Top Vented Depressurization Experiments," Proceedings of this Conference.

6. Ginsberg, T., Jones, Jr., O. C., and Chen, J. C., "Flow Behavior of Volume Heated Boiling Pools: Implications with Respect to Transition Phase Accident Conditions," Nuclear Technology, Vol. 46, No. 3, pp. 391-398, 1979.

7. Wallis, G. B., One Dimensional Two-Phase Flow, McGraw Hill Book Company, 1969, (see pg. 255, Eq. (9.34).

8. Zuber, N. and Findlay, J. A., "Average Volumetric Concentration in Two-Phase Flow Systems," Trans. ASME J. Heat Transfer, Vol. 87, Ser. C., pg. 453, 1965.

9. Condiff, D. W., Epstein, M., and Grolmes, M. A., "Transient Volumetric Pool Boiling with Foaming," AIChE Symp. Ser. No. 164, Vol. 73, p. 86, 1977.

10. Slifer, B. C., "Loss of Coolant Accident and Emergency Core Cooling Models for General Electric Boiling Water Reactors," GE Report, NEDO-10329, April 1971.

11. Wilburn, N. P., "Void Fraction Profile in a Nuclear Reactor Vessel During Cooling Blowdown," AEC Research and Development Report- Battelle Northwest Laboratories, April 1970.

Multi-Phase Flow and Heat Transfer III. Part A: Fundamentals
edited by T.N. Veziroğlu and A.E. Bergles
Elsevier Science Publishers B.V., Amsterdam, 1984 — Printed in The Netherlands

AN EVALUATION OF INCOMPLETE VAPOR PHASE SEPARATION IN FREON-12 TOP VENTED DEPRESSURIZATION EXPERIMENTS

Michael A. Grolmes and Hans K. Fauske
Fauske & Associates, Inc.
16W070 West 83rd Street
Burr Ridge, Illinois 60521

ABSTRACT

The Freon-12 (R-12) depressurization experiments reported in Ref. [1] represent unique experimental data for the two-phase discharge from a top vented container. In Ref. [1], the observed results were only related to single-phase gas flow calculations. The analysis described in this paper is shown to reproduce the two-phase as well as the single-phase discharge behavior observed in the tests.

1. INTRODUCTION

Partial vapor-liquid phase separation has been discussed in the design of emergency relief systems and the interpretation of small scale experimental results by Fauske, et al., Ref. [2]. Vapor phase superficial velocity in the vessel is an important parameter in determining the liquid phase discharge. Methods have been proposed in Ref. [2] for describing the vapor-liquid phase ratio in the two-phase discharge from a top vented vessel. The R-12 depressurization experiments reported by Guhler, Hanneman, and Sallet, Ref. [1] provide direct data for comparison with such analysis. In these experiments the vent orifice was varied such that the superficial vapor velocity in the test apparatus was changed by an order of magnitude. This variation clearly illustrates the effect of vapor superficial velocity on liquid carry out. The data and their interpretation in this report illustrate the various facets of the analytical development used to describe the two-phase discharge process.

2. EXPERIMENTAL DESCRIPTION

The data reported in Ref. [1] featured two experimental depressurization events with two different vent areas of an orifice type. The experimental apparatus was basically a transparent lucite vessel, of right cylindrical configuration, as shown in Fig. 1. The vessel was filled with R-12. The saturation pressure of R-12 at approximately ambient temperature of 297 K (75°F) corresponds to about 635 kPa. Relevant experimental details are illustrated in Table 1.

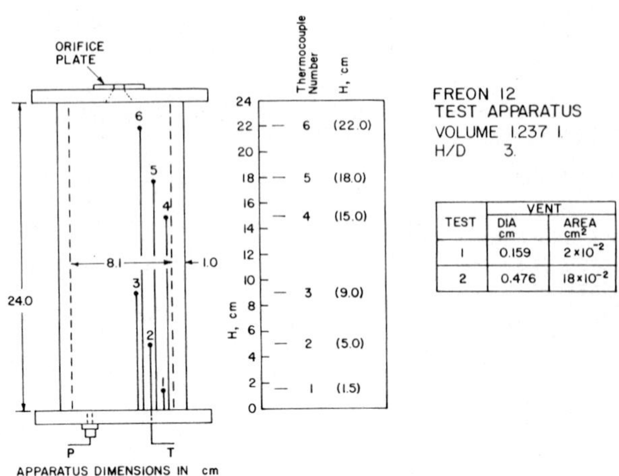

Fig. 1. Test Apparatus for R-12 Depressurization Experiments of Ref. [1]. Fluid Loss was Measured by a Strain Gage Located on a Cantilever Support Arm not Shown in Figure

Table 1. Depressurization Test Conditions

Vessel Volume		1.23×10^{-3} m^3	
Vessel i d		8.1×10^{-2} m	
Vessel Height		24.0×10^{-2} m	
Test Fluid (Mass)		R-12 (1.6 kg full)	
Test Conditions:			
		Test No. 1	Test No. 2
Vent Diameter	D	1.59×10^{-3} m	4.76×10^{-3} m
Vent Area	A	2×10^{-6} m^2	18×10^{-6} m^2
Initial Pressure	P_o	655 kPa	655 kPa
Initial Temperature	T_o	297 K	297 K
Saturation Pressure at T_o		635 kPa	635 kPa

Vapor Superficial Velocity at Initial Conditions j_{gs}	2.28 cm/s*	20.43 cm/s*

	Fluid Properties at T_o
Liquid Specific Heat	937 j/kg
Latent Heat	139 kj/kg
Liquid Density	1318 kg/m^3
Vapor Density	36.2 kg/m^3
Surface Tension	10^{-2} N/m

*Assuming C_d = .71.

The experiments will be referred to as test No. 1 or No. 2 or by the size of the discharge orifice. The 1.59 millimeter diameter discharge orifice is significant from the point of view that the discharge event had an entirely different character than the discharge event with the 4.76 millimeter diameter orifice. With the former, very little two-phase venting, if any, was observed. The pressure time history, to be discussed later, showed dramatically different characteristics than observed with the 4.76 millimeter orifice. The discharge event with the larger orifice produced two-phase flow for approximately 7 sec as reported in Ref. [1]. Pressure decreased more rapidly in comparison with the small diameter orifice discharge event. The initial rate of mass loss for Test 2 was two to three times larger than the mass loss for all vapor venting.

In order to explain the observed differences, the effect of the orifice diameter on the vapor generation rate within the bulk fluid in the vessel and the associated level swell must be considered. The vapor-liquid phase ratio entering the orifice is also important in order to correctly interpret the two-phase flow through the orifice. Finally, the appropriate two-phase vent flow model must be employed. The difference between the stagnation pressure in the reservoir and the ambient pressure is such as to insure either choked two-phase or choked single-phase flow condition. It should be noted that in Ref. [1] the data were only related to single-phase compressible flow analytical comparisons. These comparisons showed essential agreement with each discharge event in a range where single-phase flow discharge was actually occurring. We set out here to show how the methods outlined in Ref. [2] are supported by the experimental observations of Ref. [1].

3. ANALYTICAL ELEMENTS

The rate of depressurization of a vessel containing a saturated liquid is determined by the size of the discharge opening and the mass of its contents.

The vessel contents are treated as a one dimensional lumped entity for which a general energy equation can be written as,

$$\frac{d}{dt}(Mi) = -GA\,e_v \tag{1}$$

where the specific energy terms are,

$$e_v = (i + Pv + KE + PE)_v \tag{2}$$

$$i = i_f + x i_{fg} \tag{3}$$

and other terms are defined in the nomenclature. The subscript v designates conditions at the discharge opening which may be different from the vessel average or bulk conditions. Also the kinetic energy, (KE), and potential energy, (PE), terms can usually be neglected. Equation (1) can be reduced to

$$C^* \frac{dT}{dt} + i_{fg} \frac{dx}{dt} = - \frac{GA}{M} (e_v - i) \tag{2}$$

where C* is a heat capacity term,

$$C^* = \frac{di_f}{dT} + x \frac{di_{fg}}{dT} \tag{3}$$

In order to evaluate the term dx/dt one proceeds from a statement that total volume V remains constant,

$$\frac{d}{dt} (Mv) = 0 \tag{4}$$

In the expansion of (4) the term ϕ is defined as,

$$\phi = x \frac{dv_g}{dT} + (1 - x) \frac{dv_f}{dT} \tag{5}$$

From (4) one can find that,

$$\frac{dx}{dt} = \frac{v}{v_{fg}} \frac{GA}{M} - \frac{\phi}{v_{fg}} \frac{dT}{dt} \tag{6}$$

Equations (6) and (2) can be combined along with the Clapeyron equation,

$$\frac{dT}{dP} = \frac{Tv_{fg}}{h_{fg}} \tag{7}$$

to yield an expression for the rate of depressurization,

$$\frac{dP}{dt} = \frac{- \left[x_v + \frac{v_f}{v_{fg}}\right] \frac{h_{fg}\ G\ A\ \beta}{C^*\ M\ (1 + \beta)}}{\frac{Tv_{fg}}{h_{fg}} \left[1 - \frac{\phi P}{C^* \beta}\right]} \tag{8}$$

In (8) $\beta = [h_{fg}/(Pv_{fg}) - 1]^{-1}$.

Equation (8) describes the vessel pressure response to the discharge flow rate. Ancillary relations for x_v and G are needed. A sufficient discussion of the necessary relations for the mass flux G can be found in Refs. [3] and [4]. The value of the vapor mass fraction entering the discharge opening x_v may be between 0 and 1. All vapor and all liquid discharge correspond to x_v

= 1 and x_v = 0 respectively. For a top located discharge opening, $x < x_v < 1$ represents the general limits of x_v where x is the vessel average or bulk vapor mass fraction.

In order to adequately describe the experimental observations of Ref. [1], vapor phase separation in the vessel must be considered. The development is based on a vapor holdup correlation of the form,

$$j_g = \frac{2\,\alpha\,U_\infty}{1 - C_o\alpha} \tag{9}$$

where α is the average void fraction in the vessel, U_∞ is the rise velocity of discrete bubbles, and C_o is a correlating parameter related to radial nonuniformities. The rise velocity U_∞ is given by, (see Ref. [5]),

$$U_\infty = 1.53\left\{\frac{\sigma g(\rho_f - \rho_g)}{\rho_f^2}\right\}^{1/4} \tag{10}$$

Data in the literature [6] can be reasonably correlated by (9) with a C_o parameter between 1 and 2. A value of 1.5 is often a good average representation as illustrated in [7].

Partial vapor separation follows from a relation of continuity of vapor flow from the vessel to the break,

$$\rho_g\, j_g\, A_x = x_v\, G\, A \tag{11}$$

From vapor drift flux considerations it can be established for the churn turbulent flow regime that,

$$j_g = \lambda U_\infty/(1 - \lambda\alpha) \tag{12}$$

where $\lambda = \alpha_v/(1 - \alpha_v)$; $a = (1 - x_v)/x_v R$ and $R = v_g/v_f$.

The value of α_v is not conveniently known, but for the ideal churn turbulent flow regime in an open system it can be shown that,

$$\frac{\alpha_v}{1 - \alpha_v} = \frac{2\,\alpha}{1 - \alpha} \tag{13}$$

where α is the average void fraction in the vessel. It is taken as a reasonable assumption that a similar relation holds for the churn turbulent flow regime in a closed system with vapor and liquid discharge such that,

$$\lambda = \frac{2\,\alpha}{1 - C_o\alpha} \tag{14}$$

Equations (11) and (12) can be combined with the appropriate definitions in the following form,

$$\frac{x_v\, v_g\, G\, A}{\lambda\, U_\infty\, A_x} = \frac{x_v\, R}{x_v(R + \lambda) - \lambda} \tag{15}$$

Equations (15) and (8) along with an ancillary relation for vent flow G, see Ref. [3] are sufficient to describe both the flow and rate of change of vessel pressure for an adiabatic depressurization.

The nature of the solution is that for nearly full containers and (α small) and large openings (A) the vapor mass fraction x_v will be low (i.e., these are the required conditions for a low quality two-phase discharge). As an event proceeds, assuming an initial two-phase discharge, the mass loss will increase the average void fraction in the vessel α. The vapor mass fraction increases until $x_v \rightarrow 1$. This describes the change from two-phase vent flow to single-phase vent flow. This condition is equivalent to the quasi-steady state swelled height of the two-phase region in the vessel being less than the vessel height.

The maximum superficial vapor volume flux in the vessel is determined by the discharge opening and the all vapor mass flux,

$$j_g = v_g \, G \, A/A_x \tag{16}$$

where $G \simeq 0.62 \, P/\sqrt{R_g T}$ for R-12. Equation (9) describes a relation between j_g and vessel average void fraction. The void fraction α which yields the same j_g as Eq. (16) is the vessel void fraction required for complete vapor phase separation inside the vessel. A vessel void fraction less than this value leads to two-phase vent flow, in a manner similar to that discussed in [2] and [7].

4. ANALYSIS OF THE R-12 DATA OF REF. [1]

Table 1 indicates that the maximum superficial vapor velocity for the R-12 test according to (16) varies from 2.5 cm/s to 20.5 cm/s. These values can be related to an average void fraction or level swell in Fig. 2. For Test 1, this corresponds to 2-1/2% to 8% void fraction. The latter value gives consideration to the bubbly flow region not considered likely in these experiments. For the purpose of this analysis we infer that the initial liquid R-12 inventory was such as to leave a free board volume of $\leq$ 5%. For test 1, the maximum superficial vapor velocity would not lead to a two-phase discharge assuming the churn turbulent flow regime. This is consistent with observations.

However, for Test 2, Fig. 2 suggests that a two-phase discharge would continue from time 0 until the vessel void fraction exceeded 32%. This would correspond to a loss in the initial liquid of ≃ 485 g*. This is very consistent with the experiment observation as will be discussed later.

Thus, on the basis of two-phase holdup correlation for the churn turbulent regime, one can infer a significant difference between Test 1 and Test 2 that is consistent with observations. Further good agreement with the data is found by comparing the transient analysis (8) and (15) with the actual test results. For Test T2, the Henry-Fauske [8] critical flow model is used because of the short nozzle or orifice geometry. For other discharge geometry the homogeneous equilibrium critical flow model may be required [4].

The essential equations describing a vessel depressurization event are able to handle automatically either a single-phase flow from time zero or an event which precedes initially as a two-phase discharge, but because of deple-

---

*This number could be between 410 g and 510 g depending on assumptions of initial inventory.

tion of liquid inventory within the vessel, undergoes a transition to all vapor venting. Figure 3 shows a comparison of the large diameter orifice and the small diameter orifice pressure data under discussion here, along with analytical calculations for the same events. It can be seen, by examination of Fig. 3, that the analytical comparison in both cases is quite satisfactory. A discharge coefficient of .71 was used in both comparisons based on a best fit to Test 1 which showed all vapor flow.

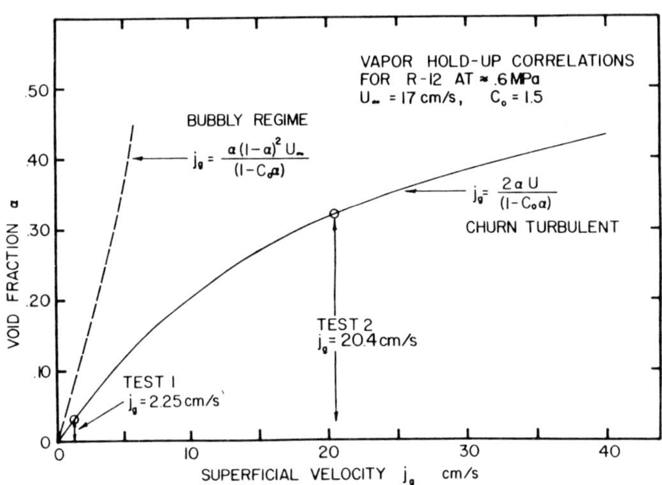

Fig. 2. Vapor-Liquid Holdup Correlation with Indications of Relative Position of R-12 Test 1 and Test 2

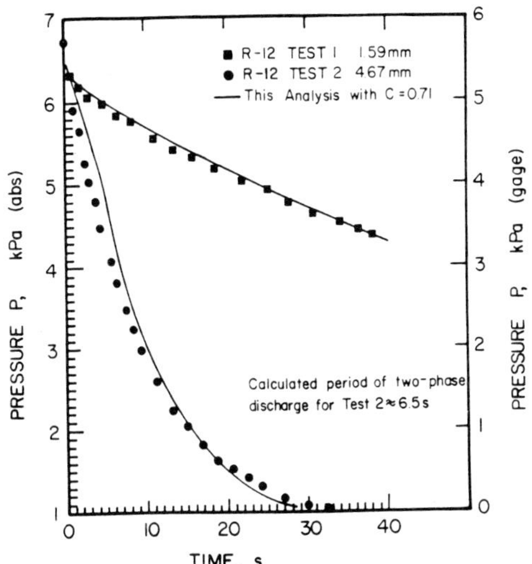

Fig. 3. Comparison of Pressure Transient Data and Analysis of R-12 Test 1 and Test 2

Now we focus attention on the more rapid depressurization associated with the larger diameter orifice and the ability of the calculations to track this behavior. Several points contained in the calculation will be illustrated by further reference to Fig. 4. The calculation indicates a transition from two-phase venting to all vapor flow at about 6.5 secs. This corresponds to a vessel average void fraction of approximately 32% which was anticipated in the earlier discussion. The observations in [1] indicated two-phase discharge for a period of 7 sec.

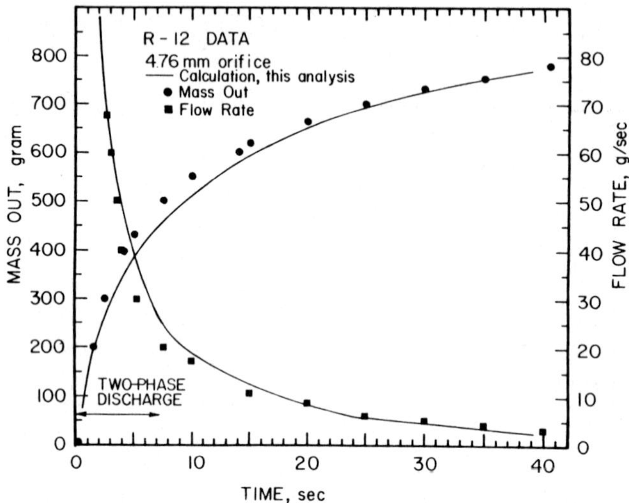

Fig. 4. Comparison of Mass Discharge Transient Data and Analysis of R-12 Test 2

Figure 4 shows the comparison of calculated mass flow rate vs time with the experimental data for the 4.76 millimeter orifice. The ability of the analysis to track the mass flow rate as well as the integral mass lost from the vessel is quite satisfying from the point of view again of integrating vessel behavior with partial vapor-liquid disengagement and coupling this behavior with an appropriate two-phase critical flow model. Again we note the agreement at 7 sec (the point observed as the cessation of two-phase flow). Data indicates ≃ 500 g mass loss. The goodness of the comparison supports the basic assumptions in the postulated partial vapor disengagement model.

5. SUMMARY

This analysis of the R-12 data of Ref. [1] represents a very useful piece of experimental support for the modeling of partial vapor-liquid disengagement and the ability to couple from flow regime within the vessel and discharge the vessel. The transition from two-phase discharge to single-phase discharge which is determined by the orifice dimensions for these experiments is accommodated satisfactorily in the analysis.

NOMENCLATURE

A	Discharge Area	P	Pressure

C^*	Heat capacity defined by (3)	PE	Specific potential energy
C_o	Parameter in (9)	R	Specific volume ratio v_g/v_f
e	Specific energy	R_g	Gas constant
G	Mass flux	T	Temperature
g	Gravity (9.8 m/s)	t	Time
h	Enthalpy	U_∞	Defined by (10)
i	Internal energy	V	Volume
j	Volume flux	v	Specific volume
KE	Specific kinetic energy	x	Vapor mass fraction
M	Mass of fluid in vessel		

SUBSCRIPTS

f	Fluid	v	Refers to conditions entering discharge opening
fg	Difference between vapor and liquid properties	x	Refers to vessel
g	Gas or vapor		

GREEK SYMBOLS

α	Volume fraction of vapor	ρ	Density
β	Defined under (8)	σ	Surface tension
λ	Defined under (12)	ϕ	Defined by (5)

REFERENCES

1. Guhler, M., Hanneman, R. J., and Sallet, D. W., "Unsteady Two-Phase Blowdown of a Flashing Liquid from a Finite Reservoir," Two-Phase Flows and Reactor Safety, Momentum Heat and Mass Transfer in Chemical Process and Energy Engineering Systems, Vol. 2, Ed. by F. Durst, et al., Hemisphere Publishing Company.

2. Fauske, H. K., Grolmes, M. A., and Henry R. E., "Emergency Relief Systems Sizing and Scale-Up," AIChE Plant Operations Progress, Vol. 2, No. 1, pp. 27-29, January 1983.

3. Grolmes, M. A. and Leung, J. C., "Scaling Considerations for Two-Phase Critical Flow," Paper in Proceedings of this Conference.

4. Fauske, H. K., "Multiphase Flow Considerations in Sizing Emergency Relief Systems for Runaway Chemical Reactions," Paper in proceedings of this Conference.

5. Wallis, G. B., One Dimensional Two-Phase Flow, McGraw Hill Book Company, 1969, (see pg. 255, Eq. (9.34)).

6. Ginsberg, T., Jones, Jr., O. C., and Chen, J. C., "Flow Behavior of Volume Heated Boiling Pools: Implications with Respect to Transition Phase Accident Conditions," Nuclear Technology, Vol. 46, No. 3, pp. 391-398, 1979.

7. Grolmes, M. A., "A Simple Approach to Transient Two-Phase Level Swell," Paper in proceedings of this Conference.

8. Henry, R. E., Fauske, H. K., "The Two-Phase Critical Flow of One Component Mixtures in Nozzles, Orifices, and Short Tubes," ASME Journal of Heat Transfer, pp. 179-186, May 1971.

Multi-Phase Flow and Heat Transfer III. Part A: Fundamentals
edited by T.N. Veziroğlu and A.E. Bergles
Elsevier Science Publishers B.V., Amsterdam, 1984 — Printed in The Netherlands

SCALING CONSIDERATIONS FOR TWO-PHASE CRITICAL FLOW

Michael A. Grolmes and Joseph C. Leung
Fauske & Associates, Inc.
16W070 West 83rd Street
Burr Ridge, Illinois 60521, U.S.A.

ABSTRACT

Similarity or scaling parameters for various two-phase critical flow models are developed which demonstrate that pressure and other fluid property dependencies can be largely eliminated. Dimensionless graphs can be developed to serve as a guide for arbitrary fluids. The homogeneous equilibrium, the homogeneous nonequilibrium, and the slip equilibrium models are considered.

The scaling considerations discussed in this paper refer to property scaling as opposed to geometric scaling. Geometric or dimensional scale effects can and do influence model selection, (an important aspect not covered here). However, the consideration of pressure and property scaling can indicate improved methods for data reduction and model evaluation.

The objective of the paper is to develop a rational non-dimensional basis for evaluation of several often used critical flow models. This information enables a quick assessment of two-phase choked flow rates to be made in instances where only limited fluid property data are available.

1. INTRODUCTION

It is useful to retain, as far as possible, a similarity with single phase gas dynamic choked flow. Consider the expansion of a compressible gas through an ideal nozzle which satisfies the energy equation,

$$G = \sqrt{2(h_o - h_1)/v_1^2} \qquad (1)$$

For a reversible isentropic process, (1) can also be written:

$$G = \frac{1}{v_1}\left\{2\int_{P_o}^{P_1} v\,dP\right\}^{1/2} \tag{2}$$

For a gas obeying the relation Pv^γ = *constant*, (2) results in,

$$G^* = \left[\frac{2\gamma}{\gamma - 1}\left(1 - \eta^{\frac{\gamma - 1}{\gamma}}\right)\eta^{\frac{2}{\gamma}}\right]^{1/2} \tag{3}$$

where:

$$G^* = G/\sqrt{P_o/v_o}$$

and

$$\eta = P_1/P_o$$

Equation (3) has a well known maximum when

$$\eta = \left[\frac{2}{\gamma + 1}\right]^{\gamma/(\gamma - 1)} \tag{4}$$

Any local condition of choking satisfies

$$G = [- dv/dP]^{-1/2} \tag{5}$$

which for a gas obeying the relation Pv^γ = *constant*, reduces to:

$$G^* = \left[\gamma\,\eta^{\frac{\gamma + 1}{\gamma}}\right]^{1/2} \tag{6}$$

The condition of gas dynamic choking simultaneously satisfies (3), (4), and (6) as shown in Fig. 1. From either (3) and (4) or (4) and (6) and the ideal gas law, $Pv = R_gT$, one can write the gas dynamic choked flow rate as

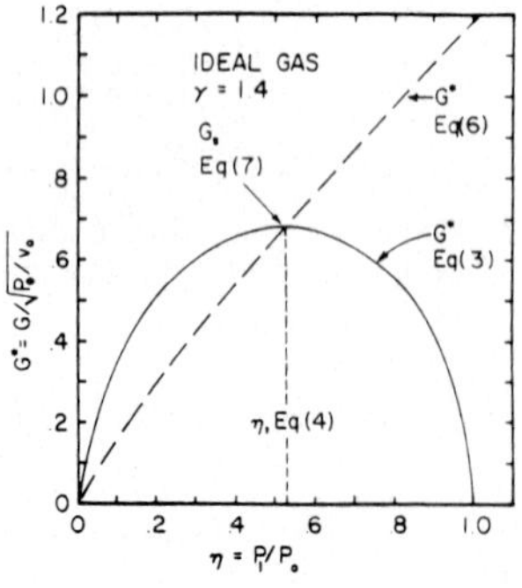

Fig. 1. Illustration of the Simultaneous Solution of (3), (4), and (6) for an Ideal Gas.

$$G_s = \frac{P_o}{\sqrt{R_g T_o}} \hat{F}(\gamma) \tag{7}$$

where

$$\hat{F}(\gamma) = \left\{ \gamma \left[\frac{2}{\gamma + 1} \right]^{\frac{\gamma + 1}{\gamma - 1}} \right\}^{1/2} \tag{8}$$

From this development, the pressure and property dependence for single-phase gas choking are evident. For a two-phase flow the results are not so readily apparent. We find that a two-phase mixture retains much underlying similarity with single-phase gas flow. For a vapor rich mixture, the vapor phase compressibility determines the similarity considerations, while for vapor lean mixtures both the liquid density and the rate of vapor formation (flashing) determine the similarity considerations. This will be illustrated for the various models.

2. HOMOGENEOUS EQUILIBRIUM MODEL

The homogeneous equilibrium (HE) critical flow model can be shown to have strong similarities with the previous gas dynamic model. Consider the thermodynamic equilibrium expansion of a two-phase compressible mixture through an ideal nozzle which satisfies the energy equation (1), where:

$$h = h_f + x h_{fg} \tag{9a}$$

and

$$v = v_f + x v_{fg} \tag{9b}$$

The maximization of (1) with (9) evaluated for an isentropic path is often cited as a working definition of the HE critical flow model. It has also been found that extreme care must be used to avoid false maxima arising from tabulated property data. We find that false maxima can be avoided by considering (2) with (9b),

$$G = \frac{1}{v_1} \left[2 \int_{P_o}^{P_1} v dP \right]^{1/2} \tag{10}$$

again valid for a reversible isentropic two-phase expansion. Equation (10) is identical to (2) except that v is the two-phase homogeneous specific volume, (9b). Using the Clapeyron equation,

$$\frac{dT}{dP} = T\, v_{fg}/h_{fg} \tag{11}$$

and the relation for vapor mass fraction x for an isentropic expansion assuming the liquid phase to be incompressible

$$x = x_o \frac{h_{fgo}}{h_{fg}} \frac{T_o}{T} + \frac{CT}{h_{fg}} \ln (T_o/T) \tag{12}$$

Equation (10) can be evaluated,

$$G = A^{-1} \sqrt{P_o/v_f} \left[2(1 - \eta) + 2B \left(\left(1 + \frac{x_o h_{fgo}}{C\ T_o} \right) (1 - \zeta) + \zeta \ln \zeta \right) \right]^{1/2} \tag{13}$$

where

$$A = [xR + (1 - x)]$$

$$B = C\ T_o/(P_o\ v_f)$$

$$R = v_g/v_f$$

$$\zeta = T/T_o$$

In the same manner as single-phase gas dynamics, any local condition of choking satisfies (5) with a two-phase homogeneous specific volume. This leads to,

$$G = \sqrt{P_o/v_f} \left\{ \frac{xR}{\gamma \eta} + \psi \left(1 - \frac{xh_{fg}}{CT} \right) \right\}^{-1/2} \tag{14}$$

where

$$\psi = v_{fg}^2\ CT\ P_o/(h_{fg}^2\ v_f)$$

The intersection of (13) and (14) determine the HE choked flow rate in an unambiguous manner that is not affected by false maxima in (13). The intersection also implicitly determines the critical pressure ratio η_s. An example for steam-water at 1.1 MPa is shown in Fig. 2.

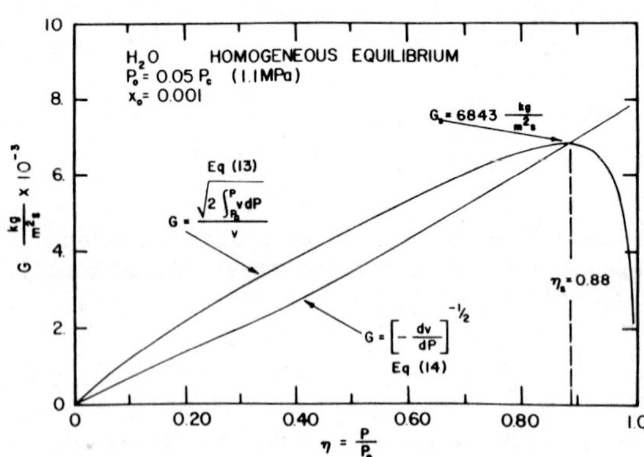

Fig. 2. Homogeneous Equilibrium Two-Phase Critical Flow for Steam-Water at 1.1 MPa Satisfies (13) and (14) at $\eta_s = 0.88$.

The homogeneous equilibrium model, evaluated with saturated steam-water properties, is shown in dimensionless form in Figs. 3a and 3b. Curves of similar reduced pressure are illustrated. Figure 3a covers the high quality vapor continuous region. For this region, the dimensionless flow parameter is

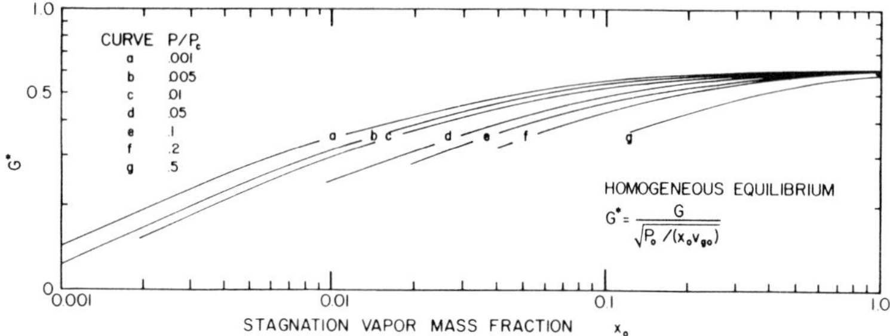

Fig. 3a. Homogeneous Equilibrium Critical Flow Model Evaluated with Steam-Water Properties. Curves at Various Reduced Pressures Cover Vapor Continuous, High Void Fraction Region $\alpha_0 > 0.6$.

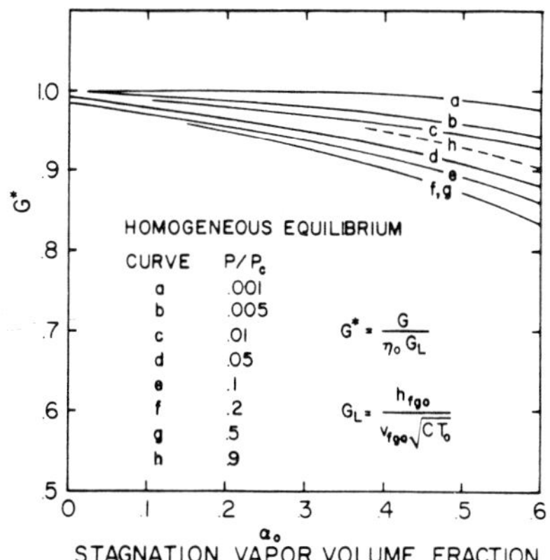

Fig. 3b. Homogeneous Equilibrium Critical Flow Model Evaluated with Steam-Water Properties. Curves at Various Reduced Pressure Cover Liquid Continuous Low Quality Region $\alpha_0 < 0.6$.

$$G^* = G/\sqrt{P_0/(x_0\, v_{go})} \tag{15}$$

Curves in Fig. 3a are terminated at a stagnation void fraction $\alpha_0 = 0.6$ for the respective reduced pressures. The high quality vapor-continuous flow regime is dominated by the compressible effect of the vapor density as represented by the first term in brackets on the RHS of (14). The value of G* as $x \to 1$ is $\simeq 0.63$ which is similar to that of gas dynamic behavior. (Recall (7) and Fig. 1).

In the low quality region, the compressibility effect of the "flashing term" becomes significant. This is represented by the second term in brackets on the RHS of (14). The dimensionless flow parameter most appropriate for the low quality regime is

$$G^* = G/[\eta_o G_L] \tag{16}$$

where G_L signifies a limiting flow as $x \to 0$ from (14)

$$G_L = h_{fgo}/[v_{fgo}\sqrt{C\ T_o}] \tag{17}$$

The rational for the denominator in (16) is geometric in nature as illustrated in Fig. 4.

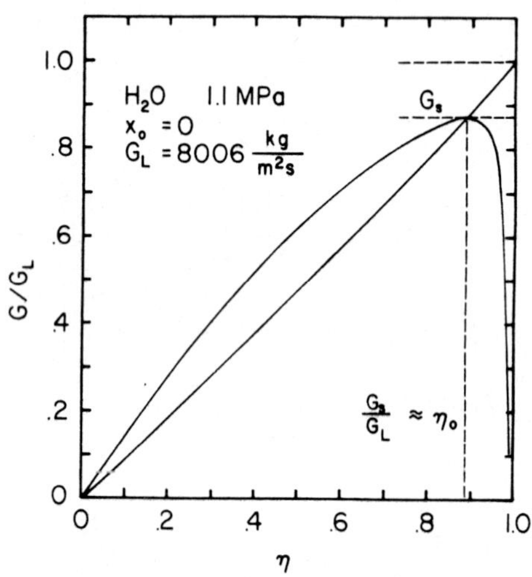

Fig. 4. Geometric Illustration of the Homogeneous Equilibrium Critical Flow Solution at x = 0. Based on Triangular Similarity, $G_s \simeq \eta_o\ G_L$.

The grouping of curves of various reduced pressure in Fig. 3b remains tightly spaced and is only slowly varying from unity over the liquid continuous flow regime, here generously extended to $\alpha_o = 0.6$. The significance of the dimensionless grouping is that η_o and G_L can be determined as general function of a few usually obtainable properties and upon so doing, reduced pressure scaling can then be used to make Fig. 3a and 3b applicable to any fluid.

It is first noted that G_L as defined by (17) is a simple function of reduced pressure except in the near critical region as shown in Fig. 5. Equally important, is the <u>parallel</u> behavior of most fluids also indicated in Fig. 5. Figure 5 can be reduced to a single line of the form,

$$\frac{G_L}{G_{L|P_r} = 0.01} = a\ P_r^b \tag{18}$$

where a = 43.56 and b = 0.82. Values of G_L at P_r = 0.01 for the fluids shown in Fig. 5 are tabulated below.

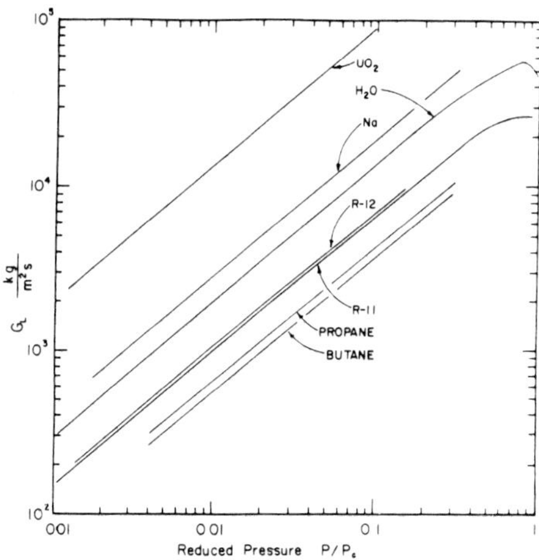

Fig. 5. Values of G_L for Various Fluids Illustrate Linear and Parallel Behavior (on Logarithmic Scale) as a Function of Reduced Pressure.

Table 1. Values of G_L at P_r - 0.01

Fluid	P_c (MPa)	$G_L\|P_r = 0.01$ $kg\ m^{-2}s^{-1}$
Butane	3.74	560
Propane	4.2	650
R-11	4.4	1000
R-12	4.02	1060
H_2O	22.09	2000
Na	40.8	2890
UO_2	≃ 140	13500

The above should be sufficient to indicate that one only needs property data at one pressure to be able to evaluate G_L and, as will be shown, G at other conditions.

One next needs to evaluate η_o. In a somewhat related application, Epstein [1] has shown that for an isentropic expansion from saturated liquid conditions, pressure and the two-phase specific volume can be related by,

$$v = \frac{a_1}{P} - b_1 \tag{19}$$

where

$$a_1 = P_o \, v_{fo}/\beta$$

and

$$b_1 = v_{fo} \, (1 - \beta)/\beta$$

The parameter β is a constant for a given pressure and will be shown also to be easily correlated with reduced pressure. Equation (19) in the above form should only be applied to a process starting at $x = 0$. However, for this special case, (19) can be used directly in (10) leading to an analytic expression for $G|_{x=0}$, which can be maximized. The HE model critical pressure ratio for $x = 0$ stagnation conditions satisfies the relation

$$2 \ln \eta_o + (\eta_o b_3)^2 - 4b_3\eta_o + (1 + 2b_3) = 0 \tag{20}$$

where

$$b_3 = (1 - \beta)$$

and

$$\eta_o = \eta_s \text{ for } x = 0$$

The curve of Eq. (20) is shown in Fig. 6 as a function of β only.

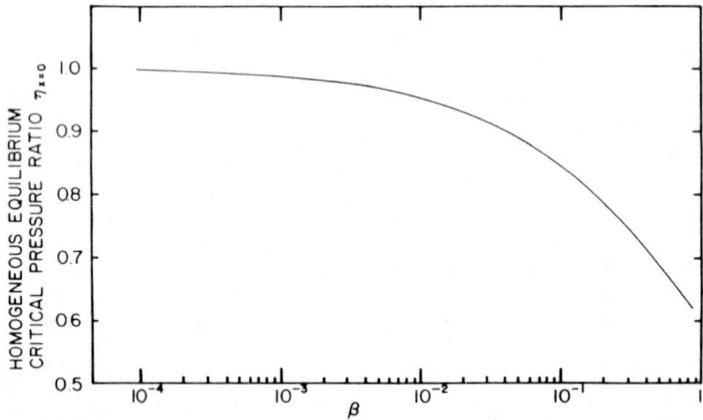

Fig. 6. A Generalized Relation for the Critical Pressure Ratio η_s for Saturated Liquid ($x = 0$) Stagnation Conditions as a Function of the Parameter β. [$\beta = v_{fo} \, G_L^2/P_o$].

Equation (19) can be manipulated with the help of (11) and (12) in such a manner that the constant β can be related to physical properties:

$$\beta \simeq h_{fgo}^2 \, v_{fo}/(P_o \, CT_o \, v_{fgo}^2) \tag{21}$$

or

$$\beta = v_{fo} \, G_L^2/P_o \tag{21a}$$

Thus for any fluid G_L, β, and η_o can be evaluated with only limited properties. A two step process is required to use Fig. 3b, for a fluid other than water, to evaluate a G in the liquid continuous regime ($\alpha_o < 0.6$). First evaluate G_L, β, and η_o for fluid "x". With either Fig. 5 or (18) find the reduced pressure P_r^o for water which yields the same value of G_L. This reduced pressure will then indicate the desired value of G* in Fig. 3b for fluid "x". For the vapor continuous region ($\alpha_o > 0.6$) simple reduced pressure equivalence is sufficient to relate Fig. 3a to fluids other than water. This methodology has been found to be surprisingly accurate for a rather large selection of fluids by comparison with direct calculation by use of (13) and (14). Often use of Figs. 3a and 3b in the generalized sense reproduces exact calculations within 5%. Differences up to 20% have been found in select cases. However, the development here points to the underlying thermodynamic and gas dynamic similarities in the HE model.

3. HOMOGENEOUS NONEQUILIBRIUM MODEL

The homogeneous nonequilibrium (HNE) critical flow model of Henry and Fauske [2] can be simply considered as follows. The HNE flow is the intersection of the homogeneous frozen Bernoulli equation and a nonequilibrium form of the local choking condition. For the former, (10) is evaluated with $v = x_o v_g + (1 - x_o) v_{fo}$ leading to,

$$G = \frac{\sqrt{\dfrac{P_o}{v_f}}\left\{2(1-\eta) + 2x_o\left[\dfrac{\gamma R_o(1-\eta^e)}{1-\gamma} - (1-\eta)\right]\right\}^{1/2}}{[1 + x_o\,(R_o\eta^{-1/\gamma} - 1)]} \tag{22}$$

where

$$e = (\gamma - 1)/\gamma$$

For the latter, (5) takes the form

$$G = \sqrt{P_o/v_f}\left\{\frac{x_o\ R}{n\ \eta} + N\,\psi\left(1 - \frac{xh_{fg}}{CT}\right)\right\}^{-1/2} \tag{23}$$

where

$$n = (x_o + \varepsilon)/(x_o/\gamma + \varepsilon)$$

$$\varepsilon = (1 - x_o)\ C/C_g$$

and

$$N = \left(\frac{dx}{dP}\right)_{nonequilibrium} \div \left(\frac{dx}{dP}\right)_{equilibrium}$$

In Ref. [2] it was proposed that

$$N = 7x \text{ for } x < 0.14$$

$$N = 1 \text{ for } x > 0.14$$

The solution of (22) and (23) for water at 1 MPa and x_o = .001 is illustrated in Fig. 7. The intersection does not occur at the maximum in (22). The philosophical argument proposed in [2] is that the flow accelerates according to (22) in a frozen manner (i.e., no phase change). The extent of the expansion, however, is limited by compressible effect embodied in the rate term (second term in RHS of (23)) of the local choking condition. N is thus a correlating parameter for the nonequilibrium effect observed in nozzle flow.

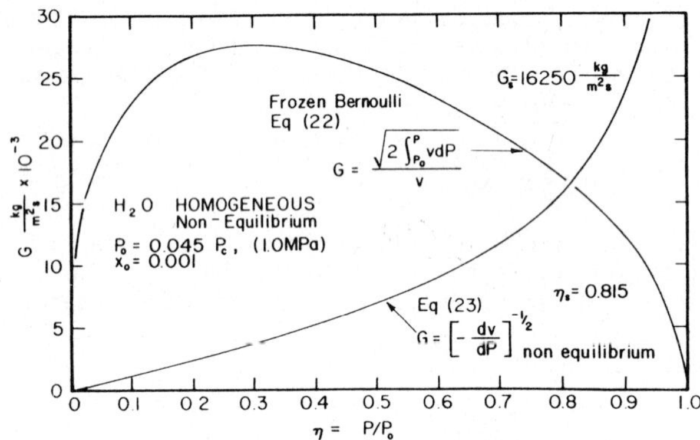

Fig. 7. Illustration of the Homogeneous Nonequilibrium Critical Flow Model as the Intersection of the Frozen Bernoulli Equation (22) and the Nonequilibrium Local Choking Condition (23) for Steam-Water at 1.0 MPa and x_o = 0.001.

The general behavior of the flow model can be compacted into very closely grouped set of curves that can be approximated as a single line in Fig. 8. Figure 8 is based on water property data. The right hand side of Fig. 8 represents the vapor continuous region (α > .5) and is made dimensionless according to

$$G^* = G/\sqrt{P_o/(x_o\, v_{go})} \tag{24}$$

The left hand side of Fig. 8 represents the liquid continuous flow regime and is made dimensionless according to

$$G^* = G/\sqrt{P_o/v_{fo}} \tag{25}$$

For this model, the use of stagnation void fraction α_o as the ordinate, seems to collapse the information into a single line where as this is not the case if x_o is used as the ordinate. Other fluids can be calculated and compacted in a similar manner. R-11 is shown in Fig. 9. The slight shift in

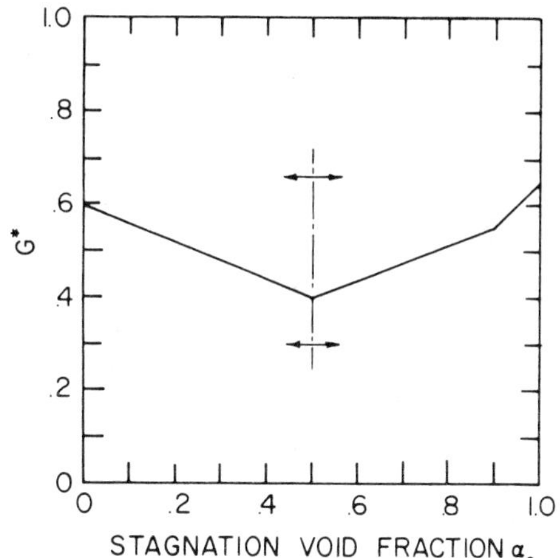

Fig. 8. Dimensionless Representation for HNE Model. (H_2O Properties Used. All Pressures Represented by 1 Line to within ± 10%).
For α_o < 0.5; $G^* = G/(P_o/v_{fo})^{1/2}$;
For α_o > 0.5; $G^* = G/(P_o/x_o v_{go})^{1/2}$.

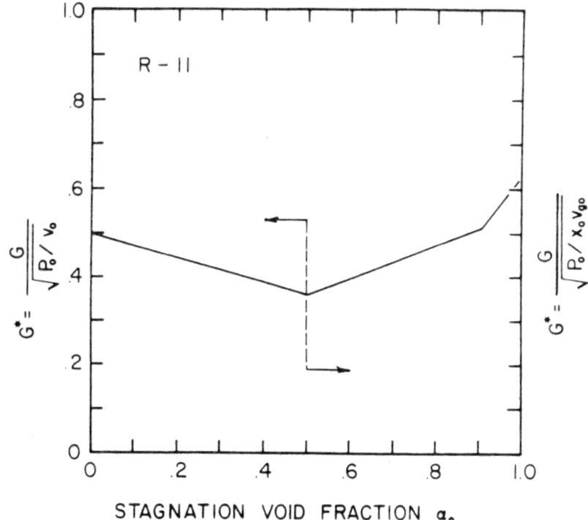

Fig. 9. Dimensionless Representation for HNE Model. (R-11 Properties Used. All Pressures Represented by 1 Line to within ± 10%).
For α_o < 0.5; $G^* = G/(P_o/v_{fo})^{1/2}$;
For α_o > 0.5; $G^* = G/(P_o/x_o v_{go})^{1/2}$.

the α_o = 0 intercept relative to water could be adjusted by employing a different N value for R-11.

The utility of the nonequilibrium parameter N in this model should not be minimized, see discussion in Ref. [3]. However, N merely plays the role of determining the intersection of (22) and (23). It should not be unexpected that N may take on different values for different conditions than the data base upon which the value was established in [2]. Data and interpretations are now appearing which shed additional information on this matter (4) and (5).

However, the purpose here is to show how the model can be compressed into a convenient form, Figs. 8 and 9. In this form pressure effects are scaled in a unifying manner. Data comparisons can be more easily made in this format. Note that in the vapor continuous regime flow scales with pressure to the first power. (This assumes the ideal gas law to eliminate v_g). This holds as well for the HE model. In the liquid continuous region, mass flux scales as pressure to the 1/2 power for the HNE model and with pressure to the 0.82 power for the HE model.

4. SLIP EQUILIBRIUM MODEL

The slip equilibrium (SE) model after Moody [6] can be treated in a similar manner. The SE model can be viewed as the mass flux satisfied by the energy equation (1) and the local choking condition (5) with special care used in the definition of a specific volume consistent with an energy equation formulation. The actual formulation used here proceeds from (2) where v under the integral is the homogeneous specific volume $v = xv_g + (1 - x)\,v_f$ as required by the equivalence between dh and vdP under isentropic conditions. However, in evaluating (2), v_1 outside the integral becomes v_E where,

$$v_E = [x\,v_g + (1 - x)\,v_f S]\,[x + (1 - x)/S^2]^{1/2} \tag{26}$$

where the slip or vapor to liquid velocity ratio is

$$S = R^{1/3}$$

The evaluation again proceeds using (11) and (12) to yield,

$$G = \frac{\sqrt{\dfrac{P_o}{v_f}\left\{2(1 - \eta) + 2B\left[\left(1 + \dfrac{x_o h_{fgo}}{C\,T_o}\right)(1 - \zeta) + \zeta \ln \zeta\right]\right\}^{1/2}}}{(xR + (1 - x)S)\left(x + (1 - x)/S^2\right)^{1/2}} \tag{27}$$

The local choking condition consistent with the energy equation for the SE model is

$$G = \left\{-\frac{v_E}{v}\frac{dv_E}{dP}\right\}^{-1/2} \tag{28}$$

where v_E is the energy specific volume having the form of (26). Equation (28) evaluated becomes:

$$G = \frac{S}{I}\sqrt{\frac{P_o}{v_f}}\left\{\frac{\eta\,\gamma\,A}{J\left[KL\left(1 - \dfrac{xh_{fg}}{CT}\right) + M + N - 1/3\right]}\right\}^{1/2} \tag{29}$$

where

$$I = XR + (1 - x)S$$

$$J = 1 + x(S^2 - 1)$$

$$K = \gamma P v_{fg} \, CT/h_{fg}^2$$

$$L = \frac{R - S}{I} + \frac{S^2 - 1}{J}$$

$$M = \frac{xR + (1 - x)S/3}{I}$$

$$N = xS^2/(3J)$$

Figure 10 illustrates the intersection of (27) and (29) leading to the SE mass flux and critical pressure ratio for water at 1 MPa and x_o = .001. The model fully evaluated for water over all pressures and stagnation qualities is shown in Fig. 11. The best compaction of the information is obtained by making the flow dimensionless according to:

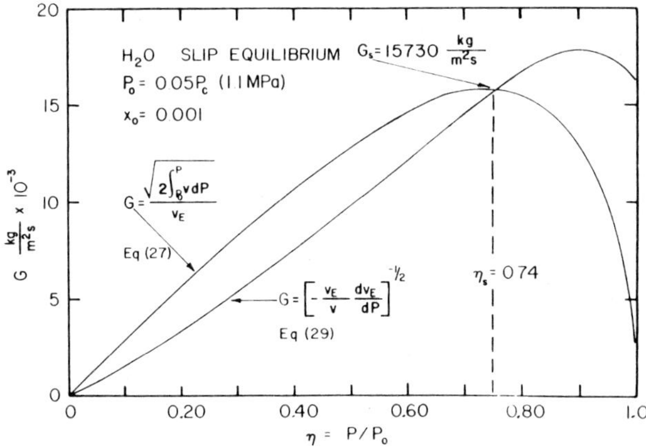

Fig. 10. Slip Equilibrium Two-Phase Critical Flow Model for Steam-Water at 1.1 MPa Satisfies (27) and (29).

$$G^* = G/\sqrt{P_o/v_{Eo}} \tag{30}$$

where v_{Eo} is the two-phase slip specific volume defined by (26) evaluated at P_o and x_o. For this model use of Fig. 11 as a generalized plot for other fluids can be done on the basis of equivalent reduced pressure with reasonable accuracy.

5. SUMMARY

The development presented here illustrates a unifying similarity of two-phase choked flow calculations with gas dynamic methods. Two-phase flow behavior however is influenced by two major considerations, gas phase compressibility and phase change or flashing. This gives rise to different pressure scaling, depending upon whether the liquid continuous or vapor continuous flow regime is important. In general, flow scales with pressure to the power m as indicated in Table 2.

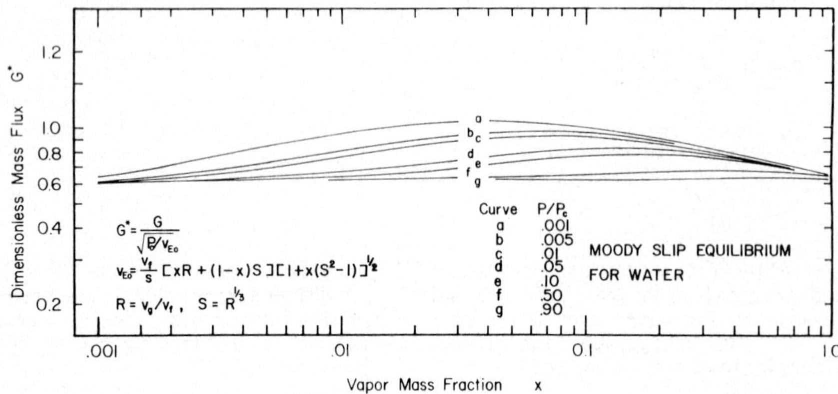

Fig. 11. Slip Equilibrium Critical Flow Model Evaluated for Steam-Water Over Indicated Reduced Pressure Range. V_{Eo} is Evaluated at Stagnation Property Conditions. Other Fluid Scale to this Figure on a Reduced Pressure Basis.

Table 2. Pressure Scaling of Two-Phase Critical Flow $G \propto P^m$

Model	Liquid Continuous	Vapor Continuous
HE	m = 0.82	m = 1
HNE	m = 1/2	m = 1
SE	m = 1/2	m = 1

Of incidental concern is the model prediction of the critical pressure ratio as shown in Fig. 12 for water as typical of all fluids.

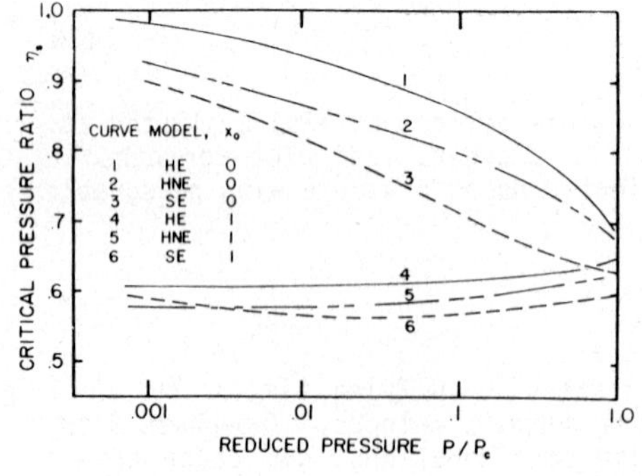

Fig. 12. Illustration of the Relationships of the Calculated Critical Pressure Ratio from the Various Models Using Steam-Water Properties.

Of significant concern is the choice of model for a calculational basis in a given application. It will be found that the HNE and SE models will give approximately the same values for mass flux, but higher than the HE model. Data can be found to agree with both. Some current thoughts on model selection are found in [5]. This development can however, be used effectively to sort out and compare data particularly when it is desired to eliminate or suppress pressure and fluid property variations.

NOMENCLATURE

a Defined under (18)

a_1 Parameter group defined under (19)

A Dimensionless homogeneous specific volume defined under (13)

B Dimensionless parameter group defined under (13)

b Defined under (18)

b_1 Parameter group defined under (19)

b_3 Defined under (20)

C Specific heat capacity of liquid phase

C_g Specific heat capacity of gas phase at constant pressure

e $(\gamma - 1)/\gamma$ used in (22)

$\hat{F}$ Function of γ defined by (8)

G Mass flux as defined by equation in context

G^* Dimensionless mass flux as defined by equation in context

h Enthalpy

I Dimensionless term defined under (29)

J Dimensionless term defined under (29)

K Dimensionless term defined under (29)

L Dimensionless term defined under (29)

M Dimensionless term defined under (29)

N Dimensionless term defined under (29)

n Defined under (23)

N Defined under (23)

P Pressure

P_r Reduced pressure relative to thermodynamic critical value

R_g Gas constant

R Specific volume ratio v_g/v_f

S Slip ratio defined under (26)

T Temperature

v Specific volume

x Vapor mass fraction

α Vapor volume fraction

β Parameter used in defining two-phase specific volume pressure relation, see (19) and (21)

γ Exponent in Pv^{γ} = *constant*

ε Defined under (23)

η Pressure ratio, P/P_o

ζ Temperature ratio, T/T_o

ψ Dimensionless property group defined under (14)

SUBSCRIPTS

c Thermodynamic critical condition

E Refers to energy specific volume of form (26)

f,g, fg Refers to fluid, vapor, and difference between vapor and fluid property value

L Limit value

r Reduced relative to thermodynamic critical value

s Refers to flow dynamic choking condition

0 Refers to properties evaluated at stagnation conditions - or refers to critical pressure ratio evaluated at x = 0

1 Refers to lower pressure state relative to stagnation condition

REFERENCES

1. Epstein, M., Henry, R. E., Midvidy. W., and Pauls, R., "One-Dimensional Modeling of Two-Phase Jet Expansion and Impingement," Proceedings of the 2nd Int. Topical Meeting on Nuclear Reactor Thermodynamics, Vol. II, pp. 769-777, Santa Barbara, January 1983.

2. Henry, R. E., Fauske, H. K., "The Two-Phase Critical Flow of One Component Mixtures in Nozzles, Orifices and Short Tubes," ASME Journal of Heat Transfer, pp. 179-186, May 1971.

3. Wallis, G. B., "Critical Two-Phase Flow," Int. Journal of Multiphase Flow, Vol. 6, pp. 97-112, February/April, 1980.

4. Abdollahian, D., et al., "Critical Flow Data Review and Analysis," EPRI Final Report NP 2192, January 1982.

5. Fauske, H. K., "Multiphase Flow Considerations in Sizing Emergency Relief Systems for Runaway Chemical Reactions," Paper in proceedings of this conference.

6. Moody, F. J., "Maximum Flow Rate of a Single Component Two-Phase Mixture," ASME Journal of Heat Transfer, pp. 134-142, February 1965.

Multi-Phase Flow and Heat Transfer III. Part A: Fundamentals
edited by T.N. Veziroğlu and A.E. Bergles
Elsevier Science Publishers B.V., Amsterdam, 1984 — Printed in The Netherlands

TRANSIENT TWO-PHASE FLOW DISCHARGE OF FLASHING LIQUIDS FOLLOWING A BREAK IN A LONG TRANSMISSION PIPE LINE

Michael A. Grolmes, Joseph C. Leung, and Hans K. Fauske
Fauske & Associates, Inc.
16W070 West 83rd Street
Burr Ridge, Illinois 60521, U.S.A.

ABSTRACT

This paper considers the transient discharge of a flashing liquid, following a postulated break in a long pipe line. A key feature of the resulting discharge is the important role of frictional effects in limiting the rate of discharge. The problem has been evaluated under several different model assumptions which lead to a best assessment utilizing a multi-node slip flow approach.

1. INTRODUCTION

Volatile and flammable liquids are transmitted in long pipelines with the distance between isolation valves often being several tens of kilometers with length to diameter ratios of the order of 10^5. In the event of an accidental line break the discharge from the pipe can lead to the formation of a hazardous vapor cloud. Its formation and abatement are dependent upon the discharge flow rate as well as the atmospheric dispersion processes. This paper focuses on a method of assessment of the discharge flow rate (source term) that is not overly simplistic - such as treating the long line as a one node volume with or without internal resistance - nor overly tedious such as would be encountered in a method of characteristics solution approach. References [1 and 2] encounter one or more of these types of difficulties.

The fluid carried in the pipe line might be a single component fluid such as propane or a multicomponent mixture of the hydrocarbon constituents of liquid natural gas or other similar products. A characteristic of the problem is that the material is assumed to be transmitted in the liquid state under sufficient pressure to prevent vapor formation. However, flashing to the vapor state will occur under a depressurization following a postulated line break.

2. MODEL

The modeling approach can have a significant effect on the resulting vapor source term. Because of the extreme length involved, a method of characteristics approach is likely to be inefficient, and treating the pipe as a single volume is likely to ignore the significance of frictional pressure loss.

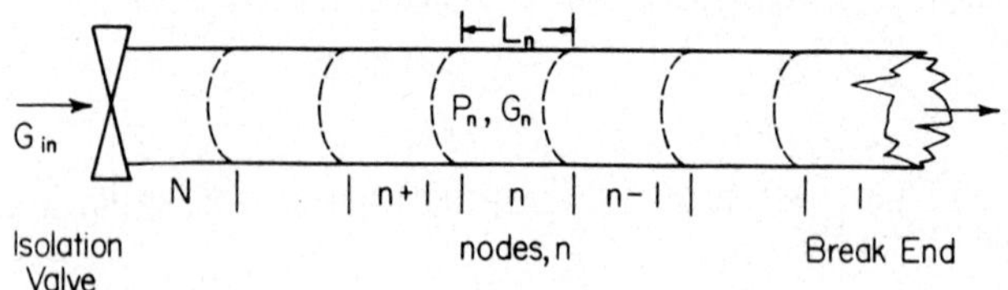

Fig. 1. Multi-Node Representation of a Pipe Line

A transient one dimensional pipe model can be formulated with the aid of Fig. 1 as follows. An unsteady momentum equation for element n is

$$\dot{G}_n = L_n^{-1}\left\{\Delta P_n - \frac{1}{2} G_n^2 v_f \left(F_n^* + \frac{2\Delta v_{2,n}}{v_f}\right)\right\} \tag{1}$$

where L_n is the length of node n, ΔP_n is the pressure drop across node n, G_n is the mass flux in node n and $\dot{G}_n$ is the rate of change of G, F_n^* is a dimensionless friction term for node n, and $\Delta v_{2,n}/v_f$ is a dimensionless two-phase specific volume change over node n (i.e., the momentum pressure loss term).

When the transient $\dot{G}$ goes to zero the terms in brackets on the RHS of (1) represent the steady-state two-phase momentum balance. The pressure change ΔP_n is defined

$$\Delta P_n = \frac{P_{n+1} + P_n}{2} - \frac{P_n + P_{n-1}}{2} \tag{2}$$

The term F_n^* is a friction loss defined by

$$F_n^* = 4\, f \frac{L_n}{D} \phi_n^2 \tag{3}$$

where ϕ_n^2 is the two-phase friction multiplier for node n. Explicitly,

$$\phi_n^2 = 1 + \bar{x}_n (\bar{R}_n - 1) \tag{4}$$

for homogeneous flow where $R = v_g/v_f$, and

$$\phi_n^2 = (1 - \bar{x}_n)^2/(1 - \bar{\alpha}_n)^2 \tag{5}$$

for slip flow.

The two-phase specific volume for the nth node is

$$v_{2,n} = \frac{x^2}{\alpha} v_g + \frac{(1 - x)^2}{1 - \alpha} v_f \tag{6}$$

where the subscript n is omitted on the RHS of (6) and all quantities are assumed to be average values. The necessary relation between quality x and void fraction α is given by,

$$\frac{x}{1 - x} = \frac{\alpha}{1 - \alpha} \frac{v_f}{v_g} \frac{u_g}{u_f} \tag{7}$$

For homogeneous flow $u_g/u_f = 1$, while for slip flow we use the Thom [3] correlation

$$\frac{u_g}{u_f} = R^m \tag{8}$$

where m is an exponent which varies between 1/3 and 1/5.

At the break, the exit flow is prescribed by,

$$G_1 = G(P_1, \bar{x}_1, P_a) \tag{9}$$

Equation (9) represents a general relation for two-phase compressible flow in the absence of friction effects. Initially the break end pressure P_1 exceeds the ambient pressure P_a by a sufficient margin such that choked flow conditions prevail. Various critical flow models such as discussed in [4] can be incorporated. In this model we are consistent such that as P_1 is reduced, subsonic flow is automatically calculated.

At the closed end either pressure or flow may be specified as a boundary condition. The remaining required equations are:

$$\dot{M}_n = \Delta G\ A \tag{10}$$

and

$$\dot{P}_n = \frac{\frac{h_{fg}}{Tv_{fg}} \left[\Delta(Gh) - i\Delta G + \frac{i_{fg}v}{v_{fg}} \Delta G \right] \frac{A}{MC}}{\left[1 - \frac{i_{fg}}{v_{fg}} \frac{B}{C} \right]} \tag{11}$$

where

$$B = \frac{dv_f}{dT} + x \frac{dv_{fg}}{dT} \tag{12a}$$

and

$$C = \frac{di_f}{dT} + x \frac{di_{fg}}{dT} \tag{12b}$$

Again subscripts are omitted on the RHS of (10) and (11) and Δ quantities are defined in the same manner as (2). Equation (10) is a continuity equation and Eq. (11) is an energy equation for node n similar to that developed in [5]. The derivatives for (12a) and (12b) are evaluated along a saturation line for the flash process.

The formulation can be expanded to any number of nodes. For a one node calculation all momentum equations containing friction terms drop out leaving only a discharge flow condition. For the purpose of calculations illustrated in the next section, a logarithmic nodal spacing was assumed with the close spacing biased toward the break end. The above set of ordinary differential equation was solved using the method of Gear [6].

3. MODEL CALCULATIONS

Two pipe line examples are illustrated with three fluids to represent various aspects of the model results. The conditions for the problem are indicated in Table 1.

Table 1. Model Examples and Calculations

PROBLEM NO.	PIPE LENGTH	DIAMETER	VOLUME
1	8 km	0.305 m	579 m^3
2	36 km	0.287 m	2342 m^3

FLUIDS AND INITIAL CONDITIONS

Fluid	Pressure	Temperature
H_2O	4 MPa	523.5 K
R-11	4 MPa	464.4 K
Propane	0.85 MPa	294.4 K

MODELS (General Designation)

xn, YP, ( )

where,

xn = number of nodes

yP = number of phases, i.e., 1 single-phase
2 two-phase

( ) = designates model

V - all vapor

H - two-phase homogeneous equilibrium

S - two-phase slip equilibrium

First, results are considered for an 8 km pipe line with water properties used to illustrate effects with well known fluid properties. Figure 2 compares the discharge mass flux for a 1 node and a 10 node two-phase homogeneous model. Of greater interest however, is the vapor source rate, which accounts for the full flashing to 1 atm pressure even though the pressure at the break may be greater than ambient. Figure 3 compares the same model as Fig. 1 in terms of vapor discharge. The differences between the 1 node and 10 node calculation are several. The 1 node calculation sustains a high vapor discharge rate over most of the pipe line depressurization. The break flow is sonic over most of the time. The break flow for the 10 node calculation is subsonic after about 50 sec and remains so for the duration of the discharge which lasts about 100 minutes. The vapor discharge rate after 50 sec is reduced by a factor of 3 or more and continues for about a factor of 3 longer in time than the 1 node calculation. Of course, the integral over time of the vapor release is the same in both calculations.

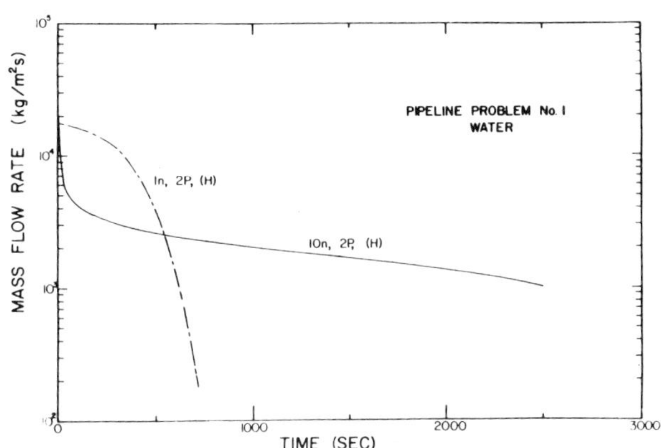

Fig. 2. Mass Flux from the Break End of an 8 km Pipe Comparing a One and Ten Node Homogeneous Equilibrium Flow Model using Water Properties

The homogeneous and slip flow models are compared in Fig. 4. The example is again an 8 km pipe line but with R-11 properties which are similar to many organic fluids in terms of critical pressure, latent heat, and specific heat. The differences between the 1 node and 10 node calculation are the same as previously discussed. The differences between the homogeneous and slip flow models are more noticeable in the 1 node calculation than the 10 node calculation. This is because in the 1 node calculation the break flow is choked and in the 10 node calculation the break flow is subsonic after ≃ 40 sec. Fc subsonic flow the differences between the two models becomes small.

The calculated pressure distribution in the pipeline for the R-11 example is shown in Fig. 5. The sensitivity of the mass loss as a function of time, to the number of nodes is indicated in Fig. 6 for the same example.

The differences between a 1 node and a multi-node calculation are more dramatic for the 36 km pipe line example with propane saturated at .85 MPa. Figure 7 compares number of nodes and slip vs homogeneous models. The difference between the slip flow model and the homogeneous model at early time is greater than in the previous illustration because of the lower reduced pressure. However, the significant aspect of comparison is the dramatic reduction in vapor release after about 100 seconds for the multi-node model. The vapor flow rate is a factor of 10 to a factor of 20 lower than the 1 node calculation, but continues for a similar increase in time. After the discharge rate is reduced by friction effects the sensitivity to models (slip vs homogeneous) is relatively minor.

5. SUMMARY

This analysis indicates that for a long transmission pipe line break the discharge flow is quickly dominated by friction effects. A one node model approximation would appear to underestimate the time interval over which the volatile material is released. While the differences between a slip flow model and a homogeneous flow model are small later in time, the slip flow model yields a higher vapor release at time zero than the homogeneous model. This would be expected. The results indicate that a multi-node flow model can be formulated to provide a reasonable assessment of the pipe break problem. The sensitivity to number of nodes becomes unimportant between 10 and 20 nodes. The model can be readily extended to multicomponent mixtures of real fluids, or to handle most upstream boundary and initial conditions.

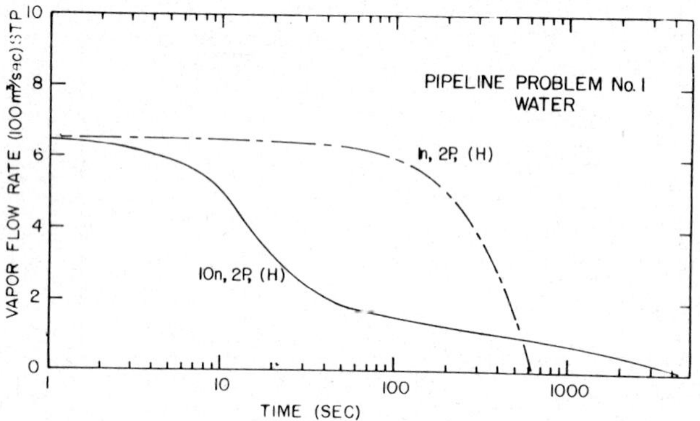

Fig. 3. Vapor Volume Release Rate for an 8 km Pipe with Water Properties Comparing a One and Ten Node Homogeneous Equilibrium Flow Model

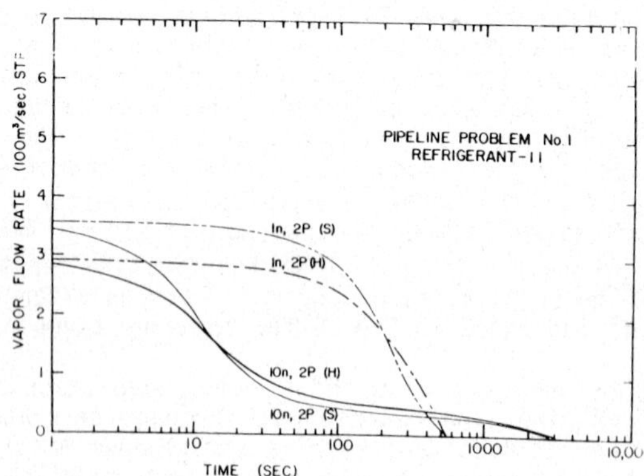

Fig. 4. Comparison of Homogeneous and Slip Flow Models for an 8 km Pipe using R-11 Fluid Properties.

NOMENCLATURE

A Area

B Specific volume derivative defined in (12a)

C Specific heat defined in (12b)

α Vapor volume fraction

ϕ^2 Two-phase friction multiplier

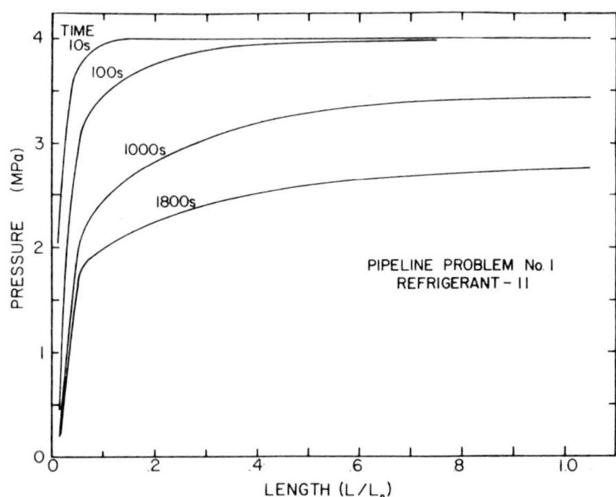

Fig. 5. Pressure Distribution in Pipe Line as Calculated for the 10 n 2P Slip Model using R-11 Properties, L_0 = 8 km

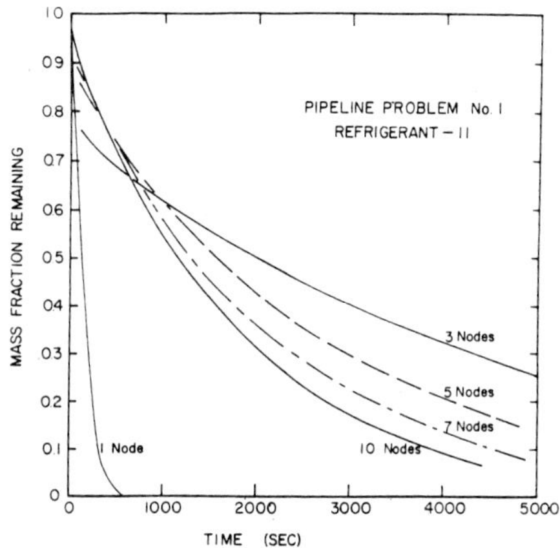

Fig. 6. Sensitivity of Mass Depletion Histories to Number of Nodes for a Two-Phase Slip Model using R-11 Properties in an 8 km Pipe Line Problem

D Diameter

f Friction factor, 0.005 used in this study

F* Dimensionless friction term

G Mass flux

h Enthalpy

i Internal energy

L Length

M Mass

m Exponent in (8)

P Pressure

R Specific volume ratio, v_g/v_f

T Temperature

u Velocity

v Specific volume

x Vapor mass fraction

SUBSCRIPTS

g,f,fg Vapor, liquid, and difference between vapor and liquid node n

n node n

2 Two-phase

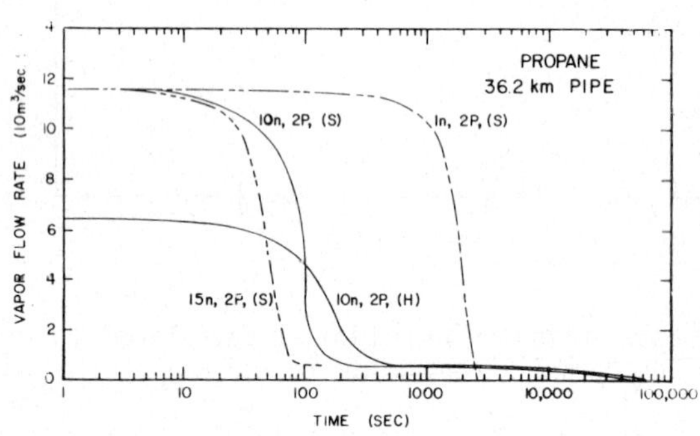

Fig. 7. Comparison of One Node vs Multi-Node and Homogeneous vs Slip Flow Models using Propane Properties in a 36 km Pipe Break Example.

REFERENCES

1. Morrow, T. B., Bass, R. L., and Lock, J. A., "An LNG Pipe Line Break Flow Model," ASME J. Energy Resources Technology, 1982.

2. Sabnis, J. Simmons, W., and Teuscher, L., "Two-Phase Blowdown from High Pressure Liquid Pipe Lines," I. Chem. E. Symposium Series, No. 71.

3. Thom, J. R. S., "Prediction of Pressure Drop During Forced Circulation Boiling of Water," Int. J. Heat and Mass Transfer, Vol. 7, pg. 709, 1964, also see Handbook of Heat Transfer, Ed. by W. M. Rohsenow and J. P. Hartnett, Section 14, p. 5, Fig. 2, McGraw-Hill, 1973.

4. Grolmes, M. A. and Leung, J. C., "Scaling Considerations for Two-Phase Critical Flow," Paper in Proceedings of this Conference.

5. Grolmes, M. A. and Fauske, H. K., "An Evaluation of Incomplete Vapor Phase Separation in Freon-12 Top Vented Depressurization Experiments," Paper in Proceedings of this Conference.

6. Gear, C. H., "Numerical Initial Value Problems in Ordinary Differential Equations," Prentice Hall, Englewood Cliffs, New Jersey, 1971.

Multi-Phase Flow and Heat Transfer III. Part A: Fundamentals
edited by T.N. Veziroğlu and A.E. Bergles
Elsevier Science Publishers B.V., Amsterdam, 1984 — Printed in The Netherlands

DISCHARGE OF A SATURATED LIQUID THROUGH A RELIEF VALVE

Joel Weisman, Zuyao Chen* and Rakesh Govind
Department of Chemical & Nuclear Engineering
University of Cincinnati
Cincinnati, Ohio 45221, U.S.A.

ABSTRACT

A series of experiments, in which heated liquid refrigerant 113 was discharged from a relief valve, have been conducted. The heated fluid was contained in a piston type accumulator and then rapidly discharged across a relief valve into a large blowdown tank. A glass viewing section assured that the fluid was single phase upstream of the valve.

The tests showed the relief valve to act as an orifice with a variable area. Vaporization apparently does not occur until the fluid leaves the valve. The fluid flow rate is proportional to the square root of the pressure difference across the valve and data for both heated and cold refrigerant lie on the same curve.

1. INTRODUCTION

Most systems which contain a fluid under pressure are protected by one or more safety relief valves. While such valves usually discharge vapor when they are called upon to operate, there are a number of circumstances when they discharge a two-phase mixture or saturated liquid. Despite the relative frequency in which two-phase or saturated liquid service is required, essentially no information could be found in the literature on relief valve behavior under these conditions. The present study represents the first phase of an experimental investigation undertaken to provide some of the needed data. A second phase, using a two-phase mixture upstream of the relief valve, is planned.

A simple schematic of the key elements of a typical relief valve is shown in Fig. 1. It may be seen that the spring holds the disc down on the seat. When the pressure below the disc exceeds the set or lift pressure, the disc rises slightly. The fluid then flows outward through the narrow passageway between the disc and the seat. The narrow passageway between the disc and seat is shown in greater detail in the insert of Fig. 1. It may be seen that there is a narrow gap in the seat area followed by a longer and slightly wider gap in the tapered region which follows. The gap would appear to be physcially somewhere between an orifice and a nozzle. Thus, at the outset of the investigation, it was not clear whether the valve would behave as an orifice or nozzle.

Henry [1] reports that most investigators have found that the discharge of a saturated liquid across an orifice may be predicted by

---

*Visiting Scholar from Shanghai Institute of Mechanical Engineering, Peoples Republic of China

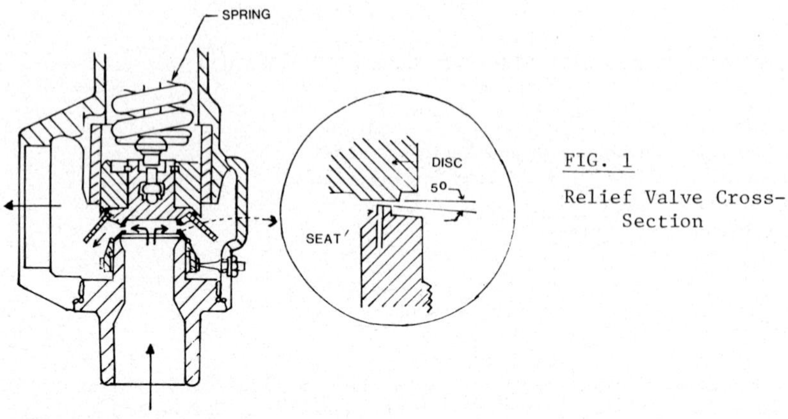

FIG. 1

Relief Valve Cross-Section

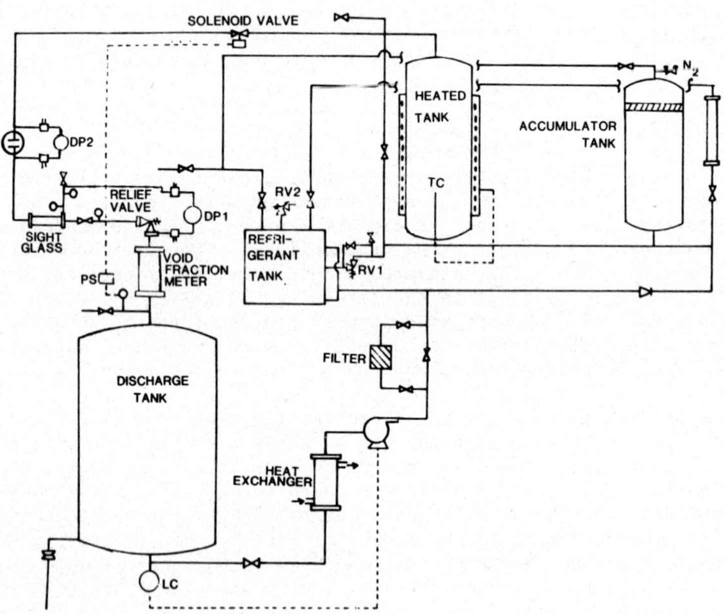

FIG. 2 - Flow Diagram of Blowdown Apparatus

$$G = 0.61 \sqrt{2g_c \rho_1 \Delta P} \tag{1}$$

Henry points out that the foregoing behavior indicates that most fluids are metastable and do not begin to flash until they have left the orifice region. However, some investigators (e.g. Bottomley [2]) have indicated that under certain conditions a discharge coefficient in excess of unity is obtained. This would indicate some flashing in the orifice region.

Flashing in the discharge region is more typical of nozzle behavior. However, here too metastable behavior is observed. Burnell [2] reports that, in a nozzle throat, flashing does not begin until the pressure at the nozzle throat is well below the saturation pressure of the liquid. Indeed, the pressure reached in the nozzle throat may be below the back pressure downstream of the nozzle. The discharge rate is thus independent of back pressure. Burnell [4] correlated this behavior by

$$G_c = \{2g_c \rho_1 [P_{in} - (1-C)P_{sat}]\}^{1/2} \tag{2}$$

The coefficient "C" was related to the surface tension (in dynes/cm) by

$$C = (0.764\ \sigma)/\sigma_s \tag{3}$$

where σ_s = 49.2 dynes/cm

For small values of σ (as in the present case), Equation (2) approaches

$$G_c = \{2g_c \rho_1 [P_{in} - P_{sat}]\}^{1/2} \tag{4}$$

Henry et al [4] present a slightly different approach to the calculation of critical flow through nozzles. For the case of zero quality at the inlet, they relate the critical flow rate, G_c, to the critical pressure ratio, η, by

$$\eta = 1 - \frac{G_c^2}{2\rho_1 g_c P_{in}} \tag{5}$$

with η being the minimum of 0.85 or

$$\eta = P_{sat}/P_{in} \tag{6}$$

When η is given by Equation (6), Equation (5) reduces to Equation (4).

Zaloudek [5] observed that the critical flow of a flashing, saturated liquid in a short pipe was at some times governed by Equation (1) and at other conditions governed by a modified version of Equation (3). However, the configuration of a standard relief valve is not very close to that of a short pipe and this composite behavior was not anticipated.

Recently Celata et al [6] made photographic observations of the discharge of a hot liquid through an orifice. Their observations confirmed the presence of a metastable liquid jet downstream of the orifice. However, they found that a fluid packet remained in this metastable state only for the order of 10^{-4} seconds. When the present relief valve tests were undertaken, it was not clear whether a metastable jet could persist long enough to clear the restricted region at the relief valve discharge location.

2. EXPERIMENTAL PROGRAM

In order to determine relief valve behavior during discharge of a saturated fluid, a typical small relief valve (1 in nominal inlet pipe connection, 1 1/2 in nominal outlet pipe connection) was obtained from the Dresser Industrial Valve & Instr. Div. The general configuration of the valve has been shown in Fig. 1. The valve was then incorporated into a blowdown apparatus.

The experimental approach used was to contain refrigerant 113 under pressure in two high pressure vessels. After heating the refrigerant, the fluid was allowed to discharge through the relief valve into a large blowdown tank. After cooling, the liquid in the blowdown tank was pumped back to the high pressure storage tanks.

Refrigerant 113 (molecular weight 187.4) was chosen for use in these tests because of its low heat of vaporization (63.5 Btu/lb at atm. pressure), low boiling point (117.6°F) and high vapor density. One can obtain significant downstream qualities with only moderate upstream temperatures. In addition, high vapor densities can be obtained at moderate pressures. Since refrigerant 113 is a liquid at room temperature and pressure, fluid handling is uncomplicated.

A schematic of the experimental system is shown in Fig. 2. The apparatus consists of an accumulator tank with a piston to pressurize the system. Pressurized nitrogen, monitored by a pressure gauge and controlled by a standard regulator and a backpressure regulator, is used to move the piston in the accumulator.

The accumulator is connected to a heating vessel equipped with electric band heaters. The top and bottom of the heating vessel are connected by a 1 1/2" pipe to facilitate the heating process by natural convection. The heat input is controlled by the temperature of freon in the vessel which is measured by a long stem thermocouple installed through the bottom of the tank. The solenoid valve, mounted on the exit pipe of the heating vessel, is kept closed during the heating process.

Note that separate vessels were used for heating and pressurization. Previous experience had shown that direct heating of the piston type accumulator vessel damanged the piston seal rings. By separating the two functions, this problem could be avoided.

A test run is begun by opening the solenoid valve on the exit pipe. The liquid then flows out of the heating vessel and across the relief valve. The outlet flow rate from the heating vessel is measured by an orifice plate connected to a differential pressure (DP) cell. A sightglass tube after the orifice plate allows the flow entering the relief valve to be observed.

Only runs in which no vapor bubbles were observed were considered. The nitrogen pressure in the accumulator maintained a nearly constant pressure, which was above saturation, on the fluid upstream of the relief valve.

Refrigerant 113 was discharged through the relief valve into a discharge tank with a volume of about 50 cubic feet. A high pressure switch on the discharge tank was used to close the solenoid valve automatically in the event of a pressure excursion. After a cooling period at the end of a run, the liquid from the discharge tank is pumped back into the accumulator through a water-cooled heat exchanger. A drier-filter, which removes water and contaminants, was present in the return line. A low liquid-level sensor in the discharge tank is used to trip the pump automatically and prevent cavitation. Possible backflow from the accumulator into the discharge tank is prevented by a check valve in the discharge line.

In the event of an abnormally high pressure in the accumulator or the heating tank, the safety relief valve, RV1, is used to discharge the freon into the Freon storage tank. Another relief valve, RV2, set at 160 psi,(the maximum allowable pressure in the system) is used to protect the freon storage tank from overpressurization.

The following orifice measurements were made during each experimental run:

(a) Exit flowrate from the heating vessel was obtained by measuring the orifice ΔP by a 0-500 in of water, strain gauge, differential pressure (DP) cell from Honeywell. The output signal, in the 0-10MA range, was amplified and recorded.

(b) Absolute pressure upstream of the relief valve is measured by a strain gauge with a range of 0-2000 in. of water.

(c) Pressure drop across the relief valve was measured by a 0-500 in. DP cell from Honeywell and its signal is amplified and recorded.

(d) Void fraction on the exit side of the relief valve was measured by a capacitance type void sensor.

(e) Fluid temperatures at the inlet and exit of the relief valve are measured by iron-constantin thermocouples.

The thermocouple outputs were measured on a standard recorder. All of the other signals were transmitted to a multi-channel light-beam oscillograph.

Prior to conducting the relief valve tests, a series of blowdown runs were made using a fixed orifice in place of the relief valve. The runs clearly showed the metastable behavior of the test fluid. The flow rates were proportional to $\sqrt{\Delta P}$ and data for hot and cold fluid fell on the same curve.

Relief valve tests followed the orifice study. Before a run began the piping upstream and downstream of the relief valve was at a low pressure and there was a low pressure difference across the valve. When the run was begun by opening the upstream solenoid, fluid flowed into the upstream pipe causing a fairly rapid rise in pressure upstream of the valve. When the upstream pressure reached the valve set pressure, the valve opened and the pressure differential across the orifice rose rapidly. Although the valve was discharging, the pressure drop across the valve continued to rise until the

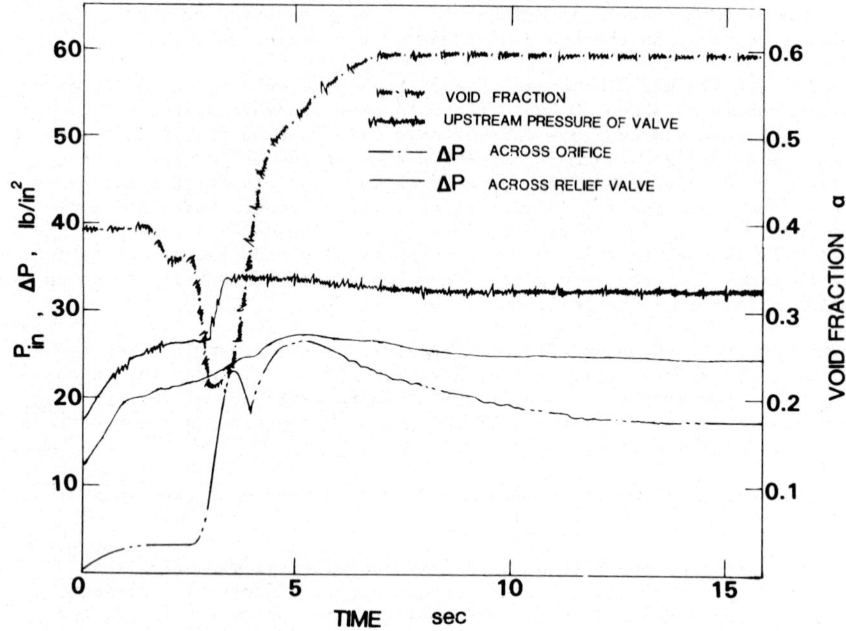

FIG. 3 - Data From Typical Oscillograph Trace

discharge rate exceeded the inflow. There is thus a period when outflow exceeds inflow and the pressure drop across the valve falls until an equilibrium is reached. For the remainder of the run, the upstream pressure and pressure drop across the orifice and relief valve remain essentially constant. The pressure data obtained from the oscillograph traces of a typical run are shown in Fig. 3.

The void fraction downstream of the relief valve is also shown in Fig. 3. At the beginning of the run, there is condensed liquid from the previous run in the void sensor so that a relatively low void fraction is observed. When flow begins, the refrigerant first reaching the void sensor was in the cold piping and the void fraction drops to a low value. As the hot refrigerant enters the void sensor, the void fraction rises and then remains at a nearly constant value.

Tests were carried out with 4 relief valve springs designed for nominal set pressures of 20, 30, 40 and 50 psig. With each spring, the valve was adjusted so that it lifted at close to the design pressure. An additional set of runs was made with the 30 psig design spring set to lift at 40 psig. For each relief setting, the upstream liquid temperature was varied over the range of about 20 to 95°C with upstream pressure held constant. The procedure was then repeated at least one other upstream pressure. In all runs, the sightglass was observed to see that only liquid was present upstream.

3. ANALYSIS OF STEADY-STATE DISCHARGE RATES

The discharge rates during the relatively long period where flows, pressure drops and pressures remained essentially unchanged were considered first. It is believed that the behavior during this steady-state period was most typical of what would be seen in an actual blowdown.

Initial examination indicated that the discharge flow rates for a given set pressure depended on the pressure differential across the relief valve. However, a different relationship was noted for each upstream pressure used. This implied orifice-like behavior but that the orifice discharge area was dependent on the upstream pressure.

The variation in orifice area with upstream pressure is not unreasonable since an increased force on a relief valve spring will cause it to compress and increase the valve opening. When the valve just lifts, the force generated by the set pressure, P_s, acting on the valve disc overcomes the spring force and starting friction. The valve then opens to some area, a_o. Any increase in area beyond a_o is due to forces generated by upstream pressures above P_s and the small force due to the right angle change in direction of the flowing stream beneath the disc. This additional force, F_a, is then given by

$$F_a = A\left[(P_{in} - P_s) + \frac{\rho u^2}{g_c}\right] \qquad (7)$$

Note, that in the foregoing both P_{in} and P_s are gauge pressures as atmospheric pressure is acting on the top of the disc.

Since the valve spring may be expected to be linear, the additional discharge area produced may be expected to be proportional to F_a. Hence the total discharge area, a, would then be given by

$$a = a_o + K_1(P_{in} - P_s + \frac{\rho u^2}{g_c}) \qquad (8)$$

However, if we consider a valve which is to operate with several different springs (several set pressures), we may expect that K_1 will vary inversely with the set pressure. We then have

$$a = a_o + \frac{K_1'}{P_s}(P_{in} - P_s + \frac{\rho u^2}{g_c}) \qquad (9)$$

As both starting friction and spring stiffness will vary with the set pressure, it seems reasonable to assume that a_o is essentially independent of the set pressure.

If the relief valve behaves as an orifice, then Equation (1) applies with "a" being determined by Equation (9). After rearrangement, we obtain

$$w/(g_c\rho\Delta P)^{1/2} = \frac{K_2}{P_s}(P_{in} - P_s + \frac{\rho u^2}{g_c}) + K_3 \qquad (10)$$

where

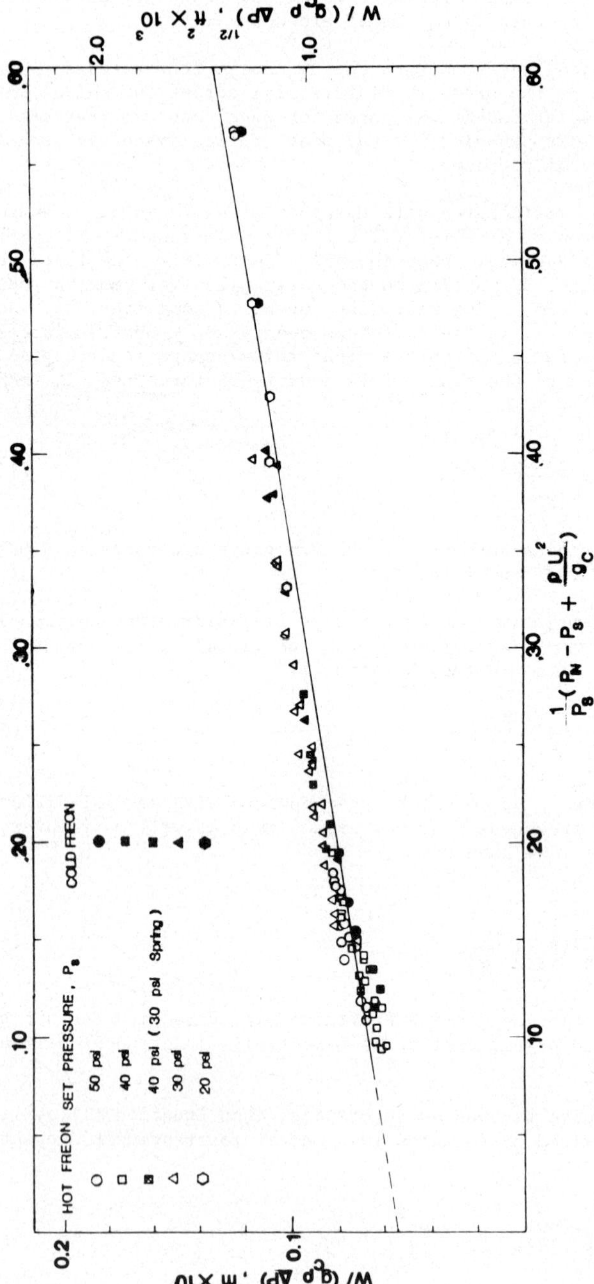

FIG. 4 - Correlation of Discharge Flow Rates

$$K_2 = C'\sqrt{2}\, K_1'$$

$$K_3 = C'\sqrt{2}\, a_o$$

The applicability of Equation (10) is illustrated in Fig. 4 where $w/(g_c \rho \Delta P)^{1/2}$ is plotted against $\frac{1}{P_s}(P_{in} - P_s + \frac{\rho u^2}{g_c})$. It may be seen that the data is well fitted by a straight line with $K_2 = 1.38\text{x}10^{-4}\text{m}^2$ and $K_3 = .55\text{x}10^{-4}\text{m}^2$ as determined by a least squares fit. It should also be noted that the data for cold (room temperature) and heated (temperatures up to 215°F) refrigerant fall on the same curve.

In Fig. 5, $w/[\frac{K_2}{P_s}(P_{in} - P_s + \frac{\rho u^2}{g_c}) + K_3]$ is plotted vs. $(g_c \rho \Delta)$ using the aforementioned values of K_2 and K_3. This representation clearly shows that the relief valve discharge rate is proportional to $(\Delta P)^{1/2}$. Slightly more scatter is observed in Fig. 5 than Fig. 4 as $(g_c \rho \Delta P)$ is used rather than its square root. An alternative attempt to represent the data in terms of Equation (4) was unsuccessful. It must therefore be concluded that the relief valve behaves as an orifice with a variable area.

The relationship between the temperature and pressure measurements just downstream of the relief valve is shown in Fig. 6. The vapor pressure curve for refrigerant 113 is also shown for comparison. It may be seen that, at the lower temperatures, the pressures in the downstream line are above the vapor pressure curve. It should be noted that at room temperature (66°F)

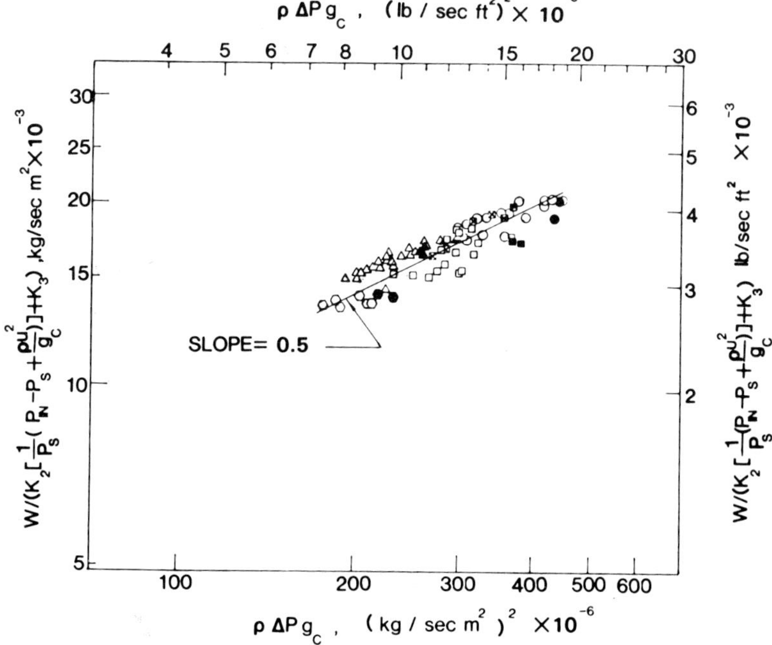

FIG. 5 - Effect of Pressure Drop Across Relief Valve on Discharge Flow

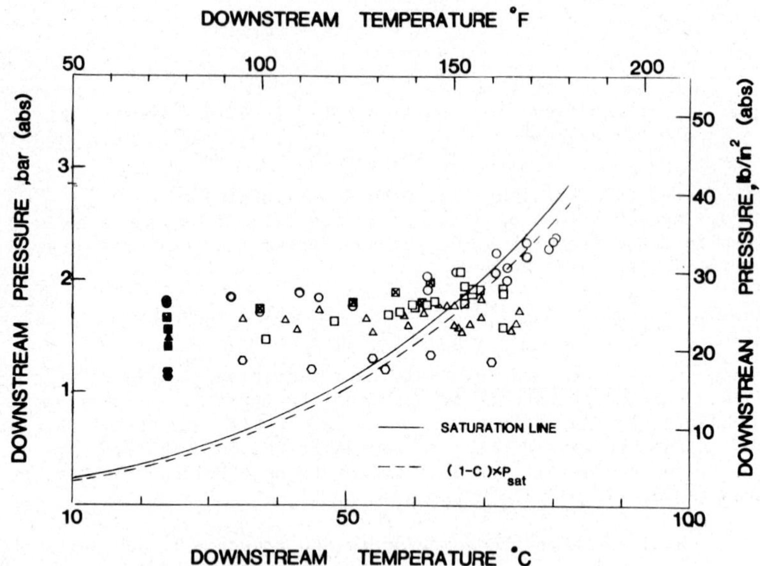

FIG. 6 - Relationship Between Downstream Pressure and Temperature

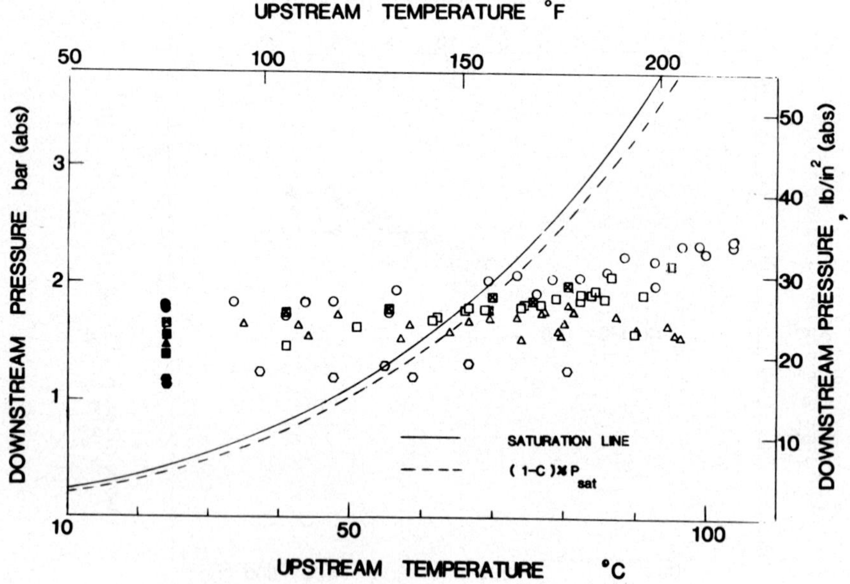

FIG. 7 - Relationship Between Downstream Pressure and Upstream Temperature

the vapor pressure of refrigerant 113 is 5.0 psia. At low temperatures the pressure in the line downstream of the relief valve is only slightly higher than the 19.7 psia which is the sum of the room temperature vapor pressure and atmospheric pressure. At temperatures above about 65°C (150°F), the pressure measurements scatter around the vapor pressure curve.

Fig. 7 presents a plot of the liquid temperature upstream of the valve vs. the downstream pressure. Both the vapor pressure curve and (1-C) times the vapor pressure curve (C is Burnell's coefficient given by Equation (3)) are shown for comparison. Note that, at the higher temperatures, the observation lie below $[(1-C)P_{sat}]$. If Burnell's equation (Equation 2) were to apply, one might expect that the downstream pressure would always be above $[(1-C)P_{sat}]$.

At the lower liquid inlet temperatures, the back pressure in the blowdown tank varies very little with time after initial transient. The pressure just downstream of the relief valve, which also shows little variation with time, is then the tank back pressure plus the pressure drop in the blowdown line. At high inlet liquid temperatures, such as present in the run illustrated in Fig. 3, there is also very little variation in the pressure downstream of the valve (Both P_{in} and the relief valve of ΔP are constant and hence the downstream pressure is constant). Measurements of the tank pressure with a Bourdon gage during discharge of such high temperature liquid showed appreciable increases in the blowdown tank pressure during the run. The near invariance of the pressure just downstream of the valve indicates that the critical flow rate of the flashing two-phase mixture was reached at a restriction in the line further downstream of the relief valve. In actual application of a relief valve, care should be taken to see that critical flow in the discharge line does not limit the valve discharge rate. This criterion is apparently commonly used in the sizing of such lines [7].

4. DISCHARGE FLOW DURING TRANSIENT PORTION OF BLOWDOWN

The discharge rates during the transient period which preceeded steady-state operation were examined for several typical runs. In the run illustrated by Fig. 3, the valve began to open at about 2.5 seconds. Approximately 1 second was allowed for the valve to open fully and hence the data for the transient between about 3.5 to 10 seconds was examined.

A series of data points were computed for the run illustrated in Fig. 3 and two other typical runs at the same set pressure. These data points are plotted in Fig. 8, as

$$w/(\rho \Delta p g_c)^{1/2} \text{ vs } \frac{1}{P_s}\left[P_{in} - P_s + \frac{\rho u^2}{g_c}\right]$$

which is the same procedure used for the steady-state data (Fig. 4). The straight line shown in Fig. 8 is the least square line from Fig. 4. The data all lie very close to this line. The slight deviation of the transient points from this line is similar to the deviation seen for these same runs on the steady-state plot. One may therefore conclude that, once the relief valve has fully opened, discharge rates may be computed from Equation (10) even though pressures have not reached their steady-state levels.

5. CONCLUSIONS

The present tests have clearly shown that, when discharging saturated or subcooled liquid, a typical relief valve acts as an orifice with a variable area. Providing upstream and downstream pressures are known, the discharge rate can be predicted on the basis of data obtained with a cold liquid. Further studies which consider relief valve behavior in the presence of a two-phase mixture upstream of the valve would be desirable.

ACKNOWLEDGEMENT

The financial assistance provided by the University Research Council of the University of Cincinnati is gratefully acknowledged. We would also like to thank Mr. S. Khajenejefi for modifying an existing experimental apparatus to enable these tests to be conducted. The generosity of the Dresser Industrial Valve & Inst. Div. in furnishing one of its relief valves for testing is appreciated.

NOMENCLATURE

a = discharge area

a_o = discharge area when $(P_{in} + \frac{\rho u^2}{g_c}) = P_s$

A = disc area exposed to flowing stream

C = Burnell's coefficient (Eq. 3) dimensionless

C' = discharge coefficient, dimensionless

g_c = gravitational conversion constant

G = mass flow rate, (mass/area time)

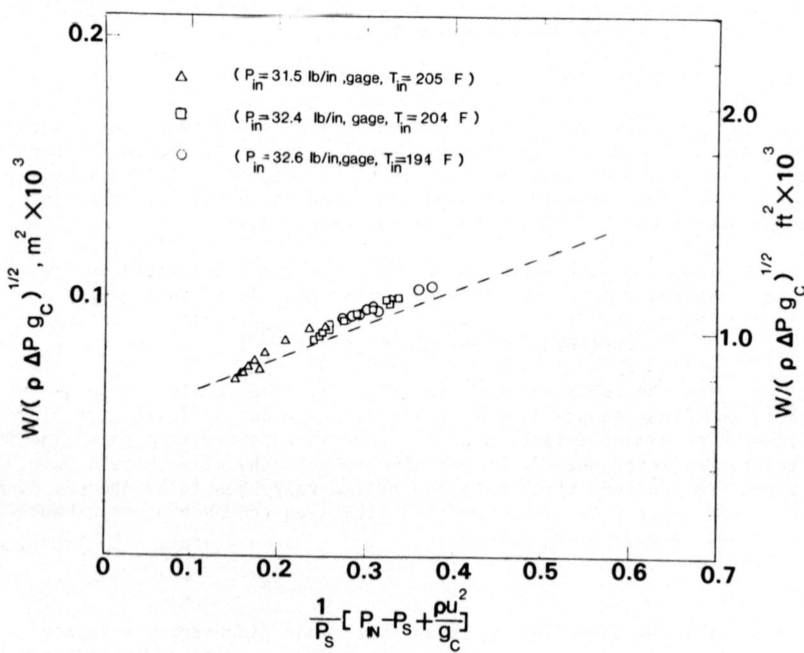

FIG. 8 - Correlation of Transient Discharge Data

G_c = critical mass flow rate (mass/area time)

K_1, K_1, K_2, K_3 = constants

P_{in} = inlet gauge pressure (force/area)

P_s = gauge pressure at valve set (or lift) point

P_{sat} = saturation pressure at T_{in} unless otherwise specified (force/area)

T_{in} = inlet fluid temperature

u = velocity of fluid in passage below valve seat

W = discharge flow rate, (mass/time)

ΔP = pressure drop across relief valve (force/area)

ρ_1 = liquid density (mass/volume)

σ = surface tension (force/length)

σ_s = surface tension at standard conditions (force/length)

REFERENCES

1. Henry, R., "Calculational Techniques for Two-Phase Critical Flow", in Two-Phase Flow Dynamics , A. Bergles and S. Ishigai, editors, Hemisphere Washington, D.C. (1980).

2. Bottomley, W. T., "The Flow of Boiling Water Through Orifices and Pipes", Trans. North-East Coast Inst. of Engrs. and Shipbuilders, 43, 65 (1936)

3. Burnell, J. G., "Flow of Boiling Water Through Nozzles, Orifices and Pipes", Engineering 164, 572 (1947)

4. Henry, R. E., M. A. Grolmes and H. K. Fauske, "Pressure Drop and Compressible Flow of Cryogenic Liquid-Vapor Mixtures", in 'Heat Transfer at Low Temperatures , W. Frost, ed., Plenum Press, N.Y. & London (1975)

5. Zaloudek, F. R., "The Low Pressure Discharge of Steam-Water Mixtures from Pipes", AEC Report HW-68934, Rev., Hanford Works, Richland, Washington (1961)

6. Celata, G. P., M. Cumo, G. E. Farello, P. C. Incalaterra and A. Naviglio, "Thermodynamic Disequilibrium in Critical Flow of Subcooled Liquids", Nuc. Tech. 60, 137 (1983)

7. Richter, H., "Size Relief Valve Systems for Two-Phase Flow", Hydrocarbon Processing, July, 1968

Multi-Phase Flow and Heat Transfer III. Part A: Fundamentals
edited by T.N. Veziroğlu and A.E. Bergles
Elsevier Science Publishers B.V., Amsterdam, 1984 — Printed in The Netherlands

MICROLAYER EVAPORATION IN SUBCOOLED NUCLEATE BOILING

L. D. Koffman
School of Mechanical Engineering
Georgia Institute of Technology
Atlanta, Georgia 30332, U.S.A.

ABSTRACT

The role of microlayer evaporation as a heat transfer mechanism in highly subcooled nucleate boiling is considered. The model of Plesset and Prosperetti is extended to account for microlayer profile. The measurements of Koffman and Plesset provide a reasonable estimate for the initial microlayer profile. The experiments of Gunther and Kreith are used as a basis for the analysis and conclusions. The contribution of microlayer evaporation is found to represent about 40% of the total heat transfer per bubble. Furthermore, this result does not appear to be very sensitive to the estimate of the initial microlayer profile.

1. INTRODUCTION

The problem of concern in the present study is the role of microlayer evaporation as a heat transfer mechanism in highly subcooled nucleate boiling. Essentially two mechanisms, microlayer evaporation and microconvection, have been advanced to explain the very high heat transfer rates observed in highly subcooled nucleate boiling. Both mechanisms are associated with the growth and collapse of vapor bubbles at the solid surface. As a vapor bubble grows at a solid surface a thin film of liquid, the microlayer, is left beneath the hemispherical bubble. This microlayer, which is on the order of a few micrometers in thickness, may evaporate during the bubble lifetime and transfer substantial energy through the latent heat of vaporization. Microconvection refers to an increased convective heat transfer due to the local fluid motion at the solid surface associated with the bubble dynamics. The relative roles of these two mechanisms in nucleate boiling have been studied by numerous investigators but most studies have dealt with near saturated and slightly subcooled boiling. Few studies have considered the case of highly subcooled nucleate boiling and the available conclusions are contradictory.

The experiments of Gunther and Kreith [1] and Gunther [2] have been used as the basis for analysis of the heat transfer mechanisms in highly subcooled nucleate boiling. These investigators report single bubble radius as a function of time as well as bubble population, bubble frequency, nominal wall temperature and heat flux, and liquid subcooling. These data can be used to analyze the role of a single vapor bubble in the heat transfer. On the basis of observed vapor bubble volume Gunther and Kreith attributed only 1-2% of the heat transfer to the latent heat of vaporization requirement and hence they argued that microconvection was the dominant heat transfer mechanism. A few years later the postulated existence of the microlayer and the proposed microlayer evaporation mechanism [3] led to a reexamination of the Gunther and Kreith data. Bankoff [4-6] considered microlayer evaporation as well as possible condensation over the bubble cap extending into the subcooled liquid (so that latent heat transport could be significantly greater than observed vapor volume). He suggested that the role of latent heat transport was an order of magnitude greater than previously estimated by Gunther and Kreith. Snyder and Robin [7-9] proposed an elaborate model from which they attributed nearly 100% of the heat transfer to microlayer evaporation. Recently, Plesset and Prosperetti [10] have proposed a direct model for the analysis of microlayer evaporation in the Gunther and Kreith data. They came to the conclusion that microlayer evaporation could account for only 10-20% of the heat transfer which is somewhat surprising since in recent years it has often been assumed that at high subcooling microlayer evaporation would be the dominant mechanism.

In the present study the model of Plesset and Prosperetti is extended to include some features which may impact on their conclusion. The primary extension is the inclusion of the microlayer profile for which the recent measurements of Koffman and Plesset [11] may be used. Plesset and Prosperetti used an approximate integral technique to solve the moving boundary problem and we have compared this solution to a more accurate finite difference solution. Finally, the kinetic theory expression for evaporative mass flux at an interface has been the subject of recent controversy and we consider the impact of a modified expression. The preliminary results reported here indicate that microlayer evaporation may account for 30-40% of the heat transfer in the Gunther and Kreith data.

2. FORMULATION OF THE MODEL

In the following the basic model of Plesset and Prosperetti is retained and the extensions and differences of the present study are noted. The microlayer is formed during rapid bubble growth but, as Cooper and Lloyd [12] have pointed out, because the microlayer is so thin it may be assumed that the viscous forces rapidly damp out any radial motion. Hence the microlayer evaporation may be treated as a transient heat conduction moving boundary problem. Furthermore, since the microlayer is thin compared to the radius of the bubble it is reasonable to assume that conduction in the radial direction is negligible compared to conduction normal to the wall. Hence a one dimensional form of the heat equation is sufficient to calculate microlayer evaporation at a given radial location,

$$\frac{\partial T}{\partial t} = D \frac{\partial^2 T}{\partial z^2} \quad ,$$

where D is the thermal diffusivity of the liquid. For the initial condition we assume that the microlayer is initially at the wall temperature T_w. The Gunther and Kreith data were taken on stainless steel strips and Plesset and Prosperetti have shown that a constant wall temperature may be assumed since stainless steel has a high thermal conductivity compared to water.

The proper choice of an interfacial boundary condition presents some difficulty. In many moving boundary problems the interfacial temperature is set equal to the saturation temperature. However, for the Gunther and Kreith data the wall is superheated about 30-35°C and nonequilibrium at the interface may be quite important. For this type of situation Plesset [13] has suggested use of the Hertz-Knudsen formula from kintic theory which gives the evaporative mass flux J as

$$J = \alpha(2\pi RT_b)^{-1/2} \; [p^e(T_b) - p_v] \tag{1}$$

where α is the accommodation coefficient for evaporation, R is the universal gas constant divided by the molecular weight of the vapor, T_b is the liquid interfacial temperature, $p^e(T_b)$ is the equilibrium vapor pressure at temperature T_b, and p_v is the pressure in the vapor at the interface. For the case of bubble dynamics Plesset and Prosperetti [14] have shown that p_v in (1) may be taken to be the internal pressure in the bubble, $p_i(t)$. The value of the accommodation coefficient is not well known but is taken to be unity since the microlayer is a freshly formed surface. The mass flux J can be used in an energy balance at the interface,

$$-k \frac{\partial T}{\partial z} = LJ \quad ,$$

where L is the latent heat of vaporization.

While the Hertz-Knudsen formula (1) has often been used, it was suggested by Schrage [15] some time ago that the expression for J should actually be larger by a factor of about 2 to account for the effect of a mean flow. More recent kinetic theory calculations suggest the correction should be 1.665 although the theory is still controversial [16]. Since there seems to be some agreement on the functional form in (1) and disagreement only in the numerical coefficient, the effect of a correction of 1.665 is considered in the present study.

A needed input in the problem formulation is the initial microlayer thickness δ_0 which can be regarded as a function of the bubble growth rate. Few measurements of microlayer thickness are available, especially for the case of water. The only detailed measurements of microlayer profile available are due to Cooper and Lloyd [12] for toluene and isopropyl alcohol, Voutsinos and Judd [17] for methylene chloride, and Koffman and Plesset [11] for water and ethanol. These latter measurements were unavailable to Plesset and Prosperetti and they chose to assume that the microlayer maintained a uniform thickness $\delta(t)$ over the base of the bubble; they then ran a sensitivity study to the initial microlayer thickness δ_0. In the present study the measurements of Koffman and Plesset are used to provide a more accurate input for the initial microlayer profile.

The remaining unknown in the problem is the internal pressure in the bubble, $p_i(t)$, which appears in (1) for the mass flux J. The internal bubble pressure couples the microlayer evaporation to the bubble dynamics. Gunther and Kreith observed that the vapor bubbles maintained a hemispherical shape while they grew and collapsed at the solid surface. In this case it is reasonable to apply the Rayleigh-Plesset equation which governs spherical bubble growth,

$$R \frac{d^2R}{dt^2} + \frac{3}{2} \left(\frac{dR}{dt}\right)^2 = \frac{1}{\rho} \left[p_i(t) - p_\infty - \frac{2\sigma}{R}\right] \quad , \tag{2}$$

where R(t) is the bubble radius, σ is the surface tension, p_∞ is the ambient pressure, and ρ is the liquid density. Plesset and Prosperetti used the data of Gunther and Kreith for R(t) to compute $p_i(t)$ from (2). With $p_i(t)$ known the problem is closed.

We summarize the model used by Plesset and Prosperetti:

- uniform thickness microlayer $\delta(t)$ with initial thickness δ_0 given
- transient heat conduction in the microlayer $\frac{\partial T(z,t)}{\partial t} = D \frac{\partial^2 T(z,t)}{\partial z^2}$ (3)
- initial condition (microlayer at wall temperature) $T = T_w$ at $t' = 0$ (4)
- wall boundary condition $T = T_w$ at $z = 0$ (5)
- interfacial boundary condition $-k \frac{\partial T}{\partial z} = LJ$ at $z = \delta(t)$ (6)
- kinetic theory mass flux $J = (2\pi RT_b)^{-1/2}[p^e(T_b) - p_i(t)]$ (7a)

  where $T_b(t) = T(\delta(t),t)$
- microlayer evaporation (moving boundary) $\frac{d\delta}{dt} = \frac{-J}{\rho}$ (8)
- $p_i(t)$ known from R(t) data used in (2).

In order to evaluate the contribution of microlayer evaporation, the total latent heat extracted from the microlayer, Q, is compared to the heat transfer per bubble reported by Gunther and Kreith. The evaporating area of the bubble base as a function of time must be accounted for in determining Q. For the uniform thickness microlayer Plesset and Prosperetti add to the above model the equation

$$\frac{dQ}{dt} = \pi R^2 LJ \quad . \tag{9}$$

Figures 1 and 2 are presented as an example of such a calculation by Plesset and Prosperetti. In Fig. 1 the R(t) data of Gunther and Kreith are shown along with the parabolic curve fit used in (2) to determine $p_i(t)$. The total heat extracted from the microlayer, Q, is plotted as a function of the assumed initial microlayer thickness δ_0 in Fig. 2. The total energy per bubble from Gunther and Kreith is also shown in Fig. 2. These results will be discussed subsequently when compared to the present results.

In the present study the basic model of Plesset and Prosperetti, equations (3)-(8), is retained. The correction to the kinetic theory mass flux may be included by replacing (7a) with

$$J = 1.665\ (2\pi RT_b)^{-1/2}[p^e(T_b) - p_i(t)] \quad . \tag{7b}$$

The primary extension of the model is the inclusion of a microlayer profile rather than assuming a uniform thickness profile. Since radial conduction is assumed negligible compared to conduction normal to the plate we may use the above model at various radial positions from which a picture of the overall microlayer evaporation can be obtained. We note that the calculation at each radial position depends on the time at which the bubble growth reaches that position which in turn couples the microlayer evaporation to the existing internal bubble pressure.

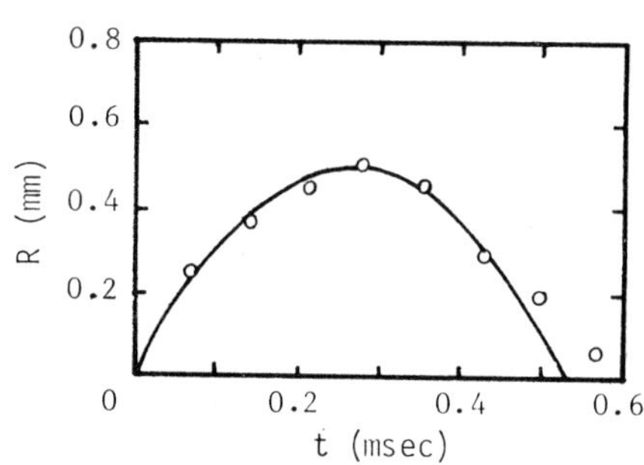

Fig. 1. The circles are observed values for bubble radius, R(t), from Gunther and Kreith [1]. The line is the fit to the data used in the analysis of Plesset and Prosperetti [10]. Experimental conditions: T_W = 132.2°C, T_∞ = 36.7°C, q = 3.26 MW/m^2, bubble density = 43 bubbles/cm^2, bubble frequency = 1000 bubbles/sec.

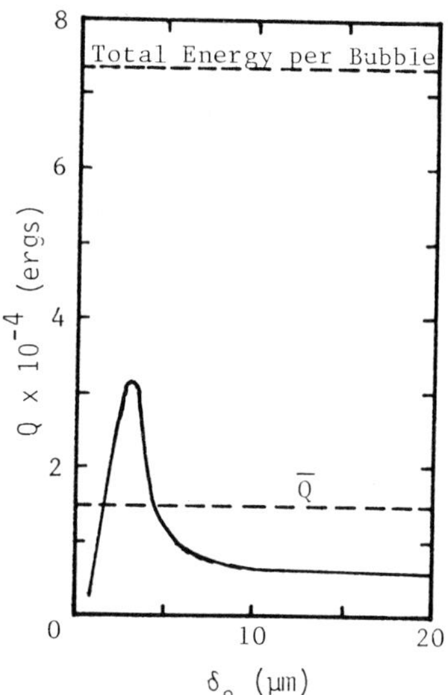

Fig. 2. The calculated total heat, Q, extracted from the microlayer over a single bubble lifetime is shown as a function of the initial thickness δ_0. (From Plesset and Prosperetti [10].)

3. COMMENTS ON THE SOLUTION TECHNIQUE

Equations (3)-(8) with given δ_0 and given $p_i(t)$ represent a well defined problem. The governing heat equation is a partial differential equation in time and one space variable and a numerical solution is required for this nonlinear problem. Plesset and Prosperetti chose to use an approximate integral method in which a parabolic profile is assumed for the temperature. The spatial dependence could then be eliminated and the set of ordinary differential equations in time are solved with a Runge-Kutta method. Although the approximate integral method often gives good results, the parabolic profile appears to be fairly crude for this problem. We felt that the accuracy of this method should be checked with a finite difference solution.

Several methods for solving moving boundary problems can be found in the literature. For the one-dimensional problem considered here, a fixed grid variable time step implicit finite difference formulation was chosen. This approach is simple to formulate, the position of the moving boundary is tracked directly, and the implicit formulation provides good accuracy and stability with a relatively course grid so that computing time is minimized. A fixed uniformly spaced grid is used and an iteration is performed on the time step so that the interface moves precisely to the next nodal point. The basis for the iteration is the energy balance at the interface given by (6). The evaporative mass flux J is very sensitive to the liquid interfacial temperature T_b so that a Newton-Raphson method must be used to obtain convergence to the time step.

The method described above has been successfully developed to solve the microlayer evaporation model; the details of the calculation are presented elsewhere [18]. Fig. 3 shows the calculation of the microlayer evaporation for an initial microlayer thickness of δ_0 = 1μm and Fig. 4 for δ_0 = 10μm. In each figure the results of the approximate integral method of Plesset and Prosperetti and the present finite difference method are shown as applied to equations (3)-(8). In addition, a finite difference solution is given for the model in which the interface is taken to be at saturation temperature; in this case (7) is replaced by the condition $T = T_{sat}$ at $z = \delta(t)$ and the energy balance (6) is still used as the basis for the iteration on the time step.

For the thin microlayer in Fig. 3 it is apparent that the integral method agrees well with the finite difference method. However for the thick microlayer in Fig. 4 the integral method differs considerably from the finite difference method. We note that the solution shapes differ near the beginning of the calculation but are similar near the end for both figures. Plesset and Prosperetti broke the solution into two parts: the solution before the wall effect is felt and the solution after the thermal layer reaches the wall. It appears that the profile in the first part may be too crude and the error from this part becomes more pronounced as the initial microlayer thickness increases.

The comparison of the T_{sat} boundary condition with the evaporative flux boundary condition is interesting. For the thick microlayer in Fig. 4 there is little difference between the two solutions. This is not surprising since the wall effect is small and the interface rapidly reaches equilibrium. However, for the thin microlayer in Fig. 3 nonequilibrium at the interface is significant and the T_{sat} solution overestimates the rate of evaporation. It seems clear that the nonequilibrium condition should be used to correctly model

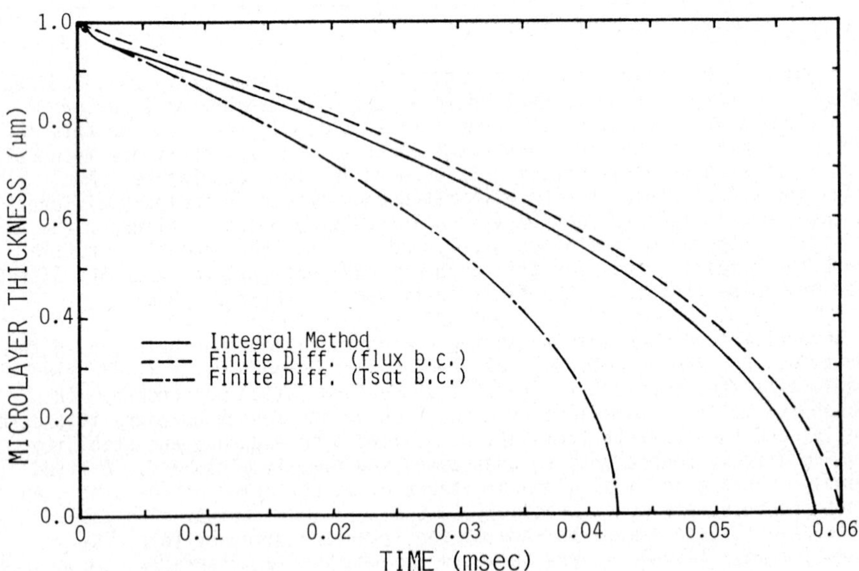

Fig. 3. Calculation of microlayer evaporation for an initial microlayer thickness of δ_0 = 1 μm.

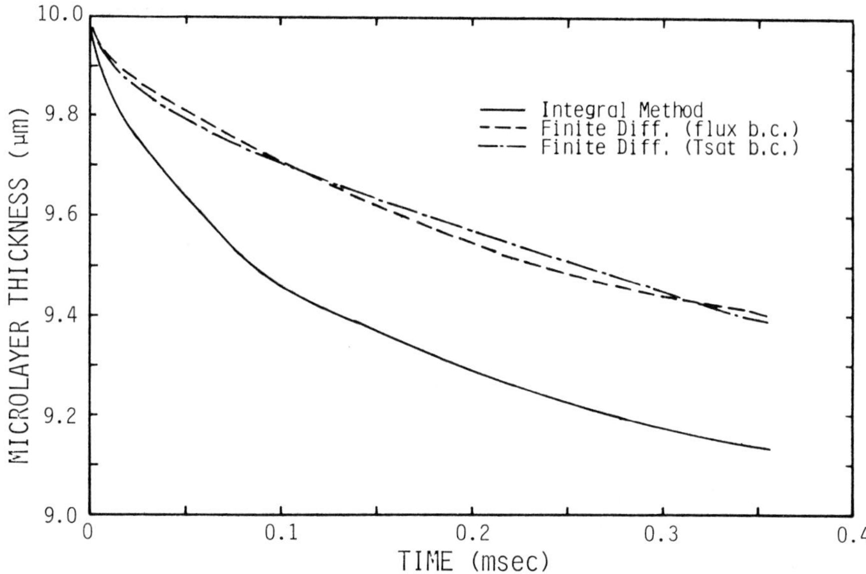

Fig. 4. Calculation of microlayer evaporation for an initial microlayer thickness of $\delta_0 = 10$ μm.

the thin portion of the microlayer. This point may be quite important in trying to compare detailed measurements of microlayer evaporation with a model.

4. RESULTS

We have considered the same bubble shown in Fig. 1 in the following calculations. The initial microlayer thickness profile is taken from the measurements of Koffman and Plesset [11] shown in Fig. 5. In this figure the microlayer profile is shown at sequential instants of time. This data gives the initial microlayer thickness and the microlayer evaporation. We note that the experimental conditions in Fig. 5 are quite different than those in Fig. 1. However, the bubble growth rates are similar and since microlayer formation is thought to be a function of bubble growth rate we may reasonably assume that the initial microlayer profiles are similar. With the initial microlayer profile known, we solve equations (3)-(8) at various radial locations from which we can form a plot similar to Fig. 5. The result of this calculation is shown in Fig. 6. We have used Plesset and Prosperetti's integral method in constructing Fig. 6 but since the microlayer is thin we can expect reasonable results as discussed in the last section.

In Fig. 6 the initial microlayer profile is shown as a dashed line. Equations (3)-(8) were solved at twenty evenly spaced radial positions. At each radial position we plotted the position of the microlayer at sequential instants of time taken in increments of 0.015 msec. We then connected the points corresponding to each instant of time to give a picture of the microlayer profile at sequential instants of time. The similarity between the calculation in Fig. 6 and the experimental measurement in Fig. 5 is quite encouraging although no direct comparison can be made because of the widely

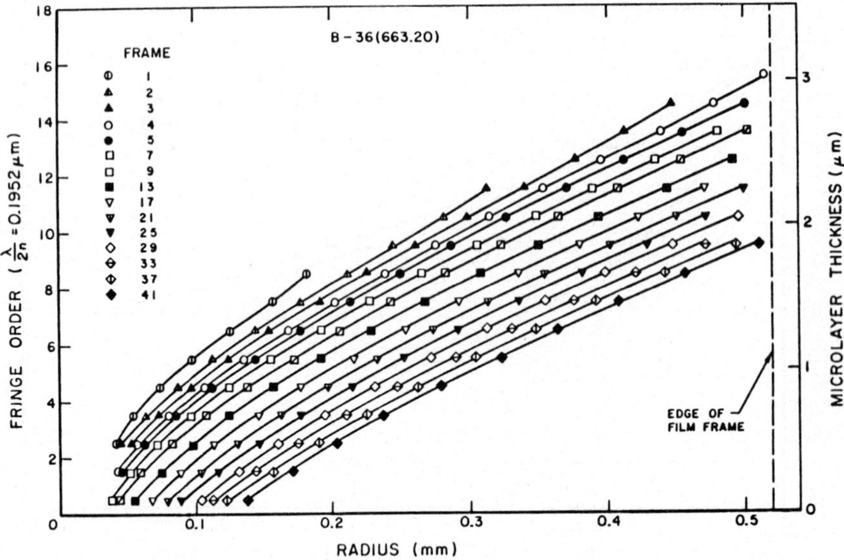

Fig. 5. Observed microlayer profiles for a water vapor bubble at atmospheric pressure with a mean heat flux of 204 kW/m² and a subcooling of 21.7°C; 0.066 msec/frame. (From Koffman and Plesset.)

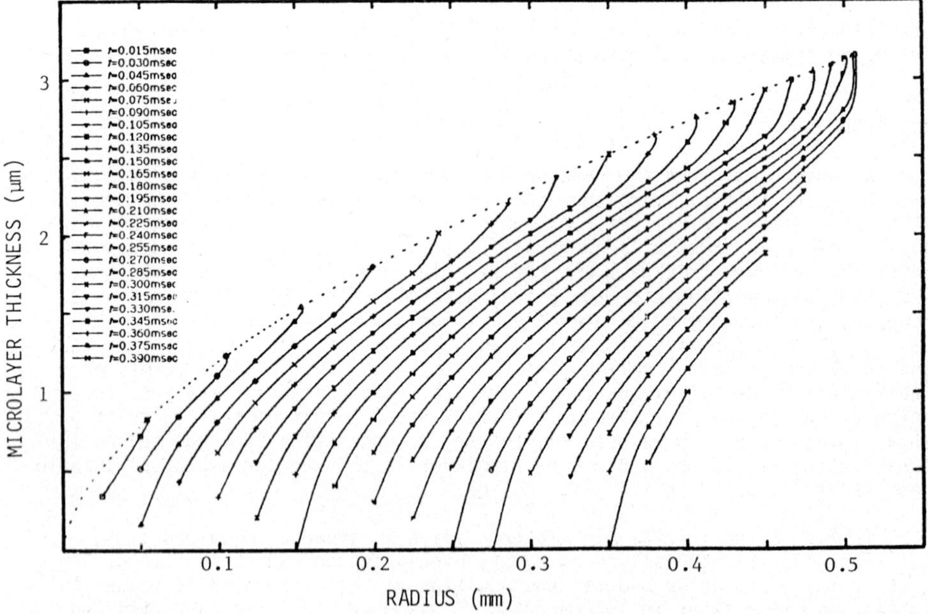

Fig. 6. Calculated microlayer profiles for the bubble growth, R(t), shown in Fig. 1 and for the initial microlayer profile given by the dashed line. The solid lines depict the microlayer profile at sequential instants of time in increments of 0.015 msec.

different experimental conditions.

From Fig. 6 we can determine the total evaporation and hence the total latent heat extracted from the microlayer. The resulting value of 2.9×10^4 ergs can be compared to Fig. 2. We see that this value represents approximately 39% of the total heat transfer per bubble. Plesset and Prosperetti estimated the order of magnitude of the initial microlayer thickness to be 9μm and thus concluded from Fig. 2 that the microlayer contribution was only about 9% for their uniform thickness model. They also performed a calculation to estimate the effect of a profile and this result is denoted as $\bar{Q}$ in Fig. 2; this calculation gave a microlayer contribution of 21%. We see that the present calculation gives a significantly higher contribution of microlayer evaporation.

We note that had Plesset and Prosperetti assumed an initial microlayer profile of 3μm that their conclusion from Fig. 2 would be quite different and more in line with the present result. A difficulty with their uniform thickness model is that the microlayer thickness is decreasing as the bubble base radius increases. Since this is counter to the experimental observation it is not clear how their initial microlayer thickness should be compared to experimental measurements.

Another feature of Fig. 2 is that the result for Q is very sensitive to the choice of initial microlayer thickness. Because we are estimating the initial microlayer profile it is of interest to consider the sensitivity of our result to changes in the initial microlayer profile. To consider this sensitivity we have modified the curve fit to the initial microlayer profile used in Fig. 6 as

$$\delta_0 = C\ (0.00188\ R^{0.6}) \tag{10}$$

where R and δ_0 are both in centimeters. The value of C = 1 corresponds to the curve shown in Fig. 6 and by varying C we can consider thicker and thinner initial microlayer profiles. The result of the calculations for various values of C is given in Fig. 7. We see that Q is much less sensitive to the initial profile than is indicated in Fig. 2. For values of C in the range $0.6 < C < 3.0$ we see that the calculated contribution of microlayer evaporation is 30-40% of the total heat transfer per bubble.

We have considered the correction to the kinetic theory mass flux by using (7b) in place of (7a) in the previous calculations. The results are given in Fig. 8 and can be compared to Fig. 7. We see that there is some effect of the correction but it is not dramatic. The calculated Q is even less sensitive to C in this case and the contribution of microlayer evaporation is seen to be about 40% of the total heat transfer per bubble.

5. CONCLUSIONS

The model of Plesset and Prosperetti used to analyze microlayer evaporation for the Gunther and Kreith data has been extended to include the microlayer profile. Using data for one case from Gunther and Kreith and using the measured initial microlayer profile of Koffman and Plesset, we have found that the contribution of microlayer evaporation represents about 40% of the total heat transfer per bubble. This is to be compared to Plesset and Prosperetti's estimate of 21%. Furthermore, while Plesset and Prosperetti found their results to be very sensitive to the initial microlayer thickness, we have found that inclusion of the microlayer profile reduces this sensitivity.

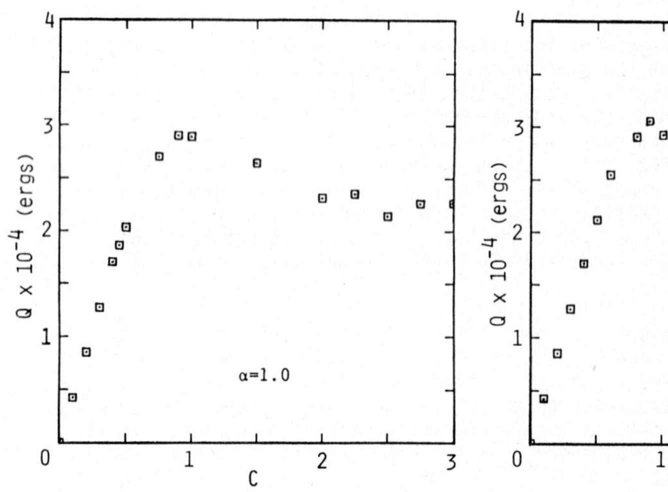

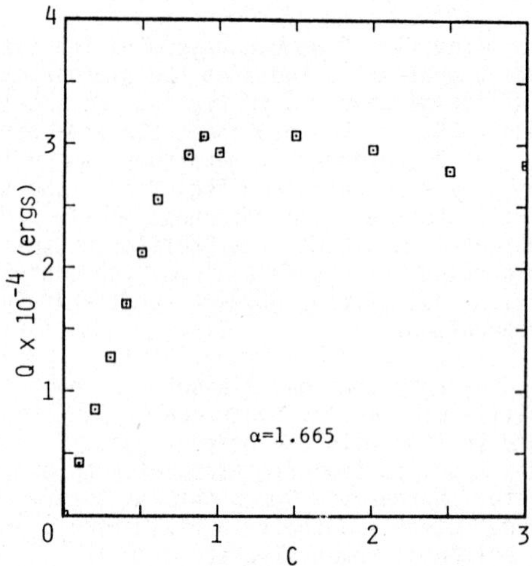

Fig. 7. The calculated total heat, Q, extracted from the microlayer as a function of the thickness parameter, C, used in Eqn. (10). The Hertz-Knudsen formula, Eqn. (7a), is used.

Fig. 8. The calculated total heat, Q, extracted from the microlayer as a function of the thickness parameter, C, used in Eqn. (10). The modified kinetic theory mass flux, Eqn. (7b), is used.

ACKNOWLEDGEMENTS

The author wishes to thank Mr. David Kemp for his assistance in the numerical computations. Mr. Kemp was supported by an Olin Corporation Summer Project Grant.

The research reported here is supported by the Heat Transfer Program of the National Science Foundation under Grant No. MEA81-05526.

REFERENCES

1. Gunther, F. C., and Kreith, F., "Photographic Study of Bubble Formation in Heat Transfer to Subcooled Water," Jet Propulsion Lab., Progress Report No. 4-120, March 1950.

2. Gunther, F. C., "Photographic Study of Surface-Boiling Heat Transfer to Water with Forced Convection," Transactions ASME, Vol. 73, 1951, pp. 115-123.

3. Snyder, N. W., and Edwards, D. K., "Summary of Conference of Bubble Dynamics and Boiling Heat Transfer Held at the Jet Propulsion Laboratory," JPL Memo No. 20-137, 1956.

4. Bankoff, S. G., and Mikesell, R. D., "Bubble Growth Rates in Highly Subcooled Nucleate Boiling," Chemical Engineering Progress Symposium Series, Vol. 55, No. 29, 1959, pp. 95-102.

5. Bankoff, S. G., and Mason, J. P., "Heat Transfer from the Surface of a Steam Bubble in a Turbulent Subcooled Liquid Stream," AIChE Journal, Vol. 8, No. 1, March 1962, pp. 30-33.

6. Bankoff, S. G., "A Note on Latent Heat Transport in Nucleate Boiling," AIChE Journal, Vol. 8, No. 1, March 1962, pp. 63-65.

7. Snyder, N. W., and Robin, T. T., "Mass-Transfer Model in Subcooled Nucleate Boiling," Journal of Heat Transfer, Vol. 91, No. 3, Aug. 1969, pp. 404-412.

8. Robin, T. T., and Snyder, N. W., "Theoretical Analysis of Bubble Dynamics for an Artifically Produced Vapor Bubble in a Turbulent Stream," International Journal of Heat and Mass Transfer, Vol. 13, 1970, pp. 523-536.

9. Robin, T. T., and Snyder, N. W., "Bubble Dynamics in Subcooled Nucleate Boiling Based on the Mass Transfer Mechanism," International Journal of Heat and Mass Transfer, Vol. 13, 1970, pp. 305-318.

10. Plesset, M. S., and Prosperetti, A., "The Contribution of Latent Heat Transport in Subcooled Nucleate Boiling," International Journal of Heat and Mass Transfer, Vol. 21, 1978, pp. 725-734.

11. Koffman, L. D., and Plesset, M. S., "Experimental Observations of the Microlayer in Vapor Bubble Growth on a Heated Solid," Journal of Heat Transfer (to appear).

12. Cooper, M. G., and Lloyd, A. J. P., "The Microlayer in Nucleate Pool Boiling," International Journal of Heat and Mass Transfer, Vol. 12, 1969, pp. 895-913.

13. Plesset, M. S., "Note on the Flow of Vapor Between Liquid Surfaces," Journal of Chemical Physics, Vol. 20, No. 5, May 1952, pp. 790-793.

14. Plesset, M. S., and Prosperetti, A., "Flow of Vapor in a Liquid Enclosure," Journal of Fluid Mechanics, Vol. 78, Part 3, 1977, pp. 433-444.

15. Schrage, R. W., A Theoretical Study of Interphase Mass Transfer, Columbia University Press, New York, 1953.

16. Koffman, L. D., "An Investigation of the Theory of Evaporation and Condensation," Ph.D. Thesis, Part II, California Institute of Technology, 1980.

17. Voutsinos, C. M., and Judd, R. L., "Laser Interferometric Investigation of the Microlayer Evaporation Phenomenon," Journal of Heat Transfer, Vol. 97, No. 1, Feb. 1975, pp. 88-92.

18. Karschner, D. W., "Finite Difference Solution for a Moving Boundary Problem with a Nonequilibrium Interfacial Boundary Condition," M.S. Thesis, Georgia Institute of Technology (in preparation).

Multi-Phase Flow and Heat Transfer III. Part A: Fundamentals
edited by T.N. Veziroğlu and A.E. Bergles
Elsevier Science Publishers B.V., Amsterdam, 1984 — Printed in The Netherlands

EFFECT OF SURFACE ROUGHNESS ON POOL BOILING OF REFRIGERANT 115 AT BURNOUT POINT

Mohamed Ibrahim Salem
Ain Shams University
Faculty of Engineering
Energy and Auto. Department
Abbassia - Cairo - Egypt

ABSTRACT

The heat transfer coefficient and the heat flux at the burnout point from a horizontal copper flat plate with different surface roughness is theoretically calculated for refrigerant R 115 in a wide range of saturation pressure between 2 to 95 percent of the critical pressure P_c.

An analytical expression is presented which permits the prediction of both the maximum nucleate heat flux (q_{max}) and the corresponding coefficient of heat transfer (α_{max}) in pool boiling of saturated refrigerants which also includes the effect of surface conditions.

The surface condition has been taken into consideration using the distribution function N(r) for the size of stable vapor bubble in active nucleation sites. The theoretical relations have been developed applying the transient conduction model of heat transfer within the range of influence of active sites. The calculated maximum nucleate heat flux (q_{max}) and the heat transfer coefficient (α_{max}) by this simple idealized approach are in qualitative as well as quantitative good agreement with measured results available in literature. In this work, it is proved theoretically, as it was early proved experimently in literature, that the coefficient of heat transfer at the burnout point in independent on the surface conditions, while the critical heat flux is very dependent on such surface conditions.

1. INTRODUCTION

With the increased use of cryogenic equipment, nuclear reactors, and rocket engines, design engineers have expressed increased interest in the accurate prediction of boiling behavior in the nucleate boiling region. The maximum heat flux holds special interest for the designer, since very high heat transfer coefficients may be obtained as the burnout point is approached; however, severe equipment damage may be sustained if film boiling is initiated. Because designers need a reliable maximum heat flux correlation, a great deal of attention and investigation has been devoted to development of a reliable maximum heat flux correlation. To date, however, there has not been a truly general maximum heat flux correlation which can be used with confidence in all situations. Perhaps one of the most significant reasons that the problem has not been completely solved is that combined effect of thermophysical, surface roughness, and geometric variables is not thoroughly understood. In order to successfully predict maximum heat flux conditions, one must be able to eliminate these variables which can not, as yet, be explained. Therefore, the purpose of this paper is to present a new approach which may be helpful in solving the problem of accurately correlating maximum heat flux (q_{max}) and heat transfer coefficient at the burnout point (α_{max}).

2. PREVIOUS WORK

As pointed out, there has been renewed interest in the heat transfer conditions at the burnout point. A reliable equation for predicting these conditions is essential, but to date no such equation has been developed. However, many investigators have adopted the view that the peak nucleate heat flux marks a transition from one regime to another (just as the critical Reynolds modulus marks the transition from laminar to turbulent flow) and proposed equations which do not derive from the behaviour of nucleate region alone. Such contributions are to be found in the work by Kutateladze [1] , Borishanskii [2] , and Rohsenow and Griffith [3] . According to this view the change from nucleate to transition boiling accures when the velocity in the vapour phase reaches a critical value. Starting from the nonlinear Euler equation of motion and the energy equation, it has been derived by Kutateladze [1] using dimensional analysis an equation for the peak nucleate heat flux in pool boiling of liquids at saturation in the form :

$$q_{max} = k \cdot \sqrt{\rho''} \cdot \Delta h_v \sqrt[4]{g\,(\rho' - \rho'')} \qquad (1)$$

The constant k was determined from experiments and its value was found to be k = 0.16 for horizontal flat plate surfaces.

Rohsenow and Griffith [3] presented Equation (2) based on data for organic liquids and water :

$$q_{max} = 143\, \rho'' \cdot \Delta h_v \left(\frac{\rho' - \rho''}{\rho''} \right)^{0.6} \qquad (2)$$

Considering the limiting stabaility of two-phase flows, Zuber [4] developed Equation (3) :

$$q_{max} = \frac{\pi}{24} \cdot \sqrt{\rho''} \cdot \Delta h_v \sqrt[4]{g(\rho' - \rho'')} \sqrt{\frac{\rho'}{\rho' + \rho''}} \qquad (3)$$

Chichelli and Bonilla [5] empirically correlated their experimental data for a large number of organic liquids by Equation (4) :

$$\frac{q_{max}}{p^*} = k \;\; f(p^*) \qquad (4)$$

Where k is equal to 1 for clean surfaces and to 1.15 for dirty surfaces.

Data of literature [6] , [7] indicate that the maximum nucleate boiling flux varies for a given fluid on surfaces of different chemical and physical properties. Cummings and Smith [8] report a correlation between (q_{max}) for helium at 1 bar and the physical properties of the soild involved as :

$$q_{max} = \Psi \left(\sqrt{(\rho \lambda c)_{su}} \cdot \sqrt{\left(\frac{\rho c}{\lambda}\right)_{su}} \cdot \sqrt{f} \right) \qquad (5)$$

Where (Ψ) is an empirical function.

These equations, presented above, can only be used as rough approximations, and in many cases they predict values which have an error of several hundred percent. An examination of the above equations leads to three important facts:

- Since these equations are based on thermodynamic properties, they are suitable to correlate the burnout point for only thermodynamically similar fluids boiling on the SAME heat transfer surface.

- They do not include any parameters describing to effect of surface roughness which vary from one investigation to another as long as the model suggested is purely a hydrodynamic model.

- The hydrodynamic approach helps only to predict the maximum heat flux but fails to predict the coefficient of heat transfer at the burnout point.

3. DERIVATION OF EQUATIONS FOR BURNOUT POINT

Although nucleate boiling does not form the focus of this paper, it is desirable to drive theoretical relations to predict both of the peak nucleate heat flux (q_{max}) and its corresponding coefficient of heat transfer (α_{max}) starting from the point of view that the burnout is brough about because of the mutual interference of many bubbles. It seems reasonable that if a quantitative understanding of the change from nucleate to film boiling is desired, and if the maximum heat flux is to be predicted from an equation for nucleate portion of the boiling curve, the bubble population has to be taken into acount. Therefore, the nucleate boiling characteristics of the surface under consideration represented by the size distribution N(r) for active nucleate sites should firstly be known.

3.1. Size Distribution N(r) for Active Nucleation Sites

The literature has shown conclusively that the surface conditions plays the dominant role in determining the nucleate boiling charactristics, and the results of preceding papers [9] , [10] indicate, however that the influence of surface roughness on the coefficient of heat transfer (α) cannot be described by a single roughness parameter for an extended pressure range and for surfaces with different forms of treatment. It seems, however that it would be possible to describe pool boiling heat transfer data for different surface properties by a three parametric distribution function of the size of stable bubbles in active nucleation sites of the heated surface. It has been shown in the above mentioned papers that the most suitable distribution, which correlates the relation between the number of active sites per unit area (N/A) and the radius of the minimum active cavity (r), is an exponential function with a power term.

$$\ln \frac{N}{A} = \left(\ln \frac{N_o}{A} \right) . \left[1 - \left(\frac{r}{r_s} \right)^m \right] \qquad (6)$$

Where

N_o The maximum value of N (r = 0)

r_s The maximum value of r (which corresponding to the beginning of nucleation (N=1))

m An exponent.

The three parameters (N_o, r_s and m) can be only determined if there are enough measuring data on the heat transfer coefficient during pool boiling in a wide range of pressures for the heating surface under test [10] .

Starting from these considerations, in this paper, theoretical relations for the heat transfer coefficient (α_{max}) and the maximum heat flux (q_{max}) are obtained combining the transient conduction model of heat transfer [11] within the range of influence of active nucleation sites with the assumption that :

- The three surface parameters (N_o, r_s and m) are known for the heating surface under consideration.
- As heat flux is increased in nucleate boiling, more nucleation sites become active. The maximum nucleate heat flux is brought about because of the mutual interference of a critical "bubble packing". This critical number of active nucleation sites will be denoted by (N_{max}).

3.2. Heat Transfer Coefficient at the Burnout Point

The total heat flow rate (Q) leaving the heated surface in pool boiling consists of the heat flow rate by natural convection (Q_{nc}), and the heat flow rate induced by the nucleation process (Q_{nuc}) :

$$Q = Q_{nc} + Q_{nuc} \qquad \text{or}$$

$$Q = q_{nc} \cdot A_{nuc} + q_{nuc} \cdot A_{nuc} \tag{7}$$

Where (q_{nc}) and (q_{nuc}) are the corresponding heat fluxes and (A_{nc}), (A_{nuc}), the pertaining areas on the heated surface. Dividing by the total area (A) and the excess temperature (ΔT) of the heated surface and rearranging yields :

$$\frac{\alpha - \alpha_{nc}}{\alpha_{nuc} - \alpha_{nc}} = \frac{A_{nuc}}{A} \tag{8}$$

According to Han and Griffith the area of influence of a single active site is πD_b^2 (D_b= diameter of the bubble when leaving the heated surface)[12] thus we obtain for all the (N) active sites

$$A_{nuc} = N \cdot \pi D_b^2 \tag{9}$$

Within the range of influence of active nucleation sites, the transient conduction model is applied [10] , thus

$$\alpha_{nuc} = \frac{2}{\sqrt{\pi}} \cdot \sqrt{\rho' c' \lambda'} \cdot \sqrt{f} \tag{10}$$

The critical heat flux may be taken place when the heated surface is covered with the maximum number of vapor bubbles (N_{max}); thus we obtain from eqs.(8) and (9) an expression for the heat transfer coefficient at the burnout point (α_{max}) as follows :

$$\frac{\alpha_{max} - \alpha_{nc}}{\alpha_{nuc} - \alpha_{nc}} = \frac{N_{max}}{A} \cdot \pi D_b^2 \tag{11}$$

Since ($\alpha_{nc} \ll \alpha_{nuc}$) and ($\alpha_{nc} \ll \alpha_{max}$) thus eq. (11) becomes :

$$\alpha_{max} = \alpha_{nuc} \cdot \frac{N_{max}}{A} \cdot \pi D_b^2 \tag{12}$$

With $\frac{N_{max}}{A} \cdot \pi D_b^2 = C$, then (12a)

$$\alpha_{max} = \frac{2C}{\sqrt{\pi}} \cdot \sqrt{\rho' c' \lambda'} \cdot \sqrt{f} \quad (13)$$

3.3. Maximum Heat at the Burnout Point

The excess pressure (ΔP) necessary for the existance of stable bubble with radius (r) is given by the well known Thomson equation

$$\Delta P = \frac{2\sigma}{r} \left(\frac{\rho'}{\rho' - \rho''} \right) \quad (14)$$

Where (σ) is the surface tension. With the assumption of thermodynamic equilibrium between superheated liquid and vapor bubble, (ΔP) can be replaced by (ΔT) calculated from the vapor pressure curve. In order to obtain an approximate explicit correlation the Clausius Clapeyron equation :

$$\frac{dp}{dT} = \frac{\Delta h_v}{T (v'' - v')} \quad (15)$$

is integrated with the assumption that the right side properties of eq. (15) are independent of pressure and temperature within the range of excess temperature (ΔT) .

$$\frac{\Delta P}{\Delta T} = \frac{\Delta h_v}{T (v'' - v')} \quad (15a)$$

Where Δh_v = Specific latent heat of vaporization,

v = " volume and $T = T_{st} + (\Delta T/2) = T_m$

The bubble radius (r) can be calculated from eqs. (14) & (15a) as:

$$r = \frac{2 \sigma T}{\rho'' \cdot \Delta h_v} \cdot \frac{1}{\Delta T} \quad (16)$$

is equal to the minimum radius (r_c) of a stable bubble in a nucleation center activated at a given excess temperature (ΔT), while the number (N) of active sites at the given (ΔT) comprises all nucleation sites with bubble $r > r_c$

The excess temperature of the heated surface at the burnout point (ΔT_{max}) may be obtained from eq. (16) as

$$\Delta T_{max} = \frac{2 \sigma T}{\rho'' \Delta h_v} \quad \frac{1}{r_{max}} \quad (16a)$$

Where (r_{max}) is the minimum radius of a stable bubble just before the occuring of burnout.

From the definition of the heat transfer coefficient $\alpha = q/\Delta T$ and eqs. (13), (16a) the maximum heat flux at the burnout point becomes :

$$q_{max} = \frac{4C}{\sqrt{\pi}} \cdot \frac{\sqrt{\rho' c' \lambda'} \cdot \sqrt{f} \cdot T}{\rho'' \Delta h_v} \cdot \frac{1}{r_{max}} \quad (17)$$

(r_{max}) can be obtain from eqs. (6) & (12a) as

$$r_{max} = r_s \left[\left(\ln \frac{N_o \cdot \pi D_b^2}{A.C} \right) / \left(\ln \frac{N_o}{A} \right) \right]^{1/m} \tag{18}$$

From eqs. (17) & (18) (q_{max}) becomes :

$$q_{max} = \frac{4C}{\sqrt{\pi}} \cdot \frac{\sqrt{\rho' c' \lambda'} \cdot \sqrt{f} \cdot T}{\rho'' A h_v . r_s \left[\left(\ln \frac{N_o . \pi D_b^2}{A.C} \right) / \left(\ln \frac{N_o}{A} \right) \right]^{1/m}} \tag{19}$$

At the burnout point the coefficient of heat transfer (α_{max}) and the peak heat flux (q_{max}) can be calculated from eqs. (13) & (19) respectively from the thermophysical properties of the boiling liquid and the three surface parameters (N_o, r_s, and m), if additional data on the bubble frequency (f) and the bubble diameter (D_b) ara available. These data are taken from equation of Mc-Fadden and Grassman for f [13] and of Fritz [14] for D_b

$$f^2 . D_b = 0.314 \, g \frac{\rho' - \rho''}{\rho'} \tag{20}$$

for refrigerants :

$$D_b = 0.0144 \; \beta \sqrt{\frac{2 \sigma}{g (\rho' - \rho'')}} \tag{21}$$

Where g is the acceleration of gravity

β is the contact angle (35° for refrigerants [15])

4. RESULTS AND DISCUSSIONS

The coefficient of heat transfer at the burnout point (α_{max}) predicted by eq. (13) which is based on the previous theoretical assumptions, is indepentent of the surface parameters (N_o, r_s, and m) but depends only on the thermophysical properties of the boiling refrigerant and the constant (C). This constant (C) indicates the mutual interference of effective areas which take place between the neighbouring active bubbles just before the burnout point. At the burnout point, i.e. at the moment at which the maximum possible interference between the effective areas of the active bubbles happens, the value of (C) should be constant and independent on the surface roughness. (C=number of active bubbles just before the burnout point (N_{max}) x effective area of each bubble (πD_b^2)). The constant (C) is difficult to be estimated theoretically, therefore, (C) should be determined experimentally to fit the experimental heat transfer data at the burnout point from different treated flat plate heat transfer surfaces. The available heat transfer data in literature [16] for boiling of R 115 at the burnout point from four different treated copper flat plate heat transfer surfaces of known surface parameters were used to determine the value of the constant (C) in eqs. (13) & (19), (C) was found to be equal 19.

In Fig. (1) the coefficient of heat transfer (α_{max}) at the burnout point predicted by equation (13) with C = 19 are compared with experimental data for R 115 from four different treated copper flat plate surfaces by Tanes [16] and from rubbed surface copper tube of 8 mm diameter by Bier et al [17] . It appears from the comparison that this simple idealized approach is qualitatively and quantitatively in good agreement with experimental results. It is seen from experimental data that the coefficients of heat transfer at the burnout point are independent of the surface roughness. It should be noted this result is proved by eq. (13) in which (α_{max}) is independent of the surface parameters (N_o, r_s and m) but depends only on the thermophysical properties of the boiling refrigerant. Therefore, the coefficient of heat transfer at the burnout point could be correlated for corresponding states of fluids boiling on different heat transfer surfaces.

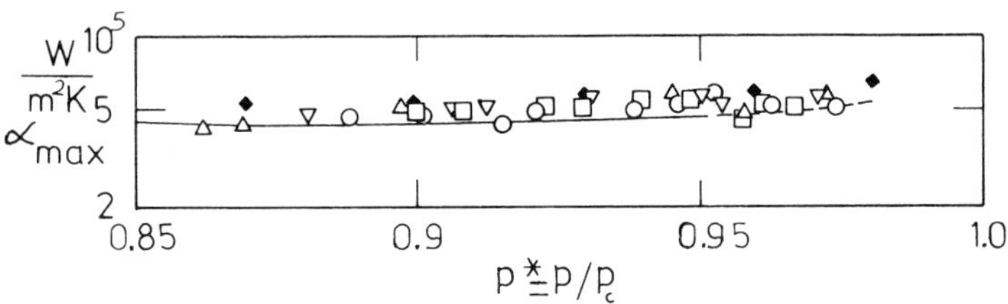

Fig. (1) : Comparison between measured heat transfer coefficient α_{max} and that calculated from eq. (13) with C= 19 (symbols used are shown in table (1)),--- extrapolation.
◆ Rubbed copper tube with 8mm. diameter [17].

Fig. (2) shows the effect of surface conditions on both (q_{max})&(α_{max}) as a function of (P*). The three-parametric constants (N_o, r_s, and m) for the different surfaces under test were obtained in previous papers [9], [10] and are given in table (1) :

Table (1)
Treatment and Three-Parametric Surface Constants of the Investigated Plates.

Plate No.	Symbol	Surface Treatment	N_o m^{-2}	r_s μm	m
1	△	Chemically etched (fine)	1.6×10^9	3.31	0.42
2	▽	" " (rough)	1.6×10^9	3.31	0.42
3	○	Turned .	3.18×19^9	12.09	0.265
4		Rubbed with emery paper	3.65×10^8	4.86	0.59
5	□	Polished to mirror finish	1.69×10^{10}	11.62	0.21

As noted above, the coefficient of heat transfer at the burnout point is independent of the surface conditions and remains nearly constant at about 4600 W/m^2K for a wide range of pressures (from the atmospheric to near the critical pressure). This result means (from the definition of the heat transfer coefficient) that the temperature difference between the plate surface and the saturated refrigerant at the burnout point changes with the pressure in similar manner as the peak heat flux. Contrary to that, the heat flux at the burnout point depends strongly on both of the pressure and the surface conditions. For a specified surface, the maximum heat flux changes with pressure and has a maximum at certain (P*) which varies from one surface to another (for the surfaces under test the maximum of the peak heat flux lies at (P*) between 0.065 and 0.08). Near the critical pressure the heat flux at the burnout point decreases strongly as the pressure increases and tends to zero as (P^*) tends to one. It is seen also that the surface conditions play a dominant role in determining the peak heat flux. For a certain pressure, (q_{max}) changes from one surface to another, the mirror finished surface shows the biggest value of (q_{max}) while the rubbed surface with emery paper shows the smallest value. The value of (q_{max}) for the turned and the chemically etached surfaces lies between (q_{max}) of the mirror and the emery rubbed surfaces. To compare the present equation for (q_{max}) with C = 19, the correlations of Rohsenow and Griffith [3] and Kutateladze [1] were plotted on the same graph.

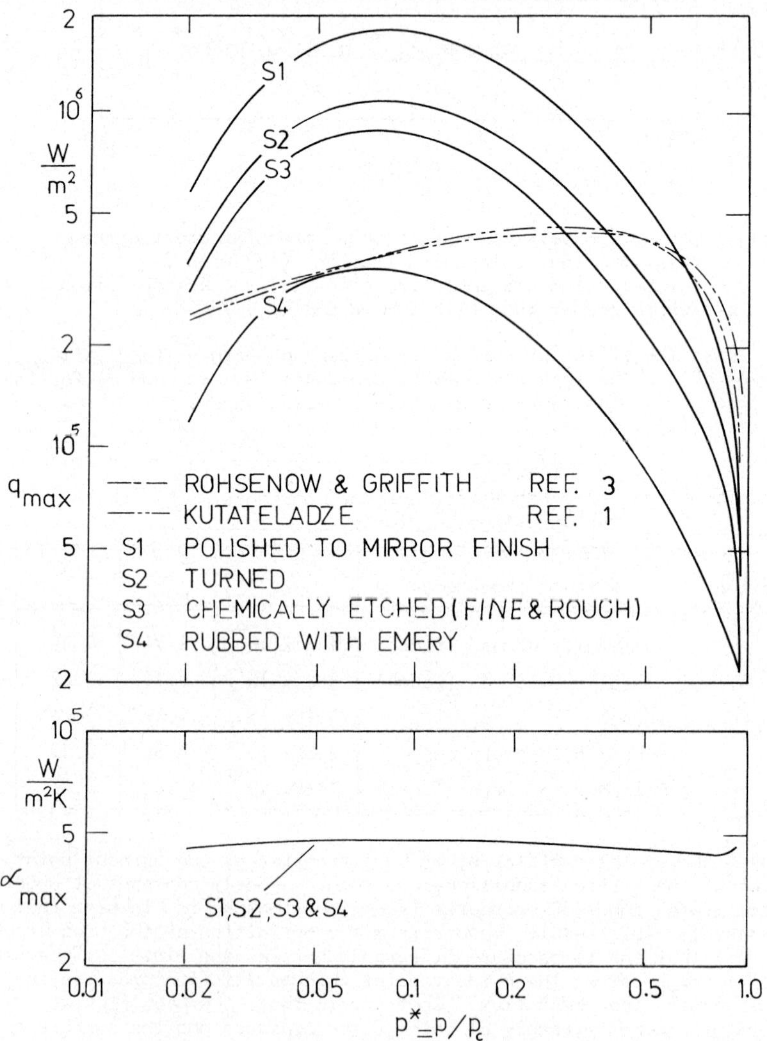

Fig. (2): Effect of surface roughness on the heat flux q_{max} (upon) and the coefficient of heat transfer α_{max} (down) at burnout point from different treated surfaces.

Fig. (3), Fig. (4) and Fig. (5) show a comparison between the measured values of both (q_{max}) and (α_{max}) for R 115 on horizontal copper flat plate surfaces by Tanes [16] with that calculated by eqs. (13) & (19) with C = 19 and the three-parametric surface constants (N_o, r_s, and m) from table (1) for the turned, chemically etched and the polished to mirror finish surfaces. The correlation of Rohsenow and Griffith [3] and Kutateladze [1] were plotted in Fig. (3), Fig. (4) and Fig. (5) in the upper graph to compare with the present work. It is seen, that the present eq. (19) is capable to correlate the experimental data of (q_{max}) for R 115 from different treated heat transfer surface better than the previous correlations. Since these previous correlations were based on purely hydrodynamic model, they contain no surface parameters but only thermophysical properties of the boiling refrigerant, therefore, the values of (q_{max}) depend only on the pressure. Eq. (13) with C = 19 correlates the experimental results of (α_{max}) in Fig. (3), Fig. (4), and Fig. (5) in the lower graph from five different treated surfaces with diversity of ± 10%. It should be noted that there are no correlation in the literature to correlate (α_{max}) to use it for comparison with the present work, eq. (19).

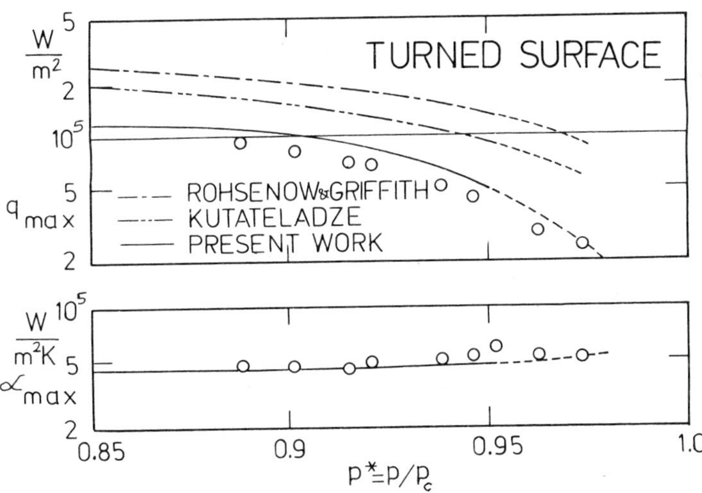

Fig. (3) : Comparison between measured q_{max} and α_{max} with that calculated from eqs. (13) and (19) with C = 19 for turned surface --- extropolation.

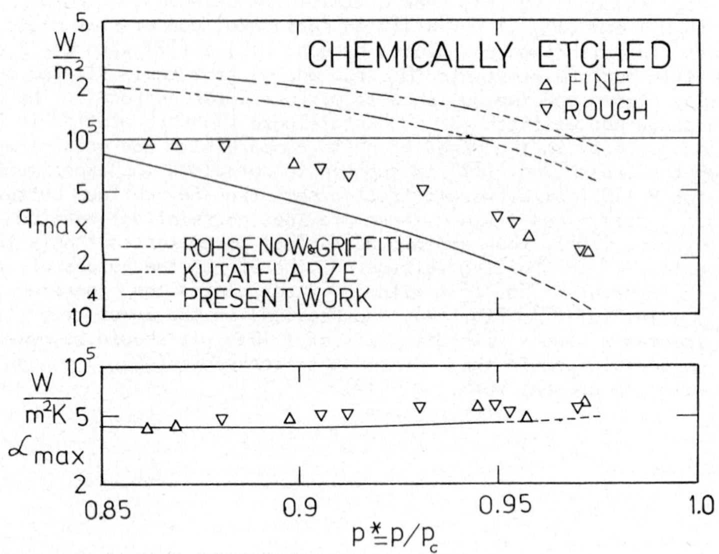

Fig. (4): Comparison between measured q_{max} and α_{max} with that calculated from eqs. (13)&(19) with C = 19 for chemically etched --- extrapolation.

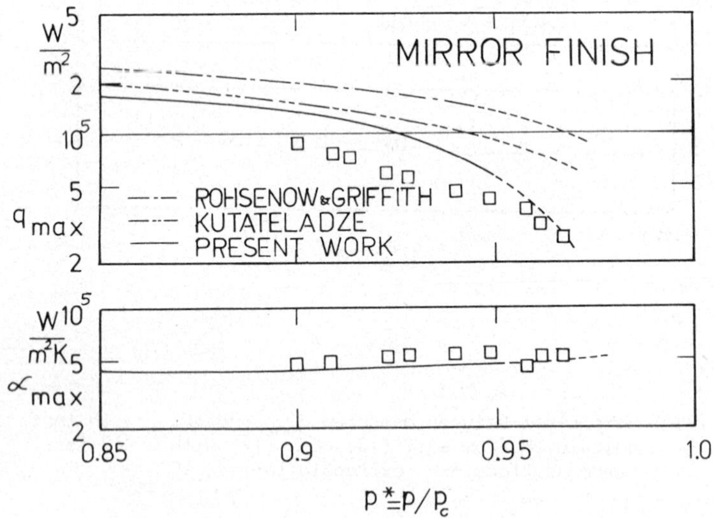

Fig. (5): Comparison between measured q_{max} and α_{max} with that calculated from eqs.(13)&(19) with C=19 for mirror finish --- extrapolation.

5. CONCLUSIONS

The influence of the surface roughness on pool boiling of R 115 on copper flate plate heat transfer surfaces at the burnout point was theoretically studied. The approach used above will possibly lay the foundations for a study to determine exactly the effect of surface roughness on pool boiling at the burnout point. However, more experimental data are needed in a wide range of pressure from surfaces of different roughness in order to establish the limitation of the present approach. The agreement of the analytical predictions, based on the discribed simple idealized model, with experimental data was shown and leads to the following conclusions:

1. In as much as the transient conduction model was sucessfully used to correlate the nucleate boiling heat transfer data [9] and [10], it has also proved (in the present work) to be capable of describing the burnout point as a maximum limit of nucleate boiling of comparison with available experimental data by Tanes [16].

2. The calculated values of (q_{max}) and (α_{max}) from eqs. (13) & (19) are qualitatively and quantitatively in good agreement with measured values, although only approximate expressions for the bubble frequency (f) and bubble diameter (D_b) were used in the theoretical analysis.

3. The maximum heat flux (q_{max}) is very dependent on the surfece conditions as well as the saturation pressure of the boiling refrigerant.

4. The coefficient of heat transfer at the burnout point (α_{max}) for a horizontal copper flat plate heat transfer surface is independent of the surface conditions, but depends only on thermophysical properties of the boiling refrigerant, such a result was deduced experimentaly by Tanes [16], and was proved theoretically in the present work. Therefore, the coefficient of heat transfer at the burnout point may be correlated for corresponding states of fluids such as refrigerants.

5. The coefficient of heat transfer at the burnout point (α_{max}) remains nearly constant in a wide range of pressure from the atmospheric to near the critical pressure (most of the refrigerants have $P^* - 0.02$ at the atmospheric pressure). It may be concluded, therefore, that the temperature difference between the plate surface and the saturated refrigerant changes with pressure in similar manner as the maximum heat flux (q_{max}).

6. All the variables which affect the nucleate boiling heat transfer, affect the burnout heat transfer in the same way. Such variables are surface material, surface roughness, geometry and thermophysical properties of the boiling refrigerant.

6. NOMENCLATURE

A	Surface area	m^2
C	Constant	
D	Diameter	m
N	Number of active nucleation sites	
Q	Total heat flow rate	W
T	Temperature	K
ΔT	Excess temperature of the heated surface	K
c	Specific heat capacity	J/kg K
f	Bubble frequency	1/s

g	Acceleration of gravity	m/s^2
Δh_v	Specific latent heat of vaporation	J/kg
p	Saturation pressure	bar
Ap	Excess pressure of stable bubble	bar
q	Heat flux	W/m^2
r	Bubbles radius	m
v	Specific volume	m^3/kg
α	Heat transfer coefficient	W/m^2K
β	Contact angle	o
λ	Thermal conductivity	W/mK
ρ	Mass density	kg/m^3
σ	Surface tension	N/m

SUBCRIPTS

b	Bubble at departure from heated surface
c	At the critical state
m	Mean value
max	At the burnout point
nc	At natural convection
nuc	With range of influence if nucleation site
su	Heating surface
st	Saturation condition

SUPERSCRIPTS

`	Saturated liquid
``	Saturated vapor
*	Property normalized by its value at the critical state

REFERENCES

1. Kutateladze, S.S., "A Hydrodynamic Theory of Changes in the Boiling Process Under Free Convection Conditions", Izv. Akad. Nauk. SSR, Otd. Nauk, No. 4,p. 529 (1951).

2. Borishanskii, V.M., "An Equation Generalizing Experimental Data on the Cessation of Bubble boiling in a Large Volume of Liquid", Zhurn. Tekh. Fiz, vol. 25, P. 252 (1956) (See Soviet Physics - Technical Physics, Vol 1, No. 2, p. 438, American Inst. of Physics).

3. Rohsenow, W.M., and Griffith, P., "Correlation of Maximum Heat Flux Data for Boiling of Saturated Liquids", Preprint No. 9, ASME - AICHE Heat Transfer Symposium, Louisville, Ky., March (1955).

4. Zuber, N., "On The Stability of Boiling Heat Transfer", Trans. Am. Soc. Mech. Engrs, Vol. 80, p. 711 (1958).

5. Cichelli, M. T., and Bonilla, C.F., "Heat Transfer to Liquids Boiling Under Pressure", Trans. Am. Inst. Chem. Engrs., Vol. 41. p. 755 (1945).

6. Lyon, D.N., Int. J. Heat Mass Transfer, Vol. 7, p. 1097 (1964).

7. Carne, M., and Charlesworth, D.H., Chem. Eng. Progr. Symposium, No. 64, Vol. 62, p. 24 (1966).

8. Cummings, R.D., and J.L. Smith, paper presented at Int'l Inst. Refrig.

Commission I Meeting, Boulder, Colo. (June, 1966); Cummings, R.D., Sc. D. Thesis, Mass. Inst. Technol., Cambridge (Nov. 1965).

9. Salem, M.I., "Wärmeübergang Beim Blasensieden und Grössenverteilungen von Stabilen Blasenkeimen sowie von Rauhigkeitsverteilungen der Heizfläche", Diss., Universität Karlsruhe (TH) (1979).

10. Bier, K., Gorenflo, D., Salem, M.I., and Tanes, Y., "Pool Boiling Heat Transfer and size of Active Nucleation Centers for Horizontal Plates with Different Surface Roughness", 6th Int. Heat Transfer Conf., Toronto Vol. 1 (1978).

11. Mikic, B.B., and Rohsenow, W.M., "A New Correlation of Pool- Boiling Data Including the Effect of Heating Surface Characteristics"., J. of Heat Transfer, Vol. 91, p. 245 (1969).

12. Han, C. - Y., and Griffith, P., "The Mechanism of Heat Transfer in Nucleate Pool Boiling", Parts I and II, Int. J. Heat Mass Transfer, Vol. 8, p. 887 (1965)

13. McFadden, P.W., and Grassmann, P., "The Relation between Bubble Frequency and Diameter During Nucleate Boiling", Int. J. Heat Mass Transfer, Vol. 5, p. 185 (1970).

14. Fritz, W., "Berechnung des Maximalvolumens von Dampfblasen", Phys. Zeitschrift, Vol. 36, p. 379 (1935).

15. Stephan, K., "Mechanismus und Modellgesetz des Wärmeübergangs bei der Blasenverdamfung", Chem. Ing. Tech., Vol. 35, p. 775 (1963). "Beitrag zur Thermodynamik des Wärmeübergangs beim Seiden", Abh. d. Deutschen Kältetechn. Vereins, No. 18, Karlsruhe (1964)

16. Tanes, Y.M., "Zum Einfluss der Oberflächenbeschaffenheit der Heizfläche auf den Wärmeübergang beim Blasensiden", Diss., Universität Karlsruhe (TH) (1976).

17. Bier, K., Engelhorn, H.R. and Gorenflo, D., "Heat Transfer at Burnout and Leidenfrost Points for Pressures up to Critical", Progress in Boiling Heat Transfer, Hemisphere Publ. Corp., Washington, D.C., p. 85 (1977).

Multi-Phase Flow and Heat Transfer III. Part A: Fundamentals
edited by T.N. Veziroğlu and A.E. Bergles
Elsevier Science Publishers B.V., Amsterdam, 1984 — Printed in The Netherlands

EXPERIMENTAL STUDIES OF TRANSIENT POOL BOILING HEAT TRANSFER ON A VERTICAL TUBE

Shi-Chune Yao, Associate Professor
Department of Mechanical Engineering
Carnegie-Mellon University
Pittsburgh, Pennsylvania 15213, U.S.A.

Yung Chang
Bell Laboratories
P. O. Box 1008
Indianapolis, Indiana 46206, U.S.A.

ABSTRACT

An Experimental study has been performed for the transient boiling heat transfer on a vertical tube in a Freon-113 pool. Liquid tin at a high temperature was poured into the tube to heat the surface through a temperature-controlled transient. Local heat flux and surface temperature were obtained by implicit numerical calculations from the measurements of two embedded thermocouples at different depths in the tube wall. Strong hysteresis of the boiling curve has been observed between the transient heat-up and cool-down processes. The hysteresis increases with the increasing heat-up rate and can be related with the transition time which was introduced in the quenching studies in reference [1]. The mechanisms for the transient boiling heat transfer and the boiling hysteresis are suggested.

1. INTRODUCTION

The knowledge of the transient pool boiling heat transfer is of great importance to the transient heat transfer analysis of the nuclear reactor, the steam generator, and the cryogenic cooling of superconductors. Although the heat transfer of steady pool boiling has been studied extensively in the past two decades, the understanding of transient boiling heat transfer is still insufficient to handle many practical engineering problems satisfactorily.

The heat transfer of transient heat-up processes has been studied for the nucleate boiling conditions [2,3,4,5]. It has been observed that the incipient boiling may occur at a higher superheat than that of a steady state. The transient critical heat flux can reach a value of as much as 5 to 10 times the steady state critical heat flux, especially under sub-cooled conditions. Most experiments were performed on thin narrow ribbons or horizontal wires or tubes, and the heat was provided by electric Joule heating on the testing surface as a heat flux controlled heating process. Therefore, these experiments could not be performed beyond the critical heat flux because at this condition the heat flux controlled process would cause the test section to burn out.

The temperature controlled transient has only been studied by Sakurai and Shiotsu [6] for thin ribbons at a slow transient rate of $2^{o}K/s$ at the condition of mixed boiling. There is little knowledge about the boiling heat transfer beyond the critical heat flux at a higher rate of temperature controlled transient.

This paper describes a unique method of studying the transient boiling heat transfer through all of the nucleate, transition, and film boiling regimes. Boiling occurred on a vertical tube which was situated in a Freon-113 liquid pool. Transient temperature-controlled heating was provided by pouring very hot liquid tin in the inside of the tube. Then the transient cool-down boiling followed. This method of heating had been tried before at this laboratory [7]; however, more systematical studies were performed [8] presently. An improved method was employed here to measure accurately the local transient heat flux and temperature on the tube surface. Interesting results of this unique experiment will be presented and discussed.

2. EXPERIMENTAL SET-UP

As an exploratory study, the experiment was performed in a liquid pool of Freon-113. The schematic of the set-up is shown in Figure 1. The vessel was made of pyrex-glass with 15.2 cm inside diameter and 60 cm height. The vessel was operated at atmospheric pressure. The pool temperature was maintained at saturation level with a rod heater.

The test section was made of stainless steel 304 with 28.9 mm O.D. and 19.05 mm I.D. The bottom of the tube was closed, and the tube was immersed into the pool for 25 cm. There was an inner funnel and a guide tube to transfer the molten tin into the tube without splashing. The molten tin was prepared in an oven in an inert argone gas environment. Its temperature was measured before discharging into the test section. The solidified tin after each experiment was remelted in another oven for repeated use. Then the inner wall of the test section was cleaned with a steel brush to remove any residual tin.

In order to evaluate the instantaneous local surface temperature and heat flux, two thermocouples were embedded in the wall of the test section as shown in Figure 2. The thermocouples are of ungrounded Copper-Constantan type with stainless steel sheath. The O.D. of the thermocouples are 0.813 mm. The thermocouples were located in the test section wall at the radial positions 10.3 mm and 13.8 mm from the center line, respectively. (These thermocouple locations were measured after cutting apart the test section when all the experiments were finished.) The elevation of the thermocouples was at 12.7 cm below the liquid surface and they were located 10 degrees apart circumferentially to avoid the interference of heat conduction between each other.

In embedding the thermocouples, two grooves were cut with different depths along the tube as shown in Figure 2. After the sheathed thermocouples were laid in the grooves, the grooved wall was filled with Nickro-braze-30 whose physical properties are close to those of stainless steel.

Before each experiment, the Freon-113 pool was degassed by boiling for 20 minutes using the rod heater. Boiling was also maintained at the tube surface for 20 minutes by blowing hot air through the test tube. After each experiment, the test section surface was polished with #320 sandpaper to assure the reproducibility of surface condition. Aging of the test section was also performed through repeated try-runs of the new test section before taking the first set of data.

3. DATA ANALYSIS

During the experiment the data was collected every 0.010 second. After the first 20 seconds the time interval was changed to 0.1 second during the film boiling. The instantaneous surface temperature and heat flux were calcu-

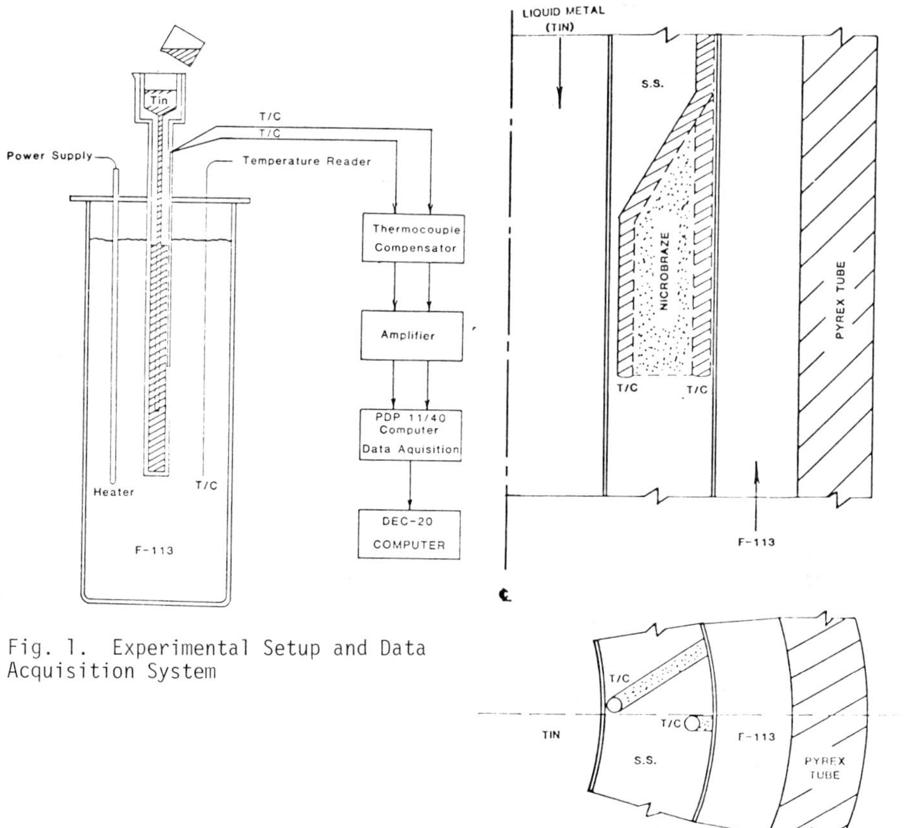

Fig. 1. Experimental Setup and Data Acquisition System

Fig. 2. Details of the Thermocouple Installations and the Schematic for the Data Reduction

lated numerically. Since the transient temperature at both of the thermocouple locations were available, the temperature profile between these two thermocouples were calculated by a one-dimensional transient conduction, using an implicit numerical scheme. Having the calculated instantaneous heat flux and temperature at the outer thermocouple location, the instantaneous temperature profile between the outer thermocouple and the surface of test section were calculated numerically. This part of the computation was an inverse heat conduction calculation.

In the data analysis, special attention was paid to the time constant of the embedded thermocouples. With the nickro-braze, the thermocouple would have an intimate contact with the surrounding solid and the contact resistance could be very small. The time response of the thermocouple was studied separately by dipping it suddenly into a molten pot of woodsmetal and observing the temperature transient of the thermocouple reading. The result is shown in Figure 3. The time constant which appeared from this curve is 21.5 ms; however, this is not the time constant which has been used for the data analysis. In

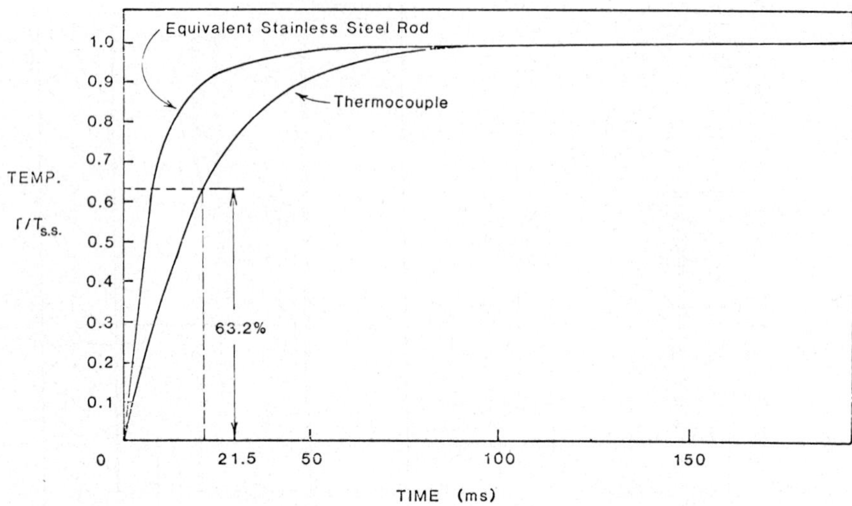

Fig. 3. Transient Temperature Responses of a Thermocouple and an Equivalent Stainless Steel Rod

the data analysis the thermocouple was considered a point located at the centerline of the thermocouple with all the surrounding space filled with stainless steel. Therefore, the real time constant which should be considered in analysis was the difference between the time response of the above test and that at the center line of a solid stainless steel rod with the diameter same as the thermocouple. For comparison, the temperature history at the center point of the equivalent stainless steel rod was also plotted in Figure 3. The difference of the time response was in the order of 15 ms. Since the error induced by this time difference in the data reduction calculation appeared to be negligible, the time delay effect was not considered.

4. RESULTS AND DISCUSSION

This experiment gives accurate local heat transfer information at transient boiling. However, due to the particular nature in heating up the test section (by pouring the hot molten tin into the tube), the surface temperature transient bears a character which is system dependent. The rate of heating-up may be varied by adjusting the initial temperature of tin poured into the test section, but its functional form cannot be varied with a definite control. Nevertheless, this novel type experiment offers us a unique opportunity to study the heat up transient boiling heat transfer through all the regimes, especially at beyond the critical heat flux temperature.

4.1. Hysteresis

During the experiment, the surface temperature of the test section rose up to a peak and cooled down. For a typical condition when the Freon-113 was at the saturation temperature and the initial liquid metal temperature (ILMT) was 977°K, the temperature histories of the two thermocouples are shown in Figure 4.

After the reduction of data the overall boiling curve of this test and its uncertainty are obtained as shown in Figure 5. In the range A to B the transient nucleate boiling occurs. The point B is the transient critical heat flux. In the range BCD, the heat flux reduces with the increase of wall temperature. This is traditionally called transition boiling regime. The heat flux drops rapidly from B to C in one second and then from C to D in 12.5 seconds. After the point D the wall temperature starts to decrease along a cool-down boiling curve. It takes 95 seconds to cool down from D to E in film boiling and passes through the quenching curve in a short duration as shown in Figure 4.

From an overview of this boiling curve, thermal hysteresis is observed. The heat-up transient critical heat flux is about twice that of the transient cool-down critical heat flux. In the heat-up process the transition boiling extends into the temperature range where the film boiling occurs during the cool-down process.

4.2. Effect of Initial Liquid Tin Temperature

The experiments were performed with different initial liquid tin temperatures. The heat-up boiling curves of various initial liquid tin temperatures are shown in Figure 6. The higher the initial liquid tin temperature is, the higher the rate of heat-up transient will be produced. It appears from Figure 6 that with the higher initial liquid tin temperature the heat-up boiling curves shift upwards with higher critical heat flux. The explanation of this phenomena will be provided in a later section.

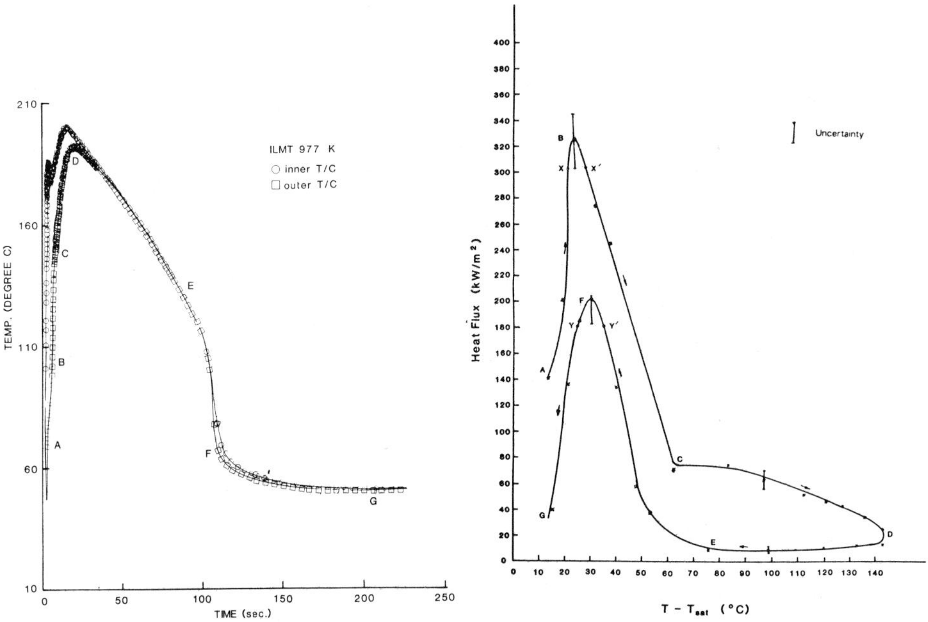

Fig. 4. Typical Temperature Histories of Thermocouples

Fig. 5. Typical Result for Heat-up and Cool-down Boiling Curves

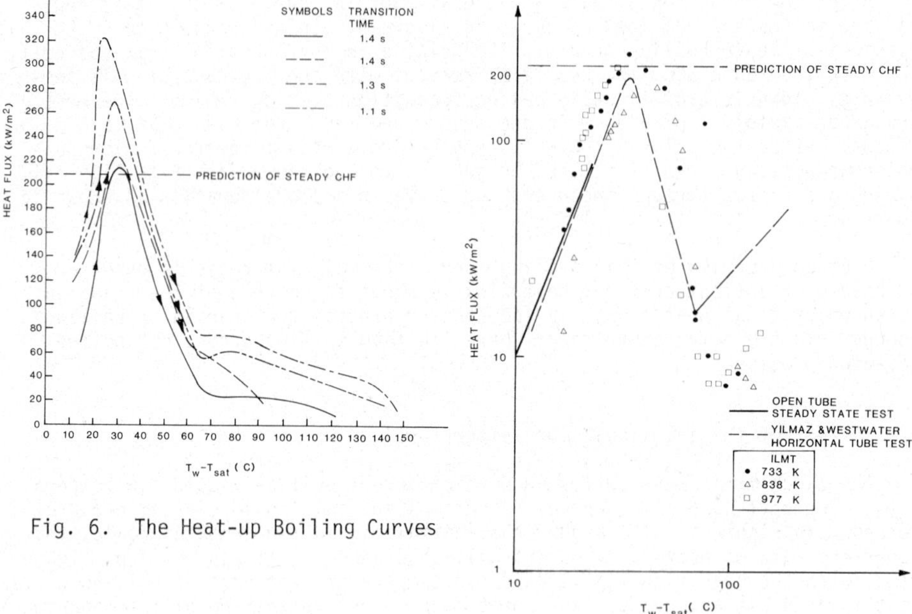

Fig. 6. The Heat-up Boiling Curves

Fig. 7. The Cool-down Boiling Heat Transfer Data

The data of the cool-down boiling curves are shown in Figure 7. For the tests with different initial liquid metal temperatures the cool-down boiling data are about the same. This is because all the cool-down processes start from a similar quasi-steady film boiling irrespective to their heat-up rates. In Figure 7, the data of cool-down nucleate boiling are compared with our steady-state pool boiling data of a vertical tube in the same Freon-113 pool. That thin wall tube has an O.D. 25.4 mm, and is heated with direct electric current. The critical heat flux is not reached in this steady-state experiment due to the limitation of electric power supply; however, the nucleate boiling data of present cool-down study appear to be very close to the steady state results. The pool boiling of a horizontal tube in Freon-113 has been studied by Yilmaz and Westwater [9]. Their boiling curve is also plotted in Figure 7. Although the orientation and the surface condition of that tube are different from those of the present study, the present cool-down data is not very far from their steady-state results. The predicted steady CHF with the convection for vertical tube geometry [10,11] is 207KW/m^2 which is very close to the present results.

4.3. Transient Boiling Mechanics

The transient nucleate boiling in a liquid pool may be contributed from two major heat transfer mechanisms. The transient process of boiling may give a different heat transfer than that of steady-state boiling. The boiling in the pool also induces transient recirculating flow in the pool which enhances the heat transfer by the effect of the convection. In the present experiments, the boiling occurs on a vertical tube. Compared with the transient boiling on

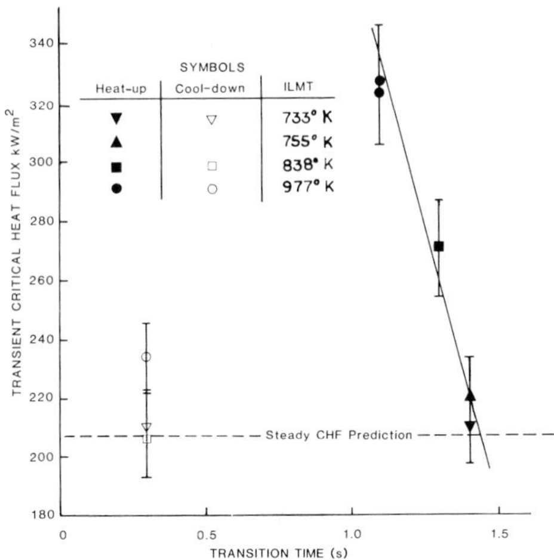

Fig. 8. The Transient Critical Heat Flux vs. the Transition Time

horizontal plates, it is likely that the transient boiling will influence the heat transfer on a vertical tube more significantly due to the stronger induced transient flow in the pool. The boiling on the vertical tube below a measuring location will seriously affect the fluid flow near the measuring point. It is important to point out that all the reported transient pool boiling heat transfer in literature includes both of these two mechanisms and cannot be separated easily due to their coupled relationship. Therefore, the boiling heat transfer on a vertical tube is a strong function of the system and the location of measurement. The reported results of this exploratory study will be considered as typical and local information.

A characteristic time scale, called the transition time in [1], for the transient critical heat flux in a temperature controlled system can be defined as the time for the boiling to transpass the 90 percent of the critical heat flux. As indicated in Figure 5, this is the time to travel between x to x′ or y′ to y. This time will be infinite for a steady-state boiling process. This transition time denotes the characteristic time for the boiling mechanism to shift from nucleate boiling to transition boiling in a heat-up transition, or from transition boiling to nucleate boiling in a cool-down transient.

The transient critical heat flux of the present experimental studies are plotted in Figure 8 with respect to the transition time. The initial liquid tin temperature of each test is also denoted in this Figure. For a heat-up boiling process the transition time is about a constant of 1.4 second when the initial liquid tin temperature is below 755°K. The transition time becomes shorter when the initial liquid tin temperature is higher than 755°K. On the other hand, all the cool-down boiling processes have about the same transition time of 0.3 second irrespective to their initial liquid tin temperature. This is because all the cool-down processes start from a quasi-steady film boiling which is not influenced by the initial heating characteristics of the transients.

As pointed out previously, for the cool-down processes the transition times are about 0.3 second and the critical heat flux is very close to the steady-state critical heat flux, (207 K/M^2). On the other hand, the transient heat-up critical heat flux are also at about 207 KW/m^2 when the transition time is at 1.4 seconds. If the heat-up transition time is shorter than 1.4 seconds, the transient critical heat flux increases with the decreasing of the transition time. Therefore, it can be temporarily concluded that the natural transition time for the heat-up critical heat flux to maintain at its steady state value is about 1.4 seconds in the present study. A minimum time of 1.4 seconds is required to change naturally from the nucleate boiling into transition boiling. A faster heat-up process with a shorter transition time will cause a higher transient critical heat flux.

It is interesting to notice that the natural transition time during heat-up transient is about 1.4 seconds; however, during the cool-down process a transition time of 0.3 second gives the critical heat flux about the same value as the steady-state. It is possible that these two transient processes may have very different mechanisms. The transient heat-up process passes through the liquid-riched nucleate boiling to the vapor-riched transition boiling. The rapid heat-up from nucleate boiling may leave the liquid residual on the heating surface and causes higher critical heat flux due to the extra storage of liquid on surface to supply the evaporation. At this time, the liquid in the pool is already unable to get down to the surface due to the hydrodynamic limitation posed by the departing vapor. To evaporate this liquid residue would require a rather long duration of time, the transition time for heat-up process is as long as 1.4 seconds in this study. On the other hand, the cool-down process passes through the vapor-riched transition boiling to the liquid-riched nucleate boiling. Rapid cool-down from transition boiling to nucleate boiling may leave more vapor residue on the heating surface which makes it hard for the liquid to come to the surface immediately. Therefore, the resulting transient critical heat flux will be lower than that of a steady-state process. In fact, the lower CHF at very rapid cooling process has been observed in [1]. Since the removal of the vapor residue on the surface is a much easier process than the evaporation of the liquid residue in a heat-up process, the cool-down transition time of 0.3 second in the present study may not be short enough to cause a visible reduction of the transient CHF.

The transition boiling is traditionally defined as the boiling regime in which the heat flux decreases with the increase of wall temperature. Therefore, a negative slope is observed on the boiling curve. In the present study with the initial liquid tin temperature 733°K the transition boiling curve is close to that of the steady-state. With higher initial liquid metal temperature, the transition boiling curves show two distinctly different slopes. In a heat-up process, the transition boiling curve with a severa slope near the critical heat flux covers about 1 second of duration. The boiling in this subregime is assumed to be the nucleate-boiling-like transition boiling where most of the surface is rather wet. It takes about one second to change from the wet surface subregime to the regime with a rather dry surface. The other part of the transient transition boiling is characterized by a smaller slope. This film-boiling-like transition boiling covers a duration more than 5 seconds. The heat flux is higher than that of steady film boiling at the same wall superheat. The higher heat flux may be attributed to the extra disturbance of the liquid and vapor near the surface induced by the agitation of the pool during the transient. The details of these mechanisms are not fully identified in this exploratory experiment.

5. CONCLUSION

1. The boiling curve of Freon-113 on a vertical tube show strong hysteresis between the processes of transient heat-up and cool-down. The hysteresis

increases with the increasing heat-up rate.

2. The transition time (which is defined as the time required for transient boiling to pass the 90% of the critical heat flux) characterizes the behavior of the transient boiling. In the heat-up transient of this study, if the transition time is shorter than 1.4 seconds the transient critical heat flux increases as the transition time decreases. The increasing of CHF is possibly due to the effect of liquid residue on the surface. In cool-down transient, all the observed transition times are about 0.3 second. The CHF at this condition is very close to that of steady-state process. When the transition time is very short the transient critical heat flux at cool-down process will decrease due to the effect of the vapor residue on the heating surface. The natural period of transition time during heat-up could be much longer than that during cool-down because the time required to evaporate the liquid residue during heat-up is much longer than the time required to remove the vapor residue during a cool-down transient.

3. The transition boiling curve at transient heat-up process shows two different slopes. The transient transition boiling curve extends into the high temperature range where the film boiling usually occurs at steady boiling conditions.

ACKNOWLEDGEMENT

The support for this research from the National Science Foundation, Heat Transfer Program under the agreement ENG-7818333 is greatly appreciated by the authors.

REFERENCES

1. PeyayopanKul, W. and Westwater, J.W., "Evaluation of the Unsteady Quenching Method for Determining Boiling Curve," International Journal of Heat Mass Transfer, Vol. 21, pp. 1437-1445, 1978.

2. Rosenthal, M.W. and Miller, R.L., "An Experiment Study of Transient Boiling," U.S. Atomic Energy Commission, Report ORNL-2294, 1957.

3. Johnson, H.A., "Transient Boiling Heat Transfer to Water," 4th Int. Heat Transfer, Conf., Paris, Paper B3.1, 1970.

4. Hall, W.B. and Harrison, W.G., "Transient Boiling of Water at Atmospheric Pressure," Proc. 3rd Int. Heat Transfer Conf., Chicago, p. 186, 1966.

5. Sakurai, A. and Shiotsu, M., "Transient Pool Boiling Heat Transfer," Part 1,2, J. Heat Transfer, ASME, 99, pp. 547-560, Nov. 1977.

6. Sakurai, A. and Shiotsu, M., "Studies on Temperature-Controlled Pool-Boiling Heat Transfer," Proc. of 5th Int. Heat Transfer Conf., Heat Transfer 1974, paper No. B.3.

7. Baum, A., "A Study of Transition and Film Boiling from Vertical Surfaces," Ph.D. Thesis, Carnegie-Mellon University, Pittsburgh, 1977.

8. Chang, Y., "The Effects of Confinement and Transient to Pool Boiling on Vertical Tubes," Ph.D. Thesis, Carnegie-Mellon University, Pittsburgh, 1981.

9. Yilmaz, S. and Westwater, J.W., "Effect of Velocity on Heat Transfer to Boiling Freon-113," Journal of Heat Transfer, Trans ASME, Vol. 102, pp. 26-31, Feb. 1980.

10. Zuber, N., "On the Stability of Heat Transfer," Trans. ASME, Vol. 80, pp. 711-720.

11. Lienhard, J.G. and Dhir, V.K., "Hydrodynamic Prediction of Peak Pool-Boiling Heat Fluxes from Finite Bodies," J. of Heat Transfer, Trans. ASME, Sec. C., Vol. 95, pp. 152-158, 1973.

Multi-Phase Flow and Heat Transfer III. Part A: Fundamentals
edited by T.N. Veziroğlu and A.E. Bergles
Elsevier Science Publishers B.V., Amsterdam, 1984 — Printed in The Netherlands

VOID DISTRIBUTION AND VAPOR HOLDUP IN A VESSEL WITH AN IMMERSED HORIZONTAL HEATER

A.M.C. Chan, M. Shoukri and M. Martins
Research Division, Ontario Hydro
800 Kipling Ave, Toronto, Ontario
Canada M8Z 5S4

ABSTRACT

Void distribution and vapor holdup in nucleate pool boiling has been investigated experimentally. Results obtained using a single horizontal heater in a vessel were presented. Different heat fluxes were used. The local void fraction was found to go through a minimum when it was traversed vertically from the heater to the water free surface. The entrained void or vapor holdup in the vessel can be obtained directly from a LVDT-float unit or indirectly by integrating the void distribution curve. Some initial efforts in data reduction were also attempted.

1. INTRODUCTION

Void entrainment and heat transfer characteristics in nucleate pool boiling situations with horizontal tubes and tube bundles are of direct interest to the operation and control of kettle reboilers as well as a large variety of heat exchange equipment.

The analysis of void entrainment involves estimating the amount of steam contained in the two-phase coolant in the shell-side of a reboiler or steam generator. The amount of entrained void will affect the coolant level and hence the operation of the reboiler. In the event of an upset condition, such as the loss of heat source in the tube-side, the void in the shell-side will collapse and the coolant level change rapidly. This rapid change in coolant level must be predictable to be controlled. The change in the coolant level may result in some of the upper tubes in the reboiler being left exposed, ie, above the shell-side coolant level. These exposed tubes are cooled by the rising steam and to some extent, by the splashing coolant from below. The heat transfer characteristics are quite different from the tubes below the coolant surface.

Also, understanding of the basic characteristics of nucleate boiling within horizontal tube bundles is of interest, especially the effect of boiling around neighboring tubes in a tube bundle geometry. In fact, enhancement in heat transfer in the upper tubes of a bundle has been observed/1/. However, the mechanisms causing that need to be identified and correlation to describe this effect need to be developed.

An extensive experimental program is being conducted in the Research Division of Ontario Hydro to investigate systematically the void entrainment and heat transfer characteristics in a simulated reboiler geometry. This paper presents some initial results using a single immersed heater. Only the void entrainment characteristics are reported. The experimental facilities are described and some initial efforts in data reduction are attempted.

2. EXPERIMENTAL FACILITY

2.1. Experimental Setup

A schematic diagram of the experimental facilities is shown in Figure 1. The setup consists of the following components:

(i) The boiling vessel - The test vessel is a rectangular tank measured 546 mm long, 184 mm wide and 745 mm high. Two Lexan windows (305 mm x 610 mm) were fitted to the front and back of the tank for visual observation. The vessel is large enough to house up to 100 heaters in a rectangular pitch. Only one heater was used in the present experiment. It was mounted horizontally 270 mm from the bottom of the tank. The heating element consisted of a specially designed high heat flux tubular heater 12.5 mm O.D. with 520 mm heated length and an overall length of 686 mm. An aluminum tube having proper tolerance was heated and press fitted over the tubular heater. Thin axial slots were machined on the surface of the aluminum tube to house the thermocouples for surface temperature measurements. A thin walled stainless steel tube (19.05 mm O.D.) was then swaged over the aluminum tube with the thermocouples. The process was carefully performed to ensure good thermal contact and thus good surface temperature measurements. Two auxiliary heaters were located at the bottom of the tank to maintain a uniform water temperature.

(ii) The condenser - A water cooled condenser with a designed heat removal capacity of about 70 kW was used.

(iii) The collection tank - A cylindrical aluminum vessel (168 mm O.D. x 910 mm long) was used to collect the condensate from the condenser and feed it back to the boiling vessel by gravity.

(iv) Power supply - The heater was heated by a 600 volt, 200 Amp ac power supply. The power input to the heater was controlled using a manual variac.

2.2. Instrumentation

The parameters measured during the experiments included:

(i) void distribution in the vessel above the heater; (ii) average void or vapor holdup in the vessel; (iii) power input to

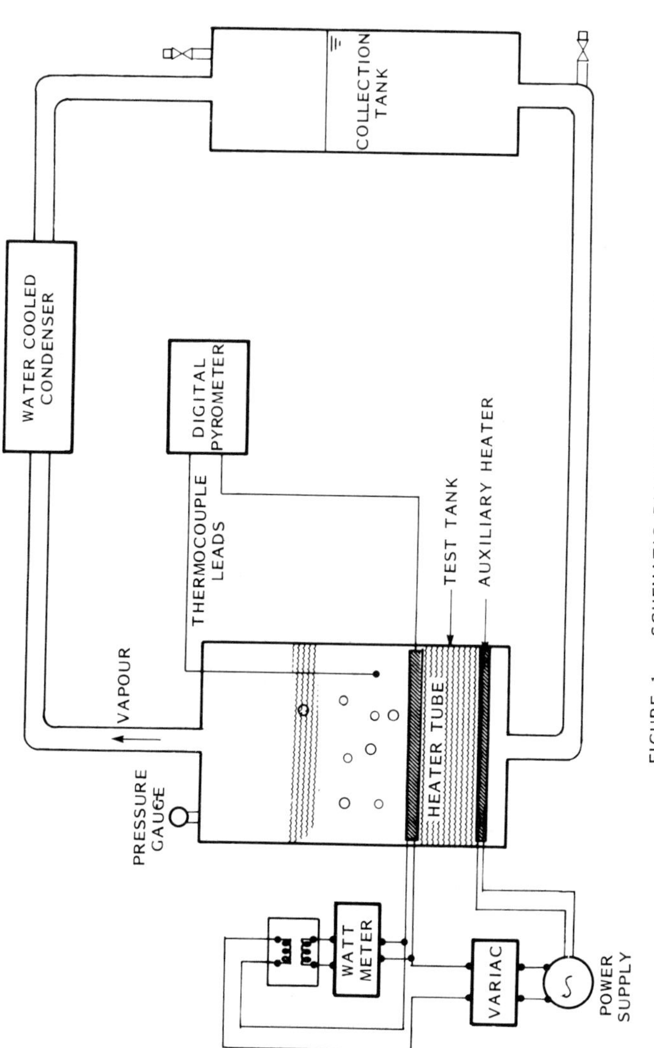

FIGURE 1 SCHEMATIC DIAGRAM OF TEST RIG

the heater; (iv) heater surface temperatures and (v) water temperature in the tank. The instrumentation used are described below.

(i) Void distribution: A traversing gamma densitometer was used to obtain the local void fraction measurements above the heater. Details of the design and operation of gamma densitometers can be found in/2/. The system used in the present experiment is shown schematically in Figure 2. A 50 mCi Cobalt-57 source (12.5 mm O.D. and 50.8 mm long) was used. The energies of the dominant gammas are 122 keV (85.2%) and 136.5 keV (11.1%). The gamma beam was collimated using steel blocks providing a rectangular beam 50.8 mm wide and 12.5 mm high. The transmitted beam was detected by a large (762 mm dia and 762 mm long) thallium activated sodium iodine, NaI(Tℓ) crystal. The transmitted gammas were then processed and counted using a standard pulse counting system/2/. The radioactive source and the scintillator were mounted in opposite sides of the boiling vessel (Figure 2) on a platform which could be moved in the vertical direction by a variable speed motor. Provisions were also made to traverse the densitometer in the horizontal direction. However, this was not done in the present study. Instead, the movement of the densitometer was confined to a vertical plane in the middle of the vessel.

(ii) Average void or vapor holdup: This was obtained by integrating the void distribution curve in a vertical plane, assuming the same void distribution curve along the length of the heater or tank. As an independent measurement, a LVDT-float system was used (Figure 2). The float was used to follow the movement of the water free surface in the vessel. It was attached to the armature of a linear variable differential transformer (LVDT). Depending on the armature displacement in the core, the LVDT produces a digital readout which was calibrated to yield the position of the water free surface. The average void or amount of entrained vapor was thus determined.

(iii) Power input: The power input to the heater was monitored using a Weston Model 432 Wattmeter. It was connected to the load through a current transformer as shown in Figure 1.

(iv) Heater surface temperature: The heater surface temperature was obtained using exposed junction stainless steel sheath type-K thermocouples (0.254 mm O.D.). As many as sixteen thermocouples were used to obtain a detailed heater surface temperature distribution both axially and circumferentially.

(v) Water temperature: Two stainless steel sheath, grounded junction type-K thermocouples were used to measure the water temperature in the tank. The thermocouples could be traversed to take measurements at different elevations in the tank.

3. EXPERIMENTAL PROCEDURES

The experiments were performed using distilled water as the working fluid. When the boiling tank was first filled with water, it was allowed to boil for at least two hours in order to degass the system. The collection tank was opened to the atmosphere (Figure 1). Therefore, the operating pressure in the boiling vessel was close to atmospheric.

To obtain the void entrainment and heat transfer data, the following procedures were followed:

(i) The power to the heater and the auxiliarly heaters was switched on to boil the water in the vessel.

(ii) After the water was boiled for some time, the power was switched off and the saturated water level was recorded after all the void in the system has been collapsed.

(iii) The power input was then switched on again and adjusted to the required level.

(iv) After steady state operation was achieved, heater surface temperatures were recorded using a rotary switch and a digital pyrometer.

(v) The transmitted gamma fluxes at different elevations were then obtained by traversing the gamma densitometer. A counting period of about two minutes is sufficient to get good statistics at each location.

(vi) The water temperature in the vessel was also recorded.

(vii) The power input to the heater was then changed and steps (iv) to (vi) were repeated.

4. RESULTS AND DISCUSSION

4.1. Void Distribution

Local void fractions were obtained from the transmitted gamma flux measurements. Typical void distribution curves for different heat fluxes are shown in Figure 3. Plotted are the local void fraction at different elevations from the top of the heater to the water free surface. The saturated single phase water level was 254 mm above the heater.

From Figure 3, it can be seen that all the void curves go through a minimum. This minimum tends to appear at lower elevations as the heat flux increases. The void fraction then increases rapidly as the water free surface is approached except for very low heat fluxes (eg, $Q/A=5.0\ w/cm^2$). These results were also evident from visual observations as shown in Figure 4. The pictures in Figure 4 were obtained for two

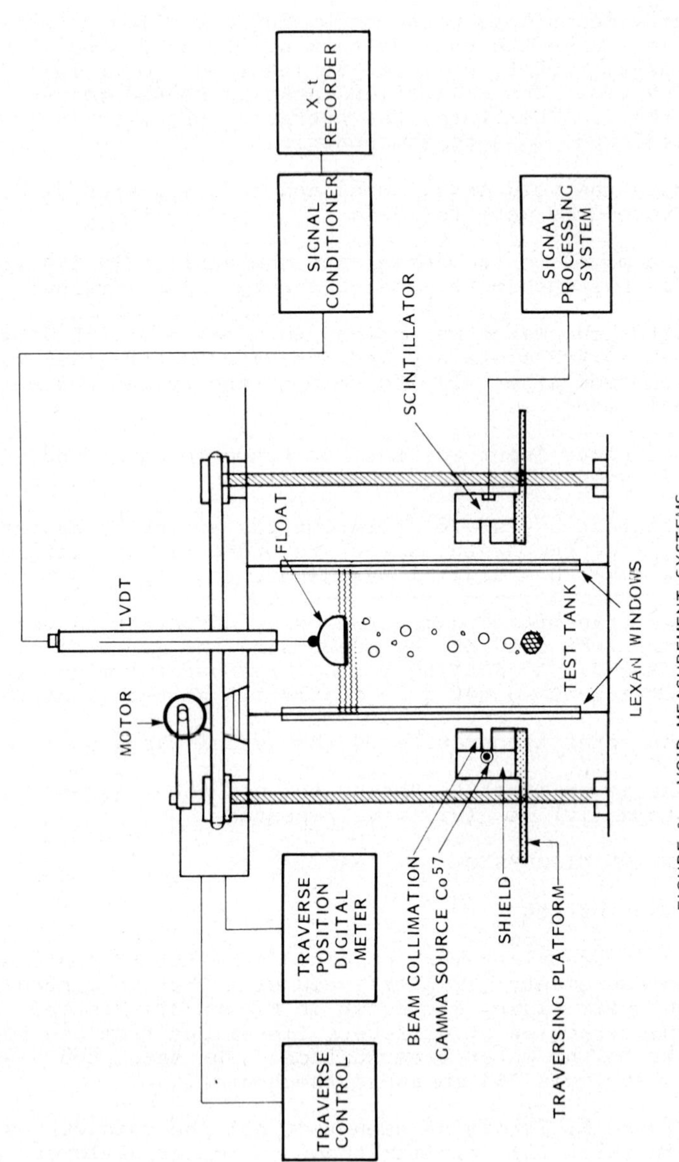

FIGURE 2 VOID MEASUREMENT SYSTEMS
(i) GAMMA DENSITOMETER
(ii) LVDT - FLOAT UNIT

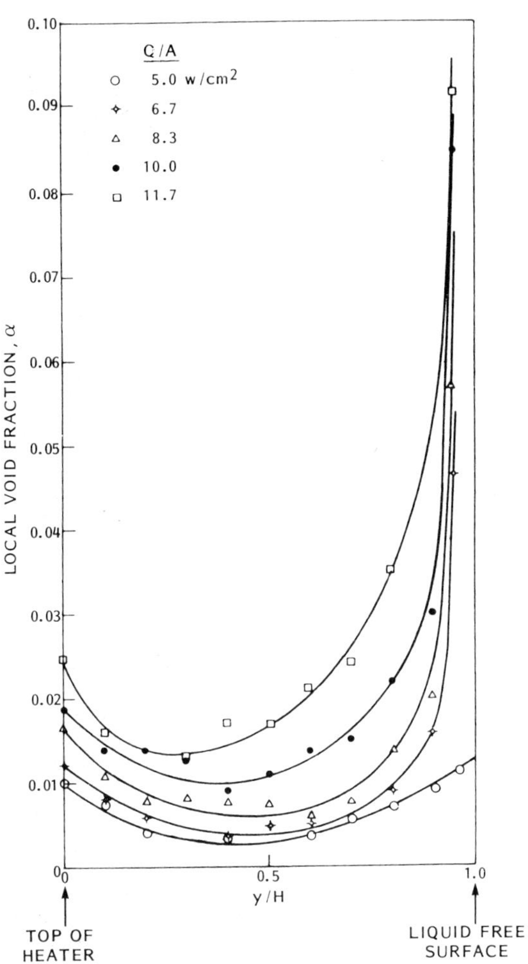

FIGURE 3 VOID DISTRIBUTION ABOVE THE HEATER

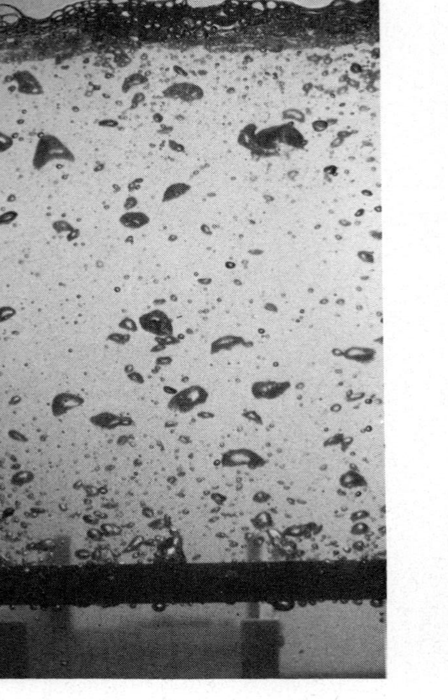

(a) $Q/A = 4\ w/cm^2$

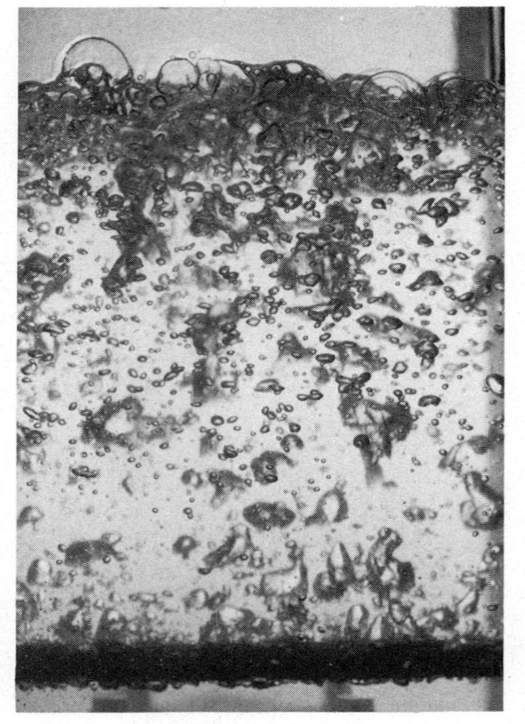

(b) $Q/A = 12\ w/cm^2$

FIGURE 4
BOILING CHARACTERISTICS

different heat fluxes. The decrease in void fraction in the central regions of the tank can be clearly seen in both cases.

Since there is no condensation in the liquid bulk, by vapor continuity, a minimum in the void distribution curve implies a maximum in vapor velocity. Physically, when bubbles detach from the heater surface and rise toward the free surface, their velocity increases due to buoyancy and the recirculating flow of the continuous phase. In general,

$$v_b \simeq v_\infty + v_{flow} \tag{1}$$

where v_∞ is the bubble terminal velocity in a stagnant mixture. The effect of coalescence of smaller bubbles to form larger ones can be accounted for via the terminal velocity.

It should be noted that a bubble can achieve its terminal velocity at a relatively short distance, the vapor rise velocity is thus more dependent on the magnitude of the recirculation flow (v_{flow}) in the vessel. In general, the recirculation flow in a vessel is stronger in the central region. Its magnitude decreases as it approaches the free surface. Therefore, the vapor rise velocity will also achieve a maximum in the central region and then decreases towards the free surface. This will be discussed later. The rapid increase in local void fraction near the free surface for higher heat fluxes (Figure 3) is believed to be a surface effect. When large number of bubbles reach the free surface, if the total interfacial area of the bubbles is greater than the free surface area, a blockage is created. A high void layer is thus formed underneath the free surface (Figure 4b). This layer of bubbles becomes thicker for higher heat fluxes and reduces the effective flow path of the recirculating flow. The strength of the recirculation is thus reduced. This will be discussed further in Section 4.3.

4.2 Vapor Holdup

The entrained void or vapor holdup in the vessel was obtained directly from the LVDT-float unit and indirectly by integrating the void distribution curve shown in Figure 3. The results are compared in Figure 5. Good agreement was obtained. This implies that either technique can be used to predict the amount of vapor holdup. It should be noted that the agreement tends to be better at lower heat flux. The discrepancy between the two measurement techniques in the higher heat flux region can be attributed to two reasons:

(i) The local void fraction is not well defined close to the free surface at high heat flux (Figure 3).

(ii) For the LVDT-float unit, the displacement of the float depends on the density of the fluid it is floating on. For higher heat flux, the mixture density in the surface region tends to be lower because of the high void layer, thus the float will tend to sink a little deeper with respect to the true water surface.

4.3. Vapor Rise Velocity in the Vessel

As discussed previously, the vapor rise velocity in the vessel changes as the bubbles rise to the free surface. It is given by Eqn (1). In order to determine v_b, we need to know v_∞ and v_{flow}. A lot of work has been done on the terminal velocity of single bubbles and a number of correlations are available/3/. The circulation velocity, v_{flow}, however, is not readily known. It is believed to be a function of local void, heat flux, as well as the vessel geometry. If we can assume the circulation to be two-dimensional in a vertical plane perpendicular to the heater, the flow characteristics in the vessel will be very close to the well known gas bubble driven circulation flows or recirculation flows in gas-agitated reactors/4,5/ for which numerical solutions are possible. However, direct local measurements of v_{flow} are difficult, especially in violent boiling situations.

The local vapor rise velocity can also be determined indirectly from the void fraction measurements in the present study. By conservation of vapor, it can be shown that

$$v_b(y) = \frac{Q}{\rho_g \Delta h_{fg} A_s} \frac{1}{\alpha} \tag{2}$$

where Q is the input power, ρ_g is the vapor density, Δh_{fg} is the heat of vaporization, A_s is the cross-sectional area of the vessel and α is the local void fraction. Using the void fraction results shown in Figure 4, the vapor rise velocity can be determined. They are plotted in Figure 6. As discussed previously, the vapor rise velocity increases from the top of the heater, reaches a maximum and decreases towards the free surface. It is interesting to note that the velocity decreases as the input power increases. An obvious explanation is the vapor blockage created near the water free surface at higher heat flux as discussed in Section 4.1. This will increase the flow resistance and reduces the effective flow path of the recirculating flow.

It should be noted that the vapor rise velocity as shown in Figure 6 is many times larger than the bubble rise velocity in stagnant liquid. The difference is expected to be the velocity of the recirculating flow in the vessel, v_{flow}. Therefore, using the v_b as derived and assuming a realistic v_∞ based on an average bubble size, we can determine approximately the strength of the recirculation flow in the vessel.

5. CONCLUSION

Void distribution and vapor holdup in a vessel has been investigated experimentally. Results obtained using a single horizontal heater were presented. It was found that the local void fraction decreases in the vertical direction as bubbles detach and rise from the heater surface. The trend is reversed at higher elevations, resulting in a minimum in the void distribution curve. The local void fraction then increases rapidly as it approaches the water free surface. This interesting observation can be explained in terms of changes in vapor rise velocity in the vessel. By conservation of mass, an increase in vapor rise velocity implies a decrease in local void fraction.

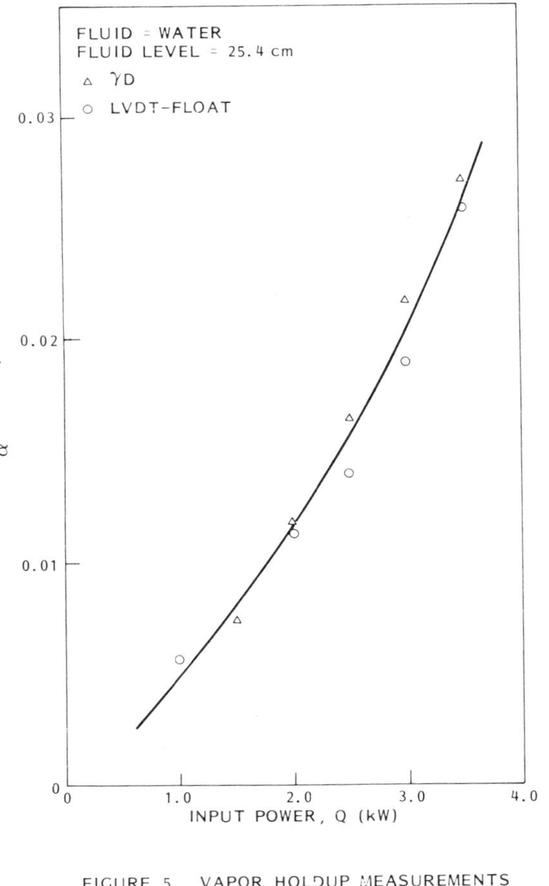

FIGURE 5 VAPOR HOLDUP MEASUREMENTS

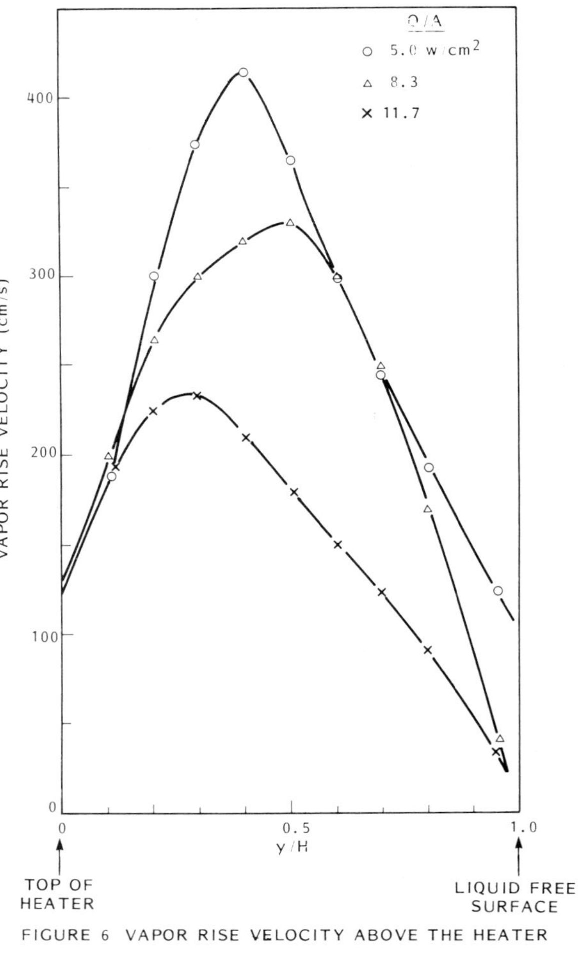

FIGURE 6 VAPOR RISE VELOCITY ABOVE THE HEATER

The vapor holdup in the vessel can be obtained by integrating the void distribution curve. The results were found to be in good agreement with independent measurements obtained using a LVDT-float unit.

The vapor rise velocity in the vessel was obtained from the local void fraction measurements. It was found to decrease as the input power was increased. From the vapor rise velocity, it is also possible to estimate the recirculation flow velocity in the vessel if a realistic terminal velocity can be assumed.

REFERENCES

1. K. Cornwell and R.B. Schuller, "A Study of Boiling Outside a Tube Bundle Using High Speed Photography", Int. J. Heat Mass Transfer, Vol 25, pp 683-690, 1982.

2. A.M.C. Chan and S. Banerjee, "Design Aspects of Gamma Densitometer For Void Fraction Measurements in Small Scale Two-Phase Flow", Nuclear Instruments and Methods, Vol 190, pp 135-148, 1981.

3. G.B. Wallis, One-Dimensional Two-Phase Flow, McGraw Hill, 1969, Chapter 9.

4. J.H. Grevet, J. Szekely and N. El-Kaddah, "An Experimental and Theoretical Study of Gas Bubble Driven Circulation Systems", Int. J. Heat Mass Transfer, Vol 25, pp 487-497, 1982.

5. T. Deb Roy, A.K. Majumdar and D.B. Spalding, "Numerical Prediction of Recirculation Flows with Free Convection Encountered in Gas-agitated Reactors", Appl Math Modelling, Vol 2, pp 146-150, 1978.

Multi-Phase Flow and Heat Transfer III. Part A: Fundamentals
edited by T.N. Veziroğlu and A.E. Bergles
Elsevier Science Publishers B.V., Amsterdam, 1984 — Printed in The Netherlands

AN INVESTIGATION OF SUBCOOLED FLOW BOILING

Z. H. Lin
Xian Jiao-Tong University
Xian, Shaanxi Province
The People's Republic of China

T. N. Veziroğlu
Clean Energy Research Institute
University of Miami
Coral Gables, Florida 33124, U.S.A.

ABSTRACT

The experimental data for low pressure subcooled flow boiling of Freon-11, under steady state conditions, are reported. A new relationship for the heat transfer coeffcient, under the above conditions, is obtained from the experimental data. The proposed relationship is compared with the experimental data for the low pressure subcooled boiling of Freon-11, and the existing Chen and Pokhalov correlations. The new relationship is found to give better agreement, especially at the limits. It is also easier to use.

NOMENCLATURE

Latin Letters

c_p Specific Heat at Constant Pressure; J/kg.K

d Inside Diameter; m

H Enthalpy; J/kg, kJ/kg

Δh_v Latent Heat of Vaporization; J/kg

m Constant

P Pressure; bar

q Heat Flux Density; W/m^2

T Temperature; K

u Velocity; m/s

Greek Letters

α Heat Transfer Coefficient; $W/m^2 \cdot K$

λ Thermal Conductivity; $W/m \cdot K$

μ Dynamic Viscosity; $kg/m \cdot s$

ρ Density; kg/m^3

Subscripts

b	Bulk Boiling
c	Critical
in	Inlet
ℓ	Saturated Liquid
o	Fluid at the Point where Wall Temperature is Equal to the Fluid Saturation Temperature
s	Saturation
v	Vapor
w	Wall

1. INTRODUCTION

When a subcooled liquid flows inside a heated tube, at some point along the tube nucleate boiling will commence from cavities at the wall, in spite of the fact that the bulk of the fluid has not reached its saturation temperature. Bubbles at first grow due to heat and mass transfer from the superheated boundary layer near the wall, later their volume grows and they move beyond the superheated boundary layer, and eventually they condense. This phenomenon is called subcooled flow boiling.

Subcooled flow boiling range may be divided into two: the partial subcooled boiling region and fully developed subcooled boiling region. In the partial boiling region, once boiling has been initiated only comparatively few nucleation sites are operating. As the surface temperature increases, the number of bubble sites also increases and thus the area for single phase heat transfer decreases. In this region the heat is transferred by both forced convection single phase flow and subcooled boiling simultaneously. As the fluid moves further downstream and the wall temperature increases, the whole surface becomes covered by bubble sites, boiling becomes fully developed, and the single phase component of the heat transfer becomes negligible. This region is called the fully developed subcooled boiling region. As the fluid bulk temperature reaches the saturation temperature, the fully developed nucleate boiling region starts.

Subcooled flow boiling heat transfer is very important for the design of heat generating systems with high power densities. This knowledge is widely used in oil fired boilers, chemical reactors, nuclear reactors and refrigeration systems.

Up to the present, several studies [1-5]* have been reported in the literature about the subcooled flow boiling heat transfer. For predicting the heat transfer coefficient, a number of empirical correlations have been proposed. However, most of the studies were conducted with water. Experimental data for subcooled flow boiling of Freon-11 are quite rare. Most of the existing correlations are

*Numbers in square brackets refer to references at the end of the paper.

valid for water only. A few correlations, such as the Chen correlation [6] and the Pokhvalov correlation [7], have been suggested to be valid for fluids other than water, as well. However, their validity for Freon-11 had not been examined.

The objective of the present investigation was to study the subcooled flow boiling of Freon-11, compare the data with the correlations given in the literature, and propose a new correlation.

2. EXPERIMENTS

A schematic diagram of the experimental system is shown in Figure 1. Test fluid, Freon-11, is supplied from the main tank pressurized by nitrogen gas. A cooling unit provides an inlet temperature range of 233 K to 385 K, with a control accuracy of 1.0 K. The test section is an electrically heated, vertical, smooth Nichrome tube, having an inside diameter of 7.49 mm, wall thickness of 1.02 mm, and a heated length of 605 mm. Following the test section is a recovery system consisting of a condenser and a collector tank. Condensed liquid is stored in the collector tank under a constant pressure. In order to eliminate the heat losses, a radiation guard with a compensating heater is built around the heated tube, and the whole assembly is then housed in a vacuum jacket.

The experimental system has been instrumented for temperature, pressure, flow rate and heat input measurements. Temperature measurements are made by standard copper-constantan thermocouples. Pressures are measured at five different points by Bourdon-type pressure gauges. A strain gauge-type pressure transducer is used to record pressure variations at the heater inlet. Flow rates are measured by a calibrated rotameter and a differential pressure transducer. Outputs of the pressure transducers and thermocouples are fed into a four-channel amplifier-recorder unit for recordings. Heat input is obtained by measuring the current and the voltage drop across the heater.

During the test, in order to obtain steady flow, the surge tank was kept full of Freon-11. For a given heat input different sets of experiments corresponding to various flow rates were conducted. Following each adjustment of flow rate, measurements of temperatures, pressures, flow rate and heat input were taken. From the recorded outside wall temperatures of the heater, the known heat source strength and the heater wall properties, the inside wall temperatures were calculated. Using the flow rate, temperature and pressure data, the fluid temperatures and properties were calculated as a function of location within the heater. Then, the local wall superheats, heat fluxes and heat transfer coefficients were computed. The range of flow parameters covered were as follows: heat flux density 0.69×10^4 - $2.8 \times 10^4 W/m^2$; mass flow rate 0.97×10^{-3} - $20.17 \times 10^{-3} kg/s$; 1.7 - 2.7 bar; fluid inlet temperature 293-297 K.

3. RESULTS AND DISCUSSION

Figure 2 presents a typical plot of heat flux versus wall

superheat for three different mass fluxes. As can be seen from the figure, at low heat fluxes the wall superheat strongly depends on the mass flux--increasing with decrease in mass flux. This corresponds to the forced convection and the partial subcooled boiling regions. As the heat flux increases, the curves corresponding to three different mass fluxes join, indicating that in this high heat region the mass flux has little or no effect on the wall superheat. This is the fully developed subcooled boiling region.

Figure 3 presents heat transfer versus fluid enthalpy for three different mass flux and heat flux combinations. The circles indicate the experimental points. H_o is the liquid enthalpy when the wall temperature is equal to the liquid saturation temperature, and H_s the saturated liquid enthalpy. An examination of Figures 3(a), 3(b) and 3(c) show that between H_o and H_s (i.e., the subcooled boiling region) heat transfer coefficient increases as enthalpy increases; and for enthalpies greater than H_s, the heat transfer coefficient is more or less constant. Also, as the heat flux to mass flux ratio (q/G) increases, the heat transfer coefficient increases.

In the pure liquid region (i.e., enthalpy $\leq H_o$), the applicable heat transfer relationship is the Dittus-Boelter correlation, via.,

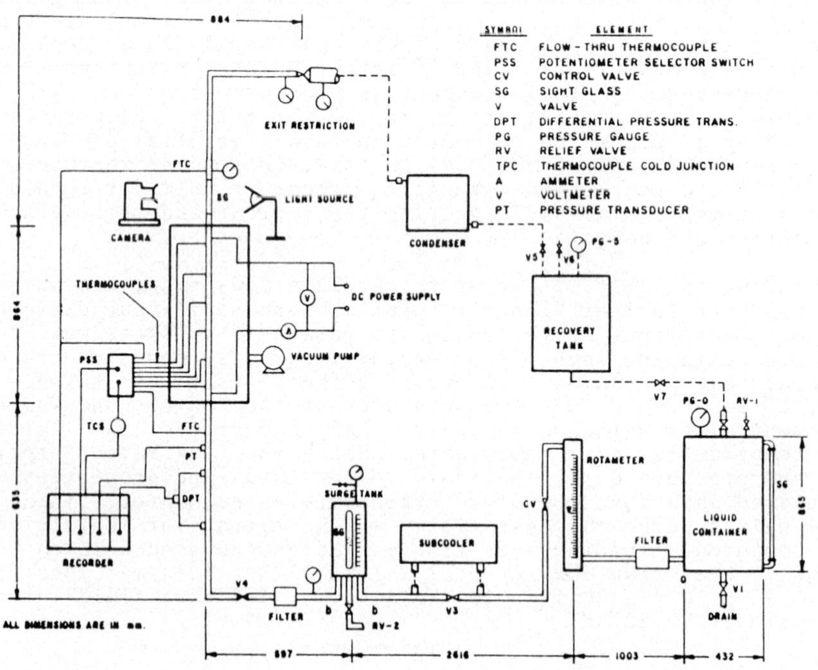

Fig. 1 - Schematic Diagram of Experimental System.

$$\frac{\alpha d}{\lambda} = 0.023(\frac{du\rho}{\mu})^{0.8}(\frac{\mu c_p}{\lambda})^{0.4} \quad (1)$$

For the saturated boiling region (i.e., enthalpy $\geq H_s$), the authors obtained the following correlation, using the experimental results and the method outlined in an earlier paper [8]:

$$\frac{\alpha d}{\lambda_\ell} = 190(\frac{P}{P_c})^{0.25}(\frac{dq}{\Delta h_v \mu_\ell})^{0.8} \quad (2)$$

Calculations show that equation (2) agrees with the current data with an accuracy better than ±15%. J. G. Collier [9] has suggested that the heat transfer coefficient for the fully developed boiling (or the saturated boiling) region is not markedly different from that obtained for the saturated pool boiling. Such a pool boiling heat transfer coefficient relationship has been proposed by Stephan and Abdelsalam [10].

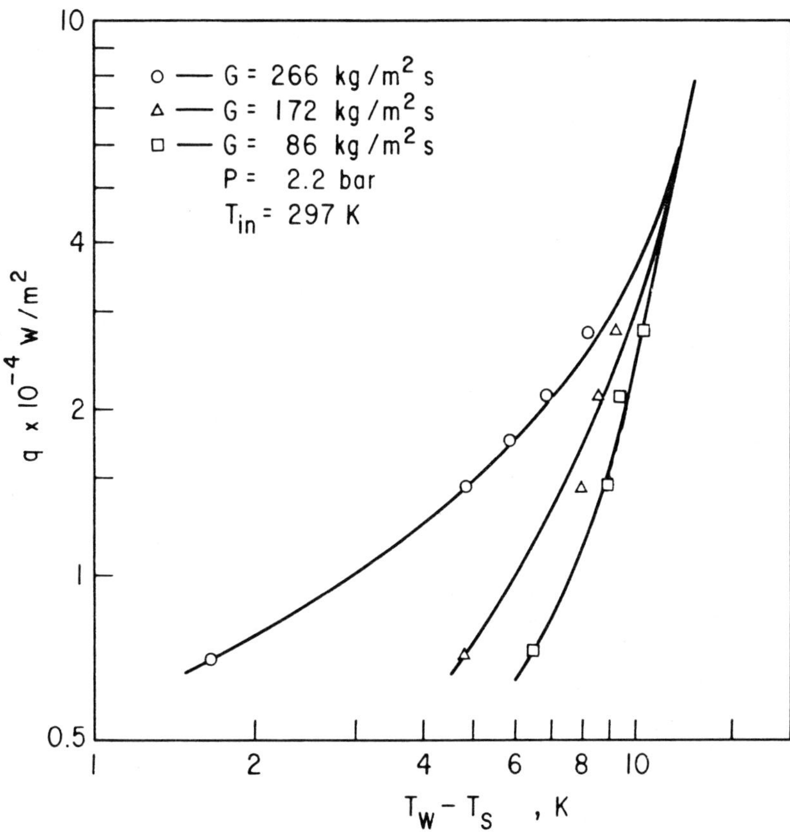

Fig. 2 - Local Heat Flux versus Wall Superheat for Three Different Mass Fluxes.

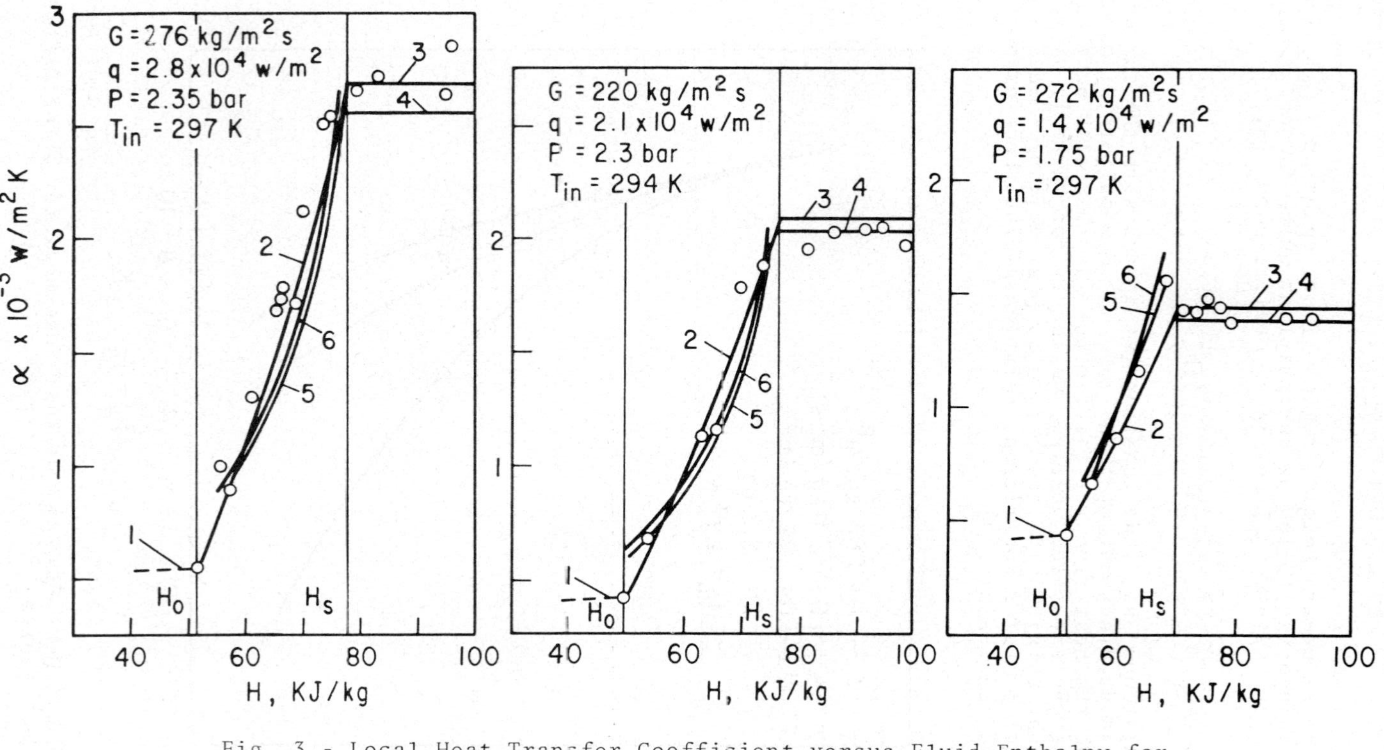

Fig. 3 - Local Heat Transfer Coefficient versus Fluid Enthalpy for Three Different Heat Flux and Mass Flux Combinations. (1: Dittus-Boelter Correlation; 2: Equation (4); 3: Equation (3); 4: Stephan Correlation; 5: Chen Correlation; 6: Pokhvalov Correlation)

When the complete subcooled boiling region (i.e., $H_o \leq$ Enthalpy $\leq H_s$) is considered, it is clear that the heat transfer relationship must agree with equation (1) when the fluid enthalpy is equal to H_o and with equation (2) when the fluid enthalpy is equal to H_s. Because of the foregoing, the following relationship is proposed for the complete subcooled boiling region (i.e., the partial subcooled boiling region plus the fully developed subcooled boiling region):

$$\alpha = \alpha_o + \left\{\frac{H - H_o}{H_s - H_o}\right\}^m (\alpha_b - \alpha_o) \tag{3}$$

where H is the fluid enthalpy, α_o the heat transfer coefficient for $H = H_o$ given by equation (1), and α_b the heat transfer coefficient for $H = H_s$ given by equation (2). As can be seen from equation (3), it satisfies the boundary conditions, viz., $\alpha = \alpha_o$ when $H = H_o$ and $\alpha = \alpha_b$ when $H = H_s$. A regression analysis of the present experimental data has resulted as the best value of the power m to be 1.2. With this value of m, the heat transfer coefficients calculated from equation (3) fall within ±18% of the experimental values.

The heat transfer correlations proposed in this paper and earlier, together with the experimental results, are plotted in Figures 3(a), 3(b) and 3(c). The curve 1 (eq. 1) gives the heat transfer coefficient in the single phase forced convection region, the curve 2 (eq. 3) in the subcooled boiling region, and the curve 3 (eq. 2) in the saturated boiling region. In addition, Stephen and Abdelsalam correlation (curve 4), Chen correlation (curve 5) and Pokhvalov correlation (curve 6) have been plotted, for comparison purposes. A study of Figure 3 shows that the proposed heat transfer coefficient correlation for the forced convection subcooled Freon-11 boiling (eq. 3 in conjunction with eq. 2) agrees better with the experimental data and with the boundary conditions, than the other proposed correlations.

4. CONCLUSIONS

A new heat transfer coefficient correlation has been obtained for the forced convection subcooled Freon-11 boiling, as well as a correlation for the saturated boiling flows. The subcooled boiling flow correlation satisfies the boundary conditions completely, and agrees with the experimental data better than those proposed earlier.

ACKNOWLEDGEMENTS

The authors gratefully acknowledge the support of the National Science Foundation. The assistance of Aykut Menteş, İlker Gürkan and Donna Pressley, all of the University of Miami Clean Energy Research Institute, is also gratefully acknowledged.

REFERENCES

1. Jens, W. H. and Lottes, P. A., U.S.A.E.C. Report ANL-4627, May, 1951.

2. Rohsenow, W.M., Trans. ASME, Vol. 74, p. 969, 1952.

3. Bergles, A.E. and Rohsenow, W.M., Paper 63-HT-22, 6th National Heat Transfer Conference ASME/AIChE, 1963.

4. Thom, J.R.S., Walker, W.M., Fallow, T.A. and Reising, G.F.S., Paper 6, the Symposium on Boiling Heat Transfer in Steam Generating Units and Heat Exchangers, Manchester, 1965.

5. Bjorge, R. W., Hall, G. R, and Rohsenow, W. M., Int. J. Heat Mass Transfer, Vol. 25, pp. 753-757, 1982.

6. Chen, J. C., Paper 63-HT-34, the 6th National Heat Transfer Conference ASME/AIChE, 1963.

7. Pokhvalov, Yn. E., Kronin, G. H. and Kunganova, I. V., Teploenergetika, Vol. 13, No. 5, pp. 63-68, 1966.

8. Lin, Z. H., Veziroğlu, T. N., Kakaç, S., Gürgenci, H., Menteş, A., Paper TF-24, Proceedings of the 7th International Heat Transfer Conference, Vol. 5, pp. 331-336, 1982.

9. Bergles, A. E., Collier, J. G., Delhaye, J. M., Hewitt, G. F. and Mayinger, F., Two Phase Flow and Heat Transfer in the Power and Process Industries, Hemisphere Publishing Corporation, pp. 226-255, 1981.

10. Stephan, K. and Abdelsalam, M., Int. J. Heat Mass Transfer, Vol. 23, pp. 73-87, 1980.

Multi-Phase Flow and Heat Transfer III. Part A: Fundamentals
edited by T.N. Veziroğlu and A.E. Bergles
Elsevier Science Publishers B.V., Amsterdam, 1984 — Printed in The Netherlands

CRITICAL HEAT FLUX MAPPING

J. E. Dahlquist and F. S. Gunnerson
Department of Mechanical Engineering and Aerospace Sciences
University of Central Florida
Orlando, Florida 32816, U.S.A.

R. A. Nelson
EG & G Idaho, Inc.
P.O. Box 1625
Idaho Falls, Idaho 83415, U.S.A.

ABSTRACT

Prediction of the critical heat flux (CHF) over a wide range of thermal hydraulic conditions, is essential for accurate nuclear reactor modeling. In spite of the very large amount of work that has been done on natural and forced convective CHF, there is not yet a complete understanding of the physical mechanisms involved. Numerous correlations have evolved over the years to model the CHF behavior, unfortunately, most are severely limited to a narrow range of thermal hydraulic conditions and specific geometries. Many CHF correlations must be incorporated to model the CHF behavior over a wide range of conditions. Advanced computer codes currently used for Light Water Reactor (LWR) modeling are limited in this area and are often inept in CHF predictions for certain accident scenarios.

In this study the process of developing a CHF map is presented and may be used to identify specific regions of the CHF mechanisms and establish applicability limits for some of the current CHF correlations. These correlations are applied, with some minor modifications, to give continuous CHF predictions over a wider range of thermal hydraulic conditions. Emphasis is placed on low pressure and flow conditions. The resultant model is then visually presented by means of a 3-D surface in mass flux-quality-CHF space.

1. INTRODUCTION

In all phases of nuclear reactor design and operation, one major concern is the prediction of a safety margin to the critical heat flux (CHF). There is considerable disparity in nomenclature for the CHF condition. The term CHF is often used to define the condition for which a small increase in the surface heat flux, or a small decrease in the coolant mass flux, gives rise to inordinate deterioration in the heat transfer. It is often considered synonymous with other terms such as "dryout," "burnout," "departure from nucleate boiling (DNB)," "boiling crisis," or "boiling transition." The term "critical heat flux" is used herein as an encompassing term to describe the phenomenon of deteriorated heat transfer. It is chosen since it has possibly the widest international recognition.

There are many credible single and coincidental events that may give rise to the CHF. In general, these events may be classified as coolant deficiency events or overpower events. Coolant deficiency may result from a breach in the primary system where coolant inventory is lost; commonly referred to as the small or large break loss of coolant accidents (LOCA). In addition, equipment malfunctions, flow blockages, or secondary system malfunctions that result in decreased heat removal from the primary coolant system may likewise spawn the CHF. Overpower events could result from malpositioned or unplanned withdrawal of control rods, a decrease in soluble poison concentration, fuel assembly enrichment errors or a host of other reactivity initiated accident *RIA) scenarios. The wide range of events between the LOCA and RIA are often referred to as power-cooling-mismatch (PCM) events. This paper addresses the problem of predicting the CHF over a wide range of thermal hydraulic conditions which may be encountered during normal accident operations.

2. CHF BEHAVIOR AND PREDICTION

A nuclear rod bundle is a very complex system for CHF analysis due to the large number of thermal, hydraulic and geometric parameters involved. To avoid these difficulties, experimental efforts have been directed toward test conditions that closely resemble reactor channels with respect to internal construction and toward tests conducted at or near the operating conditions of the reactor. Since no universal analytical treatments are available for CHF prediction, most of the available literature is confined to reporting experimental data and a wide variety of corresponding empirical or semi-empirical correlations. Correlations of an empirical nature make no assumptions whatever about the mechanisms involved in the CHF condition, but solely attempt a functional relationship between the CHF and the independent variables. Semi-empirical correlations arise from attempts to model, in a general sense, the hydrodynamic and heat transfer processes occurring in the heated channel and relate these to the CHF.

Two different phenomenological approaches are generally used to model the CHF behavior with respect to forced convective conditions. Under low void fraction or low quality conditions, the CHF is generally considered to be governed by the local bubble population near the heating surface. This behavior is represented by the thermally limited and pool boiling regions in Figure 1.

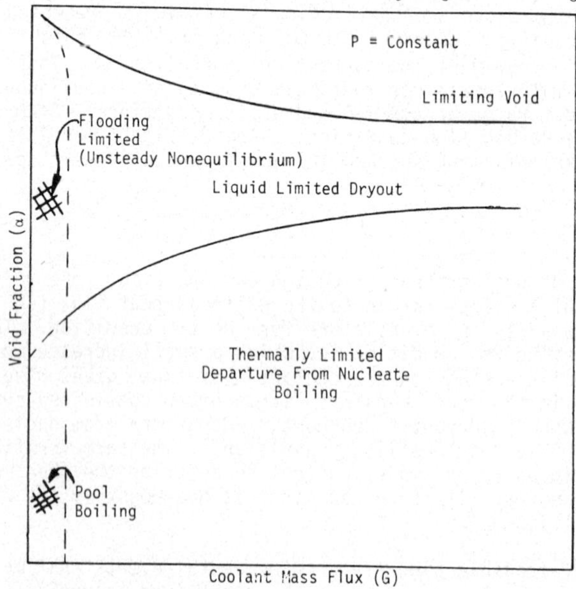

FIG. 1 REPRESENTATIVE CRITICAL HEAT FLUX MAP FOR STEADY STATE CONDITIONS

When the void fraction becomes sufficiently large, an annular flow pattern may develop and the CHF behavior is phenomenologically liquid limited. The hot surface in the annular pattern is covered with a liquid film, disturbed by deposition and entrainment of droplets within the vapor core. Essentially all the heat has to be transferred through the film and consequently moderate wall superheats may be expected. Under certain critical conditions, however, the film departs or vanishes from the heating surface and the surface assumes direct contact with the vapor. This liquid limited CHF behavior is illustrated on Figure 1 and is often referred to as "dryout" and results in a deterioration of the heat removal with a corresponding rise of the heater temperature. Repeated dryout and rewetting could lead to thermal shock which can be more dangerous than high wall temperature. The CHF behavior at high void fractions or high qualities is of particular interest in the assessment of accident consequences in nuclear systems. The limiting void line in Figure 1 represents a group of operating conditions where the heat flux versus void curve is nearly independent of heat flux. The region of flooding limited CHF (Figure 1) has not yet been included within this study and the corresponding mode of CHF prediction in this region is relegated to future studies.

The lines or boundaries illustrated on Figure 1 correspond to mechanistic behavioral changes in the CHF, and may be viewed as corresponding to changes in the hydraulic flow pattern. The onset of annular flow, for example, is known to induce CHF behavior under certain conditions. Prediction of the onset of an annular flow pattern, however, is not yet possible for all conditions and geometries. As discussed by Saito [1] and by Mishima and Ishii [2], considerable uncertainty may be expected with predicted boundaries.

There are numerous correlations available for CHF prediction. Table I summarizes several of the common correlations used within the initial phase of this study. Collectively, they span a wide range of thermal-hydraulic conditions, and are commonly used within the nuclear industry. The parametric data base ranges for each of the correlations is given, but this does not mean that the correlations are directly applicable over these entire ranges. Even for apparently identical thermal-hydraulic and geometry conditions, the CHF values predicted by the correlations often span a wide range, with differences or uncertainties of over 100% not being unusual.

Figures 2 and 3 illustrate the pressure-mass flux and quality-mass flux ranges, respectively, of the correlation data bases. Again, this does not imply that the respective correlations are valid for the entire spectrum of its parameters. From these figures, certain problem areas in CHF predictions can be identified. For example, the Modified-Barnett and Westinghouse-3 correlations are used in subcooled CHF predictions, and as shown in Figure 2 and 3, become undefined in the medium pressure ($\sim$ 5 to 7 MPa) range. Similarly, at very low pressures and mass fluxes, the inverse mass flux dependence of the Biasi, et al. correlation gives rise to unrealistically high CHF values ($CHF \rightarrow \infty$ as $G \rightarrow 0$). Such a trend is due, in part, to the limited amount of experimental data in this area. The Modified-Barnett correlation, which is not applicable at elevated pressures ($P > 5$ MPa) leaves a predictive gap at low mass flux and low quality when pressure exceeds its correlation limit, as shown in Figure 3.

3. COMBINING CHF CORRELATIONS

The CHF correlations given in Table 1 are the backbone of several large computer codes used within the nuclear industry [3,4,5]. Individually, the correlations are limited to somewhat narrow parametric ranges and, therefore, must be combined to adequately model the CHF behavior over a wide range of conditions.

TABLE I

CRITICAL HEAT FLUX CORRELATIONS

Westinghouse Company, W-3, [3,4]

$$q''_{CHF} = 1.\times 10^6\,[2.022 - 4.302\times 10^{-4}P + (0.1722 - 9.84\times 10^{-5})\exp((18.177 - 4.129\times 10^{-3}P)X)]\times [1.157 - 0.869X][(0.1484 + X(-1.596 + 0.1729\,ABS(X)))G' + 1.037]\times[0.8258 + 7.94\times 10^{-4}(H_f - H_{IN})]\times[0.2664 + 0.8357\exp(-3.151\,D_{HE})]$$

where

$G' = G/10^6$. $1000 < P\ (psia) < 2400$

$-.2 < D_H(inches) < 0.7$ $10^6 < G\ (lbm/ft^2\cdot hr) < 5(10)^6$

$10 < L\ (inches) < 144$ $-0.15 < X < 0.15$

Modified Barnett [5]

$$q''_{CHF} = 10^6\left[\frac{A + B(H_f - H_{in})}{C + L}\right]$$

where

$A = 73.71\,D_{HE}^{0.052}G'^{0.663}\ 1.0 - 0.315\,e^{(-11.34D_{HY}G')}$

$B = 0.104\,D_{HE}^{1.445}G'^{0.691}$

$C = 45.44\,D_{HY}^{0.0817}G'^{0.5866}$

$G' = G/10^6$

$0.395 < D(inches) < 0.543$

$32.9 < L(inches) < 174.8$

$150 < P(psia) < 725$

$0.03(10)^6 < G(lbm/ft^2\cdot hr) < 1.7(10)^6$

Inlet subccoling: 6 to 373 Btu/lbm

Biasi Correlation [6]

$$q''_{CHF} = \left[\frac{1.883\times 10^3}{D_H^a\,G^{1/6}}\right]\left[\frac{Y(p)}{G^{1/6}} - X_o\right]$$

and

$$q''_{CHF} = \left[\frac{3.78\times 10^3}{D_H^a\,G^{0.6}}\right)h(P)]\,(1 - X_o)$$

where

$a = 0.4$ for $D_H > 1cm$

$a = 0.6$ for $D_H < 1cm$

$y(p) = 0.7249 + 0.099\,P\cdot\exp(-0.032\cdot P)$

$h(p) = -1.159 + 0.149\,P\cdot\exp(-0.019\cdot P) + \dfrac{8.0\cdot P}{10 = P^2}$

$0.3 < D_H\ (cm) < 3.75$

$20 < L\ (cm) < 600$

$2.7 < P\ (atm) < 140$

$10 < G(gm/cm^2\cdot s) < 60$

$X_o = 1/(1 + \rho_f/\rho_g)$

Limiting Quality [7]

$$X_1 = [0.39 + 3.53\,(P/P_{CR}) - 10.3\,(P/P_{CR})^2 + 7.62\,(P/P_{CR})^3]\times(G/1000)^{-0.5}\,(8/d)^{0.15}$$

where

P/P_{CR} = reduced pressure X_1 = limiting quality

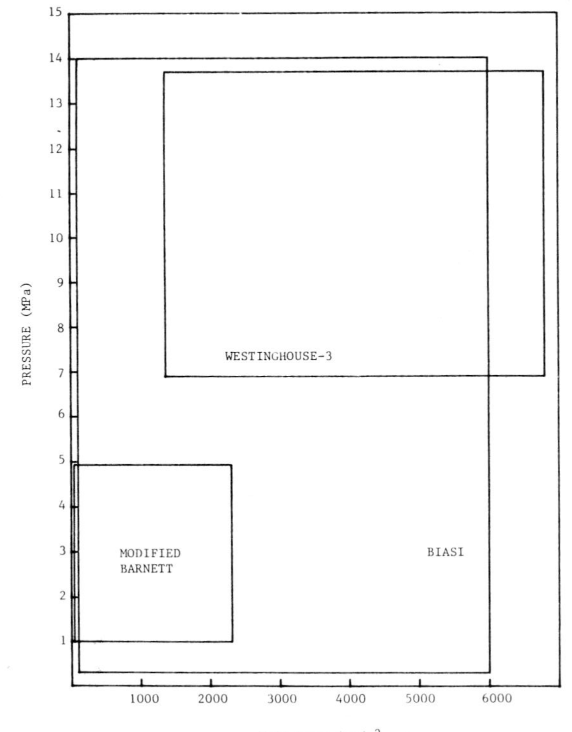

FIG. 2 DATA RANGES FOR CORRELATIONS USED IN THE PRESSURE MASS FLUX PLANE

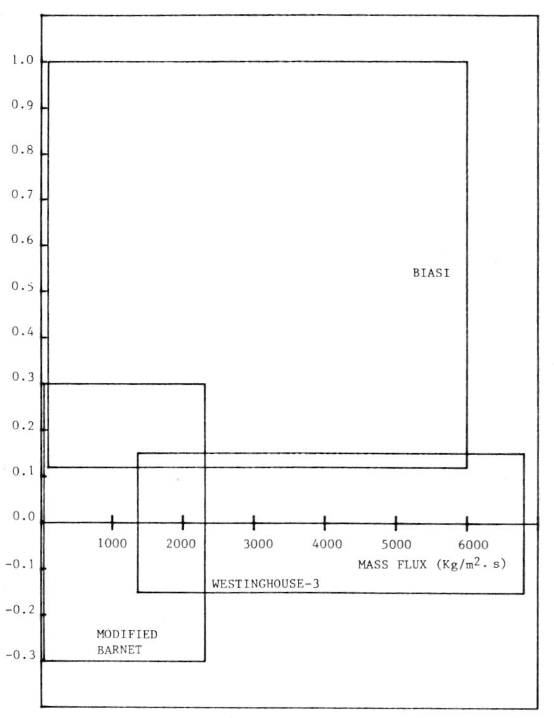

FIG. 3 DATA RANGES FOR THE CORRELATIONS USED IN THE MASS FLUX QUALITY PLANE

Considerable uncertainty, however, arises under certain conditions when going from one correlation to another. To combine the CHF correlations together in a congruent manner it is assumed that no major discontinuities should exist in total CHF behavior without an understanding for its existance. This means that the "CHF-surface" should be smooth with no abrupt changes, except perhaps at the onset of the limiting quality phenomenon. For a given set of thermal-hydraulic conditions and fixed geometry, the value of the CHF should be unique. To support this presumption, it is considered that the simultaneous existance of two or more flow patterns at a given location (annular and churn for example) for a given set of conditions is unlikely.

Figure 4 is a flow chart indicating how the CHF correlations (listed in Table I) are initially combined for this study. As shown, system pressures are first considered as high (P > 6.8 MPa), medium (5.1 < P < 6.8 MPa) or low (P < 5.1 MPa) ranges. Quality follows as the next qualifying parameter which, in turn, prescribes the CHF correlation. In essense, the CHF correlations are slightly modified under certain conditions to insure continuous CHF behavior. The linear interpolation, initially used, between the W-3 and Biasi correlations prevents excessively large predictions of the CHF by the Biasi correlation at low flows and qualities. This interpolation will be illustrated in Figure 12. At voids greater than α = 0.96 current CHF correlations are inept and additional analysis is required in this area. Therefore as a preliminary estimate within this study CHF is considered to be zero beyond α = 0.96.

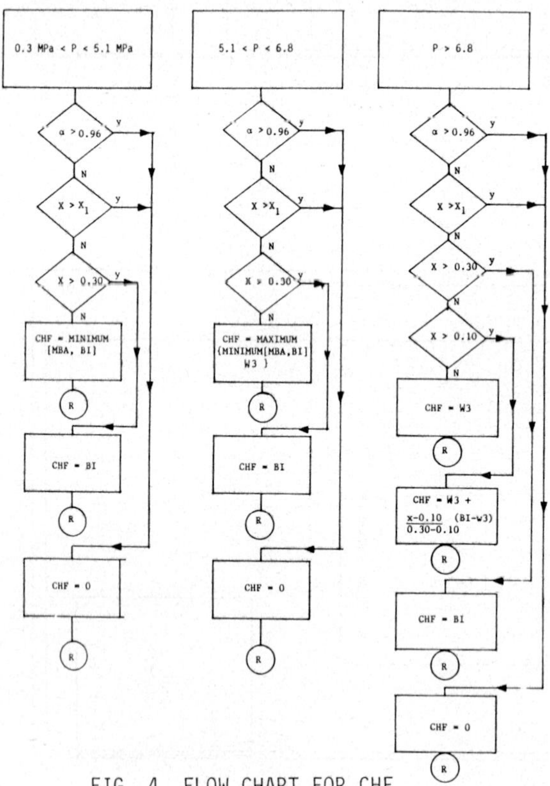

FIG. 4 FLOW CHART FOR CHF PREDICTION

4. COMPARISON WITH EXPERIMENTAL DATA

To effectively combine the CHF correlations together in a continuous manner, justification must be given for suggested modifications. Since the correlations are empirically based, a better comparison with data provides the justification.

An extensive CHF data base, with over 8,000 datum points is used for comparison.* The data base is currently used in nuclear reactor safety assessments and the original sources for these data are given by Groeneveld and Snoer [11]. The distribution of the experimental data are illustrated in Figures 5, 6 and 7.**Even though the data are from many different investigations, the limitations of the experimental facilities becomes evident. Note the lack of low quality, low flow data and high quality, high flow data. Such regions with limited data are expected since for any given mass flux there is a limited range of qualities where the CHF can be achieved within a given facility. Figures 5 and 6 illustrate the influence of the mass flux and quality respectively on the critical heat flux.

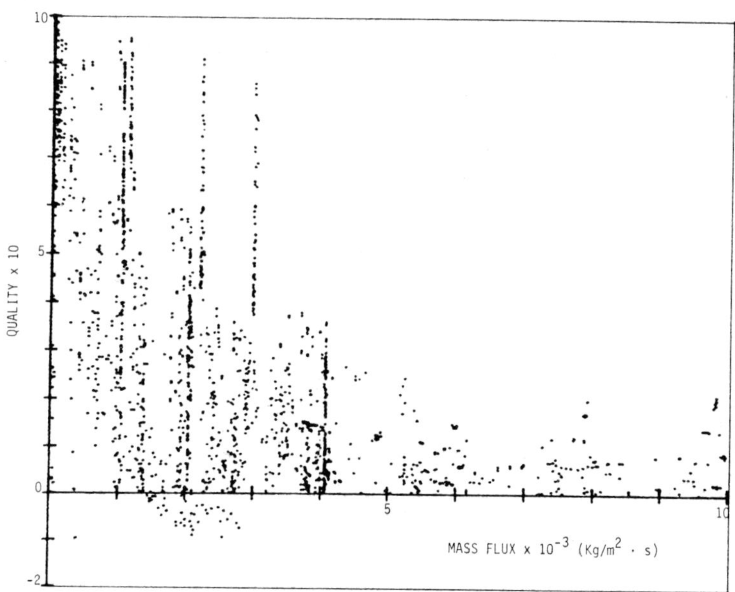

FIG. 5 REPRESENTATIVE DATA PLOTTED IN THE MASS FLUX QUALITY PLANE (2000 points)

---

* At the time this paper was written, a complete comparison with all experimental data had not yet been performed.

** Only 2000 representative data points are shown for clarity.

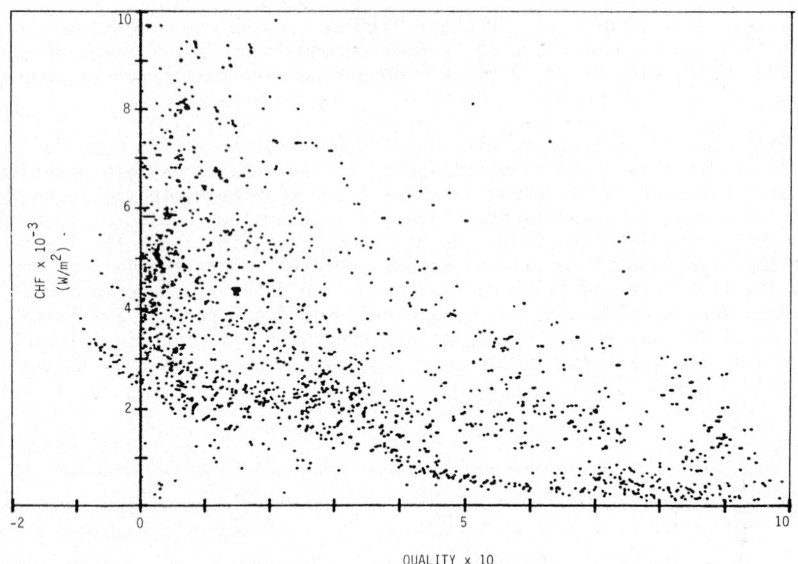

FIG. 6 REPRESENTATIVE EXPERIMENTAL DATA PLOTTED IN THE QUALITY CHF PLANE (2000 points)

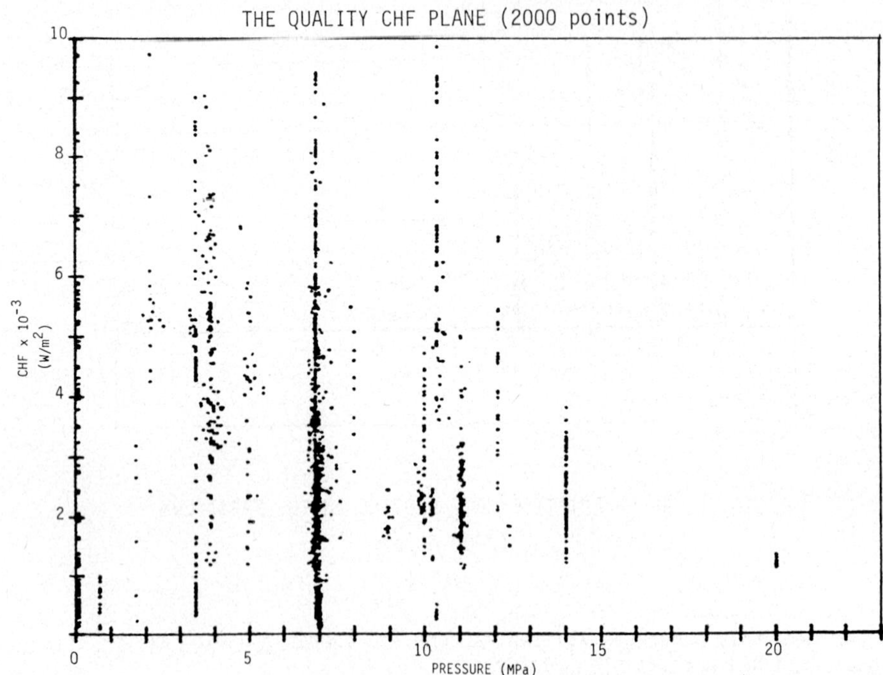

FIG. 7 REPRESENTATIVE EXPERIMENTAL DATA PLOTTED IN THE PRESSURE QUALITY CHF PLANE (2000 points)

Figure 8 is a plot of the predicted CHF versus the experimentally measured CHF. The diagonal line is the locus of points where the predicted value is identical to the measured value. Above the diagonal implies overprediction by the correlations and below the diagonal implies underprediction by the correlations. As shown, the present general trend is a slight underprediction which may be due, in part, to the inherent conservativeness of the correlations involved. The use of the limiting void correlation, which sets the CHF equal to zero beyond a certain quality, may cause underprediction of the CHF. This problem is currently being addressed.

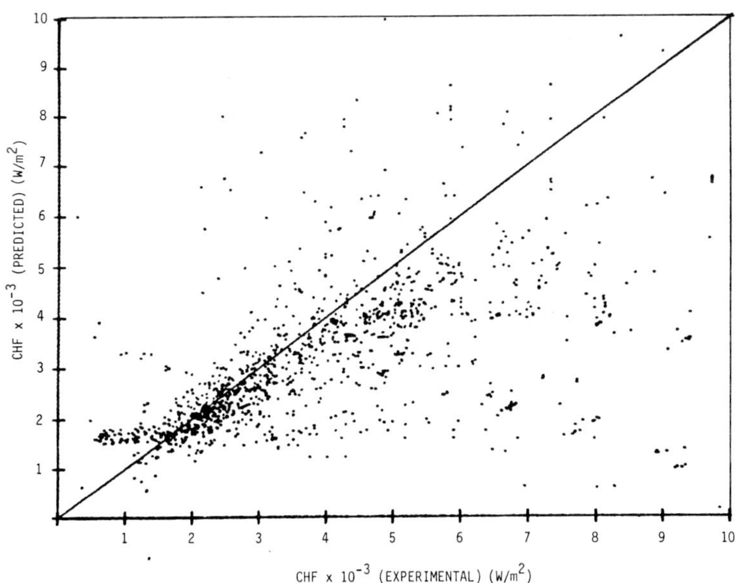

FIG. 8 MEASURED VS PREDICTED CHF FOR FLOW CHART IN FIGURE 4.

Figures 9 through 11 are residual plots for the quality, pressure and mass flux variables respectively. The residual is defined as

$$\text{Residual} = \frac{\text{Predicted CHF} - \text{Measured CHF}}{\text{Measured CHF}}$$

where the residual can have a value from negative one to positive infinity. The residual plots are useful in observing correlation trends with respect to the independent variables. The figures indicate ranges of the independent variables where there are large uncertainties in the prediction of the data. However, as shown, there does not appear to be any trends in the correlation with respect to the independent variables.

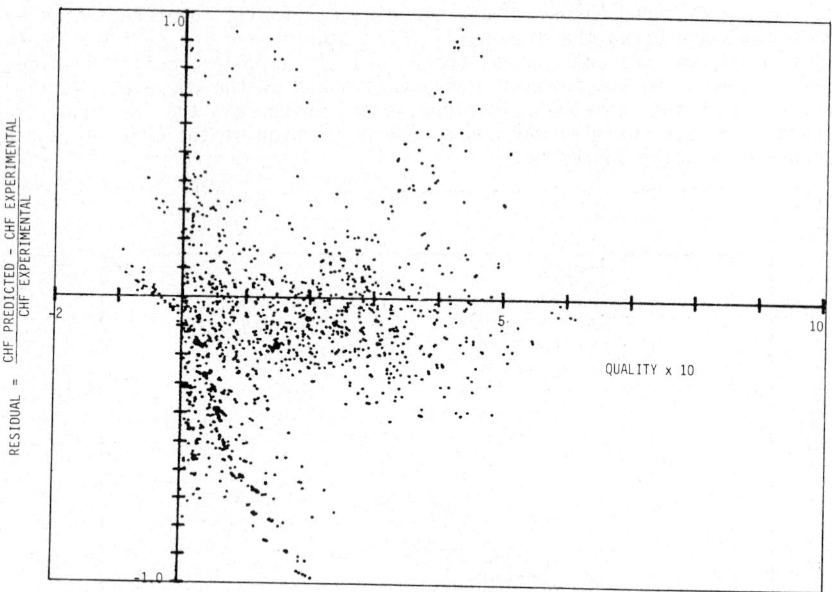

FIG. 9 RESIDUALS VS QUALITY FOR MODEL ILLUSTRATED IN FIG. 4

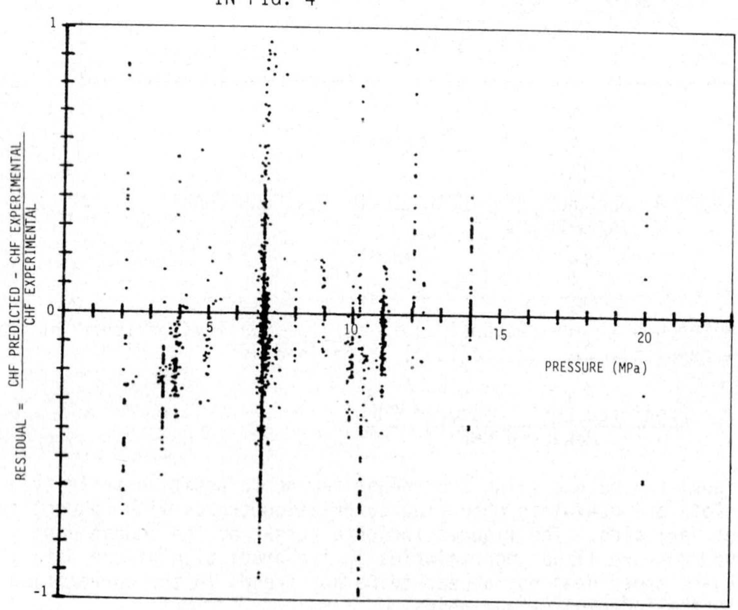

FIG. 10 RESIDUALS VS PRESSURE FOR MODEL ILLUSTRATED IN FIG. 4

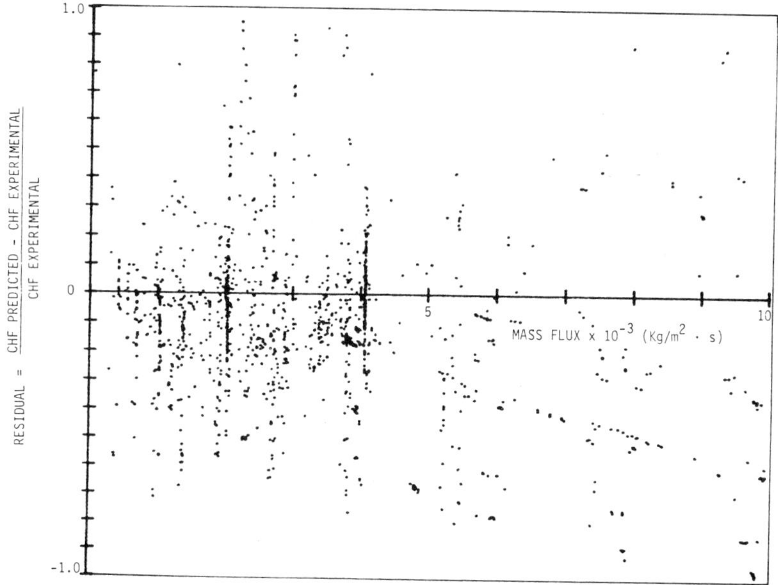

FIG. 11 RESIDUALS VS MASS FLUX FOR MODEL ILLUSTRATED IN FIG. 4

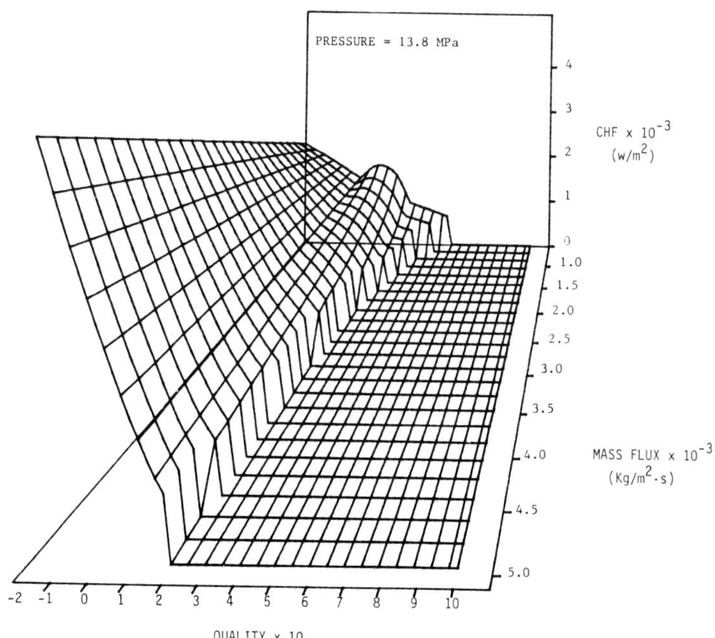

FIG. 12 A CHF SURFACE AT 13.8 MPa GENERATED USING THE CORRELATION IN FIG. 4

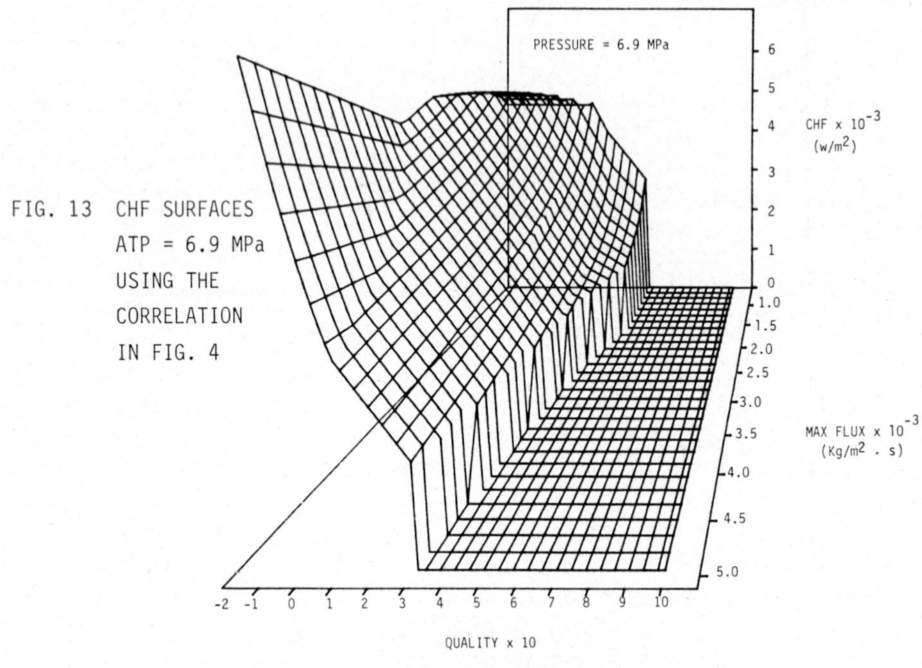

FIG. 13 CHF SURFACES ATP = 6.9 MPa USING THE CORRELATION IN FIG. 4

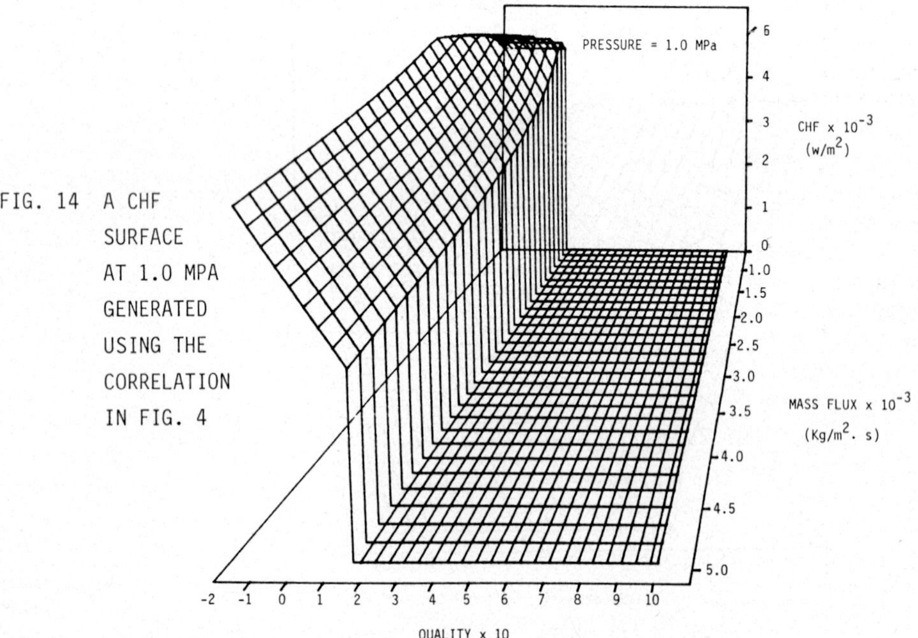

FIG. 14 A CHF SURFACE AT 1.0 MPA GENERATED USING THE CORRELATION IN FIG. 4

5. CRITICAL HEAT FLUX SURFACE

Any multi-dimensional function can be represented as a three-dimensional surface if all but two independent variables are held constant.

Figures 12, 13 and 14 illustrate three-dimensional representations of critical heat flux surfaces at high, medium and low pressures, respectively. The surfaces were developed from the CHF correlations listed in Table I at constant pressure, subcooled inlet conditions and fixed geometry. The distinct "cliff" shown on the surfaces are indicative of the limiting void phenomenon where the CHF goes to zero.

A projection of the correlation intersection in Figures 12, 13 or 14 onto the quality-mass flux plane produces a critical heat flux map similar to that illustrated in Figure 1.

6. RESULTS AND CONCLUDING REMARKS

The results of this study have not yet been fully realized at the time of this writing. Extensive comparisons with experimental data continues, and future modifications are expected.

Preliminary comparison of the model presented in Figure 4 with experimental data indicate favorable agreement at elevated pressures ($P > 6.8$ MPa), at all qualities and mass fluxes greater than about 500 Kg/m^2·s. In addition, at pressures less than 6.8 MPa and mass fluxes between 150-2500 Kg/m^2·s (all qualities) good comparison is likewise attained. The average error for all data is less than 10% with a standard deviation of several hundred percent. Specific regions are now being more thoroughly examined to reduce this large deviation in the error.

The critical heat flux behavior for a wide spectrum of thermal-hydraulic conditions can be visually illustrated as a three-dimensional surface, as shown in Figures 12, 13 and 14. From such surfaces, phenomenological trends and predictive discontinuities can be readily identified.

Future assesssments may require the incorporation of additional CHF correlations for better predictions in the regions not currently encompassed by the subject model. Inclusion of the Limiting Void Phenomenon, for example, may result in under prediction of the CHF for high quality high flow conditions.

NOMENCLATURE

BI	=	Biasi CHF prediction
CHF	=	critical heat flux
D	=	fuel rod diameter
D_H	=	hydraulic diameter
D_{HY}	=	$\sqrt{D(D + D_{HE})} - D$
D_{HE}	=	heated equivalent diameter
G	=	mass flux
H	=	enthalpy
H_{fg}	=	heat of vaporization

L = channel length

MBA = modified Barnett CHF prediction

N = No

P = pressure

P_{CR} = critical pressure

q''_{CHF} = critical heat flux

R = return

X = quality

X_1 = limiting quality

Y = yes

α = equilibrium void fraction

ρ = density

REFERENCES

1. Saito, T., "Multi-Fluid Modeling of Two-Phase Flow and Heat Transfer: Application to Critical Heat Flux Prediction for BWR Conditions," Ph.D. Dissertation, The University of Wisconsin-Madison, Nuclear Engineering, 1977.

2. Mishima, K. and Ishii, M., "Critical Heat Flux Experiments Under Low Flow Conditions in a Vertical Annulus," NUREG/CR-2647, ANL-82-6, 1982.

3. Ransom, V. H., et al., "RELAPS/MOD 1 Code Manual," NUREG/CR-1826, EGG-2070, Draft, Rev. 1, March 1981.

4. "TRAC-PD2: An Advanced Best Estimate Computer Code for Pressurized Water Reactor Loss-Of-Coolant Accident Analysis," LASL LA-8709-MS, 1981.

5. Spore, J. W., et al., "TRAC-BD1 Manual," NUREG/CR-2178, Sept. 1981.

6. Tong, L. S., "Prediction of Departure from Nucleate Boiling for an Axially Non-Uniform Heat Flux Distribution," J. Nuclear Energy, 21, 1967.

7. Tong, L. S., "Boiling Crisis and Critical Heat Flux," TID-25887, AEC Critical Review Series, U.S. Atomic Energy Commission (1972).

8. Hughes, E. D., A Correlation of Rod Bundle Critical Heat Flux for Water in the Pressure Range 150 to 725 psia, IN-1412 (July 1970).

9. Biasi, L., et al., "Studies on Burnout, Part 3-A New Correlation for Round Ducts and Uniform Heating and its Comparison with World Data," Energia Nucleare, Vol. 14/n.9/Sept. 1967.

10. Doroshchuk, V. E., "On the Burnout Mechanism in Two Phase Annular Flow," ASME Paper 73-HT-37, 1973, as cited by Kitto, J. B., "Critical Heat Flux and The Limiting Quality Phenomenon," AIChE Symposium Series 199, Vol. 76, 1980.

11. Groeneveld, D. C. and Snoer, C. W., "A Review of Current Heat Transfer Correlations Used in Reactor Safety Assessment," Proceedings 2nd International Topical Meeting on Nuclear Reactor Thermal Hydraulics, Vol. 1, pp. 20-28, Santa Barbara, California, Jan. 1983.

Multi-Phase Flow and Heat Transfer III. Part A: Fundamentals
edited by T.N. Veziroğlu and A.E. Bergles
Elsevier Science Publishers B.V., Amsterdam, 1984 — Printed in The Netherlands

PREDICTION OF TRANSITION BOILING HEAT TRANSFER FOR SUBCOOLED WATER AT LOW FLOW AND ATMOSPHERIC PRESSURE

H.S. Ragheb* and S.C. Cheng
Department of Mechanical Engineering
University of Ottawa
Ottawa, Canada K1N 6N5

ABSTRACT

This paper presents a phenomenological model using existing correlations to predict transition boiling heat transfer for water at low flow and atmospheric pressure. The model incorporates the effects of mass flux, local enthalpy (or inlet subcooling) and thermal properties of heated surface. Comparison of the predicted results with experimental data shows fair agreement in the nucleate and transition boiling regions. The proposed method establishes a general groundwork and can be easily extended after some modifications to predict transition boiling heat transfer for water as well as other fluids in the quality region at higher mass flux and higher pressure.

1. INTRODUCTION

Safety analysis of water-cooled nuclear reactors aims at the accurate prediction of fuel sheath temperature during accidents that could lead to a situation where the fuel elements are not adequately cooled. In these accidents, a very wide range of heat transfer regimes is possible, ranging from forced convection to all boiling modes: nucleate boiling, transition boiling and stable film boiling. In order to predict heat transfer coefficients and hence the fuel sheath temperature, it is necessary to calculate flow and pressure transients in the reactor core. This is accomplished by the thermohydraulic computer codes which scan a wide range of flow conditions. Once the flow and pressure are determined, the codes proceed to calculate the heat transfer coefficients using a set of correlations that covers the range of all heat transfer regimes involved.

In selecting a heat transfer correlation for a thermohydraulic computer code, several features should be sought:

(a) A correlation must have correct asymptotic trends. For example, a correlation for the transition boiling region should approach the value of film boiling heat transfer at higher wall superheats and the critical heat flux (CHF) at lower wall superheats. Also, at a very low mass flux, the correlation should approach the value of pool boiling heat transfer.

---

* Present address: Atomic Energy Control Board, Ottawa, Canada.

(b) A correlation must be based on local conditions since,in many instances, the predictions deal with flow reversal situations. Under these circumstances an inlet condition is hard to define.

(c) A correlation should be based on physical understanding so that the extension to a wider range of conditions may be possible.

It appears to be a difficult task to accommodate the above features in a transition boiling correlation. As shown in Fig. 1, the transition boiling is an intermediate heat transfer mode where the heated surface temperature is too high to maintain nucleate boiling but too low to maintain stable film boiling. At the low temperature boundary, where the transition boiling is most efficient, the heat flux can be predicted by a variety of correlations for the critical heat flux; discrepancies among them have not been resolved. At the high temperature boundary lies the minimum stable film boiling point (MSFB). Little is known about this point except that it is affected by flow, subcooling, pressure and surface conditions. Between the two boundaries there is a scarcity of experimental data and their conditions are limited. This is due to the unique nature of transition boiling in which heat flux decreases with an increase of surface temperature, such a heat transfer mode is unstable in a heat flux controlled system and usually requires a temperature controlled system.

This study presents a model to predict the transition boiling heat transfer suitable for application to problems that require thermohydraulic codes. The method of prediction has been tested in the subcooled region, at low flow and atmospheric pressure and compared with the experimental data and observa-

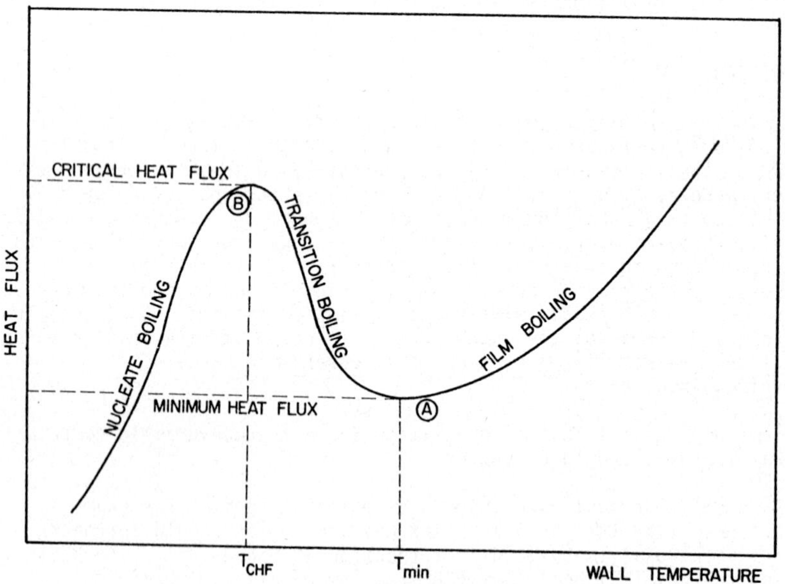

Fig. 1. Boiling Curve.

tions reported earlier |1-4|. The data have been obtained from different types of short test sections where the variation of local enthalpy along the test section is minimum. This allows more accurate prediction of local conditions in the boiling channel. A simple computer code has been developed to predict heat transfer in the transition boiling region using the present model and based on local conditions. Since most of the available experimental data are reported in terms of inlet conditions, an extra feature is included in the code to predict local conditions. Such a prediction usually requires an iterative procedure since local conditions are themselves functions of heat flux.

Application of the present model to low flow, low pressure region has become very important due to increasing studies involving heat removal from the reactor core in the absence of driving force, e.g. loss of reactor circulating pumps. Although attempts have been made recently to formulate a general predictive model for the post dryout region yet better success has generally been achieved with models limited to a smaller range of conditions.

2. ASSUMPTIONS

(a) The transition boiling mode is a combination of both unstable nucleate boiling and unstable film boiling alternately existing at any given location at the heated surface. The variation of heat transfer rate with temperature is primarily a result of the change in the fraction of time that each boiling mode exists. This postulation is based on Berenson's description |5| which has been applied to cryogenics by Kalinin et al |6|. The heat flux can, therefore, be expressed as

$$q_{TB} = q_{CHF}\, f + q_{FB}\,(1-f) \tag{1}$$

where f is the fraction of surface that is wet.

(b) The fraction of wetted area f is assumed to be equal to 1 at the CHF and approaches zero at the minimum film boiling heat flux |2|.

(c) The transition boiling is a violent heat transfer mode where intense mixing takes place between liquid and vapour phases. As a result, the heat transfer coefficient is underdeveloped. Therefore, the heat transfer coefficient is assumed to be dependent on local enthalpy and independent of heated length.

In the following section, all the terms in Eq. (1) will be determined for a given system pressure, flow and local enthalpy.

3. FORMULATION

3.1 Critical Heat Flux q_{CHF}

Basic formula. The critical heat flux is calculated based on the instability of a wave at the liquid-vapour interface using Zuber's theory |7| of burnout. The critical heat flux for saturated pool boiling is:

$$(q_{CHF})_{sat} = K\, h_{fg}\, \rho_g^{\frac{1}{2}}\, [\sigma g(\rho_f - \rho_g)]^{\frac{1}{4}} \qquad (K = 0.12\text{-}0.15) \tag{2}$$

The theory provides theoretical verification of Kutateladze's equation |8|, and determines the critical heat flux hydrodynamically which is independent of surface conditions. It also allows an inherent scattering of 22 percent of critical heat flux value due to the variation of K from 0.12 to 0.15.

Effect of local enthalpy. To account for the effect of subcooling (x<0), Kutateledze |8| proposed a semiempirical dimensionless factor that has been described by Bonilla |9|, and later modified by Ivey and Morris |10|.

$$(q_{CHF})_{x\leq 0} = (q_{CHF})_{sat} \left[1 + \left(\frac{\rho_\ell}{\rho_g}\right)^{0.75} (C_p \Delta T_{sub}/9.8\ h_{fg})\right] \tag{3}$$

in the above equation C_p was evaluated at

$$T' = T_{sat} - \left(\frac{\Delta T_{sub}}{2}\right) \tag{4}$$

Effect of mass flux. Based on the experimental data of Thompson and Macbeth |11| one can propose the following expression to account for the effect of mass flux

$$(q_{CHF})_{G,x\leq 0} = (q_{CHF})_{x\leq 0} \left(\frac{G}{G_{min}}\right)^{0.51} \tag{5}$$

G_{min} is an arbitrary mass flux value below which the CHF approaches the pool boiling CHF.

3.2 Wall Temperature at CHF T_{CHF}

Bjornard and Griffith |12| suggested that T_{CHF} can be computed by Chen's correlation |13| since it is physically based and it automatically transforms into the Forester-Zuber relation for pool boiling at low flow |14|. The Chen correlation was originally derived for saturated nucleate boiling, then extended by Butterworth and Heweitt |15| to the subcooled boiling region where the F factor in Chen's correlation was set to 1 and the S factor was evaluated from the single-phase Reynolds number, GD/μ. The resulting form of correlation which is used for the present application is:

$$q_{G,x\leq 0} = F(0.023) \frac{k_\ell^{0.6}\, G^{0.8} (1-x)^{0.8}\, C_{p\ell}^{0.4}}{\mu_\ell^{0.4}\, D_e^{0.2}} (T_w - T_\ell)$$
$$+ S(0.00122) \frac{k_f^{0.79}\, C_{pf}^{0.45}\, \rho_f^{0.49}}{\sigma^{0.5}\, \mu_f^{0.29}\, h_{fg}^{0.24}\, \rho_g^{0.24}} (T_w - T_{sat})^{1.24} (P_w - P)^{0.75} \tag{6}$$

Equating Eqs. (5) and (6), one would obtain T_w which is T_{CHF}. The solution of T_{CHF} normally requires an iterative procedure.

Effect of wall thermal properties on T_{CHF}. Magrini and Nannei |16| investigated the wall thickness and thermal properties on the heat transfer of nucleate boiling. Their experiments were conducted under conditions of saturated pool boiling of water at atmospheric pressure, resulting in the following correlation

$$T_w - T_{sat} = 0.27(q)^{0.25} \{1 + \frac{4 \times 10^4}{\sqrt{k_w \rho_w C_{pw}}} [1 - e^{-4.5 \times 10^9} (k_w s)^3]\}^{0.75} \quad (7)$$

Eq. (7) indicates that for a given heat flux q and wall thickness s, a heated surface with a smaller value of $k_w \rho_w C_{pw}$ will result in higher wall superheat. This confirms our experimental data.

They found that when the wall thickness exceeds a certain value, the effect of thickness becomes negligibly small. In fact, this is the case for our test sections |1-4|, thus the exponential term in Eq. (7) can be dropped. Furthermore, Eq. (7) can be rearranged to express the wall superheat of any surface material in terms of the wall superheat of copper surface under the same heat flux as

$$T_w - T_{sat} = 0.571(T_w - T_{sat})_{cu} \left(1 + \frac{4 \times 10^4}{\sqrt{k_w \rho_w C_{pw}}}\right)^{0.75} \quad (8)$$

where k_w, ρ_w and C_{pw} are referred to the thermal properties of any surface material. It has been found that Eq. (6) gives good prediction of wall superheat in nucleate boiling up to the CHF point (when coupled with Eq. (5)) for the copper surface. Thus for a given heat flux, the wall superheat of copper surface is obtainable from Eq. (6), and for the same heat flux, the wall superheat of any other surface material can be calculated via Eq. (8).

3.3 Fraction of Wetted Area f

Experimental observations |2| showed that the fraction of wetted area f during transition boiling decreases with increasing wall superheat. One can assume a possible relation between them as

$$f = a(T_w - b)^2 \quad (9)$$

where a and b are constants to be determined by two boundary conditions, i.e.

At $T_w = T_{CHF}$, $f = 1$;

At $T_w = T_{min}$, $f = 0$.

The expression of f then becomes

$$f = \left(\frac{T_w - T_{min}}{T_{CHF} - T_{min}}\right)^2 \quad (10)$$

Bjornard and Griffith |12| proposed the same expression as Eq. (10).

3.4 Minimum Stable Film Boiling Temperature T_{min}

The best available analysis that predicts the minimum surface temperature to support stable film boiling, is the model presented by Henry |17|. His correlation incorporates the effect of heated surface thermal properties on the heat transfer coefficient near the minimum film boiling point that was originally derived by Berenson |18|. Berenson's expression for the heat transfer coefficient is physically based and it combines Bromley's heat conduction model across a thin vapor film |19| with the hydrodynamic consideration of Zuber |20|.

The correlation by Henry is:

$$\frac{T_{min} - (T_{min})_I}{(T_{min})_I - T_\ell} = 0.42 \left[\sqrt{\frac{k_\ell \rho_\ell C_{p\ell}}{k_w \rho_w C_{pw}}} \frac{h_{fg}}{(\Delta T_{min})_I} \right]^{0.6} \tag{11}$$

where $(T_{min})_I$ is the minimum film boiling temperature for the isothermal surface predicted by Berenson |18| and can be calculated as follows:

$$(\Delta T_{min})_I = 0.127 \frac{\rho_g h_{fg}}{k_g} \left[\frac{g(\rho_\ell - \rho_g)}{\rho_\ell + \rho_g} \right]^{2/3} \left[\frac{\sigma}{g(\rho_\ell - \rho_g)} \right]^{\frac{1}{2}} \left[\frac{\mu_g}{g(\rho_\ell - \rho_g)} \right]^{1/3} \tag{12}$$

3.5 Minimum Heat Flux q_{min}

In a series of update to BE (Best Estimate) Code, Hsu et al |21| recommended Bromley's equation |19| for low void film boiling. Bromley's equation is based on the assumption that heat transfer through the vapour film is by molecular conduction only. Subcooled flow boiling experiments show that the film boiling heat flux can be much higher than values predicted by Bromley's equation. This could be due to the assumption that the vapor flow is laminar. It is suggested here that under flow boiling conditions, the moving liquid-vapor interface could cause the vapor flow to be turbulent. As a result, the assumption of linear temperature profile across the vapor film is not appropriate. Thus, the Bromley equation is used in the present model with a slight modification to the thermal conductivity of the vapor.

The heat transfer coefficient for film boiling given by Bromley is:

$$h_{Bromley} = 0.62 \left[\frac{k_g^3 \, g \, \rho_g (\rho_f - \rho_g) \, h_{fg}}{D \, \mu_g \, \Delta T_{sat}} \right]^{\frac{1}{4}} \tag{13}$$

One way to account for the eddy diffusivity, is to use the simple model of Baum et al |22|. They suggested that turbulence has the effect of making heat transfer proportional to Re raised to the power of 0.8 rather than Re_0. The modified thermal conductivity is then

$$k_m = k_g \left[1 + \left(\frac{Re - Re_0}{Re_0} \right)^{0.8} \right] \tag{14}$$

where Re_0 is the transition Reynolds number of 500. In the present application, a value of the vapor Reynolds number Re in the order of 10,000 was found to provide a reasonable trend in the transition boiling region when compared with experimental data.

The minimum heat flux q_{min} is then calculated by

$$q_{min} = h_{Bromley}(T_{min} - T_{sat}) \tag{15}$$

This expression of q_{min} can be used to replace q_{FB} in Eq. (1). This is because q_{FB} in the transition boiling region is near constant for subcooled conditions and thus it can be approximated by q_{min} |2|.

4. PREDICTION

Several sets of boiling curves predicted by the present model are presented in Figs. 2-4 along with experimental data |23|. Each of the boiling curves shown in these figures is constructed for a given set of parameters: mass flux, inlet subcooling and surface material. Since the model has been developed to predict heat flux as a function of local conditions, an iteration procedure was followed to calculate the local enthalpy which is in itself a function of heat flux. The iterative procedure was carried out at each point on the boiling curve and is incorporated in the computer code.

Figs. 2-4 show the effect of inlet subcooling, mass flux and heated surface material on boiling curves, respectively. In general, good agreement is observed between the predicted and experimental results in both nucleate and transition boiling regions.

In Fig. 4, the completed, predicted boiling curve for the copper surface is shown, this includes the film boiling region, CHF point and minimum point. Film boiling experimental data from the same surface are also added for comparison. The agreement between the predicted and experimental data in the nucleate boiling region, where Chen's correlation was used, provides more confidence in the prediction of the wall temperature at CHF. However, the present

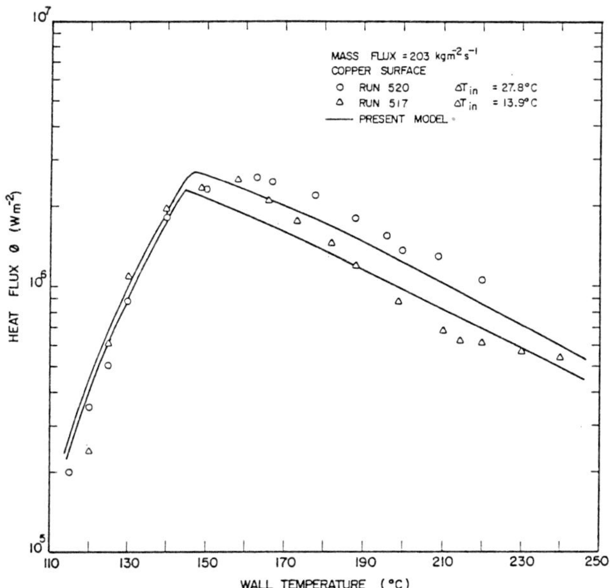

Fig. 2. Effect of Subcooling on Boiling Curves of Distilled Water, 500 Run Series at 203 kg/m^{2}s.

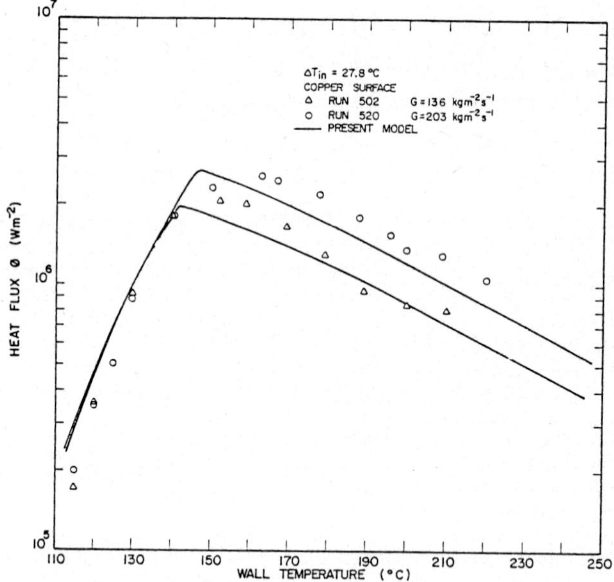

Fig. 3. Effect of Mass Flux on Boiling Curves of Distilled Water, 500 Run Series at $\Delta T_{in} = 27.8^oC$.

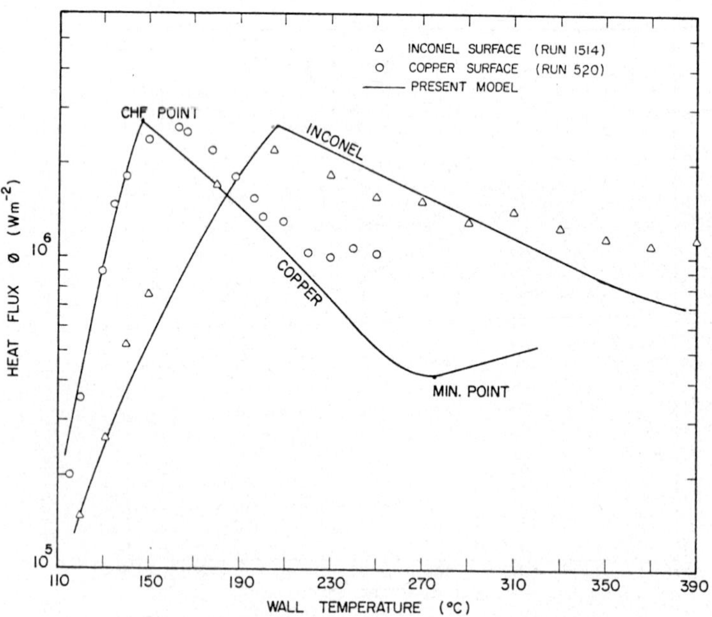

Fig. 4. Comparison of Boiling Curves for Copper Test Section and Inconel-Copper Test Section at $G = 203\ kg/m^2s$ and $\Delta T_{in} = 27.8^oC$.

model underpredicts the film boiling data by one order of magnitude. This is not entirely unexpected. From our earlier work |1|, we found that film boiling is a very strong function of upstream history which itself is partially dependent on heated length. The difference in heat flux value could be as much as several orders of magnitude. In view of our very short test section, a strong entrance effect is present which tends to enhance heat transfer. No existing correlation has taken this factor into consideration.

5. SUMMARY AND CONCLUSIONS

This paper presents a phenomenological model using existing correlations to predict transition boiling heat transfer for water at low flow and atmospheric pressure. The model incorporates effects of mass flux, local enthalpy (or inlet subcooling) and thermal properties of heated surface. Comparison of the predicted results with experimental data shows fair agreement in the nucleate and transition boiling regions.

The proposed method establishes a general ground work and can be easily extended after some modifications to predict transition boiling heat transfer for water as well as other fluids in the quality region at higher mass flux and higher pressure. For example, transition boiling at higher pressure is obtainable by multiplying a pressure factor term from atmospheric data |24|.

ACKNOWLEDGEMENTS

The authors wish to express their gratitude to the U.S. Nuclear Regulatory Commission, Argonne National Laboratory and Natural Sciences and Engineering Research Council of Canada for their financial support.

NOMENCLATURE

C_p	specific heat
CHF	critical heat flux
D	diameter
f	fraction of wetted area
g	gravitational acceleration
G	mass flux
h	enthalpy
h	heat transfer coefficient
h_{fg}	latent heat of evaporation
k	thermal conductivity
p	pressure
q	heat flux
Re	Reynolds number
Re_o	transition Reynolds number
s	wall thickness
T	temperature
x	quality
ΔT	temperature difference
μ	viscosity
ρ	density
σ	surface tension

Subscripts

CHF	critical heat flux
e	equivalent
f	saturated liquid
FB	film boiling
G	mass flux
g	vapour or saturated vapour
I	isothermal
in	inlet
ℓ	liquid
m	modified
min	minimum
sat	saturated
sub	subcooled
TB	transition boiling
w	wall
x	quality

REFERENCES

1. Cheng, S.C., Ng, W.W.L. and Heng, K.T., "Measurements of Boiling Curves of Subcooled Water Under Forced Convective Conditions", Int. J. Heat Mass Transfer 21, 1385-1392 (1978).

2. Ragheb, H.S. and Cheng, S.C., "Surface Wetted Area During Transition Boiling in Forced Convective Flow", J. Heat Transfer 101, 381-383 (1979).

3. Ragheb, H.S., Cheng, S.C., and Groeneveld, D.C., "Observations in Transition Boiling of Subcooled Water Under Forced Convective Conditions", Int. J. Heat Mass Transfer 24, 1127-1137 (1981).

4. Cheng, S.C. and Ragheb, H.S., "Transition Boiling Data of Water on Inconel Surface Under Forced Convective Conditions", Int. J. Multiphase Flow 5, 281-291 (1979).

5. Berenson, P.J., "Experiments in Pool-Boiling Heat Transfer", Int. J. Heat Mass Transfer 5, 985-999 (1962).

6. Kalinin et al, "Heat Transfer in Transition Boiling of Cryogenics", Cryogenics Eng. Conf., Queen's Univ., Kingston, Ont., Canada (1975).

7. Zuber, N., "Hydrodynamic Aspects of Boiling Heat Transfer", Ph.D. Thesis, Univ. of Calif., Los Angeles (1959).

8. Kutateladze, S.S., "A Hydrodynamic Theory of Changes in Boiling Process Under Free Convection", Izv. Akademia Nauk Otdelenie Tekh. Nauk 4, 529-536 (1951).

9. Bonilla, C.F., Nuclear Engineering, McGraw-Hill, New York, 399-431 (1957).

10. Ivey, H.J. and Morris, D.J., "On the Relevance of the Vapor-Liquid Exchange Mechanism for Subcooled Boiling Heat Transfer at High Pressure", AEEW-R137 (1962).

11. Thompson, B. and Macbeth, R.V., "Boiling Heat Transfer Burnout in Uniformly Heated Round Tubes: A Compilation of World Data with Accurate Corre-

lations", AEEW-R356 (1964).

12. Bjornand, T.A. and Griffith, P., "PWR Blowdown Heat Transfer", Symposium on the Thermal Hydraulic Aspects of Nuclear Reactor Safety, ASME, New York (1977).

13. Chen, J.C., "A Correlation for Boiling Heat Transfer to Saturated Fluids in Convective Flow", ASME Paper No. 63-HT-34 (1963).

14. Forster, H.K. and Zuber, N., "Dynamics of Vapor Bubbles and Boiling Heat Transfer", AIChE J. 1, 531-535 (1955).

15. Butterworth, D. and Hewitt, G.F., Two-Phase Flow and Heat Transfer, Harwell Series, Oxford University Press, 219 (1977).

16. Magrini, U.and Nannei, E., "On the Influence of Thickness and Thermal Properties of Heating Walls on the Heat Transfer Coefficients in Nucleate Pool Boiling", J. Heat Transfer 97, 173-178 (1975).

17. Henry, R.E., "A Correlation for the Minimum Film Boiling Temperature", AIChE Symposium Series 70, 138, 81-90 (1974).

18. Berenson, P.J., "Transition Boiling Heat Transfer from a Horizontal Surface", J. Heat Transfer 83c, 351-358 (1961).

19. Bromley, L.A., "Heat Transfer in Stable Film Boiling", Chem. Eng. Prog. Symp. Ser. 46, 221-227 (1950).

20. Zuber, N., "Stability of Boiling Heat Transfer", Trans. ASME 80, 711-720 (1958).

21. Hsu, Y.Y. and Sullivan, L., "Updating of 'Best-Estimate' Heat Transfer Recommendation for Transient CHF and Post-CHF Heat Transfer Modes", CSNI Meeting on Two-phase Flow, Los Angeles (1981).

22. Baum, A.J., Purcupile, J.C. and Dougall, R.S., "Transition and Film Boiling Heat Transfer from Vertical Surfaces", ASME Paper No. 77-HT-82 (1977).

23. Cheng, S.C., Ragheb, H.S., Ng, W.W.L., Heng, K.T. and Roy, S., "Transition Boiling Heat Transfer in Forced Vertical Flow", Final Report (1977-78), NUREG/CR-0357, ANL-78-75 (1978).

24. Cheng, S.C., Poon, K.T., Ng, W.W.L. and Heng, K.T., "Transition Boiling Heat Transfer in Forced Vertical Flow", Final Report (1979-81), Argonne Contract Nos. 31-109-38-3654 and 31-109-38-5503 (1981).

Multi-Phase Flow and Heat Transfer III. Part A: Fundamentals
edited by T.N. Veziroğlu and A.E. Bergles
Elsevier Science Publishers B.V., Amsterdam, 1984 — Printed in The Netherlands

COOLING OF A HOT PLATE WITH AN IMPINGING CIRCULAR WATER JET

Toshiaki Ochi*, Shigeyasu Nakanishi**, Masuo Kaji**, and Seikan Ishigai**
* Osaka Prefectural Technical College
Neyagawa City, Osaka Pref., 572, Japan
** Faculty of Engineering, Osaka University
Suita City, Osaka Pref., 565, Japan

ABSTRACT

Boiling heat transfer to a circular water jet impinging on a hot plate was experimentally investigated. Transient temperatures were measured on the back-face of the plate. Temperature and heat flux variations on the cooling surface were calculated from these cooling curves by solving the inverse heat conduction problem. Heat flux at the stagnation point increases with water subcooling and with impinging jet velocity divided by the nozzle diameter. It was observed that the film boiling is maintained on the whole cooling surface when the plate temperature is high. At a time when the complete film boiling ends, liquid-solid contact occurs instantaneously on the impingement zone. After that the cooling surface gets wet and the wet zone advances outwards taking with splashing of the water. The wetting velocity was defined and the factors affecting on it were discussed.

1. INTRODUCTION

The rapid cooling of a hot solid surface with an impinging water jet is applied in some technical fields such as an iron works [1], and it is a possible method for the emergency core cooling of a nuclear reactor. An experiment on the boiling heat transfer of a water jet impinging on a hot surface was carried out and the nucleate and film boiling heat flux was obtained by Copeland [2]. In this experiment, however, the jet diameter was very small and the heat flux was obtained as an average value throughout the whole cooling area. The relationship between the burnout heat flux and the cooling conditions was clarified by Monde and Katto [3]. An experiment on the boiling heat transfer of the impinging plane water jet was done and the heat flux for all boiling regions in the impingement zone was obtained by authors [4].

In this paper, first, the radial velocity distribution which dominates the heat transfer on the impingement surface was measured and the boiling heat transfer characteristic was studied for the impinging circular water jet. Cooling process from the film boiling to the nucleate boiling in the impingement and parallel flow zone was clarified.

2. EXPERIMENTAL APPARATUS AND METHOD

The experiment was carried out by a transient method. A test plate was initially heated and cooled with an impinging water jet. Temperature variations of the plate were measured and heat fluxes were determined from these temperature-time curves.

2.1. Experimental Apparatus and Procedure

The schematic diagram of the experimental apparatus is shown in fig.1. The cooling water adjusted to the prescribed temperature in the head tank /2/ was led into the calming chamber /6/, and issued from the nozzle /8/. A shutter /9/ was employed to deflect the jet from the test plate before cooling. The water flow rate was measured by an orifice /4/ and a manometer /5/ . When the water temperature in the chamber and the flow rate were stabilized, the cooling test plate /10/ was heated by A.C. power up to 1100 °C . Afte the heating was stopped, the test plate was cooled by the impingement of the water jet. The distance between the nozzle exit and the cooling surface was 25 mm. During the cooling, temperature variations of the plate were measured by thermocouples welded on the back-face of the plate and recorded by an oscillograph /16/ or a data recorder /17/. Details of the cooling test plate are shown in Fig.2. The test plate was made of stainless steel, and its cooling surface was polished with #100 emery cloth. Five chromel-alumel thermocouples of 50 μm diameter were spot welded on the back-face of the plate. Nozzle diameters were 5, 10 and 20 mm, the water subcooling ranged from 5 to 80 K and the impinging jet velocity changed from 2 to 7 m/s.

To measure the pressure distributions on the impingement surface, an acrylic test plate, which had the same shape as the cooling test plate and was equipped with pressure taps of 0.3 mm diameter, was used. The radial velocity on the surface was calculated from the pressure distribution by using Bernoulli's equation.

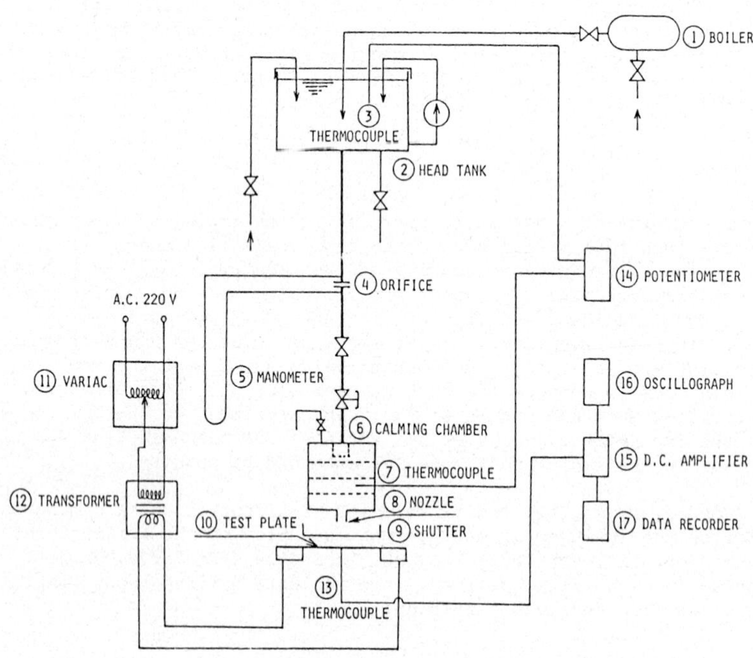

Fig.1. Schematic diagram of the experimental apparatus.

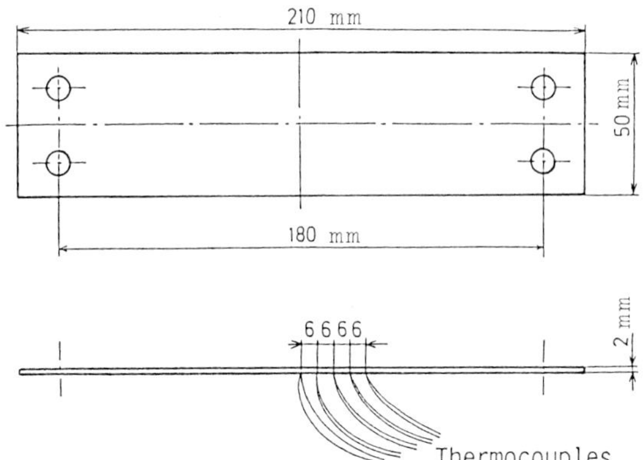

Fig.2. Test plate.

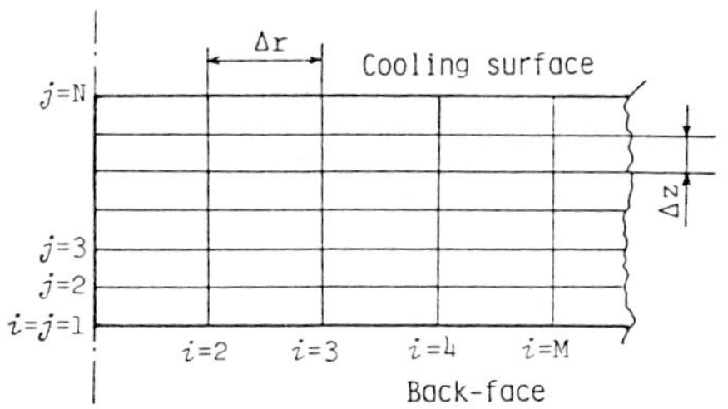

Fig.3. Nodal scheme for calculation.

2.2 Data Reduction

Temperatures and heat fluxes on the cooling surface were numerically calculated by the following method. Axisymmetric heat conduction equation,

$$\rho c \frac{\partial T}{\partial t} = \frac{\partial}{\partial r}\left(k \frac{\partial T}{\partial r} \right) + \frac{1}{r}\left(k \frac{\partial T}{\partial r} \right) + \frac{\partial}{\partial z}\left(k \frac{\partial T}{\partial z} \right) \tag{1}$$

is put into the difference equation,

$$T_{i,j+1} = \frac{(\rho c)_{i,j}}{\lambda_{i,j}} \frac{\Delta T_{i,j}}{\Delta t} \Delta z^2 - \frac{\Delta z^2}{\Delta r^2}\left(T_{i+1,j} - 2T_{i,j} + T_{i-1,j} \right)$$

$$- \frac{\Delta z^2}{2r_i \Delta r}\left(T_{i+1,j} - T_{i-1,j} \right) + \left(2T_{i,j} - T_{i,j-1} \right) \tag{2}$$

by a mesh of axial length Δz, radial length Δr and time Δt. The temperature of every node in Fig.3 is calculated under the following boundary conditions assumed. When M is the outer limit of the measured point,

$$\left.\begin{aligned} i=1 &: T_{i-1,j}=T_{i+1,j} \\ i=M &: T_{i+1,j}=2T_{i,j}-T_{i-1,j} \\ j=1 &: T_{i,j-1}=T_{i,j+1}-q_{loss,i}\frac{\Delta z}{2\lambda_{i,j}} \end{aligned}\right\} \quad (3)$$

where $q_{loss,i}$ is the heat loss from the back-face of the plate. It was preliminarily determined by experiment as a function of the temperature. Heat flux to the water jet q_i was determined from the temperatures and $q_{loss,i}$.

3. EXPERIMENTAL RESULTS AND DISCUSSIONS

3.1. Pressure and Velocity Distribution

Before cooling experiments, the pressure distribution along the surface of the plate, on which the circular water jet impinges, was measured. The pressure and the velocity along the impingement surface are nondimensionalized as, $(p-p_o)/(p_s-p_o)$ and $u_{r\infty}/V_s$ respectively, where p_s is the pressure at the stagnation point and V_s is the impingement velocity given by,

$$V_s = \sqrt{2(p_s-p_o)/\rho_\ell}$$

The results of the measurements for the nozzle diameter D = 10 mm is shown in Fig.4. The impingement zone is defined by the range where the radial velocity $u_{r\infty}$ is in proportion to the distance from the stagnation point. And the region out of the impingement zone is the parallel flow zone. In the present case, the impingement zone falls in the range $r/D \leq 1.28$.

3.2. Boiling Curve

The change of the back-face temperature recorded by the oscillograph is shown in Fig.5. In this figure, the temperature scale is shifted for every curve. Boiling curves determined from these temperature-time curves are shown

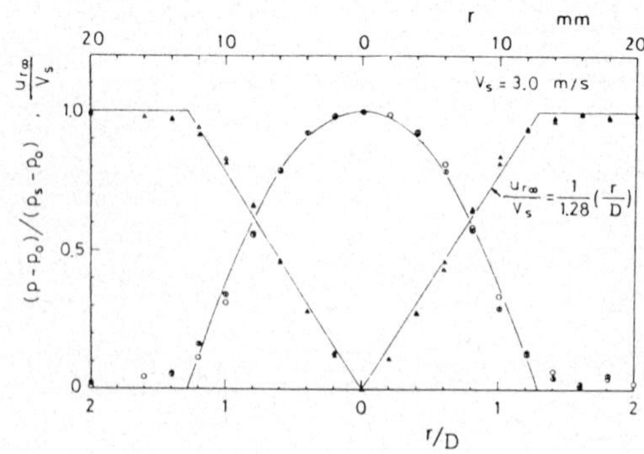

Fig.4. Pressure and velocity distributions in the radial direction on the impingement surface, D = 10 mm.

in Fig.7, and the sketches of the boiling phenomena in the cooling process are shown in Fig.6. In the cooling experiment, it was observed that the stable film boiling is maintained on the whole surface of the plate just after the beginning of the cooling (figure (a)). During the film boiling, the temperatures of the plate fall linearly with time. At the end of the film boiling, liquid-solid contact takes place on the impingement zone accompanied with the peculiar boiling noise (figure (b)). From this time the heat flux of the impingement zone increases. This boiling corresponds to the transition boiling in Fig.7. After that the cooling surface of the impingement zone gets wet completely with another boiling noise, and the wet zone advances outwards taking with the splashing of the liquid film (figure (c)). The heat transfer of the wet zone is made by nucleate boiling and nonboiling. When the nozzle diameter is equal to 20 mm, the differences of five boiling curves are small. This is due to that five measured points are in the impingement zone. On the other hand, the differences are clear when the nozzle diameter is small. Fig.8 shows the boiling curves for D = 5 mm. In this case, the heat flux at the stagnation point is higher than that for D = 20 mm. But the heat flux decreases with distance from the stagnation point, because the water film thickness decreases and the water temperature rises. When the hot surface is not completely wet (Figs.6-(a) and (b)), the whole surface is covered with liquid film. But it gets wet in the impingement zone, the liquid film out of the wet zone is splashed upwards as shown in Fig.6-(c). Then the heat flux out of the rewetting front drops. And it ascends suddenly when the rewetting front reaches the point. The splashing angle becomes larger with decreasing nozzle diameter and jet velocity.

The heat flux is the highest at the stagnation point. Comparisons of the boiling curve at the stagnation point for various conditions are shown in Figs.9 and 10. The boiling heat transfer is strongly affected by the water subcooling as shown. Increasing the water subcooling shifts the boiling curve to higher heat fluxes and higher wall superheats. When the subcooling is

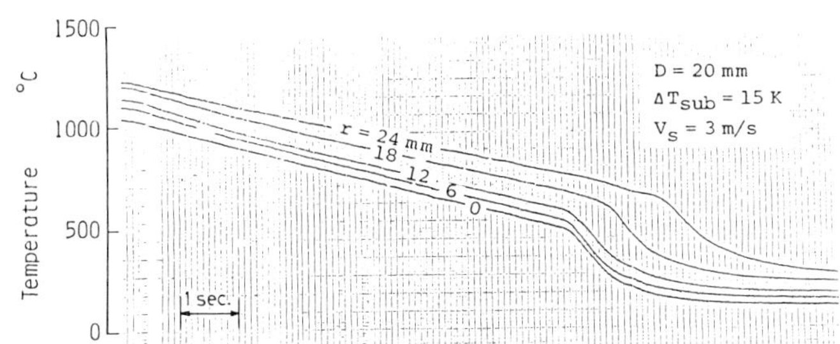

Fig.5. Typical oscillograph record of temperatures.

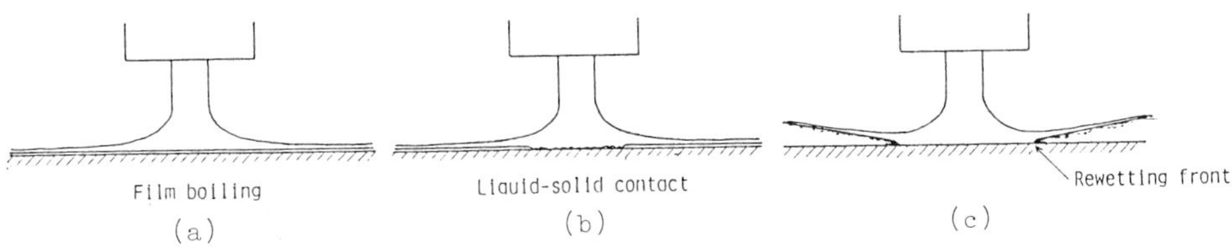

Fig.6. Sketches of boiling phenomena on the impingement surface.

raised, the film boiling ends at higher wall temperature. As a result, the temperature range of the transition boiling becomes wide and the peak of the boiling curve gets blunt. As seen from the figure, in the cases of high subcooling ($\Delta T_{sub} \geq 65$ K), the temperature of the minimum heat flux point is higher than the initial temprtature of the plate, and no film boiling appears in the present experimental range. Fig.10 shows the effect of the nozzle diameter. The heat flux increases with decreasing nozzle diameter, because the radial velocity gradient of the water on the impingement surface becomes larger.

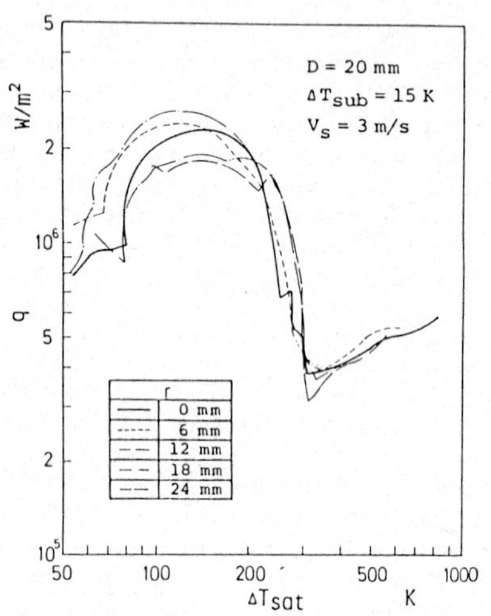

Fig.7. Boiling curves for different positions, D = 20 mm.

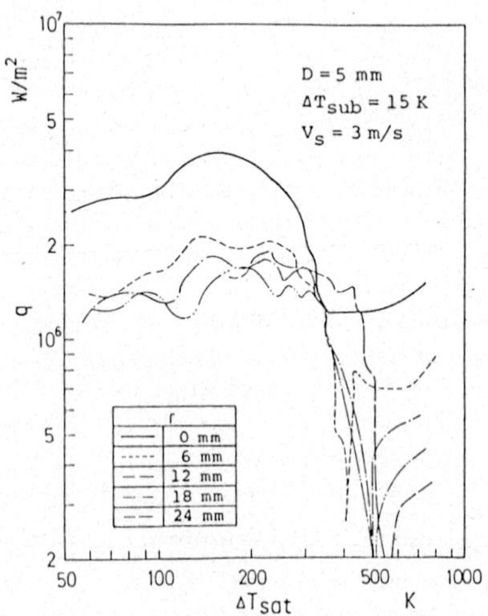

Fig.8. Boiling curves for different positions, D = 5 mm.

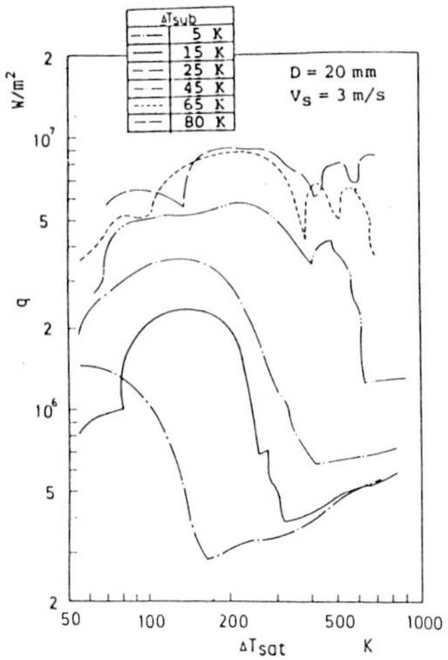

Fig.9. Comparison of the boiling curve at the stagnation point for various subcoolings.

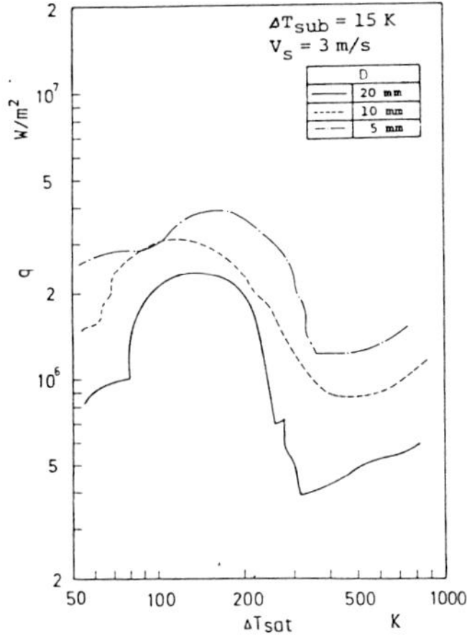

Fig.10. Comparison of the boiling curve at the stagnation point for various nozzle diameters.

This result is analytically shown by authors [4] for the film boiling of the impinging plane water jet. It can be applicable to all boiling regions in tendency. The temperature of the minimum heat flux point increases by a small amount with decreasing nozzle diameter, as the vapor film thickness of the film boiling is made thinner. For the velocity gradient becomes larger with jet velocity, increase of the jet velocity has the same effect on the heat transfer as decrease of the nozzle diameter.

Minimum heat flux points are indicated clearly in the boiling curves as shown. The effect of the water subcooling, jet velocity and nozzle diameter on the minimum heat flux is shown in Fig.11. As the radial velocity on the impingement surface dominate the heat flux of the film boiling, the jet velocity has the same effect as the reciprocal of the nozzle diameter. And the following empirical correlation equation is obained:

$$q_{min} = 3.18 \times 10^5 \cdot (1 + 0.383\,\Delta T_{sub})(V_s / D)^{0.828} \tag{4}$$

The exponent of V_s is o.607 for the plane water jet [4]. The effect of the velocity for circular water jet is more significant than that for the plane water jet. The relationship between the temperature of the minimum heat flux point and the jet velocity for three nozzle diameters is shown in Fig.12. When the nozzle diameter is small, it is not very much affected by the jet velocity, but the effect of it becomes stronger with nozzle diameter.

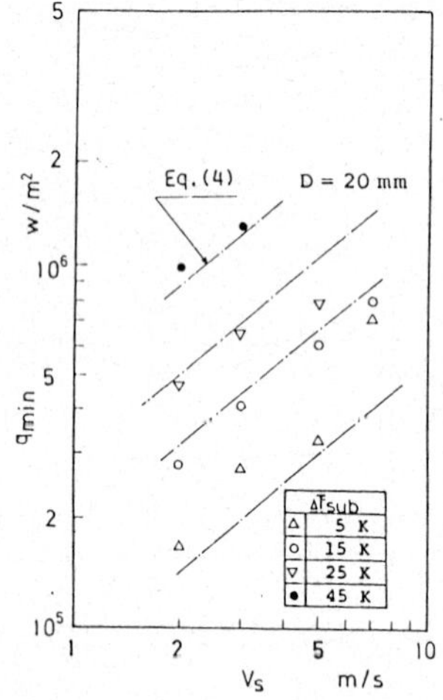

(a) Effect of jet velocity.

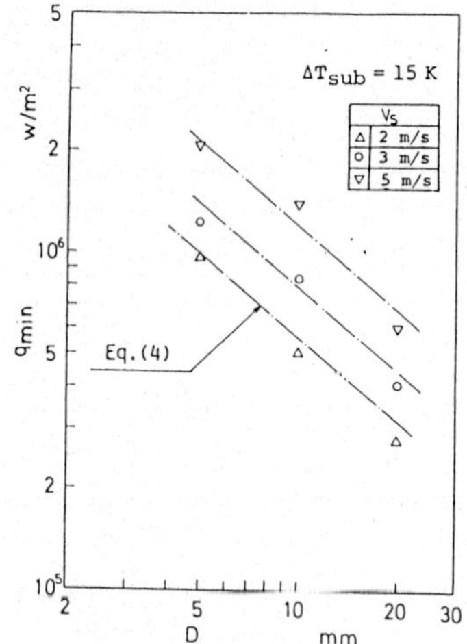

(b) Effect of nozzle diameter.

Fig.11. Variation of the minimum heat flux at the stagnation point.

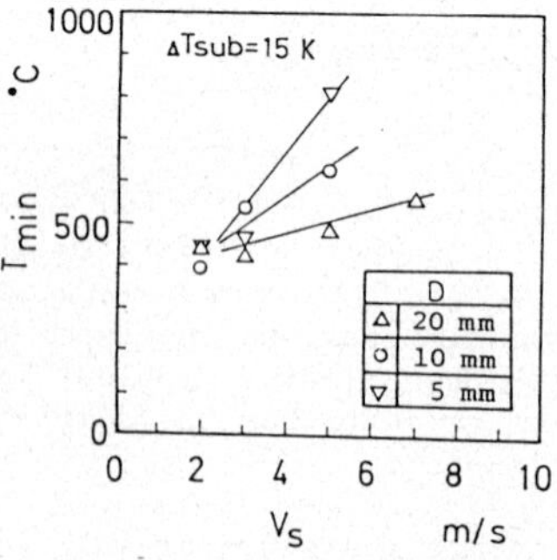

Fig.12. Variation of temperature of the minimum heat flux point with jet velocity.

3.3. Velocity of Rewetting Front

After the cooling surface in the impingement zone gets wet, the rewetting front advances gradually. When it reaches, the heat flux at the point increases suddenly. The velocity of the rewetting front V_r is defined as

$$V_r = \Delta r/\Delta t \tag{5}$$

where Δr is the distance between the adjacent thermocouples and Δt is the difference of time when the sudden increase of the heat flux begins at their points. Fig.13 shows the variation of inverse rewetting velocity along the distance from the stagnation point. The velocity defined above can be obtained also in the impingement zone. But its value indicates the velocity of the liquid-solid contact for there is no rewetting front in this region. The value of V_r^{-1} in this region is very small as shown. This means that the liquid-solid contact occurs instantaneously in the impingement zone. In the parallel flow zone, the value of V_r^{-1} is significant and it can be seen that the rewetting velocity increases with nozzle diameter. Fig.14 shows the effect of the subcooling and jet velocity on the rewetting. As seen in figure(a), the rewetting velocity becomes the minimum value when the subcooling is 15 K. This result is explained as follows: when the subcooling is higher than 15 K, the rewetting velocity increases with subcooling due to the effective cooling for high subcooling, but when the subcooling is too low (= 5 K), film boiling is kept for long time till low surface temperature, and then the rewetting front advances fast. In general, the rewetting velocity increases almost in proportion to the impinging jet velocity. The order of the velocity of the rewetting front agrees with other experiments such as for spray cooling [5] and quenching[6]. The rewetting temperature becomes higher with distance from the stagnation point. The rewetting temperature at the position of r = 18 mm for D = 10 mm is shown in figure (b). It increases linearly with subcooling. This result is in agreement with the experimental result of quenching [6].

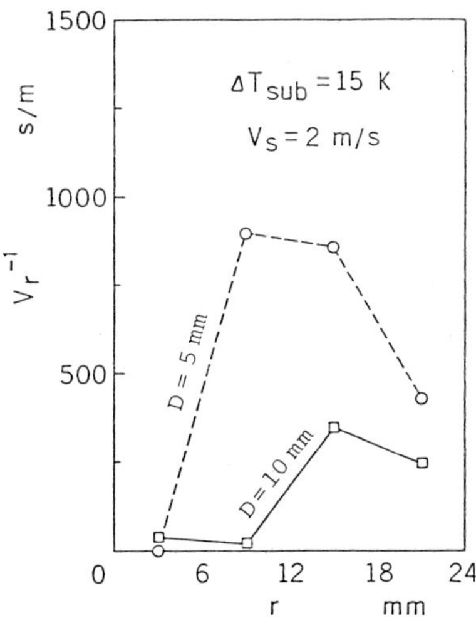

Fig.13. Variation of inverse rewetting velocity with distance from the stagnation point.

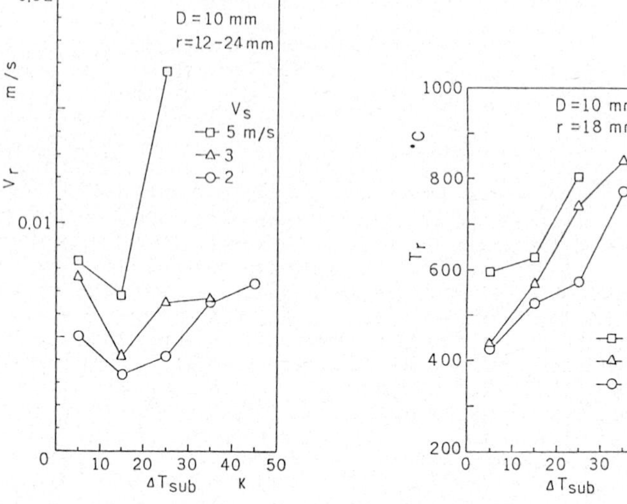

(a) Variation of rewetting velccity. (b) Variation of rewetting temperature.

Fig.14. Effect of subcooling and jet velocity on the rewetting.

4. CONCLUSIONS

Experiments on the cooling of a hot plate by an impinging circular water jet were carried out and the following results were obtained.

1. Boiling curves of the impingement zone are only slightly different each other, but, in the parallel flow zone, the heat flux decreases with radial distance from the stagnation point. The boiling curve at the stagnation point shifts to higher heat fluxes with increasing water subcooling and jet velocity, and with decreasing nozzle diameter.

2. The minimum heat flux at the stagnation point is dominated by the water subcooling and the jet velocity divided by the nozzle diameter, and an empirical correlation equation was obtained.

3. At the end of the film boiling, liquid-solid contact takes place instantaneously on the impingement zone. After that the cooling surface gets wet, and the wet zone advances taking with the splashing of the liquid film. The velocity of the rewetting front increases with nozzle diameter, jet velocity and water subcooling.

NOMENCLATURE

c	specific heat of the plate, J/kg K
D	nozzle diameter, mm
k	thermal conductivity of the plate, W/m K
p_o	atmospheric pressure, Pa
p_s	pressure at the stagnation point, Pa
q	heat flux, W/m^2

q_{min} minimum heat flux at the stagnation point, W/m^2

r distance from the stagnation point, mm

t time, s

T temperature, $^\circ C$

T_r rewetting temperature, $^\circ C$

ΔT_{sat} wall superheat, K

ΔT_{sub} water subcooling, K

$u_{r\infty}$ velocity along radial direction, m/s

V_r velocity of rewetting front, m/s

V_s impingement velocity of jet, m/s

ρ dencity of the plate, kg/m^3

ρ_ℓ dencity of water, kg/m^3

REFERENCES

1. Adcock, J. N., "The Laminar Jet System for Cooling Hot Steel Strip", Journal of the Iron and Steel Institute, Vol.200, 909-913 (1962).
2. Copeland, R. J., "Boiling Heat Transfer to a Water Jet Impinging on a Flat Surface", Ph.D. Thesis, Southern Methodist University (1970).
3. Monde, M., and Katto, Y., "Burnout in a High Heat-Flux Boiling System with an Impinging Jet", Int. J. Heat Mass Transfer, Vol.21, 295-305 (1978).
4. Ishigai, S., Nakanishi, S., and Ochi, T., "Boiling Heat Transfer for a Plane Water Jet Impinging on a Hot Surface", Sixth International Heat Transfer Conference, Toronto, Vol.1, 445-450 (1978).
5. Yamanouchi, A., "Effect of Core Spray Cooling in Transient State After Loss of Coolant Accident", Journal of Nuclear Science and Technology, Vol.5, 547-558 (1968).
6. Dhir, V. K., Duffey, R. B., and Catton, I., "Quenching Studies on a Zircaloy Rod Bundle", Journal of Heat Transfer, Vol.103, 293-299 (1981).

Multi-Phase Flow and Heat Transfer III. Part A: Fundamentals
edited by T.N. Veziroğlu and A.E. Bergles
Elsevier Science Publishers B.V., Amsterdam, 1984 — Printed in The Netherlands

FORCED CONVECTION FILM BOILING

Warren M. Rohsenow
Department of Mechanical Engineering
Massachusetts Institute of Technology
Cambridge, Massachusetts 02139, U.S.A.

ABSTRACT

The mechanism of heat transfer beyond the dryout point in forced convection boiling is described. Also phenomenological models leading to predictions of T_w and q'' are presented.

1. FILM BOILING, FORCED CONVECTION

In normal forced convection boiling the flow changes from bubbly flow nucleate boiling to annular film flow at a void fraction of around ≃ ~ 80% ($x \approx 5\%$). The annular liquid thins down due to evaporation plus entrainment minus the deposition of drops from the vapor core. As the film thins down, nucleation may be suppressed before the film completely disappears from the wall, the dryout or critical condition, Fig. 1.

Beyond this point the wall temperature rises rapidly. Heat transfer from the walls is in two parts $q_{wd} + q_{wv}$, wall to drops that approach the wall and wall to the vapor. The vapor becomes superheated beyond this point and heat is transferred from the vapor to the liquid drops, q''_{vd}. This nonequilibrium causes drops to exist well beyond the point in the tube where enough heat has been added to evaporate all of the liquid. This suggests defining an "equilibrium quality" x_{eq}, determined from the heat input as though the vapor were at saturation temperature and an "actual quality", x_a, which exists for the same upstream heat input, but with the vapor superheated. Then from an energy balance,

$$x_{eq} - x_{do} = \frac{4q''}{DGi_{\ell g}} (z - z_{do}) \tag{1}$$

$$x_a - x_{do} = \frac{4q''}{DGi_{\ell g}} (z - z_{do}) - \frac{c_p(T_g - T_{sat})}{i_{\ell g}} x_a \tag{2}$$

A measure of the extent of nonequilibrium at any z is

$$\varepsilon \equiv \frac{x_a - x_{do}}{x_{eq} - x_{do}} \tag{3}$$

which decreases from unity at the dryout point, goes through a minimum and returns to the unity at some $x_{eq} > 1.0$.

Two extreme unreal limiting operating conditions may be visualized. If one assumes that the vapor-to-drop and wall-to-drop heat transfers, q''_{vd} and q_{wd} are zero, then all of the q" goes to superheating the vapor and no additional vapor is formed, as shown in Fig. 2a. At the other extreme if q''_{vd} is sufficiently high the vapor remains close to T_{sat} and thermal equilibrium is established as shown in Fig. 2b. The actual performance for T_w and T_{ga}, lies between these two extremes as shown in Fig. 2a,b. The same performances may be shown on a plot x_a vs. x_e in Fig. 2c.

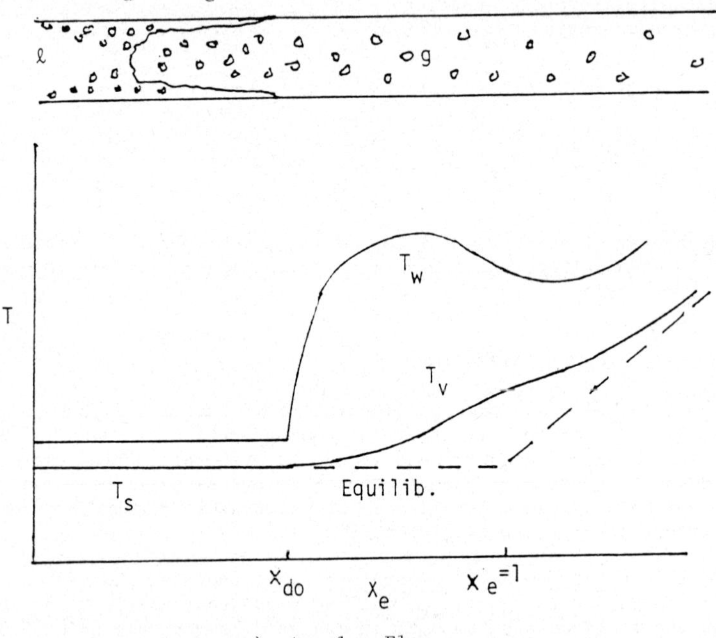

a) Annular Flow

Fig. 1. Post Dryout Temperature Distribution

2. PHENOMENOLOGICAL MODELS

A comprehensive analysis of the process shown in Fig. 1 involves accounting for the three heat fluxes q_{wv}, q_{wd}, and q_{vd}, neglecting radiation. The approximate mass, momentum and energy equations with various assumptions are integraded step-wise from the dryout point to determine all conditions, T_{ga}, D_d, ν_ℓ, ν_g, x_a, α, and T_w when q" is given.

The equations employed are a mass balance on the drops, momentum equations for the drops and for the vapor, and energy balance for the liquid, the vapor and the wall. In solving these equations various assumptions are made for heat transfer correlations, for the three heat flux terms, the drag coefficient on the drop, drop size and distribution at the dryout point.

The earliest of these analyses were by Laverty [1] and Forslund [2] and by Bennett [3]. Forslund introduced the use of a critical Weber number to establish drop break-up in flowing downstream. Bennett neglected q_{wd}. Hynek [4] extended Forslund's analysis and investigated the effect of twisted tapes to

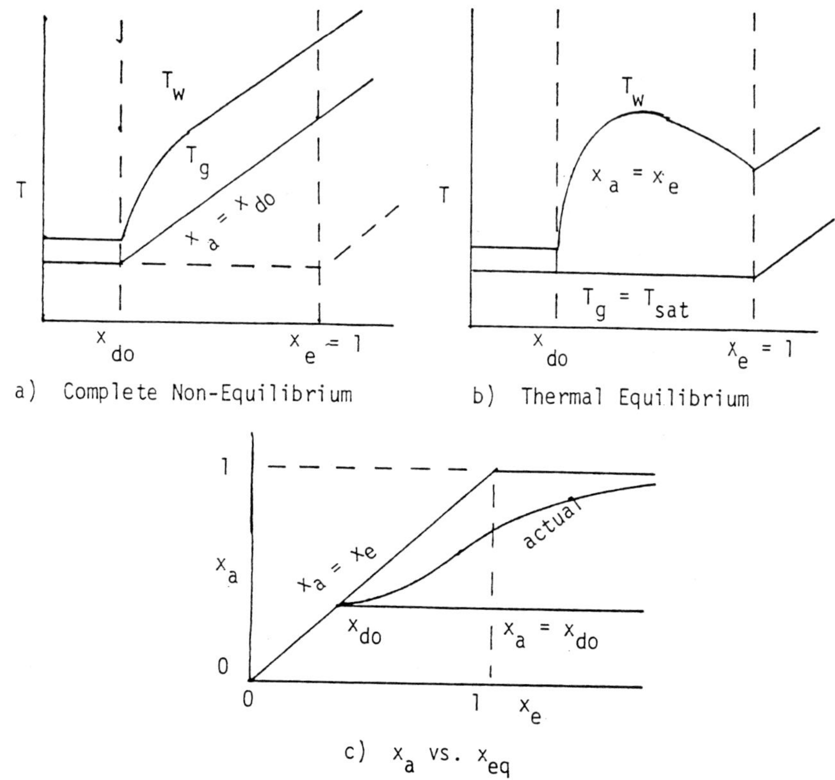

Fig. 2. Limiting Conditions for Post Dryout

induce swirl flow. Groeneveld [5] allowed pressure drop along the tube. Iloeje [6] developed a droplet-wall contact model for q_{wd} where some drops contacted the wall and others didn't. Ganic [7] introduced a droplet size distribution. In each case at least one coefficient or parameter was determined by comparison with data.

Kendall (McCarthy) [8] experimentally and analytically showed that below some Leidenfrost wall temperature the drops striking the wall essentially completely evaporate and above this temperature they bounce of the wall and less than 1% of the drop evaporates. Yoder [9] incorporated this analysis along with a prediction of average drop diameter at dryout and obtained good agreement with data without empirical coefficients.

Others have used various modifications of the above procedure. Koizumi etal [10] measured drop sizes and applied the above type of analysis to data. Yao [11] introduced the universal velocity distribution for the vapor, neglected slip and q_{wd}, and did a computer solution both radially and axially. Webb and Chen [12] performed a similar solution with different assumptions.

All of the above modified analyses give reasonable agreement with data, but require a step-wise computer solution.

Plummer [13] made simplified assumptions from the results of the step-wise calculations of the detailed model in order to arrive at a procedure for predicting the q" at any local position given, T_w, x_{do}, L_{do}, G, p, and x_e.

From the computer calculations x_a vs. x_e is actually a curve which was approximated as a straight line or ε is the constant in Eq. (3). The following is a curve fit for three different fluids:

$$\varepsilon = C_1 \ln \left[G \left(\frac{D}{\rho_g \sigma}\right)^{1/2} (1 - x_{do})^5\right] + C_2 \tag{4}$$

where

FLUID	C_1	C_2
Nitrogen	0.0821	0.290
Water	0.0700	0.400
Freon-12	0.0780	0.255

With this relation x_a is determined from Eq. (3), and T_{ga}, the superheated vapor temperature, is calculated by combining Eqs. (1) and (2):

$$\frac{x_e}{x_a} = 1 - \frac{c_{p,g}(T_{ga} - T_{sat})}{i_{\ell g}} \tag{5}$$

The slip ratio, $S \equiv V_g/V_\ell$, beyond dryout was approximated by

$$\frac{S - 1}{S_{do} - 1} = \left[1 - \frac{\varepsilon(x_e - x_{do})}{1 - x_{do}}\right]^{A/\varepsilon^B} \tag{6}$$

where

$$S_{do} - 1 = \frac{1}{2}\left[\frac{(\rho_\ell/\rho_g)^{0.205}}{(GD/\mu_\ell)^{0.016}} - 1\right] \tag{7}$$

and

FLUID	A	B
Nitrogen	5.0	0.486
Water	2.5	0.264
Freon-12	1.0	0.37

The void fraction α is

$$\alpha = \frac{x_a}{x_a + (\rho_g/\rho_\ell)(1 - x_a)S} \tag{8}$$

The length of tube beyond the dryout, L_{do}, is calculated from an energy balance:

$$L_{do} = \frac{GDi_{\ell g}}{4} \int_{x_{do}}^{x_e} \frac{dx_e}{q''} \tag{9}$$

With calculated values from the above equations for x_a, T_{ga}, α, and L_{do}, the heat flux is calculated from

$$q'' = 0.023 \frac{k_g}{4} \left(\frac{Gx_a D}{\mu_g \alpha}\right)^{0.8} Pr_g^{1/3} \left(\frac{\mu_g}{\mu_{g,w}}\right)^{0.14}$$

$$x \left[1 + 0.3 \left(\frac{D}{L_{do} + 0.1D}\right)^{0.7}\right] (T_w - T_{ga}) +$$

$$+ Ck_{gf} (1 - \alpha)\exp\left(-\frac{2D}{L_{do}}\right) (T_w - T_{sat}) \tag{10}$$

where $C = 2500\ ft^{-1}$ or $8202\ m^{-1}$.

The first term represents q''_{wv} and the second term q''_{wd}. The empirical L_{do}/D terms give better results close to dryout location.

The following are the limits for the above procedure:

$$G > 1.22 \left(\frac{\rho_\ell \rho_g}{\rho_\ell - \rho_g}\right) \left(\frac{\sigma(\rho_\ell - \rho_g) g g_c}{\rho_v^2}\right)^{1/4}$$

$0.01 < x_{do} < 0.95$

$x_e > x_{do}$, $L_{do} > 0$

Nitrogen, water, Freon12.

Yoder [9] improved this "local condition" model by combining the various equations and observing from calculated results for a variety of conditions that K in the following equations remained essentially constant along the tube.

$$K \frac{x_a^{3/4} x_e}{(1 - x_a)^{7/12}} \frac{dx_a}{dx_e} = (x_e - x_a) \tag{11}$$

where Hill [14] developed the following relation for K:

$$K = 0.0013 \frac{f \phi Pr_v^{2/3} \left(\sqrt{\frac{\rho_\ell}{\rho_g}} x_{do} + 1\right)^{5/2} \left(\frac{\mu_g}{\mu_\ell}\right)^{5/6} A_c^{3/4} Re_{do}^{4/3}}{(1 - x_{do})^{5/12} x_{do}^{7/4} \left(\frac{\rho_\ell}{\rho_g}\right)^{1/12} We_{do}^{5/4}} \tag{12}$$

where

$$A_c \equiv \frac{q''}{Gi_{\ell g}}$$

$$Re_{do} = \frac{GD_T}{\mu_g}\left[x_{do} + \frac{\rho_g}{\rho_\ell}(1 - x_{do})\right]$$

$$We_{do} = \frac{G^2 x_{do}{}^2 D_T}{\rho_g \sigma}$$

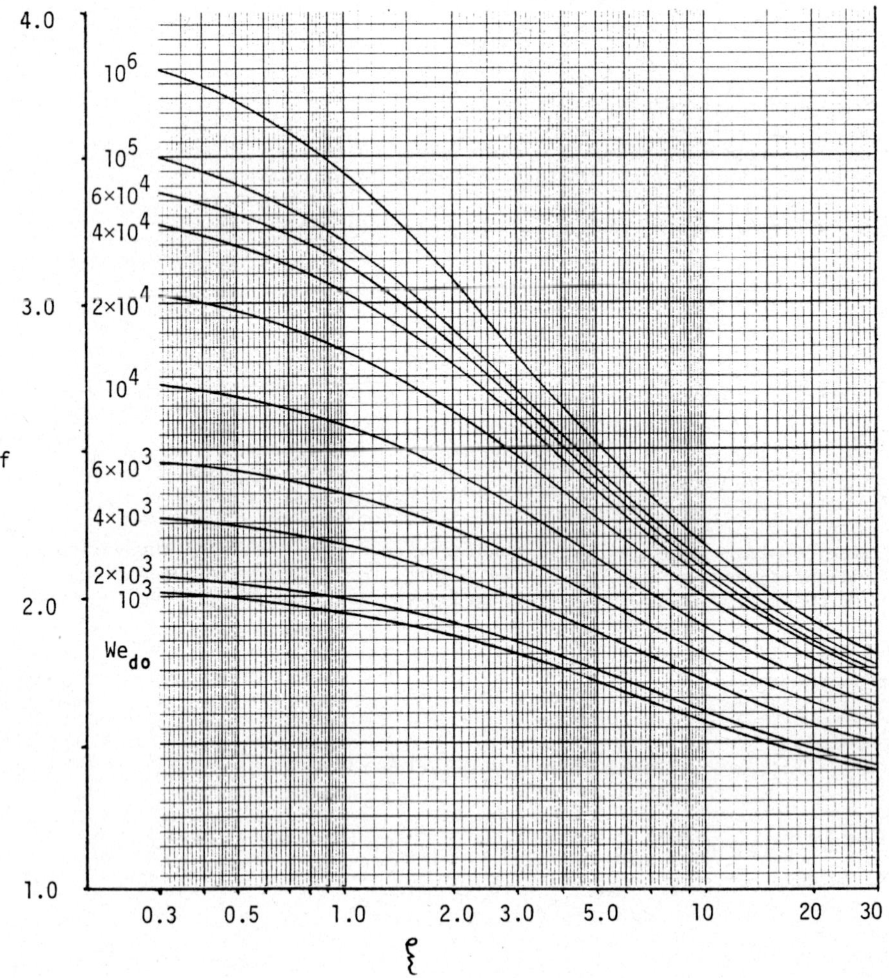

Figure 3 Distribution Factor for $x_{do} = 0.20$

The quantities f and ϕ are read from Figs. 3 and 4, where

$$\xi = \frac{0.0338 \ x_{do}}{Re_{do}^{1/8} A_c} \tag{13}$$

Equation (11) is shown plotted in Fig. 5 as x_a vs. x_e for various magnitudes of the non-equilibrium parameter K for x_{do} = 0.1. A family of such graphs may be constructed for various mangitudes of x_{do}. However, Fig. 5 may be used for any x_{do} by making a simple construction. It is noted that since at x_{do}, $x_a = x_e = x_{do}$, Eq. (11) shows dx_a/dx_e = 0. If Eq. (11) is differentiated once the radius of curvature of a constant K curve at x_{do} is:

$$\frac{1}{d^2x_a/dx_e^2} \equiv r_{do} = \frac{K \ x_{do}^{7/4}}{(1 - x_{do})^{7/12}} \tag{14}$$

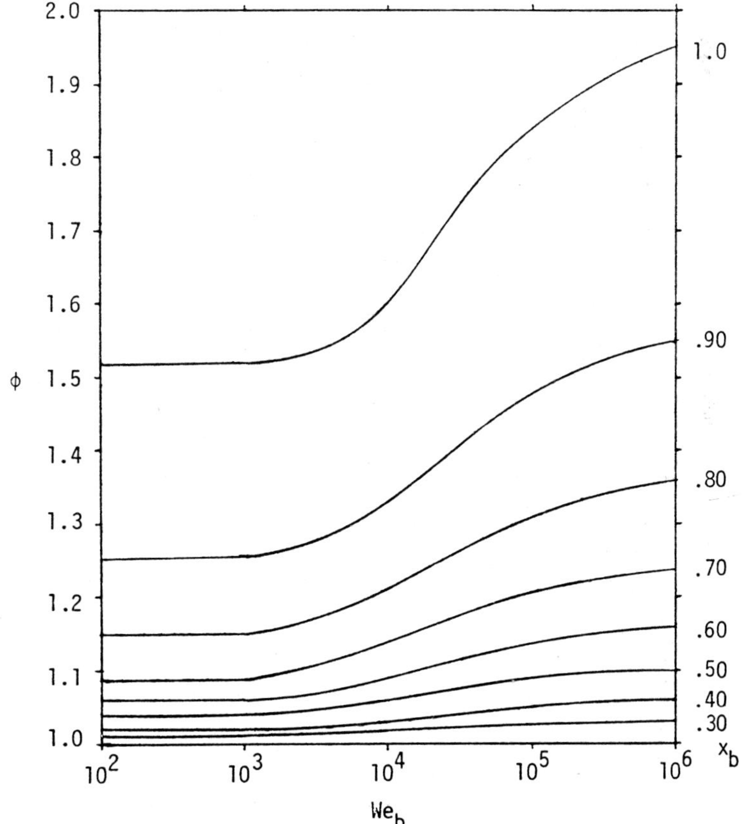

Figure 4 Distribution Factor Multiplier for Dryout Quality

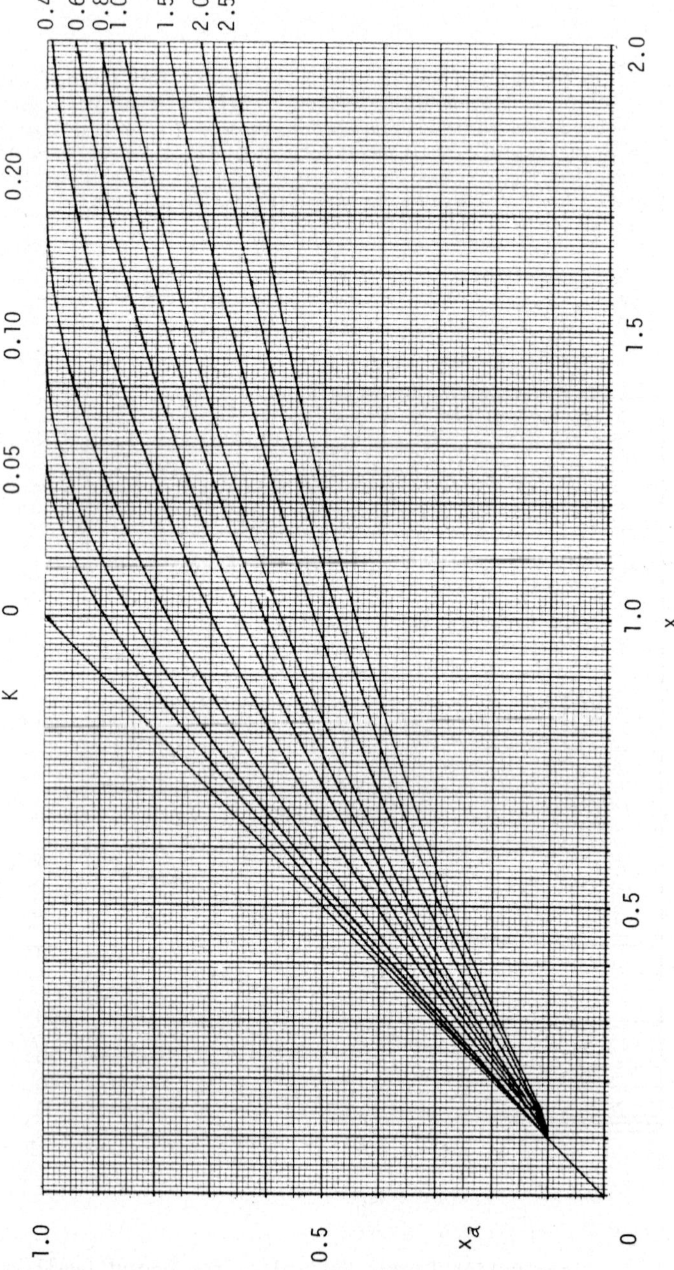

Figure 5 Actual Quality vs. Equilibrium Quality

Figure 6 shows the construction to be made on Fig. 5 to obtain the constant K curve for any $x_{do} > 0.1$. The solid curve on Fig. 6 is a constant K curve reproduced from Fig. 5. At the larger x_{do} construct an arc of radius r_{do} as shown. A stairght line (dashed) tangent to the arc and the original constant K curve is a good approximation to the curve for the same K for the larger x_{do}. The same construction applies to any other $x_{do} > 0.1$ and any magnitude of K.

The calculation of K is as follows. In Fig. 7 the known quantities are G, D_T, z_{do}, x_{do}, q" and all fluid properties. To determine the wall temperature T_w at any $z > z_{do}$:

1) Calculate A_c, Re_{do}, We_{do} and K from Eq. 12 where f and ϕ read from Figs. 3 and 4 with ξ calculated from Eq. (13).
2) Calculate x_e from Eq. (1).
3) Calculate r_{do} from Eq. (14) and make the construction of Fig. 6 on Fig. 5 for the approximate K and x_{do}. Read x_a from Fig. 6.
4) Calculate T_v from Eq. (2).
5) Calculate h from

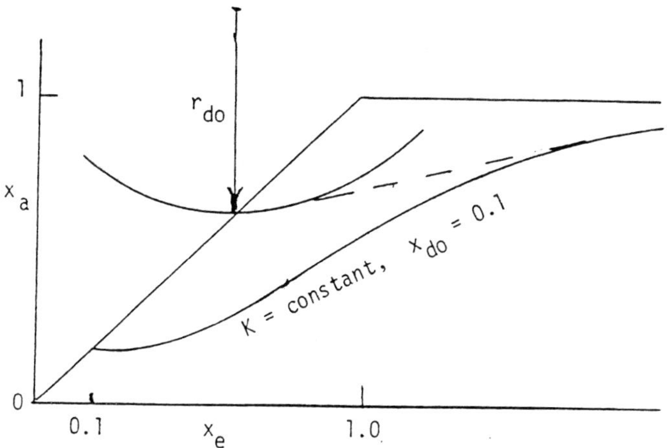

Figure 6. Construction for K = Constant for $x_{do} > 0.1$

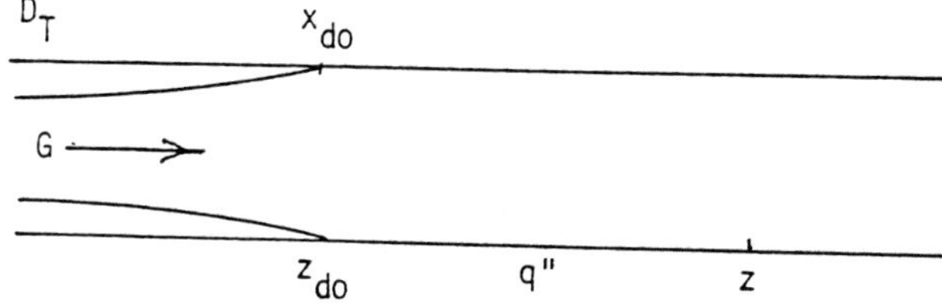

Figure 7. Local Condition Solution Given Quantities

$$\frac{hD_T}{k_{gf}} = 0.00835 \left(\frac{Gx_a D_T}{\mu_{gf}\,\alpha}\right)^{0.877} Pr_{gf}^{0.611} \tag{15}$$

where

$$\frac{1}{\alpha} = 1 + \frac{1 - x_a}{x_a} \frac{\rho_g}{\rho_\ell}$$

6) Calculate $T_w = T_v + q''/h$.

This calculation for T_w may be made at any $z > z_{do}$ without integration from z_{do}.

Figure 8 shows a comparison of this prediction and others with a representative set of data. Hill [14] has compared these various prediction equations and procedures with many test runs for water, Freon-12, and nitrogen for the case of annular flow. In general, the above procedure developed by Hill and Yoder agrees with most of the data. This analysis does not apply well near the dryout point at small $z = z_{do}$. This region is discussed in the next section.

3. CLOSE TO DRYOUT LOCATION

Just downstream of the dryout location there are very high heat transfer coefficients. This effect has been incorporated in various prediction equations empirically with a term involving exp (- D/L) as in Eq. (10). Attempts to model this region as a single phase developing boundary layer have under predicted the magnitude of this region.

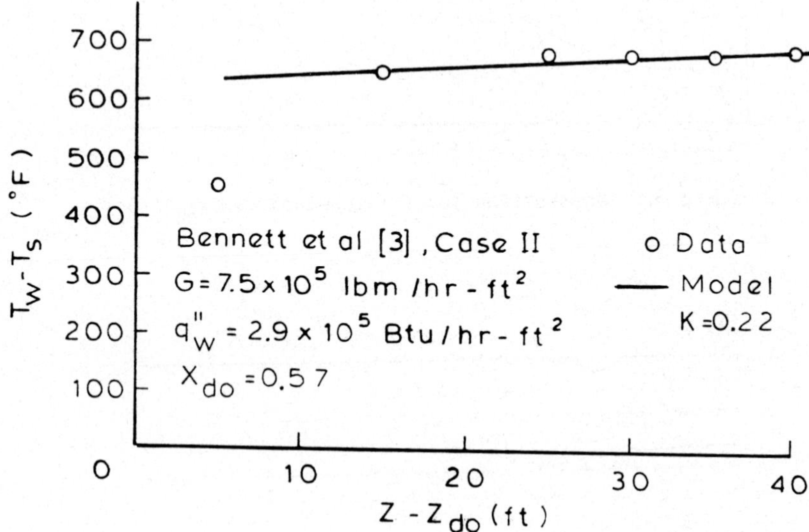

Fig. 8 Comparison with Water Data

Hull [15], acting on a suggestion by B.B. Mikic, modelled this region as a droplet laden thermal boundary as shown in Fig. 9. Essentially the droplets act like an increased specific heat of the fluid mixture resulting in higher values than for a single phase fluid.

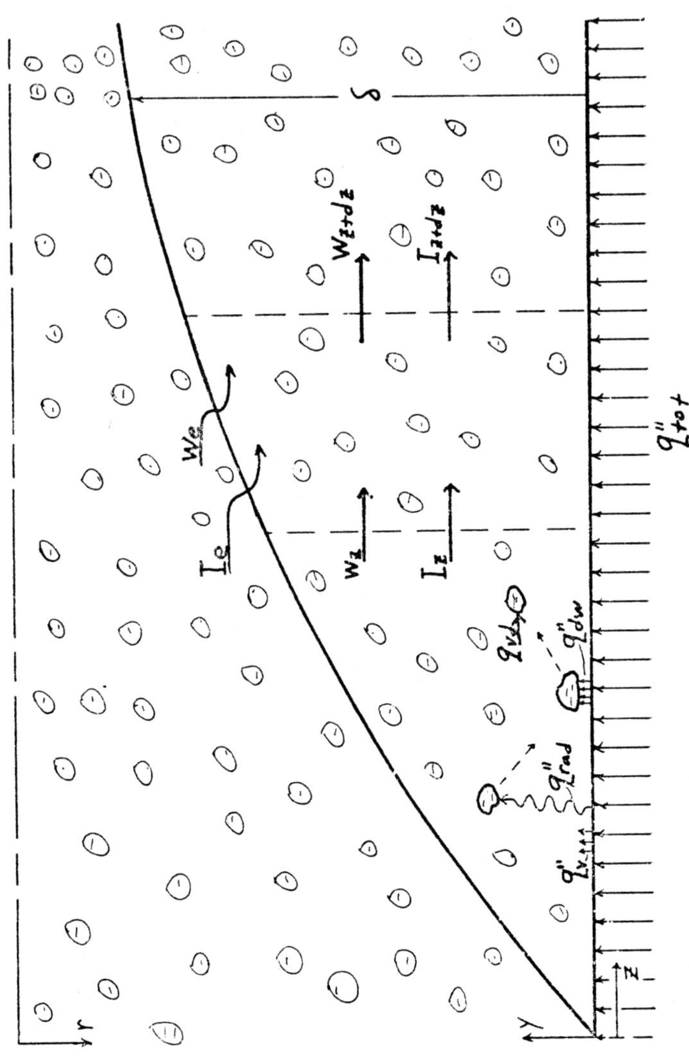

Fig. 9. Dispersed Flow Boundary Layer

The integrated equation for this boundary layer is;

$$\frac{d}{dz}\int_{r_{o-\delta}}^{r_o} \rho_v\ v\ c_{p,v}\ (T - T_s)2\pi r dr + \int_{r_{o-\delta}}^{r_o} n\pi d^2 h_{vd} 2\pi r dr = q''_{vd}\ \pi D_T \tag{16}$$

where n is the droplet density and d is droplet diameter.

Hull integrated Eq. (16) with the velocity profile unaffected by the presence of the droplet, but he derived a temperature profile which was influenced by the droplets. The droplet density was varied as the 1/7 power of distance from the wall. Slip was determined from a force balance on the drop and Nu_{vd} was determined from the resulting relative velocity. The heat transfer from wall to vapor was calculated incrementally downstream of the dryout location. This computer solution is lengthy and involved trial-and-error at three steps for each incremented downstream position.

This complex analysis was simplified assuming the turbulent velocity and temperature distributions are unaffected by the droplets, the turbulent Prandtl number is unity. This results in the following equations.

For the heat flux from the wall to vapor q''_{wv}:

$$Nu = \frac{q''_{wv} D_T}{(T_w - T_{sat})k_v} = 0.033 \left(\frac{G\,x\,D_T}{\alpha\,\mu_v}\right)^{0.78} Pr\left[\frac{\Gamma}{1 - e^{\frac{\Gamma(z-z_{do})}{D_T}}}\right]^{0.13} \tag{17}$$

where

$$\Gamma \equiv \frac{4\,q''_{w,tot}}{G\,i_{\ell g}} \frac{(1 - x_{do})^{7/12}}{K\,x_{do}^{3/4}}$$

$$\alpha = \frac{1}{1 + \frac{1-x_a}{x_a}\frac{\rho_v}{\rho_\ell}}$$

and K is given by Eq. (12).

Also, the actial quality x_a is given by

$$x_a = x_e - \frac{q''_{wv}/q''_{w,tot}}{\lambda}\left[1 - e^{-\lambda(x_e - x_{do})}\right] \tag{18}$$

where

$$\lambda \equiv \frac{(1 - x_{do})^{7/12}}{K\,x_{do}^{3/4}}$$

Assuming radiation from wall-to-drops q''_{rad} and wall-to-drop interaction q''_{wd} are negligible, $q''_{wd} = q''_{w,tot}$. For any position $z > z_{do}$, calculate x_e from Eq. (1), K at dryout point conditions, and x_a from Eq. (18). Then, calculate α and Γ and T_w from Eq. (17).

The preceding analysis leads to the following expression of the droplet-laden thermal boundary layer thickness, δ:

$$\frac{\delta}{D_T} = 0.366\ Re^{-0.277} \left[\frac{1 - e^{-\frac{\Gamma(z-z_{do})}{D_T}}}{\Gamma}\right] \quad (19)$$

where the analysis is certainly invalid beyond distance z at which $\delta = D_T/2$.

4. DROP-WALL INTERACTION

The total heat flux from the wall is:

$$q''_{w,tot} = q''_{wv} + q''_{dw} \quad (20)$$

In many cases the latter quantity is negligible. Near the dryout point at lower qualities q''_{dw} may become large enough to consider. Kendall - McCarthy [8] studied this problem leading the the following result:

$$q''_{dw} = 0.03 \frac{\rho_v}{\rho_\ell} \frac{i_{fg}\mu_v}{D_T} \varepsilon\ (1 - \alpha)\ Re_{DT}^{7/8} \quad (21)$$

Very close to dryout ε is approximately unity. The wall temperature rises in fractions of an inch to above a Leidenfrost temperature and ε becomes $\varepsilon = 0.0025$.

The analysis of the previous section is then accomplished by substituting Eqs. (17) and (21) in Eq. (20), and T_w determined for a given $q''_{w,tot}$.

The effect of radiation between the wall and drops may also be included, but is omitted here.

Hull [15] performed experiments by introducing known masses of drops of known diameter into the inlet vapor flow and measured wall temperatures at many locations along the tube. Figure 10 is one set of data showing the predictions from his complex computer solution, from the simplified solution of the previous section, and from a single phase boundary layer.

The simple solution usually predicts wall temperature slightly higher than the measured values. More work will be done to improve the simple solution for better agreement.

5. CONCLUSIONS

Detailed phenomenological models of the post dryout region of film boiling have lead to complex computer solutions prediction of wall temperature in good agreement with data. These complex solutions have been simplified to lead to "local condition" solutions for any position downstream of dryout without requiring intergration down the tube.

In Section 2 the simplified solution gives reasonable agreement with data downstream of dryout, but does not apply very close to dryout.

In Section 3 the simplified solution gives reasonable agreement with data close to the dryout locaiton, but does not apply far downstream. The two solutions intersect a short distance from the dryout point.

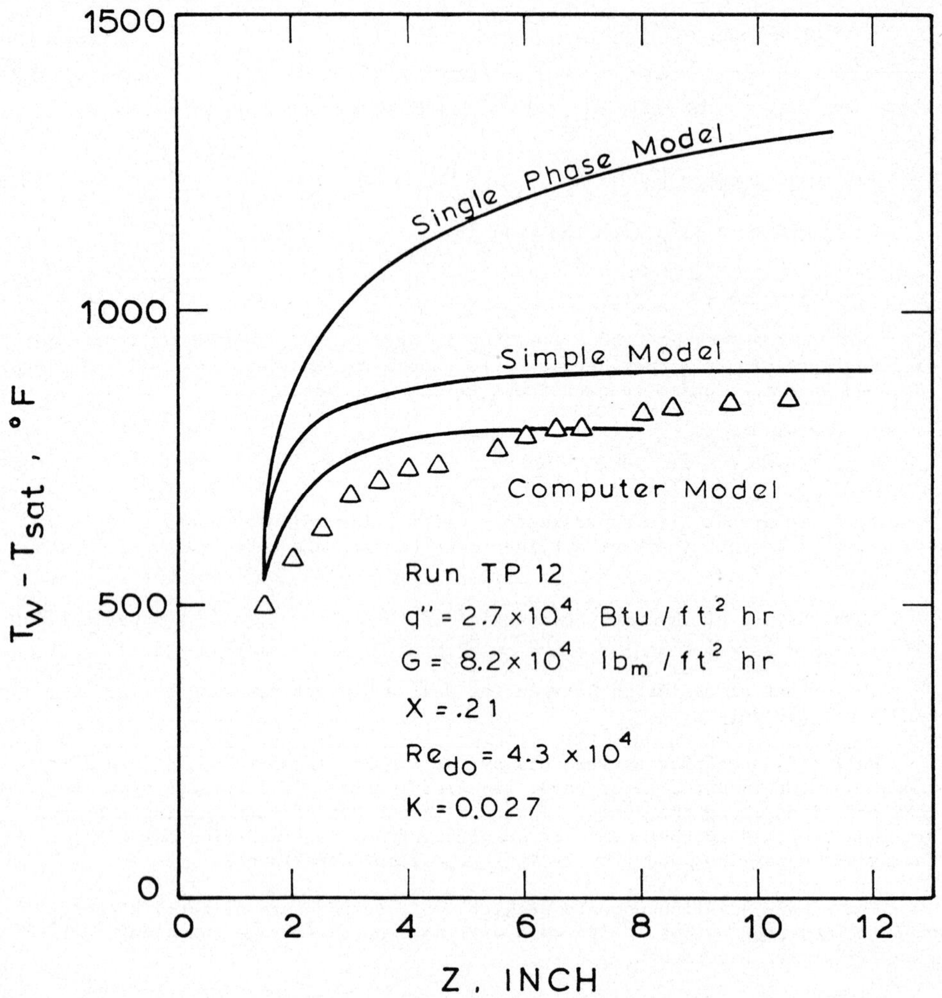

Fig. 10. Comparison of Prediction with Data Near Dryout Location (15)

ACKNOWLEDGEMENT

We gratefully acknowledge the support of the Heat Transfer Research Program of the National Science Foundation.

NOMENCLATURE

(SI and English Units)

A_c	$q''/Gi_{\ell g}$
c_p	specific heat at constant pressure, J/kg·K; Btu/lb_m °F
d	droplet diamter, m; ft
D, D_T	tube inside diameter, m; ft
G	fluid mass velocity = W/A_o, kg/m^2·s; lb_m/hr·ft^2
g	gravitational acceleration, m/s^2; ft/s^2
g_c	proportionally constant in Newton's second law of motion, 1; 32.174 lb_m·ft/lb_f·s^2
h	heat transfer coefficient, W/m^2·K; Btu/hr·ft^2°F
i	enthalpy per unit mass, J/kg; Btu/lb_m
$i_{\ell g}$	latent heat of evaporation, J/kg; Btu/lb_m
K	defined by Eq. (12)
k	thermal conductivity, W/m·K; Btu/hr·°F·ft
L	length, m; ft
n	number droplet density
Nu	Nusselt number = hD/k
Pr	Prandtl number = $\mu c_p/k$
q"	heat flux (heat transfer rate per unit of area), W/m^2; Btu/hr·ft^2
Re	Reynolds number
r_{do}	radius of curvature, Eq. (14), units of x
S	slip ratio = V_g/V_ℓ
T	temperature, °C,K; °F,R
V	velocity, m/s; ft/s
We	Weber number = $G^2x^2D_T/\rho_g\sigma$
x	mass vapor quality
z	distance along tube, m; ft

GREEK SYMBOLS

α	void fraction
ε	defined by Eq. (3); also Eq. (20)
Γ	defined with Eq. (17)
λ	defined by Eq. (18)
μ	dynamic viscosity, Pa·s; lb_m/hr·ft
ν	kinematic viscosity, m^2/s;ft^2/hr
ρ	density, kg/m^3; lb_m/ft^3
σ	surface tension, N/m; lb_f/ft
ϕ	quantity from Fig. 4 in Eq. (12)
ξ	defined by Eq. (13)

Subscripts

a	actual
do	dryout
eq	equilibrium
g	vapor
	liquid
sat,s	saturation condition
vd	vapor to drop
w	wall condition
wd	wall to drop
wv	wall to vapor
w,tot	total from wall

REFERENCES

1. Laverty, W.F., and Rohsenow, W.M., "Film Boiling of Nitrogen Flowing in Vertical Tubes", ASME J. Heat Transfer", February 1967, 89c, 1.

2. Forslund, R., and Rohsenow, W.M.,"Dispersed Flow Film Boiling", ASME J. Heat Transfer, November, 1968, 90c,4.

3. Bennett, A.W., Hewitt, G.F., Kearsey, H.A. and Keeys, R.K.F., "Heat Transfer to Steam Water Flowing in Uniformly Heated Tubes Beyond CHF", AERE-R-5373, 1967.

4. Hynek, S.J., Rohsenow, W.M., and Bergles, A.E., "Forced Convection Dispersed Flow Film Boiling", Heat Transfer Lab. Report 70586-63, M.I.T., Cambridge, 1966.

5. Groeneveld, D.C., "Thermal Behavior of a Heated Surface at and Beyond Dryout, AE CL-4309, 1972.

6. Iloeje, O.J., Rohsenow, W.M., and Griffith, P., "Three-Step Model of Dispersed Flow in Post CHF", ASME 75-WA/HT-1, July, 1975.

7. Ganic, E.N., and Rohsenow, W.M., "Dispersed Heat Transfer ", Int. J. Heat Mass Transfer, 20, 8, August, 1977, pp. 855-866.

8. Kendall-McCarthy, G.E., and Rohsenow, W.M., "Heat Transfer to Impacting Drops and Post-Critical Heat Flux Dispersed Flow", Heat Transfer Lab. Report 85694-100, M.I.T., Cambridge, March, 1978.

9. Yoder, G.L., Jr., and Rohsenow, W.M., "Dispersed Flow Film Boiling", Heat Transfer Lab. Report 95694-103, M.I.T., Cambridge, March, 1980; also ASME Paper, Session 31, 20th Nat. Heat Transfer Conference, Milwaukee, August, 1981.

10. Koizumi, Y., Ueda, T., and Tanaka, H., "Post Dryout Heat Transfer to R-113", Int. J. Heat Mass Transfer, 22, 5, May, 1978, pp. 669-678.

11. Yao, S., and Rane, A., "Numerical Study of Turbulent Droplet Flow Heat Transfer", ASME 80-WA/HT-48, November, 1980.

12. Webb, S.W., and Chen, J.C., "Numerical Model for Turbulent Dispersed Flow, ASME Paper Session, 31, 20th Nat. Heat Transfer Conference, Milwaukee, August, 1981.

13. Plummer, D., Griffith, P., and Rohsenow, W.M., "Post Critical Heat Transfer in Vertical Tubes", Trans. Canadian SME, 4, No. 3, 1976-1977, pp. 151-158.

14. Hill, W.S., and Rohsenow, W.M., "Dryout Droplet Distribution and Dispersed Flow Film Boiling", M.I.T. Heat Transfer Lab. Report No. 85694-105, Cambridge, August, 1982.

15. Hull, L.M., and Rohsenow, W.M., "Thermal Boundary Layer Development in Dispersed Flow Film Boiling", M.I.T. Heat Transfer Lab. Report No. 85694-104, Cambridge, August, 1982.

Multi-Phase Flow and Heat Transfer III. Part A: Fundamentals
edited by T.N. Veziroğlu and A.E. Bergles
Elsevier Science Publishers B.V., Amsterdam, 1984 — Printed in The Netherlands

THE EFFECTS OF ANNULAR GAP SIZE AND INCLINATION ON HYDRO-THERMODYNAMIC BEHAVIOR IN NATURAL CONVECTION BOILING IN AN ANNULAR TUBE

In Suhk Kim
Assistant Professor of Myong Ji University
Seoul, Korea
(Currently Visiting Assistant Professor of
Nuclear Engineering at University of Illinois
at Urbana-Champaign.)

ABSTRACT

This paper presents the results of an experimental investigation of the effects of gap size and angle of inclination of an annular test section on the hydro-thermodynamic behavior in natural convection boiling flow.

The liquid used is pure water at saturated temperature and atmospheric pressure. Three annular test sections, consisting of outer glass tubes with inside diameters of 16mm, 22mm, and 28mm in which there is an electically heated stainless steel pipe of 4.6mm in diameter and 250mm in length, were used. Equal angular intervals of 22.5 degrees were used to position the test section at the five angles of inclination from horizontal to vertical in this study.

1. INTRODUCTION

In natural convective boiling annular flow, the parameters such as flow patterns, oscillations, burnout heat flux and heat transfer are important for the thermal design of water-cooled nuclear reactors and high performance of heat exchangers.

Flow patterns tend to be strongly influenced by void fraction, phase velocity and shearing force as well as showing a close relation with pressure drop, flow oscillations and local heat transfer level. Hewitt and Hall-Taylor [1] carried out a detailed study of flow patterns. Hewitt and Roberts [2] developed the experimental technique of x-ray absorption for detecting flow patterns. Although various descriptive names for flow patterns have been used, there are typically only six distinct patterns for horizontal flows and five for vertical flows [3].

It is common to describe flow pattern studies in terms of flow pattern maps. Manhane [4] and Baker [5] have presented such maps.

Oscillation conditions are characterized by the correlation among heat flux, void fraction, flow rate and flow geometry. Quandt [6] carried out an analytical study of flow oscillations by means of perturbation methods. Most theoretical analyses of oscillation flows are based on system mathematical models.

In reactor core design there is a critical need to predict burnout heat fluxes in complex geometries. Whalley [7], who compared his results with experimental data of Jensen [8], estimated burnout heat flux levels in annular flows. Tong [9] suggested that such burnout heat flux values be calculated on a parameter based on the ratio of hydraulic radius-to-heating length. With an increase in aspect ratio of the heating tube, the burnout heat flux decreases. Throttling at the inlet of heating tube restrains the oscillations and the burnout heat flux increases.

B. R. Rich [10] has presented both experimental and theoretical results of heat transfer from an inclined flat plate in natural convection. A major portion of the studies for heat transfer in inclined geometry have used air or single phase flow conditions.

There are neither analytical studies nor experimental studies for annular boiling flow which take into account both channel inclination and annular gap size. For this reason, this experimental investigation was undertaken to provide reactor designers with good experimental data with which to select reactor dimensions and boiling flow systems.

2. EXPERIMENTAL APPARATUS

Figure 1 shows the natural convective circulating loop used in these studies. The liquid used is pure water at saturation temperature and atmospheric pressure. The water bath sustained the saturated temperature condition during the experiment. One face of the water bath included a visual window for observation of the flow patterns. Using a precise angle indicator, inclination of the tubes was varied in five steps from horizontal to vertical using the angular intervals; 0°, 22.5°, 45°, 67.5°, 90°. An electrical heater was used to raise the feed water temperature to the saturated level. Utilizing variable power supply, several series of experiments have been carried out using step increases in power input, in which the power was controlled in the range from 0 to 25 DC volts, and from 0 to 800 amperes.

In order to prevent instantaneous overcurrent from flowing into the standard resistance, a series resistance was inserted into the circuit. The total resistant value of the standard resistance is 4.73 mΩ. The double bridge is used to measure resistance of stainless steel tube. Both a voltmeter and a galvanometer were connected with the double bridge so as to measure the voltage of the test section. The manometer was connected with pressure taps attached to the inlet and outlet of the test section, which provided measurement of both pressure drop and flowrate. The oscillations of flowrate and pressure drop were recorded by the recorder.

2.1. Test Section

Figure 2 shows the annular test section which consists of an electrically-heated stainless steel inner tube of 4.6 mm in diameter and 250 mm in length, and interchangeable outer glass tubes of 16 mm, 22 mm, and 28 mm, respectively. The pitot tube was used to measure flowrate.

3. EXPERIMENTAL PROCEDURE

After some four hours of operation with the heater operating and circulating a flow of purified water, the system was considered ready for experimental tests. As a first step, flow pattern formations were specifically observed and, simultaneously, the flowrate and differential pressure were measured.

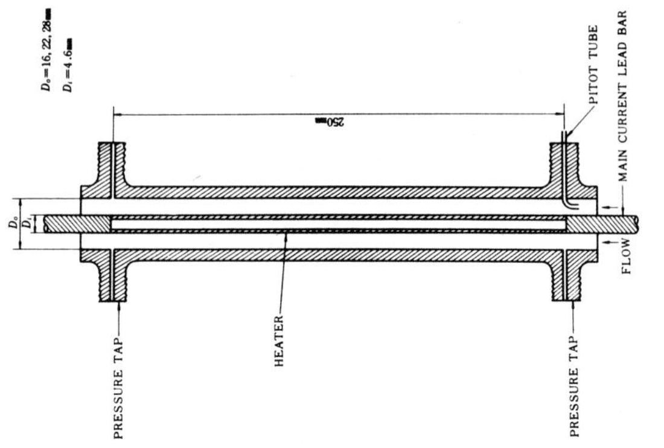

FIG. 2. TEST SECTION.

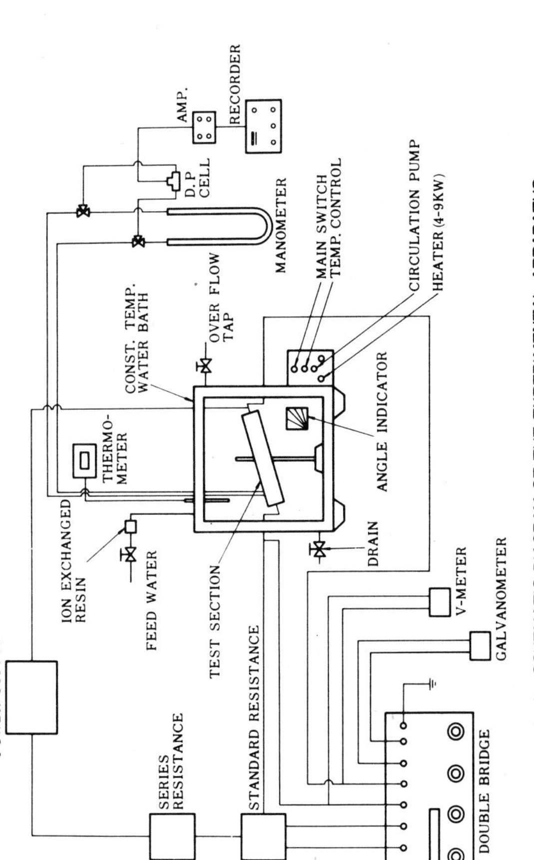

FIG. 1. SCHEMATIC DIAGRAM OF THE EXPERIMENTAL APPARATUS.

After measuring the electrical resistance of the heating tube using the double bridge, the surface temperature of it was calculated. The input power was increased in each stage after finishing all measurements and at the same time, measurement and observations were performed with the previous procedures.

When oscillations occured from a heat flux increase, the recorder was operated in order to record these oscillations for subsequent analysis.

The heat flux was increased continuously until burnout took place with full observation and monitoring of the system.

When the heat flux reached the burnout situation, there was violent oscillation in the flow channel which was used as an indication that burnout was occurring.

After completing the horizontal case, inclination of the test section was varied in five steps from horizontal to vertical, and the outer tube diameter was varied in three steps. Figure 3 shows the mean temperature measurement circuit, where R_X is the resistance of the heating element and R_S is the standard resistance. M, N, m and n are variable resistances.

It is common to use thermocouples to measure surface temperatures. But, in this experiment, it was impossible to mount thermocouples on heating element for two reasons. Firstly, the diameter of heating element is only 4.6 mm and secondly, the thermocouple would have to be mounted on the outside surface which would disturb the flow and, therefore, the reading.

As is well known in boiling flow systems, the local evaluation of the heat transfer, as would be calculated using surface temperature, does not indicate the overall heat transfer of the system which is related with void fraction, thickness of liquid film, and quality. Therefore, it is reasonable and desirable to use electrical resistance in order to calculate average heat transfer. After measuring resistance of the heating element using the double bridge, surface temperature with the correlation between resistance and temperature was calculated for each case. As a result, the temperature difference between surface temperature and mean temperature can be calculated.

4. EXPERIMENTAL RESULTS

4.1. Flow Patterns

Figure 4 shows the flow patterns associated with flow channel orientation. Six types of flow are given for horizontal flows, and five for vertical flows. It is possible to simplify the treatment of two-phase flow by classifying the types of interfacial contact between the phases and the distribution of phases in terms of their flow regimes or flow patterns. It should be stressed that this classification of flow patterns is highly qualitative.

The confusing and qualitative nature of the visual observation of flow patterns has led a number of workers to develop alternative schemes for flow pattern designation. Although a vast variety of different names has been generated in the literature to describe these patterns, five types of flow were confirmed through the experiment: bubbly flow, plug flow, stratified flow, slug flow and annular flow. In case of horizontal flow, the flow

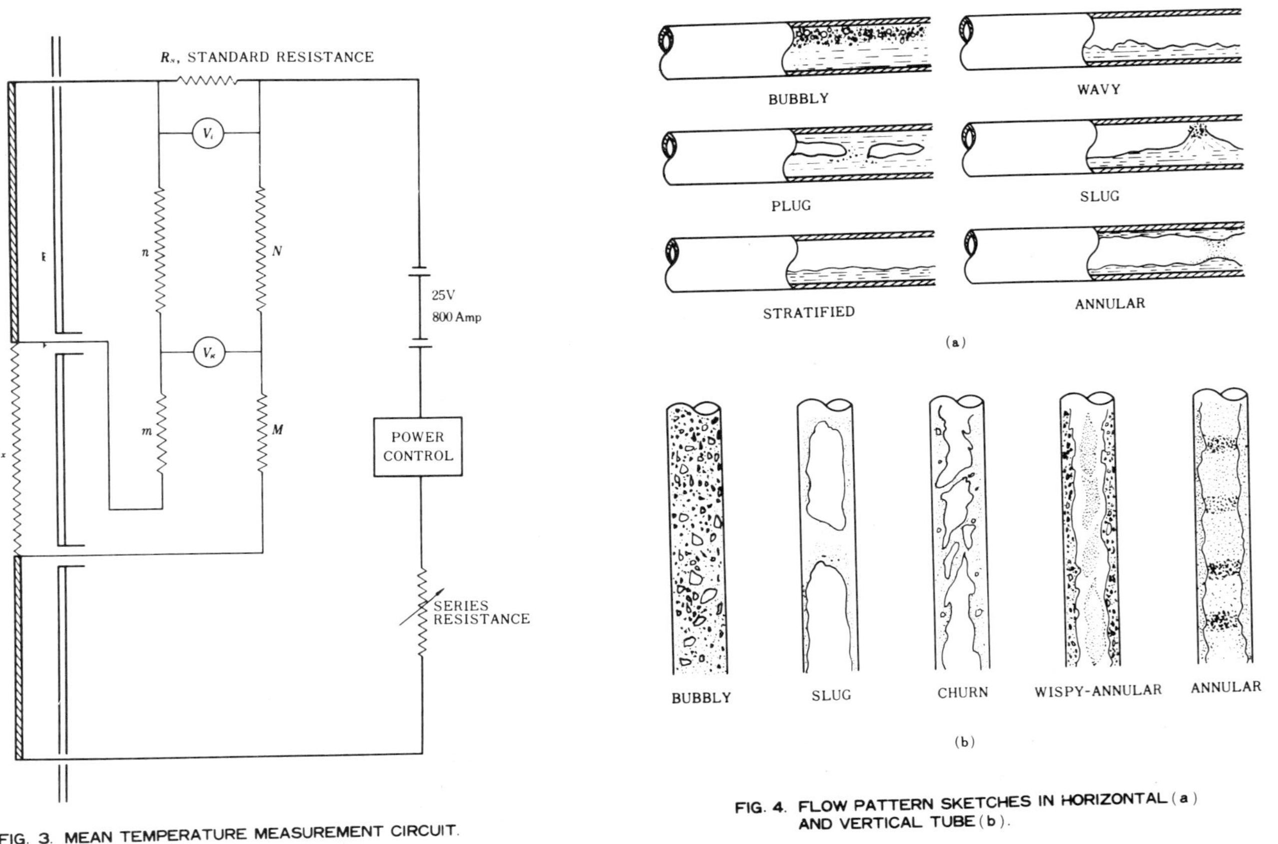

FIG. 3. MEAN TEMPERATURE MEASUREMENT CIRCUIT.

FIG. 4. FLOW PATTERN SKETCHES IN HORIZONTAL (a) AND VERTICAL TUBE (b).

patterns took place in the order of bubbly flow, stratified flow, slug flow and annular flow as the mass flow rates were increased.

Figure 5 shows the flow pattern for horizontal flow. In this case, the period of the existence of bubbly flow is very short, and the situation is complicated by the fact that gravitational forces pull the liquid phase toward the bottom of the tube, the bubbles tending by buoyancy to flow toward the top of the tube. As shown in the figure two phase counter flow occured at the middle position due to the density difference between top and bottom of the tube.

With a volumetric increase due to phase change and bubble generation along the length of the channel, bubbles in the flow induce flow in the liquid and cause it to be expelled from both ends of the horizontal tube.

Due to such counter flow in the channel, interfacial and frictional effects are significant in inhibiting the ability of the liquid to re-enter the channel and provide a distributed source of water to allow boiling to continue. Thus, the heat transfer is lower for this orientation of the channel.

Figure 6-1 shows the transient flow patterns in heat flux and flowrate correlation for $\theta=0°$.

In the case of D_0=16 mm, the friction loss between bubble phase and liquid phase was much larger than for the other outside diameters owing to violent mass exchange. Therefore, burnout was facilitated as a result of the subsequent flowrate decrease. The flowrate was about 10 Kg/m^2s at the burnout condition, which is about 1/6 of the value observed for the maximum flowrate.

Under the same heat flux, the regime of bubbly flow was extended with increasing diameter. On the contrary, the regime of stratified flow became gradually narrower.

At a heat flux of 10^5 $kcal/m^2h$, annular flow occurred in the flow channel and gradually the flowrate decreased.

Figure 6-2 shows for $\theta=22.5°$ that the regime of stratified flow becomes more narrow but slug flow is expanded as the annular thickness increases. Particularly, at D_0=28 mm, the regime of stratified flow became very narrow. This suggests that stratified flow does not form beyond certain outside diameters sizes.

For $\theta > 22.5°$, the reduction of flowrate became smaller with increasing outside diameter because of a decrease of shearing force.

Figure 6-3 shows that both slug flow and annular flow regimes contain oscillations and, therefore, burnout is delayed. This fact means that the heat transfer is improved.

Figures 6-4 and 6-5 show similar results for θ of 67.5° and 90°, respectively.

Figures 7-1 through 7-5 show the transitions of flow patterns as a function of flowrate and quality for several angles of inclination and three test section sizes. As shown previously in figures 6-1 and 6-2 for shallow angles of inclination, the regime of stratified flow became more narrow but

FIG. 5 TWO PHASE COUNTER FLOW IN THE CHANNEL WITH THE TURNING POINT AT THE MIDDLE.

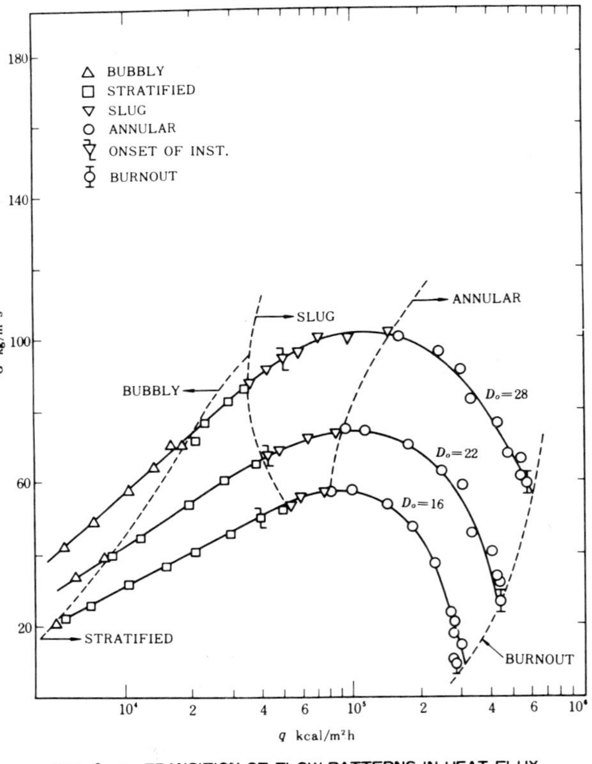

FIG. 6-1. TRANSITION OF FLOW PATTERNS IN HEAT FLUX AND FLOW RATE CORRELATION FOR $\theta = 0°$.

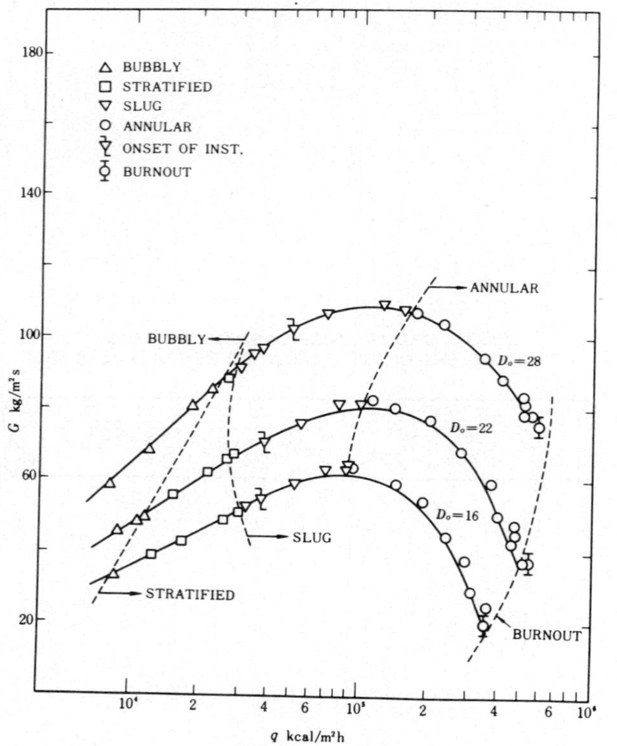

FIG. 6-2. TRANSITION OF FLOW PATTERNS IN HEAT FLUX AND FLOW RATE CORRELATION FOR $\theta=22.5°$.

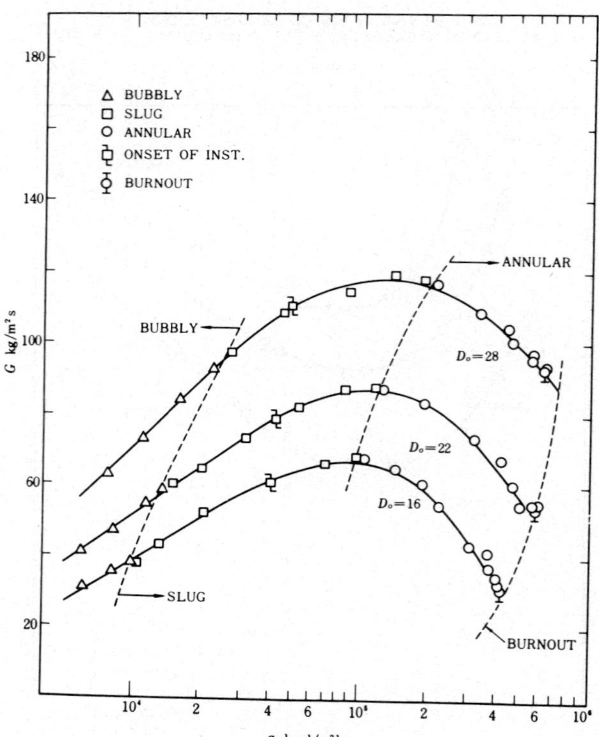

FIG. 6-3. TRANSITION OF FLOW PATTERNS IN HEAT FLUX AND FLOW RATE CORRELATION FOR $\theta=45°$.

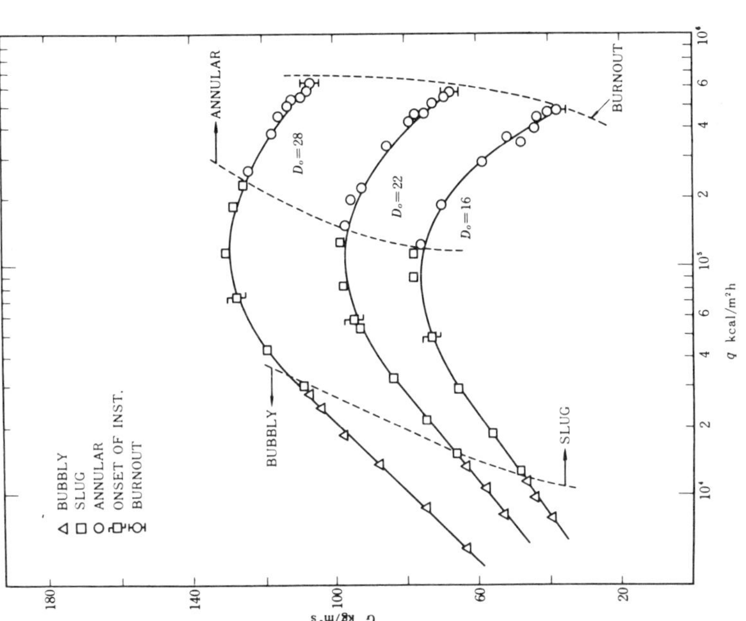

FIG. 6-4. TRANSITION OF FLOW PATTERNS IN HEAT FLUX AND FLOW RATE CORRELATION FOR $\theta=67.5°$.

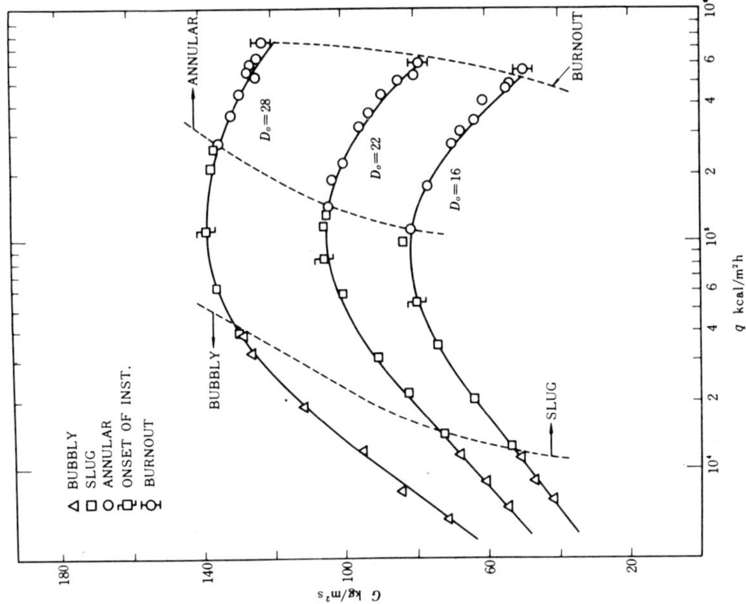

FIG. 6-5. TRANSITION OF FLOW PATTERNS IN HEAT FLUX AND FLOW RATE CORRELATION FOR $\theta=90°$.

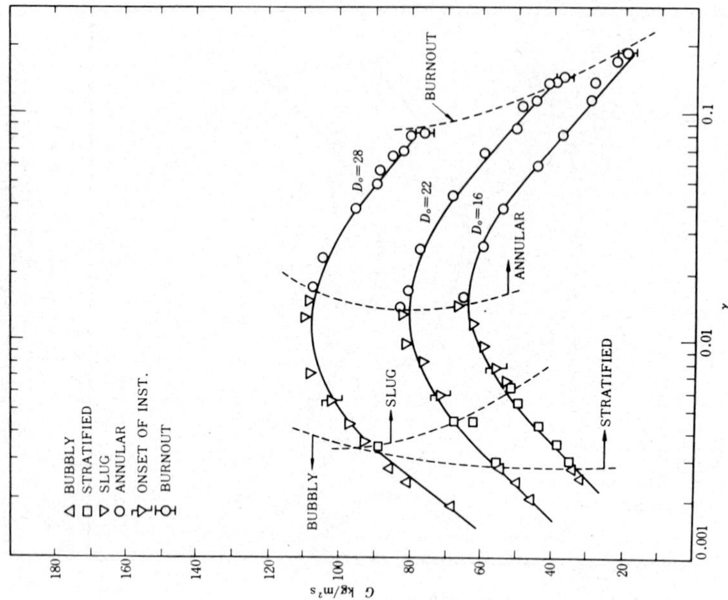

FIG. 7-2. TRANSITION OF FLOW PATTERNS IN FLOW RATE AND QUALITY CORRELATION FOR $\theta=22.5°$

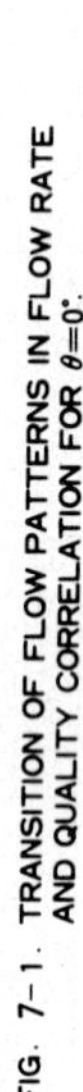

FIG. 7-1. TRANSITION OF FLOW PATTERNS IN FLOW RATE AND QUALITY CORRELATION FOR $\theta=0°$

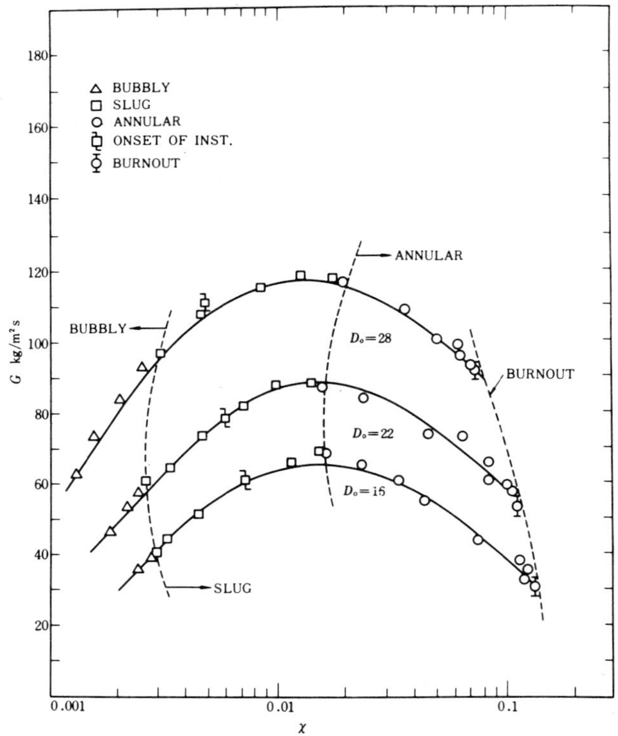

FIG. 7-3. TRANSITION OF FLOW PATTERNS IN FLOW RATE AND QUALITY CORRELATION FOR θ=45°.

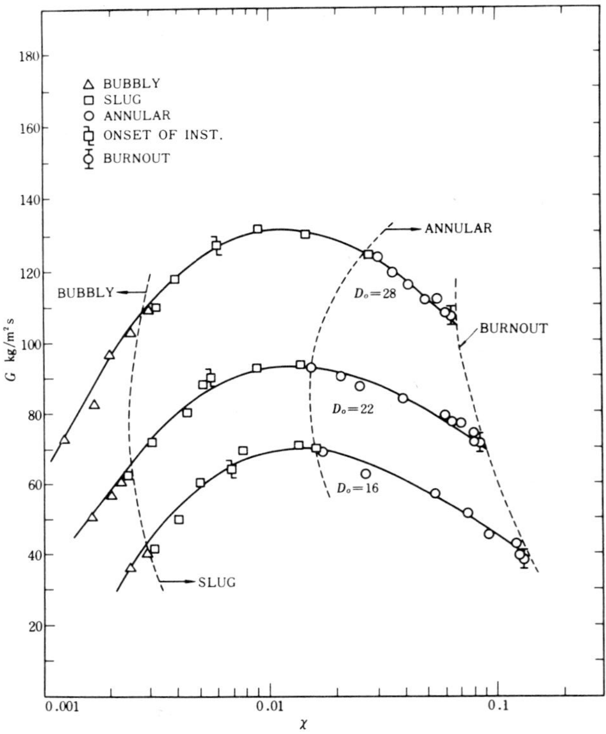

FIG. 7-4. TRANSITION OF FLOW PATTERNS IN FLOW RATE AND QUALITY CORRELATION FOR θ=67.5°.

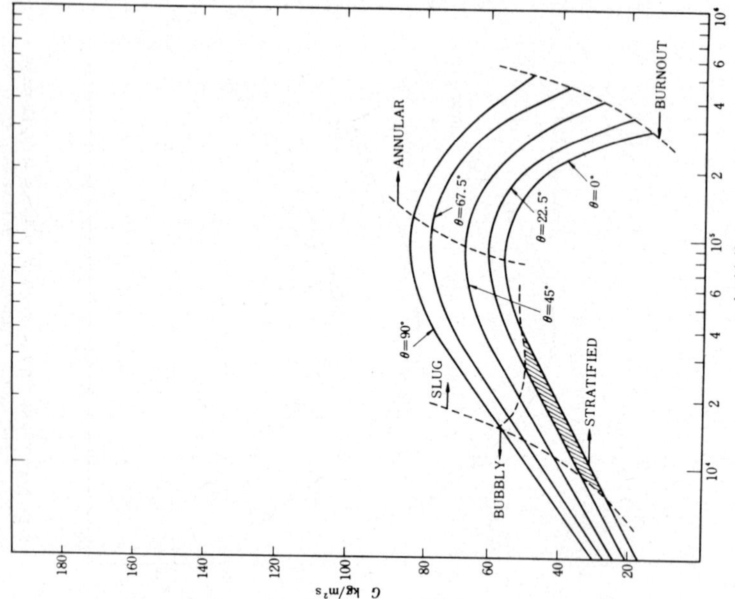

FIG. 8–1. TRANSITION OF FLOW PATTERNS IN FLOW RATE AND HEAT FLUX CORRELATION FOR $D_o = 16$mm.

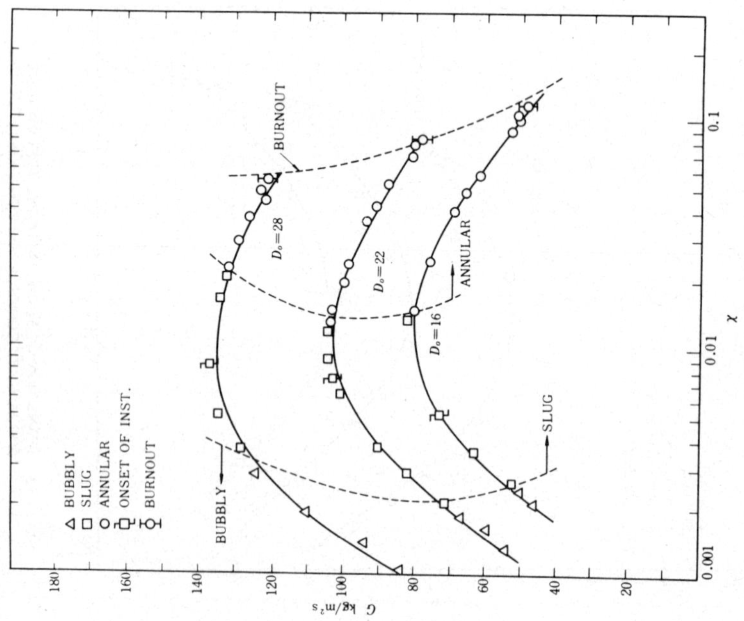

FIG. 7–5. TRANSITION OF FLOW PATTERNS IN FLOW RATE AND QUALITY CORRELATION FOR $\theta = 90°$.

the slug flow regime expanded as the test section size increased. As the angle of inclination was increased, no stratified flow regime was observed.

Under the same heat flux, the average quality was observed to decrease with increasing diameter and the associated void fraction decreased.

Figures 8-1 through 8-3 show the influences of increasing the inclination angle for three test section diameters. Each shows that, when the inclination angle is increased, the flow velocity increases for constant heat flux due to the increase of the vertical component of buoyancy force, showing the significant contribution of the buoyancy effect. The reduction of the occurrance of the stratified flow regime with increasing test section diameter and angle of inclination is very well demonstrated.

Also demonstrated in these figures is the extension of the bubbly flow regime's mass flow rate with increasing angle of inclination as well as with test section size. A similar trend is observed for the transition from slug to annular flow with increasing angle of inclination. Burnout took place in the annular flow regime in all cases.

Figure 9 shows the correlation between quality and angle of inclination at the threshold annular flow. Quality is observed to be relatively independent of the diameter or inclination. Thus, annular flow is established when a certain void fraction is exceeded in the channel.

4.2. Hydraulic Oscillations

Oscillatory behavior may be encountered in two phase flows. Among other instability phenomena, oscillations are characterized by their regularity in shape, period, and amplitude and by the fact that near the oscillation threshold, they are sine-wave-shaped with only small amplitudes. Sustained flow oscillations may cause forced mechanical vibration of components and/or system control problems. Flow instabilities also affect the local heat transfer characteristics and may induce prematurely the boiling crisis. Such conditions are of particular importance in water-cooled nuclear reactors.

Instabilities in two phase flows are caused by a variety of conditions: inlet enthalpy oscillations, pressure drop oscillations between the inlet and the outlet of the channel, and flowrate oscillations. In this experiment, the inlet enthalpy is sustained constant and, since pressure drop was dependent on flowrate, consequently, the important factor is flowrate oscillation.

A set of four controlling conditions may be distinguished generally in the test system.

(1) The inlet thermal boundary condition which imposes the enthalpy at the inlet of the test section. (2) The wall thermal boundary condition which imposes the heat transfer conditions at the wall. (3) The pressure boundary condition which imposes the pressure at some point in the system. (4) The internal hydrodynamic conditions of the fluid flow within the system.

In this experiment, system pressure is kept constant at atmospheric pressure. The fluid temperature at inlet to the test section is maintained at saturated temperature. Therefore, the internal hydrodynamic conditions

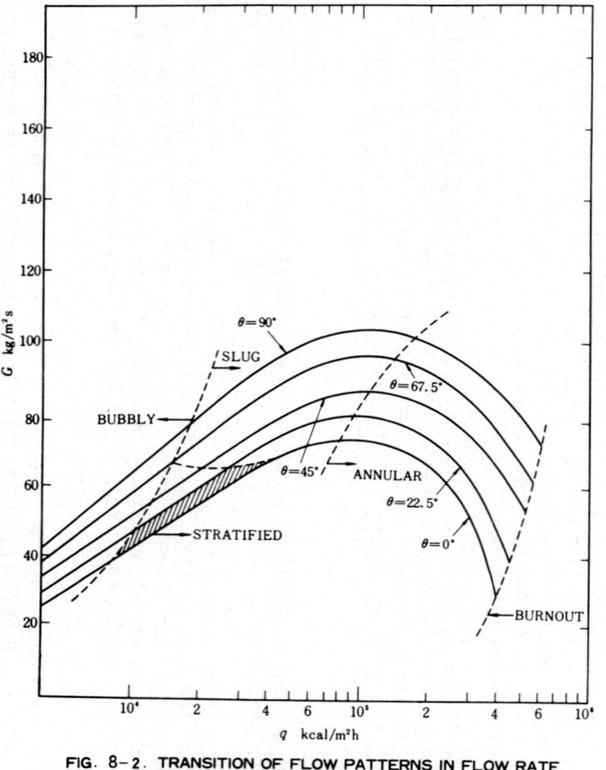

FIG. 8-2. TRANSITION OF FLOW PATTERNS IN FLOW RATE AND HEAT FLUX CORRELATION FOR $D_o = 22$mm.

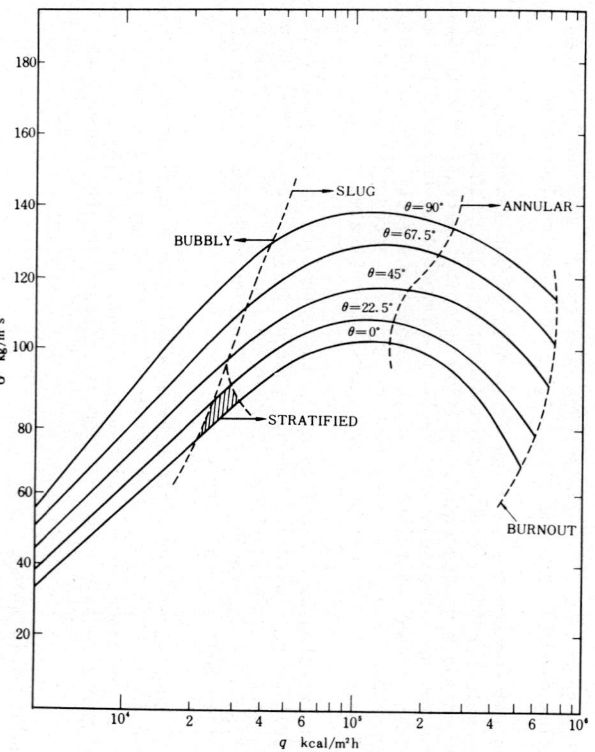

FIG. 8-3. TRANSITION OF FLOW PATTERNS IN FLOW RATE AND HEAT FLUX CORRELATION FOR $D_o = 28$mm.

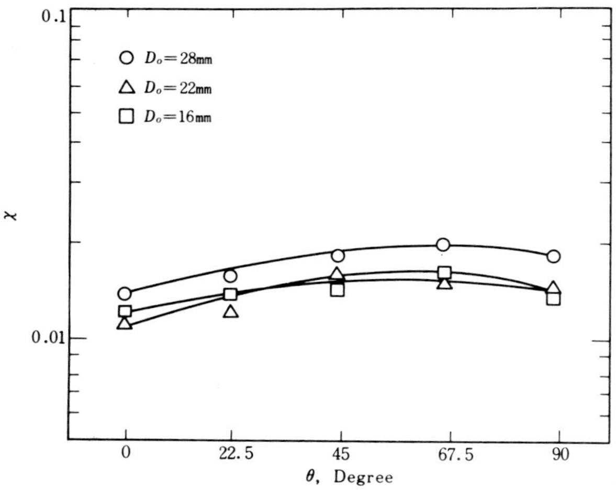

FIG. 9. DATA OF QUALITY VS. INCLINATION AT THRESHOLD ANNULAR FLOW.

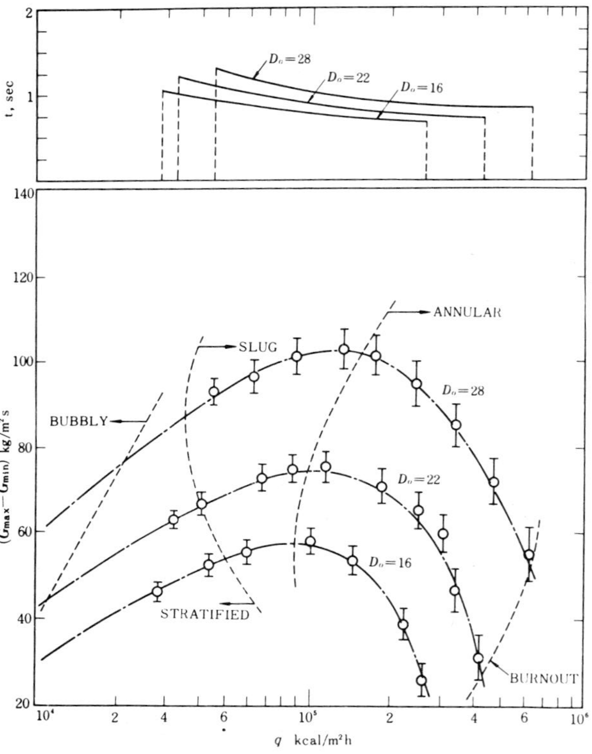

FIG. 10 - 1. VARIATION OF AMPLITUDE $(G_{max}-G_{min})$ AND FREQUENCY WITH HEAT FLUX FOR $\theta=0°$.

and the wall thermal boundary condition are the variables for the study. When flowrate oscillation sets up, feedback processes occur in the system. The oscillation is observed to be a cyclic phenomenon characterized by the periodic bubble behavior.

Figures 10-1 through 10-5 show variations of amplitude and frequency for the full set of tests performed. The onset of flow oscillations occurred in either the slug or stratified flow regime, and they continued to exist with increasing wall heat flux up to the occurance of burnout. The oscillation became more severe with both increasing annular gap size and angle of inclination of the test section. All observed burnouts occurred in the oscillatory annular flow regime.

While oscillation is delayed with increasing diameter, its associated amplitude and frequency tended to increase. The amplitude became maximum near the burnout situation in all cases. Because, the dryout condition on heating surface is facilitated by large amplitude flow oscillations, it tends to promote burnout conditions and, so, the burnout heat flux is decreased.

While the amplitude increases with increasing heat flux, the associated frequency tends to decrease with an increase of wall heat flux, the void fraction is observed to increase and, so, when the gas phase is scavenged, the entrained liquid flowrate will be decreased.

During proceeding flow pattern studies, in which phase change occurred, bubble growth and scavenging by the flow of bubbles took place. This phenomenon will be interpreted by means of the concept of an increasing fluid velocity, an associated scavenging velocity of the gas phase and a resulting increasing flow resistance.

Due to shearing force between the liquid phase and the gas phase, the magnitude of the scavenging velocity will not grow without bounds as the quality is increased. If the gas phase velocity is equivalent to the liquid phase velocity, the channel will not exhibit the scavenging action, but scavenging may be considered complete.

The residual gas phase, when incorporated with newly generated bubbles, becomes larger explosively and, consequently, the local gas phase pressure may be increased. This rapid increase in the gas phase tends to dominate the flow dynamics, in comparison to the average shearing force, and scavenges the nucleated bubbles at once. This phenomenon was observed to dominate at both $\theta=0°$ and $\theta=22.5°$. However, for $\theta \geq 45°$ the influence of bouyancy became significant and tended to dominate the flow dynamics. However, the oscillatory phenomenon is observed to be similar to the horizontal flow case.

Data show that there is a change in the scavenging characteristics of the flow with the angle of inclination of the test section. This tends to suggest that this may be related to the relative influence of the bouyancy force with the local shear force. Since oscillatory characteristics have continuously varied with the inclination angle, there is the possibility that this phenomenon may correlate closely with the release of the gas entrained in the wall region. Figure 11-1, which shows the diameter and angle of inclination of the test section do not influence strongly the quality at which the onset of instability occurs, suggests that onset of oscillation will be established in the place where void fraction reaches a certain value. However, as figure 11-2 shows, the burnout quality is significantly decreased with increasing angle of inclination of the test section.

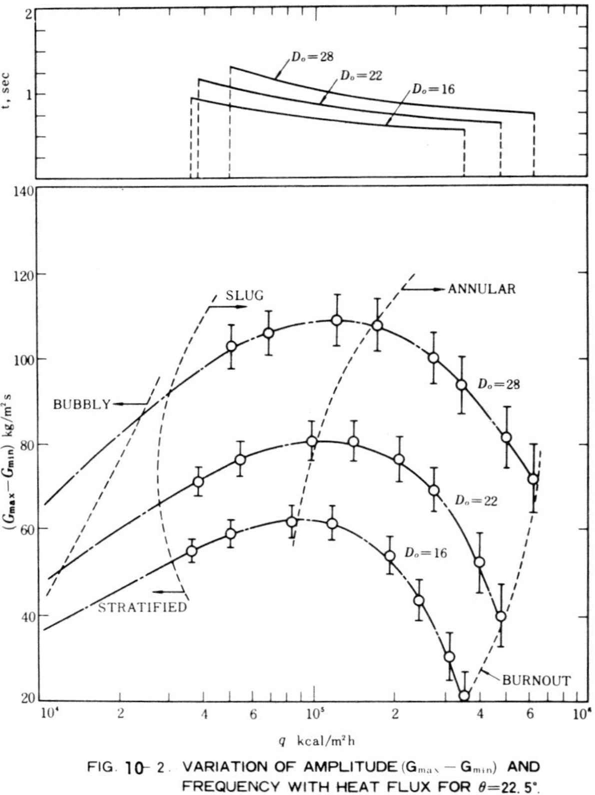

FIG. 10-2. VARIATION OF AMPLITUDE ($G_{max}-G_{min}$) AND FREQUENCY WITH HEAT FLUX FOR $\theta=22.5°$.

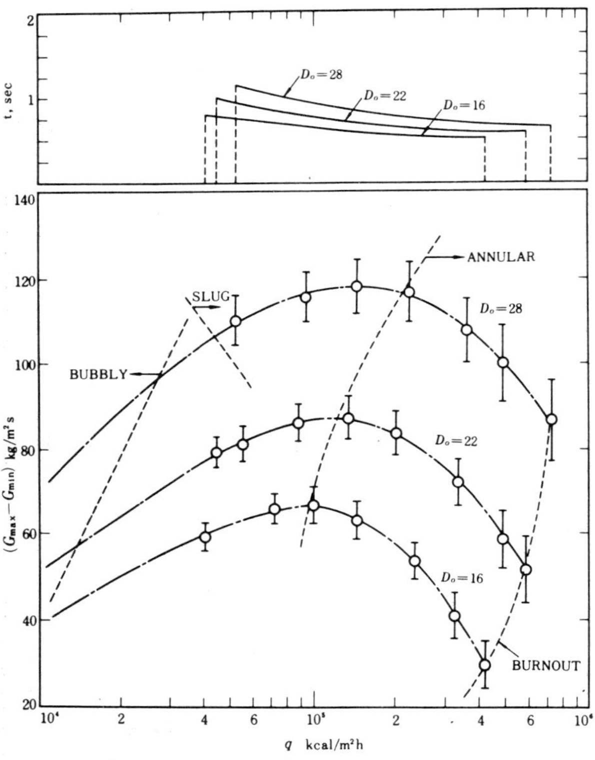

FIG. 10-3. VARIATION OF AMPLITUDE ($G_{max}-G_{min}$) AND FREQUENCY WITH HEAT FLUX FOR $\theta=45°$.

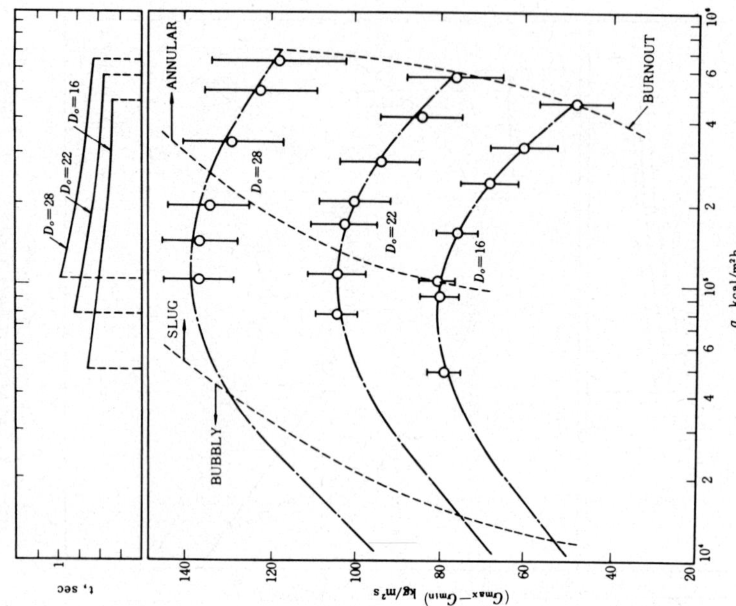

FIG. 10-5. VARIATION OF AMPLITUDE ($G_{max}-G_{min}$) AND FREQUENCY WITH HEAT FLUX FOR $\theta=90°$.

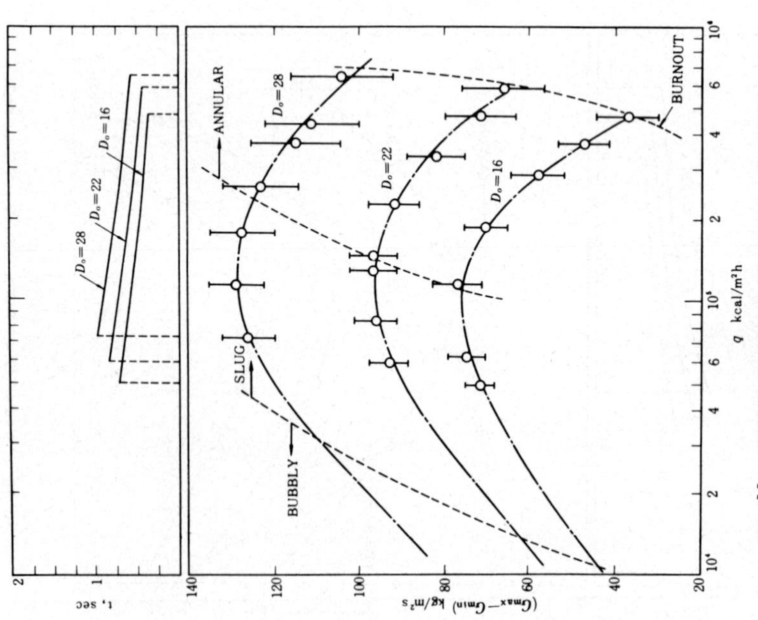

FIG. 10-4. VARIATION OF AMPLITUDE ($G_{max}-G_{min}$) AND FREQUENCY WITH HEAT FLUX FOR $\theta=67.5°$.

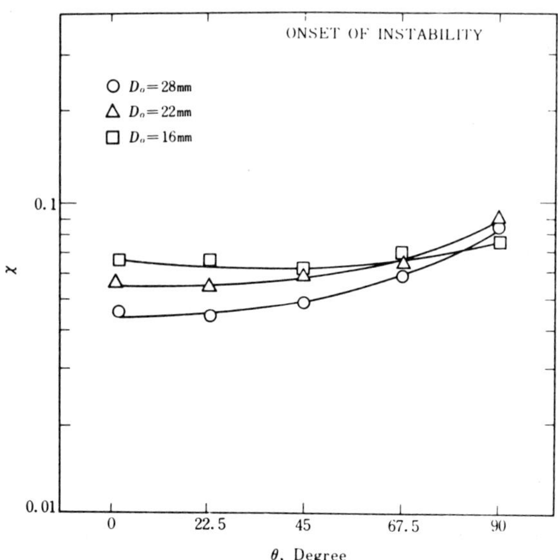

FIG. 11 - 1. INFLUENCE OF INCLINATION ON QUALITY AT INSTABILITY ONSET.

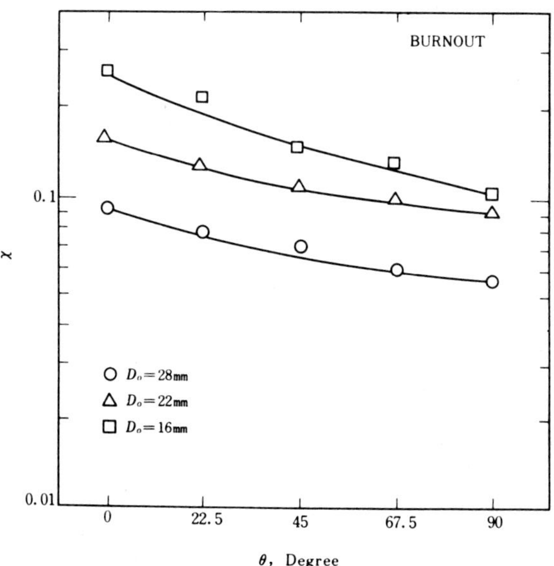

FIG. 11- 2. INFLUENCE OF INCLINATION ON BURNOUT QUALITY.

4.3. Burnout

In general, the burnout condition can be defined by the following functional equation:

$$f(p, T_{in}, q_{Bo}, X_{Bo}) = 0 \qquad (1)$$

where, p is the system pressure, T_{in} is the inlet temperature, q_{Bo} is the burnout heat flux, and X_{Bo} is the burnout quality. The flowrate at burnout can be calculated by the following thermal equilibrium equation:

$$G_{Bo} = \frac{L\ P\ q_{Bo}}{h_{fg}\ X_{Bo} + (h_f - h_i)} \qquad (2)$$

where, L is the heated length, P is the wetted parameter, h_i is the inlet enthalpy, h_f is the saturated liquid enthalpy, and h_{fg} is the latent heat. In this experiment, the pressure is constant and the inlet temperature is saturated. Therefore, the equation (1) becomes:

$$f(q_{Bo}, X_{Bo}) = 0, \; f(q_{Bo}, G_{Bo}) = 0 \qquad (3)$$

To study these functional forms the data at burnout for all the experimental tests have been plotted in figures 12-1 and 12-2 showing the correlation of both q_{Bo} and X_{Bo} with G_{Bo}. The burnout heat flux is observed to increase with increasing diameter and, though not explicitly shown, with increasing angle of test section inclination. However, the burnout quality tends to decrease with the same condition variations. This tendency is interpreted to result from the superheated vapour film overspreading the heated surface preventing heat transfer and, since the oscillatory amplitude becomes larger instantaneously with increasing diameter and inclination, preburnout takes place.

It is also instructive to examine the location along the test section at which burnout tends to occur. Generally this tends to occur more frequently at the outlet with increasing angle of inclination, and shows no decernable influence on test section diameter. There appears to be no dominant circumferential position for burnout in the test section.

Comparing the relation between burnout heat flux and system pressure drop, figure 13 shows a good collapse of all the data on a straight line.

4.4 Heat Transfer

Figures 14-1 through 14-3 show for all the data the influences of angle of inclination and diameter of the test section on heat flux and ΔT_{sat}. It is noted that the heat transfer performance is generally improved with increasing annular gap size and inclination of the test section. However, the heat transfer performance shows a decline in its increase at the burnout condition which appears to be due to an abrupt flowrate decrease.

At the occurance of the incipient nucleate boiling stage, the test section diameter shows little effect on the heat transfer performance, but strongly effects the rate of increasing heat flux. This implies the void fraction and flow patterns are different in each case. But, obviously, the heat transfer performance is improved by the diameter increase.

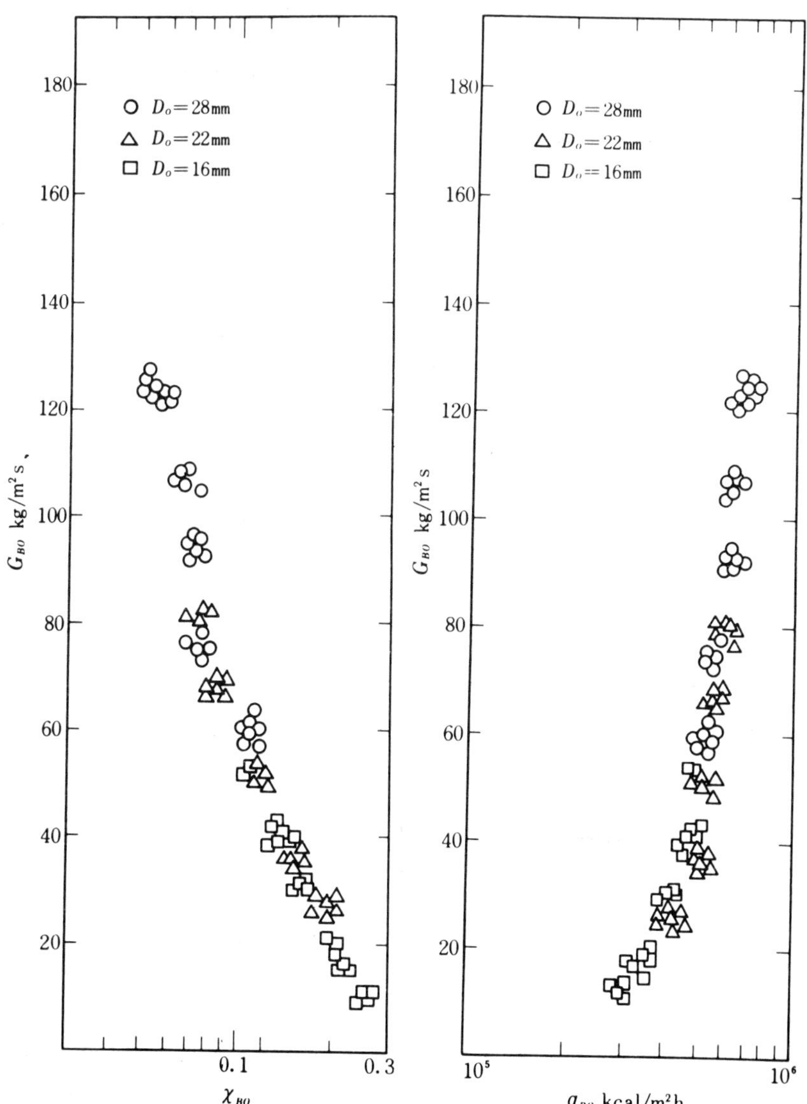

FIG.12-1. FLOW RATE AT BURNOUT VS. BURNOUT QUALITY.

FIG.12-2. FLOW RATE AT BURNOUT VS. BURNOUT HEAT FLUX.

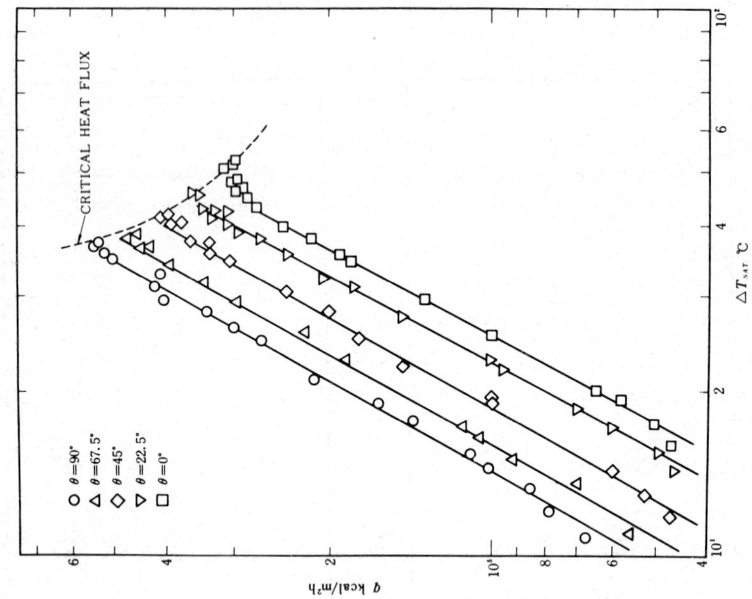

FIG. 14-1. INFLUENCE OF INCLINATION ON NUCLEATE BOILING HEAT TRANSFER TO WATER FOR D_o=16mm.

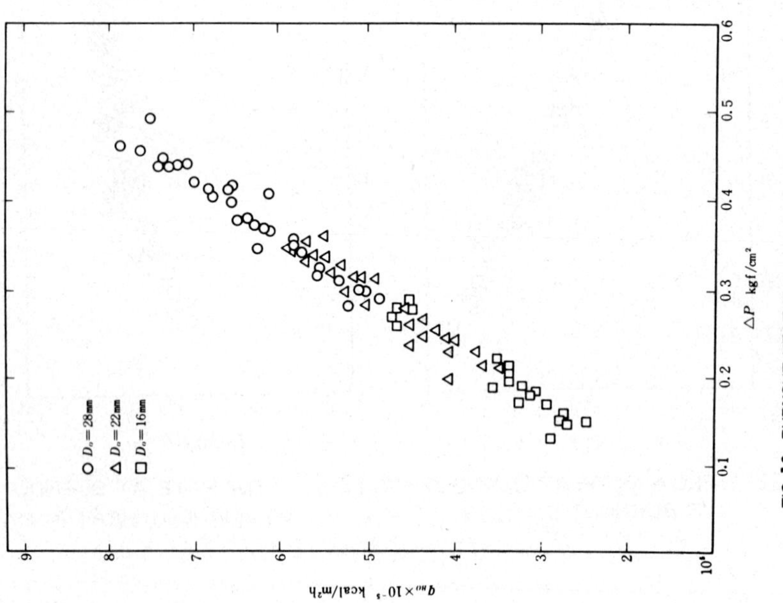

FIG. 13. BURNOUT HEAT FLUX VS. PRESSURE DROP.

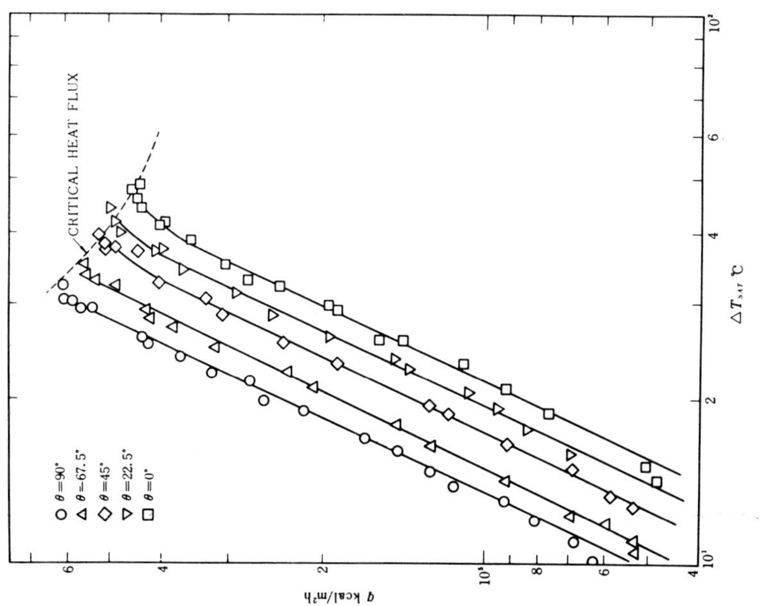

FIG. 14-2. INFLUENCE OF INCLINATION ON NUCLEATE BOILING HEAT TRANSFER TO WATER FOR D_o=22mm.

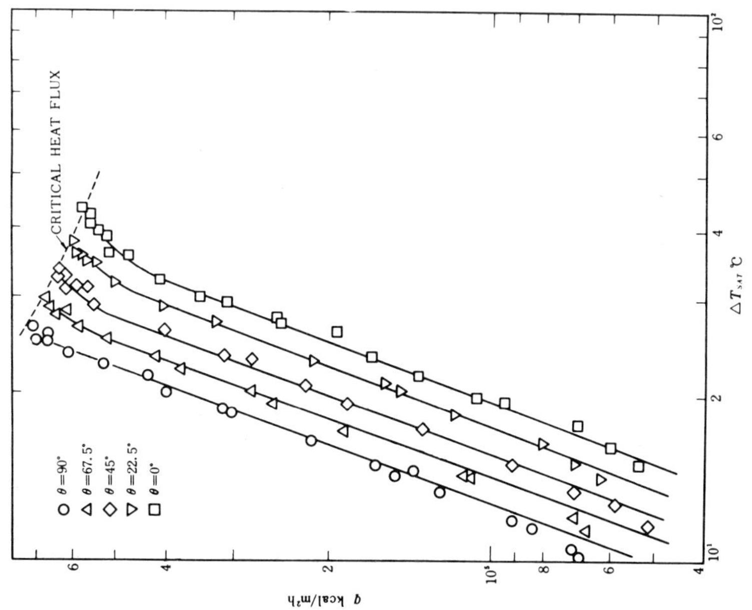

FIG. 14-3. INFLUENCE OF INCLINATION ON NUCLEATE BOILING HEAT TRANSFER TO WATER FOR D_o=28mm.

To correlate the heat flux and heat transfer coefficient for the set of experimental data an equation of the form:

$$q \text{ or } h = C\Delta T^{m}_{sat}$$

is employed. Tables 1 through 3 show the coefficients for this empirical equations for heat flux and heat transfer coefficient for each experimental case. The experimental results show that the coefficient, C, is strongly dependent upon both the angle of inclination of the test section and the annular gap thickness and the exponent, m, was strongly dependent on annular gap size but independent of the angle of inclination of the test section.

5. CONCLUSIONS

Through visual observations of boiling flow in an annulus, bubbly, slug and annular flow regimes were confirmed at all five angles of inclination. However, the stratified flow regime appeared only at zero and 22.5 degrees of tube inclination between the bubbly and slug flow regimes.

The onset of flow oscillations appeared during either the slug or the stratified flow conditions and continued up to the occurence of burnout. The oscillations became much more violent as the annular gap size and the inclination angle of the test tube were increased.

All burnout conditions occurred in an oscillatory annular flow regime, and the burnout heat flux was found to increase with annular gap size and inclination angle of the test section.

The heat transfer performance of the test section generally improved with increasing annular gap size and inclination angle of the test section. This performance was correlated using the equation, $q = C\Delta T^{m}_{sat}$. It was shown through the experimental data that the coefficient, C, was dependent upon both the angle of inclination of the tube and the annular gap size, and that the exponent, m, was determined in accordance with the annular gap size of the test section and not by the angle of test section inclination.

ACKNOWLEDGEMENTS

The author expresses his sincere appreciation to Dr. Barclay G. Jones, Professor of Nuclear Engineering at the University of Illinois, for his kind encouragement and suggestions in preparing this manuscript.

REFERENCES

1. Hewitt, G. F., and Hall Taylor, "Annular Two-Phase Flow," N.S. (1970), Pergamon Press, Oxford.

2. Hewitt, G. F., and Roberts, "Studies of Two-Phase Flow Patterns by Simultaneous X-Ray and Flash Photography," UKAEA Rpt. No. AERE-M2159, D.N. (1969), (MTFS).

TABLE 1. EMPIRICAL EQUATIONS OF HEAT FLUX AND HEAT TRANSFER COEFFICIENT FOR D_o=16mm.

Inclination \ Eq.	$q=f(\triangle T_{sat})$	$h=f(\triangle T_{sat})$
0°	$q=246.2020\,\triangle T_{sat}^{1.852446}$	$h=246.1121\,\triangle T_{sat}^{0.84186}$
22.5°	$q=265.557\,\triangle T_{sat}^{1.852446}$	$h=265.311\,\triangle T_{sat}^{0.84186}$
45°	$q=525.3476\,\triangle T_{sat}^{1.852446}$	$h=525.2075\,\triangle T_{sat}^{0.84186}$
67.5°	$q=692.1320\,\triangle T_{sat}^{1.852446}$	$h=691.8573\,\triangle T_{sat}^{0.84186}$
90°	$q=778.1651\,\triangle T_{sat}^{1.852446}$	$h=777.9045\,\triangle T_{sat}^{0.84186}$

TABLE 2 EMPIRICAL EQUATIONS OF HEAT FLUX AND HEAT TRANSFER COEFFICIENT FOR D_o=22mm.

Inclination \ Eq.	$q=f(\triangle T_{sat})$	$h=f(\triangle T_{sat})$
0°	$q=66.1780\,\triangle T_{sat}^{2.27936}$	$h=66.0862\,\triangle T_{sat}^{1.27948}$
22.5°	$q=93.4946\,\triangle T_{sat}^{2.27936}$	$h=93.5481\,\triangle T_{sat}^{1.27948}$
45°	$q=150.5078\,\triangle T_{sat}^{2.27936}$	$h=150.5073\,\triangle T_{sat}^{1.27948}$
67.5°	$q=227.6553\,\triangle T_{sat}^{2.27936}$	$h=276.4054\,\triangle T_{sat}^{1.27948}$
90°	$q=340.4957\,\triangle T_{sat}^{2.27936}$	$h=340.4957\,\triangle T_{sat}^{1.27948}$

TABLE 3. EMPIRICAL EQUATIONS OF HEAT FLUX AND HEAT TRANSFER COEFFICIENT FOR D_o=28mm.

Inclination \ Eq.	$q=f(\triangle T_{sat})$	$h=f(\triangle T_{sat})$
0°	$q=25.7727\,\triangle T_{sat}^{2.67896}$	$h=25.7584\,\triangle T_{sat}^{1.67906}$
22.5°	$q=49.7953\,\triangle T_{sat}^{2.67896}$	$h=49.7953\,\triangle T_{sat}^{1.67906}$
45°	$q=76.3475\,\triangle T_{sat}^{2.67896}$	$h=76.3135\,\triangle T_{sat}^{1.67906}$
67.5°	$q=89.4578\,\triangle T_{sat}^{2.67896}$	$h=89.4273\,\triangle T_{sat}^{1.67906}$
90°	$q=129.5956\,\triangle T_{sat}^{2.67896}$	$h=129.5956\,\triangle T_{sat}^{1.67906}$

3. Jean J. Ginoux, "Two-Phase Flows and Heat Transfer," Hemisphere Publishing Corp. pp. 3-14, (1978).

4. Manhane, T. M., Gregory, and Aziz, K., "A Flow Pattern Map for Gas-Liquid Flow in Horizontal Pipes," Int. J. of Multiphase Flow, Vol. 1, No. 4, pp. 537-554 (1974).

5. Baker, O., "Simultaneous Flow of Oil and Gas," Oil Gas J. 53, pp. 185-190 (1954).

6. Quandt, E., "Analysis and Measurement of Flow Oscillations," Chem. Eng. Progr., Symp. Ser. 57, No. 32, pp. 111-126 (1961).

7. Whalley, P. B., etc. "The Calculation of Critical Heat Flux in Forced Convective Boiling," Proc. Int. Heat Transfer Conf. 5th, Vol. 4 (1974).

8. Jensen, A., and Mannov, G., "Measurements of Burnout, Film Flow and Pressure Drop in a Concentric Annulus 3500x26x17mm with Heated Rod and Tube", Paper presented at the European Two Phase Flow Group, Harwell, England, June.

9. Tong, L. S. and F. C. Engel, USEAC Rpt. WCAP-3736, (1964).

10. B. R. Rich, "An Investigation of Heat Transfer from an Inclined Flat Plate in Free Convection," J. of Heat Transfer, Trans. ASME., Vol. 75, pp. 489-499 (1953).

Multi-Phase Flow and Heat Transfer III. Part A: Fundamentals
edited by T.N. Veziroğlu and A.E. Bergles
Elsevier Science Publishers B.V., Amsterdam, 1984 — Printed in The Netherlands

HEAT TRANSFER FROM HIGH PRESSURE AIRLESS SPRAYS

Scott R. Herman
Exxon Corporation

William S. Janna
Department of Mechanical Engineering
University of New Orleans
New Orleans, Louisiana 70148, U.S.A.

James E. A. John
School of Engineering
University of Massachusetts at Amherst
Amherst, Massachusetts 10002, U.S.A.

ABSTRACT

An airless, fan-jet pressure nozzle was used to spray hot water vertically downward into the atmosphere at back pressures to 2500 psig (17 240 kPa). Sprayed water was collected in cups into which thermocouples had been placed. The cooling efficiency was determined for a wide variety of spraying conditions and nozzles. A dimensional analysis was performed, with excellent correlation found from the equation

$$\frac{\eta}{\mathrm{Nu}} = 0.5098 \left(\frac{\rho Q}{L\mu g_c}\right)^{-0.1312} \left(\frac{L^2\sigma}{\mu Q}\right)^{-0.2903} \left(\frac{L}{D}\right)^{0.8748} \left(\frac{\theta}{\theta_r}\right)^{0.0355}$$

1. INTRODUCTION

The purpose of this study was to develop an expression for predicting the cooling efficiency of a hot liquid sprayed from an airless pressure nozzle. The controllable variables are spraying pressure, spraying distance, nozzle type and initial water temperature. The free or uncontrollable variables are the wet bulb temperature, dry bulb temperature, and properties of the sprayed liquid. Results of this study should be helpful in the development of spray cooling models for commercial use.

The need for cooling in industrial facilities and in power generation has become increasingly important in recent years due to the problem of thermal pollution. Waste heat has traditionally been rejected by one of several processes: discharge of the heated water into a nearby body of water, use of a mechanical or natural draft cooling tower, or use of a spray to reject the heat into the atmosphere.

Spraying systems are now being developed that will dissipate heat economically and require a minimum of maintenance, yet still have a high thermal efficiency comparable to more expensive systems. There are three basic nozzle types: pressure nozzles, spinning mechanical nozzles, and gas atomizing nozzles. The pressure nozzle is most commonly used for the cooling of water. The water is pumped through the nozzle forming a spray of water from the nozzle exit. Under lower pressure a sheet of water is formed at the nozzle exit; this sheet tears into ligaments, and then into droplets providing a large surface area from which heat is convected. This heat transfer takes place in two modes: latent heat transfer and sensible heat transfer. Latent heat transfer is due to vaporization of a small amount of water from the spray, the remaining heat transfer is due to the air entrained in the spray.

Pressure nozzles can be used for cooling in either a fixed pipe system, where an array of nozzles is fed by a central pumping system, or a modular system (spray canal), which is an array of moored modules, each module consisting of a pump and one to four nozzles [1]. The fixed pipe systems are currently used as emergency cooling systems for nuclear plants whereas the modular systems are used to dissipate waste heat (see Figure 1).

There are various mathematical models used to describe these systems. The manufacturer's model for a modular system [2] is based on the measured performance of a single nozzle, with varying nozzle characteristics, pressure, and flow rates. The nozzle performance is plotted in the form of curves of sprayed water temperature versus nozzle hot water temperature for a given wind speed and range of wet bulb temperatures.

The fixed pipe system version of the manufacturer's model [2] uses data which reflect the averaged performance of a large number of different spray ponds. These design data are presented in a family of curves from which the cooling range can be estimated provided wet bulb temperature and nozzle water temperature are given.

The system model [2] treats each row of modules in a spray canal as a separate channel. A fraction of the flow in each channel is pumped through each module, cooled, and mixed back with the remaining flow in the channel. This mixed flow then proceeds to the next channel. The temperature of the sprayed water is then found from the module when the hot water temperature, local wet bulb temperature, and wind speed are known.

The cellular model [2] for both the modular system and the fixed pipe system, treats the spray field as a number of identical droplets, each sur-

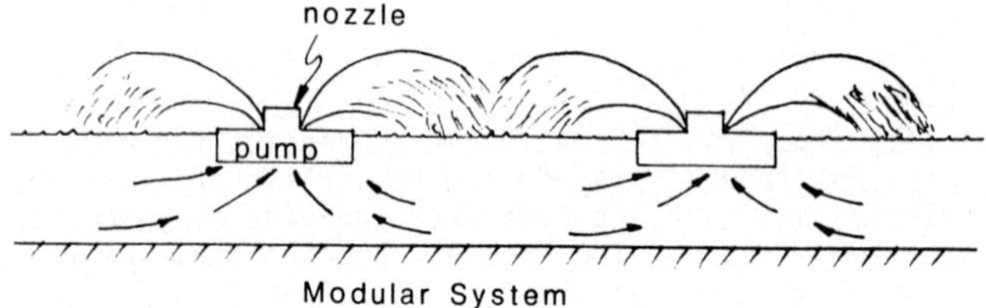

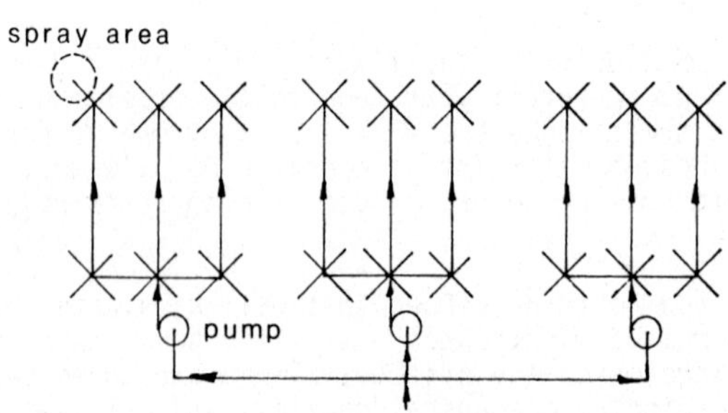

FIGURE 1. *Typical modular and fixed pipe systems.*

rounded by a cell of air. Heat and mass transfer from the droplet to the air can be predicted with published formulae. Elgawharry [16] analyzed the problem of a single droplet to evolve a computer model that includes the effects of drop size, wind speed and direction, air dry bulb temperature, air wet bulb temperature, distance of droplet travel, pressure drop through the nozzle, dimensions of the spray volume, and water flow rate through the volume. The model neglects spray geometry, however.

The ASHRAE technique [2] considers the flow of air through the spray volume and uses the wet bulb temperature of the air leaving the spray volume as the basis for performance calculations. First, a cooling range is selected and then for a given wind speed and flow area, the enthalpy increase of the air passing through the spray volume is calculated. The saturated air enthalpy leaving the volume is used to determine the exit wet bulb temperature. An empirical correction, based on air travel distance through the spray volume and the cooling range, is applied to that wet bulb temperature to determine the spray temperature. This technique does not consider drop size or nozzle type.

Chen and Trezek [3, 4, 5] defined a Spray Energy Release (SER) technique to give practical results in the field of direct contact evaporative heat transfer. The SER is defined as the actual quantity of energy released by the spray per unit total heat driving potential, based on spray distribution and a spray mixing parameter. The method relies on experimentally determined performance of a single spray unit to model the entire system. Thermal efficiency can then be determined from the computed SER, the total heat transfer factor, and the ratio of coolant spray rate to the amount of air participating in the energy transfer process.

Chen and Porter [6] developed the concept of spray cooling efficiency, which can be represented by a dimensionless ratio known as the number of transfer units (NTU). NTU can be defined as the ratio of heat rejected by a unit mass of coolant to the difference between the enthalpy of sprayed water leaving the nozzle and that of ambient air. This number can be determined, in principle, from the average dynamic and thermodynamic behavior of droplets. In practice, however, it is obtained from either experiments on a single module or by back calculations from system performance. These ideal results can then be corrected, as done by Porter et al [7] for interference allowance between adjacent modules, psychrometric corrections, a radiation correction (a modification of 1 to 2%) for non linear total heat variation, and an evaporative loss correction.

Hoffman [8] obtained an empirical curve that relates NTU to wind speed. Wilson [9] used the same data as Hoffman to find a maximum in the performance curves for wind speeds in the 9 to 12 mile per hour (4 to 5.4 m/s) range. This approach has been criticized on the grounds that NTU is a function of the total heat difference (change in enthalpy) as well as the wind speed. This drawback may explain the spread in data on plots of NTU versus windspeed found in Wilson's results.

Pita [10] reports on the study of heat transfer associated with evaporative water sprays in the atmosphere. Using the volume-surface mean diameter, experimental evaporative cooling rates are compared to those predicted by simultaneously solving the motion and transfer equations numerically. Also, an energy transfer coefficient is defined by correlating experimental data with a Reynolds number, spray dimensions, and a supply pressure.

Berger and Taylor [11] developed a mathematical model of atmospheric spray cooling without reliance upon experimental thermal performance data (NTU).

Instead, such real effects as droplet thermal resistance, buoyancy (entrainment due to the action of the spray), and wet bulb degradation were included. Comparison of these results with extracted experimental data from Rancho Seco spray ponds [15] indicates good agreement except at high wind conditions.

Jakubowski [1] experimentally investigated the interaction that takes place between adjacent flat hot water sprays for various configurations. From these data, an equation for optimum spacing between nozzles in a spray array, based on Reynolds number and the nozzle dimensions, was developed from dimensional analysis. Also defined were efficiencies for nozzle spacings greater than and less than the optimum spacing.

Prior to this investigation, spray cooling had been analyzed only at back pressures under 6 900 kPa (1000 psig). This study will present data on sprayed water cooling efficiency up to back pressures of 17 240 kPa (2500 psig) using two DeVilbiss spray caps, operating under a variety of conditions. The data are then correlated by means of dimensional analysis.

2. EXPERIMENTAL APPARATUS, TECHNIQUE AND PROCEDURE

In this study, heated water was sprayed downward under hydraulic pressure into the atmosphere from a single nozzle. Measurements of wet bulb temperature, spray back pressure and sprayed water temperature were made. A description of the experimental apparatus involved in these measurements and capabilities follows.

2.1. Experimental Apparatus

Tests were conducted in a room 4.6 m wide by 12.2 m long by 6.1 m high (15 x 40 x 20 ft). A minimum of ventilation was used to disperse the humid air. Wind speed and solar radiation were therefore neglected. A schematic of the airless spray system is shown in Figure 2.

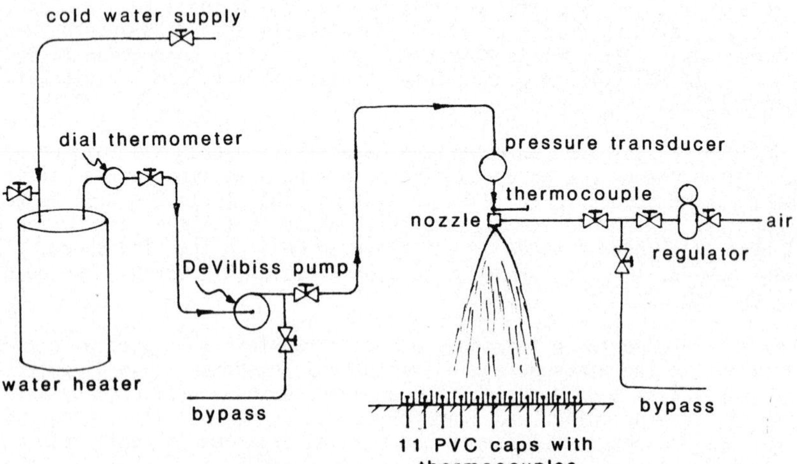

FIGURE 2. *Schematic of the liquid flow circuit.*

Hot water was supplied by a Ruud Manufacturing Company gas water heater (model number GL100-75-1) with a capacity of 0.379 m^3 (100 U. S. gallons) which produced water in an amount sufficient to obtain data on one complete run of the experiments. Hot water was pumped by a DeVilbiss Electric Airless Pump (model number QFE 504). The pump delivers a maximum of 47 x 10^{-6} m^3/s (1.66 ft^3/s), over a range of 0 to 20 690 kPa (0 to 3000 psig). Flow rate was controlled by means of a by-pass valve; spraying pressure was controlled by a pressure regulator. Connecting liquid lines were 0.64 cm, 24 800 kPa (¼-in., 3600 psi) high pressure hoses also manufactured by the DeVilbiss Company.

Pressure was sensed just upstream of the nozzle by a Setra Systems pressure transducer (model number 204, serial number 27155). This transducer had a calibrated pressure range of 0 to 34 500 kPa (0 to 5000 psig) with a proof pressure of 41 380 kPa (6000 psig). The 117 V needed to drive this device was provided by a Wall regulated power supply (model number XW211). Pressure in the outlet line was read from a Setra Systems digital pressure readout (model 300B, serial number 27156). The readout device operated on 115 VAC with a range of 0 to 34 480 kPa (0 to 4999 psig) and an accuracy of ±0.01% ± 1 digit.

Supply and collected water temperatures were sensed by chromel-alumel thermocouples, 3.05 m long (10 ft). The voltages generated by the thermocouples were read (in °F) with a Fluke digital temperature readout, model number 2175A serial number 1770058, with a range of -100 to 1000°F ± 1°F.

Collected water temperatures were measured by inserting thermocouples into eleven 1.91 cm (3/4-nominal) PVC pipe caps. These were symmetrically placed about the spray pattern as shown in Figure 3. This method [1] was selected over placing a thermocouple in the apex of a funnel or placing a thermocouple directly in the line of the spray, because the funnel proved too cumbersome and because the direct method also measured the temperature of the air entrained by the sprayed droplets.

Water was sprayed with a DeVilbiss airless spray gun (model number WVA-501) actuated by air from an Ingersoll-Rand air compressor (type 30). Pressure to the gun was regulated by a Speedair Regulator (model number 4Z028), range 0 to 1 380 kPa (0-200 psig), at 550 kPa (80 psig). Two DeVilbiss spray caps having the same equivalent orifice diameter of 2.79 mm (0.011 in.) and spray angles of 25° and 65° were used. The dimensions of these nozzles are shown in Figure 4 and Table 1.

Wet bulb and dry bulb temperatures were sensed by a Honeywell Resistance thermometer-type psychrometer with a water bottle supply (model number 12685). These temperatures were then read by a Honeywell Servoline, class 45 recorder with a range of 0 to 37.8 C (100°F) on chart paper (number 14694) on a rotation of once every 24 hrs. The accuracy is ±5% of the full scale. Mass flow rates were measured by weighing an amount of water sprayed and dividing by the corresponding time. Volume flow rate as a function of pressure is graphed in Figure 5.

2.2. Experimental Technique

For each spray cap, wet bulb temperature, dry bulb temperature, exit nozzle water temperature, and collected water temperature were recorded for each of 6 runs. One run consisted of readings at vertical distances (nozzle to cups) of 0.11 through 0.36 m (4.3125 through 14.3125 in.) at 0.051 m (2 in.) increments for one pressure reading. Six runs were made at pressures of 2 070,

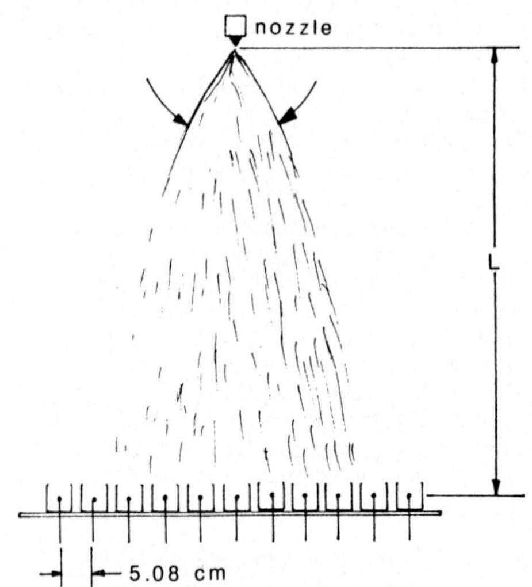

FIGURE 3. *Nozzle and collection cup arrangement.*

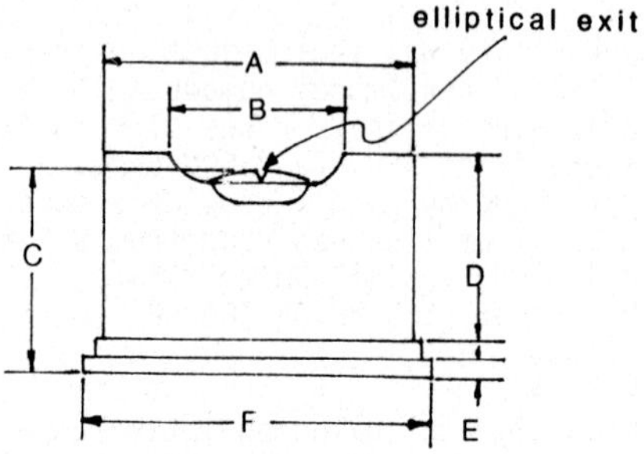

FIGURE 4. *Spray nozzle detail.*

TABLE 1. *Dimensions of spray nozzles used in this study.*

Nozzle	Spray Angle	dimensions in cm A	B	C	D	E	F	major axis	minor axis
1125	25°	1.036	0.495	0.822 5	0.480	0.241 3	1.506	0.041 7	0.027 7
1165	65°	1.036	0.495	0.822 5	0.480	0.241 3	1.506	0.044 7	0.022 1

4 830, 6 900, 10 300, 13 800, and 17 240 kPa. The spray was directed vertically downward for each case. Therefore a total of thirty-six readings were obtained for each spray nozzle. Measurements were made indoors because when conducted outdoors, drift caused little water to be collected in the PVC caps.

The collection temperature was found by mounting eleven 1.91 cm pipe caps at 5.08 cm (2 in.) intervals symmetrically about the nozzle centerline. The temperature measuring cups, labeled one through eleven , shown in Figure 3, had one thermocouple located 1.3 cm (0.5 in.) from the top of the cup. A 1.6 mm (1/16 in.) hole was drilled in the bottom of each cup so there would be a continuous flow of cooled water through the cups. This would prevent thermal stratification and in effect yield the cup-mixing temperature at each thermocouple.

2.3. Experimental Procedure

The same procedure was used for recording the temperatures of the sprayed water for all nozzle settings. For each nozzle type, the by-pass valve was opened to avoid starting against a load, and the pump was activated. The bypass was then closed and the gun was actuated by the air supply valve. The

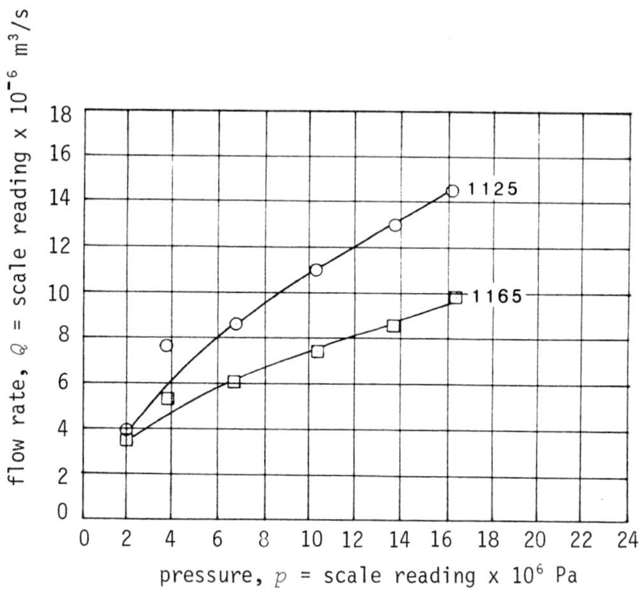

FIGURE 5. *Pressure-flow rate curves for nozzles tested.*

pressure was then set with the pressure adjusting valve. Starting at the lowest pressure and allowing time for conditions to become steady, the sprayed water temperature, nozzle water temperature, spray back pressure, dry and wet bulb temperature, and number of collecting cups covered by the spray were recorded for all six spraying distances (nozzle to cups) previously mentioned. The pressure was then raised to the next level, the conditions allowed to become steady, and all data were then recorded. This was repeated until a maximum pressure of 17 240 kPa (2500 psig) was reached. To ensure repeatability, three runs of each nozzle were made.

3. EXPERIMENTAL RESULTS AND DATA CORRELATION

Plots of factors such as equivalent orifice diameter, spray back pressure, and spray angle versus average spray cooling efficiency were made to examine the results and to determine if an obvious relationship existed between any of these which might be correlated. Dimensional analysis was then performed to determine an equation to predict the thermal efficiency once various fluid properties and ambient conditions are known.

An average spray cooling efficiency was used in this study, defined by Chen and Trezek [3] as

$$\eta = \frac{T_s - T_c}{T_s - T_{WB}} \times 100\%$$

where T_S = temperature of water at nozzle inlet
T_C = average collected water temperature used in this study
T_{WB} = wet bulb temperature

This equation expresses, in per cent, how close the sprayed water temperature approaches the wet bulb temperature of the atmosphere. The wet bulb temperature is the theoretical lower limit of the sprayed water temperature.

3.1. Data Correlation

A dimensional analysis was performed to obtain a relationship between the operating variables; specifically to correlate the average thermal efficiency of the sprayed water with the operating variables. This efficiency is a function of several variables as shown below:

$$\eta = \eta(L, Q, \theta, \theta_r, \rho, D, \sigma, \mu, h, k, g_c) \quad (1)$$

where
L = spraying length, m or ft
Q = volume flow rate, m^3/s or ft^3/s
θ = spray cap angle, rad
θ_r = reference spray angle, rad
ρ = liquid density, kg/m^3 or lbm/ft^3
D = volume median diameter of spray, m or ft
σ = surface tension, N/m or lbf/ft
μ = absolute viscosity of liquid, N-s/m^2 or lbf-s/ft^2
h = convection heat transfer coefficient, W/(m^2-K) or BTU/hr-ft^2-°F
k = conduction heat transfer coefficient, W/(m-K) or BTU/hr-ft-°F

The lengths and spray angles are included to account for the effects of spray geometry. Flow rate is also included. Although only one fluid (water) was used in this experiment, surface tension, density, and absolute viscosity were included. A convection heat transfer coefficient and a conduction heat transfer coefficient were included to account for the different cooling effects.

It should be noted that volume median rather than Sauter Mean Diameter was used here because data on volume median diameter are all that are available [13] for the high pressure sprays of this study.

The Janna-John equation [13] for predicting volume median diameter of sprays produced by airless nozzles was used to calculate typical drop sizes. These varied by ±18%, at the various pressures for each nozzle. Results of the calculations are shown in Table 2. The nozzle dimensions were measured with a microscope.

Using the Rayleigh method of dimensional analysis, the average spray cooling efficiency can be determined as

$$\eta = C_1 \left(\frac{\rho Q}{L \mu g_c}\right)^a \left(\frac{L^2 \sigma}{\mu Q}\right)^b \left(\frac{L}{D}\right)^c \left(\frac{\theta}{\theta_r}\right)^d \left(\frac{hD}{k}\right)^e$$

where hD/k is the Nusselt number. Rewriting the above equation,

$$\eta = C_1 \left(\frac{\rho Q}{L \mu g_c}\right)^a \left(\frac{L^2 \sigma}{\mu Q}\right)^b \left(\frac{L}{D}\right)^c \left(\frac{\theta}{\theta_r}\right)^d (\mathrm{Nu})^e \quad (2)$$

TABLE 2. *Drop size data for nozzles tested.*

Nozzle	Spraying Pressure, kPa	Volume Median Diameter of Spray, μm†	Terminal Velocity, m/s
1125	2 070	483	3.75
	4 830	329	3.00
	6 900	274	2.67
	10 350	199	2.14
	13 790	145	1.66
	17 240	107	1.24
1165	2 070	447	3.60
	4 830	438	3.54
	6 900	387	3.20
	10 350	292	2.59
	13 790	252	2.32
	17 240	213	2.04

†As determined from Ref. [13].

Because 25° was the smallest spray angle tested, this was selected as the reference angle.

Water inlet temperature of this experiment varied from approximately 26.7 C to 48.9 C. An average value of 37.8 C (100°F) was used to evaluate properties (Table 3). The Nusselt number for heat transfer from droplets was expressed by Ranz and Marshall [12] as

$$Nu = \frac{hD}{k} = 2.0 + 0.6Re^{0.5}Pr^{0.33}$$

where Re = Reynolds number = $\rho VD/\mu g_c$

Pr = Prandtl number

The Reynolds number was found by a trial and error procedure involving a force balance on the droplet [17]. It was assumed that the drops reached terminal velocity by the time they reached the collection cups. Terminal velocity is given in Table 2 for various spraying pressures. All variables were then inserted into the equation for average thermal efficiency and arranged in tabular form. From a simultaneous solution of the five equations for the five unknown constants or exponents and a lengthy trial and error procedure, suitable constants were found which gave reasonable results. The equation becomes

$$\frac{\eta}{Nu} = 0.509\ 8\left(\frac{\rho Q}{L\mu g_c}\right)^{-0.1312}\left(\frac{L^2\sigma}{\mu Q}\right)^{-0.2903}\left(\frac{L}{D}\right)^{0.8748}\left(\frac{\theta}{\theta_r}\right)^{0.0355} \tag{3}$$

TABLE 3. *Physical properties of water.*

Variable	Value
Surface tension, σ	0.007 1 N/m
Viscosity, μ	6.82×10^{-4} N-s/m^2
Density, ρ	999 kg/m^3

A plot of calculated versus actual points is presented in Figure 6. A summary of the calculations is provided in Table 4.

4. DISCUSSION

Although not presented here, the average spray cooling efficiency was plotted, for each nozzle, versus spray angle, wet bulb temperature, spray back pressure, and spraying length. From these figures, information regarding spray cooling can be ascertained.

The graph of efficiency versus spray angle (with constant pressure and length) indicated an increase in efficiency with corresponding increases in pressure and length. The increase with pressure is expected because a finer spray is formed at higher pressures giving a greater surface area for heat transfer. The increase with length is expected because there is more contact between the warmer sprayed water and the cooler air.

Graphs of spray angle versus efficiency indicate that there is no definite increase or decrease of efficiency with spray angle, although a relationship was expected because the drops travel through a larger distance when sprayed through nozzles with larger angles. This occurs with nozzles of equivalent orifice diameters, the different orifices produce different volume median diameters although they ideally produce equal flow rates. Because the drop diameters vary, the cooling rates and efficiencies will also vary. Efficiency was also plotted versus wet bulb temperature but provided no conclusive results.

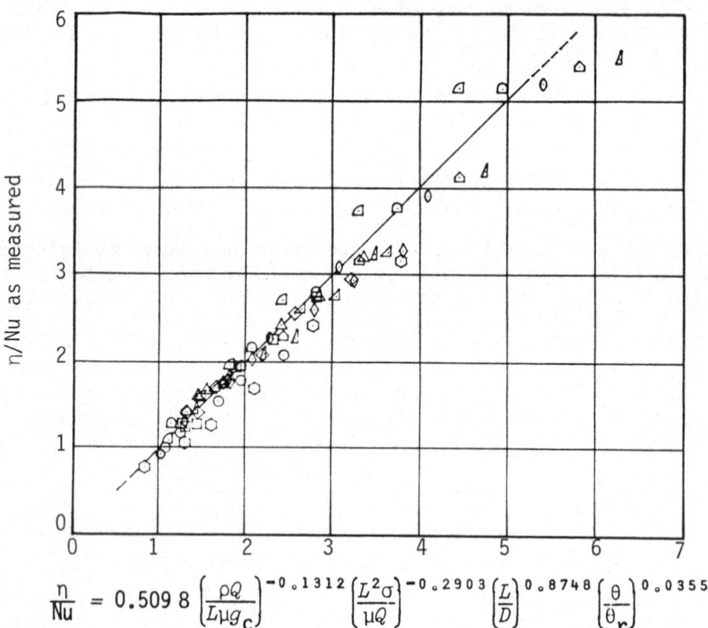

$$\frac{\eta}{Nu} = 0.509\,8\left(\frac{\rho Q}{L\mu g_c}\right)^{-0.1312}\left(\frac{L^2\sigma}{\mu Q}\right)^{-0.2903}\left(\frac{L}{D}\right)^{0.8748}\left(\frac{\theta}{\theta_r}\right)^{0.0355}$$

FIGURE 6. *Actual average spray cooling efficiency vs. predicted cooling efficiency. (See Table 4 for symbol legend.)*

TABLE 4. *A summary of the data and calculations.*

Spray nozzle	length, L	pressure, kPa (psig)	T_c,°F	T_s,°F	T_{wb},°F	D, µm	η/Nu	by equation η/Nu	% error
1125 ⬡	0.11 m 0.36 ft	2 070 (300)	80.0	92.0	62.0	483	0.756	0.859	-13.6
		4 830 (700)	88.4	106.8		329	1.04	1.28	-23.1
		6 900 (1000)	91.2	111.4		274	1.19	1.54	-29.4
		10 300 (1500)	92.2	116.2		199	1.67	2.11	-26.3
		13 800 (2000)	90.0	119.0		145	2.43	2.84	-16.9
		17 240 (2500)	91.2	120.8		107	3.20	3.78	-18.1
1125 ▱	0.16 m 0.53 ft	2 070 (300)	74.2	92.0	62.0	483	1.12	1.01	9.8
		4 830 (700)	77.8	107.2		329	1.70	1.50	11.8
		6 900 (1000)	78.3	112.2		274	1.97	1.81	8.1
		10 300 (1500)	77.3	116.4		199	1.70	2.47	-45.3
		13 800 (2000)	74.6	119.0		145	3.78	3.33	11.9
		17 240 (2500)	74.1	121.6		107	5.09	4.42	13.2
1125 ⯃	0.21 m 0.69 ft	2 070 (300)	72.1	92.2	62.0	483	1.26	1.13	10.3
		4 830 (700)	75.7	107.8		329	1.77	1.69	6.3
		6 900 (1000)	75.5	112.4		274	2.13	2.03	4.7
		10 300 (1500)	75.1	116.4		199	2.85	2.77	2.8
		13 800 (2000)	74.3	119.0		145	3.81	3.74	1.8
		17 240 (2500)	74.0	121.8		107	5.09	4.97	2.4
1125 ⬯	0.26 m 0.86 ft	2 070 (300)	70.6	92.6	62.0	483	1.36	1.24	8.8
		4 830 (700)	74.1	108.0		329	1.87	1.85	1.1
		6 900 (1000)	73.5	112.4		274	2.25	2.23	0.9
		10 300 (1500)	72.1	116.4		199	3.06	3.04	0.7
		13 800 (2000)	72.7	119.0		145	3.94	4.10	-4.1
		17 240 (2500)	72.5	121.4		107	5.24	5.44	-3.8
1125 ⌂	0.31 m 1.03 ft	2 070 (300)	69.6	93.4	62.0	483	1.43	1.33	7.0
		4 830 (700)	72.5	107.8		329	1.95	1.99	-2.1
		6 900 (1000)	72.2	112.8		274	2.33	2.40	-3.0
		10 300 (1500)	71.3	116.4		199	3.12	3.27	-4.8
		13 800 (2000)	71.3	119.2		145	4.06	4.42	-8.9
		17 240 (2500)	70.7	121.2		107	5.43	5.86	-7.9
1125 ◿	0.36 m 1.19 ft	2 070 (300)	68.8	93.6	62.0	483	1.48	1.42	4.1
		4 830 (700)	71.3	107.8		329	2.02	2.12	-5.0
		6 900 (1000)	71.3	112.6		274	2.38	2.55	-7.1
		10 300 (1500)	69.9	116.4		199	3.21	3.49	-8.7
		13 800 (2000)	69.8	119.0		145	4.19	4.70	-12.2
		17 240 (2500)	69.6	121.2		107	5.55	6.25	-12.6
1165 ○	0.11 m 0.36 ft	2 070 (300)	61.1	75.6	43.5	447	0.906	1.03	-13.7
		4 830 (700)	68.5	92.8	43.3	438	1.00	1.11	-11.0
		6 900 (1000)	70.4	100.4	43.5	387	1.20	1.26	-5.0
		10 300 (1500)	73.2	107.2	43.5	292	1.53	1.68	-9.8
		13 800 (2000)	73.1	110.6	43.5	252	1.81	1.95	-7.7
		17 240 (2500)	73.3	113.8	43.5	213	2.14	2.44	-14.0
1165 □	0.16 m 0.53 ft	2 070 (300)	55.2	76.0	43.5	447	1.28	1.29	-0.8
		4 830 (700)	62.7	94.2	43.3	438	1.26	1.30	-3.2
		6 900 (1000)	63.3	100.6	43.5	387	1.25	1.48	-18.4
		10 300 (1500)	63.3	107.4	43.5	292	1.98	1.96	1.0
		13 800 (2000)	61.6	110.8	43.5	252	2.37	2.29	3.4
		17 240 (2500)	61.6	113.0	43.5	213	2.75	2.87	-4.4
1165 ◇	0.21 m 0.69 ft	2 070 (300)	52.7	76.4	43.5	447	1.44	1.36	5.6
		4 830 (700)	58.2	94.6	43.3	438	1.44	1.46	-1.4
		6 900 (1000)	57.1	99.8	43.5	387	1.72	1.67	2.9
		10 300 (1500)	59.0	107.2	43.5	292	2.17	2.20	-1.4
		13 800 (2000)	58.2	110.4	43.5	252	2.52	2.57	-2.0
		17 240 (2500)	57.6	113.4	43.5	213	2.97	3.22	-8.4
1165 △	0.26 m 0.86 ft	2 070 (300)	50.3	76.4	43.5	447	1.59	1.49	6.3
		4 830 (700)	53.1	94.4	43.2	438	1.64	1.60	2.4
		6 900 (1000)	53.7	100.0	43.3	387	1.85	1.82	1.6
		10 300 (1500)	54.2	106.8	43.5	292	2.44	2.41	1.2
		13 800 (2000)	53.5	110.4	43.5	252	2.75	2.81	-2.2
		17 240 (2500)	53.5	113.4	43.5	213	3.19	3.34	-4.7
1165 ◿	0.31 m 1.03 ft	2 070 (300)	48.8	76.4	43.5	447	1.68	1.60	4.8
		4 830 (700)	51.3	94.2	43.3	438	1.72	1.73	-0.6
		6 900 (1000)	51.6	100.0	43.5	387	1.94	1.97	-1.5
		10 300 (1500)	52.4	107.4	43.5	292	2.52	2.59	-2.8
		13 800 (2000)	52.2	110.8	43.5	252	2.82	3.01	-6.7
		17 240 (2500)	52.5	113.4	43.5	213	3.24	3.61	-11.4
1165 ◊	0.36 m 1.19 ft	2 070 (300)	47.7	77.0	43.5	447	1.75	1.71	2.3
		4 830 (700)	50.0	94.8	43.3	438	1.77	1.84	-4.0
		6 900 (1000)	49.9	100.2	43.5	387	2.01	2.09	-4.0
		10 300 (1500)	50.7	107.0	43.5	292	2.60	2.76	-6.2
		13 800 (2000)	50.1	111.0	43.5	252	2.92	3.23	-10.6
		17 240 (2500)	50.4	113.4	43.5	213	3.35	3.84	-14.6

Equation 3 was used to predict the average thermal performance of sprayed water between 0 and 17 240 kPa (0 and 2500 psig). A linear correlation coefficient [14], which defines a perfect correlation as unity and no correlation between predicted and actual data as zero, was calculated. It was found that the correlation coefficient is

$$r = 0.989\,6$$

This result gains meaning from a probability table. For over seventy observations, there is a probability of 0.001 that a correlation coefficient greater than 0.394 will occur for unrelated variables.

5. CONCLUSIONS

This experiment was concerned with the cooling efficiency of hot water sprayed from an airless pressure nozzle over a pressure range of 2 070 to 17 240 kPa (300 to 2500 psig).

The analysis of the experimental data of the airless spray nozzles yeilded the following expression for average thermal efficiency

$$\frac{\eta}{\mathrm{Nu}} = 0.509\,8\left(\frac{\rho Q}{L\mu g_c}\right)^{-0.1312}\left(\frac{L^2\sigma}{\mu Q}\right)^{-0.2903}\left(\frac{L}{D}\right)^{0.8748}\left(\frac{\theta}{\theta_r}\right)^{0.0355}$$

valid for water sprayed under pressure through two nozzles over the pressure range 2 070 through 17 240 kPa (300 through 2500 psig).

6. FUTURE WORK

An economic analysis which could optimize electrical cost with the cost of other cooling methods could prove very useful in the marketing of spray cooling systems. Here an optimum drop size and spray back pressure could be specified.

Studying the effects of windspeed on spray cooling efficiency would also be worthwhile because this efficiency is directly proportional to windspeed. From data collected in this experiment, modifications could be made on the equation derived in this paper as well as studies on drift. Drift, which is affected by windspeed and droplet size could be measured and compared with works by Hebden and Shah [15] who defined an equation that related per cent drift and windspeed.

Because all work done in this experiment was performed indoors, effects due to radiation were neglected. This could be analyzed by running two nozzles (one indoors and one outdoors) at the same conditions (no wind) and comparing the effects on efficiency.

As stated in the introduction, there are many methods available to model a spray cooling system. A calculation of the efficiencies predicted by these models would prove useful for comparison purposes.

Lastly, each of the above methods has been tested only with water. It is useful to study the effects and efficiencies of other fluids and cooling. This could be used to validate the equation derived in this experiment.

ACKNOWLEDGEMENTS

The authors wish to express their gratitude to Mr. Paul A. D'Amico for his help and guidance in assembling the spraying system and the temperature measurement system used in this study.

REFERENCES

1. Jakubowski, Gerald S., "Hot Water Cooling from Spray Arrays", Ph.D. Dissertation, University of Toledo, 1978.

2. Meyers, D. M. and Ryan, P. J. "Spray Cooling: A Review of Thermal Performance Models", Proceedings of the American Power Conference, 1976, volume 38.

3. Chen, K. H. and Trezek, G. J., "Spray Energy Release Approach to Analyzing Spray System Performance", Proceedings of the American Power Conference, 1976, volume 38.

4. Chen, K. H. and Trezek, G. J., "The Effect of Heat Transfer Coefficient, Local Wet Bulb Temperature, and Droplet Size Distribution Function on the Thermal Performance of Sprays", *Journal of Heat Transfer*, volume 99, August 1977.

5. Chen, K. H. and Trezek, G. J., "Thermal Performance Models and Drift Loss Predictions for a Spray Cooling System", *Journal of Heat Transfer*, volume 99, May 1977.

6. Chen, K. H. and Porter, R. W., "Heat and Mass Transfer of Spray Canals", *Journal of Heat Transfer*, page 286, August 1974.

7. Porter, R. W., Yank U., and Yanik, A., "Thermal Performance of Spray Cooling Systems", Proceedings of the American Power Conference, 1976, volume 38.

8. Hoffman, D. P., "Spray Cooling for Power Plants", Proceedings of the American Power Conference, 1973, volume 35.

9. Wilson, D. E., "Thermal Performance of Powered Spray Modules", Master of Science Thesis, University of Illinois at Urbana-Champaign, 1972.

10. Pita, E. G. and John, J. E. A., "The Effect of Forced Convection on Evaporative Cooling of Sprays in Air", University of Maryland, College Park, Md, 1969.

11. Berger, M. H. and Taylor, R. E., "An Atmospheric Spray Cooling Model", Tennessee Valley Authority, Knoxville, TN, 1978.

12. Ranz, W. E. and Marshall, W. R., "Evaporation from Drops", *Chemical Engineering Progress*, volume 48, No. 3, page 141, March 1952.

13. Janna, W. S., "Drop Size Distributions of Sprays Produced by Fan Jet Pressure Nozzles", Ph.D. dissertation, University of Toledo, 1976.

14. Young, H. D., Statistical Treatment of Experimental Data, McGraw-Hill Book Company, Inc., 1962.

15. Hibden, W. E. and Shah, A. M., "Effects of Nozzle Performance on Spray Ponds", Proceedings of the American Power Conference, volume 38, 1976.

16. Elgawhary, M., "Spray Pond Mathematical Model for Cooling Fresh Water and Brine", Ph.D. Dissertation, Oklahoma State University, July 1971.

17. John, J. E. A. and Haberman, W., Introduction to Fluid Mechanics , New Jersey; Prentice-Hall, Inc. 1971, Chapter 6.

Multi-Phase Flow and Heat Transfer III. Part A: Fundamentals
edited by T.N. Veziroğlu and A.E. Bergles
Elsevier Science Publishers B.V., Amsterdam, 1984 — Printed in The Netherlands

VAPORIZATION OF DROPLETS IN AN IMPINGING JET

Luciano M. de Socio
Department of Mechanics
Politecnico di Torino
C.so Duca degli Abruzzi,24-10129 TORINO(Italy)

Giovanni Gaffuri
Department of Mathematics and Physics
Università di Camerino
62032 CAMERINO (MC)(Italy)

ABSTRACT

The characteristics of a turbulent round air jet carrying finely dispersed water droplets, between two parallel disks at different temperatures,are investigated by means of a numerical experiment, under the assumption that the liquid phase and the vapor do not perturb the air flow. The jet issues from a circular hole in the cold disk and impinges vertically on the hot disk. The evolution of a swarm of vaporising droplets is evaluated by means of a deterministic-stochastic approach leading to the calculation of the vaporization region. The effect of the latent heat delivered by the liquid phase on the heat transfer at the hot wall is eventually investigated by considering the corresponding conductive resistance.

1. INTRODUCTION

Two-phase systems of dispersed droplets in gaseous streams present noticeable interest in many technological fields. In this paper those peculiar situations will be considered which correspond to the interaction of cold two - phase jets of liquid drops in air with hot solid surfaces, as it may occur, for instance, in mist flow tempering of steel and glass plates or in turbomachinery (see,for example,Ref.1).

A particular geometry is here investigated, where an axisymmetric round misty jet issues axially from a circular hole in an upper cold disk and impinges vertically on a second lower hot disk parallel to the first one. The diameter ofthe disks will be supposed to the large when compared with the diameter of the hole and with the spacing between the disks (Fig.1).

Following an approach already adopted with success in a previous work [2]it will be assumed that the number and the size of the droplets are small enough so that the presence of the dispersed phase and of the steam originated by its evaporation does not perturb the flow and the thermal field of the air. Under this hypothesis the problem of the time-evolution of the liquid phase can be considered separately and after having solved the problem of the velocity and temperature distribution in the air. Furthermore gravity effects will be neglected in the gaseous phase whereas they will be taken into account for the liquid particles.

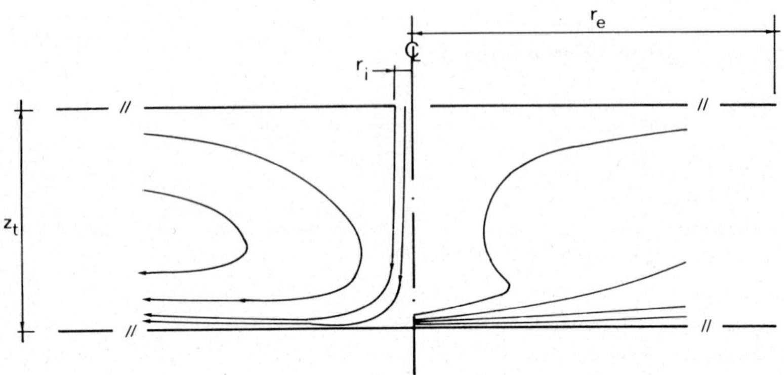

Fig.1 Geometry of the problem with streamlines (left) and isotherms (right)

When all the characteristics of the motion and thermal fields of the air are known, the time-evolution of the particles can be dealt with by means of a deterministic-stochastic procedure, corresponding to an application of the theory of differential equations with random initial conditions [3,4]. This technique, as in [2], consists of solving the basic equations of motion and vaporization rate for a finite number of drops and - at the same time - an equation for a suitable operator which translates an initial given probability density of finding some variables in an assigned state into a final unknown probability density.

2. ANALYSIS

As already said in the Introduction, the first step of the investigation is the solution of the basic equations for the continuous phase. Due to the complicated geometry of a turbulent jet confined between two finite walls the full Stokes-Navier, continuity and energy equations are to be solved through a suitable numerical procedure, together with the equations for the turbulence model. Among the various possibilities, in this paper a finite difference scheme will be adopted, as discussed later. Once the thermal and the flow fields for the air have been determined, the problem of the time-evolution of a single drop is to be solved.

In a cylindrical coordinate system, let z and r be the axial and the radial coordinate, respectively, let $\underline{u}$ and $\underline{v}$ be the local values of the velocity vectors of the air and of the liquid particle, and let ρ_a and ρ be the densities of the gas and of the liquid. Furthermore, let d be the diameter of the particle, $\underline{g}$ the acceleration of gravity, c_D the drag coefficient in the relative motion of the droplet with respect to the air, and t the time.
The deterministic state of the droplet is governed by a nonlinear system of first order ordinary differential equations corresponding to the second principle of dynamics and to the mass evolution law relative to the evaporation process.

In particular, one has

$$
\left.\begin{aligned}
dz/dt &= v_z \\
dr/dt &= v_r \\
dv_z/dt &= (1-\rho_a/\rho)g-(3\rho_a c_D/d\rho)|\underline{v}-\underline{u}|(v_z-u_z) \\
dv_r/dt &= -(3\rho_a c_D/d\rho)|\underline{v}-\underline{u}|(v_r-u_r)
\end{aligned}\right\} \quad (1a)
$$

$$
dd/dt = e \quad (1b)
$$

The rate equation (1b) follows the model proposed in [5], which assumes the form

$$
e = \frac{2}{d}\left(\frac{M}{\overline{M}}\frac{\rho_a}{\rho} D \text{ Sh } \ln\{(1-\omega_\infty)/(1-\omega_0)\}\right) \quad (2)
$$

where M is the molecular mass of the droplet, $\overline{M}$ is the average molecular mass of the air, D is the diffusion coefficient of the water vapor in air, Sh is the Sherwood number. In (2) ω_0 is the mole fraction of vapor at the droplet surface temperature and ω_∞ is the mole fraction of vapor at the local temperature of the surrouding air. The Sherwoord number obeys the Marshall-Ranz relation [6]

$$
\text{Sh} = 2 + 0.6\,\text{Re}_p^{\frac{1}{2}}\,\text{Sc}^{\frac{1}{2}}
$$

where Re_p is the Reynolds number of the relative motion, Sc is the Schmidt number $\mu/\rho_a D$, whereas μ stays for the viscosity of the continuum phase. Beard and Pruppacher's expression will be assumed for the drag coefficient

$$
c_D = (24/\text{Re}_p)(1+a\,\text{Re}_p^b)
$$

where

$a = 0$; $\text{Re}_p \leq 0.2$

$a = 0.1$; $b = 0.99$; $0.2 < \text{Re}_p \leq 2$

$a = 0.11$; $b = 0.81$; $2 < \text{Re}_p \leq 21$

$a = 0.189$; $b = 0.632$; $21 < \text{Re}_p \leq 200$

The fundamental system (1) is subjected to a proper set of initial conditions that corresponds to assigning, at $t=0$, to each particle of given initial diameter d_0, the value $z(0) = 0$ and an arbitrary value of $r(0) = r_0$, and by assuming that, at the inlet section, the velocities and the temperatures of the two phases coincide.

The basic set and related initial conditions can be put into non-dimensional form as

$$
\left.\begin{aligned}
dZ/dT &= V_Z \\
dR/dT &= V_R \\
dV_Z/dT &= A-(BF/D^2)(V_Z-U_Z) \\
dV_R/dT &= -(BF/D^2)(V_R-U_R) \\
dD/dT &= E \\
Z(0) = 0\ ,\ & R(0) = R_0\ ,\ V(0) = U(0,R_0)
\end{aligned}\right\} \quad (3)
$$

where capital letters indicate dimensionless quantities. In particular, the lengths are non-dimensional with respect to the radius of the inlet section r_i, the velocities are non-dimensional with respect to the average flow velocity at the entrance u_i and the reference time is r_i/u_i . In addition

$$A = (1 - \rho_a/\rho)r_i g/u_i^2$$

$$B = 144\ \rho_a/\rho Re_i$$

$$F = 1 + a\ Re_p^b$$

where $Re_i = \rho_a u_i r_i/\mu$ is the Reynolds number of the jet at the inlet.

Equations (3) can be solved by means of a fourth order Runge-Kutta procedure. Care must be paid to the case where a droplet hits the hot wall. If this happens, the liquid particle can either stick to the wall or bounce away, while evaporating, depending on the Weber number $We = \rho d\ v_z^2/\sigma$ at the impact (σ is the surface tension of the liquid)[7]. In particular, if $We > 30$, the particle bounces away, if $30 < We$, the droplet evaporates at the wall. In this work the Weber number of a particle hitting the wall has been calculated and, for We >30 specular reflection has been assumed.

As an example, on the axis of a jet issuing from a circular hole of radius r_i = 5 mm with u_i = 96 m/s, at an initial temperature of 303 K, a particle hitting a hot wall at 873 K and at the distance of 60 mm, bounces away and evaporates if d_0 = 110 μm, and sticks to the wall if d_0 = 115 μm. Here one considers as evaporated a drop whose mass is 10^{-6} times the initial mass. Particles of diameter d_0 smaller than 110 μm evapo rate before hitting the wall.
On the other hand, in the same jet, for d_0 = 115 μm, particles evaporate at the hot wall for $0 \leq r_0 < 2$ mm, whereas are specualr reflected for $2 < r_0 < 3$ mm and evaporate without touching the wall for $3 < r_0 \leq 5$ mm.

The system (3) can be put into the form of an evolution equation for the state of the droplet $\underline{X} \equiv \{Z,R,V_Z,V_R,D\}$:

$$d\underline{X}/dT = \underline{G}(\underline{X},T)\ ,\qquad \underline{X}_0 = \underline{X}(0) \tag{4}$$

In order to take into account the fact that the initial condition on the state of the particles can be assigned in a stochastic form, Eq.(4) must be augmented by considering the equation for the operator J which connects the initial probability density $P_0(\underline{X}_0)$ to the probability density $P(T;\underline{X})$ which corresponds to the probability associated to some state $\underline{X}$ of the object, at the time T 3 .
Since $J = |\partial \underline{X}_0^T/\partial \underline{X}|$, is the Jacobian of the transformation from $\underline{X}_0$ into $\underline{X} = \underline{h}(\underline{X}_0,T)$, then

$$P(T;\underline{X}) = |J|\ P_0\{\underline{X}_0 = \underline{h}^{-1}(\underline{X},T)\}$$

for a one to one mapping of $\underline{X}_0$ onto $\underline{X}$ and, from Liouville's theorem [4] ,

$$dJ/dT = -J\ \nabla \cdot \underline{G}\ ;\qquad J(T=0)=1 \tag{5}$$

Therefore the deterministic-stochastic problem is associated to the solution of the differential system (4,5) for a discrete number of droplets.

3. RESULTS AND CONCLUSIONS

As a practical application, the case of a turbulent jet, issuing at Re_i = 2,000 from the hole in the cold disk, such that $r_e/r_i = 50$, and for a spacing between the disks $z_t/r_i = 12$ was investigated.
Before considering the fundamental system (4,5) the equations for the thermal and flow fields were reduced to a finite difference form and were solved by an A.D.I. method, assuming a turbulence model governed by two differential equations for the dissipation of turbulent energy and for the turbulent energy itself [8].

The main characteristic of the flow field is the fact that the axisymmetric central jet, after impinging on the facing surface, spreads over the opposite

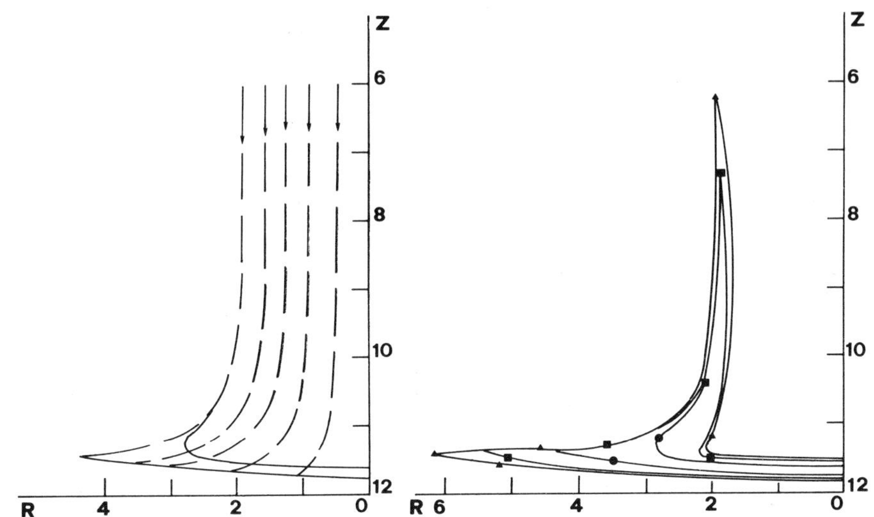

Fig.2 Droplet vaporization regions for the standard D_o = 0.002 (left) and for three different initial diameters (right)

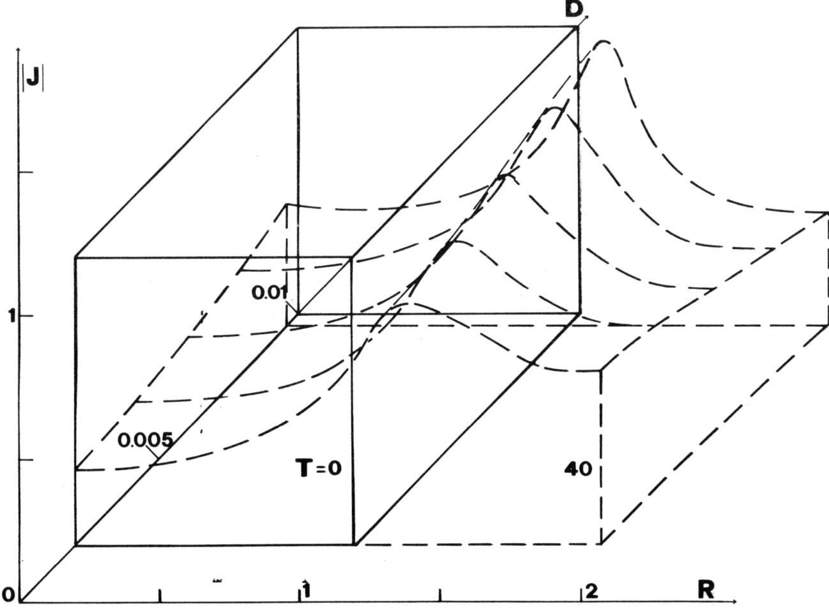

Fig.3 Evolution of the Jacobian of the transformation $P(T;\underline{X})=|J|P_o(\underline{X}_o)$

wall while driving a large vortex between the disks. This is shown at the l.h.s. of Fig.1, where some streamlines are reported, whereas some isotherms appear on the r.h.s.

Figure 2 shows on the left and side the trajectories of a number of droplets of dimensionless initial diameter $D_o = 0.002$ (10 μm). In the same figure the region is indicated where most of the evaporation process takes place. In the space, this evaporation region is bounded by the outermost trajectories, by an inner surface correspoding to the points where a sizeable evaporation is evaluated and by a third surface, close to the hot wall, where the residual mass of the droplets is considered to be negligible. In this work, it was conventionally supposed that evaporation has begun, where the mass of the droplet has lost the equivalent of the 10% of the initial mass of a standard droplet of diameter 0.002, whereas evaporation has ended where the mass of the droplets has reduced to 0.1% of the initial mass of the standard droplet.

The evaporation regions for particles of three different diameters are shown on the right hand side of Fig.2, where circles stay for $D_o = 0.002$, squares stay for $D_o = 0.005$ and triangles for $D_o = 0.01$.
The case where the initial distribution of diameters between the limits 0.002 and 0.01 is equiprobable and the initial radial position of the droplets is also equiprobable between 0 and 1 was then considered. In this situation the evolution of the Jacobian represents also the evolution of the probability density.

Figure 3 shows J versus D and R at T=0 (continuous lines) and T = 40 (dashed lines). Beyond T=40 , the Jacobian cannot be evaluated anymore, as: $\underline{X}_o \to \underline{X}$ is not a one-to-one mapping, due to the evaporation of the drops. Observation of Fig.3 enables one to realize that, at increasing time, the majority of the particles migrates toward higher R whereas the range of probable diameters is shortened in correspondence with the upper limit of the initial interval. In other words, particles of diameter close to the initial maximum value become improbable.

When the heat transfer at the hot wall is considered, the role of the latent heat of vaporization can be taken into account as follows. The overall heat transfer process between the hot wall and the two-phase system takes place approximately through a convective resistance extending from the wall to the near boundary of the evaporation region and through the couple of a convective resistance and a parallel conductive resistance from this boundary to the inlet section. In particular, the conductive resistance corresponds to the latent heat absorption.
The convective resistances can be evaluated as in the case where the particles are not present, due to the basic initial hypothesis of no interference of water vapor and water drops on the air flow and temperature. The conductive resistance R_s , following the guidelines of [9] can be approximately obtained by the expression

$$R_s = \Delta\Theta / (\dot{m}'' \pi r_i^2 L_v)$$

where $\Delta\Theta$ is the mean difference between the initial temperature of the jet Θ_j and the temperature of the surface of the vaporization region Θ_n adjacent to the hot wall, L_v is the latent heat and $\dot{m}''$ is the mass flux of the liquid at the entrance.

As an indication, the values of R_s corresponding to three different diameters are given, for $u_i = 96$ m/s ; $r_i = 5$ mm; $r_e = 250$ mm; $\Theta_j = 300$ K; $\Theta_n = 873$K; $z_t = 60$ mm:

$$d_0\,(\mu m) = \{10,30,50\} \to \frac{R_s}{\dot{m}''} \; \frac{(K\,s/J)}{(kg/m^2 s)} = \{0.16, 0.46, 0.64\}$$

ACKNOWLEDGMENT

This work was partially supported by the Italian National Ministry for Education (M.P.I.)

REFERENCES

1. Toda,S.,"A Study of Mist Cooling", Trans.J.S.M.E.,Vol.39,1973,pp.2160-2193

2. de Socio,L.M.,Gaffuri,G.and Polzonetti,A.,"Analytical Results on Two-Phase Flows in Ducts",Wärme-und Stoffübertragung,Vol.14,1980,pp.183-188

3. Soong,T.:*Random Differential Equations*,Academic Press,New York,1973

4. Bellomo,N.,Loiodice,R. and Pistone,G.,"Time Evolution of a Multidroplet System in the Kinetic Theory of Condensation-Vaporization" in *Rarefied Gasdynamics*,J.L.Potter Ed.,Progress in Astronautics & Aeronautics,Vol.51,Part II, AIAA,New York 1977,pp.1227-1239

5. Savery,C.W.,Juedes,D.L. and Borman,G.L.,"n-Heptane,Carbon Dioxide and Chlorotrifluoromethane Droplet Vaporization Measurements at Supercritical Pressures",Ind.Eng.Chem.Fund.,Vol.10,1971,pp.543-553

6. Bird,R.B.,Lightfoot,E.N. and Stewart,W.E.:*Transport Phenomena*,J.Wiley, New York,1964

7. Moureau,J.C., Bolle,L. and Giot,M.,"Influence of the Droplet Size Distribution on the Spray Cooling Efficiency", in *Two-Phase Momentum, Heat and Mass Transfer in Chemical Process and Energy Engineering Systems*,F.Durst, G.V. Tsiklauri,N.H.Afgan Eds.,Hemisphere Publ.Co.Washington,1979,pp.171-184

8. de Socio,L.M. and Polzonetti,A.,"Numerical Evaluation of the Characteristics of a Turbulent Jet Impinging on a Wall" (in Italian) Atti Ac.Sc.Torino,Giugno 1981

9. Bhatti,M.S. and Savery,C.W.,"Augmentation of Heat Transfer in a Laminar External Gas Boundary Layer by the Vaporization of Suspended Droplets"Trans. ASME,J.Heat Transfer,Vol.97,1975,pp.179-184

Multi-Phase Flow and Heat Transfer III. Part A: Fundamentals
edited by T.N. Veziroğlu and A.E. Bergles
Elsevier Science Publishers B.V., Amsterdam, 1984 — Printed in The Netherlands

EMPIRICAL FORMULA FOR HEAT TRANSFER COEFFICIENT IN TWO-COMPONENT TWO-PHASE FLOW

M. M. El-Kassaby
Department of Engineering
Purdue University Calumet
Hammond, Indiana 46323, U.S.A.

E. N. Ganic
Faculty of Mechanical Engineering
University of Sarajevo
Sarajevo 71000, Yugoslavia

ABSTRACT

Measurements of heat transfer in two-component (air-water) dispersed flow were performed. The heat transfer test section was a smooth, stainless steel circular tube, 12.95 mm I.D. and 890 mm long. The electrical power was dissipated in the test section, which acted as a resistance element with nearly uniform wall heat flux along the test section. The experimental conditions included different values for the inlet pressure, mass velocity, the inlet water droplet concentrations, and the wall heat flux. An empirical correlation for heat transfer coefficient, based on the experimental data of this study, was developed.

1. INTRODUCTION

Analysis of two-phase heat transfer in dispersed flow is of interest in many practical applications. Cryogenic machinery, steam generators, nuclear reactor systems, and wet steam turbines all involve two-phase heat transfer to some extent. Much attention has, therefore, been paid to the study of post-dryout heat transfer in one-component two-phase flows involving steam-water mixtures and cryogenic fluids.

Correlations are normally developed using data from a limited number of sources, and as such are typically limited to a range of flow conditions and one fluid. However, they are convenient to use because they do not require a computer solution and are therefore very attractive.

Many such correlations begin with an accepted equation for pure vapor, such as the McAdams or Dittus-Boelter correlation, and modify it to account for such things as non-equilibrium, slip (the ratio of vapor to liquid velocities), and entrance length effects in the flow.

Miropolski [1] developed a film boiling correlation for water, including a correction for slip in the flow. He modified the Dittus-Boelter correlation to describe the heat transfer to pure steam, and applied the resulting correlation to two-phase steam-water flows.

Groeneveld [2] presented a heat transfer correlation including a slip factor similar to Miropolski's. The equation was generalized for different flow geometries by determining a set of constants appropriate for each geometry.

Matteson et al. [3] used a regression analysis on several data sets for different flow geometries and developed a separate correlation for each geometry. His correlations included a heat flux effect and a gas conductivity to critical point conductivity ratio.

A partial list of available correlations is presented by Groeneveld [2].

From the foregoing research, there is no available correlation dealing with two-component two-phase flow. Such a correlation is the goal of the present research.

2. EXPERIMENTS ON DISPERSED FLOW HEAT TRANSFER

2.1. Experimental Apparatus and Description

A schematic of the experimental set-up used for the heat transfer studies is illustrated in Fig. (1). The air and water supply and the atomizer used for droplet generation are essentially the same as described for the deposition studies using a stainless steel tube test section. The apparatus is the same as used by Mastanaiah [4].

The heat transfer test section is a smooth, commercially drawn, stainless steel type-304 circular tube, 12.95 mm I.D., 3.04 mm thickness, and 889 mm long. Electrical power is supplied by a D-C rectifier with an A.C. input of 220 V, 3 Φ, 60 C/s and an output range of 0 - 15 V and 0 - 200 A. The electrical power is dissipated in the test section, which acts as a resistance element and insures nearly uniform wall heat flux along the test section. The inlet and outlet lines connecting the test section to the rest of the flow system are insulated electrically from the test section with asbestos washers.

The liquid film formed at the outlet of the entrance (flow development) section is removed through the extraction unit placed at the inlet of the test section, and is connected to a vacuum pump.

2.2. Experimental Technique

Measurement of wall temperature. The outer wall temperature distribution of the test section is measured by twenty-nine chromel-alumel thermocouples (.51 mm diameter) spot welded to the outer surface. The thermocouple locations on the test section are spaced 38.1 mm apart longitudinally in a zig-zag manner, and are displaced 120° apart circumferentially. Thermocouple #1 is fixed at 12.7 mm from the inlet, while thermocouple #29 is attached at 38.1 mm below the outlet of the test section. At three locations along the test section (z/D = 9.8, 36.27 and 62.74), the three thermocouples are fixed at equal distances around the circumference to check the circumferential non-uniformity of the surface temperature.

The error in the thermocouple output due to D.C. interference caused by the existence of a very small gap between the thermocouple wires on the surface has been calibrated by Mastahaiah [4] using the current shut-off method suggested by Davenport et al. [5], and rechecked by the author using the same method.

The thermocouple output is measured by a Hewlett-Packard digital multimeter (Model 4365-B) having 1.0 μv sensitivity and 0.03% accuracy in 0 - 20 mv range. The inside wall temperatures are estimated from the measured outer surface temperatures by the steady-state heat conduction solution for a hollow cylinder with uniform internal heat generation and insulated outer wall [6, 7].

Measurement of gas temperature and pressure. Thermocouples are placed at the center line, at the inlet, and at the outlet of the test section to estimate the bulk fluid temperature. A pressure measurement is made at the inlet of the test section, and the pressure drop across the test section is measured by a water manometer.

Measurement of heat flux. The electrical power input to the test section is determined from the measured voltage drop across the test section, and the resistivity of the test section material corresponding to the measured average wall temperature of the test section, as

$$q'' = \frac{V^2}{\rho_e\,(\bar{T}_{wm}) \cdot L/A} \qquad (1)$$

The electrical resistivity, ρ_e, of the test section was measured by [4] in a temperature range of 323 - 373°K, using a standard resistor (400 A, 50 MV) in the circuit. The measured resistivity was found to be within 1% of the literature for similar material [8], and, therefore, the literature values have been used in the present work for the entire temperature range.

Heat loss by natural convection from the test section insulation is evaluated using the measured outer surface temperature of the insulation, ambient temperature, and employing a standard correlation [8] at five locations along the test section (z = 10, 30, 45, 60 and 80 mm).

The heat loss is found to be less than 3% of the total power input in computing the average wall heat flux in the test section, $\bar{q}_w$. The longitudinal heat loss due to conduction at the downstream end of the test section is computed from the measured wall temperatures and the wall thermal conductivity. However, the longitudinal heat loss is found to be less than 0.3% of the total input in all present experiments, and, therefore, it can be neglected.

Local wall heat flux is obtained from the average wall heat flux and a correction factor accounting for the variation of tube metal resistivity from end to end, in conjunction with the calculated local wall temperature. This relation is given by

$$q_w''(z) = \overline{q_w''}\rho_e\ T_w(z)/\rho_e\ (\bar{T}_{wm}) \qquad (2)$$

Finally, the heat transfer coefficient can be calculated as

$$h_2(z) = \frac{q_w''(z)}{T_w(z) - T_g(z)} \qquad (3)$$

In evaluating $T_g(z)$, the model suggested by [9] was adopted here.

The test section conditions for dispersed flow heat transfer are displayed in Table 1, showing the inlet pressure, the mass velocity, the inlet water concentration to the test section, the average wall heat flux, the inlet gas mass

Run #	P_{in} bar	C Kg/m·s	$\bar{C}_{in}$x10 Kg/m^3	$\overline{q_w}$ K_w/m^2	w_gx10^3 Kg/s	d_{10} μm	Δpx10 bar	Re
1	1.22	72.26	0.7227	37.64	9.082	10.75	0.4117	48400
2	1.36	72.16	0.7189	26.97	9.077	10.75	0.3725	48380
3	1.226	71.91	0.6582	22.02	9.077	10.75	0.3039	48390
4	1.317	89.39	0.7984	43.30	11.22	8.593	0.5391	59830
5	1.299	89.21	0.7639	34.27	11.22	8.599	0.490	59770

Table 1. Test Conditions for Dispersed Flow Heat Transfer

flow rate, the inlet drop diameter, the pressure drop, and the initial value for Re.

The empirical formula for heat transfer coefficient in two-component two-phase-flow. It was found by many investigators [9, 10, 11] that the local heat transfer coefficient for two-component, two-phase flow in circular duct is a function of $\overline{q}_w$, C_{in}, Re and z/D, assuming

$$h_2(z) = \frac{q_w}{T_{in}} \; \Phi \, (\mathrm{Re}, \, z/D, \, C_{in}) \tag{4}$$

where Φ is the nonlinear function of the dimensionless parameters Re, z/D and C_{in}. C_{in} here represents the ratio of the mass flow rate of water entering the test section divided by the total mass flow rate entering the test section (kg/kg), i.e., the mass fraction upon entering the test section.

Using the least square fitting technique, the best equation to fit the experimental data was found to be

$$h_2(z) = 748.2 \, \frac{\overline{q''_w}}{T_{in}} \left(\frac{z}{D}\right)^{-0.5} \mathrm{Re}^{-0.6} \, C_{in}^{\ \alpha} \tag{5}$$

where

$$\left.\begin{aligned} \alpha &= -1.6, && \text{for } 0.06 < C_{in} < 0.08 \\ &= -1.0, && \text{for } 0.04 < C_{in} < 0.06 \\ &= -0.8, && \text{for } 0.01 < C_{in} < 0.04 \end{aligned}\right\} \tag{6}$$

The comparison of the present formula (Eq. 5) with the present experimental data is plotted in Figs. 2.a through 2.e. The figures show a good agreement, within a maximum error of $\pm$ 20%. This error decreases with the increase of z/D and also with the increase of the Reynolds number.

In order to check the validity of the obtained formula, it is important to compare it with the existing data dealing with two-component, two-phase flow.

The experimental data obtained by Ganic & Mastanaiah [10], who used the same apparatus explained here, were also compared with the obtained formula. The comparison is plotted in Figs. 3.a through 3.e. These figures show a good agreement with the same observation noticed before.

Takagi and Ogasaware [11] obtained experimental data using an air-water mist flow in a rectangular cross-section heated from one side. Hence, their experimental data are used, and the length is divided by the hydraulic diameter to get the ratio (z/D), to be able to use the present formula (Eq. 5). The comparison is plotted in Figs. 4.a through 4.d.

The figures show a large discrepancy of 50%, because they used a rectangular cross-section instead of a circular cross-section and applied heat only to one side. This suggests that the heat transfer coefficient is dependent on the configuration.

Empirical formula for T_w. T_w was also found to be dependent of q_w, inlet concentration, Re, and z/D assuming

$$T_w = \frac{q_w\text{-}L}{k_g} \phi \, [\mathrm{Re},\ z/D,\ C_{in}] \qquad (7)$$

where k_g is the thermal conductivity of the gas and ϕ is the non-linear function. As done previously, using the least square fitting, the best equation to fit the experimental data was found to be

$$T_w(z) = 0.925 \, \frac{\overline{q_w} L}{k_g} \, \mathrm{Re}^{-1.0827} \left(\frac{z}{D}\right)^{0.26201} C_{in} \qquad (8a)$$

$$0.01 < C_{in} < 0.06$$

and

$$T_w(z) = 2.9\text{x}10^{-3} \, \frac{\overline{q_w} \; L}{k_g} \, \mathrm{Re}^{-0.614} \left(\frac{z}{D}\right)^{0.1262} C_{in} \qquad (8b)$$

$$0.007 < C_{in} < 0.04$$

The comparison between Eq. (8a-b) and the present experimental data is plotted in Figs. 5a through 5 , which show a good agreement in the range $15 \leq z/D \leq 76$, with an error of $\pm$ 10%.

A comparison between eq. (8a-b) and the experimental data obtained by Ganic & Mastanaiah [10] is also plotted in Figs. 6a through 6e.

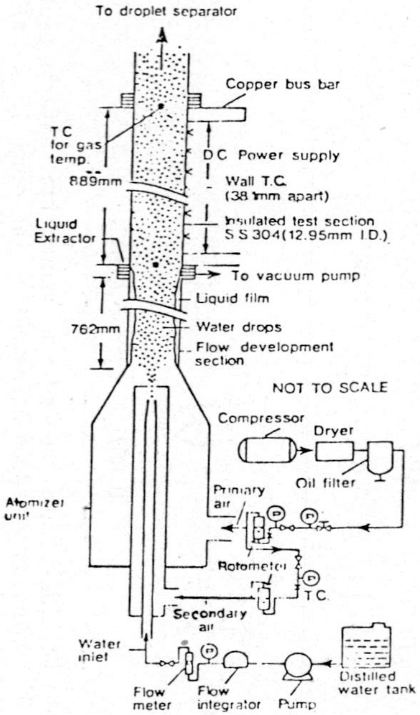

Fig. 1: Schematic of experimental set-up for dispersed flow heat transfer.

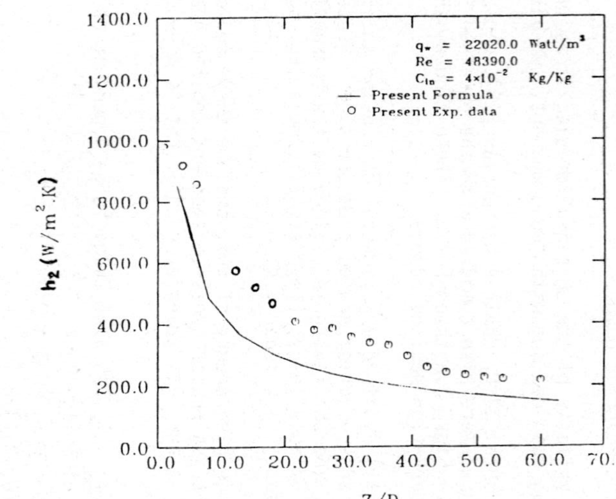

Fig. 2a: A comparison between the present correlation and the present experimental data at Re = 48390 and $\overline{q}_w$ = 22020 watt/m^2.

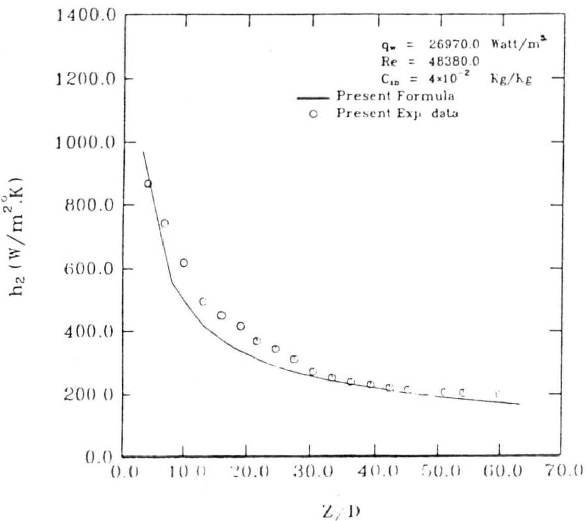

Fig. 2b: A comparison between the present correlation and the present experimental data at Re = 48380, and $\overline{q}_w$ = 26970 watt/m^2.

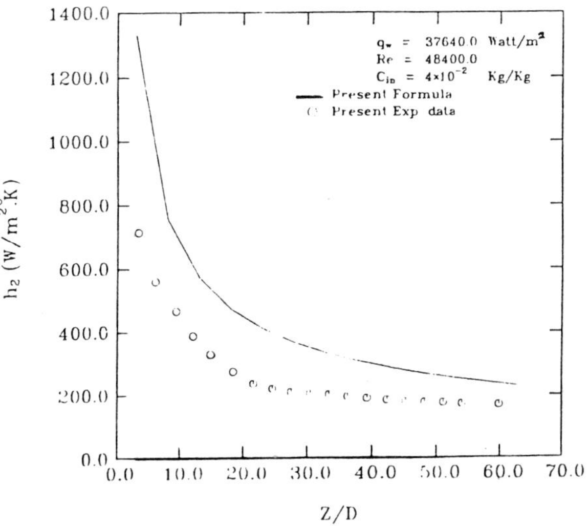

Fig. 2c: A comparison between the present correlation and the present experimental data at Re = 48400, and $\overline{q}_w$ = 37640 watt/m^2.

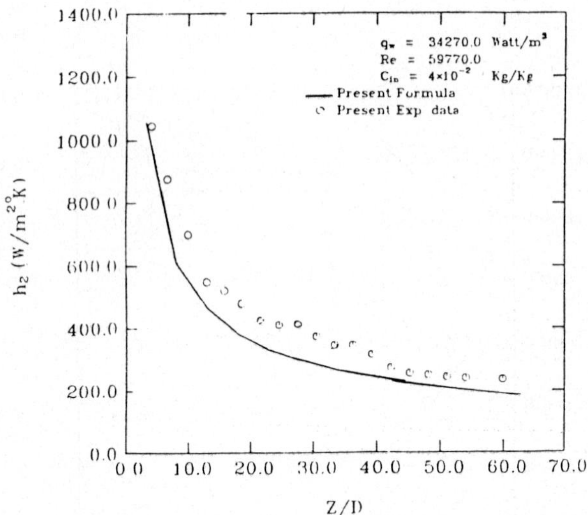

Fig. 2d: Comparison between the present correlation and the present experimental data at Re = 59770 and $\overline{q}_w$ = 34270 watt/m^2.

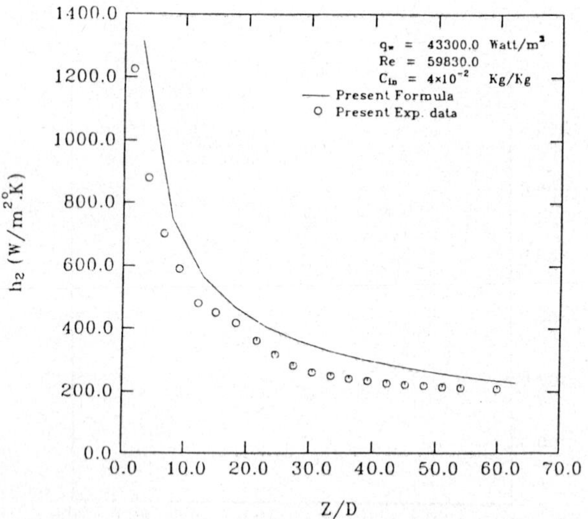

Fig. 2e: Comparison between the present correlation and the present experimental data at Re = 59830 and $\overline{q}_w$ = 43300 watt/m^2.

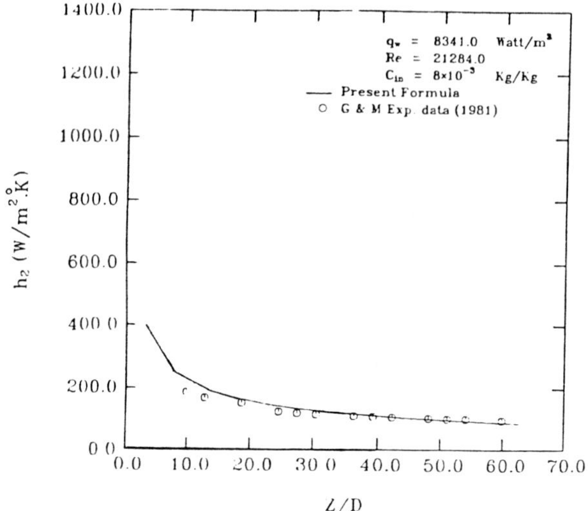

Fig. 3a: Comparison between the present formula and the experimental data by Ganic and Mastanaiah (1981).

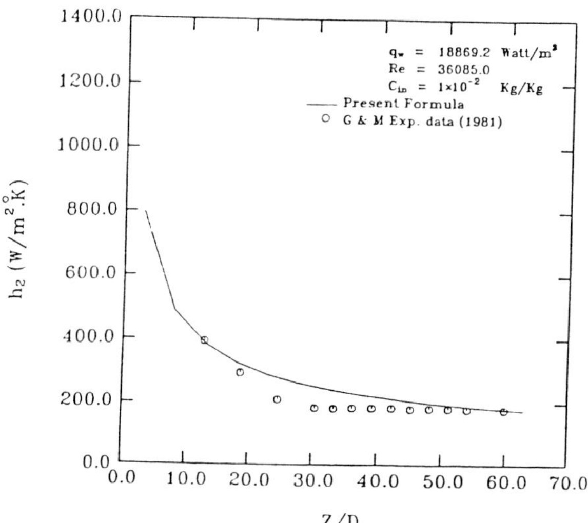

Fig. 3b: Comparison between the present formula and the experimental data by Ganic and Mastanaiah (1981).

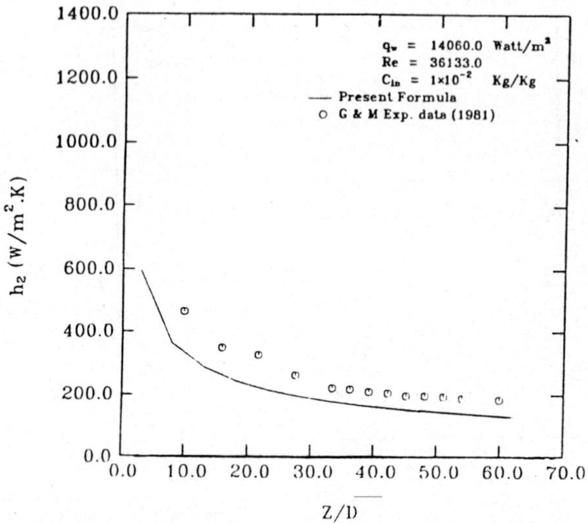

Fig. 3c: Comparison between the present formula and the experimental data by Ganic and Mastanaiah (1981).

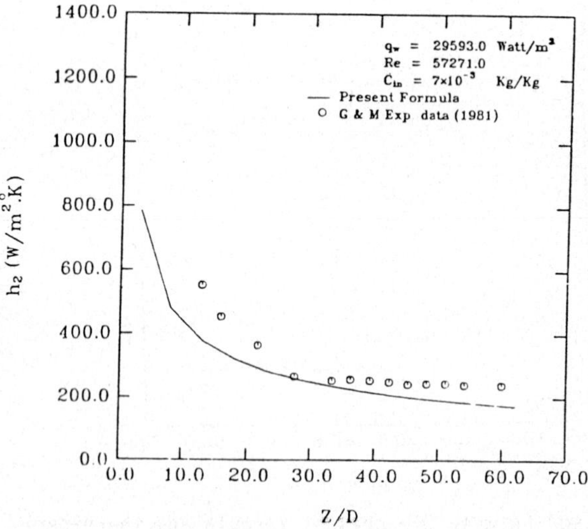

Fig. 3d: Comparison between the present formula and the experimental data by Ganic and Mastanaiah (1981).

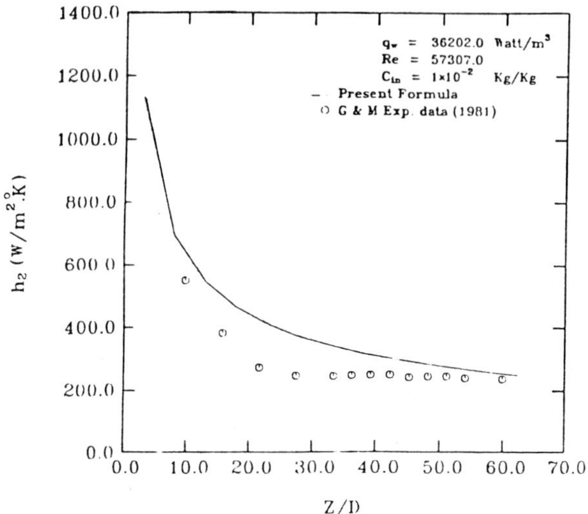

Fig. 3e: Comparison between the formula and the experimental data by Ganic and Mastanaiah (1981).

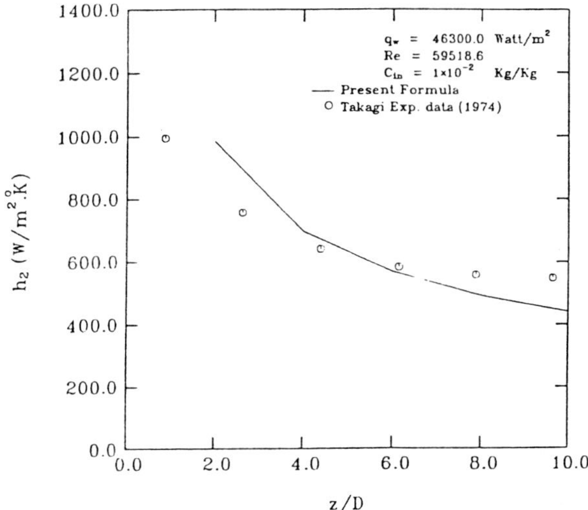

Fig. 4a: Comparison between the present formula and the experimental data by Takagi (1974).

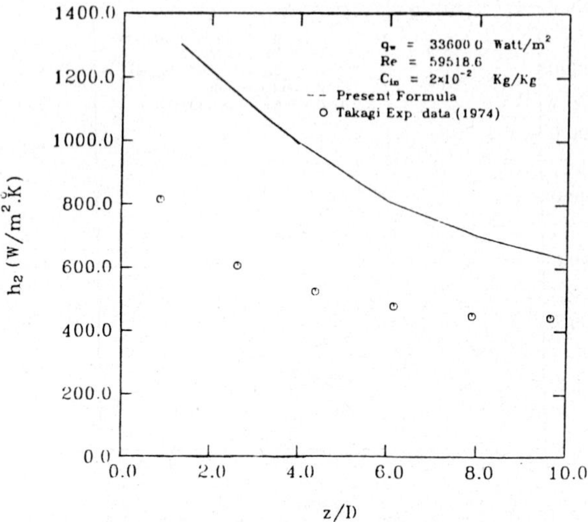

Fig. 4b: Comparison between the present formula and the experimental data by Takagi (1974).

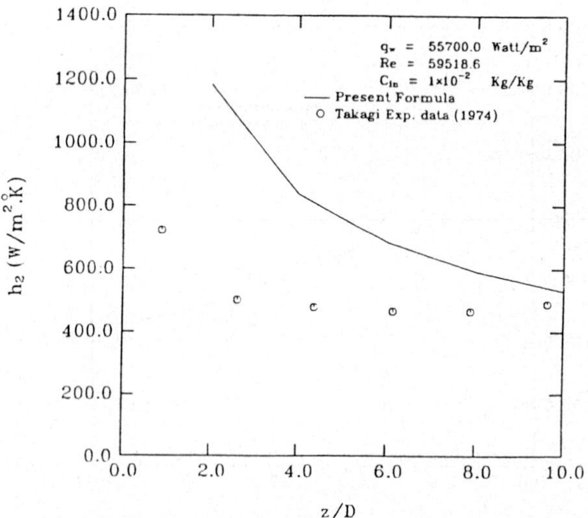

Fig. 4c: Comparison between the present formula and the experimental data by Takagi (1974).

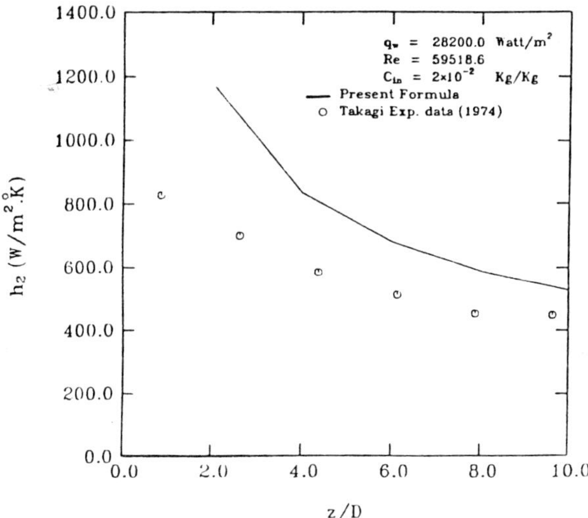

Fig. 4d: Comparison between the present formula and the experimental data by Takagi (1974).

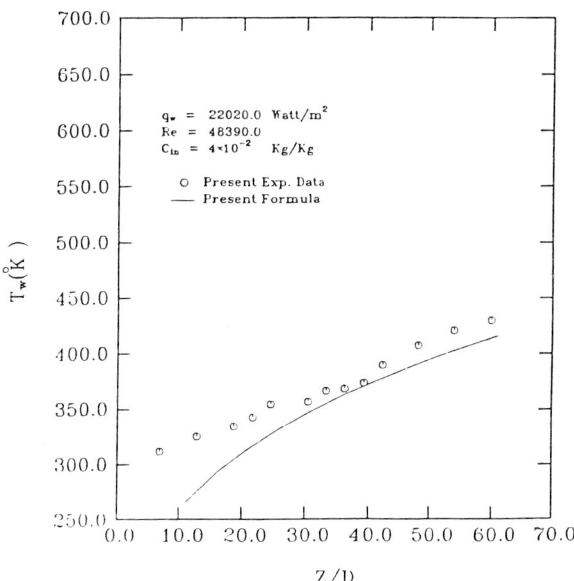

Fig. 5a: Comparison between the present formula and the present experimental data.

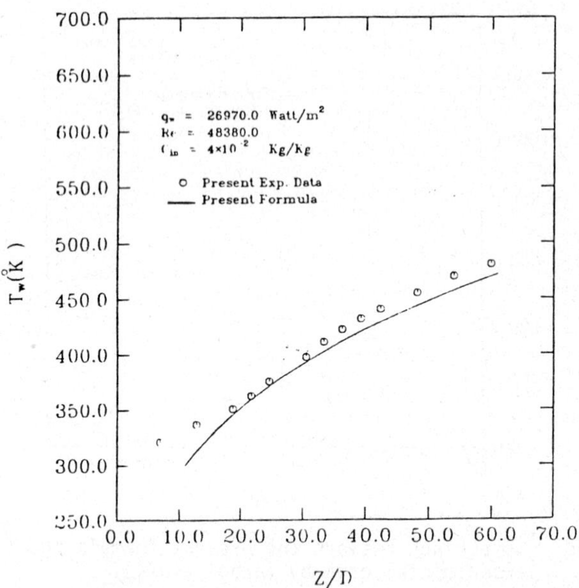

Fig. 5b: Comparison between the present formula and the present experimental data.

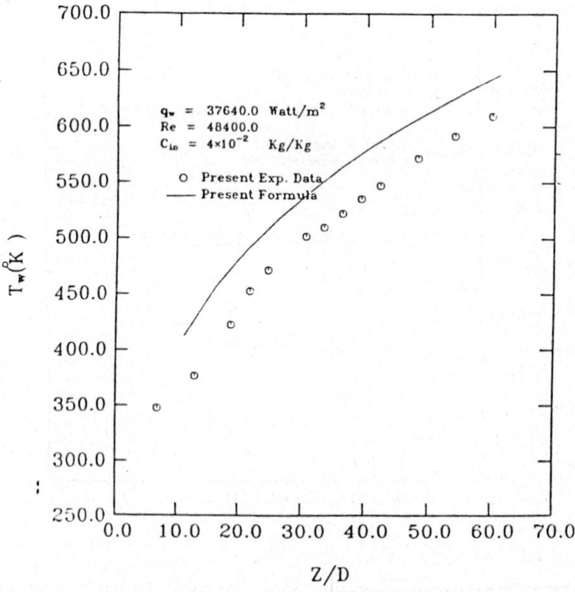

Fig. 5c: Comparison between the present formula and the present experimental data.

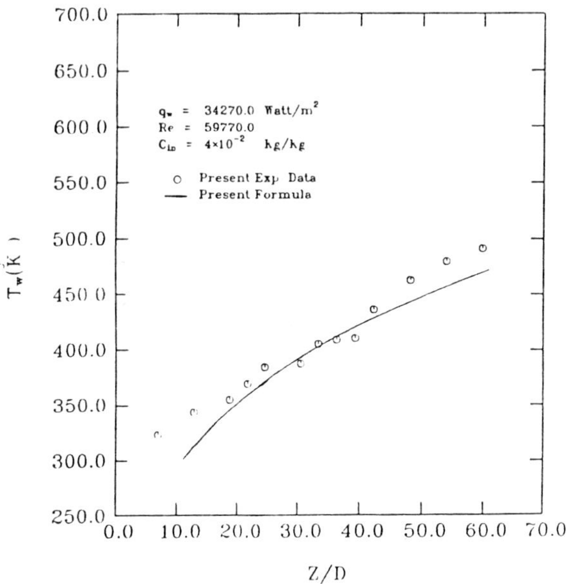

Fig. 5d: Comparison between the present formula and the present experimental data.

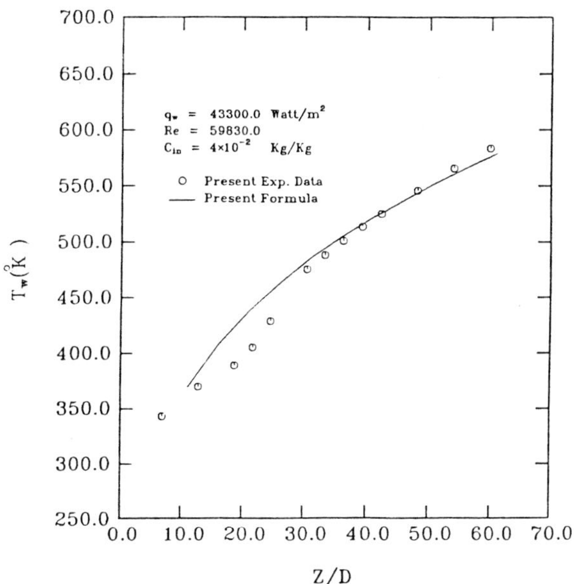

Fig. 5e: Comparison between the present formula and the present experimental data.

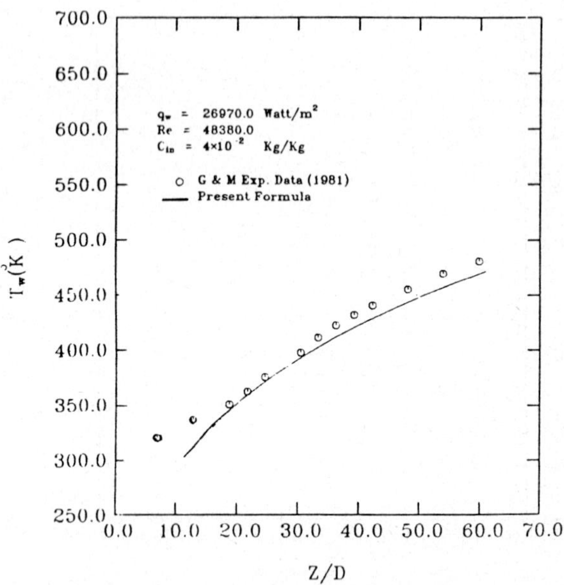

Fig. 6a: Comparison between the present formula and the exp. data by G & M (1981) for evaluating T_W.

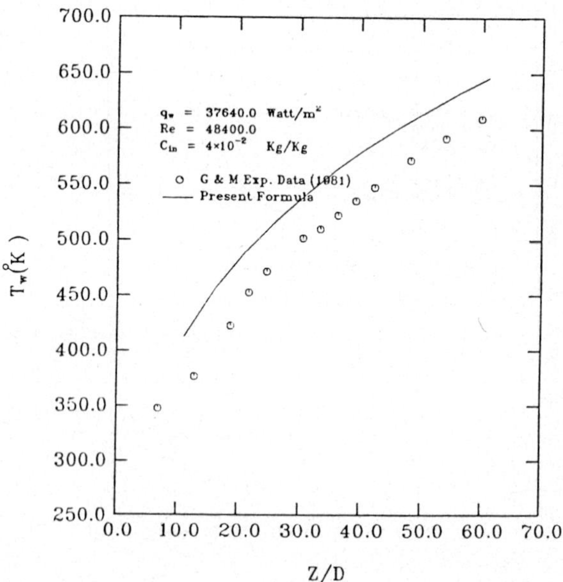

Fig. 6b: Comparison between the present formula and the exp. data by G & M (1981) for evaluating T_W.

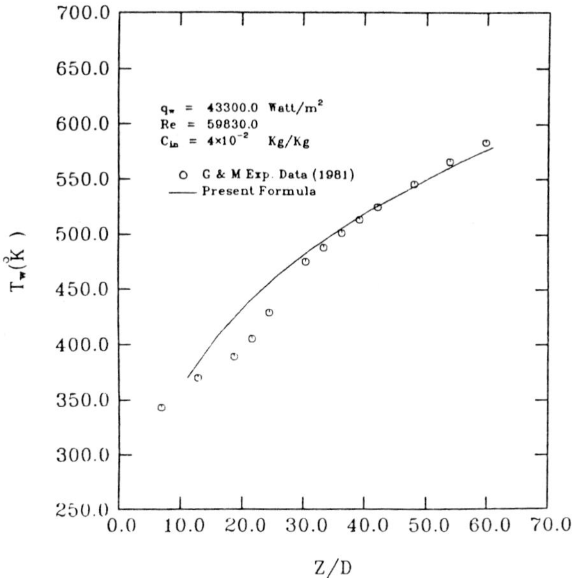

Fig. 6c: Comparison between the present formula and the exp. data by G & M (1981) for evaluating T_w.

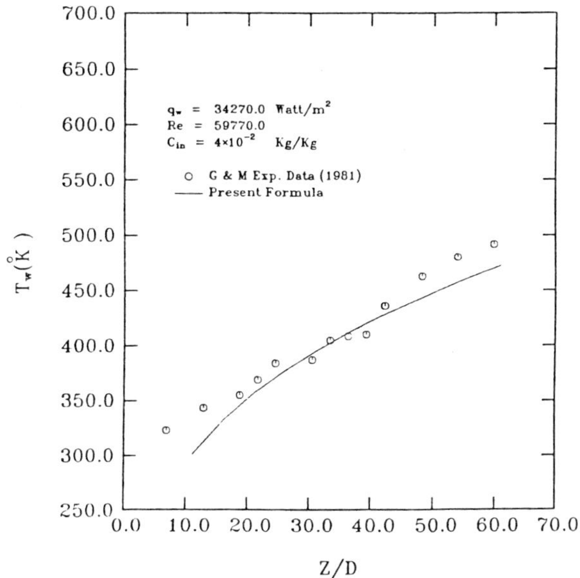

Fig. 6d: Comparison between the present formula and the exp. data by G & M (1981) for evaluating T_w.

3. CONCLUSION

Experiments have been performed on heat transfer in two-phase flow in an air-water dispersed system. A formulated correlation has been found to estimate both the heat transfer coefficient, h_2, and the wall temperature, T_w. The following conclusions can be drawn:

1. The two-phase heat transfer coefficient, h_2, decreases with an increase in wall temperature.
2. The recommended empirical formula presented by Eq. 5 for evaluating the heat transfer coefficient, h_2, is applicable with an error of $\pm$ 20%, in the specified ranges for the initial concentration given by Eq. (6). Outside of this range, one can use it with caution, i.e., the power for C_{in} may need to change.
3. The empirical formula for evaluating the wall temperature given by Eq. (8a-b) is applicable with an error of $\pm$ 10%.

ACKNOWLEDGMENT

This work was supported by NSF Grant CME-8005681.

NOMENCLATURE

A = cross-sectional area, m^2

D = tube diameter, m

k = thermal conductivity, W/(m.°K)

q = heat rate, W

Re = tube Reynolds number, $\bar{U}\ D/\nu_g$

T = temperature, °K

V = voltage drop across the test section, V

z = axial distance, m

Greek Symbols:

ρ_e = electrical resistivity, Ω - m

Subscripts:

2 = two-phase

in = inlet

m = measured

w = wall

Superscripts:

- = average sign

" = flux

REFERENCES

1. Miropolski. A.L., "Heat Transfer in Film Boiling of a Steam Water Mixture in Steam Generating Tubes," Teploenergetika 10, pp. 49-53 (1963).

2. Groeneveld, D.C., "The Thermal Behavior of Heated Surface at and Beyond Dryout," AECL 4309 (1972).

3. Mattson, R.J., Condie, K.G., Bengston, S.J. and Obenchain, C.F., "Regression Analysis of Post- CHF Flow Boiling Data," Int. Heat Trans. Conf., Tokyo, 1974.

4. Mastanaiah, K., "Experimental and Theoretical Investigation of Droplet Deposition and Heat Transfer in Air-Water Dispersed Flow," Ph.D. Thesis, Dept. of Energy Engineering, Univ. of Illinois at Chicago Circle, 1980.

5. Davenport, M.E., Magee, P.M., and Leppert, G., "Thermocouple Attachment to a Direct Current Heater," ASME J. of Heat Transfer 84, May 1962, pp. 187-188

6. Hartnett, J.P., "Experimental Determination of the Thermal Entrance Lengths for the Flow of Water and of Oil in Circular Pipes," Trans. ASME 77, pp. 1211-1220 (1955).

7. Yao, S.S., "Head Transfer and Friction Factors of Non-Newtonian Fluids in Turbulent Pipe Flow." Ph.D. Thesis, Dept. of Engng., Univ. of Illinois at Chicago Circle, 1974.

8. Dickinson, N.L. and Welch, C.P., "Heat Transfer to Supercritical Water," Trans. ASME 80, pp. 746-752 (1958).

9. El-Kassaby, M.M., "Drop Deposition and Heat Transfer in Two-Component Dispersed Flow," Ph.D. Thesis, University of Illinois at Chicago (1982).

10. Mastanaiah, K. and Ganic, E.N., "Heat Transfer in Two-Component Dispersed Flow," J. of Heat Transfer 103, pp. 300-306 (1981).

11. Takagi, T. and Agasaware, M., "Some Characteristics of Heat and Mass Transfer in Binary Mist Flow," Proc. 5th International Heat Transfer Conf., Tokyo, 1974.

Multi-Phase Flow and Heat Transfer III. Part A: Fundamentals
edited by T.N. Veziroğlu and A.E. Bergles
Elsevier Science Publishers B.V., Amsterdam, 1984 — Printed in The Netherlands

MATHEMATICAL MODELLING OF COLD DROPLETS-HOT SURFACE INTERACTION

Luciano M.de Socio
Department of Mechanics
Politecnico di Torino
C.so Duca degli Abruzzi,24-10129 TORINO(Italy)

Amilcare Pozzi
Institute of Aerodynamics
Università di Napoli
Piazzale Tecchio-80125 NAPOLI(Italy)

ABSTRACT

The heat transfer between a hot flat plate and a stream of impinging cold droplets, which vaporize completely in a very thin layer adjacent to a solid surface, is considered. A mathematical model is proposed, corresponding to the energy equation for the vapor phase and the solid plate, with a suitable matching condition at the interface, where heat of vaporization and radiation act as energy sinks. Analytic solutions are obtained by means of the Laplace transformation.

1. INTRODUCTION AND MATHEMATICAL MODEL

The problem of the interaction between hot solid bodies and swarms of impinging vaporizing droplets presents noticeable interest in various technological applications. The corresponding heat transfer process can show very different aspects depending on the widely changing characteristics of the particular physical circumstances.

In this paper one will consider the heat transfer between a hot flat plate of finite thickness and of infinite extent and a constant stream of cold impinging droplets. The droplets vaporize completely in a very thin layer adjacent to the solid surface and the resulting vapor is considered to be of constant density and moving back from the plate at a constant velocity normal to the interface. In the process, it is assumed that the liquid particles and the possible presence of a non-condensable gas do not affect the temperature and velocity fields in the vapor region. Furthermore, those situations will be investigated, where radiant here exchange between the hot surface and the ambient is present.
The radiative heat absorbed by the vapor will be neglected, in circumstances where the difference between the temperature of the invested surface and the temperature of the ambient infinitely far from the solid is relatively small with respect to the ambient temperature.

Under the hypotheses mentioned above the entire field of interested can be modelled according to the sketch reported in Fig.1.

If $\dot{m}$ is the mass flux of drops impinging per unit surface, then $v = \dot{m}/\rho_V$

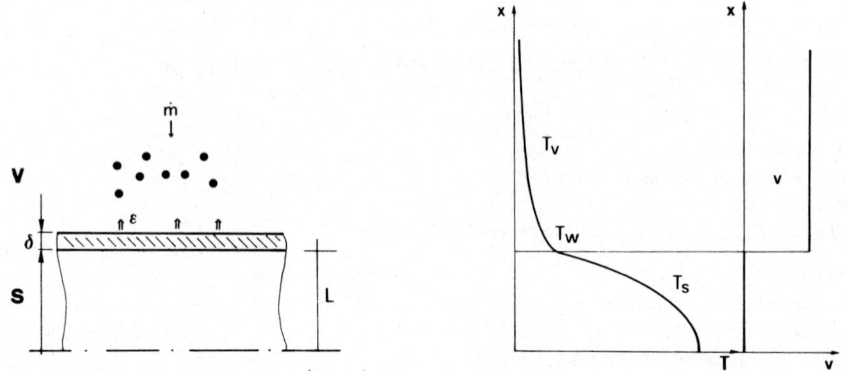

Fig.1 A sketch of the model (left) and the distributions of temperature and velocity (right)

is the velocity of the vapor, where ρ_V is its density.
In the infinitely thin layer δ there is the complete vaporization of the droplets which release a latent heat of vaporization per unit mass equal to L_V .
When the abscissa x is measured from the insulated wall of the plate of thickness L , a thin layer δ at $x=L$ corresponds to a line source of mass flux $\rho_V v$, and of energy flux $L_V\dot{m}$. In addition, at the body surface, there is a net amount of leaving radiation flux equal to $\varepsilon\sigma(T_w^4 - T_\infty^4)$, where T is the temperature, ε is the emissivity of the surface, σ is the Stefan constant and the indexes w and ∞ are relative to the wall and to the ambient at infinity, respectively.
The letters S and V indicate the solid and the vapor region and, in the following, will be appended to quantities pertaining to these two regions. The schematic diagrams on the r.h.s. of Fig.1 show the qualitative behaviour of the temperature and velocity distributions.

The plate at the initial temperature T_i , is invested, at the time $t=0$, from the side at $x=L$, by a continuous swarm of droplets at the ambient temperature T_∞. At the adiabatic wall of the flat plate at $x=0$ the condition $\partial T_S/\partial x = 0$ is to be fullfilled by the temperature profile which, at $x=L$ shows a discontinuity of the derivative caused by the localized heat sinks.

When dealing with the energy equation for the vapor region, both convection and conduction must be taken into account, whereas the energy equation for the solid region reduces to the classic conduction equation. At the interface δ the matching conditions corresponding to the continuity of the temperature and to the balance of energy fluxes will be applied.

Let the following non-dimensional quantities be introduced:

$$\theta = (T-T_\infty)/(T_i-T_\infty) \ ; \quad X = x/L \ ; \quad \tau = t/(c_S\rho_S L^2/\lambda_S) \ ;$$

$$\alpha^2 = (\rho_V c_V \lambda_S/\rho_S c_S \lambda_V) \ ; \quad \beta = \tfrac{1}{2}\, \rho_V c_V\, vL/\lambda_V\, \alpha \ ;$$

where λ is the heat conductivity, c_S is the specific heat of the plate and

c_V is the specific heat at constant volume of the vapor. Furthermore, let

$$\gamma = \lambda_V/\lambda_S \; ; \quad \eta = L\, L_V\dot{m}/\lambda_S(T_i - T_\infty) \; ; \quad \zeta = 4\varepsilon\sigma T_\infty^3 L/\{\lambda_S(T_i - T_\infty)\} \; .$$

The differential problem corresponding to the investigated mathematical model is given by the following fundamental set of basic equations together with the pertinent initial and boundary conditions:

Region S

$$\partial\theta_S/\partial\tau = \partial^2\theta_S/\partial X^2 \quad ; \quad \tau > 0 \; , \quad 0 \leqq X < 1 \tag{1}$$

$$\theta_S(0,X) = 1 \; , \quad 0 \leqq X < 1 \quad ; \quad \partial\theta_S(\tau,0)/\partial X = 0 \; , \quad \forall\tau \tag{2}$$

Region V

$$\alpha^2\partial\theta_V/\partial\tau + 2\,\beta\alpha\,\partial\theta_V/\partial X = \partial^2\theta_V/\partial X^2 \tag{3}$$

$$\theta_V(0,X) = 0 \; , \quad 1 < X \leqq \infty \quad ; \quad \theta_V(\tau,\infty) = 0 \; , \; \tau > 0 \tag{4}$$

At the interface δ

$$\theta_S = \theta_V \tag{5}$$

$$\partial\theta_S/\partial X = \gamma(\partial\theta_V/\partial X) - \eta + \zeta\,\theta_S \tag{6}$$

The order of magnitude of the dimensionless parameters appearing in the model can be realized by considering their values in the particular cases of mist tempering of a steel plate and of a glass plate. For a thickness of the plate of 4 mm, to the steel vapor couple correspond the typical values α=0.65, β=1.14, γ=0.0007, η=0.0018 ζ=0.22 $\times 10^{-4}$, whereas for the vapor glass couple one has α=0.122, β=6 , γ=0.034, η=0.09, ζ=0.21 $\times 10^{-4}$.

2. SOLUTION

Introducing the Laplace transforms on the time $F = \mathscr{L}_\tau\{\theta_S\}$ and $G = \mathscr{L}_\tau\{\theta_V\}$, from Eqs.(1) and (3) one has

$$pF - 1 = d^2F/dX^2$$

and

$$p\,\alpha^2 G + 2\,\beta\alpha\, dG/dX = d^2G/dX^2$$

which, with the proper boundary conditions, give

$$F = A\cosh(X\,p^{\frac{1}{2}}) + 1/p \tag{7}$$

$$G = C\exp\{-\alpha(X-1)P\} \tag{8}$$

where:

$$P = (\beta^2 + p)^{\frac{1}{2}} + \beta \tag{9}$$

When the conditions at the interface (5,6) are taken into account, if one puts:

$$C = F(p,1) = G(p,1) \tag{10}$$

then

$$A = (-\eta + \zeta - \mu P)/(p^{\frac{1}{2}}\tanh p^{\frac{1}{2}} - \zeta + \mu P)p \cosh p^{\frac{1}{2}} \tag{11}$$

and

$$C = 1/p + (-\eta + \zeta - \mu P)/(p^{\frac{1}{2}} \tanh p^{\frac{1}{2}} - \zeta + \mu P)p \tag{12}$$

with $\mu = \alpha\gamma$. Note that, at $\tau = 0$, these expressions give the exact values

$$\theta(0,1) = 1 - \mu_1 \quad ; \quad \theta(0,X>1) = 0 \quad ; \quad \theta(0,X<1) = 1$$

where $\mu_1 = \mu/(1+\mu)$.

To evaluate the exact inverse transforms of the expressions (7,8), with the positions (11) and (12), being hopeless, approximate forms of F and G will first be determined and then inverted. In particular, following the approximation procedure of [2] one can assume

$$P \cong p^{\frac{1}{2}} + 2\beta \quad ; \quad \tanh p^{\frac{1}{2}} \cong p^{\frac{1}{2}}/(1+p^{\frac{1}{2}})$$

from which one can write

$$G \cong \exp\{-\alpha P(X-1)\}\{1-\mu_1-(a_1 p^{\frac{1}{2}}+\nu_1 a_2)/[p+\mu_1(\nu+1)p^{\frac{1}{2}}+\nu_1]\}/p$$

where

$$a_1 = (\eta_1+\nu+1-\nu_1-\mu_1)\mu_1 \quad ; \quad a_2 = (\eta_1+\nu-\nu_1)\mu_1/\nu_1$$

with

$$\eta_1 = \eta/\mu \quad ; \quad \nu_1 = \nu\mu_1 \quad ; \quad \nu = -(\zeta-2\beta\mu)/\mu$$

Since μ_1 and ν_1 are usually small, the dimensionless temperature distribution in the vapor phase will then be expressed as

$$\theta_V = \exp\{-2\alpha\beta(X-1)\}[(1-\mu_1)\operatorname{erfc}\{\alpha(X-1)/2\tau^{\frac{1}{2}}\}-a_1H_1\{\alpha(X-1),\nu_1^{\frac{1}{2}}\}-$$
$$-a_2H_2\{\alpha(X-1),\nu_1^{\frac{1}{2}}\}]$$

where

$$H_1(a,b) = 2(\tau/\pi)^{\frac{1}{2}}\exp(-a^2/4\tau)-(a+2b\tau)\exp(ab+b^2\tau)\operatorname{erfc}(a/2\tau^{\frac{1}{2}}+b\tau^{\frac{1}{2}})$$

and

$$H_2(a,b) = \operatorname{erf}(a/2\tau^{\frac{1}{2}})-2b(\tau/\pi)^{\frac{1}{2}}\exp(-a^2\tau/4)+(2\tau b^2+ab-1)\exp(ab+b^2\tau)\times$$
$$\times\operatorname{erfc}(a/2\tau^{\frac{1}{2}}+b\tau^{\frac{1}{2}})$$

Following the same procedure, the anti-transformation of F can be carried out. One first writes the approximate expression

$$F \cong 1/p + \exp\{-(1-X)p^{\frac{1}{2}}\}[-\mu_1/p - a_1/\{p^{\frac{1}{2}}(p^{\frac{1}{2}}+\nu_1^{\frac{1}{2}})\} - a_2\nu_1/\{p(p^{\frac{1}{2}}+\nu_1^{\frac{1}{2}})\}]$$

and then the dimensionless temperature distribution in the solid phase can be evaluated as

$$\theta_S = 1-\mu_1\operatorname{erfc}\{(1-X)/2\tau^{\frac{1}{2}}\}- a_1H_1(1-X,\nu_1^{\frac{1}{2}}) - a_2H_2(1-X,\nu_1^{\frac{1}{2}})$$

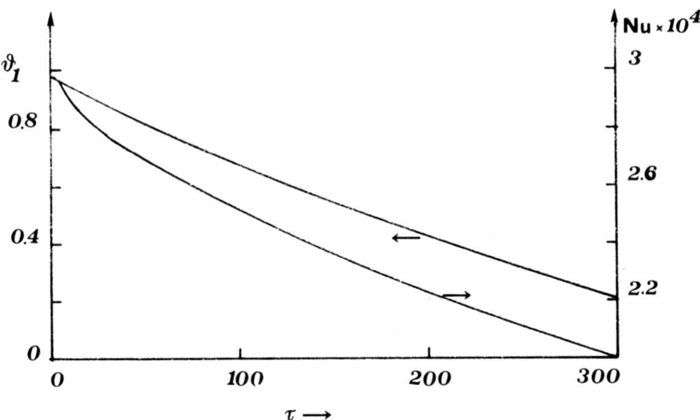

Fig.2a Temperature at the wall and Nusselt number in the case of steel tempering. $\alpha = 0.65$, $\beta = 1.$; $\gamma = 0.0007$, $\zeta = 0$, $\eta = 0.0018$

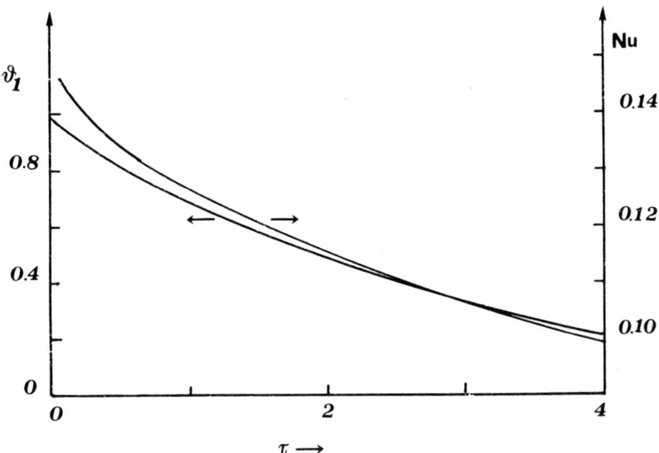

Fig.2b Temperature at the wall and Nusselt number in the case of glass tempering. $\alpha = 0.12$, $\beta = 6$, $\gamma = 0.0034$, $\zeta = 0$, $\eta = 0.009$

3. RESULTS AND CONCLUSION

The procedure exposed in the preceding chapter allows the evaluation of the complete unsteady thermal field in both the solid region and the vapor region. The heat transfer at the wall can be easily calculated by taking into account the fact that the Nusselt number at the solid wall Nu is immediately expressed as

$$Nu = |\partial\theta_S(\tau,1)/\partial X|$$

An improvement of the physico-mathematical model can be obtained for those cases where the physical characteristics of the two mediums are temperature dependent. For instance, if one considers the vapor region, the dimensionless density $\tilde{\rho}_V$ and the thermal conductivity $\tilde{\lambda}_V$ approximately satisfy the relation : $\tilde{\rho}_V\tilde{\lambda}_V = 1$. In this case, a transformation of variables, similar to Dorodnitsyn's transformation of compressible aerodynamics [3], i.e.

$$\tilde{\tau} = \tau \quad ; \quad \tilde{X} = \int_1^X \tilde{\rho}_V dX$$

leads to writing the energy equation in the same form of eq.(3).

As an application, the case of mist tempering of a steel plate and of a glass plate have been considered.
Figure 2a and Fig.2b are relative to the steel tempering and to the glass tempering, respectively, and there the temperature history at the wall, $\theta_1 = \theta(\tau,1)$ and the behaviour of the Nusselt number are reported. In particular, one has

$$Nu = \left| -\mu_1/(\pi\tau)^{\frac{1}{2}} - a_1 H_{3w} - a_2 \nu_1 H_{1w} \right|$$

where

$$H_{1w} = H_1(0, \nu_1^{\frac{1}{2}})$$

and

$$H_{3w} = 1-2(\nu_1\tau/\pi)^{\frac{1}{2}} - (1-2\nu_1\tau)\exp(\nu_1\tau)\mathrm{erfc}(\nu_1\tau)^{\frac{1}{2}}$$

Note that the reference time for the steel vapor-couple of Fig.2a is equal to 0.35 s and for the glass-vapor couple (Fig.2b) is equal to 10 s.

ACKNOWLEDGMENT

This work has been partially supported by the Italian National Research Council (C.N.R.) through grant n.CT 80.01131.07.

REFERENCES

1. Toda, S., "A Study of Mist Cooling", Trans.J.S.M.E., 39, 1973, pp.2160-2193
2. Bianchini, A., Pozzi, A. and Teodori, A.R., "A Solution with Discontinuous Derivative to the Heat Equation" (*in Italian*), Proc.XXXVII Congr.ATI, Padova 1982
3. Lagerstrom, P.A., "Theory of Laminar Flows" in *Laminar Flow Theory*, Moore, F.K. Ed., Princeton Univ.Press, Princeton 1964.

Multi-Phase Flow and Heat Transfer III. Part A: Fundamentals
edited by T.N. Veziroğlu and A.E. Bergles
Elsevier Science Publishers B.V., Amsterdam, 1984 — Printed in The Netherlands

MIXING CHARACTERISTICS OF BUBBLY FLOW

M. M. Sorour
Mechanical Engineering Department
Faculty of Engineering
Alexandria University
Alexandria, Egypt

Abstract

In this investigation the mixing characteristics of two component, two phase flow was studied. Heated water flows axially in a vertical pipe and air, at ambient temperature, was introduced under pressure to the flowing water such that heat transfer occurs between the two phases. In addition, various system parameters affecting the mixing characteristics were investigated, such as the flow rates of both phaseses, flow direction of the liquid phase, bubble size and distribution of the gas phase.

The results indicate that as the velocity and size of bubbles were increased the mixing was enhanced and that downwards flow induces mixing more efficiently than upwards flow. Furthermore, for some conditions of downwards flow temperature fluctuations were observed. These fluctuaions were damped as the number of bubbles was increased and/or the water flow rate was increased.

Introduction

Bubbly flow is that flow pattern in two phase gas liquid flow where the gas phase is distributed within a liquid continuum in discrete bubbles smaller than the characteristic dimension of the channel. Bubbly flow is the most simple and widely employed in industry. The industrial use of bubble columns has recently been reviewed by Alvarez-Cuenca(1), but recently bubbly flow was used to prevent ice formation in lakes, as barriers against saltwater intrusion in rivers, and for aeration in water purification and waste treatment plants(2).

Bubbly flow is successfully produced by various ways; injecting air from an air chamber into the flowing liquid through several holes at the tube wall is the most common method(3). Other methods to produce bubbly flow are through staggered porous tubes[4], and by using liquid gas ejectors[5]. However, the present investigation utilizes perforated thin Aluminum sheets for introducing the air from an air chamber at relatively high pressure.

Previous investigations in two phase gas liquid flow indicate that this flow is associated with intense increase in heat transfer rate compared with single phase flow under similar flow conditions.[6-8]. In addition, the heat transfer process is a function of the phase distribution properties which depends greatly on the flow conditions, such as the flow rates of both phases, bubble generation method, flow direction as well as the physical properties of fluids[9]. However, there are less investigations concerned with thermal mixing of two phases with different temperatures, which is of interest in many applications.

This investigation is an experimental study of thermal mixing of hot flowing water in a vertical pipe with air, at ambient temperature, rising through the fluid as discrete bubbles, i.e., thermal characteristics of bubble column when the two phases are initially introduced with different temperatures. In addition, the study involves the effect of gas and liquid flow rates, flow direction of the liquid phase and number and size of the holes in the distributing plate.

Experimental Facility and Procedure

The experimental facility is a modified version of a test rig described in an accompanying paper[10]. In essence, it is composed of a vertical pipe, water loop, and an air circuit. The vertical pipe is 7.5 cm diameter and composed of three parts; the main section is 150 cm long and contain 8 axial measuring stations, and two 20 cm pipe connections from both upper and lower sides. These connections have 3 holes drilled circumferentially 120° apart to allow the introduction and delivery of the water.

The hot water flows from the overhead tank through an orifice meter to the insulated cylinder through insulated piping, in addition the water is heated and its temperature controlled by 3 kw thermostatic electric heater immersed in the overhead tank. On the other hand, the air circuit is composed of a compressor, rotameter and flexible piping. To distribute the air in the test section, air is compressed to a small air chamber in the bottom of the tube through a small hole and then forced to flow through holes in a distributing plate located between the air chamber and the pipe. Two such groups were used in this investigation; the first group, A, has equal area of perforations by having different number and size of holes, viz., 3 holes (2mm) diameter, 12 holes (1 mm) and 48 holes (0.5 mm) while the second group, B, have orifices of constant size but different in number, viz., 6, 30 and 60 holes of 0.5 mm diameter.

A total of 312 experiment were performed investigating the mixing characteristics of the water, which was typically introduced at 50°C, and air typically injected at 20°C at various water and air flow rates, flow directions and distributing plates. The liquid flow rate varied in the range of $Q_w \leqslant 11 \times 10^{-3}$ kg/sec, and the gas flow rates in the range of $Q_a \leqslant 70 \times 10^{-6}$ kg/sec. In each experiment, the water flow rate was adjusted to the required value by changing the valve of the overhead tank. A specific amount of air flow rate was then introduced into the cylinder after being carefully measured by the gas meter and recorded. Temperature measurements were then assessed and recorded after steady state conditions prevailed.

Results and Discussion

The effect of air and water flow rates on the axial temperature distribution in the bubble column is illustrated in figures 1-2 for upwards and downwards flow respectively. The bubbles in this case were formed from 3 holes of 2 mm diameter. This is a typical sample of many results showing the same trend. For upwards flow the temperature distribution is characterised by an initial steep temperature gradient in the entrance section, which indicates the intensive mixing within a very small distance, followed by a nearly constant temperature section. This constant temperature section exhibits a small temperature gradient as the liquid flow rate increases, due to the increase in water velocity and the need of larger distance for the same contact time between the two phases. In addition, as the air flow rate increases the intensive mixing section is slightly extended, which is analogous to the behaviour of single phase flow. On the other hand, downwards flow temperature distribution considerably differs from that of upwards flow. Steep temperature gradients occur at the upper section of the pipe together with small temperature fluctuations in most of the pipe. These temperature fluctuations were damped as the liquid flow rate increased.

The effect of distributing plates on the temperature distribution of the bubbly flow mixture is presented in figures 3-4 for upwards and downwards flow respectively. This sample is for three distributing plates of group A when the water flow was kept constant, $Q_w = 3 \times 10^{-3}$ kg/sec, but the air flow rate was varied. It can be seen that for upwards flow the temperature distribution is not significantly changed by the variation of the hole size and number. The constant temperature section

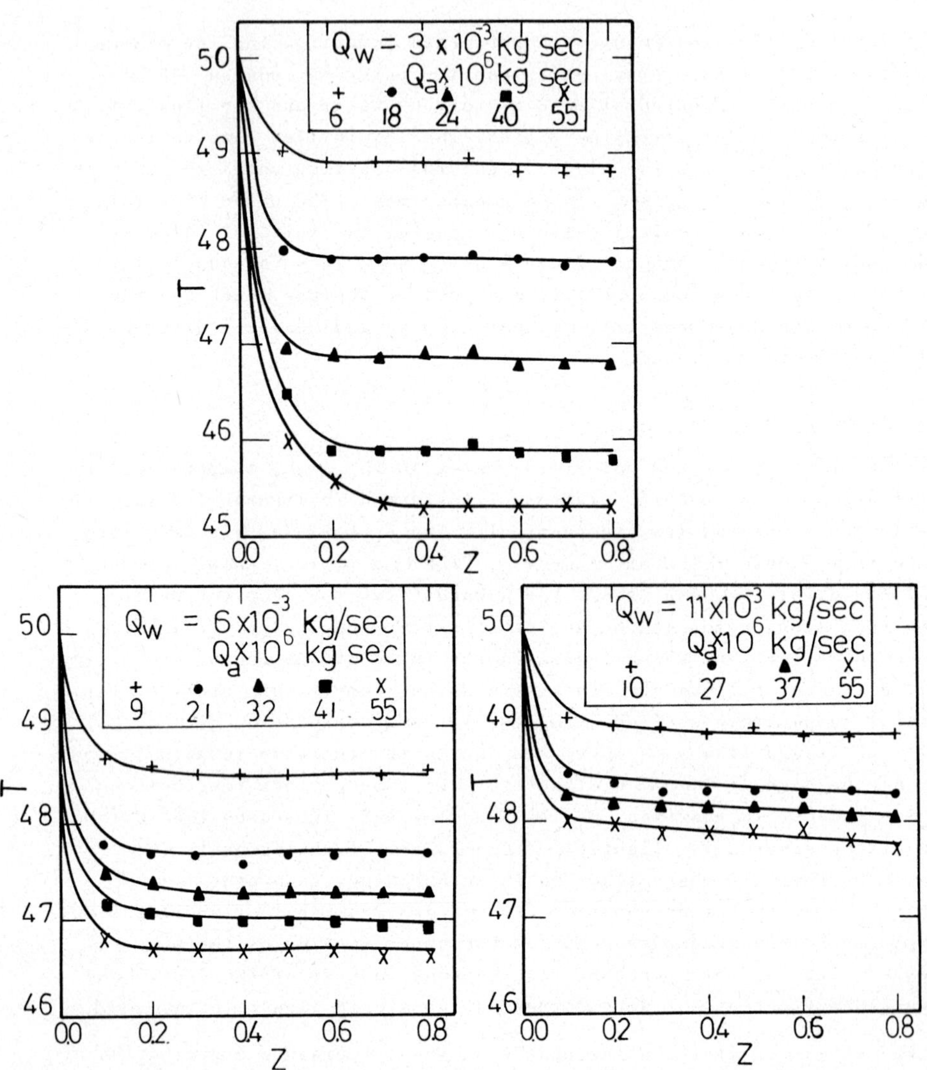

Fig.(1) EFFECT OF AIR AND WATER FLOW RATES ON THE AXIAL TEMP. DISTRIBUTION FOR UPWARDS FLOW. D.P: 3 holes 2mm diameter.

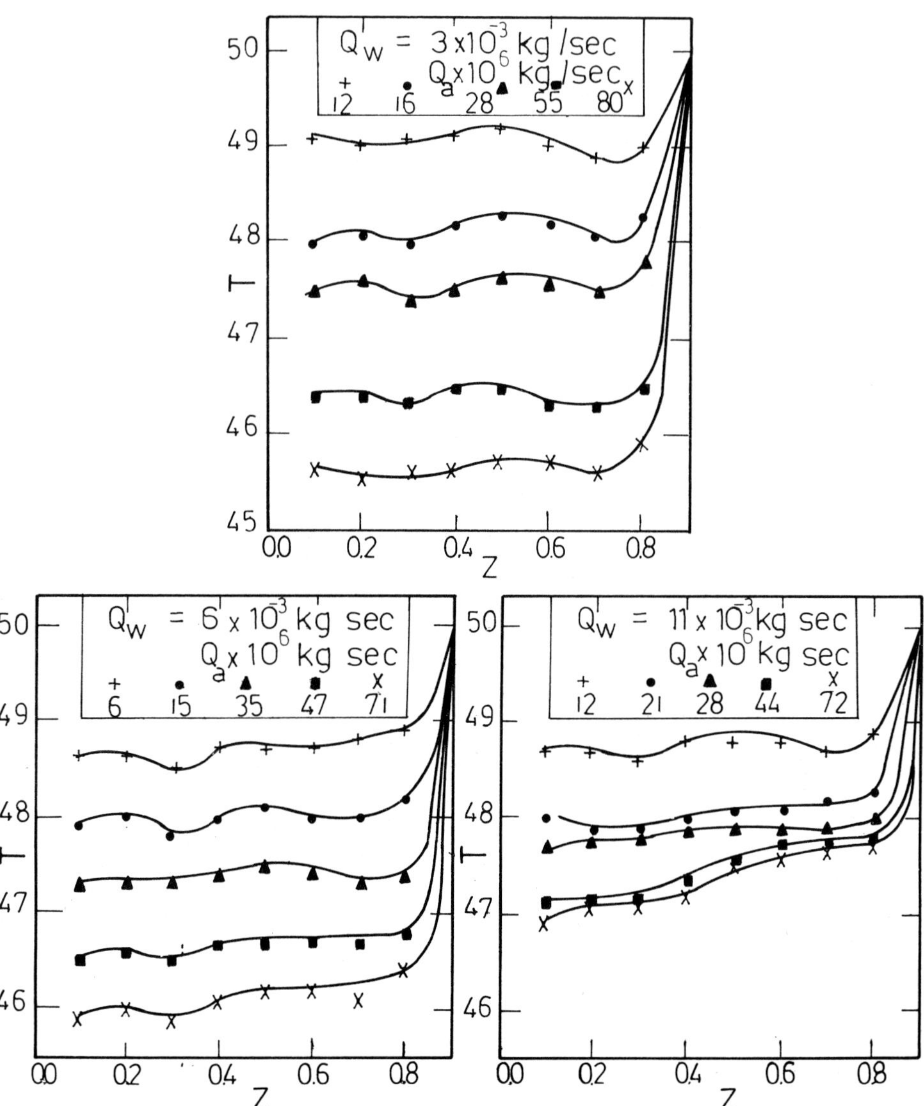

Fig.(2) EFFECT OF AIR AND WATER FLOW RATES ON THE AXIAL TEMP. DISTRIBUTION FOR DOWNWARDS FLOW. D.P : 3 holes 2mm.

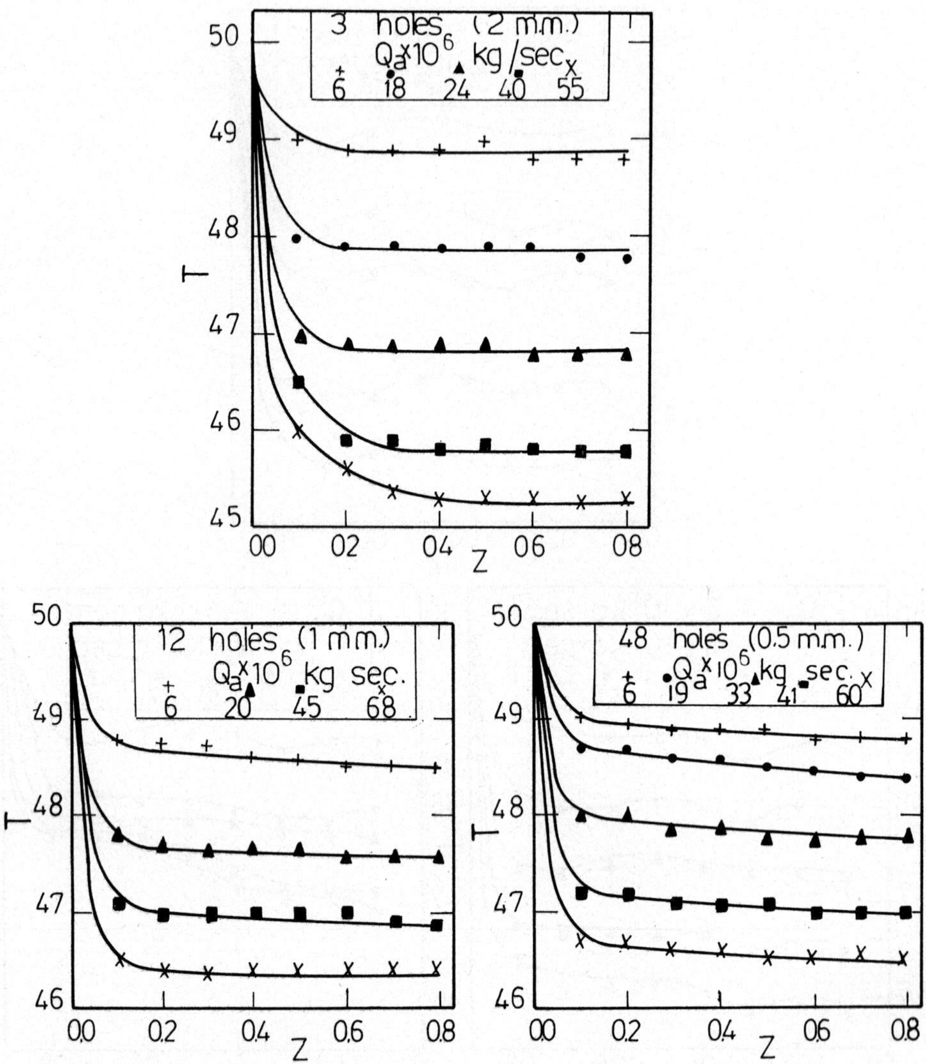

Fig.(3) EFFECT OF SIZE AND NUMBER OF HOLES ON THE AXIAL TEMP. DISTRIBUTION FOR UPWARDS FLOW FOR $Q_w = 3 \times 10^{-3}$ kg/sec.

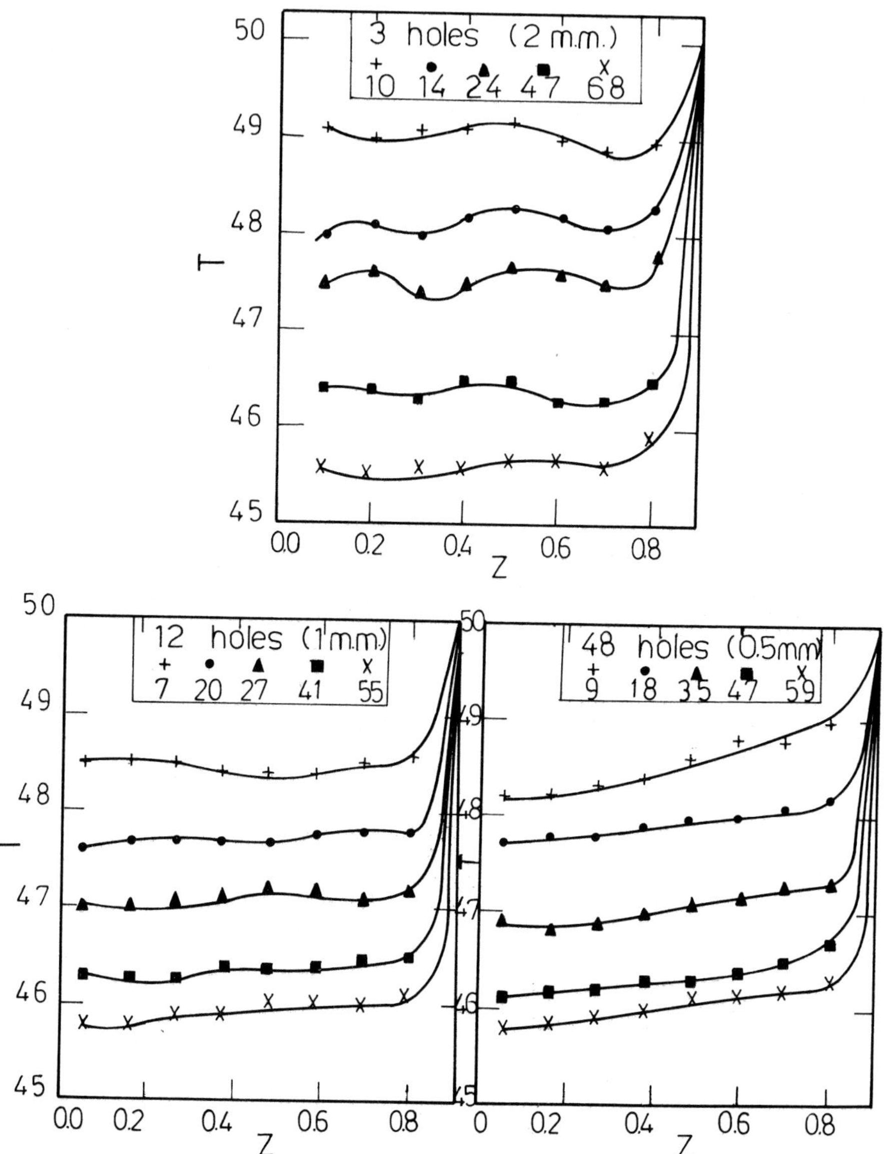

Fig.(4) EFFECT OF SIZE AND NUMBER OF HOLES ON THE AXIAL TEMP. DISTRIBUTION FOR DOWNWARDS FLOW FOR $Q_w = 3 \times 10^{-3}$ kg/sec.

which characterises the bulk of the bubble column slightly builds up a negative temperature gradient with decreasing the hole size and increasing their number. Conversely, for downwards flow the influence of the distributing plate is much more significant. As the number of holes increases and their size decreases the temperature fluctuations are damped.

The present investigation reveals that for some conditions of downwards flow, temperature fluctuations were observed. These fluctuations were damped as the number of bubbles was increased and/or the water flow rate was increased. The temperature fluctuations noted are due to the flow oscillations in the two phase mixture and not thermal oscillations due to thermal stability of the liquid phase. Flow oscillations restrict the continuity of flow in the pipe such that dead zones are formed in some axial locations, that attain different temperatures than the bulk of the mixture.

It is to be noted that in upwards flow experiments the air flows out of the tube with the water through three large circumferential holes. Conversely, for downwards flow experiments the air outlet was through a small hole in the uppermost position of the apparatus, similar to the air inlet hole. In addition, two phase, two component gas liquid flows are affected by the sensitivity of the exit volume flow characteristics to mixture density[10]. Therefore for downwards flow the instability is caused by the oscillations in the flow rate of the liquid component. This is the reason why a small number of large bubbles will effectively influence the oscillations more than a large number of small bubbles and why the oscillations were damped as the liquid flow rate increased.

Figure 5 a,b presents a sample of the total temperature drop in the bubbly column at various air and water flow rates for upwards and dowwards flow respectively. As expected, the temperature difference increases with the air flow rate but decreases with the water flow rate. However, it can be seen that downwards flow maintain higher temperature difference than upwards flow, and that the difference between the two directions decays with the number of holes and the decrease of the flow rates. Furthermore, to compare the effect of a number of holes on the temperature difference for various cases, fig. 6 presents a sample when the number of holes varies from 3 to 48 holes. It can be seen that a small number of holes produces a larger temperature difference than a large number of holes; however the difference is much more significant for small flow rates.

Table 1 presents a sample of some relevent characteristics of the

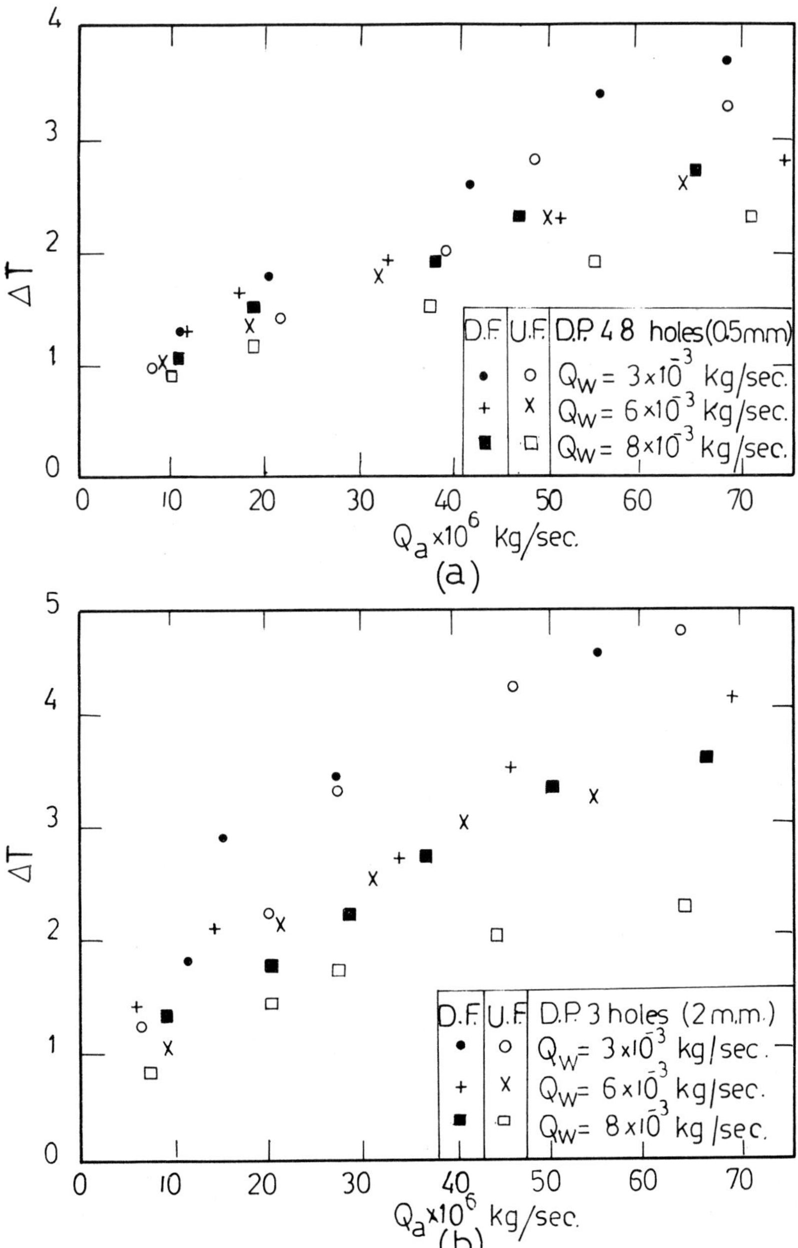

Fig.(5) TOTAL TEMPERATURE DIFFERENCE FOR VARIOUS AIR AND WATER FLOW RATES AND DIRECTIONS.

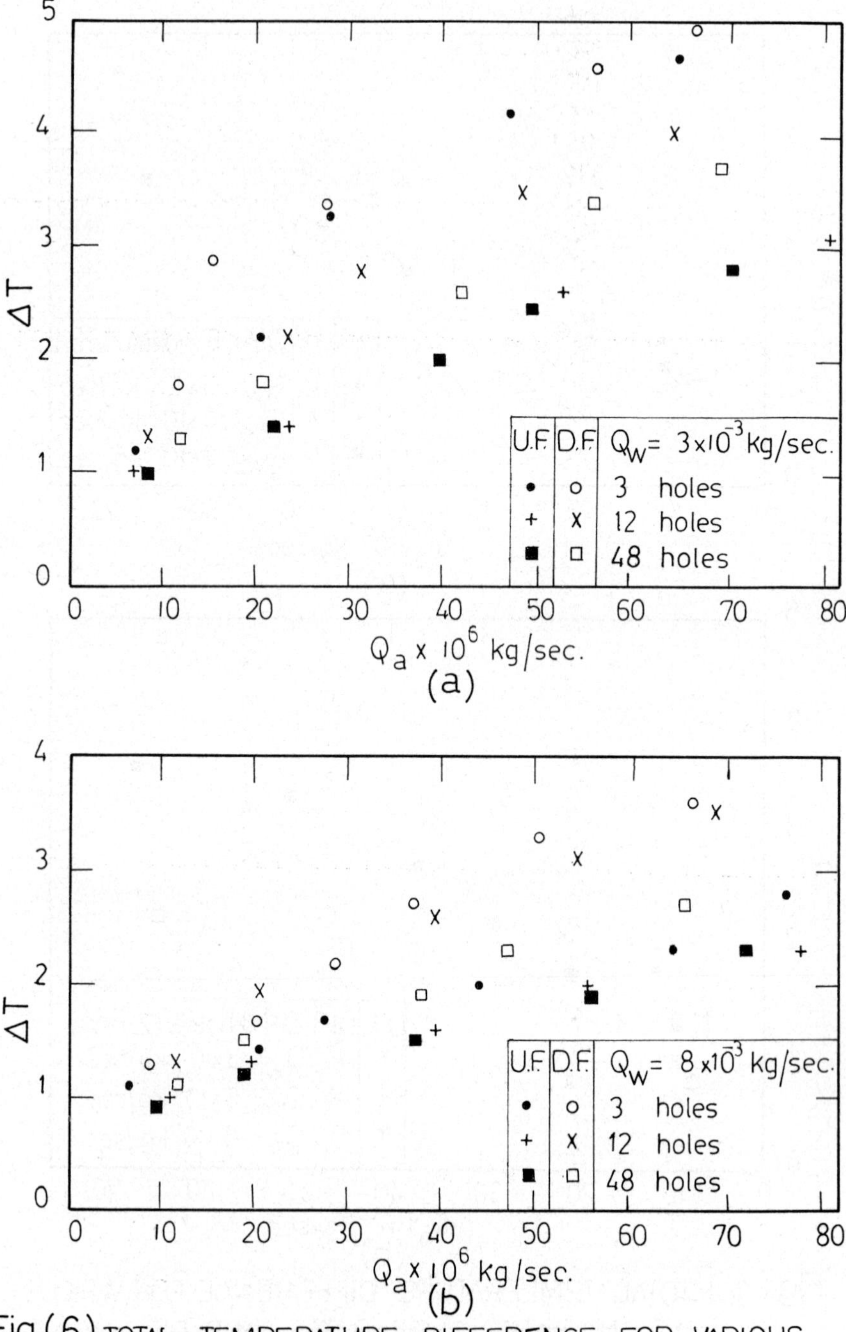

Fig.(6) TOTAL TEMPERATURE DIFFERENCE FOR VARIOUS NUMBER OF HOLES.

air bubbles as they are formed from three distributing plates having equal area of perforations for a stagnant liquid column. These calculations assume that bubbles are perfect spheres formed without coalescence. It can be seen that for any air flow rate within the experimental range, as the number of holes increase and their size decreases the generated bubbles become smaller in size (their diameter D_b decreases), and travel slower (their velocity V_g decreases), but they have a higher frequency f of generation and larger contact area between the air and the water A. It is to be noted that although the presented table is for the case of a stagnant liquid column the same trend applies when the liquid is forced to flow in the pipe.

		3 holes(2mm)	12 hole(1 mm)	48 hole(0.5 mm)
$Q_a = 6x10^{-6}$ kg/sec	D_b, m	0.004	0.0023	0.0013
	V_g, m/sec	0.14	0.106	0.079
	f, sec^{-1}	149	785	4331
	A,m^2 $x10^3$	53	122	289.8
$Q_a=24x10^{-6}$ kg/sec	D_b,m	0.0069	0.0040	0.0023
	V_g, m/sec	0.183	0.14	0.106
	f,sec^{-1}	114	595	3134
	A,m^2 $x10^3$	95	212	481
$Q_a = 72x10^{-6}$ kg/sec	D_b,m	0.0108	0.0062	0.0035
	V_g,m/sec.	0.23	0.174	0.130
	f,sec^{-1}	39	477	2672
	A,m^2 $x10^3$	130	332	790

Table 1 Comparison of the details of Bubble formation for group A.

This study reveals that few large bubbles produce a larger axial temperature drop than a large number of small bubbles. This is not to be expected from Table 1 since the contact area between air and water increases with the number of bubbles. In addition, in a previous study of a cooling pond(12) in which air bubbles, at ambient temperature were dispersed in a continuous downwards flow of a thin hot water layer, the auther indicated that as the bubbles are smaller in size and larger in number the effectiveness of the pond increases, i.e., larger contact area produces greater temperature difference.

This phenomenon may be explained by consideration of the heat transfer process in the direct contact exchangers. In general, the heat transfer depends on two factors. The first is the heat conducted between the two fluids, which depends on the thermal conductivity of both phases, the temperature difference, and the contact area between the two fluids. The second is the turbulence generated by the stirring action of the bubbles. In addition, visual observations indicate that a small number of large bubbles is more irregular and travels in a much more random motion than a large number of small bubbles that are less distorted and travel in a more rectilinear path, plate 1. Consequently, since the velocity of

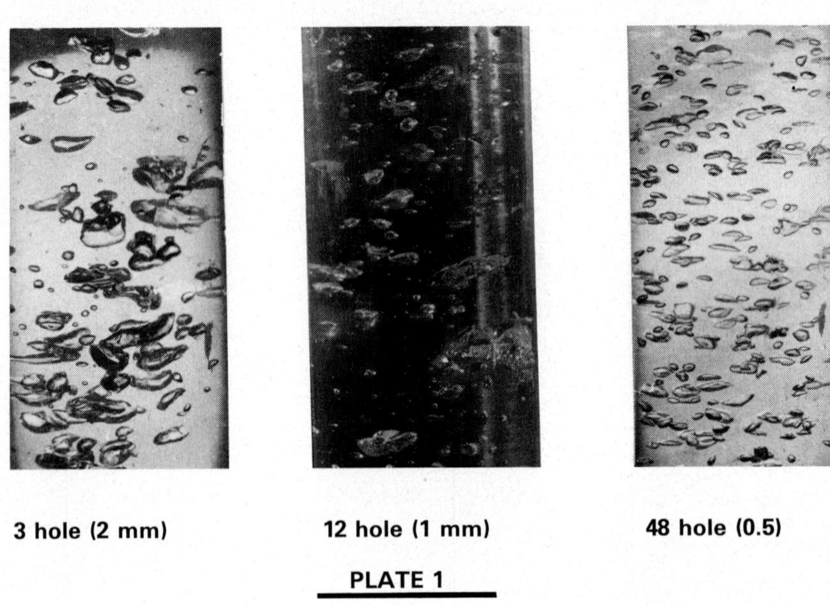

3 hole (2 mm) **12 hole (1 mm)** **48 hole (0.5)**

PLATE 1

large bubbles is higher than small bubbles, the turbulence generated from the former case is much more significant than that produced from the latter one.

Therefore, it can be concluded that, as the number of bubbles increase the contribution of conduction increases but the contribution of turbulence decreases. When the influence of motion of bubbles is negligible and the bubbles behave as solid spheres, as for example in very thin layers, the heat exchange increases when the number of bubbles increases. On the other hand, when the motion of bubbles is influential, as in the case of long cylinder, the heat exchange increases when the number of bubbles decreases. This explains why downwards flow produces larger temperature drop than upwards flow and why mixing is very much intensified in the entrance section.

Lastly, this investigation indicates that the thermal characteristics of group B showed similar trends to the presented characteristics of group A, i.e., the effect of changing the number of constant size holes was similar to the effect of changing the number of different size holes. This proves that the temperature distribution and total temperature drop vary with the number of holes regardless of their size. A comparison between groups A and B reveals that the hydrodynamic characteristics of bubbles, (D_b, V_g, f & A) for groups A and B follow the same sequence with the number of holes although the size of holes differs, Table 2. Therefore the similarity between groups A and B is not surprising.

	3 holes (2mm)	6 holes (0.5mm)	12 holes (1mm)	30 holes (0.5 mm)	45 holes (o.5 mm)	60 holes (0.5 mm)
D_b	0.0052	0.004	0.0030	0.0021	0.0017	0.0016
V_g	0.159	0.14	0.12	0.10	0.091	0.088
f	135	300	707	2061	3870	4620
A	72×10^{-3}	106×10^{-3}	166×10^{-3}	282×10^{-3}	386×10^{-3}	420×10^{-3}

Table 2. Comparison of the details of bubble formation between group A and B for $Q_a = 12 \times 10^{-6}$ kg/sec, $Q_w = 0$

Conclusions

The results of this investigation can be summarised as follows:

1. The thermal characteristics are strongly related to the hydrodynamic characteristics and the influence of the distributing plates is only to control the size of bubbles.

2. The contribution of turbulence in the mixing process is much more dominant than the contribution of conduction which is very small.

3. Temperature fluctuations were noticed in the temperature profiles for small downwards liquid flow due to flow oscillations in the tube.

References

1. Alvarez-Cuenca, M., Baker, C.G.J., Nerenberg, M.A., and Bergougnou, M.A., "Oxygen Mass Transfer in Bubble Columns Working at Large Gas and Liquid flow Rates" Symp. on fundamental Research in Heat and Mass Transfer. Annual AICHE Meeting San Francisco. CA(1979).

2. Hussam, N.A., and Narang, B. S., "Analysis of two dimensional Air-Bubble Plumes" AIChE Journal 23, 596, 1977.

3. Oya, T., "Upward liquid flow in small tubes into which air streams," Bull. J.S.M.E. 14, 1390 (1971).

4. Huir, J.F., and Eichorn, R.,"Compressible flow of an air water mixture through a vertical two dimensional converging-diverging nozzle." Proj. Squid Tech. Rept, PR-105, 1963

5. Deich, M.Y., "A method for calculating a two phase Jet injector " Heat Transfer Sov. Res., 3 (1971).

6. Kubie. J." Heat transfer in two phase gas liquid system multiphase flow systems," Strathclyde-Glasgow, Scotland 2-4 April, vol II-H1 (1974).

7. Fair, J.S., Lambricht, A.J., and Anderson, J.W., "Heat transfer and gas hold up in a sparged contactor". Ind. Eng. Chem. Process Design Develop., 1, 33 (1962).

8. Kolbel, H., Siemes, W., Mass, R., and Muller, K., "Wärmeübergang an Blasensorten," Chemie Ing. Tech., 30, 400 (1958).

9. Sato, Y., and Sadatomi, M., "On the variety of the void fraction distribution in two-phase bubble flow," 14th Symp. Heat Transfer Society of Japan, 316-318.

10. Sorour, M.M., "Experiments in two phase flow: Bubble slug transition in a vertical pipe." Published elsewhere in this volume.

11. Veziroglu, T.N., Lee, S.S., and Kakac, S.,"Fundamentals of two phase flow oscillations and experiments in single channel systems", Two phase flows and heat transfer, Vol. I, Proceedings of NATO Advanced Study Institute, August 16-27 1976, Istanbul, Turkey.

12. Sorour M.M. "A compact cooling pond" "The Bulletin of the Faculty of Engineering, Alexandria University, Vol XX, pp. 309-323, 1981.

Multi-Phase Flow and Heat Transfer III. Part A: Fundamentals
edited by T.N. Veziroğlu and A.E. Bergles
Elsevier Science Publishers B.V., Amsterdam, 1984 — Printed in The Netherlands

LAMINAR FILM CONDENSATION ON DOWNWARD VERTICAL FIN

Chao-Kuang Chen, Han-Taw Chen
Department of Mechanical Engineering
National Cheng Kung Universtiy
Tainan, Taiwan, R.O.C.

John W. Cleaver
Department of Mechanical Engineering
The University of Liverpool
Liverpool, England, U.K.

ABSTRACT

In this paper, by using the Nusselt's laminar film condensation theory, heat transfer between the saturated vapor and the downward vertical rectangular fin is analyzed.

The fourth order Runge-kutta method can be used to obtain the fin temperature distribution, fin efficiency, and the thickness of condensation film. A reasonable fin length can also be solved.

1. INTRODUCTION

Extension of the effective surface, usually in the form of fins, is one of the methods of enhancing the rate of heat exchange between some solid body and the fluid in which the body is immersed. Thus, the character of fins is widely used to reduce the size of the condenser and the manufacture cost by engineers.

In order to comprehend this method properly, one should study the performance of the fin in the film condensation. Such actual fin problems are rather complicated, hence one must invoke the following assumptions to establish the governing equations for this study, namely:

(1) Simple force and heat balances within the condensate film, and the inertia forces, energy convection, and interfacial shear force are not considered.
(2) The saturated vapor does not consist of noncondensable gases.
(3) The temperature distribution in the condensate film is assumed to be linear.
(4) The fin temperature does not vary across the fin.

The physical situation to be studied is pictured schematically in Fig. 1. A cooled vertical fin has a variable temperature, while the interface between the liquid condensation and the vapor is at the saturate temperature Ts corresponding to the local vapor pressure. Only gravity-induced flow is considered; that is, there is no forced convection in the bulk flow of the vapor. The fluid motions within the condensate film arise owing to the body forces.

The subject of this study is the film condensation on a downward vertical fin. Lienhard and Dhir (2) showed the results of this study n graphical forms, but the results are slightly less accurate and their numerical methods are slightly complicated. In the following, a more convenient method is developed for this problem; that is, the fin length is assumed to be unknown and the temperature at the fin tip is given. In accordance with the simple idea, one can easily determine the accurate or reasonable results such as the fin temperature distribution, the fin length, the fin efficiency, and the heat flux at fin base etc.

2. ANALYSIS AND NUMERICAL PROCEDURE

2.1. Physical Model and Coordinate

A schematical representation of the physical model and coordinate system is shown in Fig. 1. The positive x-axis is directed downwards; the horizontal y-axis is orthogonal to the centre plane of the fin; the common origin is at the base of the fin, the positive y-axis pointing away from the fin. A z-axis is not needed, because we shall consider the problem as a two dimensional steady state one.

2.2. The Governing Equations

In this problem we consider the vertical fin of length, L. thickness, t, thermal conductivity, k, and base temperature, T_o. Assuming that the saturated temperature Ts is higher than T_o. Thus there will be the condensate film formed along the fin from the fin base. In other words, the film thickness δ increases from its initial value, zero, at the fin base. Using these informations, the rate of mass flow M is easily determined as follows:

$$M = \frac{g\Delta\rho_c}{\nu_c}\frac{\delta^3}{3} \tag{1}$$

In accordance with the conservative theory of energy within tiny elements of the fin, we may obtain the following thermal equilibrium differential equation:

$$\frac{d^2\theta}{dx^2} = \frac{2k_c}{kt}\frac{\theta}{\delta} \tag{2}$$

where $\theta = T - T_s$, $h_c = k_c/\delta$

We equate the conduction of heat through the fin, within tiny elements, to the sum of heat liberated during the condensation of the vapor; the result is:

$$\frac{k_c\theta}{\delta} = \frac{g\Delta\rho_c h_{fg}}{\nu_c} \cdot \delta^2 \frac{d\delta}{dx} \tag{3}$$

For the obligatory normalization of the engineering problem, we select the following dimensionless variables: $\xi=\theta/\theta_o$, $\underline{X}=x/t$ and $\eta=\delta/t$. Thus, one may transcribe eq. (2) and eq. (3) into the following forms, respectively:

$$\frac{d^2\xi}{d\underline{X}^2} = \frac{2k_c}{k}\frac{\xi}{\eta} \tag{4}$$

$$\xi = \frac{g\Delta\rho_c h_{fg} t^3}{k_c \nu_c(-\theta_o)} \cdot \eta^3 \frac{d\eta}{d\underline{X}} \tag{5}$$

where $\theta_o = T_o - T_s$

The dimensionless forms of boundary conditions are given by:

$$\xi\ (0) = 0 \tag{6}$$

$$\xi\ (0) = 1 \tag{7}$$

$$\xi\ (\frac{L}{t}) = \xi_L \tag{8}$$

$$\frac{d\xi}{d\underline{x}} = 0\ , \qquad \text{at} \qquad \underline{x} = \frac{L}{t} \tag{9}$$

2.3. Numerical Procedure

Rearranging both of eq. (4) and eq. (5). We may have the following differential equation of the actual fin problem:

$$\frac{d^2\xi}{d\underline{x}^2} = \frac{2g\Delta\rho_c h_{fg} t^3}{k(-\theta_o)\nu_c} \cdot \eta^2 \frac{d\eta}{d\underline{x}} \tag{10}$$

Integration gives

$$\eta = (\frac{2g\Delta\rho_c h_{fg} t^3}{3k\nu_c(-\theta_o)})^{-\frac{1}{3}} (\frac{d\xi}{d\underline{x}} - \xi'_o)^{\frac{1}{3}} \tag{11}$$

Substituting eq. (11) into eq. (4), the second order nonlinear differential equation with dimensionless temperature within the fin is obtained as:

$$\frac{d^2\xi}{d\underline{x}^2} = 1.74716 \cdot Dch^{-\frac{1}{3}} \cdot (\frac{d\xi}{d\underline{x}} - \xi_o')^{-\frac{1}{3}} \cdot \xi \tag{12}$$

Where the parameter Dch is defined as

$$Dch = \frac{k^4(-\theta_o)\nu_c}{g\Delta\rho_c h_{fg} t^3 k_c{}^3} \tag{13}$$

Integration eq. (12) from $\underline{x} = 0$ to $\underline{x} = \frac{L}{t}$, we may obtain

$$\xi_o' = -1.53495 \cdot Dch^{-\frac{1}{7}} (1-\xi_L{}^2)^{\frac{3}{7}} \tag{14}$$

Where ξ'_L is designated as the temperature at the fin tip.

The heat flux at the fin base q_o is

$$q_o = 1.53495k(-\theta_o) \cdot Dch^{-\frac{1}{7}} \cdot (1-\xi_L{}^2)^{\frac{3}{7}} \tag{15}$$

In accordance with the Nusselt theory, we know the average coefficient $\bar{h}$ is

$$\bar{h} = 0.943\ (\frac{g\Delta\rho_c k_c{}^3 h_{fg}}{L\nu_c(-\theta_o)})^{\frac{1}{4}} \tag{16}$$

One defines the fin efficiency, the ratio of the actual fin heat transfer to the fin heat transfer that would occur if the fin were everywhere at the base temperature, the result is

$$\eta_f = \frac{q_o}{2\bar{h}L(-\theta_o)} \tag{17}$$

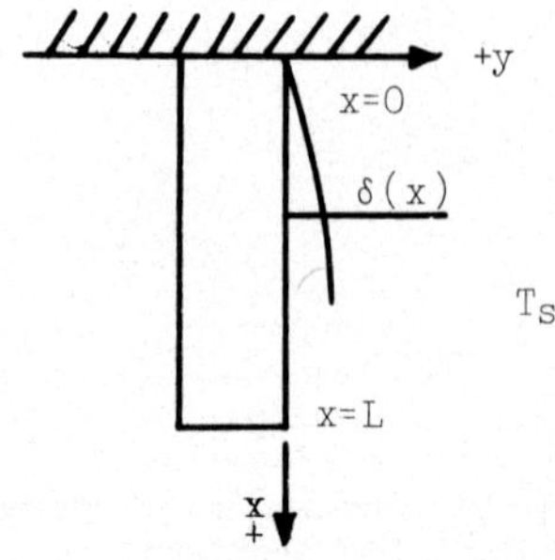

Fig. 1 Physical Model and coordinate system.

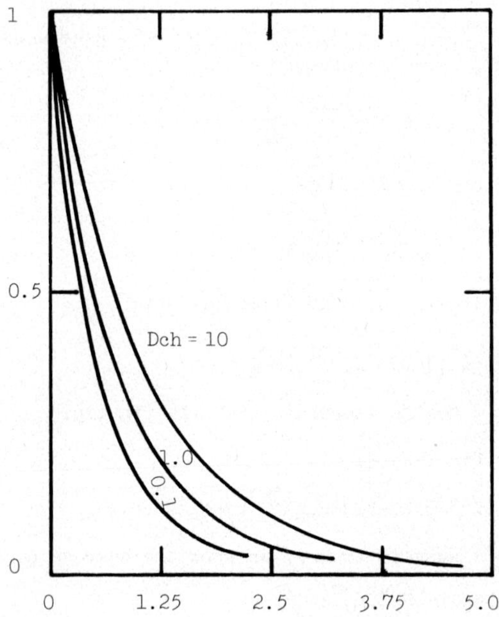

Fig. 2 Dimensionless temperature distribution along the axis of the fin.

3. RESULTS AND DISCUSSION

This theory is based solely on the governing equations and is free of emperical constants. The results are obtained by the 4-th order Runge-kutta method. It is worth emphasizing that the present theory provides a very accurate and straight-forward approach for the design of fins heated by condensation. At one time, the numerical computations requires very little computer time.

If the dimensionless tip temperature ξ_L is given, the heat flux at the fin base and the fin length will easily be obtained. Now, rearranging eq. (16) as

$$Nu = 0.943 \cdot \left(\frac{L}{t}\right)^{\frac{3}{4}} \cdot Dch^{-\frac{1}{4}} \tag{18}$$

Where $Nu = \frac{\overline{h}L}{k}$. Namely, eq. (18) shows that Nu has close relations with Dch. Substituting eq. (15) and eq. (18) into (17), we may obtain

$$\xi_f = 0.813865 \cdot Dch^{\frac{3}{28}} \cdot \left(\frac{L}{t}\right)^{-\frac{3}{4}} \tag{19}$$

Eq. (12) is again computed, by the 4-th order Runge-kutta method, to fit the boundary condition (9). The results for the downward vertical fin are plotted in Fig. 2. The curves in Fig. 2 show the dimensionless temperature variations with $\underline{X}$. According to the results in Fig. 2, we may find that the efficiency of the present fin will be predicted to be a constant value, 0.335 when ξ and $d\xi/d\underline{X}$ approach to zero.

NOMENCLATURE

Dch : D-chen number, $(k^4 \nu_c(-\theta_o))/(g\Delta\rho_c h_{fg} k_c^3 t^3)$
g : acceleration of gravity
h : heat transfer coefficient
h_{fg} : saturation enthalpy
k : thermal conductivity
L : fin length
M : mass flow rate
t : fin thickness
T : temperature
$\underline{X}$: dimensionless length
x, y : Cartesian coordinate
θ : $= T - T_s$ temperature difference
ξ : dimensionless temperature
ρ : mass density
$\Delta\rho$: $\rho_c - \rho_s$ mass density difference
ν : dynamic viscosity
δ : film thickness
η : dimensionless film thickness

SUBSCRIPTS

c : condensate
o : base
r : reference
s : saturation vapor

REFERENCE

1. C. K. Chen and H. T. Chen, "Study on Laminar Film Condensation Heat Transfer onto Fin", J. CSME Vol. 2, No. 1 (1981).

2. Dhir, V. K. and Lienhard, J. H., "Laminar Film Condensation on Nonisothermal and Arbitrary-Heat-Flux Surface, and on Fins", J. of Heat Transfer, Trans. ASME, May 1974.

Multi-Phase Flow and Heat Transfer III. Part A: Fundamentals
edited by T.N. Veziroğlu and A.E. Bergles
Elsevier Science Publishers B.V., Amsterdam, 1984 — Printed in The Netherlands

CONDENSATION ON A CIRCUMFERENTIAL FIN SURFACE

Qijie Wang* and Henping Chen**
Department of Mechanical Engineering
University of Minnesota
Minneapolis, Minnesota 55455, U.S.A.

ABSTRACT

In this article, a numerical solution is performed by a finite difference scheme to predict the temperature field in and heat flux onto vertical circumferential fins with film condensation. The condensation process onto the fin surface is strongly coupled to conduction within the fin so that it becomes a problem of considerable complexity.

An approximate model is introduced, e.g. the condensate on the upper-half base of the fin flows away only along the base tube. As a result, the problem is reduced to a two-dimensional boundary value problem and a one-dimensional initial value problem, which can be described by a partial differential equation and an ordinary differential equation. The present work employs two coordinate systems to deal with the set of equations. Polar coordinates are used for the former and cartesian coordinates are used for the latter. An alternate iteration process is carried out to solve the problem.

The computation is carried out for $r_i/r_o = 0.25$, 0.5, and 0.75, the dimensionless synthetic variable F_1 is covered from 4.425×10^{-7} to 2.023×10^{-10} The computation results are presented in dimensionless form and are compared with an isothermal fin model (Nusselt model) as well as with that of fin without condensation. Finally, the error which is caused by the approximate model is estimated and discussed.

INTRODUCTION

The use of finned condensers is quite common in many engineering fields. Therefore, the information about the condensation on a circumferential fin surface is of interest for design engineers working with heat exchanges.

There are two different patterns of circumferential fin design. In one case the fins are short and close together, the spacing ratio b/r_o-r_i being less than 0.6. In this situation surface tension has significant effects on the flow of condensate film. It causes not only additional pressure

* Permanent address: Xi'an Jiaotong University, Xi'an, PRC
** Permanent address: Shanghai Jiaotong University, Shanghai, PRC

gradient which draws condensate film towards fin base but also capillary ascension at low-part of fin tube. In the other case the fins are tall and far apart. The spacing ratio is larger than 0.6. Gravity and viscous resistance dominates the flow of condensate film while the influence of surface tension can be neglected. Most freon condensers are examples of the former and some investigations have been published [1, 2, 3]. Moisture-air cooler and moisture eliminators which are often used in air conditioning systems, mines, or pits for cooling air and eliminating the moisture from the air by condensation are examples of the latter. From a search of the published literature, it appears that only very little has been done in this area. In [4], the conventional one-dimensional fin analysis was modified by assuming that the fin was covered with a uniform film of condensate. Nader [5] dealt with a vertical fin whose lower end is maintained at a fixed temperature. Variations of the condensate film thickness and fin temperature were confined only to the vertical direction so that the problem is of the Nusselt type. S.V. Patankar and E.M. Sparrow [6] carried out a similarity solution of the problem in a limited range. Beyond that range where the similarity is valid, a finite-difference marching solution was given. However, all of the above referenced papers only dealt with a rectangular fin. Up to now, the condensation on circumferential fin surfaces which are tall and far apart has not been investigated from theory or from experiment. It should be noted that in this case an experiment would be very difficult.

In the present study, a numerical solution is performed by a finite difference scheme to predict the temperature field and heat flux onto the tall and sparse circumferential fins with film condensation.

ANALYSIS AND COMPUTATION

Figure 1 shows a vertical circumferential fin surface which is attached to a cooled horizontal tube, film condensation occurs on the fin surface. A cartesian coordinate system is used to formulate the behaviour of the condensate film. The coordinate y coincides with the gravity direction.

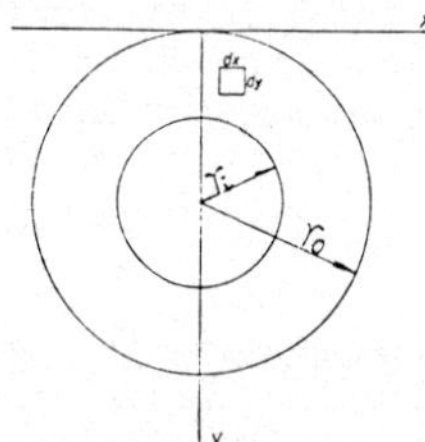

Fig. 1 Circumferencial fin

An energy balance for a typical volume element t·dx·dy (Fig. 1) involves conduction heat flows within the fin in the x and y directions and an inflow of heat from the condensate which washes the exposed surface of the element. We assume that the fin is sufficiently thin so that temperature variation across its thickness can be neglected and the condensate on the upper-half base of the fin flows away only along the base tube. This means we ignore the piling up of condensate near the base. We also assume that the respective condensate films do not interact. Pure saturated vapor temperature T_s and base surface temperature T_W which are considered uniform are given. All of Nusselt's assumptions are kept except that the fin temperature depends on both x and y.

For the given volume element, the fin energy balance can be written as

$$K_f\left(\frac{\partial^2 T(x,y)}{\partial x^2}+\frac{\partial^2 T(x,y)}{\partial y^2}\right)\cdot(dx\cdot dy\cdot t) = \frac{K_l}{\delta(x,y)}(dx\cdot dy)\cdot(T(x,y)-T_s) \tag{1}$$

The left-hand side of equation (1) is the net energy entering the element volume due to conduction within the fin, the right-hand side represents the inflow of heat from the condensate. Energy equation (1) can be recast as

$$\frac{\partial^2 T(x,y)}{\partial x^2} + \frac{\partial^2 T(x,y)}{\partial y^2} = \frac{K_\ell}{K_f\, t} \cdot \frac{(T(x,y) - T_s)}{\delta(x,y)} \tag{2}$$

In order to obtain the fin temperature distribution by solving the energy equation, $\delta(x,y)$ must first be found. In order to obtain an equation which governs the variation of δ with x and y, we will introduce a film momentum equation.

The velocity profile of condensate film on fin surface can be written as

$$w(x,y,z) = \left(\frac{g\rho_\ell}{\mu_\ell}\right) \cdot \left(z \cdot \delta(x,y) - \frac{1}{2} z^2\right)$$

The rate at which condensate passes through the element area ($\delta(x,y) \cdot dx$) is

$$\dot{m}(x,y) = \int_{z=0}^{z=\delta(x,y)} \rho_\ell \cdot w(x,y,z)\, dx\, dz = \left(\frac{g\rho_\ell^2 \delta^3(x,y)}{3\mu_\ell}\right) dx$$

The change in $\dot{m}$ in the gravity direction between x, y and x,(y + dy) can be obtained by differentiation

$$d\dot{m} = \left(\frac{\partial \dot{m}}{\partial y}\right) dy = \left(\frac{g\rho_\ell^2 \delta^2}{\mu_\ell}\right) \cdot \left(\frac{\partial \delta}{\partial y}\right) dx\, dy$$

The change in $\dot{m}$ for unit area is

$$\frac{d\dot{m}}{dx\, dy} = \left(\frac{g\rho_\ell^2 \delta^2}{\mu_\ell}\right) \left(\frac{\partial \delta}{\partial y}\right)$$

The amount of heat which is released by vapor at the liquid-vapor interface is

$$\lambda \cdot \frac{d\dot{m}}{dx\, dy} = \frac{\lambda g \rho_\ell^2 \delta^3}{\mu_\ell} \cdot \left(\frac{\partial \delta}{\partial y}\right)$$

Here we accept the assumption [6] that any transverse velocities which might exist in the film are very small. The effects of these velocities are negligible.

The amount of heat which is released by the vapor should equal that which is conducted to fin surface passing through the condensate film.

$$\frac{\lambda g \rho_\ell^2 \delta^3}{\mu_\ell} \left(\frac{\partial \delta(x,y)}{\partial y}\right) = \frac{K_\ell (T_s - T(x,y))}{\delta(x,y)}$$

or

$$\frac{\partial(\delta^4(x,y))}{\partial y} = \frac{4\mu_\ell K_\ell}{\lambda g \rho_\ell^2} \left(T_s - T(x,y) \right) \tag{3}$$

Equation (3) is the film momentum equation which relates $\delta(x,y)$ with $(T_s - T(x,y))$.

To get generalized solution, dimensionless variables are introduced as follows

$$\theta = \frac{T(x,y) - T_s}{T_w - T_s}$$

$$\Delta = \left[\frac{K_f\, t}{K_\ell L^2}\right] \cdot \delta(x,y)$$

$$X = \frac{x}{L}$$

$$Y = \frac{y}{L}$$

With these the energy and momentum equations become

$$\frac{\partial^2\theta}{\partial X^2} + \frac{\partial^2\theta}{\partial Y^2} - \frac{\theta}{\Delta} = 0 \qquad (4)$$

$$\frac{\partial\Delta^4}{\partial Y} - F_1\theta = 0 \qquad (5)$$

where F_1 is a dimensionless synthetic variable.

$$F_1 = \frac{4\mu_l k_l L\,(T_s - T_w)}{\lambda\, g\, \rho_l^2} \cdot \left(\frac{k_f t}{k_l L^2}\right)^4$$

The above two coupled differential equations (4) and (5) give the mathematical description of the condensation on a circumferential fin surface.

For a circumferential fin the boundary conditions for equation (4) are given on the inner and outer surfaces of the fin. It is, therefore, convenient to describe the equation in polar coordinates;

$$\frac{\partial^2\theta}{\partial R^2} + \frac{1}{R}\frac{\partial\theta}{\partial R} + \frac{1}{R^2}\frac{\partial^2\theta}{\partial\varphi^2} - \frac{\theta}{\Delta} = 0 \qquad (6)$$

where R represents the dimensionless ratio r/r_o, and the outer radius r_o of the fin is taken as the length dimension L. The corresponding boundary conditions are

$$\theta = 1.0 \quad \text{at} \quad R = r_i/r_o \qquad (7)$$

$$\frac{\partial\theta}{\partial R} = 0 \quad \text{at} \quad R = 1.0 \qquad (8)$$

Obviously, this is a two-dimensional boundary value problem with a source term θ/Δ. Once the dimensionless film thickness Δ is known, the numerical solution of the equation can be obtained by using a finite difference scheme. On the other hand, equation (5) is a simple ordinary differential equation. We can solve it by using a matching intergral technique so long as the dimensionless temperature θ is known. The equation (5) however, represents a one-dimensional initial value problem and its boundary conditions are given only on the part of the inner and outer circles. Therefore, the best way is to keep the equation in cartesian coordinates. The boundary conditions of equation (5) can be written as

$$\Delta = 0 \quad \text{at} \quad \sqrt{x^2+y^2} = 1.0 \quad \cap \quad Y \leq 1.0 \qquad (9)$$

$$\Delta = 0 \quad \text{at} \quad \sqrt{x^2+y^2} = r_i/r_o \quad \cap \quad Y > 1.0 \qquad (10)$$

The condensation process onto the circumferential fin surface (equation (5)) is strongly coupled to conduction within the fin (equation(6)), an alternate iteration process is used to solve the problem.

Due to the symmetry of the problem only a half of the fin surface is taken as the computational domain.

By solving equation (5) and (6) we can obtain temperature field, the profile of condensate film thickness, and the local and overall rate of heat transfer. The present computation is carried out for $r_i/r_o = 0.25$, 0.5, and 0.75. The dimensionless synthetic variable F_1 varies from 4.245×10^{-7} to 2.023×10^{-10} which comes from properties and geometry parameters of some typical moisture-air coolers. For convenience of comparison the inner radius of all fins is taken as 0.00125 m.

Numerical solution is performed by a finite difference scheme. 20 x 24 constant-interval grid points are employed to fins with $r_i/r_o \geq 0$, and 20 x 24 logarithmic-interval grid points for fins with $r_i/r_o < 0.5$. For each grid point the value Δ is computed individually by a marching integration based on equation (5). The beginning points of integration are located on the circles above the grid points where the initial values of Δ are equal to zero. The step lengthes of integration are selected such that it is always less than the grid intervals. The values of θ in equation (5) come from interpolations of the solution of equation (6). A linear interpolation is used in the circular direction and a logarithmic interpolation is used in the radical direction.

RESULTS AND DISCUSSION

The dimensionless temperature profiles are plotted with a set of isothermals on Figs. 1, 2 and 3. Most of isotherms approximately are circular. Shown in Fig. 1 are the computation results of $r_i/r_o = 0.5$. Figs. 1-a and 1-b correspond to temperature differences ($T_s - T_w$) of 30°c and 5°c respectively. It is to be noted that the temperature drop is essentially within the region which is near the fin base, for example, it is seen from Fig. 1-a that the thickness of the region in which the dimensionless temperature dropped from 1.0 to 0.1 is about 0.38 ($r_o - r_i$). A similar result is observed in Fig. 1-b. Comparing Fig. 1-a with Fig. 1-b, we can conclude that temperature differences $\Delta T = T_s - T_w$ have no significant influence on dimensionless temperature profiles, except that the temperature drop is slightly faster for small temperature differences. Same results can be seen in Figs. 3 and 4.

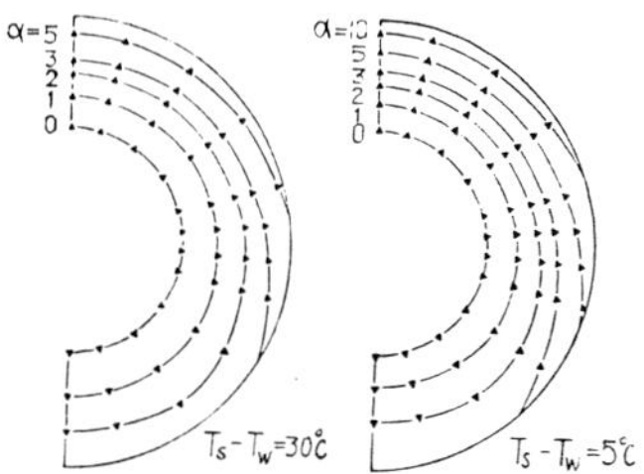

Fig. 2 Dimensionless temperature profile $\theta=10^{-\alpha}$ for $r_i/r_o=0.5$

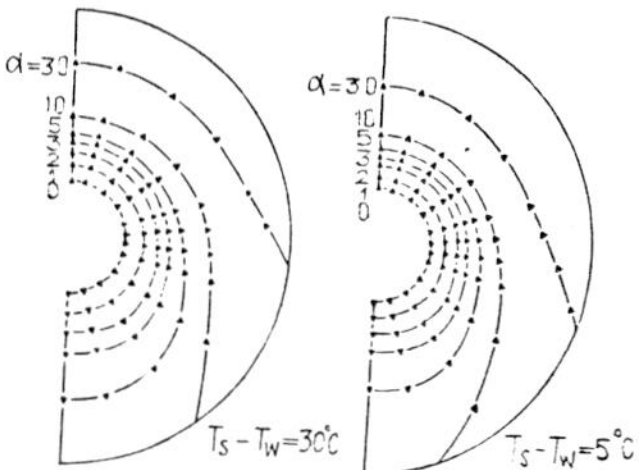

Fig. 3 Dimensionless temperature profile $\theta=10^{-\alpha}$ for $r_i/r_o=0.25$

As a reference quantity, it is appropriate to use the heat transfer corresponding to an ideal circumferential fin whose temperature is uniform everywhere and equal to T_w. Such isothermal circumferential fins have been calculated numerically.

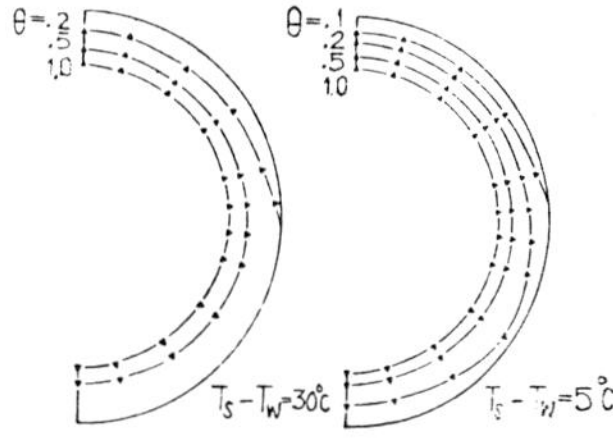

Fig. 4 Dimensionless temperature profile for $r_i/r_o=0.75$

Fig. 5 shows the comparison between an ideal fin and a real fin at three different regions. Since in Figs. 5-a and 5-d the θ of the ideal fin would always be equal to unity, they are not shown. From Fig. 5 it is noted that the use of an isothermal fin model, which is equivalent to using the simple Nusselt theory, gives rise to significant errors in the heat transfer, fin temperature, and condensate film thickness. The ratio $Q_{f\,re}/Q_{f\,id}$ of the overall rate of heat transfer from fin to base is only about 17% for $r_i/r_o = 0.5$ and $\Delta T = 30°c$. Therefore the use of the ideal fin model can lead to a significant overestimation of the heat transfer capabilities of the fin.

The overall rate of heat transfer is plotted in Fig. 6 as the function of ratio r_i/r_o. The parameter is temperature differences. It is to be noted here that the overall rates of heat transfer for the three different ratios $r_i/r_o = 0.25$, 0.5, and 0.75 are almost a constant. This fact shows us that the ratio r_i/r_o covered in practical air-cooler (from 0.25 to 0.75) has only a 2% influence on the overall rate of heat transfer. This result coincides with the fact which we indicated previously: the temperature drop is essentially within a region near the base tube. In view of this fact — we didn't consider that the condensate will pile up on the fin surface near the base. As will be discussed later, it will be expected that the overall rate of heat transfer for smaller r_i/r_o (for example 0.25) will decrease somewhat.

The overall rate of heat transfer onto a dry circumferential fin has been calculated under the donditions $\Delta T = 30^\circ c$, $r_i/r_o = 0.5$, and average heat transfer coefficient $h = 80[W/m^2 \cdot c]$, the result indicates that the fin efficiency of the dry fin is about 0.9 which is much higher than that of wet fins. The latter is only about 17%. The reason is that the condension heat transfer coefficient is much larger than that of forced convection. Therefore, it is important for an engineer who is engaged in the design of air coolers to judge

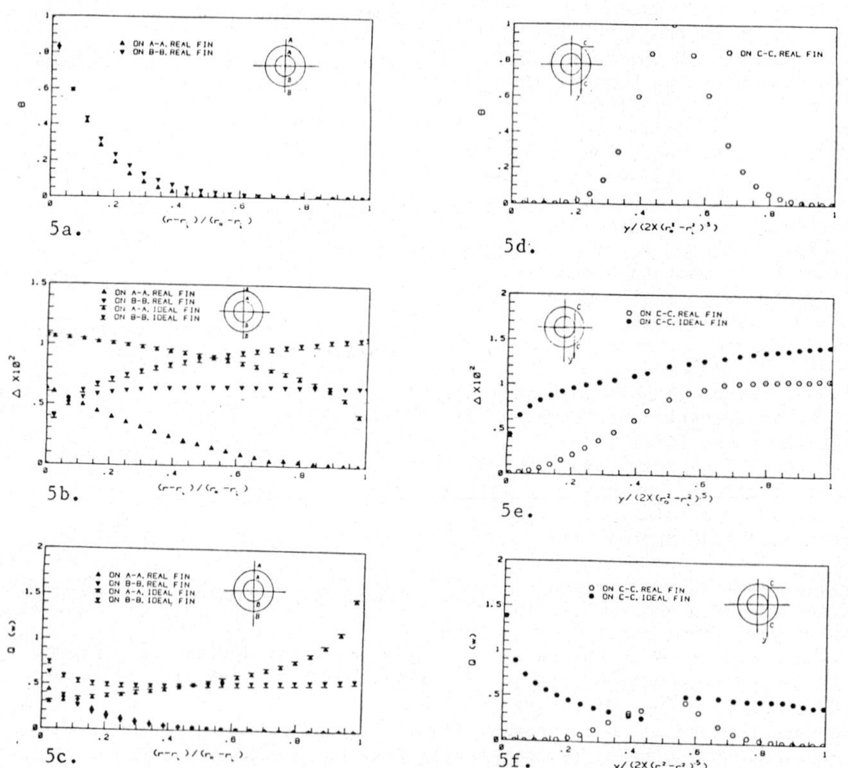

Fig. 5 Comparison between real and ideal fins in dimensionless temperature, film thickness and heat flux profiles

whether or not condensation will occur on the fin surface.

Finally, we will estimate and discuss the error which is caused by the assumptions. First, the condensate flows only along the base tube. In fact, the condensate will pile up near the upper-half base and flows only partly along the base as shown in Fig. 7. In order to estimate the error, two approximate assumptions were introduced. They are that the velocity of condensate which flows downward equals the velocity at which it flows along the base, and the shape of $\widehat{PQ}$ (Fig. 7) is a quarter round.

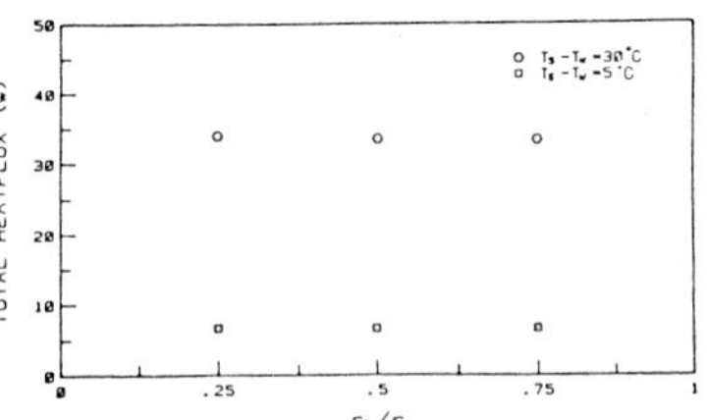

Fig. 6 Comparison of heat flux

Based on the continuity equation and the above two assumptions, an estimate of the error can be obtained. For the $r_i/r_o = 0.5, \Delta T = 30^oc$ situation the area upon which the condensate piles up is only about 7% of the total fin area and the error of overall heat transfer rate will be less than 10%. Due to the fact that the temperature drop is essentially within a region near the base tube, we can desregard the piling up of condensate on the edge of the lower-half of the fin.

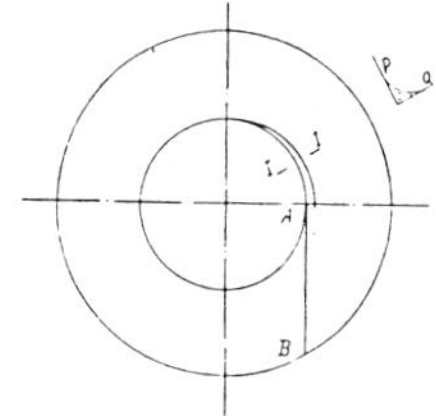

Fig. 7 Model of film collection on the fin base

Secondly, the film thickness will not be continued on two sides of $\overline{AB}$ (Fig. 6) because we ignore the surface tension and transverse velocity. Computation shows that the relative film thickness differences are only about 15% for most parts of $\overline{AB}$. In view of the following two reasons: the discontinuous region of film thickness is quite small compared with whole fin area and the influence of the discontinuous region on the temperature field will be reduced significantly by conduction of the fin-self, it can be expected that the influence of a discontinuous film on heat transfer rate can be ignored.

CONCLUSIONS

1. Numerical solution is performed by a finite difference scheme to predict the temperature field and heat flux onto vertical circumferential fins with film condensation. The computation is carried out for $r_i/r_o = 0.25$, 0.5 and 0.75, The dimensionless synthetic variable F_1 is covered from 4.425×10^{-7} to 2.023×10^{-10}.

2. Computation results are compared with an isothermal fin which is equivalent to using the simple Nusselt theory and dry fin. It is noted that the use of an isothermal fin model gives rise to significant errors in heat transfer. For example, the ratio Q_{fre}/Q_{fid} is only about 17% for $r_i/r_o = 0.5$ and $\Delta T = 30^oc$. The fin efficiency of the dry fin is much higher than that of wet fins under the above same conditions.

3. The error of the overall rate of heat transfer which is caused by ignoring the piling up of condensate on the upper-half base of the fin is about 10% for $r_i/r_o = 0.5$ and $\Delta T = 30^oc$.

4. Condensation on a circumferential fin surface is a problem of considerable complexity. In the present study we give a simplified but practical computation model. It would be worthwhile to search for a more

exact model which can be expected to improve the present solution.

ACKNOWLEDGEMENTS

The authors wish to acknowledge support received from the University of Minnesota and the Ministry of Education PRC.

NOMENCLATURE

b - interval between fins
F_1 - dimensionless synthetic variable
g - accelaration of gravity
K_1 - thermal conductivity of condensate
K_f - thermal conductivity of fin
L - characteristic dimension
$\dot{m}$ - condensate flow rate
Q - overall rate of fin
r - local radius of fin
r_i - inner radius of fin
r_o - outer radius of fin
R - dimensionless variable, r/r_o
T_w - fin base temperature
T_s - saturation temperature
T - fin temperature
t - fin half thickness
w - vertical velocity component
x - transverse coordinate
y - vertical coordinate
z - coordinate normal to fin
X - dimensionless coordinate, x/L
Y - dimensionless coordinate, y/L
Δ - dimensionless film thickness
δ - condensate film thickness
θ - dimensionless temperature
λ - latent heat of condensation
μ_l - viscosity of condensate
ϕ - circulat coordinate

REFERENCES

1. B.A. Kapxy, Engineering-Physics Magazine, U.S.S.R., T19, No. 4, p. 617, 1970.

2. C.B. Harjanof, Refrigerant Technique, U.S.S.R., No. 1, 1971.

3. P.N. Wu, Journal of Xi'an Jiaotong University, 1978.

4. Threlkeld, J.L., Thermal Environmental Engineering, second ed., Prentice-Hall, Englewood Cliffs, N.J., 1970, pp. 257-259.

5. Nader, W.K., "Extended Surface Heat Transfer With Condensation," Paper CS-5, Sixth International Heat Transfer Conference, Toronto, Canada, August 1978.

6. S.V. Patankar, E.M. Sparrow, Journal of Heat Transfer, Vol. 101, August, 1979, pp. 434-440.

Multi-Phase Flow and Heat Transfer III. Part A: Fundamentals
edited by T.N. Veziroğlu and A.E. Bergles
Elsevier Science Publishers B.V., Amsterdam, 1984 — Printed in The Netherlands

'COMPAGA' ANALYSIS OF SURFACE CONDENSERS

L. Lakshminarayana
Deputy Manager, Energy Systems Group
BHEL, Vikasnagar, Hyderabad - 500 593
INDIA

ABSTRACT

This paper presents COMPAGA (Combined macro-phenomenological and giant-average) analysis of steam-air mixture of surface condensers. Assuming the steam-air mixture to be steady, two-dimensional, non-diffusing, inviscid, irrorational and saturated, and treating it as a single phase medium, in untubed region, and as a porous medium, in tubed region, and assuming further its bulk fluctuation terms, in the latter region, as negligible, a mathematical formulation of the combined macro and giant-average state of the steam-air mixture is given, using macro-phenomenological theory, for untubed region, and giant-average theory of porous medium, for tubed region. The mathematical formulation consists of the followings: a) two sets of coupled Poisson's equations, each having three equations, and governing the three quantities, namely, mixture mass flux potential, air mass flux potential, and total pressure of each of the untubed and tubed regions; b) interfacial conditions, at the interface of untubed and tubed regions; and c) appropriate boundary conditions, at entrance, vent and shell surface. The mathematical formulation is then reduced to a set of algebraic equations, by standard Galerkin finite element method, using linear, isoparametric, quadrilateral elements. A computer code, in FORTRAN IV, is developed to solve the algebraic equations.

To demonstrate the application of the code, numerical results are obtained for an example condenser, having rectangular tube-nest and parallel-plate shell. The local and local volume-averaged values of velocity and air concentration in untubed and tubed regions, and local volume-averaged values of pressure, temperature, condensation rate and overall heat transfer coefficient, in tubed region, are presented and discussed.

1. INTRODUCTION

Surface condenser is one of the main elements of land-based thermal power stations and marine-based power plants, found in merchant and warships. Analysis of steam-air mixture of surface condenser is therefore important from the design and performance checking points of view.

The macro-thermo-hydrodynamic state of steam-air mixture, in the combined untubed and tubed region of the condenser, can be analysed numerically, in principle, by using, macro-phenomenological theory (mass diffusion equation for air, continuity equation, Navier-Stokes equations and energy equation for steam-air mixture) coupled with appropriate interfacial and boundary conditions, and finite difference or finite element method. However, it is not possible to do so, in practice, in view of very close grid to be considered in the tubed region (in order to obtain the numerical solution, with meaningful detail and tolerable accuracy, in that region), and consequent huge computation time required on any present day computer.

All the analyses of steam-air mixture till to-day, therefore, concern with its macro-thermohydro-dynamic state, only in untubed region, and with its macro-macro(giant*) - thermo-hydrodynamic state, on the other hand, in tubed region. Chisholm and McFarlane[1] were the first to present a one-dimensional analysis of the combined macro and giant state of the steam-air mixture. Wilson[2, 3] next presented a two-dimensional analysis of the combined state of the steam-air mixture. In his analysis, the number of nodes, at which the numerical solution was obtained, was twice the number of tubes in tube-nest, and its application was meant to small tube-nest condensers, having 2000 tubes or less. For large tube-nest condensers, this limitation, however, was overcome by Davidson and Rowe[4], by using giant-phenomenological theory of porous medium, for tube-nest region, and finite difference method.

The objective of this paper is to develop a COMPAGA(combined macro-phenomenological and giant-average) analysis, governing the combined macro and giant state of steam-air mixture, using macro-phenomenological theory for untubed region, giant-average theory (theory of giant state by average approach) of porous medium, for tubed region, and finite element method.

In what follows Sec.2 presents the analysis. It consists of physical model and coordinate system, mathematical formulation and Finite Element Method. Section 3 presents application of the analysis to an example condenser.

2. ANALYSIS

2.1 Physical Model and Co-ordinate system

Figure 1 shows schematic diagram and co-ordinate system of half the portion of a general symmetric surface condenser. Steam-air mixture at density $\bar{\rho}_{cen}$, at the axis of symmetry, air mass fraction $\bar{\epsilon}_{en}$ and velocity $\bar{u}_{1en}$ enters the shell through the entrance. The steam part, in the steam-air mixture, passes uncondensed in the untubed region and condenses in the tubed region. The condensate and the uncondensed steam, alongwith the non-condensible air, leaves the exit. It may be noted that the exit and

---

*'giant', whenever it appears hereafter means 'macro-macro'

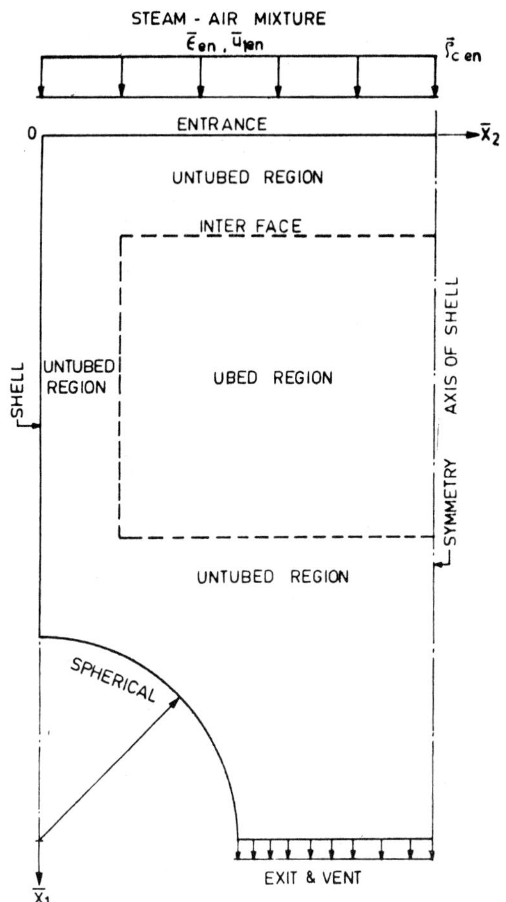

FIG. 1 SCHEMATIC DIAGRAM & CO-ORDINATE SYSTEM OF SURFACE CONDENSER

vent are one and the same in the figure, though they are different in practice, and no generality in the analysis would however get affected on this basis.

A cartesian co-ordinate system is chosen, as can be seen from the figure. O is the origin of the co-ordinate system. $\bar{x}_1$ $\bar{x}_2$ are the coordinate axes, along the straight portion of the shell and the entrance respectively.

2.2 Mathematical Formulation

Governing equations. Assuming the steam-air mixture to be steady, two-dimensional, non-diffusing, inviscid and saturated, and treating it as a single phase medium, in the untubed region, and a porous medium, in the tubed region, and further assuming its bulk fluctuation terms in the latter region, as negligible, the differential equations of state, governing its macro-state, in untubed region, and giant state, in tubed region, can be written from Lakshminarayana 5 , following tensor notation, as

$$(v \bar{\rho} \bar{u}_{\bar{i}})_{,\bar{i}} = -\bar{\dot{m}} \tag{1}$$

$$(v \bar{\epsilon} \bar{\rho} \bar{u}_{\bar{i}})_{,\bar{i}} = 0 \tag{2}$$

$$(v \bar{\rho} \bar{u}_{\bar{i}} \bar{u}_{\bar{j}})_{,\bar{j}} = v \bar{\rho} g_{\bar{i}} - (v \bar{p})_{,\bar{i}} + \bar{F}_{\bar{i}} \tag{3}$$

$$\bar{\rho}_s = \bar{\rho}_s (\bar{p} - \bar{p}_a) \tag{4}$$

$$\bar{T} = \bar{T}_s (\bar{p} - \bar{p}_a) \tag{5}$$

$$\bar{p}_a = \bar{\epsilon} \rho R_a \bar{T} \tag{6}$$

$$\bar{\rho} = \frac{\bar{\rho}_s}{(1-\bar{\epsilon})} \tag{7}$$

where v is the volume fraction of the mixture, having a value one in untubed region and a non-zero, positive value, less than 1, in tubed region; $\bar{\rho}$, $\bar{u}_{\bar{i}}$, $\bar{p}$ and $\bar{T}$ are local values, in untubed regions, and local volume-averaged values, in tubed region, of density, velocity, pressure and temperature, respectively, of the steam-air mixture; $\bar{\rho}_s$ is local value, in untubed region, and local volume-averaged value, in tubed region, of density of steam, in the steam-air mixture; $\bar{\epsilon}$, $\bar{\rho}_a$ and $\bar{p}_a$ are local values, in untubed region, and local volume-averaged values, in tubed region, of concentration (or mass fraction), density and pressure of air, in the steam-air mixture; R_a is the individual gas constant of the air; $g_{\bar{i}}$ is acceleration due to gravity; and $\bar{\dot{m}}$ and $\bar{F}_{\bar{i}}$ are respectively condensation rate of the steam and friction force, both per unit volume, and both having zero values, in tubed region, and non-zero values in untubed region.

Equations (1)-(3) are continuity equation of mixture, mass diffusion equation of air, and equation of momentum, respectively obtained from macro-phenomenological theory, for untubed region, and giant-average theory of porous medium, for tubed region. Equations (4) and (5) are equations of state for $\bar{\rho}_s$ and $\bar{T}$, and are functions of partial pressure of steam i.e. $p-p_a$ only, since the steam was assumed to be saturated. Equation (6) is the ideal gas equation of state for air. Equation (7) expresses $\bar{\rho}$ in terms of $\bar{\rho}_s$ and $\bar{\epsilon}$.

The mathematical models used for $\bar{\dot{m}}$ and $\bar{F}_{\bar{i}}$, in the tubed region, are the same as those used by Wilson [2,3], but in a form appropriate for the present analysis.. Tabular functions are used for $(\bar{p} - \bar{p}_a)$ and $\bar{T}_s(\bar{p} - \bar{p}_a)$, and are taken from [6].

Apart from the assumptions, made earlier for the steam-air mixture, we shall assume further that the flow of the mixture is irrotational i.e.

$$\epsilon_{\bar{i}\bar{j}}(\upsilon \bar{\rho} \bar{u}_{\bar{i}})_{,\bar{j}} = 0 \tag{8}$$

$$\epsilon_{\bar{i}\bar{j}}(\upsilon \bar{\rho} \bar{\epsilon} \bar{u}_{\bar{i}})_{,\bar{j}} = 0 \tag{9}$$

where $\epsilon_{\bar{i}\bar{j}}$ is two-dimensional alternating tensor. As a result, there exist, velocity and air concentration potentials ϕ and $\phi_{\bar{\epsilon}}$ respectively, such that

$$\upsilon \bar{\rho} \bar{u}_{\bar{i}} = -\bar{\Phi}_{,\bar{i}} \tag{10}$$

$$\upsilon \bar{\rho} \bar{\epsilon} \bar{u}_{\bar{i}} = \bar{\Phi}_{\bar{\epsilon},\bar{i}} \tag{11}$$

Now, with the help of Eqs. (10) and (11), Eqs. (1)-(3) can be written as

$$\bar{\Phi}_{,\bar{i}\bar{i}} = \bar{\dot{m}} \tag{12}$$

$$\bar{\Phi}_{\bar{\epsilon},\bar{i}\bar{i}} = 0 \tag{13}$$

$$\bar{H}_{,\bar{i}\bar{i}} = -F_{\bar{j},\bar{j}} \tag{14}$$

where $\bar{H}$ is the total pressure given by

$$\bar{H} = -\upsilon\left(\bar{P} + \frac{1}{2}\bar{u}_{\bar{j}}\bar{u}_{\bar{j}} + \bar{\rho}\,\Omega\right) \tag{15}$$

Here Ω is the gravity potential such that $g_{\bar{i}} = -\Omega_{,\bar{i}}$

Using $\bar{L}$, an appropriate length scale, $\bar{u}_{1en}$ and $\bar{\rho}_{cen}$ as characteristic quantities, and introducing the following non-dimensional quantities,

$$x_i = \frac{\bar{x}_i}{\bar{L}} \tag{16}$$

$$\phi = \frac{\bar{\Phi}}{\bar{\rho}_{cen}\bar{u}_{1en}\bar{L}} \tag{17}$$

$$\phi_{\bar{\epsilon}} = \frac{\bar{\Phi}_{\bar{\epsilon}}}{\bar{\rho}_{cen}\bar{u}_{1en}\bar{L}} \tag{18}$$

$$H = \frac{\bar{H} - \left\{\bar{P}_{cen} + \frac{1}{2}\bar{\rho}_{cen}\bar{u}_{1en}^{2} + \bar{\rho}_{cen}\Omega_{cen}\right\}}{\frac{1}{2}\bar{\rho}_{cen}\bar{u}_{1en}^{2}} \tag{19}$$

$$\dot{m} = \frac{\bar{\dot{m}}}{\left\{\frac{\bar{\rho}_{cen}\ \bar{u}_{1en}}{\bar{L}}\right\}} \tag{20}$$

$$F_i = \frac{\bar{F}_i}{\left\{\frac{1}{2}\ \frac{\bar{\rho}_{cen}\ u_{1en}^2}{\bar{L}}\right\}} \tag{21}$$

Eqs. (12)-(14) become

$$\phi_{,ii} = \dot{m} \tag{22}$$

$$\phi_{\bar{\epsilon},ii} = 0 \tag{23}$$

$$H_{,ii} = F_{i,i} \tag{24}$$

Interfacial conditions.

At the interface of untubed and tubed regions (see Fig.1)

$$\phi \text{ and } \phi_{,n} \text{ are continuous} \tag{25}$$

$$\phi_{\bar{\epsilon}} \text{ and } \phi_{\bar{\epsilon},n} \text{ are continuous} \tag{26}$$

$$H \text{ is continuous, and } H_{,n} = -F_* \tag{27}$$

where 'n' is the outward drawn normal and F_* is the component of F_i along n.

In Eq. (25), there are two conditions. The first one(ϕ is continuous) is an imposed one. This imposition is plausible in view of freedom available in selecting $\bar{\phi}$, differing by arbitrary constants, in untubed and tubed regions, based on its definition, Eq.(10). The second condition($\phi_{,n}$ is continuous) is the result of the application of conservation of mass at the interface. For Eq. (26), the interpretation is similar to the one given to Eq.(25). For Eq.(27), however, the first condition has interpretation similar to the first condition in Eq.(25). But the second condition($H_{,n} = -F_*$) is obtained, on the other hand, from law of conservation of momentum or by specializing Eq.(3) at the interface.

Boundary conditions.

At entrance,

$$\phi_{,n} = 1 \tag{28}$$

$$\Phi_{,n} = \bar{\epsilon}_{en} \quad (29)$$

$$H_{,n} = 0 \quad (30)$$

At vent,

$$\Phi = \Phi_{\bar{\epsilon}} = 0 \quad (31)$$

$$H_{,n} = 0 \quad (32)$$

At shell wall,

$$\Phi_{,n} = \Phi_{\bar{\epsilon},n} = H_{,n} = 0 \quad (33a,b,c)$$

At the axis of symmetry of the shell,

$$\Phi_{,n} = \Phi_{\bar{\epsilon},n} = H_{,n} = 0 \quad (34a,b,c)$$

Equations (22) and (24) are in the form of Poisson equation. For sake of generality, Eq.(23) can also be considered as Poisson equation, with non-homogenous term identically zero. Equation (23) is uncoupled from Eqs.(22) and (23), but the latter are coupled with one another, via the non-homogenous terms. These three equations, alongwith the equations of state, Eqs. (4) - (7), interfacial conditions, Eqs. (25) - (27), and boundary conditions, Eqs.(28) - (34 a,b,c) complete the formulation for Φ, $\Phi_{\bar{\epsilon}}$ and H.

2.3 Finite Element Method

Let G represent Φ or $\Phi_{\bar{\epsilon}}$ or H, and f represent correspondingly $-\dot{m}$ or 0 or $F_{i,i}$. Then Eqs. (22) - (24) can be written as

$$G_{,ii} = -f \quad (35)$$

Following standard Galerkin Finite Element Method [7], G can be approximated as

$$G \simeq \sum_{N=1}^{N_t} \Psi_N G_N \quad (36)$$

where G_N is the value of G at node N; Ψ_N is global shape function at N such that the boundary conditions, Eqs. (28) - (34 a,b,c) are satisfied; and N_t is the total number of nodes, in the untubes or tubed region, depending upon the case.

The residual R with respect to Ψ_N can be written as

$$R = \left(\sum_{N=1}^{N_t} \Psi_N G_N \right)_{,ii} + f \quad (37)$$

$$(R, \Psi_N) = 0 \qquad \text{for } N = 1, N_t \quad (38)$$

Substituting Eq.(37) into Eq.(38), we get

$$\sum_{M=1}^{N_t} A_{NM}\, G_M = F_N \qquad \text{for } N = 1, N_t \tag{39}$$

where

$$A_{NM} = \iint_D \psi_{N,i}\, \psi_{M,i}\, dD \qquad \text{for} \begin{cases} N = 1, N_t \\ M = 1, N_t \end{cases} \tag{40}$$

$$F_N = \iint_D f\, \psi_N\, dD + \int_\Gamma G_{,i}\, n_i\, \psi_N^*\, d\Gamma \qquad \text{for } N = 1, N_t \tag{41}$$

Here D and Γ represent domain and boundary respectively of the untubed region or tubed region, depending on whether G_M pertains to the former or the latter, ψ_N^* is the boundary value of the shape function ψ_N.

Equation (39) represents a set of algebraic equations coupled via the interfacial conditions, Eqs. (25) - (27). Linear, iso-parametric, quadrilateral elements are used, and a typical way of element discretisation of untubed and tubed regions, having 110 elements and 132 nodes, is shown in Fig.2. Note that the arbitrary dimensions shown for shell, tube nest and elements are guided by the numerical example to be considered in the next section. To solve the algebraic equations, a computer code, in Fortran IV, has been developed. The output from the computer program includes $\bar{u}_i$, ϵ, p, T, $\dot{m}$ and $\bar{h}_o$, over all heat transfer coefficient of tube, in the tubed region.

3. EXAMPLE

To demonstrate the application of the code, an example condenser is considered here. It is a rectangular nest, symmetric condenser (whose half nest is square), with parallel plate shell (the same as in Fig.1, with the convergence of the shell at the exit as absent). Table-1 below gives its geometric and thermo-hydrodynamic input data for the code.

The element discretisation used for the example condenser is same as that in Fig.2, but specialized by assuming the shell wall at exit as straight. The numerical results are obtained for $\bar{\epsilon}_{en}$ = 0.005 and 0.05 (as indicated in Table 1) using the computer code and the results are presented graphically in Figs. 3-8. The graphical presentation of the results, however, is confined to $\bar{\epsilon}_{en}$ = 0.005 only, in Figs. 3-7, but to both $\bar{\epsilon}_{en}$ = 0.005 and 0.05, in Fig.8.

Figures 3 and 4 present respectively u_i, the absolute $\bar{u}_i$ $\left(= \frac{\bar{u}_i}{\bar{u}_{ien}}\right)$ and $\bar{\epsilon}$, in untubed and tubed regions, for $\bar{\epsilon}_{en}$ = 0.005. As may be seen, for the case presented, all the steam gets condensed over the tube nest and an air blanket forms at the end of the nest.

Table 1. Geometric and Thermohydrodynamic Data of Example Condenser

Dimensions of shell	= 4 ft x 3.5 ft
Dimensions of the tube nest	= 3 ft x 1.5 ft
Material of the tubes	= Cupro-nickel(70:30)
S_T	= .07812 ft
S_L	= .06785 ft
d_o	= .05205 ft
d_i	= .04396 ft
velocity of coolant water	= 8.663 ft/sec
Temperature of coolant water	= 579.66°R
$\bar{L}$	= 1 ft.
$\bar{\rho}_{c\,en}$	= .009492 lbm/ft^3
$\bar{u}_{1en}$	= 35.08 ft/sec
$\bar{\epsilon}_{en}$	= 0.005, 0.05

In the above table, S_T, S_L, d_o and d_i are respectively transverse and longitudinal pitches of tube nest, and outer and inner diameters of the tubes. The quantities, third to the nineth from the top, in the table, are required to evaluate $\bar{\dot{m}}$ and $\bar{F}_i$.

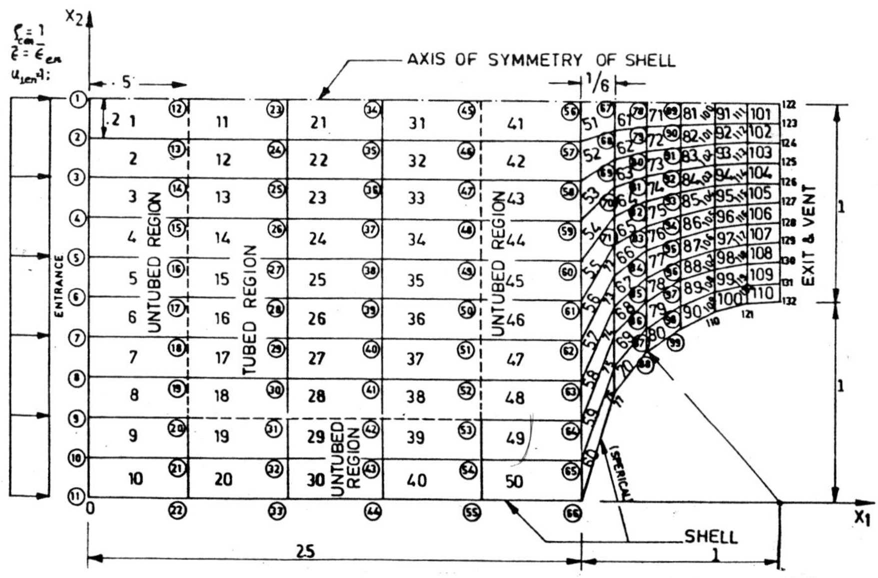

FIG. 2 SCHEMATIC DIAGRAM OF CONDENSER, NODES AND ELEMENTS

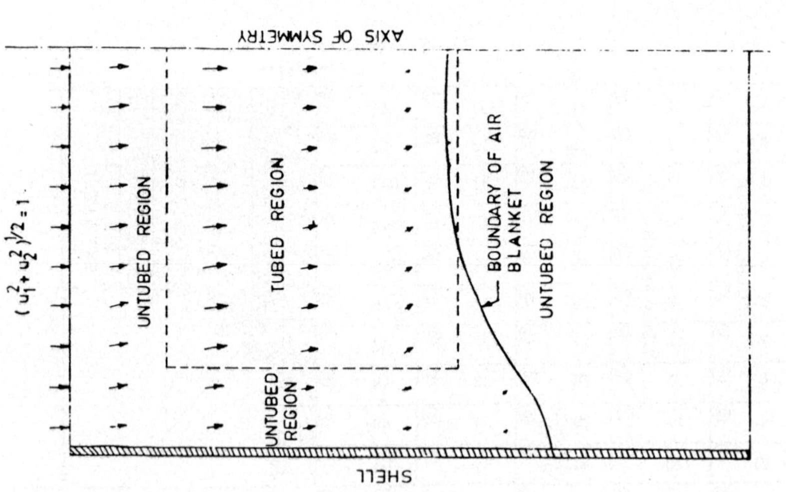

FIG. 3 NON-DIMENSIONAL VELOCITY DISTRIBUTION

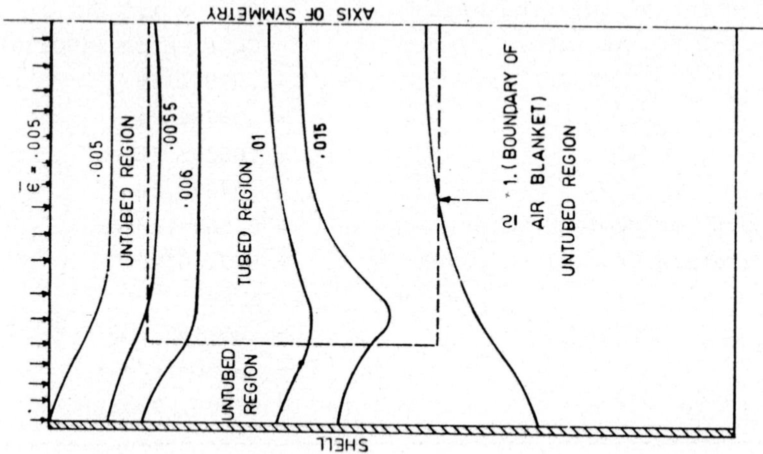

FIG. 4 AIR CONCENTRATION DISTRIBUTION

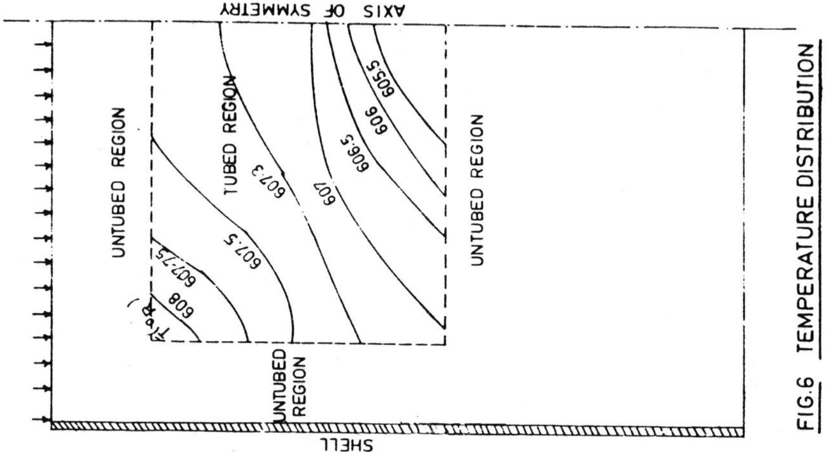

FIG.6 TEMPERATURE DISTRIBUTION

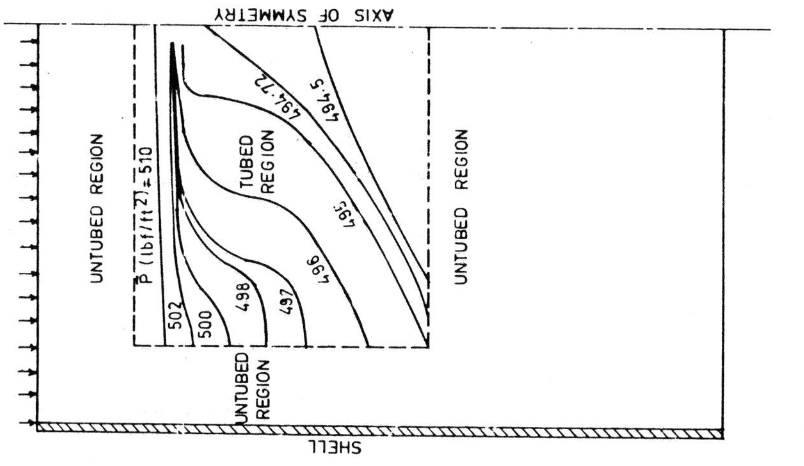

FIG. 5 PRESSURE DISTRIBUTION

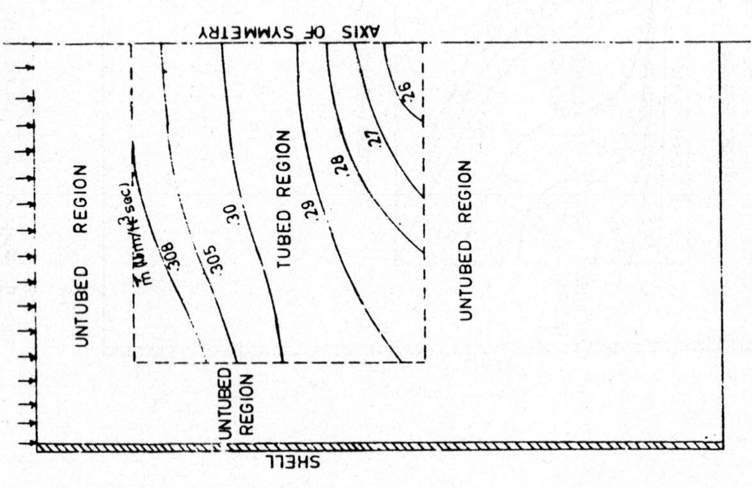

FIG. 7 CONDENSATION RATE DISTRIBUTION

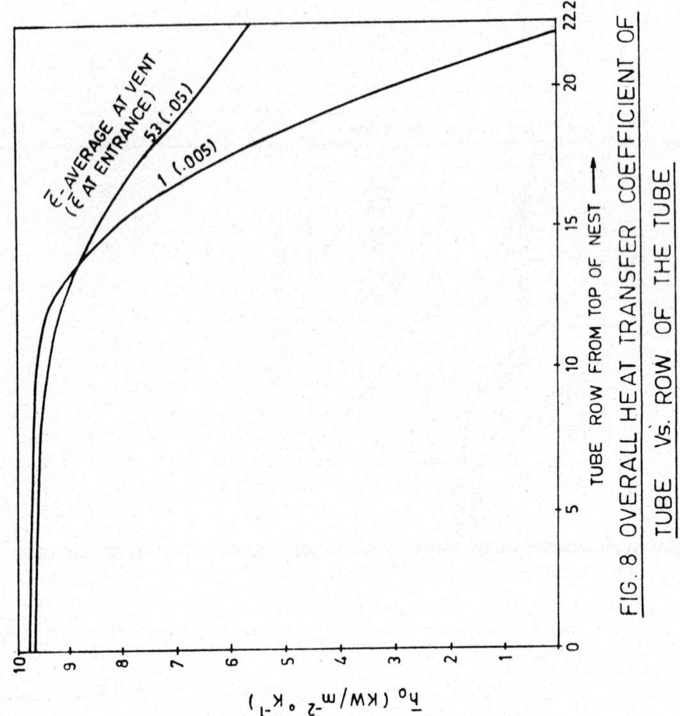

FIG. 8 OVERALL HEAT TRANSFER COEFFICIENT OF TUBE Vs. ROW OF THE TUBE

Figures 5-7 present respectively $\bar{p}$, $\bar{T}$ and $\bar{\dot{m}}$ in the tubed region. As to be expected all the three quantities fall gradually as the steam-air mixture proceeds towards vent.

Figures 8 presents overall heat transfer coefficient of tube, $\bar{h}_o$, at the axis, versus tube row, from the top of the nest, for $\bar{\epsilon}_{en}$= 0.005 and 0.05. As may be seen, $\bar{h}_o$, for the tubes at the beginning of the nest, is more or less flat and its value is higher for $\bar{\epsilon}_{en}$= 0.005 than for $\bar{\epsilon}_{en}$= 0.05. The values of h_o for the two cases are, however, reversed at the end of tube nest.

Further, in case of = 0.005, the value of $\bar{h}_o$, close to the end of the nest, is zero. This is because of the formation of air blanket, as pointed out above.

ACKNOWLEDGMENTS

The author expresses deep sense of gratitude to K.D. Udeshi, Manager, for keen interest and several discussions during the course of the work. The author also expresses gratitude to Dr.A. Gopalakrishnan, General Manager, for suggesting the problem and for several discussions on the problem. Finally, the author is grateful to Dr.A.L. Chandraker, Deputy Manager, for many fruitful discussions at different stages of the problem.

REFERENCES

1. Chisholm, D. and McFarlane, Mrs.M.W., "The Prediction of Condenser Performance using a Digital Computer", Report No.161, Sept.1964, National Engineering Lab., Glasgow.

2. Wilson,J.L., "The Design of Condensers by Digital Computers", I.Chem.E.Symposium Series, No.35, 1972.

3. Wilson, J.L., "NEL Two-Dimensional Condenser Computer Program" Report No.619, Sept.1974, pp.132-151, National Engineering Lab., Glasgow.

4. Davidson, B.J. and Rowe,M., "Simulation of Power Plant Condenser Persormance by Computational Methods: An Overview", Power Condenser Heat Transfer Technology, Edited by Marto, P.J. and Nunn, R.H., Hemisphere Publishing Corporation, Washington, pp. 17-49, 1980

5. Lakshminarayana, L., "Giant-Average Theory of Phase-Pore Medium: Its Application to Multiphase-Pore Medium", Tech. Report No.RD-49-81-01, BHEL(R&D), Hyderabad, India, Aug.1981.

6. Keenan, J.H. et al., Steam Tables , John Wiley & Sons, Inc., USA, 1969.

7. Chung, T.J., Finite Element Analysis in Fluid Dynamics , McGraw - Hill, Inc., USA, 1978.

Multi-Phase Flow and Heat Transfer III. Part A: Fundamentals
edited by T.N. Veziroğlu and A.E. Bergles
Elsevier Science Publishers B.V., Amsterdam, 1984 — Printed in The Netherlands

EXPERIMENTAL STUDY ON STEAM JET CONDENSATION IN SUBCOOLED WATER POOL

Giovanni DEL TIN, Evasio LAVAGNO, Mario MALANDRONE
Politecnico di Torino - Dipartimento di Energetica
Corso Duca degli Abruzzi, 24
10129 Torino, Italy

ABSTRACT

Pressure and temperature measurements in the surrounding zone and inside a steam jet condensing in a subcooled water pool have been performed. The steam jet shapes and the several flow regions in the surrounding subcooled water have been investigated in various operating conditions. The measured jet length to nozzle diameter ratios have been determined and a comparison has been made with theoretical predictions. Axial and radial temperature profiles in the vapour cone and in the outside regions have been obtained; the temperature profiles have been compared with the predicted ones through an analytical model, based on the A. Kudo approach, which assumes the kinetic gas theory condensation law, the Prandtl-Taylor theory for the liquid mixing zone and the integral conservation equations for the water in the main region. Axial and radial static and total pressure profiles inside the vapour cone and in the surrounding water have been determined. Dynamic pressures in the mixing and in the main regions have been measured and the evaluated fluid velocities have been compared with the predicted ones by means of the previously mentioned analytical model.

1. INTRODUCTION

The present study is concerned with the experimental investigation of the high velocity vapour jets condensing in a subcooled water pool. The investigations into the field of the direct contact condensation are very useful to improve the knowledge of the basic jet condensation mechanisms: not a great deal is known about this topic. The majority of the experimental work in this area, /1/ - /14/, does not examine thoroughly the behaviour of the parameters such as pressure, temperature, liquid entrainment and so on, inside the vapour jet and in the interface between vapour and surrounding liquid. On the other side, there are no complete theoretical models of submerged condensing vapour jets capable of treating the inside jet phenomena as well as the interface ones. Furthermore, the lack of proper experimental results does not allow to supply a suitable framework for the development of a theoretical understanding and for the checking and validation of the existing analytical models.

The primary objective of this study is to experimentally investigate the steam jet behaviour in the stable regime, where the condensing phase assumes a near conical shape. The main points of interest are the following:

- steam jet shapes and flow regions in the surrounding subcooled water;
- penetration length of vapour jets submerged in the subcooled water;
- axial and radial temperature profiles in the vapour jet and in the outside regions;
- axial and radial pressure profiles (both total and static) inside the vapour jet and in the surrounding water regions.

Some experimental results - e.g. the vapour jet length (penetration length) to nozzle diameter ratios, the axial and radial temperature profiles in the liquid regions outside the steam jet and the evaluated fluid velocities in the mixing region - are compared with the predicted ones through the A. Kudo analytical model.

Besides their scientific meaning, the steam jet condensation studies are useful in various application areas, namely: underwater propulsion, jet pumps, direct contact feedwater heaters and nuclear reactor safety systems. In the nuclear safety area the role of condensation phenomena in the so-called pressure suppression system of the Boiling Water Reactors is well known; recently, a new safety system, named Core Rescue System, has been proposed [15] as an additional safety system for the next Pressurized Water Reactors. As a component of the Core Rescue System there is a direct condenser partially filled with borated water. In emergency conditions the steam released from the primary circuit is discharged into the water pool as high speed jets which are subjected to direct contact condensation through a discharge line and steam spargers.

2. EXPERIMENTAL APPARATUS

The main apparatus components (fig. 1) are the electrically heated steam supply system which yields saturated steam at pressures ranging from 2 bar to 6 bar, the steam line with the superheater, the single straight discharge nozzle of different diameters (2-4-6 mm I.D.) and the circular section vapour suppression pool which contains 900 dm^3 of water at atmospheric pressure and temperature ranging from 20 to 75 °C. The boiler feed-water is outgassed by boiling at atmospheric pressure. The nozzles are double-walled pipes to minimize the steam condensation along the flow path. The nozzle exit submergence in the pool water is changeable, ranging from 20 to 250 mm. The steam mass velocity through the nozzle is ranging from 200 to 800 kg/(s m^2). The jet pressure measurements are performed by two different axial probes: a stainless steel 0.5 mm I.D., 0.7 mm O.D. probe for total pressure measurements, and a 0.7 mm I.D., 1 mm O.D. probe with a 0.5 mm diameter static pressure tap. The jet temperature distributions are measured by 0.25 mm and 0.5 mm chromel-alumel microthermocouples. A micrometric device allows displacements of the pressure probe and thermocouple with an accuracy better than 0.05 mm.

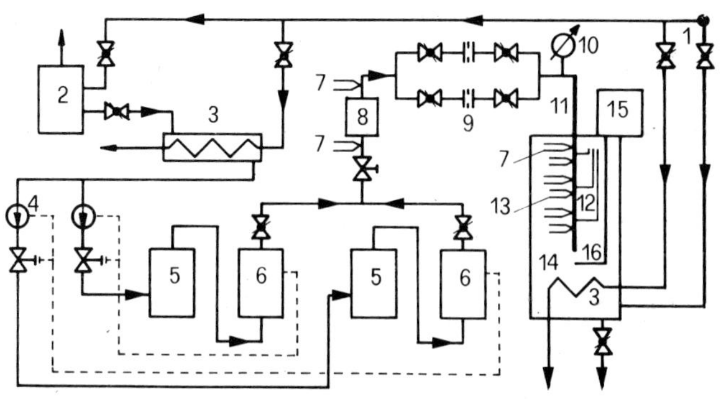

1 - water line
2 - outgassing boiler
3 - heat exchanger
4 - boiler feed-water pump
5 - feed-water heater
6 - boiler
7 - steam flow thermocouple
8 - superheater
9 - orifice meter
10 - Bourdon gauge
11 - steam discharge nozzle
12 - nozzle pressure taps
13 - nozzle wall thermocouple
14 - suppression pool
15 - micrometric device
16 - pressure and temperature probes

Fig. 1 - Experimental apparatus

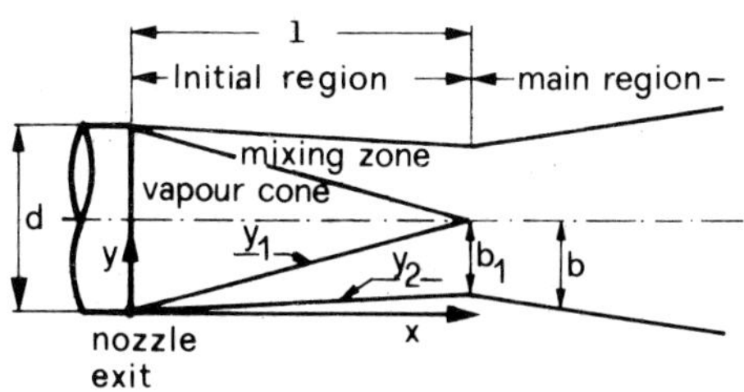

Fig. 2 - Jet model and coordinates

3. ANALYTICAL MODEL

The complexity and the lack of physical insight of jet condensation cause many difficulties to formulate mathematical models concerning the process under consideration. In this investigation a theoretical solution of the problem of steam jet condensation has been found using the Prandtl-Taylor theory following the A. Kudo [4] approach.

In the A. Kudo analysis the steam jet and the surrounding liquid regions are represented as in fig. 2. Downstream the nozzle exit the jet is subdivided in two sub-regions, namely the initial and the main region. The former is made up by the vapour cone and the mixing zone. According to A. Kudo model, the steam jet assumes a conical shape and the cone length is controlled by condensation heat transfer through the liquid-steam boundary. The heat released by condensation is transferred to the surrounding medium by turbulent diffusion, while the mixing zone formed around the vapour cone can be regarded as the liquid phase zone. Inside the vapour cone, both the steam velocity and the temperature are uniform and the latter is equal to the saturation value at the ambient pressure.

The initial region is analyzed as follows. Supposing that the theory of the boundary layer of an incompressible fluid [16] may be applied in the mixing zone of the initial region, the following two-dimensional equations can be written:

- momentum equation: $$u\frac{\partial u}{\partial x} + v\frac{\partial u}{\partial y} = c^2 x^2 \frac{\partial}{\partial y}\left[\left(\frac{\partial u}{\partial y}\right)^2\right] \tag{1}$$

- continuity equation: $$\frac{\partial u}{\partial x} + \frac{\partial v}{\partial y} = 0 \tag{2}$$

- energy equation: $$u\frac{\partial \vartheta}{\partial x} + v\frac{\partial \vartheta}{\partial y} = 2\, c^2 x^2 \frac{\partial}{\partial y}\left[\left(\frac{\partial u}{\partial y}\right)\left(\frac{\partial \vartheta}{\partial y}\right)\right] \tag{3}$$

In the above equations u, v, and ϑ are the time averaged values and c is a constant relating to the mixing length. To solve these equations, the following assumptions are made:

- the internal and external boundaries of the mixing zone y_1 and y_2 are axially linear;
- velocity and temperature distributions in the mixing zone are independent of x in the initial region;
- the temperature variation in the mixing zone is linear, owing to its small thickness.

The solutions of the above equations are given in ref. [4].

The present results, however, have been obtained introducing some modifications in the A. Kudo model, taking into account our experimental results. The topic where the model has been modified, concerns the evaluation of the temperature ϑ at the internal boundary of the mixing zone. From the molecular theory of gas, when the steam condenses, the temperature on the steam-water interface is expressed as follows:

$$\vartheta_i = \vartheta_s - \frac{\varphi_{h1}}{\Delta h_v\, \Omega} \tag{4}$$

where φ_{h1} is the heat flux transferred by turbulent diffusion at the internal boundary of the mixing zone (which is given by Taylor's free turbulence theory) and according to reference [17], Ω is given by the following expression:

$$\Omega = \frac{\sigma}{2-\sigma}\left(\frac{1}{2\pi}\right)^{\frac{1}{2}}\left(\frac{M}{R}\right)^{\frac{3}{2}}\left(\frac{p_v\,\Delta h_v}{T_v^{\frac{5}{2}}}\right) \tag{5}$$

This expression differs from A. Kudo's one. Taking into account the physical meaning of Ω (which is a heat transfer coefficient divided by a latent heat of vaporization Δh_v), the value of σ has been assumed equal to 0.5 to have a good agreement with the heat transfer coefficients obtained experimentally by the Authors [10],[12] and by Cumo and co-workers [7]. As a result of the calculations the following expression is adopted to evaluate ϑ_i:

$$\vartheta_i = \vartheta_s - \frac{a\ c_p\ \varrho_1\ Z'_{\eta=1}\ u_1\ (\vartheta_i - \vartheta_p)}{\phi_o^2\ \Delta h_v\ \Omega} \tag{6}$$

The temperature ϑ in the mixing zone is given by the following relation:

$$\vartheta = \vartheta_p + (\vartheta_i - \vartheta_p)\frac{\psi - \psi_2}{\phi_o} \tag{7}$$

The model allows also to obtain the liquid velocity profiles inside the mixing zone. The adimensional velocity u/u_1, where u_1 is the velocity at the internal boundary of the mixing zone, is given by the following equation:

$$Z = \frac{u}{u_1} = C_1\ e^{-\phi_o\eta} - C_1\ e^{\frac{\phi_o}{2}\eta}\cos(\frac{\sqrt{3}}{2}\ \phi_o\ \eta) + \sqrt{3}\ C_1\ e^{\frac{\phi_o}{2}\eta}\sin(\frac{\sqrt{3}}{2}\ \phi_o\ \eta) \tag{8}$$

where:

$$C_1 = \left[e^{-\phi_o} - e^{\phi_o/2}\cos(\frac{\sqrt{3}}{2}\ \phi_o) + \sqrt{3}\ e^{\phi_o/2}\sin(\frac{\sqrt{3}}{2}\ \phi_o)\right]^{-1} \tag{9}$$

u_1 can be calculated numerically by the following equation, being ϑ_p and u_o given:

$$\frac{u_1}{u_o}\ B^*\ \phi_o^3 = \frac{Z'_{\eta=1}\ c_p\ (\vartheta_s - \vartheta_p)\ \phi_o^2\ \Omega}{\phi_o^2\ \Delta h_v\ \Omega + a\ c_p\ Z'_{\eta=1}\ u_1\ \varrho_1} \tag{10}$$

where:

$$B^* = \frac{\varrho_v}{\varrho_1}\ (\frac{u_o}{u_1})^2\ (\frac{\psi_1}{\phi_o}) \tag{11}$$

Finally, the model allows to calculate the penetration length of the vapour jet. Therefore the ratio l/d is expressed as:

$$\frac{l}{d} = \frac{1}{2\ a\ \psi_1} \tag{12}$$

The analytical results which are given here will be compared with the experimental ones, later on.

The main region is analyzed as follows. According to the proposed model,in this region there is a single phase water jet, which is treated analytically as an axisymmetric flow. The jet width, the velocity and temperature distributions are usually evaluated from the following expressions:

- half jet width b : $$\frac{db}{dx} = 0.22 \quad (13)$$

- velocity distribution: $$\frac{u}{u_m} = \left[1 - (\frac{y}{b})^{\frac{3}{2}} \right]^2 \quad (14)$$

- temperature distribution: $$\frac{\vartheta - \vartheta_p}{\vartheta_m - \vartheta_p} = 1 - (\frac{y}{b})^{\frac{3}{2}} \quad (15)$$

Here u_m and ϑ_m are the values on the jet axis (that is at $y = 0$) which, in our case, are determined by experiments. According to [4], the condition between the two regions, initial and main, is arranged by matching each jet widths at their boundaries. The half width of the jet b_1 at $x = 1$ in the initial region, must be equal to the half width of the main region at the same position. By means of momentum and energy equations applied in the main region, taking also into account the above-mentioned temperature and velocity radial distributions, the model allows to calculate the temperature distribution on the jet axis, that is given by the following equation:

$$\vartheta_m - \vartheta_p = 9.31 \sqrt{\frac{\varrho_v}{\varrho_1}} \cdot \frac{\Delta h_v + c_p(\vartheta_s - \vartheta_p)}{c_p} \cdot \frac{d/2}{x + \frac{b_1}{.22} - 1} \quad (16)$$

This equation, which slightly differs from that given by A. Kudo, will be verified by experimental results.

4. RESULTS AND DISCUSSION

The experimental investigations are concerned with several aspects of the jet condensation phenomena involving both the vapour jet and the outside liquid region. The measured parameters, especially in the liquid region, are such that they can be compared with the theoretical ones predicted by the analytical model [4]. The present results refer to a stable jet behaviour; the jet instability thresholds and the limit lines between stable and unstable regimes have been reported in a previous work [12].

The preliminary experimental observations concern the steam jet shape; fig. 3 shows some typical shapes obtained by photographs. As it can be seen, the jet shape is strongly affected by pool temperature and it is rather complicated. At the nozzle exit, where the steam pressure is larger than the ambient one, the jet diameter increases, apparently due to the expansion of the steam leaving the nozzle; a similar jet behaviour has been observed by Wiemer at al. [3] and Kerney et al. [2]. Both the distance from the nozzle exit,

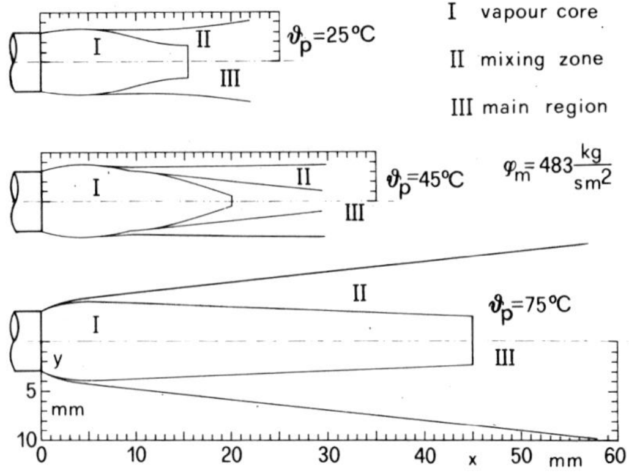

Fig. 3 - Observed jet shapes

where the diameter reaches its maximum value, and the maximum diameter itself, slightly increase with the pool temperature. The jet end is not clearly defined by photographs: a two - phase mixture can be observed, therefore the point where the vapour disappears cannot be well recognized. It is also probable that some liquid, as drops, is entrained by the high-velocity vapour: the rate of entrainment seems to be increasing with the pool subcooling. Around the steam jet a transition region can develop and extend downstream the zone where the properly said condensation takes place. In this region, near the interface with condensing vapour, there can be a small amount of vapour entrained by the liquid phase as bubbles; on the other hand, far from the interface, the observed "froth flow" can be due to air bubbles previously dissolved in the water pool.

The vapour jet lentgh, obtained by photographs, has also been measured.

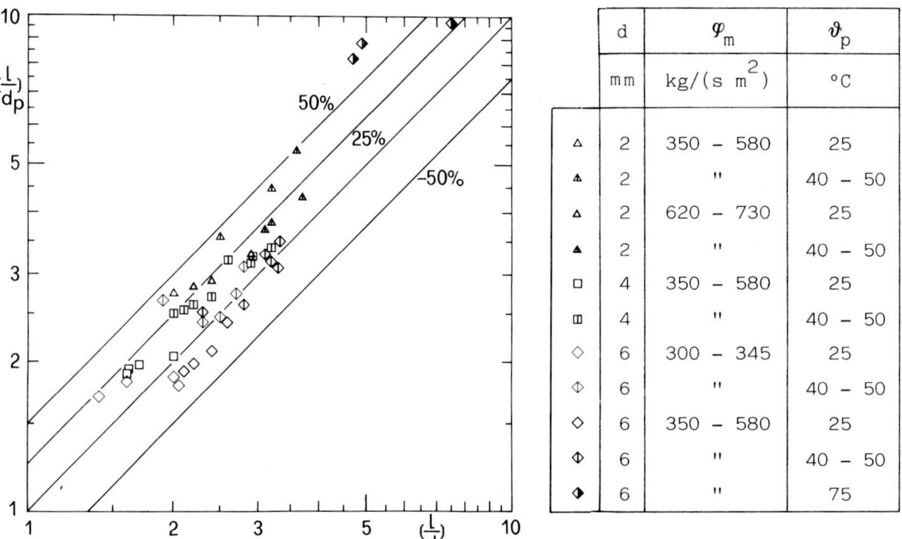

	d	φ_m	ϑ_p
	mm	kg/(s m²)	°C
△	2	350 - 580	25
◭	2	"	40 - 50
△	2	620 - 730	25
▲	2	"	40 - 50
□	4	350 - 580	25
⊞	4	"	40 - 50
◇	6	300 - 345	25
⌽	6	"	40 - 50
◇	6	350 - 580	25
◈	6	"	40 - 50
◐	6	"	75

Fig. 4 - Comparison between experimental, $(l/d)_e$, and predicted, $(l/d)_p$, penetration length to nozzle diameter ratios

The ratio between jet length and nozzle diameter, plotted in fig. 4, is a function of the nozzle diameter, the mass velocity, the steam thermodynamic conditions and the pool water subcooling. It can be observed that the experimental results are somewhat scattered. This is due to difficulties in defining precise boundaries for the turbulent two-phase flow as well as uncertainties when determining the extent of the vapour region from photographs. The penetration lengths predicted by the model are in fairly good agreement with the experimental ones at low pool temperatures. However, the model does not take correctly into account the effects of the mass velocity and of the nozzle diameter.

Several experimental data concerning the axial temperature profiles have been obtained in various operating conditions; some of these data referring to 6 mm nozzle diameter, for three different water subcoolings, are reported in fig. 5. As it can be observed, at the nozzle exit, were the jet diameter tends to increase, the steam temperature rapidly falls, then it increase again in a manner practically not dependent by pool water temperature. Subsequently the temperature decreases with a trend which is clearly dependent by pool water temperature. The sharp decrease of the temperature inside the vapour jet can be due to the subcooled liquid entrainment. The axial temperature profiles referring to the main region are compared with the experimental ones in fig. 5. The satisfactory agreement in the main region, confirms that this simple model can be reasonably applicable in these conditions. However, the best agreement is obtained when pool temperature equals 45 °C; at 75 °C the calculated temperatures are sligthly higher than experimental ones, while at 25 °C the predictions are lower than the measured data. These discrepancies can mean, among other things, that the jet width (eq. 13) is also temperature dependent. This hypothesis seems also to be confirmed by fig. 3 where it can be seen that the jet shape is strongly affected by pool temperature. Inside the vapour region, where the model assumes a constant temperature equal to the saturation value, the temperature profiles cannot be easily explained. Two subregions can be observed: the first near the nozzle exit, where the steam is subjected to a sudden expansion, is characterized by a temperature variation practically independent by pool temperatures; in the other sub-region, near the vapour jet end, the temperature profiles are instead strongly influenced by pool temperature, probably owing to the subcooled liquid entrainment.

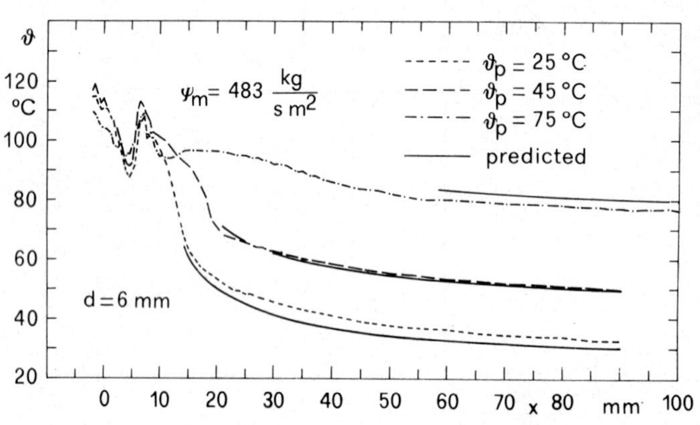

Fig. 5 - Axial temperature profiles

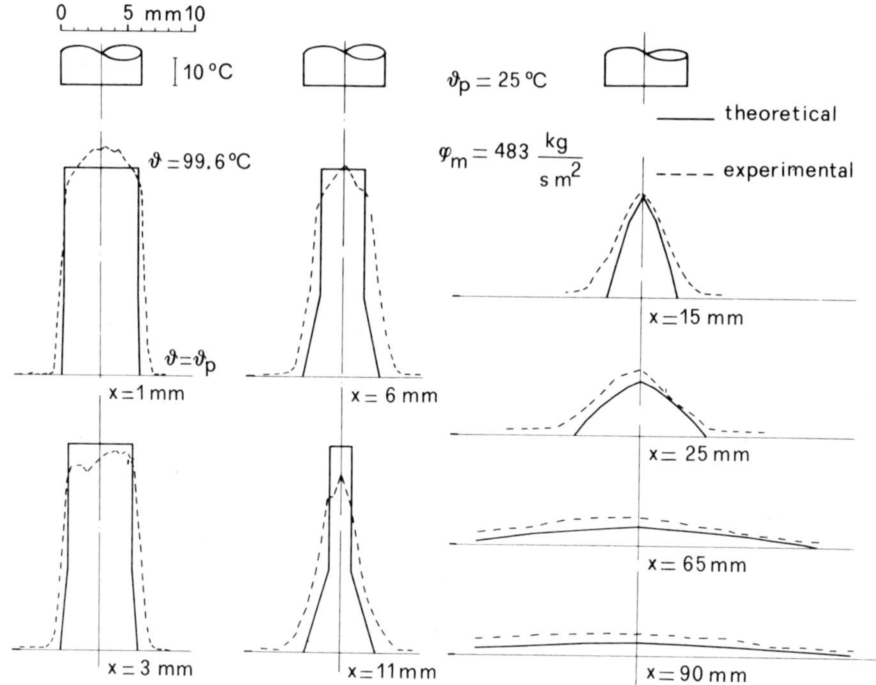

Fig. 6a - Radial temperature profiles at various axial locations, d = 6 mm

Fig. 6 shows some radial experimental and predicted temperature profiles, obtained at several axial positions along the vapour jet and in the main region (see also fig. 4), with reference to three water pool subcoolings. A brief analysis of this figure shows immediately some interesting features:

- the temperature inside the vapour zone in the initial region is not constant as assumed in the model, but it is subjected to a sensible variation in the radial direction;
- the temperature at the interface between steam and surrounding water does not show a sudden reduction as predicted by the model; this result confirms the existence of a transition zone where a two-phase mixture can be present;
- the linear variation of the temperature in the mixing zone must be assumed with some caution; however, the slope of the predicted temperature curve is very similar to the experimental one;
- the model tends to underestimate both the width of the mixing zone and of the main region; that is also due to a difference between the jet shape assumed and the real one as it results in fig. 2 and 3.
- the evolution of the radial temperature profiles along the fluid path shows a gradual transition from initial to main region: the vapour cone end is not well defined as indicated in fig. 2;

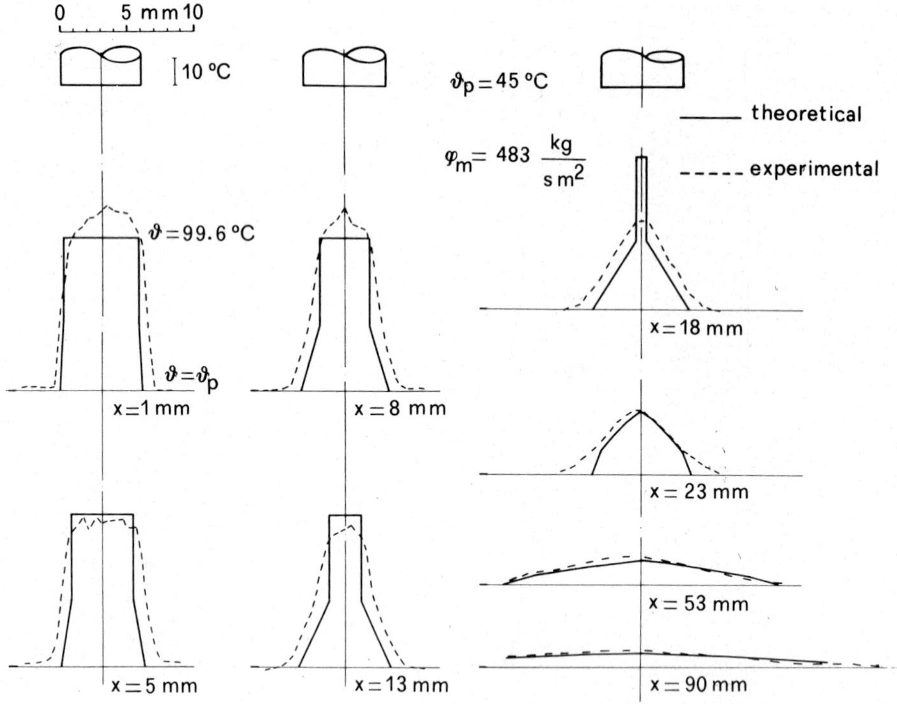

Fig. 6b - Radial temperature profiles at various axial locations, d = 6 mm

- in the main region, the temperature profiles predicted by the model are in fairly good agreement with the experimental ones.

The variations of the pressure along the jet axis and in several radial positions are also measured. Fig. 7 shows both the total and "static" axial pressure profiles; the so called static pressure is that measured by pressure taps normal to the flow direction. It can be seen that the pressure variations are very complicated and that it is difficult to explain the phenomena involved. The radial pressure profiles in fig. 8 represent typical trends at different axial locations. An attempt has been made to find the velocity profiles in the mixing zone and in the main region. In the mixing zone the pressure measurements are complicated owing to the small dimensions of this zone (2 - 5 mm thick) in comparison to the pressure probes dimensions; moreover, as it is already known, in this zone small vapour bubbles may be mixed in the liquid matrix so that the real density evaluation is rather complicated. In the present work the experimental velocity profiles have been determined with reference to liquid density at pool temperature.

Fig. 9 shows the experimental and calculated velocity profiles in the mixing region at various axial locations and at three different pool subcoolings. In this figure the point $y = 0$ is located at a distance from the

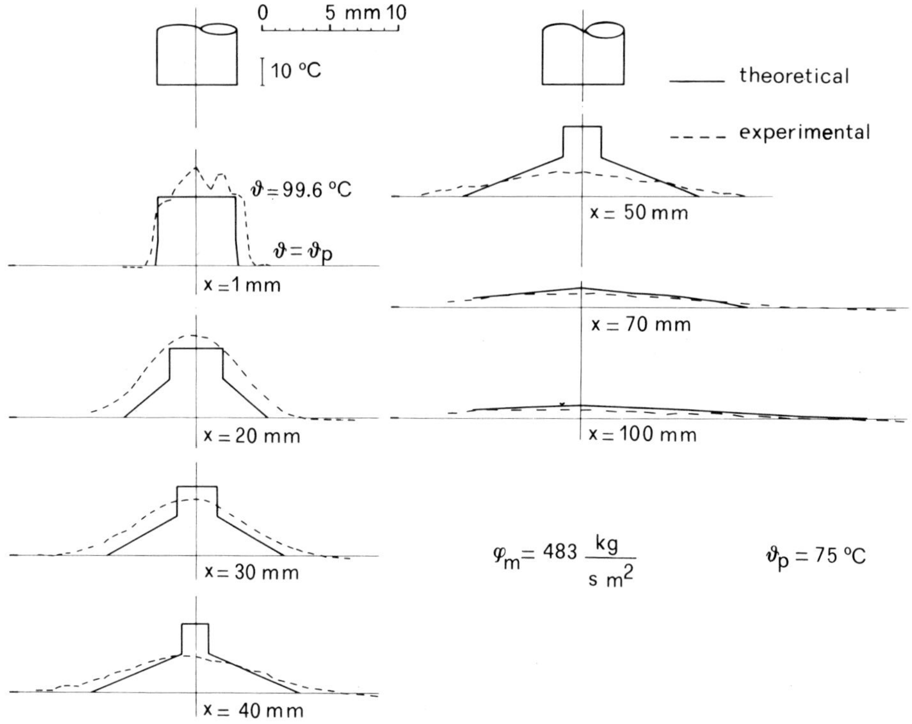

Fig. 6c - Radial temperature profiles at various axial locations, d = 6 mm

jet axis equal to the inner nozzle radius. In spite of the abovementioned experimental uncertainty and of the model limitations, the comparison shown in fig. 9 seems to confirm the substantial validity of the model in order to predict the trend and the values of the velocity in the mixing zone, especially at lower pool subcoolings. It can be seen that the predicted velocity decreases faster than the experimental one, so that the jet width becomes smaller than that experimentally observed; however this result is consistent with that concerning the predicted radial temperature profiles.

Fig. 10 shows the comparison between experimental and calculated radial velocity profiles in the main region. There the model does not calculate the velocity along the jet axis, but only its radial profile, corresponding to a given axial velocity, which in our case is an input obtained by experimental data. In this case, too, like in the mixing zone, the experimental velocity decreases at a rate lower than the calculated one; substantial discrepancies are observed for the lower subcooling.

Even if there are no predictions about the steam behaviour in the initial region, some considerations can be made observing the axial and radial pressure profiles reported in fig. 7 and 8 respectively. With reference to fig. 7,

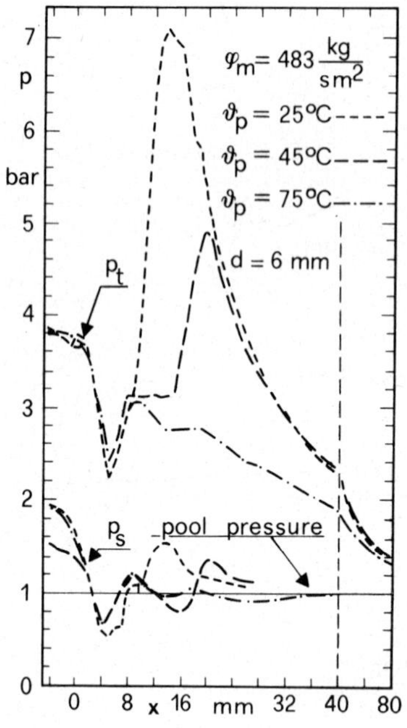

Fig. 7 - Axial pressure profiles

Fig. 8 - Radial static pressure profiles

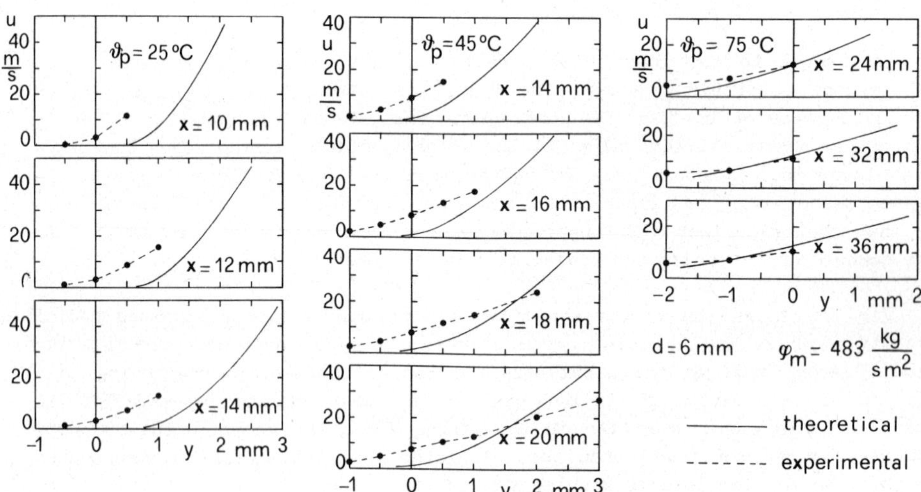

Fig. 9 - Radial velocity profiles in the mixing region at various axial locations

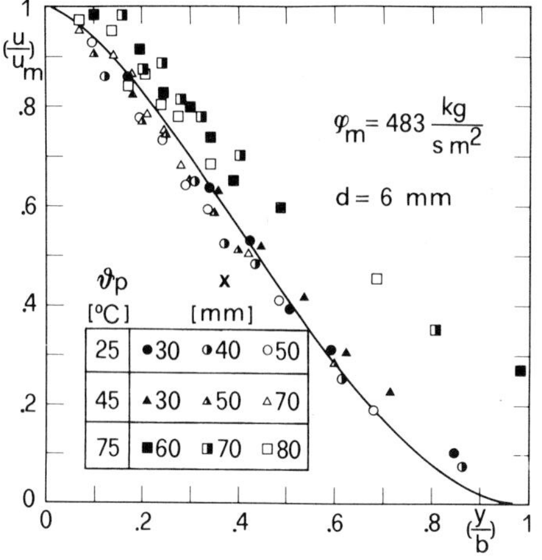

Fig. 10 - Dimensionless radial velocity profiles in the main region

where typical examples of total pressure (p_t) and "static" pressure (p_s) are reported, it can be seen that at least in our operating conditions, (steam flow at Mach number greater than unity), pressure and temperature inside the vapour region are strongly variable both axially and radially. Inside the vapour region two sub-regions can be roughly distinguished. In the first, near the nozzle exit, the vapour expands and reaches sub-atmospheric pressure; in this sub-region both total pressure and temperature decrease, while the diameter of the vapour jet increases. It can be said that the first sub-region is characterized by a vapour behaviour practically not dependent by pool subcooling. In the second sub-region the effect of pool temperature on the parameters involved in the flow process is quite evident. In this sub-region it can also be observed that, where the total pressure increases, the temperature falls at a rate depending by pool temperature. The physical meaning of the measurements performed in the vapour region must be examined thouroughly, taking into account the effect of the liquid entrainment and the probably onset of "spontaneous" or measurement devices induced shock waves.

5. CONCLUDING REMARKS

The comparison between the experimental results, concerning the mixing and the main regions, and the predicted ones, shows that the adopted model yields a reasonable description of the phenomena involved in the liquid region surrounding a condensing vapour jet.

The model describes with enough accuracy both the axial temperature profiles in the main region and the penetration lengths of vapour jets.

Temperature and velocity profiles in both mixing and main regions are correctly predicted in terms of trend and as order of magnitude; by the way, it is useful to remember that, especially in the mixing region, the temperature and velocity measurements are difficult to perform and some uncertainty on experimental data is to be expected.

However, it can be said that the model tends to underestimate the width of the flow. Additional study is needed to determine the geometry of the condensing vapour jet; a more complete model capable of treating near-injector phenomena as well as the two-phase mixture which is observed at the end of the vapour region, is believed to be necessary.

Inside the vapour jet further investigations are needed, both theoretical and experimental, to understand the observed pressure and temperature behaviour; shock phenomena can be taken into account and various flow conditions must be analyzed.

The physics of delayed condensation, droplet nucleation and droplet entrainment will be studied and the effect of vapour conditions on jet shape and interface phenomena shall be thoroughly examined.

ACKNOWLEDGEMENTS

The research reported in this paper was sponsored by the Italian Ministero della Pubblica Istruzione under contract for 1982.

NOMENCLATURE

a	constant, $a^3 = 2c^2$, $a = 0.09$
b	half width of jet
b_1	half width of jet at the position $x = l$
c	constant
c_p	specific heat of water
d	nozzle inner diameter
Δh_v	specific latent heat of vaporization
l	length of the vapour cone
M	molar mass of the water
p_s	static pressure
p_t	total pressure
p_v	pressure of steam
R	universal gas constant
t	time
T_v	thermodynamic temperature of steam
u	velocity along x
u_1	velocity at the interface between steam and water
u_o	averaged steam velocity at the nozzle exit
u_m	velocity at jet axis
v	velocity along y
x	coordinate of flow direction
y	coordinate perpendicular to flow direction
y_1	internal boundary of mixing zone
y_2	external boundary of mixing zone
Z	$= u/u_1$
η	$= (y - y_2) / (y_1 - y_2)$
ϑ	temperature
ϑ_i	temperature at the interface between steam and water
ϑ_m	temperature at jet axis
ϑ_p	pool temperature
ϑ_s	saturation temperature
ϱ_v	density of steam
ϱ_1	density of water

σ condensation coefficient

φ_m steam mass flux density in the discharge tube

φ_{h1} heat flux density at the interface between steam and water

$\emptyset_o = \psi_1 - \psi_2$

$\psi, \psi_1, \psi_2 = y / ax, = y_1 / ax, = y_2 / ax$

REFERENCES

1. Linehan, J.H., Grolmes, M.A., "Condensation of a High Velocity Vapor on a Subcooled Liquid Jet in Stratified Flow", Proc. 4th Int. Heat Transfer Conf., Paris, vol. 6, Cs 2.6, 1970.
2. Kerney, P.J., Faeth, G.M., Olson, D.R., "Penetration Characteristics of a Submerged Steam Jet", A I Ch E J., Vol. 18, No. 3, pp. 548-553, May 1972.
3. Weimer, J.C., Faeth, G.M., Olson, D.R., "Penetration of Vapor Jets Submerged in Subcooled Liquids", A I Ch E J., Vol. 19, No. 3, pp. 552-558, May 1973.
4. Kudo, A., "Basic Study on Vapor Suppression", Proc. 5th Int. Heat Transfer Conf., Tokyo, Vol. 3, pp. 221-225, 1974.
5. Chan, C. K., "Dynamical Pressure Pulse in Steam Jet Condensation", Proc. 5th Int. Heat Transfer Conf., Tokyo, Vol. 3, pp. 395-399, 1974.
6. Joung, R.J., Yang, K.T., Novotny, J.L., "Vapor-Liquid Interaction in High Velocity Vapor Jet Condensing in a Coaxial Water Flow", Proc. 5th Int. Heat Transfer Conf., Tokyo, Vol. 3, pp. 226-230, 1974.
7. Cumo, M., Farello, G.E., Ferrari, G., "Heat Transfer in Condensing Jets of Steam in Water (Pressure Suppressure Systems)", CNEN-RT/IN (77) 8, April 1977.
8. Aoki, S., Inoue, A., Kozawa, Y., Akimoto, H., "Direct Contact Condensation of Flowing Steam Onto Injected Water", Proc. 6th Int. Heat Transfer Conf., Toronto, Vol. 5, pp. 107-112, 1978.
9. Ysachenko, V.P., Solodov, A.P., Sennikov, V.V., "Heat Transfer with Condensation of a Steam Jet in a Liquid Flow", Teploenergetika, Vol. 26, pp. 20-23, 1979.
10. Del Tin, G., Giannitti, C., Lavagno, E., Malandrone, M., "Studio Sperimentale della Condensazione di Getti di Vapore in Acqua", Proc. 36th Congresso Nazionale ATI, Viareggio, Vol. 2, pp. 1065-1080, 1981.
11. Baiakleyevskiy, Yu.I., Chekhovich, V. Yu., "Condensation of a Submerged Steam Jet", Heat Transfer-Soviet Research, Vol. 13, No. 3, pp. 65-80, 1981.
12. Del Tin, G., Lavagno, E., Malandrone, M., "Pressure and Temperature Measurements in a Vapour Condensing Jet", Proc. 7th Int. Heat Transfer Conf., München, Vol. 6, pp. 159-164, 1982.
13. Chan, C.K., Lee, C.B.K., "A Regime Map for Direct Contact Condensation", Int. J. Multiphase Flow, Vol. 8, No. 1, pp. 11-20, 1982.

14. Chen, L.D., Faeth, G.M., "Condensation of Submerged Vapor Jets in Subcooled Liquids", J. Heat Transfer, Vol. 104, pp. 774-780, Nov. 1982.

15. Petrangeli, G., "Study on Safety Criteria and Systems for Next PWRs", ENEA RT/DISP(82)1.

16. Abramovich, G.N., "The Theory of Turbulent Jets", M.I.T. Press, Cambridge, Massachusetts, 1963.

17. Collier, J.G., Convective Boiling and Condensation , McGraw, Hill Book Company, 1972.

Multi-Phase Flow and Heat Transfer III. Part A: Fundamentals
edited by T.N. Veziroğlu and A.E. Bergles
Elsevier Science Publishers B.V., Amsterdam, 1984 — Printed in The Netherlands

AN ANALYSIS OF DENSITY-WAVE OSCILLATIONS IN VENTILATED BWR FUEL ROD BUNDLES

R. Taleyarkhan, M. Podowski, R. T. Lahey, Jr.
Department of Nuclear Engineering
Rensselaer Polytechnic Institute
Troy, New York 12181, U.S.A.

ABSTRACT

A mathematical model for the linear stability analysis of density-wave oscillations in systems of ventilated boiling channels [1] is discussed in this paper. This model allows for phasic slip, turbulence effects, arbitrary non-uniform axial power distributions, distributed local losses, heated wall dynamics, channel-to-channel radial power skews, and discrete or continuous lateral ventilation between the parallel channels. The model developed was applied to study the stability characteristics of ASEA/ATOM's new SVEA-type fuel [2]. The study revealed that ventilation holes can have a significant impact on stability margins.

1. INTRODUCTION

Boiling flow instabilities are an important phenomenon that must be accounted for in the design and analysis of many components in chemical process and energy production equipment. Instabilities are of a particular concern in the design and operation of Boiling Water Nuclear Reactors (BWRs). This is because sufficiently large flow and power excursions from the steady state may cause undesirable flow-induced vibration of internal components, operational problems, and degrade thermal margins. Of the various possible BWR instability mechanisims, the most important is the one due to nuclear coupled density-wave oscillations.

Comprehensive reviews of the current status of the analytical and experimental understability of density-wave oscillations have been conducted by Lahey and Drew [3], Boure, et al. [4], and Yadigaroglu [5]. These reviews indicate that Density-wave instabilities have been studied quite extensively. This instability mechanism is caused by the lag associated with movement of density-waves, which travel at the so-called kinematic wave speed. Standard feedback control techniques [6] can be used to analyze density-wave instabilities.

Most previous experimental and theoretical studies of density-wave oscillations have been conducted for one-dimensional flows, in single or parallel channels. Several experimental studies on density-wave oscillations in a system of heated channels, ventilated at discrete axial locations, have been reported by Veziroglu [7], Kakac [8], and Nakanishi [9]. However, no previous analytical model, to quantify ventilated channel-instability mechanisims has been published.

2. DISCUSSION

The purpose of this paper is to present an analytical model for the analysis of density-wave oscillations in ventilated heated channels, having discrete or continuous interconnections. This model provides a convenient tool for the stability analysis of advanced-type BWR fuels, such as ASEA-ATOM's recently introduced SVEA BWR fuel rod bundle [2].

Needless to say, there are no general solution techniques for the analysis of a set of non-linear differential equations. Solutions can be obtained in either the time domain, or if linearized, in the frequency domain. It is quite expensive to perform a systematic study of density-wave oscillations by direct numerical integration of the non-linear conservation equations. Moreover, it is often difficult to determine the instability threshold because of numerical instabilities, and truncation errors and the spatial smearing of the kinematic wave fronts. Many of these difficulities can be overcome by taking small perturbations about the steady-state, and determining stability thresholds in the frequency domain. In this case the system of equations are linearized and laplace-transformed. By doing so we can directly investigate properties of the system response to a class of initial perturbations, instead of examining each individual transient separately. Hence, this approach is chosen for the study.

Due to the presence of lateral ventillation between the parallel channels as well as the existance of non-uniform parameters (e.g., heat fluxes, heater geometries, etc.), a nodal approach was used, to arrive at the characteristic equation of the system. Location of the roots of the characteristic equation in the complex plane determines the systems stability behavior. Any root with a positive real part implies that particular operating state is linearly unstable. For convenience, we have chosen a modified Nyquist criterion [1, 6] for evaluating the position of these roots.

3. ANALYTICAL MODELING

The model described in this paper was set up to study cross-coupled density-wave oscillations between parallel channels. Typical geometries are shown in Figure-1. Note that any channel in the system can be interconnected with at most two other channels in the system. This allows an analysis of the advanced SVEA-type BWR fuel bundle shown in Figure-2. This fuel rod bundle has been partitioned into four 4 x 4 minibundles, which are separated by a water gap. These minibundles can communicate via gaps at discrete axial locations. Each minibundle communicates with two other minibundles; the similarity between the four channel case in Figure-1 and the SVEA minibundle in obvious.

The sign convention followed was that any quantity transported between channel-i, and the neighboring channels, i-1 and i+1, was positive when the flow was from channel-i to i+1 (or i-1 to i), and negative from i+1 to i (or i to i-1). Details of the mathematical model, heater wall and boiling-boundary dynamics, matrix reduction scheme, comparisons with experimental data; code verification results and parametric studies have been reported elsewhere [1, 10, 11, 12]. The purpose of this paper is to highlight the practical application of the model developed. However, for the sake of completeness, the essential aspects of the mathematical model are presented herein.

Analytical modelling of the cross-coupled system of equations is based on a subchannel analysis approach [1, 13, 14], where the flow field is arbitrarily divided into several subchannels communicating via gaps. The subchannel equations used include the transient conservation equations of mass, energy, axial momentum and transverse momentum. These account for lateral transport of

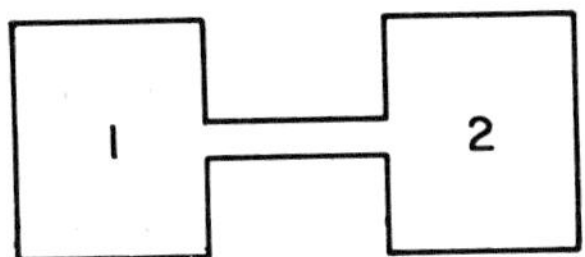

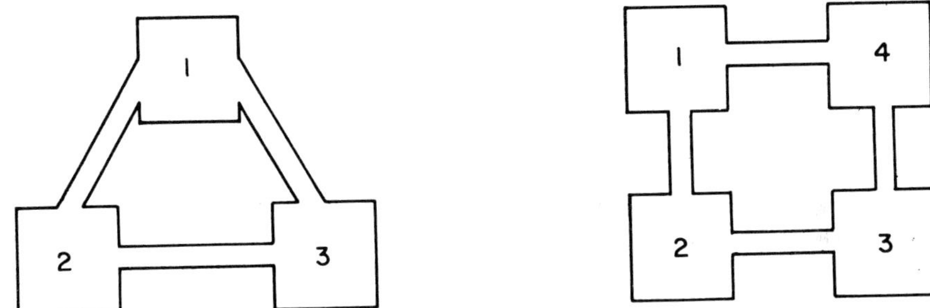

Figure 1. Schematic Diagram (top-view) for Two, Three and Four Ventilated Channels.

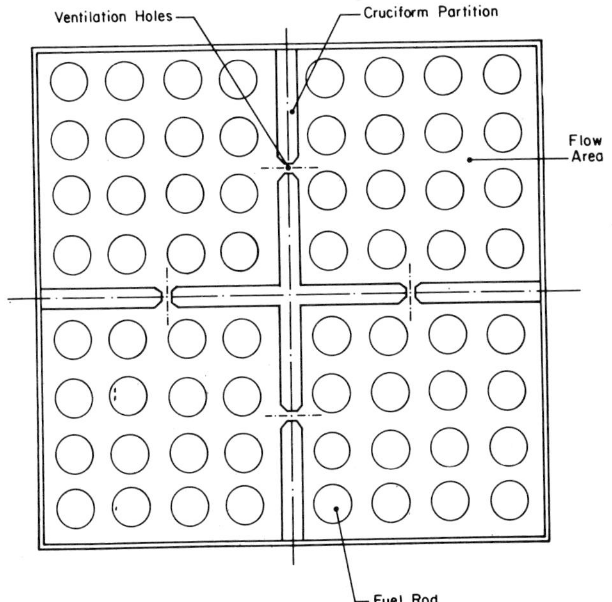

Figure 2. Schematic Representation of SVEA BWR rod bundle cross section.

mass, energy and momentum, as well as the turbulence exchanges. They are supplemented by a one-dimensional (in the radial direction) heat conduction equation for the heater, constitutive equations for heat convection, and a drift-flux model for phasic slip. The commonly used assumptions of subchannel analysis were retained [13, 14] The system pressure was taken as constant, and large in comparison to the axial drop. finally, subcooled boiling was neglected.

3.1 Discretization and Spatial Integration

The flowfield to be discretized is shown in Figure-3, for a representative two channel case. In figure 3, λ is the axial position where bulk boiling begins. Arbitrary discretization gives rise to axial nodels having moving boundaries. Possible nodal configurations are shown in Figure-4, and are differentiated from one another by use of Kronecker delta functions, defined as:

$$\delta_{m1,1} \stackrel{\Delta}{=} \begin{cases} 1, & m1=1 \quad \text{When the node has no moving boundaries within the node (Figures-4a \& 4c-4e)} \\ 0, & \text{otherwise} \end{cases}$$

$$\delta_{m2,1} \stackrel{\Delta}{=} \begin{cases} 1, & m2=1 \quad \text{When the node has fixed boundaries with a moving boundary within it (Figure-4b)} \\ 0, & \text{otherwise} \end{cases}$$

$$\delta_{m3,1} \stackrel{\Delta}{=} \begin{cases} 1, & m3=1 \quad \text{When the node has its lower boundary fixed, but the upper boundary is a function of time (Figure-4d)} \\ 0, & \text{otherwise} \end{cases}$$

$$\delta_{m4,1} \stackrel{\Delta}{=} \begin{cases} 1, & m4=1 \quad \text{When the node has its upper boundary fixed, but the lower boundary is a function of time (Figure-4e)} \\ 0, & \text{otherwise} \end{cases}$$

Note that the case shown in Figure-4c can be treated as a superposition of the cases shown in Figures-4d and 4e.

The subchannel conservation equations can now be integrated over the arbitrary node shown in Figure-5. Details of the derivation are presented elsewhere [1, 10, 12]. The final form of the nodal conservation equations are given by,

<u>Nodal Mass Conservation</u>

$$w_{i_{z+\Delta z_k}} + [\Delta\zeta_{1_i} w'_{i_z} + \Delta\zeta_{2_i} w'_{i_{z+\Delta z_k}} + \Delta\zeta_{1_i} w'_{m_{i_z}} + \Delta\zeta_{2_i} w'_{m_{i_{z+\Delta z_k}}}]$$

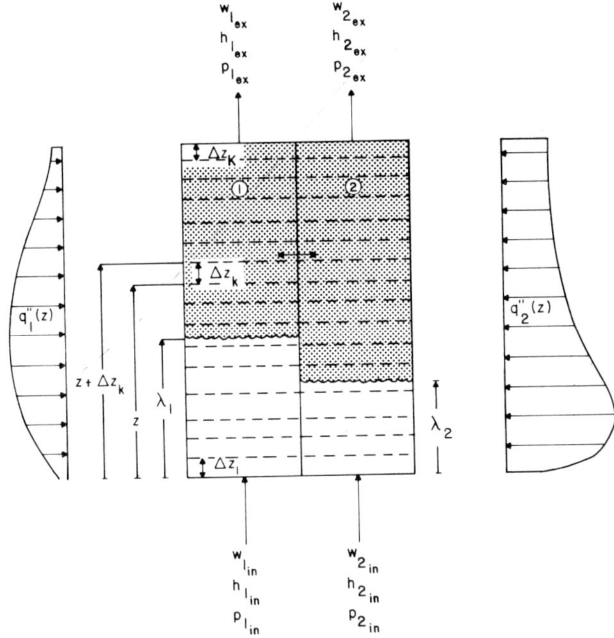

Figure 3. Nodalization of a Ventilated Two-Channel System.

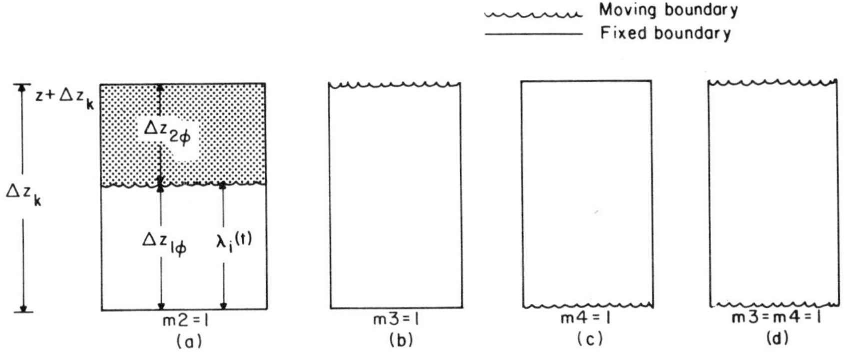

Figure 4. Possible Configuration for Fixed and Moving Boundary Nodes.

$$-w_{i_z} - [\Delta\zeta_{1_{i-1}} w'_{i-1_z} + \Delta\zeta_{2_{i-1}} w'_{i-1_{z+\Delta z_k}} + \Delta\zeta_{1_{i-1}} w'_{m_{i-1_z}} + \Delta\zeta_{2_{i-1}} w'_{m_{i-1_{z+\Delta z_k}}}]$$

$$+ \delta_{m1,1} \Delta z_k A_{x-s_i} \frac{d}{dt} \langle\bar{\rho}_i\rangle + \delta_{m2,1} \left\{ A_{x-s_i} \frac{d}{dt} [\rho_f \Delta z_{1\phi_i} + \langle\bar{\rho}_{2_i}\rangle \Delta z_{2\phi_i}] \right\}$$

$$+ \delta_{m3,1} [\langle\bar{\rho}_i\rangle - \langle\rho_i(z+\Delta z_k)\rangle] A_{x-s_i} \left.\frac{dz}{dt}\right|_{z+\Delta z_k} - \delta_{m4,1}[\langle\bar{\rho}_i\rangle - \langle\rho_i(z)\rangle] A_{x-s_i} \left.\frac{dz}{dt}\right|_z = 0 \qquad (1)$$

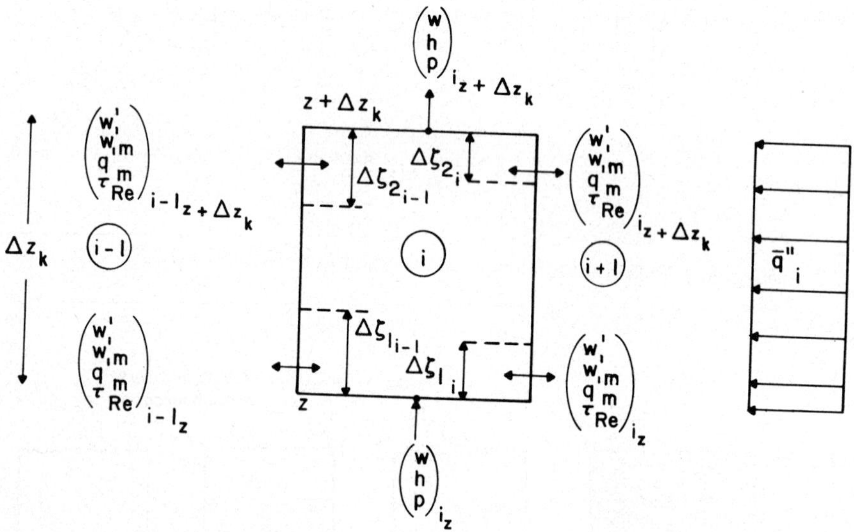

Figure 5. Typical Node Between Axial Locations z abd $z+\Delta z_k$, for Channel-i Communicating with Two Neighboring Channels.

Nodal Energy Conservation

$$(wh)_{i_{z+\Delta z_k}} + \left\{ \Delta\zeta_{1_i} w'_{i_z} [k_i^* h_{i_z} + (1-k_i^*) h_{i+1_z}] + \Delta\zeta_{2_i} w'_{i_{z+\Delta z_k}} [k_i^* h_{i_{z+\Delta z_k}} \right.$$

$$+(1-k_i^*)h_{i+1_{z+\Delta z_k}}] + \Delta\zeta_{1_i} q'_{m_{i_z}} + \Delta\zeta_{2_i} q'_{m_{i_{z+\Delta z_k}}} \Big\} - (wh)_{iz}$$

$$- \Big\{ \Delta\zeta_{1_{i-1}} w'_{i-1_z} [k_{i-1}^* h_{i-1_z} + (1-k_{i-1}^*)h_{i_z}]$$

$$+ \Delta\zeta_{2_{i-1}} w'_{i-1_{z+\Delta z_k}} [k_{i-1}^* h_{i-1_{z+\Delta z_k}} + (1-k_{i-1}^*)h_{i_z}] + \Delta\zeta_{1_{i-1}} q'_{m_{i-1_z}}$$

$$+ \Delta\zeta_{2_{i-1}} q'_{m_{i-1_{z+\Delta z_k}}} \Big\} - \bar{q}''_i P_{H_i} \Delta z_k + \delta_{m1,1} \Delta z_k A_{x-s_i} \frac{d}{dt} (\overline{<\rho_i>h}_{cm_i})$$

$$+ \delta_{m2,1} A_{x-s_i} [\frac{d}{dt}(\rho_f \bar{h}_{1_i} \Delta z_{1\phi_i} + \overline{<\rho_{2_i}>h}_{cm_i} \Delta z_{2\phi_i})] + \delta_{m3,1} [\overline{<\rho_i>h}_{cm_i}$$

$$- <\rho_i(z+\Delta z_k)> h_{cm_i}(z+\Delta z_k)] A_{x-s_i} \frac{dz}{dt}\Big|_{z+\Delta z_k} - \delta_{m4,1} [\overline{<\rho_i>h}_{cm_i}$$

$$- <\rho_i(z)> h_{cm_i}(z)] \quad A_{x-s_i} \frac{dz}{dt}\Big|_z = 0 \qquad (2)$$

where,

$\overline{<\rho>h}_{cm} = \rho_\ell \bar{h}_\ell$, for single-phase liquid flow

<u>Nodal Axial Momentum Conservation</u>

$$(wU_p)_{i_{z+\Delta z_k}} + [\{\Delta\zeta_{1_i} w'_{i_z} [k_i^* U_{p_{i_z}} + (1-k_i^*) U_{p_{i+1_z}}] + \Delta\zeta_{2_i} w'_{i_{z+\Delta z_k}}$$

$$[k_i^* U_{p_{i_{z+\Delta z_k}}} + (1-k_i^*) U_{p_{i+1_{z+\Delta z_k}}}]\} + g_c S_{g_i} (\Delta\zeta_{1_i} \tau_{Re_{i_z}} + \Delta\zeta_{2_i} \tau_{Re_{i_{z+\Delta z_k}}})]$$

$$-(wU_p)_{i_z} - [\{\Delta\zeta_{1_{i-1}} w'_{i-1_z} [k_{i-1}^* U_{p_{i-1_z}} + (1-k_{i-1}^*) U_{p_{i_z}}] + \Delta\zeta_{2_{i-1}} w'_{i-1_{z+\Delta z_k}}$$

$$[k_{i-1}^* U_{p_{i-1_{z+\Delta z_k}}} + (1-k_{i-1}^*) U_{p_{i_{z+\Delta z_k}}}]\} + g_c S_{g_{i-1}} (\Delta\zeta_{1_{i-1}} \tau_{Re_{i-1_z}}$$

$$+ \Delta\zeta_{2_{i-1}} \tau_{Re_{i-1}}{}_{z+\Delta z_k})] + \delta_{m1,1}\, \Delta z_k \frac{d}{dt}(\bar{w}_i) + \delta_{m2,1} \frac{d}{dt}(\Delta z_{1\phi_i} \bar{w}_{1_i} + \Delta z_{2\phi_i} \bar{w}_{2_i})$$

$$+ \delta_{m3,1} (\bar{w}_i - w_{i_{z+\Delta z_k}}) \frac{dz}{dt}\Big|_{z+\Delta z_k} - \delta_{m4,1} (\bar{w}_i - w_{i_z}) \frac{dz}{dt}\Big|_z$$

$$= g_c A_{x-s_i} (p_{i_z} - p_{i_{z+\Delta z_k}}) - \delta_{m1,1} (g \operatorname{Sin}\theta_i A_{x-s_i} \Delta z_k \langle\bar{\rho}_i\rangle + \frac{f_i \Delta z_k}{2\rho_f A_{x-s} D_{H_i}} \overline{w_i^2 \phi_{\ell o_i}^2})$$

$$- \delta_{m2,1} \Big\{ g \operatorname{Sin}\theta_i\, A_{x-s_i} (\Delta z_{1\phi_i} \rho_f + \Delta z_{2\phi_i} \langle\overline{\rho_{2_i}}\rangle) + \frac{f_i}{2\rho_f A_{x-s_i} D_{H_i}} [\Delta z_{1\phi_i} (\bar{w}_i^{\,2})_1$$

$$+ \Delta z_{2\phi_i} (\overline{w_i^2 \phi_{\ell_{o_i}}^2})_2] \Big\} - \frac{K_{sp_{i_z}}}{2\rho_f A_{x-s_i}} w_{i_z}^2 \, \Phi_{i_z} - \frac{K_{sp_{i_{z+\Delta z_k}}}}{2\rho_f A_{x-s_i}} w_{i_{z+\Delta z_k}}^2 \, \Phi_{i_{z+\Delta z_k}} \tag{3}$$

Nodal Transverse Momentum Conservation

$$\frac{d}{dt} [w_i' \ell]_{z+\Delta z_k} + \frac{\ell_I}{\Delta z_k} \Big\{ w'_{i_{z+\Delta z_k}} [k_i^* U_{p_{i_{z+\Delta z_k}}} + (1-k_i^*) U_{p_{i+1_{z+\Delta z_k}}}]$$

$$- w'_{i_z} [k_i^* U_{p_{i_z}} + (1-k_i^*) U_{p_{i+1_z}}] \Big\}$$

$$= \frac{-K_{G_i}}{2 S_{g_i} \rho_f} \Big\{ |w_i'|\, w_i' \,[k_i^* \Phi_i + (1-k_i^*) \Phi_{i+1}] \Big\}_{z+\Delta z_k} + S_{g_i} g_c (p_i - p_{i+1})_{z+\Delta z_k} \tag{4}$$

where, ℓ and ℓ_I have units of length. For continuously ventilated channels ℓ_I can be taken equal to ℓ. However, we have introduced the notation ℓ_I, to distinguish between the length over which momentum storage in the transverse direction occurs (ℓ), and that length over which cross-flow inertia effects are significant (ℓ_I). The remaining notation is explained in the section on nomenclature. It should be noted that the bars over the quantities, $\langle\rho\rangle$, w, h, h_{cm} and $\phi^2_{\ell o}$ represent nodal-averaged quantities. Assuming uniform heat flux over any node length, and small lateral transport of mass, energy and momentum, the variation of enthalpy and mass flow rate over the nodal length can be assumed to be linear. Hence, the above nodal-averaged parameters are evaluated at the nodal linear averaged enthalpy and mass flow rate.

Equations (1) to (4) represent the nodalized system of equations for a general node with moving boundaries. For the case of fixed boundary nodes, the nodalized system of equations can be obtained by setting m1=1 and m2=m3=m4≠1.

Equations (1) to (4) can now be linearized and Laplace-transformed. During this process the terms $\delta\hat{q}''$ and $\delta\hat{\lambda}$ arise. These represent the dynamics of the heater wall and the boiling boundary, respectively. The algebraic details involved in the linearization, Laplace-transformation and modelling of $\delta\hat{q}''$ and $\delta\hat{\lambda}$ are presented elsewhere [1, 10, 12]. The resulting set of algebraic equations can be written as,

$$\Big\{\sum_{j\varepsilon D_1}[C^{(\ell)}_{w_{j_z}}(s)\,\delta\hat{w}_{j_z} + C^{(\ell)}_{h_{j_z}}(s)\,\delta\hat{h}_{j_z} + C^{(\ell)}_{w_{j_{z\Delta+z_k}}}(s)\,\delta\hat{w}_{j_{z+\Delta z_k}} + C^{(\ell)}_{h_{j_{z+\Delta z_k}}}\,\delta\hat{h}_{j_{z+\Delta z_k}}] + \sum_{j\varepsilon D_2}[C^{(\ell)}_{w_{j_z}}(s)\,\delta\hat{w}'_{j_z} + C^{(\ell)}_{w_{j_{z+\Delta z_k}}}(s)\,\delta\hat{w}'_{j_{z+\Delta z_k}}] + \sum_{j\varepsilon D_3}[C^{(\ell)}_{p_{j_z}}(s)\,\delta\hat{p}_{j_z} + C^{(\ell)}_{p_{j_{z+\Delta z_k}}}(s)\,\delta\hat{p}_{j_{z+\Delta z_k}}] + C_z(s)\,\delta\hat{z} + C^{(\ell)}_{z+\Delta z_k}(s)\,\delta(z+\hat{\Delta} z_k)\Big\}_k = 0 \qquad (5)$$

$j = 1,2,..,N;\ \ell = 1,2,3,4;\ k = 1,2,..,K$

where N is the total number of parallel channels in the system and K is the total number of axial nodes in each channel. The set indicators D_1, D_2, D_3 are defined as:

$$D_1 \triangleq \left\{ \begin{array}{ll} i-1,i,i+1 & \text{for, } i=2,\ldots, N-1 \\ N-1,\ N,1 & \text{for, } i=N \\ N,1,2 & \text{for, } i=1 \end{array} \right\}$$

$$D_2 \triangleq \left\{ \begin{array}{ll} i-1,\ i & \text{for, } i=2,\ldots.,N \\ N,\ 1 & \text{for, } i=1 \end{array} \right\}$$

$$D_3 \triangleq \left\{ \begin{array}{ll} i,\ i+1 & \text{for, } i=1,\ldots.,N-1 \\ N,\ 1 & \text{for } i=N \end{array} \right\}$$

3.2 Matrix Foundation of Nodal Equations

The generic set of nodal equations for any node -k, given by Equation (5) can be conveniently written in matrix form as,

$$\underline{\underline{A}}_k\ \underline{\delta\hat{y}}_{k-1} + \underline{\underline{B}}_k\ \underline{\delta\hat{y}}_k = \underline{\delta\hat{z}}_k \tag{6}$$

where,

$$\underline{\delta\hat{y}} = [\underline{\delta\hat{w}},\ \underline{\delta\hat{h}},\ \underline{\delta\hat{w}}',\ \underline{\delta\hat{p}}]^T$$

and,

A_k and B_k are square 4Nx4N matrices, consisting of complex elements, corresponding to the coefficients of the sub-vectors $\underline{\delta\hat{w}}$, $\underline{\delta\hat{h}}$, $\underline{\delta\hat{w}}'$, and $\underline{\delta\hat{p}}$, at the inlet and exit of node-k. Each subvector consists of N components, for example: $\delta\hat{\underline{w}} = [\delta\hat{\underline{w}}_1,\ \delta\hat{\underline{w}}_2,\ldots.,\ \delta\hat{\underline{w}}_N]^T$. Equation [7], written for all nodes k=1,2,...,K, forms a block diagonal system as shown in Figure-6.

3.3 SYSTEM TRANSFER FUNCTION

We can notice from Figure-6 and Equation (6) that the node exit variable for node-k becomes the inlet variables for node-k+1. Taking advantage of this property, and a special matrix reduction scheme [1, 11], we can obtain the exit vector, $\underline{\delta\hat{y}}_k$, in terms of the system's inlet vector, $\underline{\delta\hat{y}}_o$, and the nodal internal heat generation rate, $\delta\hat{\underline{z}}_k$, (k=1,2,....,k). The result is given by,

$$\delta\hat{\underline{y}}_k = [\prod_{k=K}^{1}(-\underline{\underline{B}}_k^{-1}\ \underline{\underline{A}}_k)\ \delta\hat{\underline{y}}_o + \sum_{k=1}^{k}\ \{\delta_{k,K}\ \underline{\underline{B}}_k^{-1} + (1-\delta_{k,K})\ [\prod_{k'=k}^{k+1}\ (-\underline{\underline{B}}_{k'}^{-1}\ \underline{\underline{A}}_{k'})]\ \underline{\underline{B}}_k^{-1}\}\ \delta\hat{\underline{z}}_k \tag{7}$$

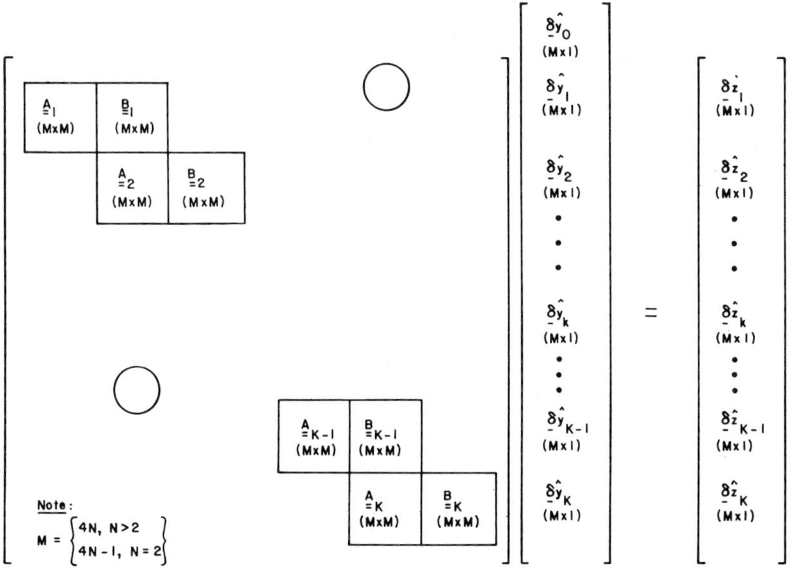

Figure 6. System of Nodalized Conservation Equations in Matrix Form for Flow in an N-Ventilated Channel System.

where,

$$\delta_{k,K} \overset{\Delta}{=} \begin{cases} 1, & k=K \\ 0, & \text{otherwise} \end{cases}$$

Out-of-core density wave oscillations can be investigated by setting $\delta\hat{\underline{w}}'_{in}$, $\delta\hat{\underline{h}}_{in}$, and $\delta\hat{\underline{z}}_k$ (k=1,....,K) to zero in Equation (7) to obtain,

$$\delta\hat{\underline{P}}_{ex} - H_1(s)\delta\hat{\underline{P}}_{in} = H_2'(s)\ \delta\hat{\underline{w}}_{in} \tag{8}$$

where, $H_1(s)$ and $H_2(s)$ are N x N square matrices. Further, assuming all channels have common upper and lower plena, and assuming incompressibility of the liquid in the lower plenum, we can write,

$$\delta\hat{P}_{i_{ex}} = \delta\hat{P}_{ex} \quad (i=1,2,...,N) \tag{9}$$

$$\delta\hat{P}_{i_{in}} = \delta\hat{P}_{in} \quad (i=1,2,...,N) \tag{10}$$

$$\delta\hat{w}_T = \sum_{i=1}^{N}\ \delta\hat{w}_{i_{in}} \quad (i=1,2,...,N) \tag{11}$$

where P_{in}, P_{ex}, and w_T are the common inlet plenum pressure, exit plenum pressure, and total mass flow rate entering the lower plenum, respectively.

Multiplying Eq. (8) by $H_2^{-1}(s)$ and summing the resultant equation we obtain,

$$\delta\hat{\Delta p}_T = G(s)\ \delta\hat{w}_T \qquad (12)$$

where ΔP_T is the plenum-to-plenum pressure drop, which is assumed to be the same for all channels, and G(s) is the characteristic equation of the system.

Equation (12) has been used to investigate parallel channel stability behavior for various operating conditions and system geometries.

The analytical model described above was encoded as a digital computer code, MAZDA-4F [1]. This code has been compared with exact analytical results [1] and good agreement was achieved. The steady-state was evaluated using a separate code, MAZDA-4S [1], which is based on classical subchannel techniques [13, 14].

4. STABILITY CHARACTERISTICS OF SVEA-BWR TYPE FUEL BUNDLES

4.1 Description of SVEA Geometry

Recently ASEA/ATOM has introduced SVEA-type BWR fuel. As can be seen in Figure-2, this fuel is intended to fit within the same channel dimensions as previous BWR fuel designs, but consists of four 4x4 mini-bundles which are ventilated at discrete axial locations. In principle, the SVEA fuel design allows for better power flattening, due to the cruciform water gap region in the interior of the bundle. The discrete ventilation holes were apparently installed to achieve adequate stability characteristics. In this paper we investigate the effectiveness of these ventilation holes on bundle stability.

The SVEA-type BWR rod bundle was treated as a system of four discretely ventilated channels. Since the exact number of ventilation paths was not known, we divided each flow channel into 52 axial nodes. The first and last nodes coincide with the channel inlet and exit. Thus ventilation paths were set up at each of the 50 interior nodes.

The full details of the SVEA bundle geometry and operating parameters have not yet been published. Nevertheless, since this fuel is intended for reload into existing BWRs, these parameters could be estimated. The key parameters used for the investigation (power, mass flux, etc) are indicated on the diagrams showing the results (Nyquist loci).

4.2 Results

As indicated earlier a modified Nyquist method was used to evaluate system stability. For the present purpose the distance (Δ) of the point where the Nyquist focus crosses the real axis (for $\omega > o$) and the origin is a measure of the system stability margin. That is, tne greater the distance (for $\Delta > o$), the greater the system stability, and vice versa for $\Delta < o$.

The first set of runs made were to test the effect of cross ventilation. This is shown in Figure-7. As can be seen, the four channel system with no power skewing (Curve-1) is more unstable than the one with power skewing

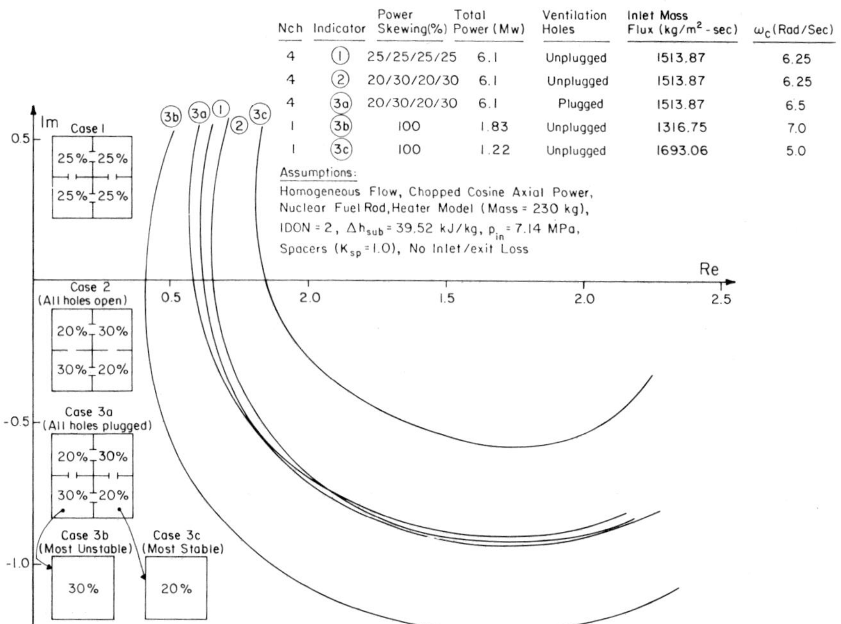

Figure 7. Effect of Ventilation on Stability of SVEA Type Rod Bundle.

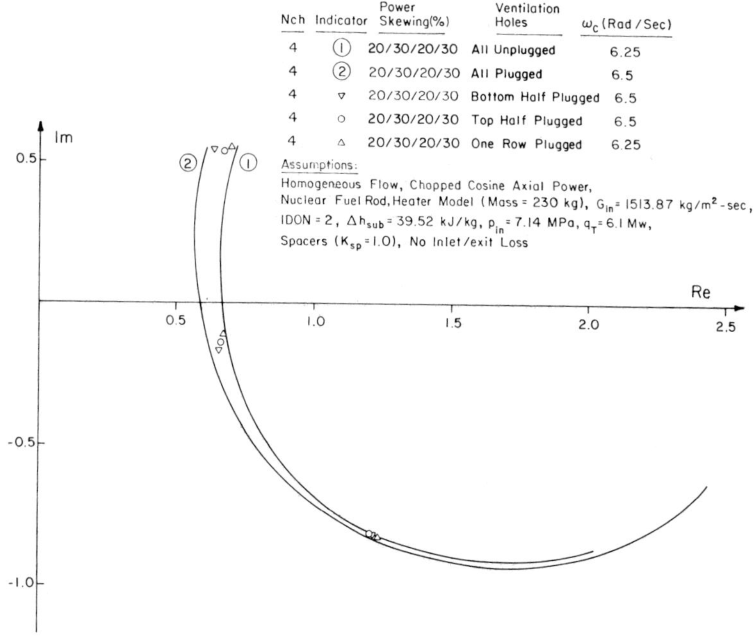

Figure 8. The Effect of Selectivity Plugging Ventilation Holes in SVEA Type Rod Bundles.

(Curve-2). Curve-3a shows how system stability would be affected if all the flow holes were plugged. As seen, plugging all the holes results in the system being more unstable than the one with no power skewing.

Curves-3b and -3c are the Nyquist loci for the least and most stable channels in the four unventilated channel system (Curve-3a) if they operated by themselves. As can be seen, the effective four unventilated channel system is more stable than the most unstable channel in the system, but less stable than the least unstable channel.

Next, the relative impact of selectively plugging up certain holes was investigated. Specifically, the effect of plugging up all the holes in the upper half, lower half and the effect of plugging up all holes in between one pair of adjacent channels was investigated. The results of this investigation are shown in Figure-8. As can be seen, for a checker-board 20/30 power skew, plugging of the holes in the lower half between all channels is the most destabilizing, followed by the effect of plugging all holes in the top half of the bundle. Nevertheless, in all cases the trend was the same. That is, reducing the number of ventilation paths leads to a more unstable system.

Since the difference between the stability margins for the various degrees of ventilation "looks" relatively small, it was decided to quantify it in terms of total bundle power. That is, we sought to answer the question, "By how much would we have to change the total bundle power to produce the observed difference between the stability margins of loci-2 and 3a in Figure-7?" As can be seen in Figure-9, lowering the bundle power in the plugged bundle by 10%, from the minimal operating power of 6.2 Mw, gives approximately the same stability margin as obtained when all holes are unplugged. Conversely, increasing the total bundle power by about 10% from the nominal operating power gives approximately the same stability margin as that due to power skewing with all holes plugged. This is a significant result, and clearly indicates the benefit of ventilation.

5. SUMMARY AND CONCLUSION

The study has outlined the mathematical modeling of density-wave oscillations in ventilated boiling channels. An analysis of the stability characteristics of SVEA-type BWR fuel has been performed, to study the effect of power skewing and ventilation. It has been shown that radial power skews, such as those used in this investigation, can have a substantial effect on a system stability. Moreover, lateral ventilation between parallel channels can also have a significant impact on the predicted stability margin. Overall, these results strongly support the need for a multidimensional stability code, such as MAZDA-4F, for the assessment of stability margins in SVEA-type fuel.

NOMENCLATURE

Latin Symbols

A_{x-s} = Cross-section flow area

D_H = Hydraulic diameter

f = Darcy-Weisbach friction factor

$\langle F \rangle$ = Cross-section average of a function, F

g_c = $32.17\ lb_m\text{-}ft/lb_f\text{-}sec^2 = 1.0\ kg\text{-}m/N\text{-}sec^2$

g = Gravitational acceleration

Δh_{sub} = $h_f - h_{in}$, inlet subcooling

h_{fg} = Latent heat of evaporation

h_{cm} = Center-of-mass enthalpy $\stackrel{\Delta}{=} \frac{1}{\langle\rho\rangle} [\rho_\ell h_\ell (1-\alpha)+\rho_g h_g \alpha]$

h = Specific two-phase enthalpy $\stackrel{\Delta}{=} [h_\ell(1-x)+h_g x]$

i = Channel number

k^* = 1.0, donor cell method used, and w'>0
0.0, donor cell method used, and w'>0
0.5, average-of-channels method used

K = Local loss coefficient

L = Length

ℓ = Transverse control volume length

M = Mass

p = Static pressure

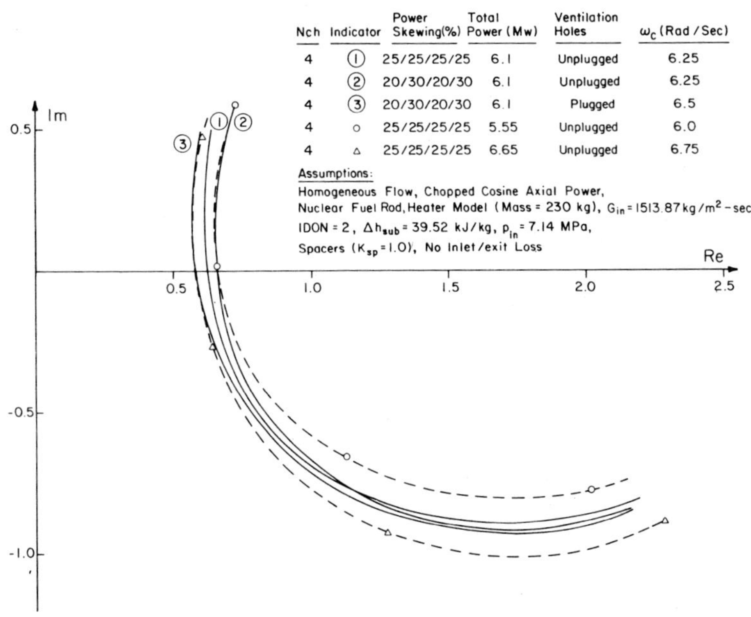

Figure 9. Determining the Effect of Ventilation Holes in Terms of Power Margin, for SVEA Typed Rod Bundles.

P_H	=	Heater perimeter
q	=	Heat rate
q'	=	Heat rate transfer unit axial length due to diverse cross flow
q''	=	Heat flux
q'''	=	Volumetric heat generation rate
q'_m	=	Heat transfer per unit length due to turbulent mixing
s	=	Laplace transform parameter
S_q	=	Gap width
t	=	Time
T	=	Temperature
ΔT	=	Temperature difference
U_p	=	$w/(\rho A_{x-s})$ = velocity of the center-of-momentum
w_m	=	Lateral mass flow rate due to turbulent mixing
w	=	Mass flow rate
w'	=	Diversion cross flow per unit axial length
x	=	Flow quality
z	=	Axial location

Greek Symbols

α	=	Void fraction
β	=	Mixing Stanton Number
Θ	=	Inclination of channels from vertical
$\gamma(t)$	=	Temporal part of internal heat generation rate
ρ_k	=	Density of phase-k
$\langle\rho\rangle$	=	Two phase density $\stackrel{\Delta}{=} \rho_f(1-\alpha) + \rho_g \alpha$
ρ'	=	Momentum density $\stackrel{\Delta}{=} [\frac{(1-x)^2}{(1-\alpha)\rho_f} + \frac{x^2}{\alpha\rho_g}]^{-1}$
μ	=	Dynamic viscosity
ε	=	Eddy diffusivity
$\phi^2_{\ell 0}$	=	Two-phase friction multiplier

ϕ = Two-Phase local loss multiplier

κ = Multiplier to assess importance of two-phase turbulence (equal-volume-exchange model)

λ = Non-boiling length (i.e., the boiling boundary)

ω = Angular frequency

Δ = Difference

τ = Stress

δ = Perturbation

δ_{ij} = $\begin{cases} 1, & \text{if } i=j \\ 0, & \text{if } i\neq j \end{cases}$ = Kronecker delta function

ζ = Length related to discrete ventilation path

Subscripts

ex = Exit

f = Saturated liquid

fg = Difference between saturated vapor and liquid properities

g = Saturated vapor

H = Heater

HT = Heat transfer

I = Inertia

in = Inlet

k = Node number

ℓ = Liquid (subcooled phase)

sat = Saturation

T = Total

V = Vapor phase

x-s = Cross-section

w = Wall

1ψ = Single-phase

2ϕ = Two-phase

Re = Reynolds

m = Turbulent

REFERENCES

[1] Taleyarkhan, R.P., "An Analysis of Density-Wave Oscillations in Ventilated Channels," NUREG/CR-2972 (1983).

[2] Nylund, O., Fredin, Bo and Johnson, Anders, "SVEA-A New BWR Fuel Assembly Concept for Improved Fuel Utilization," Proceedings of ANS Topical Meeting, Rhode Island (1981).

[3] Lahey, R.T. and Drew, D., "An Assessment of the Literature Related to LWR Instability Modes," NUREG/CR-1414 (1980).

[4] Boure, J.A., Bergles, A.E., and Tong, L.S., "Review of Two-Phase Flow Instability," Nuclear Engineering and Design, 25, 165-192 (1973).

[5] Yadigaroglu, G., "Two-Phase Flow Instabilities and Propagation Phenomenon," Von Karman Institute for Fluid Dynamics, Lecture Series (1978).

[6] Shinners, S.M., "Control System Design", John Wiley & Sons (1964).

[7] Veziroglu, T.N. and Lee, S.S., "Sustained and Transient Boiling Flow Instabilities in a Cross-Connected Parallel Channel Upflow System", AEC Oak Ridge National Laboratory Subcontract No. 2975, Final Report (1971).

[8] Kakac, S., Veziroglu, T.N. and Berkol, O., "Sustained and Transient Boiling Flow Instabilities in a Cross-Connected Four-Parallel Channel Upflow System", Proc. of the Fifth International Heat Transfer Conference, Tokyo (1974).

[9] Nakanishi, S., Ozawa, M. and Ishigai, S., "The Modes of Flow Oscillation in Multi-Channel Two-Phase Flow Systems", NATO Advanced Research Workshop: "Advances in Two-Phase Flows and Heat Transfer", Spitzingsee, F.R.G. (1982).

[10] Taleyarkhan, R.P., Podowski, M. and Lahey, R.T., Jr., "Ventilated Channel Instability Analysis", ASME Preprint 82-EA/HT-S, WAM, Phoenix, Arizona (1982).

[11] Taleyarkhan, R.P., Podowski, M. and Lahey, R.T., Jr., "The Development of a Nodal Method for the Stability Analysis of Ventilated Boiling Channels", Second International Topical Meeting on Nuclear Reactor Thermal-Hydraulics, Santa Barbara, California (1983).

[12] Taleyarkhan, R.P., Podowski, M. and Lahey, R.T., Jr., "The Effect of Heater Wall Dynamics on Density-Wave Oscillations in a Boiling Channel", Proceedings of the 7th International Heat Transfer Conference, Munich (1982).

[13] Rowe, D.S., et al., "COBRA-IV-I": An Interim Version of COBRA for Thermal-Hydraulic Analysis of Rod Bundle Nuclear Fuel Elements and Cores", BNW4-1962, PNW4 (1976).

[14] Lahey, R.T., Jr. and Moody, F.J., "The Thermal-Hydraulics of a Boiling Water Nuclear Reactor", ANS Monograph (1977).

Multi-Phase Flow and Heat Transfer III. Part A: Fundamentals
edited by T.N. Veziroğlu and A.E. Bergles
Elsevier Science Publishers B.V., Amsterdam, 1984 — Printed in The Netherlands

STABILITY OF A TWO-PHASE NATURAL CIRCULATION LOOP WITH FIGURE-OF-EIGHT SYMMETRY

K. H. Ardron, V. S. Krishnan
Atomic Energy of Canada Limited
Whiteshell Nuclear Research Establishment
Pinawa, Manitoba ROE 1LO, Canada

ABSTRACT

A series of tests has been performed in the Whiteshell Nuclear Research Establishment RD-12 pressurized water loop to study the stability of two-phase natural circulation. RD-12 has the symmetrical figure-of-eight geometry characteristic of the primary circuit of a CANDU* nuclear reactor. Tests show that the natural circulation flow can become oscillatory when the system void fraction exceeds a threshold value.

A simple linearised model is developed, using density-wave propagation theory to explain the observed oscillations. The theory suggests that net vapour condensation in the hot-legs is needed for stable two-phase natural circulation in RD-12. This corresponds to the experimental situation.

The model predicts that the oscillations can be characterised using three non-dimensional groups representing the heat input, the heater inlet subcooling, and the condensation rate in the hot-legs; a simple equation is obtained expressing the stability threshold in terms of these variables. The oscillation period is predicted to be equal to the vapour transit time in the hot-leg. Reasonable quantitative agreement between experiment and theory is found.

1. INTRODUCTION

For some postulated upsets in CANDU nuclear reactors, core heat removal is effected by natural circulation (thermosiphoning) of the primary coolant. In some of these upset situations, voidage can appear in the primary heat-transport system (PHTS) because of either coolant shrinkage, or leakage of coolant into the containment [1].

To provide information on the nature of two-phase thermosiphoning in a CANDU-type loop geometry, tests have been carried out in the RD-12 pressurized water loop at WNRE [2]. The RD-12 facility consists of two active boilers, two pumps, and two heated sections arranged in the symmetrical figure-of-eight configuration characteristic of the CANDU PHTS. The objective of the thermosiphoning tests was to investigate flow stability as voidage was gradually introduced into the loop, starting from the fully liquid condition. This paper briefly discusses the experimentally observed

---

* CANada Deuterium Uranium

AECL-7958

loop stability characteristics, and describes in detail the development of a simple linear stability model, based on density-wave propagation, to explain the observed oscillations.

2. EXPERIMENTAL

2.1 Experimental Facility And Procedure

Figure 1 is a schematic diagram of the RD-12 facility, described in detail in reference [2]. The loop, which has a maximum operating pressure of 10 MPa, contains two 4-metre long horizontal heated sections, each containing a seven-element assembly of electrically heated tubes. The loop has two active U-tube boilers, two pumps and four headers, all connected in a symmetrical figure-of-eight configuration. The elevation difference between the heated sections and the top of the boiler U-tubes is 7.7 m, and the loop mass inventory when liquid-filled, is $\sim$265 kg.

Loop instrumentation consists of 3-beam gamma-ray densitometers for fluid density measurements, differential and gauge pressure transducers, thermocouples and resistance temperature detectors. Flow measurements are made with turbine meters or orifice meters.

The experimental procedure for the present test series was to first establish stable, single-phase, natural-circulation (thermosiphoning) cooling in the loop for a given heater power and secondary-side (boiler) pressure. Voidage was then introduced by controlled draining of water from the outlet of one of the heated sections, with the secondary-side pressure and heater power held constant. In most tests, divergent oscillations in the thermosiphoning flow commenced when a threshold loop mass inventory was reached. Tests were usually terminated when the oscillation amplitude was sufficient to cause periodic dryout of the heater rods.

2.2 Experimental Results

A detailed description of the test results is given in reference [3]. Trends in data are discussed briefly below.

In all tests, it was found that stable thermosiphoning could be achieved for single-phase conditions. However, as voids were introduced into the loop by draining, a threshold condition was reached after which the loop flow rate began to show divergent oscillations. Figure 2 shows the volumetric flow-rate in the two half-loops for a typical test. (Flow measurements were made using turbine meters located at the pump outlets, where subcooled liquid conditions exist). Oscillations are seen to commence at $\sim$1100 s after about 4.7% of the mass inventory has been drained. The oscillations are characterised by a dominant large-amplitude, low-frequency mode (the period in the test shown is 66 s). Flow oscillations in the two half-loops are out-of-phase by almost exactly 180°.

Figure 3 shows the period of the dominant flow oscillations, as measured in the various tests, for different values of heater input power and secondary-side pressure. Periods vary between 50 s and 220 s over the experimental range.

Figures 4 and 5 show the primary circuit pressure and the steady thermosiphoning flow rate, respectively, as a function of heater power at the

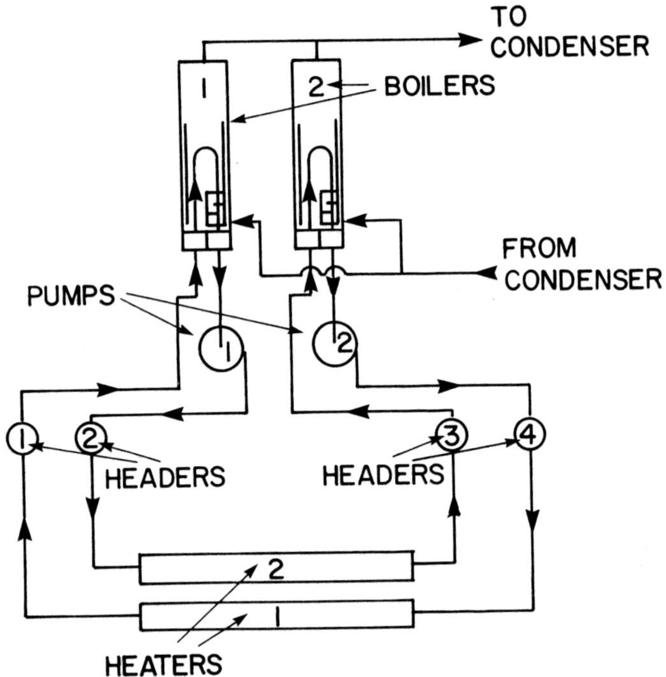

Figure 1: Schematic diagram of RD-12 primary circuit

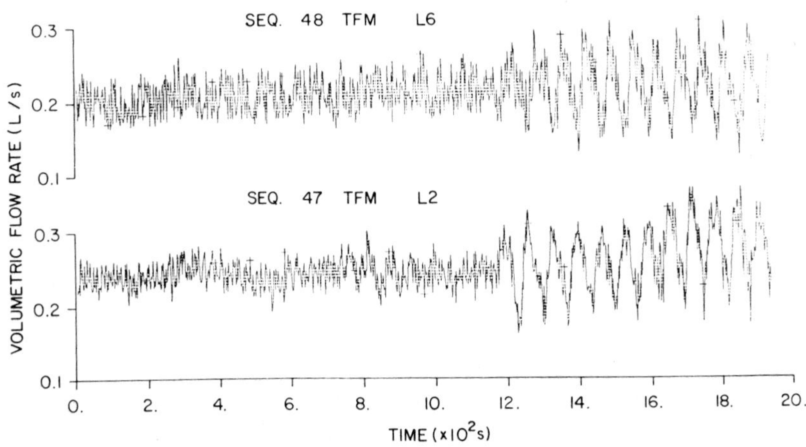

Figure 2: Flowrate transients in test T8124 (q_1 = 47 kW/test section; p_{sec} = 0.6 MPa)

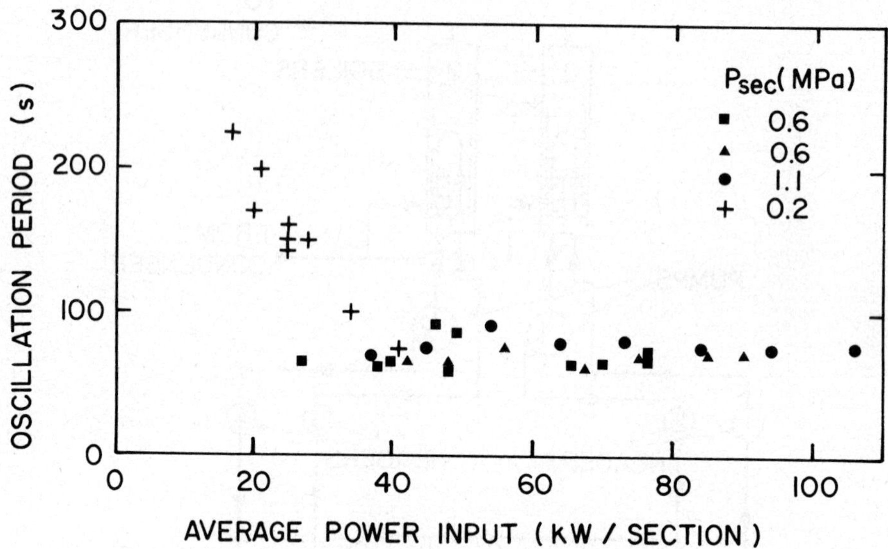

Figure 3: Oscillation periods versus heat power

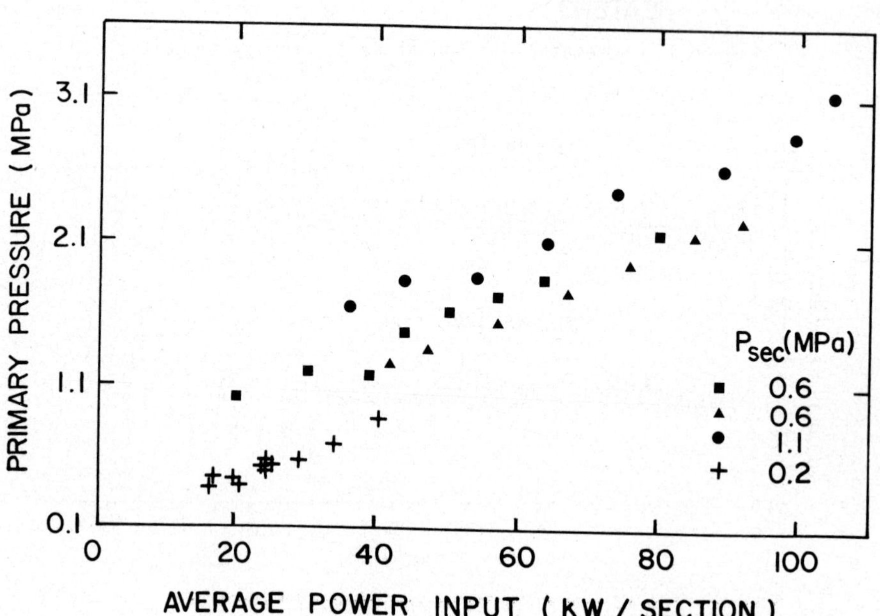

Figure 4: Primary circuit pressure versus heater power at stability threshold

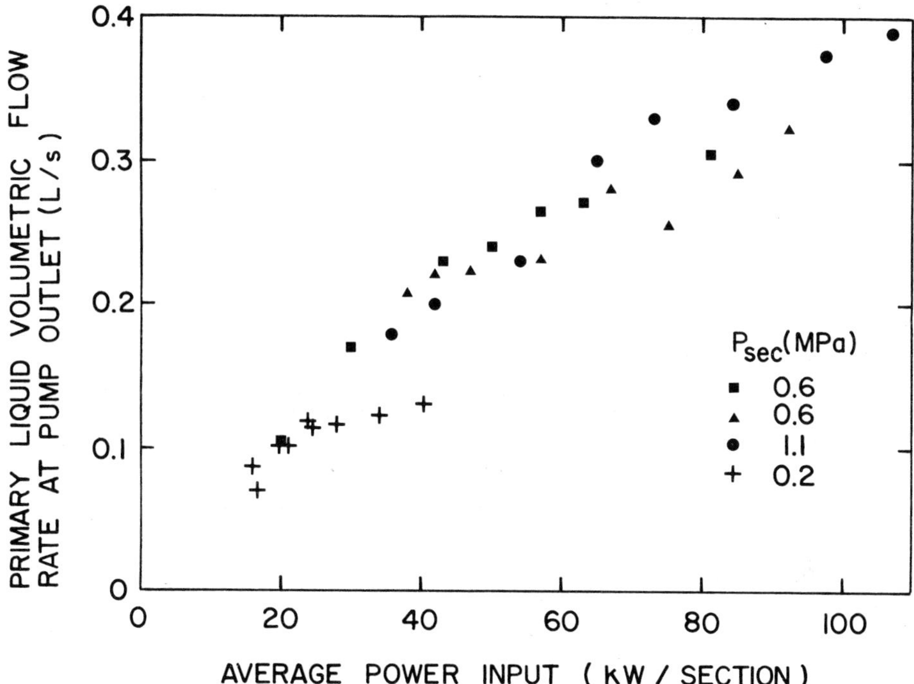

Figure 5: Flow rate versus heater power at stability threshold

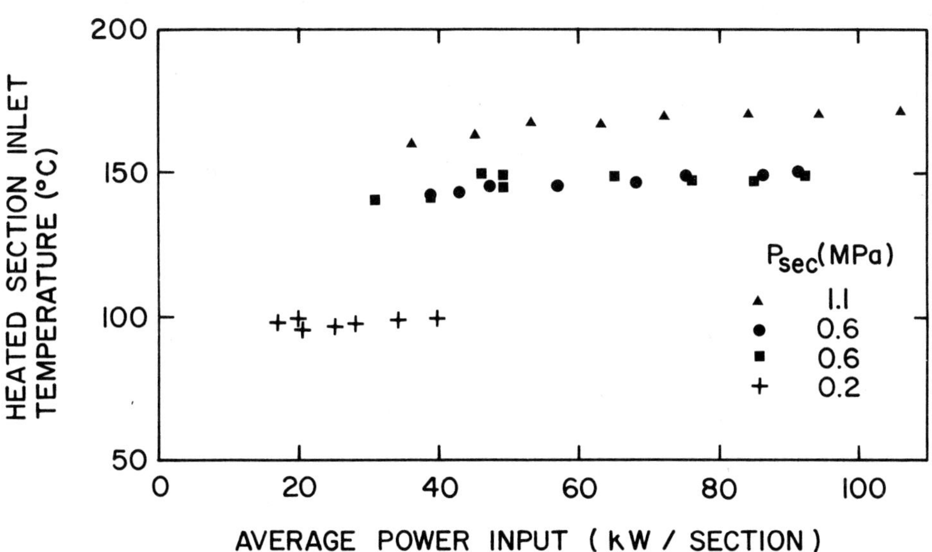

Figure 6: Heated section inlet temperature at onset of oscillations

oscillation threshold, for different secondary-side conditions. Both the pressure and flow rate are seen to increase with power, as expected. Subcooled liquid temperatures at the heater inlet for the same tests are shown in Figure 6. The inlet temperatures are seen to be almost constant for a given secondary pressure (lying $\sim$10-20°C below the appropriate secondary-side saturation temperature). This suggests that in thermosiphoning cooling, the boiler exit flow is essentially in thermal equilibrium with the secondary-side fluid.

The measured temperature distribution around the circuit shows that in each half-loop the two-phase flow is confined to the line between the heater outlet and boiler inlet plenum (referred to as the hot-leg). Table 1 gives the cross-sectional average void fraction inferred from the gamma-ray densitometers, at various locations in the hot-legs, at the onset of oscillations for one of the test series. It can be seen that the void fraction first decreases in the direction of the header, but then increases to approach quite a high value in the (horizontal) boiler inlet line. The void distribution is governed by heat loss, pressure drop and elevation change effects, with heat loss tending to decrease the void fraction and the latter two effects tending to increase it. It is difficult to account for the high void fraction observed at the boiler inlet using simple homogeneous equilibrium flow theory. The most likely explanation is the presence of an elongated gas bubble (steam or non-condensibles) trapped in the horizontal inlet line to the boiler, where the densitometer is located.

3. LINEAR STABILITY ANALYSIS

3.1 Characteristic Equation

The long period of the oscillations (50 - 220 s) suggests that the observed instabilities are driven by density-wave propagation [4]. Density-wave instabilities are commonly encountered in boiling systems, and result from delays in propagation of void fraction disturbances through the two-phase part of the system. A detailed review of density-wave instability is given by Boure et al. [5].

To calculate the stability characteristics, the loop is represented by the simple idealization shown in Figure 7. Heaters and boilers are treated as point heat sources and sinks, connected by uniform pipes. Two-phase conditions are assumed to exist in the hot-legs (H_1B_1, H_2B_2), and subcooled liquid conditions in the cold-legs (B_1H_2, B_2H_1), in line with observations. Treatment of the boilers as point sinks is justified because, in natural circulation operation, voids are observed to collapse in the boiler inlet plena. The point-source heater approximation is used because the heater length is <20% of the two-phase length; thus, it seems reasonable, to a first approximation, to neglect fluid transport delays in the heaters.

Frequencies and amplification coefficients of the flow oscillations in this symmetrical system can be obtained from symmetry arguments [6]. Consider the loop geometry shown in Figure 7(a) operating in steady-state natural circulation with flow rate $\bar{w}$. A perturbation $\delta w_1(t)$ applied to the inlet flow rate to heater H_1 will induce a perturbation in the inlet flow rate to H_2 of magnitude $\delta w_2(t)$. Denoting the Laplace transform of $\delta w_1(t)$ by $\delta w_1(s)$, and assuming $\delta w_1, \delta w_2 << \bar{w}$, the perturbations will be related by the equation

$$\delta w_2(s) = g(s)\, \delta w_1(s) \tag{1}$$

TABLE 1

Void Fraction Distribution at Onset of Oscillations

Experiment Number	q (kW/ Section)	p_{sec} (MPa)	Void Fraction[a]					
			Heated Section 1 Outlet[b]	Header 1 Inlet	Boiler 1 Inlet	Heated Section 2 Outlet	Header 2 Inlet	Boiler 2 Inlet
T8121	85	0.6	0.15	0.12	0.42	0.25	0.05	0.60
T8211	75	0.6	0.15	0.10	0.50	0.25	0.08	0.67
T8124	48	0.6	0.15	0.10	0.50	0.25	0.05	0.70
T8129	91	0.6	0.15	0.15	0.45	0.25	0.10	0.60
T8131/32	38	0.6	0.20	0.15	0.55	0.20	0.05	0.70
T8133	42	0.6	0.15	0.15	0.50	0.20	0.08	0.85

Notes: (a) Cross-sectional average values
(b) Component locations shown in Figure 1

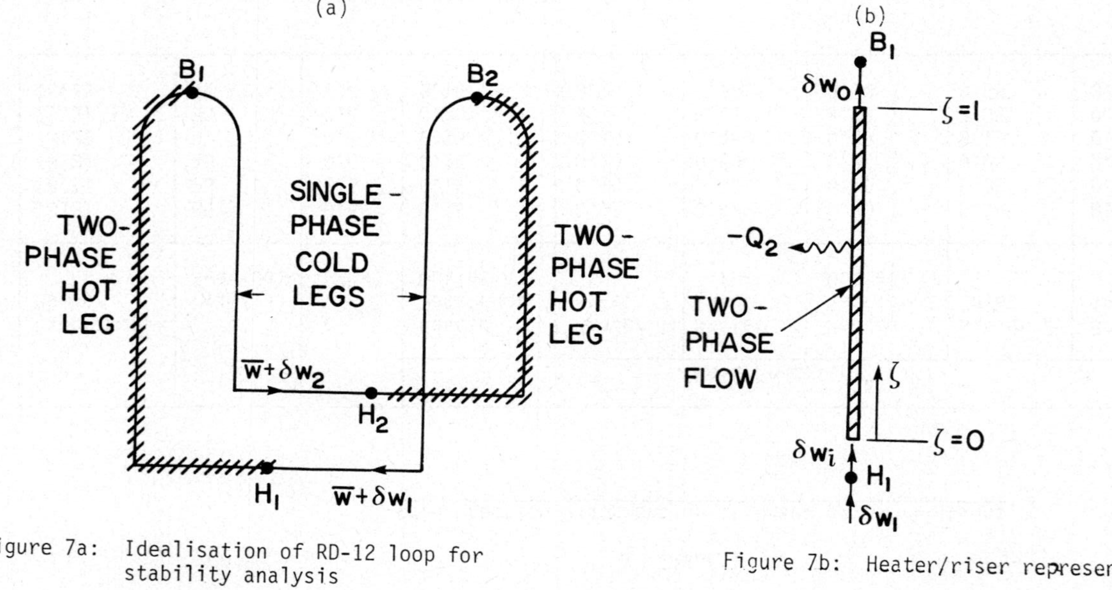

Figure 7a: Idealisation of RD-12 loop for stability analysis

Figure 7b: Heater/riser representation

where g(s) is a response function for the half-loop $H_1B_1H_2$. Now, since the second half-loop $(H_2B_2H_1)$ is identical to the first, δw_1 must be related to δw_2 by the same response function. Thus, we can also write

$$\delta w_1(s) = g(s)\ \delta w_2(s) \tag{2}$$

Comparing equations (1) and (2), we see that

$$g(s) = \pm 1 \tag{3}$$

The roots of equation (3) are a series of complex numbers which define the frequencies and amplification coefficients of possible flow oscillations. Occurrence of roots with positive real parts indicates growing oscillations, and unstable operation [7].

It can be seen that there are two families of oscillatory modes, corresponding to solutions of the equations g(s) = +1 and g(s) = -1, respectively (termed s_+ modes and s_- modes). Equation (1) shows that in an s_+ mode the flow-rate perturbations in the two half-loops are exactly in-phase, while in an s_- mode they are out-of-phase by a half-cycle.

Now, if the vapour volume in half-loop $H_1B_1H_2$ is denoted by V_1, a relationship exists between δV_1 and δw_1 of the form

$$\delta V_1(s) = f(s)\ \delta w_1(s) \tag{4}$$

where f is another response function. Similarly, for half-loop $H_2B_2H_1$ we have

$$\delta V_2(s) = f(s)\ \delta w_2(s) \tag{5}$$

But if liquid density changes are ignored, mass conservation requires that the total volume of vapour in the loop be constant (= the volume of liquid drained from the liquid-full loop). Thus,

$$\delta V_1 + \delta V_2 = 0 \tag{6}$$

It follows from equations (4) and (6) that

$$\delta w_1(s) = -\ \delta w_2(s)$$

Hence, mass conservation is only satisfied for the <u>anti-phase oscillations</u> (s_- modes). It can similarly be shown that the s_- modes are the only ones for which momentum is conserved, since for these oscillations, perturbations in the frictional, gravitational, and accelerational pressure losses in the two half-loops cancel in pairs. We thus conclude that, for the symmetric figure-of-eight loop, only anti-phase oscillations are possible, the natural frequencies of which are solutions of the characteristic equation

$$g(s) = -1 \tag{7}$$

A similar conclusion was reached by Kahn [8].

3.2 Calculation of the Response Function g(s)

To calculate g(s) for linear perturbations we assume that:

(i) the phases are incompressible;

(ii) the hot-legs are pipes of uniform area;

(iii) there is a uniform constant evaporation/condensation rate Γ_g per unit volume within the hot-leg, determined by the competing effects of flashing due to elevation change and condensation due to external heat losses;

(iv) the heater inlet enthalpy is fixed (this assumption is justified because in natural circulation operation it is observed experimentally that the boiler outlet temperature is approximately equal to the secondary-side temperature, which is held constant);

(v) the local cross-sectional average void fraction in the hot-leg is related to the volumetric flow rate of steam and water by a drift flux equation [9] of the form

$$u_g = \frac{j_g}{\alpha_g} = C_o j + u_{gj} \tag{8}$$

where C_o and u_{gj} are constants. The term $C_o j$ describes the effect of the velocity and concentration profiles on the vapour velocity (profile slip); u_{gj} describes the effect of vapour drift due to buoyancy.

With these assumptions the desired response function is equal to the ratio between flow-rate perturbations at the hot-leg exit and the heater inlet, i.e.,

$$g(s) = \frac{\delta w_o}{\delta w_1} \tag{9}$$

The problem is thus reduced to calculating the flow-rate response of the hot-leg, treated as a constant-area straight pipe with uniform vapour generation (condensation) rate per unit volume. The equivalent problem is shown in Figure 7b.

To calculate the hot-leg response we use void propagation theory [10, 11]. In this model, interphase slip is incorporated via the drift-flux equation (8). We introduce the following dimensionless variables:

$$\zeta = z/\ell \qquad \theta = t/t_{ref} \qquad \tau = t_{tr}/t_{ref}$$

$$U_k = u_k/(u_{ref}C_o) \qquad J_k = j_k/u_{ref} \qquad V_{gj} = u_{gj}/(u_{ref}C_o)$$

$$W = w/\bar{w} \qquad \alpha^* = \alpha_g C_o v_{g\ell}/v_g \qquad Q_2 = \Gamma_g (\ell v_{g\ell}/u_{ref})$$

where u_{ref} and t_{ref} are defined by

$$u_{ref} = \bar{w} v_\ell/A \qquad t_{ref} = \ell/(u_{ref}C_o)$$

With the assumptions (i)-(v)above , the governing equations for the transient flow rate in the hot-leg are [10]

$$\partial J/\partial \zeta = Q_2 \tag{10}$$

$$d\alpha^*/d\theta = Q_2(1-\alpha^*) \tag{11}$$

where equation (11) is satisfied along the <u>characteristic line</u> whose equation is

$$d\zeta/d\theta = U_g = J(\zeta,\theta) + V_{gj} \tag{12}$$

Equation (12) is the path of a vapor particle; equation (11) is referred to as the void propagation equation.

The vapor source term, Q_2, is assumed to be time-invariant (assumption (iii) above), which is a conventional approximation in applying the void propagation model to engineering systems. With this assumption, equations (10) and (11) can be integrated at once to give

$$J(\zeta,\theta) = J_i(\theta) + Q_2\zeta \tag{13}$$

$$1 - \alpha^*(\zeta,\theta + \tau) = [1 - \alpha_i^*(\theta)] \exp(-Q_2\tau) \tag{14}$$

Here τ is the dimensionless transit line between 0 and τ of a vapour particle located at ε=0 at time θ. (Subscript i denotes conditions at the hot-leg inlet, ζ = 0).

If τ is known, equations (13) and (14) completely define the transient void and velocity distribution in the hot-leg. In practice, $\tau(\zeta,\theta)$ can be found by substituting equation (13) into (12) and integrating the resulting differential equation for ζ (using the initial condition for the vapour particle path, ζ = 0 at time θ). The general solution is

$$\zeta(\tau,\theta) = \int_0^{\tau} U_{gi}\ (\tau' + \theta) \exp [Q_2\ (\tau - \tau')]\ d\tau' \tag{15}$$

Given the time-varying conditions at the hot-leg inlet, equations (13)-(15) can be used to find α^* and J at all downstream locations and times.

The total transit time of vapour in the hot-leg at θ, $[\tau_f(\theta)]$, is given by setting ζ = 1 in equation (15):

$$1 = \int_0^{\tau_f} U_{gi}\ (\tau' + \theta) \exp [Q_2\ (\tau_f - \tau')]d\tau \tag{15a}$$

<u>Steady State</u>. For steady-state conditions equations (15) and (15a) can be integrated directly to give:

$$\zeta = \bar{U}_{gi}(e^{Q_2\tau} - 1)/Q_2$$

$$e^{Q_2\tau_f} = 1 + Q_2/\bar{U}_{gi} \qquad (16)$$

The steady-state mean void fraction in the hot-leg is defined by:

$$\langle\bar{\alpha}^*\rangle = \int_0^1 \bar{\alpha}^* \, d\zeta$$

Using equations (14) and (16), it is readily shown that <u>for the case where voids exist throughout the hot-leg</u>

$$1 - \langle\bar{\alpha}^*\rangle = (1 - \bar{\alpha}^*_i) \, \tau_f \bar{U}_{gi}$$

It follows, by considering mass conservation and the drift-flux equation (8) only, that for steady conditions

$$\tau_f = (1 - \langle\bar{\alpha}^*\rangle)/(1 + V_{gj}) \qquad (17)$$

<u>Response to Fluctuations</u>. We consider the response in the hot-leg outlet flow rate to small sinusoidal perturbations in the heater-inlet flow rate about the steady state to be of the form

$$\delta W_1 = \varepsilon e^{s\theta}$$

where s is a complex non-dimensional frequency and $\varepsilon \ll 1$. Since, by definition,

$$W_o = J_{\ell o} + J_{go} \, (v_\ell / v_g)$$

and $J = J_\ell + J_g$, it follows that

$$g(s) = \frac{\delta W_o}{\delta W_1} = \lambda_1 (1 - \bar{\alpha}^*_o) \left[\frac{\delta J_o}{\delta J_i}\right] + \lambda_2 \bar{U}_{go} \left[\frac{\delta\alpha^*_o}{\delta\alpha^*_i}\right] \qquad (18)$$

where λ_1 and λ_2 describe the response of heater exit conditions to perturbations in the subcooled inlet flow:

$$\lambda_1 = \delta J_i / \delta W_1 \; ; \; \lambda_2 = - \delta\alpha^*_i / \delta W_1$$

The λs can be obtained from an energy balance for the heater (treated as a point source). This leads to the equations:

$$J_{gi} = (Q_1 - Ja\ W_1)\ v_g/v_\ell$$
$$J_i = Q_1 + (1-Ja)\ W_1 \qquad (19)$$

Here, Q_1 and Ja are dimensionless parameters representing the power input and the inlet liquid subcooling, respectively. Using equation (8) to relate α_i^* to J_{gi} and J_i it follows, by perturbing (19) (noting that Ja is assumed constant), that

$$\lambda_1 = (1 - Ja)$$
$$\lambda_2 = U_{gi}^{-1}\ [Ja + \bar{\alpha}_i^*\ (1-Ja)] \qquad (20)$$

It remains to calculate the factors in square brackets in equation (18). We consider only the case where voids exist throughout the hot-leg ($\alpha^*>0$). Perturbing equation (13), with $\zeta=1$, we see immediately that $[\delta J_o/\delta J_i] = 1$. The factor $[\delta\alpha_o^*/\delta\alpha_i^*]$ can in turn be calculated by perturbing equation (14) (with $\zeta=1$, $\tau=\tau_f$), and then eliminating $\delta\tau_f$ by perturbing equation (15a). Using these substitutions, and using equation (20) for the λs, the response function can be expressed, after some manipulation, in the form

$$g(s) = (1-Ja)(1-\bar{\alpha}_i^*)e^{-Q_2\tau_f} \left\{ 1 + \frac{Q_2\tau_f}{s\tau_f - Q_2\tau_f} \left[1 - e^{(Q_2-s)\tau_f}\right] \right\}$$
$$+ \left[Ja + (1 - Ja)\ \bar{\alpha}_i^*\right] e^{-s\tau_f} \qquad (21)$$

g(s) can also be written in terms of the three dimensionless parameters Q_1, Ja, and Q_2 characterising the electrical power input, heater inlet subcooling and vapour volumetric generation rate in the hot-legs, respectively. The transformation is effected by eliminating $\bar{\alpha}_i^*$ and τ_f, using the relationships

$$\bar{\alpha}_i^* = (Q_1 - Ja)/(Q_1 - Ja + 1 + V_{gj}) \qquad (22)$$

$$\tau_f = \frac{1}{Q_2} \ln \left[1 + \frac{Q_2}{Q_1 - Ja + 1 + V_{gj}}\right] \qquad (23)$$

Equations (22) and (23) are derived using equations (8), (19), and (16).

Note that equation (21) is applicable only for the case where voids exist throughout the hot-leg.

3.3 Numerical Solution

The characteristic equation (7) [with g(s) from equation (21)] was solved numerically by a Newton-Raphson method for various values of Q_1, Q_2, and Ja. The first six or so roots were calculated ($s_n = \sigma_n + i\omega_n$) for increasing values of ω_n. The ω_n are the angular frequencies of possible (linear) oscillations of the system, and the σ_n are the corresponding amplification coefficients. It was found that the oscillation frequencies are given to a very close approximation by the formula

$$\tau_k \omega_n = 2\pi n \qquad n = 0,1,2,\ldots.. \tag{24}$$

For $Q_2>0$ (corresponding to positive net vapour generation in the hot-leg), all the σ_n were found to be positive, indicating unstable operation. However, for $Q_2<0$ (corresponding to net vapour condensation in the hot-leg), the oscillating modes were found to be stable+ ($\sigma_n<0$) until a small finite void fraction is reached at the hot-leg outlet. This corresponds to the experimental situation in RD-12 (see Section 4). The implication is that, for oscillations in RD-12, some voidage must be present at the boiler inlet, and that the mean voidage must exceed some threshold value, which is in line with experimental observations. A typical predicted stability map is shown in Figure 8.

Parametric calculations of the threshold for oscillatory instability were carried out by varying Q_1, Ja, and Q_2 ($Q_2<0$) for the particular case $C_o=1.2$, $V_{gj}=0$. For this choice of drift parameters the instability threshold was found to be given to a good approximation by the convenient equation

$$Q_1 - Ja = -Q_2 + 0.25\ Ja\ [1 - \exp\ (0.25\ Q_2)] \tag{25}$$

The period of the fundamental (lowest frequency) oscillation mode (n=1) is given from equation (24) by

$$(2\pi/\omega_1) = \tau_f \tag{26}$$

Thus, the predicted period of oscillation is simply the void transit time in the hot-leg, as given by equation (17).

4. COMPARISON WITH EXPERIMENTAL DATA

To compare the calculated instability threshold with the experimental data, it is necessary to estimate the condensation rate, $-Q_2$, in the hot-leg for the test conditions.

It is shown in the Appendix that, for a steady low-quality, low-velocity, thermal-equilibrium flow in a uniform pipe with uniform heat losses and elevation change, the non-dimensional condensation rate is given by

$$-Q_2 = Q_L - Q_H \tag{27}$$

Here, Q_L and Q_H represent contributions to the vapour source term from heat losses and elevation change. They are given by the equations

---

+ The zero frequency mode $\omega=0$ was generally found to be unstable. The physical meaning of this result is not clear, and in applying the model the existence of the $\omega=0$ mode is ignored.

$$Q_L = k\,(T_{SAT} - T_\infty)\, v_{g\ell} / (h_{g\ell}\, v_\ell\, \bar{w})$$

$$Q_H = g\, v_{g\ell}\, \Delta y\, (dh_\ell / dp)_{SAT} / (v_\ell\, h_{g\ell})$$

where k is a heat-loss coefficient for the hot-leg as a whole, and T_∞ the ambient temperature.

The hot-leg heat-loss coefficient k is somewhat uncertain. For the comparisons we used a value of k = 20 W/K which was found to give reasonable agreement between measured and predicted stability limits over the entire pressure range. For $T_{SAT} = 200^\circ C$, a value of k = 20 W/K corresponds to a heat loss of ∿4 kW, which compares reasonably well with the experimental value of 6-10 kW deduced by calorimetry for these conditions [3].

For the experimental simulations we took $C_o = 1.2$, which is a usual value for turbulent flow [9]. The drift parameter V_{gj} was taken as zero, since 75% of the hot-leg length is formed by horizontal pipe, for which $V_{gj} = 0$.

Using the experimental values of $\bar{w}$, p, ΔT_{SUB} and q_1 at the instability threshold, and the geometrical characteristics of the loop, values of Q_1, Q_2 and Ja can be calculated at the threshold. These can be compared directly with equation (25), which is applicable for $C_o = 1.2$, $V_{gj} = 0$. Comparisons are shown in Figure 9, and agreement is seen to be reasonably good.

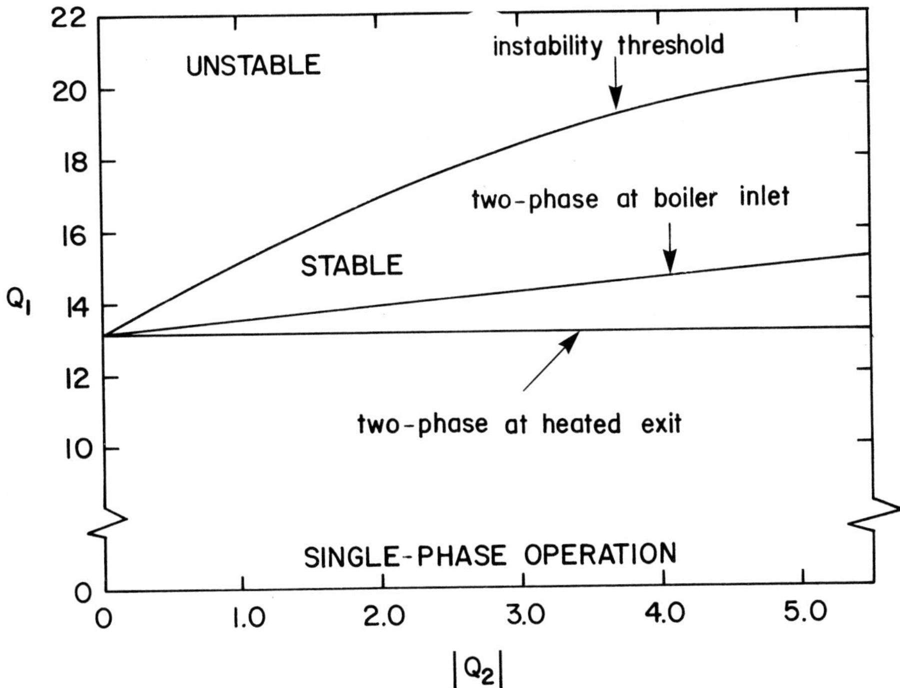

Figure 8: Stability map predicted by linearised model for Ja = 13

The comparisons can also be displayed in dimensional coordinates. Figure 10 shows the primary pressure and heater power at the oscillation threshold for different secondary side pressures. The curves are predictions of equation (25) obtained by using the experimental values of Q_2 and Ja to calculate Q_1 at threshold, and deducing a value for q_1. (Note that one predicted point is obtained for each test; the curves represent smoothed lines between the predicted points). Plotted in this way, the agreement is very good.

Figure 11 shows the oscillation periods for all tests, predicted using equations (26) and (17), with $\langle\alpha^*\rangle$ deduced from the experimentally observed loop inventory at threshold (assuming the void is confined to the hot-legs). Agreement with experiment is very good, providing convincing evidence that the oscillations are indeed due to propagation of density waves in the hot-legs.

Lastly, Figure 12 shows the measured mean voidage in the loop at threshold (% inventory drained) versus the value calculated from equation (25). Predictions were made by determining Q_2 and Ja from experimental data, and then using equation (25) to predict Q_1 at threshold; $\langle\alpha^*\rangle$ was deduced using equations (17) and (23). It is seen that the mass drained at threshold is considerably <u>over-predicted</u> by the model, although theory and experiment are in qualitative agreement in predicting instability for loop mass inventory reductions of a few percent. The discrepancies are believed to be due to

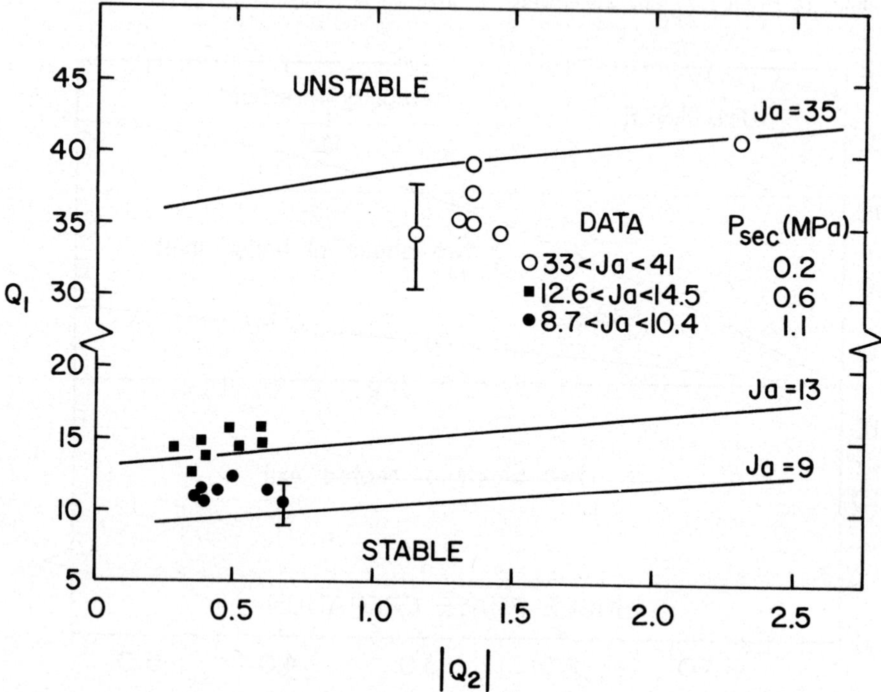

Figure 9: Comparison between measured and predicted stability thresholds. Curves denote calculated boundaries between the stable (lower) and unstable (upper) regions

the assumption in the model that in steady flow α_g decreases monotonically with distance along the hot-leg (see equations (14) and (16)). In reality, the voidage has a complex axial distribution, caused by interaction between local heat losses, elevation changes, and slip-velocity variations (see Section 2.2). Failure to represent $\bar{\alpha}_g(z)$ correctly probably results in errors in the calculated dynamic response of the system, and hence in the prediction of $\langle\alpha_g\rangle$ at threshold.

The agreement obtained with our simple model provides convincing evidence that the instabilities observed in RD-12 are driven by density-wave propagation in the two-phase region of the loop. An interesting observation is that the characteristic equation for the oscillation frequencies is independent of momentum effects, being derived solely from continuity considerations. One consequence is the prediction that a change in the flow resistance of the loop would leave the relationship between the oscillation frequencies and amplification coefficients and the characteristic dimensionless parameters Q_1, Q_2 and Ja unaltered. (In fact, it is predicted that an increase in loop flow resistance would tend to destabilise the system by decreasing $\bar{w}$ and hence increasing Q_1.) This is in complete contrast with usual parallel-channel type oscillations, which, though also driven by void-propagation effects, depend explicitly on momentum losses in the single- and two-phase regions.

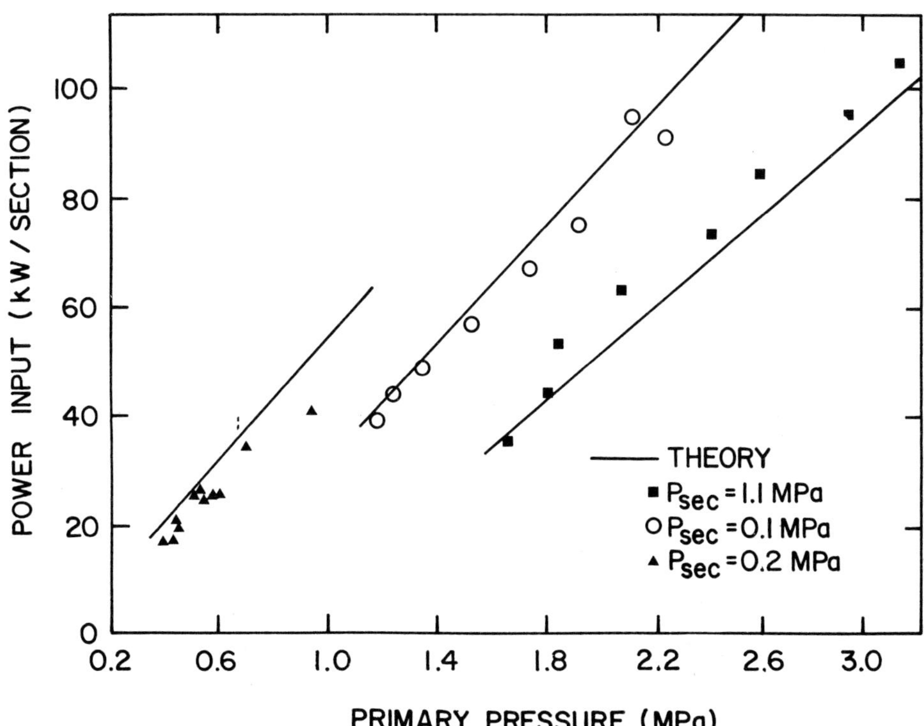

Figure 10: Comparison between measured and predicted stability thresholds

Gulshani [12] previously developed a linear stability model for RD-12 based on the idea that the driving force for the instability is an oscillation in the pressure in the two-phase region. However, this theory was unable to predict loop stability in the limit $\langle\alpha_g\rangle$ 0. The present work has shown that a simpler model based on density-wave effects (in which pressure changes are ignored) is adequate to explain most observations, and also properly predicts stability behaviour in the low void-fraction limit.

5. CONCLUSIONS

Experiments have been carried out in the RD-12 loop at WNRE to investigate the stability of two-phase natural circulation (thermosiphoning) flow in a symmetrical figure-of-eight loop geometry. Results show that the flow rate becomes oscillatory when the loop mass inventory falls below a certain threshold level.

A simple, linearised model has been developed to explain the oscillations, based on examination of propagation of density perturbations in the two-phase part of the loop (hot-legs). The model is derived solely from continuity considerations. Momentum losses, which are usually important in controlling the oscillation characteristics in two-phase systems, do not enter the analysis explicitly.

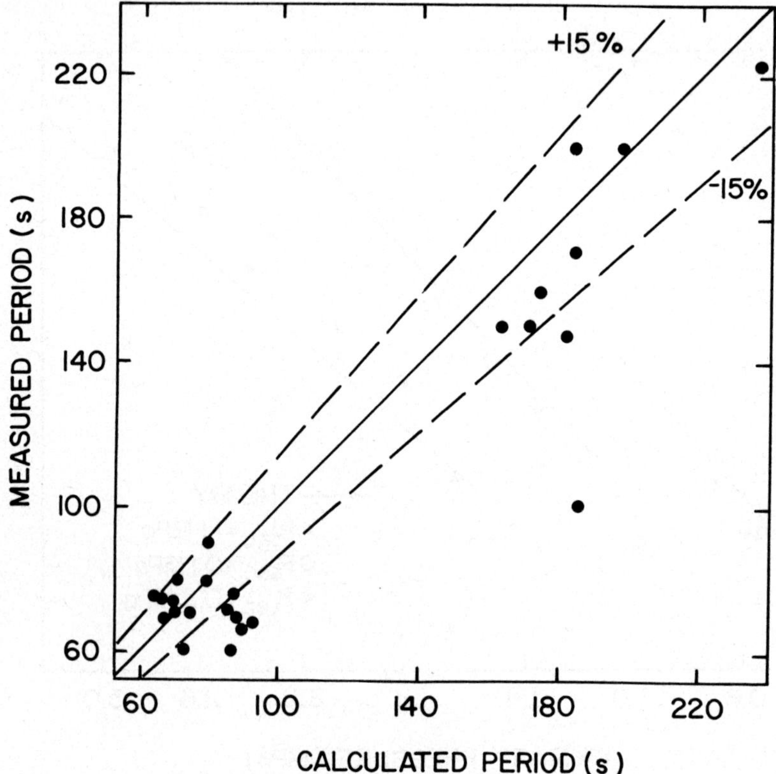

Figure 11: Comparison between measured and predicted oscillation periods

The model suggests that the oscillation frequencies and stability thresholds are governed principally by three non-dimensional groups representing the heat input, liquid subcooling at the heater inlet, and condensation rate in the hot-legs. An approximate equation has been derived for the stability threshold in terms of these three variables.

The linearised model predicts that net vapour condensation in the hot-legs is needed for stable two-phase thermosiphoning operation of RD-12. When net condensation is present, it is predicted that an instability occurs only after a small finite void fraction is reached at the boiler inlets. This corresponds to the experimental situation.

Predicted stability thresholds and oscillation frequencies have been found to be in reasonable quantitative agreement with the experimental data.

ACKNOWLEDGEMENTS

The experiments described in this paper were carried out as part of an AECL/Ontario Hydro joint research program.

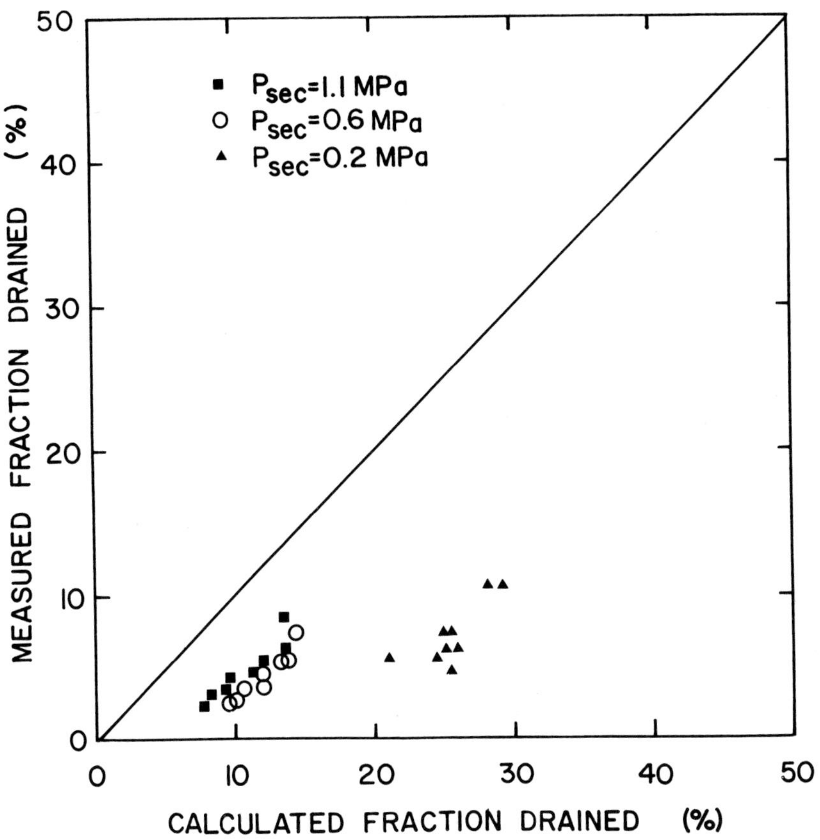

Figure 12: Fraction of loop inventory drained at threshold

REFERENCES

1. Feyginberg, Y., Sergejewich, P., and Midvidy, W.I., "A Method for Assessing Reactor Core Cooling Without Forced Circulation", in Proc. of 2nd Int. Mtg. on Nuclear Thermal-Hydraulics, Santa Barbara, CA, 11-14 January 1983. American Nuclear Society, La Grange Park, ILL, 1983, p.808.

2. Barclay, F.W., Hawley, E.H., and Nieman, R.E., "RD-12 Facility Description", Unpublished Whiteshell Nuclear Research Establishment Report, WNRE-496, December 1980.

3. Hawley, E.H., Krishnan, V.S., Borgford, T.A., Findlay, J.W., Henschell, R.M., Spitz, K.O., and Thomson, R.G., AECL unpublished data.

4. Wallis, G.B., and Heasley, J.H., "Oscillations in Two-Phase Flow Systems", Trans. ASME, J. Heat Transfer, 83, p. 363-369, 1961.

5. Boure, J.A., Bergles, A.E., and Tong, L.S., "Review of Two-Phase Flow Instability", Nucl. Eng. Des. 25, p. 165-192, 1973.

6. Hinds, H.W., McMorran, P.D., and Tseng, C.M., "Stability Studies of a CANDU Heat Transport Circuit", IMACS World Congress on System Simulation and Scientific Computation, Montreal, 8-13 August 1982, Vol. 4, p. 53-56.

7. Friedly, J.C., Dynamic Behaviour of Processes , Prentice-Hall Inc., New Jersey, 1972.

8. Kahn, A.R., "Heat Transport System Instability", Atomic Energy of Canada Limited, Engineering Company. Private communication, April 1980.

9. Zuber, N., and Findlay, J.A., "Average Volumetric Concentration in Two-Phase Flow Systems", Trans. ASME J. Heat Transfer 87, p. 453-408, 1965.

10. Zuber, N., and Staub, F.W., "An Analytical Investigation of the Transient Response of the Volumetric Concentration in a Boiling Forced-Flow System", Nucl. Sci. Eng. 30, p. 268-278, 1967.

11. Zuber, N., and Staub, F.W., "The Propagation and the Wave Form of the Vapor Volumetric Concentration in Boiling Forced Convection System Under Oscillatory Conditions", Int. J. Heat Mass Transfer 9 pp. 871-895, 1966.

12. Gulshani, P., "A Figure-of-Eight Thermosiphoning Stability Model and its Verification", in Proc. of 2nd Int. Topical Mtg. on Nuclear Reactor Thermal-Hydraulics, Santa Barbara, CA, 11-14 January 1983, American Nuclear Society, La Grange Park, IL, 1983, p. 840.

APPENDIX

Calculation Of Vapour Generation Rate In Hot-Leg

Assume a steady, low-velocity, low-quality, homogeneous, thermal-equilibrium two-phase flow in a uniform-area pipe, with elevation change. The momentum and energy equations can be written

$$(u/v)\ du/dz + dp/dz = -g\ \cos\alpha/v - \tau_w$$
$$(u/v)\ dh/dz - u\ dp/dz = q + u\tau_w \tag{A1}$$

where

$$h = Xh_g + (1 - X)\ h_\ell$$

and q, τ_w, cosα and X denote the heat-addition rate per unit flow volume, frictional force per unit flow volume, pipe inclination, and flow quality, respectively.

For cases of present interest, $X \simeq 10^{-3}$, so we can write

$$dh/dz \simeq h_{\ell g}\ dX/dz + (dh_\ell/dp)_{SAT}\ (dp/dz) \tag{A2}$$

Also, at the low flow rates encountered in thermosiphoning, the frictional and accelerational pressure drops account for less than 1% of the total pressure drop in the hot-leg. Thus, we have

$$dp/dz \simeq -\ g\ \cos\alpha/v \tag{A3}$$

Using equations (A1) to (A3), we obtain the following equation for the vapour generation rate per unit flow volume

$$\Gamma_g = (u/v)\ dX/dz = q/h_{g\ell} + (wg\ \cos\alpha/vA)\ (dh_\ell/dp)_{SAT} \tag{A4}$$

Defining the hot-leg heat-loss coefficient k by the equation

$$q = -k\ (T_{SAT} - T_\infty)/A\ell$$

Equation (A4) can be written in non-dimensional form

$$Q_2 = \Gamma_g\ (\ell V_{g\ell} A/wv_\ell) = -Q_L + Q_H \tag{A5}$$

where

$$Q_2 = k(T_{SAT} - T_\infty)\ v_{g\ell}/(h_{g\ell}\ v_\ell w)$$

$$Q_H = gv_{g\ell}\ \Delta y\ (dh_\ell/dp)_{SAT}/(v_\ell h_{g\ell})$$

$$\Delta y = \ell \cos\alpha = \text{elevation change in hot-leg.}$$

In deriving (A5), we have taken $v \simeq v_\ell$.

NOTATION

A	hot-leg flow area
C_o	drift flux parameter
h	specific enthalpy
j(J)	volumetric flow rate (non-dimensional volumetric flow rate)
Ja	sub-cooling number = $(h_{\ell\,SAT} - h_{\ell 1})\ v_{g\ell}/(h_{g\ell} v_\ell)$

ℓ	length of hot-leg
p	pressure
q_1	heat input rate
Q_1	non-dimensional heat input rate to heater
	$1 = q_1 V_{g\ell}/(h_{g\ell} V_\ell \bar{w})$
Q_2	non-dimensional vapour generation rate in hot-leg
Q_L, Q_H	non-dimensional vapour generation rate due to heat loss and elevation change [see eq. (27)]
s	complex non-dimensional frequency
t	time
t_{tr}	vapour transit time
T	temperature
u(U)	velocity (non-dimensional velocity)
$u_{gj}(V_{gj})$	drift flux parameter (non-dimensional drift flux parameter)
v	specific volume
w(W)	mass flow rate (non-dimensional mass flowrate)
z	distance from hot-leg inlet
α	volumetric concentration
α^*	modified void fraction (Section 3.2)
Δy	elevation change between heater outlet and boiler inlet
Γ_g	vapour generation rate
σ	amplification coefficient
θ	non-dimensional time
τ	non-dimensional vapour transit time
τ_f	non-dimensional vapour transit time in hot-leg
ζ	non-dimensional distance from hot-leg inlet
ω	non-dimensional oscillation frequency

<u>Subscripts and Superscripts</u>

i	value at hot-leg inlet
o	value at hot-leg outlet
g	gas property
ℓ	liquid property
$g\ell$	difference between gas and liquid property
1	property of inlet flow to heater H_1
SAT	saturation property
sec	property of system secondary side
ref	reference value for non-dimensionalization
—	property of steady flow

AUTHOR INDEX